Lecture Notes in Computational Science and Engineering

Volume 156

This series contains monographs of lecture notes type, lecture course material, and high-quality proceedings on topics described by the term "computational science and engineering". This includes theoretical aspects of scientific computing such as mathematical modeling, optimization methods, discretization techniques, multiscale approaches, fast solution algorithms, parallelization, and visualization methods as well as the application of these approaches throughout the disciplines of biology, chemistry, physics, engineering, earth sciences, and economics.

Barry Koren · Dirk Roose
Editors

Numerical Analysis and Scientific Computing: an Anthology

Dedicated to the 50th Anniversary of the Woudschoten Conference

Springer

Editors
Barry Koren
Mathematics and Computer Science
Eindhoven University of Technology
Eindhoven, The Netherlands

Dirk Roose
Department of Computer Science
Katholieke Universiteit Leuven
Leuven, Belgium

ISSN 1439-7358 ISSN 2197-7100 (electronic)
Lecture Notes in Computational Science and Engineering
ISBN 978-3-032-33427-5 ISBN 978-3-032-33428-2 (eBook)
https://doi.org/10.1007/978-3-032-33428-2

This Springer imprint is published by the registered company Springer Nature Switzerland AG
The registered company address is: Gewerbestrasse 11, 6330 Cham, Switzerland

Foreword

Since its inception in 1976, the *Woudschoten Conference on Scientific Computing* has been an institution for the numerical analysis and scientific computing community in The Netherlands and (somewhat later) Flanders, Belgium. The community has united once per year for three days, listened to experts from abroad, shared meals and drinks, and discussed developments both within the field and within the local scientific landscape. In October 2026 the Dutch-Flemish Scientific Computing Society organized its annual conference for the 50th time. (The conference planned for 2020 was postponed to 2021 due to measures to mitigate the COVID-19 pandemic.)

I asked emeritus professor Pieter van der Houwen for his recollections of the origins of the conference:

> *What I remember is that Marc [Spijker] came up with the idea of the Woudschoten Conference at one of the meetings Van der Sluis organized a couple of times per year in Utrecht. Those were attended by numerical mathematicians from the Dutch universities and CWI. Usually it was a rather small group, at most 15 to my estimate. At one of those meetings Marc mentioned that statisticians and stochasticians organized a yearly meeting at Woudschoten. . . . Marc proposed the numerical mathematicians do that as well, with as keynote speakers leading numerical mathematicians from abroad on a predetermined topic. Marc's proposal was enthusiastically embraced by those present. This would have been around 1975. The* Werkgemeenschap Numerieke Wiskunde *(in Dutch, literally "working community of numerical mathematics") was also established in the mid-1970s. . . . Shortly after the meeting in Utrecht a Woudschoten Committee was formed. . . . The secretary was always someone from the Numerical Mathematics department at CWI . . . From the first few conferences, I remember the accommodation being a bit primitive: for example, the rooms could not be locked, to the consternation of the foreign speakers.*

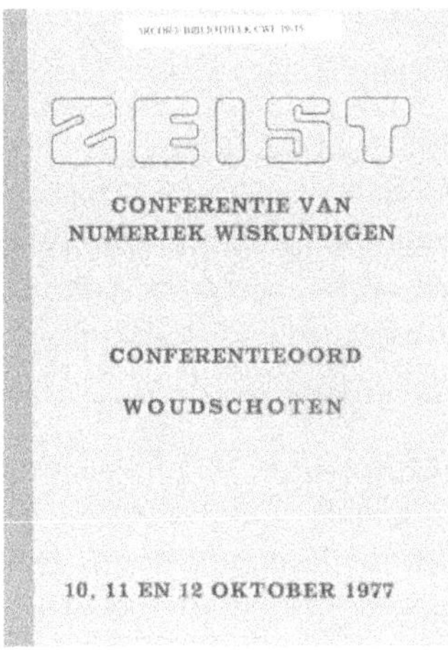

Program booklet of the second Woudschoten Conference, 10–12 October, 1977. [CWI Repository]

As national institute, CWI was the designated 'clerk' to handle the administrative organization of the conference. Eldert Slagt was the first secretary followed in succession by Jan Verwer, Willem Hundsdorfer, Ben Sommeijer, Jan Kok, Walter Lioen, Margreet Nool and Martine Anholt-Gunzeln.

Critical to the long success of the meetings, in my opinion, are the format and the location of the conference. As envisioned by Spijker (who, sadly, passed away this year), the conference has been organized around two (later three) preselected topics. Six internationally leading speakers were invited to give two talks of about one hour, the first of which of a tutorial nature suitable for graduate students new to the topic. In this way, the conference was really intended to inspire and invigorate the research of the Dutch-Flemish scientific computing community. The location at Woudschoten in Zeist, where—a bit isolated from their home institutes—participants can take long walks in the woods, is conducive to community-building.

Of course, the format only worked if those internationally leading scientists accepted the invitation. A glance at the list of keynote speakers over the last 50 years leaves a deep impression. Speakers include Claude Brezinski, John Butcher, Peter Deuflhard, Gene Golub, Bertil Gustafsson, Wolfgang Hackbusch, Peter Henrici, Nick Higham, Herbert Keller, David Keyes, Tamara Kolda, Volker Mehrmann, Bill Morton, John Ockendon, Stanley Osher, Carola Schönlieb, Josef Stoer, Andrew Stuart, Nick Trefethen, Henk van der Vorst, Mary Wheeler, James Wilkinson and Barbara Wohlmuth, to name a few. It is rare that speakers have been invited more than once. Ernst Hairer and Gerhard Wanner hold the record with three invitations each. Those who have participated two times often shared close ties with members of the local community, among others: Owe Axelsson, Paul Van Dooren, Peter Forsyth, Bram van Leer, Beresford Parlett, Sebastian Reich, Yousef Saad, Chus Sanz-Serna and Christoph Schwab. The youngest speaker was Aretha Teckentrup in 2018, two of the oldest were Phil Roe and Cleve Moler in 2015.

The topics of the 1976 conference were simply 'Approximation' and 'Discretization', but the topics have become increasingly more specialized through the years. In the mid-1970s numerical analysis was still primarily developing as a discipline. Through the years numerical methods have gained ubiquity as part of the toolbox of applied analysis and stochastics and more and more play a critical but sometimes less obvious role in complex applications, for example in data assimilation and inverse problems. Recently, the advancement of neural networks synergizes with numerical analysis, both under the hood: i.e., universal approximation and stochastic optimization methods, and at a meta-level: i.e., scientific machine learning. Time will tell how the conference evolves further.

This volume collects invited papers from a number of speakers from recent years. I would like to express my gratitude to Barry Koren and Dirk Roose for taking the initiative to organize and supervise the editing of this anthology.

Amsterdam,
2026

Jason Frank,
Chair of the Dutch-Flemish Scientific Computing Society

Preface

The Woudschoten Conference is the flagship of the Dutch-Flemish Scientific Computing Society. The unification of The Netherlands and Flanders, albeit in numerical analysis and scientific computing only, is one of the fine aspects of the conference. The cooperation between the Dutch and Flemish numerical mathematicians is excellent. The same was true for us, as editors, during the composition of this book.

The organizers of the Woudschoten Conference have always been active in making the conference broad content-wise. Lecturers have been invited to cover the spectrum from very fundamental to very applied. In composing this book, we have striven to show this diversity by inviting former Woudschoten Conference lecturers from the above-mentioned full spectrum.

To make the book rather forward-looking, rather than backward-looking, we invited 21 lecturers from the Woudschoten Conferences from 2017 up to and including 2024 to contribute a chapter to the book. Within the sharply prescribed time frame, 14 of them could write a chapter, most of them with co-authors.

We have left the authors free in choosing the character of their contribution. Hence, the chapters vary from relatively short, innovative research chapters to large, encyclopedic review chapters. The shortest chapter counts 16 pages, the two longest 56 pages. The number of references in the bibliographies varies strongly as well, from 24 to 245. Given the diversity of the chapters, we imposed rather light editorial guidelines, enabling the authors to write in their own 'comfort zone'. As a consequence, you may observe some editorial differences between the chapters here and there.

All chapters had to fulfill high standards of scientific quality. All have been reviewed by external experts, one to four per chapter. As editors, we have both reviewed each chapter as well. We were pleased by the willingness of the authors to improve their contributions, where necessary.

In the book, the chapters are ordered from fundamental first, followed by more and more applied. By having chosen this ordering, we do not suggest that fundamental research always comes first and is leading. Fundamental research often finds its origin in questions that have been posed by people working on real-life problems. In ordering the book chapters from fundamental to more applied, we clustered them in the following eight subareas: ***discretization, computability, time integration, linear algebra, high-performance computing, optimal control, scientific machine learning, applications***. We proceed by giving an overview of the contents of the 14 chapters, following the above from-fundamental-to-applied subarea structure.

Discretization

The first chapter is entitled **Metric-Driven Numerical Methods** and has as authors *Patrick Henning* (Ruhr-University Bochum, Woudschoten lecturer in 2021), *Laura Huynh* (Ruhr-University Bochum) and *Daniel Peterseim* (University of Augsburg). In the chapter, a metric-driven discretization approach is presented as a tool for solving multiscale partial differential equations. A key aspect of the approach is the minimization of functionals, in particular the idea that gradients can be represented in different metrics to accelerate convergence. The approach is applied to the Gross–Pitaevskii equation and to equations related to ground states of Bose-Einstein condensates.

The next chapter, the second, is entitled **Concentrated Real-Pole Uniform-in-Time Approximation of the Matrix Exponential** and has as authors *Stefan Güttel* (Woudschoten lecturer in 2024) and *Shuai Shao* (both from the University of Manchester). In the chapter, a theorem is provided for the asymptotically optimal choice for the single real pole in the rational approximation of the time-dependent exponential function.

The third chapter is entitled **Adaptive Robust Interior Penalty Discontinuous Galerkin Methods** and has as authors *Zhaonan Dong* (Inria and Institut Polytechnique de Paris), *Emmanuil Georgoulis* (Heriot-Watt University and National Technical University of Athens, Woudschoten lecturer in 2019) and *Natalia Kopteva* (University of Limerick). The chapter presents an interior penalty method for strongly elliptic partial differential equations. In the chapter, the focus is on a robust a posteriori error estimator, particularly for partial differential equations with discontinuities in the diffusion operator and strong mesh-size variations.

Computability

The fourth chapter is entitled **Complexity Theory Meets Ordinary Differential Equations** and is authored by *Adalbert Fono, Noah Wedlich* (both Ludwig-Maximilians-Universität München), *Holger Boche* (Technical University of Munich) and *Gitta Kutyniok* (Ludwig-Maximilians-Universität München and three more affiliations, Woudschoten lecturer in 2019). In the chapter, the computational complexity of linear ordinary differential equations on digital computers is investigated. A characterization is given of computational complexity blowup for ordinary differential equations of arbitrary order. The authors show that for any linear input-output system, there exists a polynomial time input signal such that its associated output signal is not computable in polynomial time.

Time integration

The fifth chapter is on the still important topic of numerical time integration. The chapter is entitled **Some Recent Advances in Time Integration Methods for Kinetic Equations with Time-Scale Separation**. It is authored by *Emil Løvbak* (Karlsruhe Institute of Technology), *Vince Maes* (KU Leuven) and *Giovanni Samaey* (KU Leuven, Woudschoten lecturer in 2018). The chapter gives an overview of advancements in numerical time integration over the past 30 years. The diversity of methods presented is large and the chapter has a very extensive bibliography. Challenges addressed in the chapter are high dimensionality and the presence of multiple time scales. Various approaches are considered: deterministic, stochastic, asymptotic-preserving, multilevel and adjoint optimization. The chapter focuses on kinetic equations arising in photon and neutron transport and in plasma dynamics.

Linear algebra

The sixth chapter is entitled **Randomized Orthogonalization and Krylov Subspace Methods: Principles and Algorithms** and is authored by *Jean-Guillaume de Damas* (Sorbonne University), *Laura Grigori* (École Polytechnique Fédérale de Lausanne and Paul Scherrer Institute, Woudschoten lecturer in 2023), *Igor Simunec* (École Polytechnique Fédérale de Lausanne) and *Edouard Timsit* (Académie de Versailles). The chapter gives an overview of randomized orthogonalization techniques and of their application to subspace methods. The methods use sampling to obtain approximate solutions of several problems, at a computational and communication cost that is lower than that of corresponding deterministic methods. The chapter discusses in detail randomized Cholesky and Householder QR decompositions, randomized Gram–Schmidt orthogonalization, and applications such as the solution of linear systems and eigenvalue problems.

The seventh chapter is entitled **Randomized Algorithms for Low-Rank Matrix and Tensor Decompositions**. It is authored by *Katherine Pearce* (University of Texas at Austin) and *Per-Gunnar Martinsson* (University of Texas at Austin, Woudschoten lecturer in 2021). The chapter reviews randomized solution algorithms for decompositions of vectors and matrices. The chapter starts with an overview of classical numerical linear algebra methods that can be accelerated by randomized dimension reduction methods, such as the singular-value decomposition. Several randomized dimension-reduction techniques are discussed, including new matrix sampling techniques. The chapter has an extensive bibliography.

High-performance computing

The eighth chapter, entitled **Parallel Sparse and Data-Sparse Factorization-Based Linear Solvers** is authored by *Sherry Li* (Woudschoten lecturer in 2018) and *Yang Liu* (both from Lawrence

Berkeley National Laboratory). The chapter presents a survey of parallel solution methods for linear systems of equations that have a sparse structure. A survey is presented of basic methods for parallelization, communication reduction by data replication, and sparse LU decomposition and triangular system solving. An overview is given of rank-structured matrices of different types. The chapter's bibliography is very rich.

Optimal control

The ninth chapter is entitled **Ensemble Kalman–Bucy Filtering for Nonlinear Model Predictive Control** and is single-authored by Sebastian Reich (Universität Potsdam, Woudschoten lecturer in 2023 and also in 2009). In the chapter, nonlinear optimal control under uncertainty is considered, covering nonlinear continuous time systems with linear controls and quadratic control cost. Uncertainty is also considered. It enters via an unknown initial state and via noise on nonlinear state-measurement functions. This is a research chapter combining an extension of the Ensemble Kalman Filter in its continuous-time form (the Ensemble Kalman-Bucy Filter) with model predictive control in the context of partially observed Markov decision processes.

The next chapter, the tenth, has the title **Computing Stationary Distribution via Dirichlet-Energy Minimization by Coordinate Descent** and is authored by *Konstantin Avrachenkov* (Centre Inria d'Université Côte d'Azur), *Lorenzo Gregoris* (Eindhoven University of Technology) and *Nelly Litvak* (Eindhoven University of Technology, Woudschoten lecturer in 2017). The chapter is about fast numerical methods for solving large Markov chains with application to network problems such as PageRank computation. Keywords of this chapter are: Markov chains, quadratic optimization, gradient methods, coordinate descent and symmetrized Laplacians.

Scientific machine learning

The eleventh chapter has the title **Bayesian Perspectives in Scientific Machine Learning: Methods, Challenges, and Opportunities** and as authors, both from the Scuola Internazionale Superiore di Studi Avanzati: *Dario Coscia* and *Gianluigi Rozza* (Woudschoten lecturer in 2019). The chapter exploits the fact that Bayesian learning provides probabilistic predictions, quantifying uncertainty and supporting continuous adaptation as new data becomes available. The chapter gives an overview of Bayesian methods in scientific machine learning (SciML) and outlines promising research directions at the intersection of Bayesian methods and SciML. It describes, among other things, how Bayesian methods can enhance uncertainty quantification. The chapter has a rich bibliography.

Applications

The twelfth chapter is entitled **Numerical Methods for Optimal Decumulation of a Defined Contribution Pension Plan** and is authored by *Peter Forsyth* (Woudschoten lecturer in 2023 and also in 1996) and *George Labahn* (both University of Waterloo). The chapter presents a framework for solving the optimal decumulation problem in pension plans to which the contribution is known. The retirement income problem is formulated as a stochastic control problem with annual impulsive decisions. A Fourier method is developed to solve the resulting parabolic partial integro-differential equation. The method combines a monotonicity preserving discretization, a smart control mechanism and dynamic programming. The method can handle jump-diffusion dynamics of stocks and bonds, borrowing cost, withdrawal constraints and more.

The thirteenth chapter has the compact title **Computational Optics**. Its authors are *Martijn Anthonissen* (Woudschoten lecturer in 2024), *Pieter Braam, Lisa Kusch, Koondanibha Mitra, Jan ten Thije Boonkkamp* (all from Eindhoven University of Technology), and *Wilbert IJzerman* (Eindhoven University of Technology and Signify, The Netherlands). In the chapter, a unified description is given of several problems in computational optics, leading to various types of nonlinear partial differential equations, including the Monge-Ampère and generated Jacobian equations. Several optics models are reviewed in a generic manner. The key problem addressed is to find shapes of optical surfaces that transform a given light source into a desired target distribution. For this, a mathematical model is presented that is based on the principles of geometrical optics and formulated in terms of optical maps.

The fourteenth and last chapter is entitled **Computing for Life: Modeling and Simulation of Electrodiffusion in Dense Reconstructions of Cerebral Tissue** and is authored by *Halvor Her-*

lyng, Marius Causemann (both Simula Research Laboratory, Norway), *Gaute Einevoll* (Norwegian University of Life Sciences), *Ada Ellingsrud* (Simula Research Laboratory, Norway) *Geir Halnes* (Norwegian Artificial Intelligence Research Consortium) and *Marie Rognes* (Simula Research Laboratory and K.G. Jebsen Centre for Brain Fluid Research, Norway, Woudschoten lecturer in 2024). The chapter describes the construction of multi-scale, high-resolution simulations of cerebral tissues, and it presents the development of a detailed, anatomically correct model of neurons and astrocytes and their extracellular space from microscopic data. The chapter contains a wealth of information and an extensive bibliography.

As editors, we hope that you can use this book as a reference on one or more of the aforementioned eight subareas in numerical analysis and scientific computing, or as a source of inspiration for follow-up research, or even for teaching.

To conclude, a word of thanks to all who contributed to the realization of this book, first and foremost all authors for their diligence and patience in writing and revising their chapters. We also want to thank all external experts, who reviewed one or sometimes even two chapters, for their help in further increasing the scientific quality of the book. At publisher Springer we want to thank *Martin Peters* and *Leonie Kunz* for professionally and smoothly guiding the entire publication process. We also thank external free-lance LaTeX-expert *Barnaby Sheppard*, brought in by Springer, to help in further improving the LaTeX-type-setting of the book.

Eindhoven and Leuven,
2026

Barry Koren
Dirk Roose

Contents

Metric-Driven Numerical Methods

Patrick Henning, Laura Huynh and Daniel Peterseim

Abstract In this work, we explore the concept of *metric-driven numerical methods* as a powerful tool for solving various types of multiscale partial differential equations. Our focus is on computing constrained minimizers of functionals – or, equivalently, by considering the associated Euler–Lagrange equations – the solution of a class of eigenvalue problems that may involve nonlinearities in the eigenfunctions. We introduce metric-driven methods for such problems via Riemannian gradient techniques, leveraging the idea that gradients can be represented in different metrics (so-called *Sobolev gradients*) to accelerate convergence. We show that the choice of metric not only leads to specific metric-driven iterative schemes, but also induces approximation spaces with enhanced properties, particularly in low-regularity regimes or when the solution exhibits heterogeneous multiscale features. In fact, we recover a well-known class of multiscale spaces based on the *Localized Orthogonal Decomposition* (LOD), now derived from a new perspective. Alongside a discussion of the metric-driven approach for a model problem, we also demonstrate its application to simulating the ground states of spin-orbit-coupled Bose–Einstein condensates.

1 Introduction

Solving partial differential equations (PDEs) with a complex underlying structure, such as multiscale variations or nonlinear features, can be challenging. Standard numerical approaches often require extremely fine computational meshes to resolve all structures, hence resulting in high-dimensional approximation spaces. At the same time, iterative

---

Patrick Henning and Laura Huynh
Department of Mathematics, Ruhr-University Bochum, DE-44801 Bochum, Germany, patrick.henning@rub.de, laura.huynh@rub.de

Daniel Peterseim
Institut für Mathematik & Centre for Advanced Analytics and Predictive Sciences (CAAPS), University Augsburg, DE-86159 Augsburg, Germany, daniel.peterseim@uni-a.de

B. Koren and D. Roose (eds.), *Numerical Analysis and Scientific Computing: an Anthology*, Lecture Notes in Computational Science and Engineering 156,
https://doi.org/10.1007/978-3-032-33428-2_1

solvers for handling the nonlinearities can suffer from very slow convergence. In such cases it is possible to take the structure of the PDE into account to construct problem-specific numerical methods.

For example, classical approaches for multiscale PDEs are based on enriching the shape functions of standard approximation spaces by multiscale features. This can be achieved by solving local problems based on the differential operator. The corresponding local solutions naturally contain structural information about the PDE and can then be used to correct/enrich the original shape functions. The resulting approximation spaces are called multiscale spaces, they are typically of low dimension and have very good approximation properties with respect to the solution of the underlying PDE. There is a vast landscape of different approaches how this can be realized, where we exemplarily mention the heterogeneous multiscale method (HMM) [1, 23], the multiscale finite element method (MsFEM) [42, 25], the generalized MsFEM (GMsFEM) [19, 24], the multiscale spectral generalized finite element method (MS-GFEM) [8, 9, 49, 50], gamblet-based methods [55, 56], Variational Multiscale Methods (VMM) [43, 44, 47, 60] or the localized orthogonal decomposition (LOD) [4, 51, 53]. In this contribution, we shall especially focus on the LOD methodology where we will elaborate on why this class of multiscale methods can be seen as a metric-driven approach, obtained by an "energy metric" which is induced by the differential operator.

Besides the construction of approximation spaces, problem-specific information can also be used for the design of iterative methods. A natural application is the use of gradient methods to find minimizers of functionals E. Gradient methods are based on taking a step, from a current location u^0, into the direction of the (classical) steepest descent $-E'(u^0)$ to get closer to the unknown minimizer. However, the notion of the steepest descent heavily depends on the metric in which we measure "steepness". Even though this is often not explicitly mentioned, classical gradient methods are based on the canonical L^2-metric. On the contrary, alternative metrics can accelerate the gradient descent in various applications, cf. [6, 18, 20, 21, 33, 63, 65], where the representation of gradients in different metrics is formalized by the concept of Sobolev gradients [54]. By choosing the metric adaptively depending on the operator $E'(v)$ (or linearized versions of it), additional problem-specific information can be incorporated to select an improved descent direction. This strategy of a metric-driven gradient descent was first proposed in [38].

So far, both paths of metric-driven numerical methods (i.e., the construction of problem-dependent approximation spaces and iterative solvers) were treated independently in the literature as essentially two disjoint fields. In this work we want to establish new links between the seemingly disconnected paths. Starting from a (constrained) minimization problem, we discuss how a particular energy metric in the spirit of [38] gives simultaneously rise to a metric-driven gradient method and a corresponding metric-driven approximation space. As for the approximation space, we recover the aforementioned LOD construction, but derived from a new perspective. We first describe the construction for a simple quadratic minimization problem and then discuss its generalization to a more complicated minimization problem of Gross–Pitaevskii type. Finally, to show that the ideas are not restricted to model problems, we also present an application in quantum physics where we compute the ground states of so-called pseudo-spin-1/2 Bose–Einstein condensates.

Before sketching the outline of this paper, it is important to distinguish our notion of metric-driven methods from the concept of "metric-based upscaling" [57] for elliptic PDEs $\mathcal{A}u := -\nabla \cdot (A\nabla u) = f$. The latter is based on representing the PDE in an $\mathcal{A}$-harmonic coordinate system in which heterogeneous solutions become smooth. This is not related to our approach.

Outline. In Section 2 we introduce the concept of metric-driven discretizations for a quadratic minimization problem with an L^2-normalization constraint. The generalization to minimization problems with higher order polynomial contributions is given in Section 3. Finally, in Section 4 we present the application to spin-orbit coupled Bose–Einstein condensates where we formulate and analyze a corresponding metric-driven iterative method which is afterwards illustrated in corresponding numerical experiments.

2 The Concept of Metric-Driven Discretizations

To introduce the concept of metric-driven discretizations, we consider a simple model problem that seeks the smallest eigenvalue and a corresponding eigenfunction of a linear elliptic differential operator, potentially involving rapidly varying multiscale coefficients. This simplified setting allows us to illustrate the basic ideas of metric-driven schemes before turning to more involved nonlinear problems later in the chapter.

2.1 A Model Problem

In this section we will briefly introduce the setting of the model problem and why it is suitable in this context.

2.1.1 Motivation

The model problem considered in this section serves as a simple setting in which the main ideas of metric-driven schemes can be illustrated transparently. Many PDE problems of practical interest admit a variational formulation in which a suitable energy functional is minimized under an L^2-normalization constraint. This is the case, for instance, in the computation of principal eigenpairs of elliptic operators arising in models of diffusion, wave propagation, and other physical or engineering applications.

From an algorithmic point of view, such problems are typically solved by iterative methods. Two fundamental design choices then arise: first, how the infinite-dimensional problem is discretized in space, and second, how the resulting minimization problem is solved iteratively. The key observation underlying metric-driven methods is that both aspects are strongly influenced by the choice of the metric in which gradients are represented. While the minimizer itself is independent of this choice, the metric determines the descent direction and therefore the geometry and efficiency of the iteration. In the following subsections we illustrate this idea for a simple elliptic model problem.

2.1.2 Analytical Setting

In this subsection we introduce the analytical setting of the model problem. For that, let $\mathcal{D} \subset \mathbb{R}^d$ ($d \in \{1, 2, 3\}$) denote a bounded Lipschitz domain with polygonal boundary and let

$$A \in L^\infty(\mathcal{D}, \mathbb{R}^{d\times d}) \qquad \text{and} \qquad V \in L^\infty(\mathcal{D}, \mathbb{R}_{\geq 0})$$

be two given coefficient functions, where the matrix-valued coefficient $A(x)$ is assumed to be symmetric for almost all $x \in \Omega$ and uniformly elliptic, i.e., $\sup_{\xi\in\mathbb{R}^d\setminus\{0\}} \frac{A\xi\cdot\xi}{|\xi|^2} \geq \alpha > 0$ for some positive constant α.

With this, we consider the functional $E : H^1_0(\mathcal{D}) \to \mathbb{R}_{\geq 0}$ given by

$$E(u) := \frac{1}{2}\int_{\mathcal{D}} |A^{1/2}\nabla u|^2 + V|u|^2 \,\mathrm{d}x$$

and we seek a minimizer u^* of E on $H^1_0(\mathcal{D})$ under the normalization constraint $\|u^*\|_{L^2(\mathcal{D})} = 1$. For brevity, we define the corresponding constraint manifold (L^2-sphere in $H^1_0(\mathcal{D})$) by

$$\mathbb{S} := \{v \in H^1_0(\mathcal{D}) \mid \|v\|_{L^2(\mathcal{D})} = 1\}$$

such that we can write the relevant minimizers compactly as

$$u^* = \arg\min_{v\in\mathbb{S}} E(v). \tag{1}$$

It is well-known (cf. [14]) that the minimizer exists, that it is unique up to sign and that it is either strictly positive or strictly negative in the interior $\mathcal{D}$. Problems of this type arise in many applications involving elliptic operators, including diffusion and conductivity problems, vibration analysis, and various models in physics and engineering.

The problem (1) can be equivalently expressed as an eigenvalue problem. In fact, by introducing a Lagrange multiplier $\lambda \in \mathbb{R}$ for the constraint $u^* \in \mathbb{S}$, the corresponding Euler–Lagrange equations $E'(u^*) = \lambda\,(u^*, \cdot)_{L^2(\mathcal{D})}$ take the form of a linear eigenvalue problem. Computing the Fréchet derivative $E'(u)$ in a straightforward manner, the resulting eigenvalue problem hence seeks an L^2-normalized eigenfunction $u^* \in H^1_0(\mathcal{D})$ and corresponding eigenvalue $\lambda \in \mathbb{R}$ such that

$$(A\nabla u^*, \nabla v)_{L^2(\mathcal{D})} + (Vu^*, v)_{L^2(\mathcal{D})} = \lambda\,(u^*, v)_{L^2(\mathcal{D})} \qquad \text{for all } v \in H^1_0(\mathcal{D}). \tag{2}$$

In strong form, the eigenvalue problem can be written as

$$\mathcal{L}u^* := -\nabla \cdot (A\nabla u^*) + Vu^* = \lambda u^*,$$

as an identity of L^2-functions. Since $\mathcal{L}$ represents a linear elliptic differential operator, it is well-known that an eigenfunction to the *smallest eigenvalue* λ (of $E'(u^*) = \lambda (u^*, \cdot)_{L^2(\mathcal{D})}$, or equivalently, $\mathcal{L}u^* = \lambda u^*$) is precisely a minimizer to (1). Furthermore, the eigenvalue λ is positive and simple.

Finding a minimizer of E, or respectively the smallest eigenvalue of the linear operator E', is a classical problem that can be solved with various standard techniques. We will not discuss these standard techniques here since the whole purpose of this section is an introduction to the concept of metric-driven schemes in a simple setting. For that, we have to distinguish two aspects of the discretization of (2): an iterative solver for computing a relevant eigenpair and a suitable spatial discretization of each iteration step. These two aspects are addressed individually in the next two subsections.

2.2 Metric-Driven Steepest Descents and Riemannian Sobolev Gradients

Recalling that problems (1) and (2) are equivalent for the smallest eigenvalue, we can take any of the two perspectives. In this subsection, the energy minimization perspective will be deployed.

To find a constrained minimizer of E in the sense of (1), it is possible to consider a corresponding *Riemannian* gradient flow on the (Hilbert) manifold $\mathbb{S}$. Such a gradient flow follows the direction of the Riemannian steepest descent on $\mathbb{S}$. Usually, steepest descent is measured with respect to the L^2-metric. However, alternative metrics can lead to significantly faster energy decay, which can be exploited in the construction of efficient numerical methods.

Before introducing Sobolev gradients (which depend on the chosen metric), let us briefly clarify the notation for derivatives of functionals. For a functional $E : H_0^1(\mathcal{D}) \to \mathbb{R}$, we write $E'(u)$ for its Fréchet derivative at u. Thus $E'(u)$ is a continuous linear functional acting on perturbations $v \in H_0^1(\mathcal{D})$, and the expression $\langle E'(u), v\rangle$ denotes the first variation of E at u in direction v, i.e., the canonical duality pairing between the functional $E'(u)$ and the perturbation v.

In classical optimization in $\mathbb{R}^n$, the derivative of a function E at a point x can be identified with a vector, namely the gradient $\nabla E(x)$, which points in the direction of steepest ascent. Consequently, $-\nabla E(x)$ points in the direction of steepest descent. In infinite-dimensional settings this identification is no longer automatic: the derivative $E'(u)$ is a linear functional and therefore does not directly define a descent direction. To obtain a direction of steepest descent, one has to represent this functional by an element of a Hilbert space using a suitable inner product. The choice of this inner product determines how steepness of the energy landscape is measured and therefore influences the resulting descent direction. We will make this precise in the following subsection.

2.2.1 Sobolev Gradients

The role of the metric can be understood through Sobolev gradients ∇_X of functionals E, cf. [54]. Loosely speaking, Sobolev gradients allow to characterize what we mean by steepest descent (or steepest ascent) on the graph of E when "steepness" is measured in different metrics. This is a relevant practical aspect of gradient descent methods since the choice of the metric can have a major influence on how fast the gradient descent approaches a minimizer. In particular, the metric can be viewed as a built-in preconditioner: it does not change the minimizer itself, but it may significantly change the geometry of the descent path and hence the efficiency of the iterative method.

Let us now make the definition of Sobolev gradients more precise. Given a Hilbert space X with some inner product $(\cdot,\cdot)_X$ (which represents a selected metric) the X-Sobolev gradient $\nabla_X E(u)$ of the functional E in some point $u \in H^1_0(\mathcal{D})$ is given by the Riesz-representation of the "regular gradient" $E'(u)$ in the space X, that is $\nabla_X E(u) \in X$ solves

$$(\nabla_X E(u), v)_X = \langle E'(u), v\rangle \qquad \text{for all } v \in X. \tag{3}$$

At first glance, the above definition only makes sense if $X \subset H^1_0(\mathcal{D})$ such that $\langle E'(u), v\rangle$ is well defined for $v \in X$. However, note that this can be compensated by the smoothness of u and/or A. For example, if $u \in H^1_0(\mathcal{D})$ is such that $\nabla \cdot (A\nabla u) \in L^2(\mathcal{D})$, then $X = L^2(\mathcal{D})$ is admissible and we obtain (by integration by parts) the identification

$$\nabla_{L^2} E(u) \;=\; -\nabla \cdot (A\nabla u) + Vu \;\in\; L^2(\mathcal{D}).$$

There are various choices for X, where we refer to [20, 21, 33, 46] for corresponding examples and numerical studies. A particular choice was suggested in [38], motivated by the idea to select the metric in an optimal way such that the Sobolev gradient becomes the identity, i.e., such that $\nabla_X E(u) = u$. For general functionals E, there is no fixed metric $(\cdot,\cdot)_X$ that fulfills this property for arbitrary points $u \in \mathbb{S}$, which is why such a metric needs to adaptively change depending on the location on the manifold $\mathbb{S}$. We will elaborate more on this aspect in Section 3. However, in the simplified setting of this section, E' is a linear operator and a suitable universal metric on $H^1_0(\mathcal{D})$ is canonically given by the inner-product

$$(u, v)_X \;:=\; \langle E'(u), v\rangle \;=\; (A\nabla u, \nabla v)_{L^2(\mathcal{D})} + (V u, v)_{L^2(\mathcal{D})}. \tag{4}$$

Indeed, by definition we have $(\nabla_X E(u), v)_X = \langle E'(u), v\rangle = (u, v)_X$ for all $v \in H^1_0(\mathcal{D})$ and hence $\nabla_X E(u) = u$ for all $u \in H^1_0(\mathcal{D})$. As we will exploit the above metric repeatedly, we introduce the corresponding solution operator $\mathcal{L}^{-1} : H^1_0(\mathcal{D}) \to H^1_0(\mathcal{D})$ as $\mathcal{L}^{-1}u \in H^1_0(\mathcal{D})$ solving

$$(\mathcal{L}^{-1}u, v)_X = (u, v)_{L^2(\mathcal{D})} \qquad \text{for all } v \in H^1_0(\mathcal{D}), \tag{5}$$

which effectively gives us $E'(\mathcal{L}^{-1}u) = u = \nabla_X E(u)$.

2.2.2 Riemannian Sobolev Gradients

Since we are not interested in the general steepest descent of E in $u \in \mathbb{S}$, but only in the steepest descent among all admissible directions that keep us on the manifold $\mathbb{S}$, we need to consider Riemannian Sobolev gradients, i.e., the projection of Sobolev gradients onto the tangent space of the manifold given by

$$T_u\mathbb{S} = \{v \in H^1_0(\mathcal{D}) \mid (u, v)_{L^2(\mathcal{D})} = 0\}.$$

While the metric-dependency already enters through the Sobolev gradient, it also determines how this gradient is projected onto the tangent space via the X-orthogonal projection $P_{u,X} : X \to T_u\mathbb{S}$, that is, the projection $P_{u,X}(v) \in T_u\mathbb{S}$ that fulfills

$$(P_{u,X}(v), w)_X = (v, w)_X \qquad \text{for all } w \in T_u\mathbb{S}. \tag{6}$$

This projection ensures that the descent direction remains feasible, i.e., it preserves the constraint $u \in \mathbb{S}$ while retaining the steepest descent property with respect to the X-metric. With this, the Riemannian X-Sobolev gradient of E in $u \in \mathbb{S}$ is given by $(P_{u,X} \circ \nabla_X E)u$ and, consequently, $-(P_{u,X} \circ \nabla_X E)u$ defines the Riemannian steepest descent of E in u with respect to the X-metric.

We can now apply these considerations to the particular choice of X defined in (4). For $u \in \mathbb{S}$, the projection $P_{u,X}$ is then explicitly given by

$$P_{u,X}(v) = v - \frac{(u, v)_{L^2(\mathcal{D})}}{(\mathcal{L}^{-1}u, u)_{L^2(\mathcal{D})}} \mathcal{L}^{-1}u.$$

In fact, $(P_{u,X}(v), u)_{L^2(\mathcal{D})} = 0$ and $(P_{u,X}(v), w)_X = (v, w)_X$ for all $w \in T_u\mathbb{S}$ because of $(\mathcal{L}^{-1}u, w)_X = (u, w)_{L^2(\mathcal{D})} = 0$. Furthermore, recalling $\nabla_X E(u) = u$, we obtain the corresponding Riemannian X-Sobolev gradient as

$$(P_{u,X} \circ \nabla_X E)u = u - (\mathcal{L}^{-1}u, u)^{-1}_{L^2(\mathcal{D})} \mathcal{L}^{-1}u. \tag{7}$$

2.2.3 Metric-Driven Gradient Flow

We are now ready to formulate a Riemannian Sobolev gradient flow for the minimization problem (1). It is obtained by following the Riemannian steepest descent in the X-metric, i.e., it is of the form

$$u'(t) = -(P_{u(t),X} \circ \nabla_X E)u(t) \qquad \text{for } t \geq 0$$

and some suitable initial value $u(0) = u_0 \in \mathbb{S}$. For the choice $(u, v)_X = \langle E'(u), v\rangle$ as in (4) we obtain with the representation (7) that the gradient flow $u \in C^1(0, \infty; \mathbb{S})$ fulfills

$$u'(t) = -u(t) + (\mathcal{L}^{-1}u(t), u(t))^{-1}_{L^2(\mathcal{D})} \mathcal{L}^{-1}u(t). \tag{8}$$

Well-posedness and energy dissipation $E(u(t)) \leq E(u(s))$ for any $s \leq t$ can be established as in [38]. Furthermore, for all suitable initial values $u_0 \in \mathbb{S}$ (which typically have to be non-negative), the flow converges exponentially fast to a minimizer u^* of E, i.e., for all $\delta > 0$ and all sufficiently large times $t \geq t_\delta$

$$\|u^* - u(t)\|_{H^1(\mathcal{D})} \leq c_\delta \exp(-(1 - \tfrac{\lambda_1}{\lambda_2} - \delta)t), \tag{9}$$

where $c_\delta > 0$ is a generic constant that depends on δ and where $\lambda_1 = \lambda > 0$ is the smallest and $\lambda_2 > \lambda_1$ the second smallest eigenvalue of (2).

2.2.4 Metric-Driven Riemannian Gradient Method

Motivated by the fast decay of the gradient flow (8) to a minimizer of E on $\mathbb{S}$, it is natural to build numerical methods based on corresponding time integrations. Note that t is just a pseudo-time and that we are not interested in the time evolution but only in the limiting state for $t \to \infty$. Consequently, a simple forward Euler time integration of (8) is sufficient. For that, we let $u^0 \in \mathbb{S}$ denote a starting value with $u_0 \geq 0$ and $\tau_n > 0$ a sequence of (pseudo) time step sizes. For $n \geq 0$, the iterations read

$$\begin{aligned} u^{n+1}_{\text{prel}} &:= (1 - \tau_n)\, u^n + \tau_n\, (\mathcal{L}^{-1}u^n, u^n)^{-1}_{L^2(\mathcal{D})}\, \mathcal{L}^{-1}u^n, \\ u^{n+1} &:= \frac{u^{n+1}_{\text{prel}}}{\|u^{n+1}_{\text{prel}}\|_{L^2(\mathcal{D})}}, \end{aligned} \tag{10}$$

where we normalize after each iteration step to respect the constraint $u^{n+1} \in \mathbb{S}$, which is not automatically fulfilled by a pure Euler step. Note that the iterations can be equivalently written as

$$u^{n+1} = \frac{u^n - \tau_n (P_{u^n,X} \circ \nabla_X E) u^n}{\|u^n - \tau_n (P_{u^n,X} \circ \nabla_X E) u^n\|_{L^2(\mathcal{D})}}, \tag{11}$$

where $(P_{u^n,X} \circ \nabla_X E)u^n$ denotes the Riemannian Sobolev gradient in the X-metric from (4). Hence, the scheme has a natural interpretation as a Riemannian Sobolev gradient method if the step size is chosen such that the energy reduction per iteration is maximized, that is,

$$\tau_n := \operatorname*{arg\,min}_{0<\tau\leq 2} E\left(\frac{(1-\tau)\, u^n + \tau\, (\mathcal{L}^{-1}u^n, u^n)^{-1}_{L^2(\mathcal{D})}\, \mathcal{L}^{-1}u^n}{\|(1-\tau)\, u^n + \tau\, (\mathcal{L}^{-1}u^n, u^n)^{-1}_{L^2(\mathcal{D})}\, \mathcal{L}^{-1}u^n\|_{L^2(\mathcal{D})}} \right).$$

The constraint $\tau_n < 2$ in the minimization is analytically justified by the observation that the iterations cannot decrease the energy if $\tau_n \geq 2$. On the contrary, it can be proved, that the iterations are guaranteed to reduce the energy for all sufficiently small step sizes τ_n and that they converge to the unique positive minimizer of problem (1). We summarize these properties in the following theorem and refer to [38] for a proof.

Theorem 1 *In the general setting of this section, let $u^0 \in \mathbb{S}$ denote a nonnegative starting value and consider the iterations given by* (10). *Then there exists a step size interval* $[\tau_{\min}, \tau_{\max}] \subset (0, 2)$ *such that for all* $\tau_n \in [\tau_{\min}, \tau_{\max}]$

$$E(u^{n+1}) \le E(u^n) \quad \text{for all } n \ge 0 \qquad \text{and} \qquad \lim_{n\to\infty} \|u^n - u^*\|_{H^1(\mathcal{D})} = 0,$$

where $u^* \in \mathbb{S}$ is the unique positive minimizer of problem (1). *The minimum bound $\tau_{\min}$ for the step size can be arbitrarily close to zero, it is just important that the step sizes do not degenerate.*

The theorem fully justifies the applicability of the metric-driven scheme to the linear model problem (10). However, there is another interesting aspect, which relates the scheme to a classical solution algorithm for linear eigenvalue problems: the inverse power iteration. In fact, selecting uniformly $\tau_n = 1$ in (10), we recover the scheme

$$u^{n+1} = \frac{\mathcal{L}^{-1}u^n}{\|\mathcal{L}^{-1}u^n\|_{L^2(\mathcal{D})}}. \tag{12}$$

The above inverse iteration converges, for any nonnegative starting value, globally to an L^2-normalized eigenfunction u^* belonging to the smallest eigenvalue λ of $\mathcal{L}u^* = \lambda u^*$, i.e., eigenvalue problem (2). As mentioned before, this eigenfunction coincides with the unique positive minimizer of the energy E on $\mathbb{S}$ and we recover the global convergence predicted in Theorem 1. This shows that the gradient flow perspective and the eigenvalue problem perspective are closely related, also in terms of numerical methods. In fact, in the light of the above discussion, the gradient method (10) can be viewed as an inverse power iteration with damping, where τ_n takes the role of a damping parameter.

Furthermore, we can recall a classical result for the inverse power iteration, which states that the convergence rate is linear and depends on the first spectral gap of the operator $\mathcal{L}$. Applied to our setting, we obtain that the iterates (12) fulfill

$$\|u^* - u^n\|_{H^1(\mathcal{D})} \lesssim |\tfrac{\lambda_1}{\lambda_2}|^n \, \|u^* - u^0\|_{H^1(\mathcal{D})}, \tag{13}$$

where $u^0 \in \mathbb{S}$ is a nonnegative starting value and u^* the positive L^2-normalized eigenfunction to the smallest eigenvalue $\lambda_1 = \lambda$ of $\mathcal{L}$. As before, $\lambda_2 > \lambda_1$ denotes the second smallest eigenvalue of $\mathcal{L}$. Hence, the larger the spectral gap between λ_1 and λ_2, the faster the convergence.

The estimate (13) does not only predict that the convergence of the X-Sobolev gradient method is heavily influenced by the size of the spectral gap $\frac{\lambda_1}{\lambda_2}$, but we also recover the same rate for the continuous gradient flow $u(t)$ given by (8) which approaches u^* asymptotically with the rate $\exp(-(1 - \frac{\lambda_1}{\lambda_2})t)$ (cf. (9)).

The real potential of metric-driven steepest descents unfolds for more general energy minimization problems for which the Euler–Lagrange equations become nonlinear and standard approaches are no longer applicable. Before turning towards such more complicated problems, we stay in the linear setting and consider a second important aspect of a practical numerical method, that is, the spatial discretization of the ideal iterations (10). This is addressed in the next subsection.

2.3 Minimization in Metric-Driven Approximation Spaces

In the next step, we want to discuss a spatial discretization of the gradient method (10) in a way that it respects the particular metric-dependent structure.

In the following, we let $\mathcal{T}_H$ denote a quasi-uniform and shape-regular triangulation of the polygonal computational domain $\mathcal{D} \subset \mathbb{R}^d$. On $\mathcal{T}_H$, we introduce the classical $\mathbb{P}^1$-Lagrange finite element space of H^1-conforming and $\mathcal{T}_H$-piecewise linear functions by

$$V_H \;:=\; \{v_H \in C^0(\overline{\mathcal{D}}) \mid v_H|_T \in \mathbb{P}^1(T) \text{ for all } T \in \mathcal{T}_H\}. \tag{14}$$

2.3.1 Derivation of a Metric-Driven Approximation Space

Rather than using a standard finite element space, such as V_H given in (14), whose approximation properties typically rely on additional smoothness of the minimizer u^*, we aim to construct an approximation space that is adapted to the problem through the metric

$$(u, v)_X \;:=\; (A\nabla u, \nabla v)_{L^2(\mathcal{D})} + (V u, v)_{L^2(\mathcal{D})}.$$

Recall that the corresponding metric-driven gradient descent (10) computes, as a first step,

$$u^1_{\text{prel}} \;=\; (1-\tau)u^0 \;+\; \tau\,(\mathcal{L}^{-1}u^0, u^0)^{-1}_{L^2(\mathcal{D})}\, \mathcal{L}^{-1}u^0,$$

followed by normalization $u^1 := \frac{u^1_{\text{prel}}}{\|u^1_{\text{prel}}\|_{L^2(\mathcal{D})}}$. Choosing $\tau = 1$ and taking $u^0 \in V_H$ as a classical finite element yields

$$u^1 \;=\; \frac{\mathcal{L}^{-1}u^0}{\|\mathcal{L}^{-1}u^0\|_{L^2(\mathcal{D})}}. \tag{15}$$

Thus, the iterate u^1 incorporates problem information through the metric $(\cdot,\cdot)_X$, since the update involves the induced solution operator $\mathcal{L}^{-1}$. It also coincides with the result of the first step of the inverse iteration (12). According to the error estimate (13), the approximation of the minimizer by u^1 therefore improves compared to the approximation by u^0 by a factor λ_1/λ_2.

An even better estimate can be achieved for a suitable choice of the initial function u^0, as can be seen from a block inverse iteration argument. Let $\{u_j\}_{j\geq 1}$ denote an $L^2(\mathcal{D})$-orthonormal basis of eigenfunctions of $\mathcal{L}$ with the corresponding eigenvalues $0 < \lambda_1 < \lambda_2 \leq \lambda_3 \leq \ldots$. For any $u^0 \in V_H$ with spectral expansion

$$u^0 = \sum_{j=1}^{\infty} c_j u_j,$$

the update (15) yields, up to normalization,

$$u^1 = \mathcal{L}^{-1}u^0 = \sum_{j=1}^{\infty} \frac{c_j}{\lambda_j} u_j.$$

For $c_1 \neq 0$, which is for example fulfilled for any non-negative $u^0 \not\equiv 0$, we can factor out the leading coefficient to obtain

$$u^1 = \frac{c_1}{\lambda_1}\left(u_1 + \sum_{j=2}^{\infty} \frac{c_j}{c_1}\frac{\lambda_1}{\lambda_j} u_j\right) = \frac{c_1}{\lambda_1}\left(u^* + \sum_{j=2}^{\infty} \frac{c_j}{c_1}\frac{\lambda_1}{\lambda_j} u_j\right).$$

After appropriate scaling, the common prefactor c_1/λ_1 cancels and we obtain

$$u^1_{\text{scaled}} - u^* = \sum_{j=2}^{\infty} \frac{c_j}{c_1}\frac{\lambda_1}{\lambda_j} u_j, \tag{16}$$

where $u^1_{\text{scaled}} := \frac{\lambda_1}{c_1}u^1$ differs from u^1 only by a scalar factor. In particular, both functions yield the same L^2-normalized approximation, i.e.

$$\frac{u^1_{\text{scaled}}}{\|u^1_{\text{scaled}}\|_{L^2(\mathcal{D})}} = \frac{u^1}{\|u^1\|_{L^2(\mathcal{D})}}.$$

Hence, in the light of (16), all higher modes $j \geq 2$ represent the error with respect to u^*, and their contribution is damped relative to the leading mode by the factor λ_1/λ_j. In particular, the smallest index $j \geq 2$ with $c_j \neq 0$ determines the dominant error contribution.

In general, we can choose $u^0 \in V_H$ such that its spectral expansion satisfies $c_j = 0$ for $2 \leq j \leq N_* - 1$, where $N_* \simeq N := \dim V_H$ is of the order of the dimension of the finite element space. For such a choice, the dominant error contribution after one inverse iteration stems from the N_*-th eigenfunction and, therefore,

$$\|u^* - u^1\|_{H^1(\mathcal{D})} \lesssim \frac{\lambda_1}{\lambda_{N_*}}.$$

In summary, among all possible starting values $u^0 \in V_H$ there exists one $u^0 \in \mathbb{S}$ with $(u^0, u^*)_{L^2(\mathcal{D})} > 0$ and

$$\|u^* - u^1\|_{H^1(\mathcal{D})} \lesssim \frac{\lambda_1}{\lambda_{N_*}},$$

where the hidden constant only depends on the angle $(u^0, u^*)_{L^2(\mathcal{D})}$. Weyl asymptotics for uniformly elliptic second-order operators,

$$\lambda_{N_*} \simeq \lambda_N \simeq N^{2/d} \quad \text{for large } N,$$

suggests the scaling

$$\|u^* - u^1\|_{H^1(\mathcal{D})} \lesssim N^{-2/d}.$$

Since $N \simeq H^{-d}$ for quasi-uniform meshes, this formally yields

$$\|u^* - u^1\|_{H^1(\mathcal{D})} \lesssim H^2.$$

We refer to Theorem 2 below for a rigorous non-asymptotic error bound. In other words, the metric-driven approximation space

$$V_H^{\mathcal{L}} := \mathcal{L}^{-1} V_H$$

approximates u^* uniformly well, independently of additional smoothness assumptions in u^* (whereas extra smoothness $u^* \in H^{1+s}(\mathcal{D})$, $0 < s \leq 1$, would improve the rate to H^{2+s}). Thus, $V_H^{\mathcal{L}} = \mathcal{L}^{-1} V_H$ can be interpreted as an enrichment of the standard space V_H by problem-specific information incorporated in $\mathcal{L}$. Indeed, for any $v_H \in V_H$, the function $\mathcal{L}^{-1} v_H$ is obtained as the solution of an elliptic problem with v_H as a right-hand side and therefore reflects the fine-scale structure of the operator. This explains why the space can capture multiscale features even though its dimension coincides with that of V_H.

In conclusion, the above considerations suggest a spatial discretization in which we do not minimize the energy over the standard FE space V_H, but rather the metric-driven space $V_H^{\mathcal{L}}$ to obtain

$$u_H^{\mathcal{L}} = \underset{v \in V_H^{\mathcal{L}} \cap \mathbb{S}}{\arg\min}\; E(v) \tag{17}$$

as an approximation to an exact minimizer $u^* \in \mathbb{S}$.

2.3.2 Connections to LOD and Approximation Properties

The construction above shows that the application of the metric-induced solution operator naturally transforms a standard finite element space into an enriched space that incorporates problem-specific information of the operator $\mathcal{L}$. This observation provides a conceptual link between metric-driven gradient methods and multiscale approximation spaces.

In fact, the space

$$V_H^{\mathcal{L}} = \mathcal{L}^{-1} V_H$$

coincides with a well-established construction in numerical multiscale analysis, namely the Localized Orthogonal Decomposition (LOD) space [51, 4, 53]. While LOD spaces are typically derived via orthogonal decompositions and corrector problems, it was observed in [31] that they admit the equivalent representation $\mathcal{L}^{-1} V_H$, which is exploited there for the design of suitable strategies for computing localized basis functions of the

LOD space. From the present perspective, this representation reveals that LOD spaces can be interpreted as metric-driven approximation spaces.

For completeness, we briefly sketch the classical LOD representation of $V_H^{\mathcal{L}}$ in the following lemma and afterwards discuss the practical implications.

Lemma 1 *Let $P_H : H_0^1(\mathcal{D}) \to V_H$ denote the L^2-projection given by*

$$(P_H(u), v_H)_{L^2(\mathcal{D})} = (u, v_H)_{L^2(\mathcal{D})} \qquad \text{for all } v_H \in V_H$$

and define the kernel of P_H as the detail space

$$W := \{ v \in H_0^1(\mathcal{D}) \mid P_H(v) = 0 \}.$$

With this, the corrector operator $C : V_H \to W$ is given, for $v_H \in V_H$, by $Cv_H \in W$ with

$$(Cv_H, w)_X = (v_H, w)_X \qquad \text{for all } w \in W,$$

where $(v, w)_X = (A\nabla v, \nabla w)_{L^2(\mathcal{D})} + (V\, v, w)_{L^2(\mathcal{D})}$ (cf. (4)). The following holds

$$V_H^{\mathcal{L}} = (I - C)V_H \;=\; \{ v_H - Cv_H \mid v_H \in V_H \}.$$

Proof. The proof adopts the arguments from [31]. First of all, we note that W is a closed subspace of $H_0^1(\mathcal{D})$ because the kernel of the L^2-projection can be equivalently expressed as the kernel of an H^1-stable quasi-interpolation operator [17, 51]. Since W is closed, Lax–Milgram ensures existence and uniqueness of Cv_H for each v_H. Since $W \cap V_H = \{0\}$ we conclude that $\dim (I - C)V_H = \dim \mathcal{L}^{-1}V_H$. Hence, it only remains to show the inclusion $\mathcal{L}^{-1}V_H \subset (I - C)V_H$ to conclude that the spaces are identical. For this, consider $\mathcal{L}^{-1}v_H$ for arbitrary $v_H \in V_H$. By the L^2-orthogonality of V_H and W and the definition of $\mathcal{L}^{-1}$,

$$(\mathcal{L}^{-1}v_H, w)_X \;=\; (v_H, w)_{L^2(\mathcal{D})} \;=\; 0 \qquad \text{for all } w \in W.$$

Now, define $\tilde{v}_H := P_H(\mathcal{L}^{-1}v_H)$, then $P_H(\mathcal{L}^{-1}v_H - (I - C)\tilde{v}_H) = 0$ and hence we have that $\mathcal{L}^{-1}v_H - (I - C)\tilde{v}_H \in W$. Since both $(\mathcal{L}^{-1}v_H, w)_X = 0$ and $((I - C)\tilde{v}_H, w)_X = 0$ for any $w \in W$ we conclude that $\mathcal{L}^{-1}v_H = (I - C)\tilde{v}_H$ and therefore $\mathcal{L}^{-1}V_H \subset (I - C)V_H$. □

The classical LOD characterization of $V_H^{\mathcal{L}}$ as $(I - C)V_H$ is important to understand how to obtain a practical basis for the space. In fact, it can be proved that the Green's function associated with the corrector operator C shows an exponential decay in units of the mesh size H. This implies the existence of quasi-local basis functions that can be cheaply computed and used in practical computations. For example, consider a mesh vertex z and a standard nodal hat function $\phi_z \in V_H$, then a corresponding shape function in $V_H^{\mathcal{L}}$ is given by $(I - C)\phi_z$. This shape function shows an exponential decay outside of the nodal patch $\omega_z := \operatorname{supp} \phi_z$, which justifies to truncate it on a small neighborhood of ω_z. The set of all nodal shape functions $(I - C)\phi_z$ forms an ideal (global) basis of $V_H^{\mathcal{L}}$ and the set of suitably truncated shape functions forms an approximate basis that is localized. We call such a basis an LOD basis. For details on the practical realization of the truncation and the efficient algorithmic computation of LOD basis functions using correctors, we refer to [27]. Alternative superlocalization strategies that do not require correctors but instead exploit optimization techniques are presented in [31].

With the insight that $V_H^{\mathcal{L}}$ can be used in practice by suitable localization strategies, we now turn to its approximation properties, which were extensively studied in the context of multiscale problems. The following approximation result can be found for linear eigenvalue problems in [52] and in generalized form in [35, 36]. It is a rigorous extension of the informal derivation given in Section 2.3.1.

Theorem 2 *Let $u^* \in \mathbb{S}$ denote the exact minimizer of* (1) *and $u_H^{\mathcal{L}} \in V_H^{\mathcal{L}} \cap \mathbb{S}$ the corresponding minimizer in the metric driven space $V_H^{\mathcal{L}}$ according to* (17)*. Further assume that the signs of the two minimizers are consistent, i.e., $(u^*, u_H^{\mathcal{L}})_{L^2(\mathcal{D})} \geq 0$. Then, under the minimal regularity assumptions $A \in L^\infty(\mathcal{D}, \mathbb{R}^{d\times d})$ and $V \in L^\infty(\mathcal{D}, \mathbb{R}_{\geq 0})$,*

$$\|u^* - u_H^{\mathcal{L}}\|_{L^2(\mathcal{D})} + H\,\|u^* - u_H^{\mathcal{L}}\|_{H^1(\mathcal{D})} \lesssim H^3 \qquad \textit{and} \qquad |E(u^*) - E(u_H^{\mathcal{L}})| \lesssim H^4, \tag{18}$$

where the hidden constants only depend on the maximum values of A and V. If A is additionally Lipschitz-continuous, that is $A \in W^{1,\infty}(\mathcal{D}, \mathbb{R}^{d\times d})$, then we have $u^* \in H^2(\mathcal{D})$ and the estimate improves to

$$\|u^* - u_H^{\mathcal{L}}\|_{L^2(\mathcal{D})} + H\,\|u^* - u_H^{\mathcal{L}}\|_{H^1(\mathcal{D})} \lesssim H^4 \qquad \textit{and} \qquad |E(u^*) - E(u_H^{\mathcal{L}})| \lesssim H^6, \tag{19}$$

where the hidden constant now also depends on $\|A\|_{W^{1,\infty}(\mathcal{D})}$.
Note that the eigenvalue λ to the eigenfunction u^* is given by $\lambda = 2\,E(u^*)$, such that the energy estimates translate into equivalent estimates for the eigenvalue error.

Recalling that the dimension of the LOD space $V_H^{\mathcal{L}}$ and the standard $\mathbb{P}^1$-finite element space V_H are equal, we observe that the metric-driven choice increases the approximation properties tremendously. Let us discuss the estimates (18) and (19) individually, starting with the latter one. If $A \in W^{1,\infty}(\mathcal{D}, \mathbb{R}^{d\times d})$ such that the exact solution u^* admits full H^2-regularity, then (19) ensures that the metric-driven approximation is superconvergent with order $O(H^3)$ in the H^1-norm and even with order $O(H^4)$ in the L^2-norm. The energy/eigenvalue is even approximated with the rate $O(H^6)$. On the contrary, a classical $\mathbb{P}^1$ approximation $u_H \in V_H \cap \mathbb{S}$ will only achieve the standard convergence rates with

$$\|u^* - u_H\|_{L^2(\mathcal{D})} + H\,\|u^* - u_H\|_{H^1(\mathcal{D})} \lesssim H^2 \qquad \text{and} \qquad |E(u^*) - E(u_H)| \lesssim H^2,$$

which are significantly worse.

The improvements become even more pronounced when considering estimate (18). If the coefficients, especially A, are rough and rapidly varying (i.e., multiscale coefficients), then H^2-regularity for u^* is not available and rates for the classical $\mathbb{P}^1$-finite element method can degenerate to an arbitrarily slow convergence. Additionally, meaningful approximations in V_H are only obtained if the mesh size resolves all the fine-scale variations of A, which can impose a very strong restriction and requires very fine computational meshes. For that reason, solving the problem with a standard approach can become prohibitively expensive. On the other hand, the rates (18) obtained in the space $V_H^{\mathcal{L}}$ are not only still quadratic in the H^1-error (and cubic in the L^2-error), but they also do not require that the mesh size resolves the variations of the multiscale coefficients A and V. In fact, information about these variations is naturally built into the space $V_H^{\mathcal{L}}$ and coarse mesh sizes H can be used to get meaningful approximations.

We can conclude that the metric-driven space $V_H^{\mathcal{L}}$ overcomes the issues of classical finite element approximation spaces, especially for problems with multiscale coefficients and low regularity.

2.3.3 Combined Metric-Driven Gradient-Descent

In order to formulate a fully-discrete method that combines the metric-driven gradient descent with the corresponding metric-driven approximation space $V_H^{\mathcal{L}}$ we introduce a discrete approximation of $\mathcal{L}^{-1}$ from (5) by the operator

$$\mathcal{L}_H^{-1} : V_H^{\mathcal{L}} \to V_H^{\mathcal{L}}$$

with

$$(\mathcal{L}_H^{-1} u_H^{\mathcal{L}}, v_H^{\mathcal{L}})_X = (u_H^{\mathcal{L}}, v_H^{\mathcal{L}})_{L^2(\mathcal{D})} \qquad \text{for all } v_H^{\mathcal{L}} \in V_H^{\mathcal{L}}. \tag{20}$$

With this, we can formulate the metric-driven descent (10) directly in $V_H^{\mathcal{L}}$.

Definition 1 (Metric-driven discretization) Given $u_H^{\mathcal{L},n} \in V_H^{\mathcal{L}} \cap \mathbb{S}$ the next iterate $u_H^{\mathcal{L},n+1} \in V_H^{\mathcal{L}} \cap \mathbb{S}$ is given by

$$u_H^{\mathcal{L},n+1} := \frac{u_{H,\text{prel}}^{\mathcal{L},n+1}}{\|u_{H,\text{prel}}^{\mathcal{L},n+1}\|_{L^2(\mathcal{D})}},$$

where $u_{H,\text{prel}}^{\mathcal{L},n+1} \in V_H^{\mathcal{L}}$ is defined by

$$u_{H,\text{prel}}^{\mathcal{L},n+1} = (1-\tau_n)\, u_H^{\mathcal{L},n} + \tau_n\, (\mathcal{L}_H^{-1} u_H^{\mathcal{L},n}, u_H^{\mathcal{L},n})_{L^2(\mathcal{D})}^{-1}\, \mathcal{L}_H^{-1} u_H^{\mathcal{L},n}, \tag{21}$$

where $\mathcal{L}_H^{-1} u_H^{\mathcal{L},n}$ is the approximate Riesz representation of $u_H^{\mathcal{L},n}$ in the X-metric according to (20).

The above construction combines the two key ingredients of the metric-driven approach: on the one hand, the choice of the metric induces a problem-adapted approximation space $V_H^{\mathcal{L}}$, and on the other hand, it determines the form of the iterative scheme through the corresponding Sobolev gradients. In this sense, the discretization is fully consistent with the underlying metric structure, as both the search space and the descent direction are derived from the same operator-induced geometry.

From a practical point of view, this leads to an iterative scheme that can be interpreted as a projected inverse iteration within the multiscale space $V_H^{\mathcal{L}}$. Compared to classical approaches, where discretization and iteration are designed independently, the metric-driven formulation provides a unified framework in which both aspects are intrinsically coupled.

Each iteration of the scheme (21) requires the computation of a Riesz representative with respect to the X-metric restricted to $V_H^{\mathcal{L}}$, which corresponds to solving a linear problem in the multiscale space. Due to the LOD structure of $V_H^{\mathcal{L}}$, this step can be realized efficiently using localized basis functions, making the overall scheme computationally feasible.

3 Metric-Driven Discretizations for the Gross–Pitaevskii Equation

In the previous section, a quadratic minimization problem was considered, resulting in the approximation of an eigenfunction to the smallest eigenvalue of a linear eigenvalue problem. In this section, we want to consider a more complicated setup that involves a nonlinear eigenvalue problem and we want to sketch how our previous considerations generalize to such a setting and what we can expect in terms of approximation properties.

In the following we consider the so-called Gross–Pitaevskii functional which takes an important role in many physical applications, especially when describing the ground states (i.e., lowest energy states) of quantum systems [59]. To introduce the functional we make the same assumptions as in the previous section, i.e., $\mathcal{D} \subset \mathbb{R}^d$ (for $d \in \{1, 2, 3\}$) is a polygonal Lipschitz-domain, $A \in L^\infty(\mathcal{D}, \mathbb{R}^{d \times d})$ is an almost everywhere symmetric and uniformly elliptic coefficient, $V \in L^\infty(\mathcal{D}, \mathbb{R}_{\geq 0})$ is a nonnegative potential term and, additionally, $\beta \geq 0$ is a constant parameter (which usually models the strength of repulsive particle interactions).

With this, the Gross–Pitaevskii functional $E : H_0^1(\mathcal{D}) \to \mathbb{R}_{\geq 0}$ is given by

$$E(u) := \frac{1}{2} \int_{\mathcal{D}} |A^{1/2} \nabla u|^2 + V|u|^2 + \frac{\beta}{2}|u|^4 \, \mathrm{d}x$$

and we are again concerned with finding an L^2-normalized minimizer

$$u^* = \arg\min_{v \in \mathbb{S}} E(v). \tag{22}$$

The analytical properties of the problem do not change much compared to the linear setting in Section 2. In particular, the minimizer u^* in (22) is still unique up to sign and either strictly positive or strictly negative in the interior $\mathcal{D}$, cf. [14]. From the Euler–Lagrange equations $E'(u^*) = \lambda\, (u^*, \cdot)_{L^2(\mathcal{D})}$ and the formula

$$\langle E'(u), v \rangle = (A \nabla u, \nabla v)_{L^2(\mathcal{D})} + (V u, v)_{L^2(\mathcal{D})} + (\beta |u|^2 u, v)_{L^2(\mathcal{D})}, \tag{23}$$

we obtain the nonlinear eigenvalue problem known as the *Gross–Pitaevskii equation* (GPE). The GPE seeks the smallest eigenvalue $\lambda \in \mathbb{R}_{>0}$ and corresponding eigenfunction $u^* \in \mathbb{S}$ such that

$$(A \nabla u^*, \nabla v)_{L^2(\mathcal{D})} + (V u^*, v)_{L^2(\mathcal{D})} + (\beta |u^*|^2 u^*, v)_{L^2(\mathcal{D})} = \lambda\, (u^*, v)_{L^2(\mathcal{D})} \tag{24}$$

for all $v \in H^1_0(\mathcal{D})$. The fact that the smallest eigenvalue λ of (24) belongs to a minimizer of (22) is not trivial. A proof is given in [14]. Also note that the choice of the normalization (i.e. $u^* \in \mathbb{S}$ or equivalently $\|u^*\|_{L^2(\mathcal{D})} = 1$) is important in the nonlinear setting and that different normalizations would change the eigenvalues of (24) as well as the structure of the eigenfunctions.

Next, let us again discuss how metric-driven schemes can be applied to solve the problem. As in the linear setting, we discuss the construction of iterative solvers and the choice of the spatial discretization separately.

3.1 Metric-Driven Steepest Descents for the GPE

Since the eigenvalue problem (24) is nonlinear, a direct application of linear eigenvalue solvers (such as the inverse iteration) is not possible anymore. However, we can still apply the abstract framework from Section 2.2 by exploiting the concept of Riemannian gradient methods in combination with metric-driven Sobolev gradients.

To avoid unnecessary repetitions, we skip the introduction of a preliminary (metric-driven) gradient flow and directly turn to the formulation of a gradient method. Following the general descriptions from the linear setting and in particular (11), a Riemannian gradient method in a given X-metric (induced by an inner product $(\cdot,\cdot)_X$) is of the form

$$u^{n+1} = \frac{u^n - \tau_n (P_{u^n,X} \circ \nabla_X E) u^n}{\|u^n - \tau_n (P_{u^n,X} \circ \nabla_X E) u^n\|_{L^2(\mathcal{D})}}, \tag{25}$$

where we recall $(P_{u^n,X} \circ \nabla_X E) u^n$ as the Riemannian Sobolev gradient composed from the Sobolev gradient $\nabla_X E(u) \in X$ and its X-orthogonal projection $P_{u^n,X}$ into the tangent space $T_u\mathbb{S}$ given by (6). As in the linear setting, we can ask whether we can select the metric $(\cdot,\cdot)_X$ in such a way that $\nabla_X E(u)$ becomes the identity. In the linear setting, we could find a universal inner product that yields this property, but in the nonlinear setting of the GPE, the inner product needs to depend on the location $u \in \mathbb{S}$ where we evaluate the regular gradient $E'(u)$ given by (23). For a linearization point $u \in \mathbb{S}$, we therefore define the linearized gradient $\mathcal{L}_u$ such that

$$\langle \mathcal{L}_u v, w \rangle := (A\nabla v, \nabla w)_{L^2(\mathcal{D})} + (Vv, w)_{L^2(\mathcal{D})} + (\beta |u|^2 v, w)_{L^2(\mathcal{D})}. \tag{26}$$

Obviously, in the light of (23), $E'(u) = \mathcal{L}_u u$ holds as an identity in $H^{-1}(\mathcal{D})$. We can therefore use the inner product $(\cdot,\cdot)_{X_u} := \langle \mathcal{L}_u \cdot, \cdot \rangle$ in $H^1_0(\mathcal{D})$ to obtain a Sobolev gradient with the desired feature $\nabla_{X_u} E(u) = u$. This identity can be verified by exploiting the definition of $\nabla_{X_u} E(u)$ in (3), which yields $\nabla_{X_u} E(u) \in H^1_0(\mathcal{D})$ as the solution to

$$(\nabla_{X_u} E(u), v)_{X_u} = \langle E'(u), v \rangle = \langle \mathcal{L}_u u, v \rangle = (u, v)_{X_u}$$

for all $v \in H^1_0(\mathcal{D})$ and hence $\nabla_{X_u} E(u) = u$. Again, as in the linear setting, the corresponding projection P_{u,X_u} into the tangent space can be computed explicitly as

$$P_{u,X_u}(v) = v - \frac{(u,v)_{L^2(\mathcal{D})}}{(\mathcal{L}_u^{-1}u,u)_{L^2(\mathcal{D})}} \mathcal{L}_u^{-1}u,$$

where $\mathcal{L}_u^{-1} : H_0^1(\mathcal{D}) \to H_0^1(\mathcal{D})$ is the u-linearized solution operator such that for any $z \in H_0^1(\mathcal{D})$ the image $\mathcal{L}_u^{-1}z \in H_0^1(\mathcal{D})$ fulfills

$$\langle \mathcal{L}_u (\mathcal{L}_u^{-1}z), v\rangle = (z,v)_{L^2(\mathcal{D})} \qquad \text{for all } v \in H_0^1(\mathcal{D}). \tag{27}$$

Together with $\nabla_{X_u} E(u) = u$ we obtain the following realization of the metric-driven Riemannian gradient method (25):

Definition 2 (Metric-driven Riemannian gradient method for the GPE) For a starting value $u_0 \in \mathbb{S}$ with $u_0 \geq 0$ and a sequence of (pseudo) time step sizes $\tau_n > 0$ and for $n \geq 0$, the iterations of the metric-driven Riemannian gradient method are given by

$$u^{n+1}(\tau) := \frac{(1-\tau)\,u^n + \tau\,(\mathcal{L}_{u^n}^{-1}u^n, u^n)^{-1}_{L^2(\mathcal{D})}\, \mathcal{L}_{u^n}^{-1}u^n}{\|(1-\tau)\,u^n + \tau\,(\mathcal{L}_{u^n}^{-1}u^n, u^n)^{-1}_{L^2(\mathcal{D})}\, \mathcal{L}_{u^n}^{-1}u^n\|_{L^2(\mathcal{D})}} \tag{28}$$

and we set $u^{n+1} = u^{n+1}(\tau^n)$ for the optimal step size

$$\tau_n = \operatorname*{arg\,min}_{0<\tau\leq 2} E\left(u^{n+1}(\tau)\right).$$

The scheme can be interpreted as a generalized inverse iteration with damping parameter τ. This is seen by noticing that for $\tau = 1$, the scheme (28) reduces to

$$u^{n+1} := \frac{\mathcal{L}_{u^n}^{-1}u^n}{\|\mathcal{L}_{u^n}^{-1}u^n\|_{L^2(\mathcal{D})}},$$

i.e., a direct generalization of the inverse iteration to nonlinear eigenvalue problems.

The gradient method in Definition 2 is well-studied in the literature, cf. [5, 7, 32, 38, 40, 66] for various related settings, and can be even generalized to conjugate gradients [2]. The following result can be found in [38, Theorem 5.1].

Theorem 3 (Global convergence) *Consider the metric-driven Riemannian gradient method in* (28) *for arbitrary step size τ_n and a starting value $u_0 \in \mathbb{S}$ which satisfies $u_0 \geq 0$. There exists a step size interval $[\tau_{\min}, \tau_{\max}] \subset (0,2)$ such that for all $\tau_n \in [\tau_{\min}, \tau_{\max}]$*

$$E(u^{n+1}) \leq E(u^n) \quad \textit{for all } n \geq 0 \qquad \textit{and} \qquad \lim_{n\to\infty} \|u^n - u^*\|_{H^1(\mathcal{D})} = 0,$$

where $u^* \in \mathbb{S}$ is the unique positive minimizer of (22). *In particular, the method is globally convergent if the optimal step size τ_n is selected.*

The local convergence rates of the Riemannian gradient method depend on the choice of the step size and again on the size of spectral gaps. To present an analogous result to the linear case in (13), let us consider again the choice $\tau = 1$, where the method (28) becomes a direct generalization of the inverse iteration. The following theorem can be found in [32].

Theorem 4 (Local convergence) *Consider the metric-driven method in* (28) *for a uniform step size* $\tau_n = 1$ *and let* $u^* \in \mathbb{S}$ *be the positive global minimizer to* (22). *Then, for any* $\varepsilon > 0$ *there exists an open neighborhood* $U_\varepsilon \subset H^1_0(\mathcal{D})$ *of* u^*, *such that for any starting value* $u_0 \in U_\varepsilon \cap \mathbb{S}$,

$$\|u^* - u^n\|_{H^1(\mathcal{D})} \lesssim C_\varepsilon \,|\tfrac{\lambda_1}{\lambda_2} + \varepsilon|^n \, \|u^* - u^0\|_{H^1(\mathcal{D})}. \tag{29}$$

Here, $C_\varepsilon > 0$ *is a generic, but* ε*-dependent constant and* $\lambda_1 = \lambda$ *and* $\lambda_2 > \lambda_1$ *are the smallest and the second smallest eigenvalue of the linearized GPE that seeks the eigenpairs* $(u_i, \lambda_i) \in \mathbb{S} \times \mathbb{R}_{>0}$ *to*

$$\langle \mathcal{L}_{u^*} u_i, v\rangle = \lambda_i \,(u_i, v)_{L^2(\mathcal{D})} \qquad \textit{for all } v \in H^1_0(\mathcal{D}).$$

For $\beta = 0$, when the GPE collapses into the linear setting, Theorem 4 just recovers the well-known rate (13) from the inverse iteration. However, in contrast to the linear setting, the rate in (29) is not sharp for $\beta > 0$ and can be refined. A more precise characterization of the asymptotic convergence rates (and for arbitrary τ) is more technical and we refer the interested reader to [32, 40, 29].

Remark 1 (Alternative approaches) Besides Sobolev gradient-based approaches discussed in this subsection, the Gross–Pitaevskii problem (24) is commonly approached by a variety of numerical methods, including self-consistent field (SCF) iterations [15, 16, 61], Newton-type methods [7, 62, 64], L^2-gradient flow inspired schemes [12, 28, 30], as well as nonlinear variants of inverse iteration [45, 3, 32]. We refer to [33] for a recent overview, including a quantitative comparison.

In contrast to these approaches, the metric-driven formulation is not only designed as an alternative standalone solver, but rather as a framework in which both the iterative scheme and the underlying approximation space are derived from a common operator-dependent metric. In particular, the use of Sobolev gradients can be interpreted as an intrinsic preconditioning strategy that is naturally adapted to the problem. This distinguishes the approach from classical methods, where preconditioning and space enrichment are typically introduced as separate design components. Our focus here is on the conceptual role of the metric in the joint design of the iteration and the approximation space, rather than on a detailed algorithmic or performance-oriented comparison of different solution strategies.

3.2 Minimization of the GPE Functional in Metric-Driven Approximation Spaces

Next, we will turn to the construction of a metric-driven approximation space for the GPE, where we adopt the notation from Section 2.3. In particular, $\mathcal{T}_H$ is a quasi-uniform and shape-regular triangulation of $\mathcal{D} \subset \mathbb{R}^d$ and the $\mathbb{P}^1$-Lagrange finite element space on $\mathcal{T}_H$ is denoted by $V_H \subset H^1_0(\mathcal{D})$.

For the derivation of the approximation space, we can essentially proceed analogously to the linear setting in Section 2.3. However, this would yield a sequence of spaces of the form $\mathcal{L}^{-1}_{u_H^{\mathcal{L},n}} V_H$, where $u_H^{\mathcal{L},n}$ is a previous approximation obtained in an old approximation space $\mathcal{L}^{-1}_{u_H^{\mathcal{L},n-1}} V_H$. The old space $\mathcal{L}^{-1}_{u_H^{\mathcal{L},n-1}} V_H$ itself is based on another previous approximation $u_H^{\mathcal{L},n-1} \in \mathcal{L}^{-1}_{u_H^{\mathcal{L},n-2}} V_H$ and so on. This kind of recursive construction of the approximation space involves the repeated computation and update of basis functions and is therefore impractical and not economical. For that reason, we use the linearization of $\mathcal{L}_u$ around the point $u = 0$ to construct the approximation space. It turns out that this is sufficient to obtain the optimal approximation order known from the linear setting. We fix the construction in the following definition.

Definition 3 (Metric-driven approximation space) Let $\mathcal{L}_0$ denote the elliptic operator defined in (26) for $u = 0$ (or, equivalently, $\beta = 0$) and $\mathcal{L}_0^{-1} : H^1_0(\mathcal{D}) \to H^1_0(\mathcal{D})$ the corresponding solution operator according to (27). We define the metric-driven approximation space by

$$V_H^{\mathcal{L}_0} := \mathcal{L}_0^{-1} V_H, \tag{30}$$

where V_H is the $\mathbb{P}^1$-FEM space from (14). The metric-driven approximation is now defined as a normalized minimizer of E on $V_H^{\mathcal{L}_0}$, i.e.,

$$u_H^{\mathcal{L}_0} = \underset{v \in V_H^{\mathcal{L}_0} \cap \mathbb{S}}{\arg\min}\; E(v). \tag{31}$$

Corresponding minimizers can be numerically found by performing the iterations (28) in $V_H^{\mathcal{L}_0}$, analogously to the linear case.

Again, the metric-driven space in (30) coincides with the LOD space for the GPE and accordingly we have the following approximation results proved in [35, 36].

Theorem 5 *Let $u^* \in \mathbb{S}$ denote the exact minimizer of* (22) *and $u_H^{\mathcal{L}_0} \in V_H^{\mathcal{L}_0} \cap \mathbb{S}$ the corresponding minimizer in* (31) *such that $(u^*, u_H^{\mathcal{L}_0})_{L^2(\mathcal{D})} \geq 0$.*

If merely $A \in L^\infty(\mathcal{D}, \mathbb{R}^{d\times d})$ and $V \in L^\infty(\mathcal{D}, \mathbb{R}_{\geq 0})$, then

$$\begin{aligned} \|u^* - u_H^{\mathcal{L}_0}\|_{L^2(\mathcal{D})} + H\,\|u^* - u_H^{\mathcal{L}_0}\|_{H^1(\mathcal{D})} &\lesssim H^3 \quad \textit{and} \\ |E(u^*) - E(u_H^{\mathcal{L}_0})| &\lesssim H^4, \end{aligned} \tag{32}$$

where the hidden constants only depend on β and the maximum values of A and V.

If A additionally fulfills $A \in W^{1,\infty}(\mathcal{D}, \mathbb{R}^{d\times d})$, then we have $u^* \in H^2(\mathcal{D})$ and the improved estimates become

$$\begin{aligned} \|u^* - u_H^{\mathcal{L}_0}\|_{L^2(\mathcal{D})} + H\,\|u^* - u_H^{\mathcal{L}_0}\|_{H^1(\mathcal{D})} &\lesssim H^4 \quad \textit{and} \\ |E(u^*) - E(u_H^{\mathcal{L}_0})| &\lesssim H^6, \end{aligned} \tag{33}$$

for a hidden constant that additionally depends on $\|A\|_{W^{1,\infty}(\mathcal{D})}$.

As we can see, despite using a fixed approximation space $V_H^{\mathcal{L}_0}$ that only uses the linear part of the Gross–Pitaevskii operator $E'(u)$, we still obtain the full convergence orders from the linear setting as presented in Theorem 2. As before, the results in Theorem 5 show an enormous potential in multiscale and low-regularity applications. Similar findings can be made for the time-dependent Gross–Pitaevskii equation [39, 58, 22] or other types of semi-linear problems [34].

A numerical illustration for the approximation properties of $V_H^{\mathcal{L}_0}$ is given in the next section for a more advanced application.

4 Application to Ground States of Spin-Orbit Coupled Bose–Einstein Condensates

In this section, we give an advanced example of how a metric-driven approach can be extended to more complicated models. For that, we investigate the numerical computation of ground states of two-component Bose–Einstein condensates (BECs) that are spin-orbit coupled (SO-coupled).

A Bose–Einstein condensate is a state of matter that is realized when bosonic gases are cooled to temperatures near absolute zero [13, 26, 59]. In this state of matter, the atoms are no longer distinguishable from one another and instead form a type of "superatom", i.e., all the atoms behave as if they were only a single atom, as they all occupy the same quantum state. We will consider two-component SO-coupled BECs, meaning that by means of so-called Raman lasers, the spin of an atom is tied to its momentum. Our goal is to identify ground states of the SO-coupled BEC. These ground states represent stable stationary states that minimize the energy of the condensate. The stability of a ground state allows us to examine interesting quantum phenomena. Our numerical experiments will show that, in the considered configuration, the SO-coupled BEC can exhibit a *super solid phase*, which means that it possesses properties that are both inherent to solids (i.e., crystalline structure) and superfluids (i.e. frictionless flow), see [48].

Notation. For a proper introduction of the problem we first need to specify the relevant function spaces. In the following, we denote by $\mathbf{L}^2(\mathcal{D})$ the space

$$\mathbf{L}^2(\mathcal{D}) := L^2(\mathcal{D},\mathbb{C}^2) \;\; := \;\; \{\, \mathbf{v} = (v_1, v_2) \mid v_i \in L^2(\mathcal{D},\mathbb{C}), \;\; i = 1,2 \,\},$$

equipped with the $\mathbb{R}$-inner product

$$(\mathbf{v},\mathbf{w})_{\mathbf{L}^2(\mathcal{D})} := \operatorname{Re}\int_{\mathcal{D}} v_1\overline{w_1}\,\mathrm{d}x + \operatorname{Re}\int_{\mathcal{D}} v_2\overline{w_2}\,\mathrm{d}x,$$

where $x = (x_1, x_2)$ for $\mathcal{D} \subset \mathbb{R}^2$ and $x = (x_1, x_2, x_3)$ for $\mathcal{D} \subset \mathbb{R}^3$. Similarly, we let $\mathbf{H}^1_0(\mathcal{D}) := H^1_0(\mathcal{D},\mathbb{C}^2)$ and define

$$(\mathbf{v},\mathbf{w})_{\mathbf{H}^1(\mathcal{D})} := \operatorname{Re}\int_{\mathcal{D}} \nabla v_1 \cdot \overline{\nabla w_1}\,\mathrm{d}x + \operatorname{Re}\int_{\mathcal{D}} \nabla v_2 \cdot \overline{\nabla w_2}\,\mathrm{d}x,$$

which is an inner product by the Poincaré inequality. For readability in the proofs, we sometimes also denote $(\nabla \mathbf{v}, \nabla \mathbf{w})_{\mathbf{L}^2(\mathcal{D})} := (\mathbf{v}, \mathbf{w})_{\mathbf{H}^1(\mathcal{D})}$.

4.1 Mathematical Setting and Problem Formulation

We now turn to the problem formulation. As before, we are concerned with the minimization of a functional E on a constraint manifold $\mathbb{S}$. In this section, the manifold is used to impose a normalization constraint for the mass of a BEC that consists of two components u_1 and u_2. Consequently, it is given by

$$\mathbb{S} := \{\mathbf{u} = (u_1, u_2) \in \mathbf{H}_0^1(\mathcal{D}) \mid \int_{\mathcal{D}} |u_1|^2 + |u_2|^2 \,\mathrm{d}x = 1\}.$$

The functional $E : \mathbf{H}_0^1(\mathcal{D}) \to \mathbb{R}$ in the case of SO-coupled BECs describes the total energy of a system and is defined as

$$\begin{aligned} E(\mathbf{u}) := \frac{1}{2}\int_{\mathcal{D}} \sum_{j=1}^{2} \Big(\frac{1}{2}|\nabla u_j|^2 + V_j(x)|u_j|^2\Big) + \frac{\delta}{2}\Big(|u_1|^2 - |u_2|^2\Big) + \Omega \operatorname{Re}(u_1\overline{u}_2) \qquad (34) \\ + \mathrm{i}k_0\big(\overline{u}_1\partial_1 u_1 - \overline{u}_2\partial_1 u_2\big) + \frac{\beta_{11}}{2}|u_1|^4 + \frac{\beta_{22}}{2}|u_2|^4 + \beta_{12}|u_1|^2|u_2|^2 \,\mathrm{d}x, \end{aligned}$$

for a tuple $\mathbf{u} = (u_1, u_2) \in \mathbf{H}_0^1(\mathcal{D})$ that describes the quantum states of the two components of the BEC. The functions $V_1(x)$ and $V_2(x)$ represent real-valued external trapping potentials that confine the system. The differential operator ∂_1 refers to the partial derivative with respect to the first component, i.e., $\partial_1 = \partial_{x_1}$. The constants $\beta_{ij} \geq 0$ (for $1 \leq i, j \leq 2$) characterize the interatomic interactions and depend on the type of particles. Furthermore, the constant δ denotes the detuning parameter associated with the Raman transition, Ω is the effective Rabi frequency describing the strength of the Raman coupling and k_0 is the wave number of Raman lasers which models the SO-coupling strength (cf. [10]). With this, a ground state (lowest energy state) $\mathbf{u} = (u_1, u_2) \in \mathbb{S}$ of the SO-coupled BEC is defined as a minimizer of E on $\mathbb{S}$, i.e.,

$$\mathbf{u} = \arg\min_{\mathbf{v}\in\mathbb{S}} E(\mathbf{v}). \qquad (35)$$

The physical quantity of interest is the (real-valued) density of the components, which is given by $|u_1|^2$ and $|u_2|^2$ respectively.

4.1.1 Well-Posedness

In order to compute a ground state with a metric-driven approach and to analyze the resulting method, we make a set of assumptions:

(A1) $\mathcal{D} \subseteq \mathbb{R}^d$ is a bounded Lipschitz-domain for $d = 2, 3$.
(A2) $V_j \in L^\infty(\mathcal{D}, \mathbb{R}_{\geq 0})$ fulfill $V_j(x) - \frac{|\delta|+|\Omega|+2k_0^2}{2} \geq 0$ for $j = 1, 2$.
(A3) $\beta_{11}, \beta_{12}, \beta_{22} \geq 0$ are real-valued and positive (i.e., all interactions are repulsive).

Assumption (A1) is a natural assumption on the computational domain and (A3) is a restriction on the particle types. Assumption (A2) ensures that the trapping potentials are non-negative and sufficiently strong compared to other effects. Ultimately, this guarantees that the energy remains positive. Note that (A2) is in any case uncritical, since we can always shift the energy by an arbitrary constant (such as $\frac{|\delta|+|\Omega|+2k_0^2}{2}$) without changing the minimizers of E on $\mathbb{S}$. This exploits that all minimizers satisfy the L^2-normalization constraint.

Mathematically, we require (A2) to select a suitable linearization of $E'(u)$ that induces an inner product/metric in the spirit of Sections 2 and 3. To make the linearization precise, we define the operator $\mathcal{L}_{\mathbf{u}} : \mathbf{H}_0^1(\mathcal{D}) \to \mathbf{H}_0^1(\mathcal{D})^*$ for a linearization point $\mathbf{u} \in \mathbf{H}_0^1(\mathcal{D})$ by

$$
\begin{aligned}
\langle \mathcal{L}_{\mathbf{u}}\mathbf{v}, \mathbf{w}\rangle &= \mathrm{Re} \int_{\mathcal{D}} \sum_{j=1}^{2} \Big(\frac{1}{2}\nabla v_j \cdot \nabla \overline{w}_j + V_j v_j \overline{w}_j\Big) + \frac{\delta}{2}\big(v_1\overline{w}_1 - v_2\overline{w}_2\big) + \frac{\Omega}{2}\big(v_1\overline{w}_2 + v_2\overline{w}_1\big)\,\mathrm{d}x \\
&+ \mathrm{Re} \int_{\mathcal{D}} \mathrm{i}k_0\big(\overline{w}_1\partial_1 v_1 - \overline{w}_2\partial_1 v_2\big) + \beta_{11}|u_1|^2 v_1\overline{w}_1 + \beta_{22}|u_2|^2 v_2\overline{w}_2 + \beta_{12}\big(|u_2|^2 v_1\overline{w}_1 + |u_1|^2 v_2\overline{w}_2\big)\,\mathrm{d}x
\end{aligned} \tag{36}
$$

for $\mathbf{v}, \mathbf{w} \in \mathbf{H}_0^1(\mathcal{D})$. Computing the Fréchet derivative of E confirms that $\mathcal{L}_{\mathbf{u}}$ is indeed a linearization of $E'(\mathbf{u})$ as

$$\langle E'(\mathbf{u}), \mathbf{v}\rangle = \langle \mathcal{L}_{\mathbf{u}}\mathbf{u}, \mathbf{v}\rangle.$$

If $\beta_{11} = \beta_{12} = \beta_{22} = 0$, then E' is linear and coincides with the linear operator $\mathcal{L}_{\mathbf{0}}$. Hence, the linearization $\mathcal{L}_{\mathbf{u}}$ is natural. Also note that due to the presence of the real part in front of the integral in the definition of $\mathcal{L}_{\mathbf{u}}$, the operator is self-adjoint and therefore induces a symmetric bilinear form.

As in the setting of the previous sections, constrained minimizers fulfill the corresponding Euler–Lagrange equations which take the form of a (nonlinear) eigenvalue problem. Exploiting that $E'(\mathbf{u}) = \mathcal{L}_{\mathbf{u}}\mathbf{u}$, a minimizer $\mathbf{u} \in \mathbb{S}$ to (35) is an eigenfunction with eigenvalue $\lambda \in \mathbb{R}_{>0}$ to

$$\langle \mathcal{L}_{\mathbf{u}}\mathbf{u}, \mathbf{v}\rangle \;=\; \lambda\,(\mathbf{u}, \mathbf{v})_{\mathrm{L}^2(\mathcal{D})} \tag{37}$$

for all $\mathbf{v} \in \mathbf{H}_0^1(\mathcal{D})$. Using the definition of $\mathcal{L}_{\mathbf{u}}$, the strong form of the eigenvalue problem (37) reads

$$
\begin{aligned}
\lambda u_1 &= \Big[-\frac{1}{2}\Delta + V_1(x) + \mathrm{i}k_0\partial_1 + \frac{\delta}{2} + \big(\beta_{11}|u_1|^2 + \beta_{12}|u_2|^2\big)\Big] u_1 + \frac{\Omega}{2}u_2, \\
\lambda u_2 &= \Big[-\frac{1}{2}\Delta + V_2(x) - \mathrm{i}k_0\partial_1 - \frac{\delta}{2} + \big(\beta_{12}|u_1|^2 + \beta_{22}|u_2|^2\big)\Big] u_2 + \frac{\Omega}{2}u_1,
\end{aligned}
$$

for $\mathbf{u} = (u_1, u_2)$. Note that the eigenvalue λ that belongs to a ground state (i.e. a global minimizer fulfilling (35)) is not necessarily the smallest eigenvalue of (37). This is an important difference from the settings in the previous sections.

Another difference compared to the previous settings is that existence and uniqueness of ground states are no longer obvious. The question was investigated in [10] and we have the following well-posedness result (which relies on the non-negativity of the parameters $\beta_{11}, \beta_{12}, \beta_{22}$):

Theorem 6 (Existence of ground states of SO-coupled BECs) *Assume (A1)–(A3), then there exists at least one minimizer* $\mathbf{u} \in \mathbb{S}$ *to* (35). *Ground states are at most locally unique up to complex phase shifts* $\exp(\mathrm{i}\omega)$ *for any angle* $\omega \in [-\pi, \pi)$. *This means that if* $\mathbf{u} \in \mathbb{S}$ *denotes an arbitrary ground state, then* $\exp(\mathrm{i}\omega)\mathbf{u} \in \mathbb{S}$ *shares the same density* $|u_i|^2 = |\exp(\mathrm{i}\omega)u_i|^2$ *(for* $i = 1, 2$*) and the same energy level* $E(\mathbf{u}) = E(\exp(\mathrm{i}\omega)\mathbf{u})$. *Hence, for any* $\omega \in [-\pi, \pi)$ *we have that* $\exp(\mathrm{i}\omega)\mathbf{u}$ *is another (though physically equivalent) ground state.*

4.1.2 First and Second Order Conditions for Minimizers

Classically, any global minimizer $\mathbf{u} \in \mathbb{S}$ of E must fulfill the first order and second order optimality conditions, cf. [41]. If λ denotes the corresponding Lagrange multiplier given by $\lambda := \langle \mathcal{L}_{\mathbf{u}}\mathbf{u}, \mathbf{u} \rangle$, then the *first order condition* is just the GPE (37) and reads

$$\langle E'(\mathbf{u}), \mathbf{v} \rangle - \lambda\, (\mathbf{u}, \mathbf{v})_{\mathbf{L}^2(\mathcal{D})} = 0 \qquad \text{for all } \mathbf{v} \in \mathbf{H}^1_0(\mathcal{D}),$$

whereas the necessary *second order condition* reads

$$\langle E''(\mathbf{u})\mathbf{v}, \mathbf{v} \rangle - \lambda\, (\mathbf{v}, \mathbf{v})_{\mathbf{L}^2(\mathcal{D})} \geq 0 \qquad \text{for all } \mathbf{v} \in T_{\mathbf{u}}\mathbb{S},$$

where $T_{\mathbf{u}}\mathbb{S}$ denotes the tangent space at $\mathbf{u}$ given by

$$T_{\mathbf{u}}\mathbb{S} = \{\, \mathbf{v} \in \mathbf{H}^1_0(\mathcal{D}) \mid (\mathbf{u}, \mathbf{v})_{\mathbf{L}^2(\mathcal{D})} = 0 \,\}.$$

Due to the invariance of the energy under complex phase shifts $\exp(\mathrm{i}\omega)$, the following must hold

$$\langle E''(u)(\mathrm{i}\mathbf{u}), \mathbf{v} \rangle - \lambda\, (\mathrm{i}\mathbf{u}, \mathbf{v})_{\mathbf{L}^2(\mathcal{D})} = 0 \qquad \text{for all } \mathbf{v} \in T_{\mathbf{u}}\mathbb{S}. \tag{38}$$

This is seen by considering the energy curve $\omega \mapsto E(\exp(\mathrm{i}\omega)\mathbf{u})$, which is constant. Hence, the first and the second derivative of the curve are zero. Computing these derivatives just yields (38). Therefore, λ is always the smallest eigenvalue of $E''(\mathbf{u})$ with corresponding eigenfunction $\mathrm{i}\mathbf{u}$. If all other eigenvalues of $E''(\mathbf{u})$ are strictly larger than λ, then $\mathbf{u}$ is called *locally quasi-isolated* [41, 40].

A straightforward calculation shows that the second derivative of E is given by

$$\langle E''(\mathbf{u})\mathbf{v},\mathbf{w}\rangle$$
$$= \langle \mathcal{L}_{\mathbf{u}}\mathbf{v},\mathbf{w}\rangle + 2\mathrm{Re}\int_{\mathcal{D}}\sum_{j=1}^{2}\beta_{jj}\mathrm{Re}(u_j\bar{v}_j)u_j\overline{w}_j + \beta_{12}\big(\mathrm{Re}(u_1\bar{v}_1)u_2\overline{w}_2 + \mathrm{Re}(u_2\bar{v}_2)u_1\overline{w}_1\big)\,\mathrm{d}x.$$

This can be used in practice to verify a posteriori whether a computed state $\mathbf{u}\in\mathbb{S}$ is indeed a local minimizer of E by calculating the smallest eigenvalues of $E''(\mathbf{u})|_{T_{\mathbf{u}}\mathbb{S}}$ which have to be in agreement with the above conditions.

4.1.3 Energy-Adaptive Metric

Since we want to use the metric induced by the linear operator $\mathcal{L}_{\mathbf{u}}$, we need to verify that it is elliptic. For this, we note that $\mathcal{L}_{\mathbf{u}}$ can be written as

$$\langle \mathcal{L}_{\mathbf{u}}\mathbf{v},\mathbf{w}\rangle = \langle \mathcal{L}_{\mathbf{0}}\mathbf{v},\mathbf{w}\rangle + \sum_{j=1}^{2}\beta_{jj}\mathrm{Re}\int_{\mathcal{D}}|u_j|^2 v_j\overline{w}_j\,\mathrm{d}x + \beta_{12}\mathrm{Re}\int_{\mathcal{D}}|u_1|^2 v_2\overline{w}_2 + |u_2|^2 v_1\overline{w}_1\,\mathrm{d}x, \tag{39}$$

where $\mathcal{L}_{\mathbf{0}}$ denotes the $\mathbf{u}$-independent part of $\mathcal{L}_{\mathbf{u}}$. With the convention $\beta_{21} := \beta_{12}$ we obtain

$$\langle \mathcal{L}_{\mathbf{u}}\mathbf{v},\mathbf{v}\rangle = \langle \mathcal{L}_{\mathbf{0}}\mathbf{v},\mathbf{v}\rangle + \sum_{i,j=1}^{2}\beta_{ij}\int_{\mathcal{D}}|u_i|^2|v_j|^2\,\mathrm{d}x \;\ge\; \langle \mathcal{L}_{\mathbf{0}}\mathbf{v},\mathbf{v}\rangle.$$

Hence, $\mathcal{L}_{\mathbf{u}}$ is elliptic for any $\mathbf{u}\in\mathbf{H}_0^1(\mathcal{D})$ as long as $\mathcal{L}_{\mathbf{0}}$ is elliptic. It is therefore sufficient to prove the following result.

Lemma 2 *Assume (A1)–(A2), then $\mathcal{L}_{\mathbf{0}}$ is coercive (elliptic), i.e., there is a constant $\alpha > 0$ such that*

$$\langle \mathcal{L}_{\mathbf{0}}\mathbf{v},\mathbf{v}\rangle \;\ge\; \frac{1}{4}\|\mathbf{v}\|^2_{\mathbf{H}^1(\mathcal{D})} \qquad \text{for all } \mathbf{v}\in\mathbf{H}_0^1(\mathcal{D}). \tag{40}$$

Proof. According to (36), $\langle \mathcal{L}_{\mathbf{0}}\mathbf{v},\mathbf{v}\rangle$ is given by

$$\langle \mathcal{L}_{\mathbf{0}}\mathbf{v},\mathbf{v}\rangle = \int_{\mathcal{D}}\sum_{j=1}^{2}\frac{1}{2}|\nabla v_j|^2 + V_j|v_j|^2 + \frac{\delta}{2}\big(|v_1|^2 - |v_2|^2\big) + \Omega\,\mathrm{Re}\big(v_1\bar{v}_2\big)$$
$$+\,\mathrm{Re}\int_{\mathcal{D}} \mathrm{i}k_0\big(\bar{v}_1\partial_1 v_1 - \bar{v}_2\partial_1 v_2\big)\,\mathrm{d}x.$$

To find a lower bound, we start with the last term, where we use Young's inequality for $\varepsilon > 0$ to obtain

$$|\mathrm{Re}\big(ik_0 \int_{\mathcal{D}} \bar{v}_1\partial_1 v_1 - \bar{v}_2\partial_1 v_2)|\,\mathrm{d}x \leq \frac{k_0}{2}\int_{\mathcal{D}} \frac{|v_1|^2+|v_2|^2}{\varepsilon} + \varepsilon(|\nabla v_1|^2 + |\nabla v_2|^2)\,\mathrm{d}x.$$

With the choice $\varepsilon^{-1} = 2k_0$ and by using the Cauchy–Schwarz inequality, this yields

$$\langle \mathcal{L}_0\mathbf{v},\mathbf{v}\rangle \geq \frac{1}{4}\|\nabla\mathbf{v}\|^2_{\mathbf{L}^2(\mathcal{D})} + \sum_{j=1}^{2}\int_{\mathcal{D}} (V_j(x) - \frac{|\delta| + |\Omega| + 2k_0^2}{2})|v_j|^2\,\mathrm{d}x.$$

By assumption (A2) the result follows. □

Recalling that $\mathcal{L}_\mathbf{u}$ is self-adjoint, we conclude from Lemma 2 that $\langle\mathcal{L}_\mathbf{u}\mathbf{v},\mathbf{w}\rangle$ defines an inner product. We fix this in the following conclusion.

Conclusion 1. *Assume (A1)–(A3) and let $\mathbf{u} \in \mathbf{H}^1_0(\mathcal{D})$ be arbitrary. Then $\langle\mathcal{L}_\mathbf{u}\mathbf{v},\mathbf{w}\rangle$ is symmetric and*

$$\langle\mathcal{L}_\mathbf{u}\mathbf{v},\mathbf{v}\rangle \;\geq\; \frac{1}{4}\|\mathbf{v}\|^2_{\mathbf{H}^1(\mathcal{D})} \qquad \text{for all } \mathbf{v}\in\mathbf{H}^1_0(\mathcal{D}).$$

Hence, $\langle\mathcal{L}_\mathbf{u}\mathbf{v},\mathbf{w}\rangle$ is an inner product on $\mathbf{H}^1_0(\mathcal{D})$.

Now that we verified that $\langle\mathcal{L}_\mathbf{u}\mathbf{v},\mathbf{v}\rangle$ defines an inner product that is linked to the energy, we will use it in the next step as an adaptively changing metric in a Riemannian gradient method analogously to what we did in Section 3.1.

4.2 Metric-Driven Steepest Descents for SO-Coupled BECs

Since the arguments that we developed in Section 3.1 are generally applicable, the metric-driven steepest descent in the adaptive $\langle\mathcal{L}_\mathbf{u}\mathbf{v},\mathbf{w}\rangle$-metric for computing minimizers of (35) is given as follows.

Definition 4 (Metric-driven Riemannian gradient method for SO-coupled BECs) For a starting value $\mathbf{u}_0 \in \mathbb{S}$ and a sequence of (pseudo) time step sizes $\tau_n > 0$, the iterations are given (for $n \geq 0$) by

$$\mathbf{u}^{n+1}(\tau) := \frac{(1-\tau)\,\mathbf{u}^n \;+\; \tau\,(\mathcal{L}^{-1}_{\mathbf{u}^n}\mathbf{u}^n,\mathbf{u}^n)^{-1}_{\mathbf{L}^2(\mathcal{D})}\;\mathcal{L}^{-1}_{\mathbf{u}^n}\mathbf{u}^n}{\|(1-\tau)\,\mathbf{u}^n \;+\; \tau\,(\mathcal{L}^{-1}_{\mathbf{u}^n}\mathbf{u}^n,\mathbf{u}^n)^{-1}_{\mathbf{L}^2(\mathcal{D})}\;\mathcal{L}^{-1}_{\mathbf{u}^n}\mathbf{u}^n\|_{\mathbf{L}^2(\mathcal{D})}} \tag{41}$$

and we set $\mathbf{u}^{n+1} = \mathbf{u}^{n+1}(\tau^n)$ for the optimal step size

$$\tau_n = \underset{0<\tau\leq 2}{\arg\min}\; E\left(\mathbf{u}^{n+1}(\tau)\right). \tag{42}$$

Note that each $\mathbf{u}^n \in \mathbf{H}^1_0(\mathcal{D})$ consists of two components $\mathbf{u}^n = (u_1^n, u_2^n)$.

Since the application of the metric-driven approach to SO-coupled BECs is new, it remains to verify that the method is indeed strictly energy-diminishing (for all sufficiently small step sizes τ_n, and in particular the optimal step size) and that the method converges, for any starting value, to a critical point of the energy. For that, we transfer the techniques previously developed in e.g. [38, 40] to the new setting to obtain the following convergence result.

Theorem 7 *Assume (A1)–(A3). Consider the iterations* (41) *for some given (not necessarily optimal) step sizes τ_n and a starting value $\mathbf{u}_0 \in \mathbb{S}$. Then there exists a step size interval $[\tau_{\min}, \tau_{\max}] \subset (0,2)$ such that if $\tau_n \in [\tau_{\min}, \tau_{\max}]$ for all n, then the iterates $\mathbf{u}^n \in \mathbf{H}^1_0(\mathcal{D})$ in* (41) *have the following properties:*

(i) The energy is diminished in each iteration. In particular, there is a $C_\tau > 0$ such that $E(\mathbf{u}^n) - E(\mathbf{u}^{n+1}) \geq C_\tau \|\mathbf{u}^{n+1} - \mathbf{u}^n\|^2_{\boldsymbol{H}^1(\mathcal{D})}$.

(ii) The energy decays to a limit energy, i.e., $\lim\limits_{n\to\infty} E(\mathbf{u}^n) = E_0 \in \mathbb{R}_{\geq 0}$.

(iii) The sequence of iterates possesses a subsequence $\{\mathbf{u}^{n_j}\}_{j\in\mathbb{N}}$ that converges strongly to a critical point $\mathbf{u} \in \mathbb{S}$ of E, i.e., $\lim\limits_{j\to\infty} \|\mathbf{u}^{n_j} - \mathbf{u}\|_{\boldsymbol{H}^1(\mathcal{D})} = 0$ and $\mathbf{u} \in \mathbb{S}$ fulfills the equation $E'(\mathbf{u}) = \lambda\,(\mathbf{u}, \cdot)_{\mathbf{L}^2(\mathcal{D})}$ for the eigenvalue $\lambda := \langle E'(\mathbf{u}), \mathbf{u}\rangle = \lim\limits_{j\to\infty} (\mathcal{L}^{-1}_{\mathbf{u}^{n_j}} \mathbf{u}^{n_j}, \mathbf{u}^{n_j})^{-1}_{\boldsymbol{L}^2(\mathcal{D})}$.

(iv) If the limit of the subsequence in (iii) is a locally quasi-isolated ground state in the sense of Section 4.1.2, then the entire sequence of density iterates $|\mathbf{u}^n|^2$ converges to the corresponding ground state density $|\mathbf{u}|^2$, i.e.

$$\lim_{n\to\infty} \|\, |\mathbf{u}^n|^2 - |\mathbf{u}|^2 \,\|_{\boldsymbol{L}^2(\mathcal{D})} = 0.$$

The proof of the theorem is given in Section 4.3. Before presenting it, let us make some final remarks. First, we note that Theorem 7 covers in particular the metric-driven Riemannian gradient method with optimal step size according to (42). Second, note that in (iii), we can only guarantee convergence up to subsequences due to the energy functional being invariant under complex phase shifts. The iteration does not take this into account, hence the iteration itself might continuously change the phase, whereas the density $|\mathbf{u}^n|$ remains convergent. This is because the modulus itself is unaffected by phase shifts.

4.3 Convergence Proof

In this section we give a proof of Theorem 7 using the same techniques as developed in [38, 40] for single-component BECs. To keep the notation compact, we define, for a given previous iterate $\mathbf{u}^n$ and a step size τ_n, the preliminary iterate as

$$\boldsymbol{u}^{n+1}_{\text{prel}} := (1-\tau_n)\,\mathbf{u}^n + \tau_n\,\gamma_{\mathbf{u}^n}\,\mathcal{L}^{-1}_{\mathbf{u}^n}\mathbf{u}^n, \qquad \text{where } \gamma_{\mathbf{u}^n} := (\mathcal{L}^{-1}_{\mathbf{u}^n}\mathbf{u}^n, \mathbf{u}^n)^{-1}_{\mathrm{L}^2(\mathcal{D})}. \quad (43)$$

We also define $\boldsymbol{u}^{n+1}_{\text{prel}} = (u^{n+1}_{\text{prel},1}, u^{n+1}_{\text{prel},2})$ and the new iterate is $\mathbf{u}^{n+1} := \boldsymbol{u}^{n+1}_{\text{prel}} / \|\boldsymbol{u}^{n+1}_{\text{prel}}\|_{L^2(\mathcal{D})}$. Furthermore, the energy norm induced by $\mathcal{L}_0$ is denoted by

$$|||\mathbf{v}||| = \langle \mathcal{L}_0 \mathbf{v}, \mathbf{v}\rangle^{1/2}.$$

We now turn to the proof of Theorem 7, which requires a series of lemmas that will be proved first. All these auxiliary results will silently assume that we are in the setting of Theorem 7.

We start by noting that the preliminary iterations increase the mass.

Lemma 3 *If* $\boldsymbol{u}^{n+1}_{\text{prel}} \neq \mathbf{u}^n$, *then*

$$\|\boldsymbol{u}^{n+1}_{\text{prel}}\|_{\boldsymbol{L}^2(\mathcal{D})} - \|\mathbf{u}^n\|_{\boldsymbol{L}^2(\mathcal{D})} > 0.$$

In particular, since $\mathbf{u}^n \in \mathbb{S}$, *we have* $\|\boldsymbol{u}^{n+1}_{\text{prel}}\|_{L^2(\mathcal{D})} \geq 1$ *for all n.*

Proof. Denote for brevity the Riemannian gradient by $\mathbf{d}^n := -\mathbf{u}^n + \gamma_{\mathbf{u}^n} \mathcal{L}^{-1}_{\mathbf{u}^n} \mathbf{u}^n$ and recall $\gamma_{\mathbf{u}^n} := (\mathcal{L}^{-1}_{\mathbf{u}^n}\mathbf{u}^n, \mathbf{u}^n)^{-1}_{\mathbf{L}^2(\mathcal{D})}$. This yields

$$(\mathbf{u}^n, \mathbf{d}^n)_{\boldsymbol{L}^2(\mathcal{D})} = -\|\mathbf{u}^n\|^2_{\boldsymbol{L}^2(\mathcal{D})} + \gamma_{\mathbf{u}^n}(\mathbf{u}^n, \mathcal{L}^{-1}_{\mathbf{u}^n}\mathbf{u}^n)_{\boldsymbol{L}^2(\mathcal{D})} = 0 \tag{44}$$

and we conclude

$$\begin{aligned} \|\boldsymbol{u}^{n+1}_{\text{prel}}\|^2_{\boldsymbol{L}^2(\mathcal{D})} &= \|\mathbf{u}^n\|^2_{\boldsymbol{L}^2(\mathcal{D})} + 2\tau_n(\mathbf{u}^n, \mathbf{d}^n)_{\boldsymbol{L}^2(\mathcal{D})} + \tau_n^2\|\mathbf{d}^n\|^2_{\boldsymbol{L}^2(\mathcal{D})} \\ &= \|\mathbf{u}^n\|^2_{\boldsymbol{L}^2(\mathcal{D})} + \tau_n^2\|\boldsymbol{u}^{n+1}_{\text{prel}} - \mathbf{u}^n\|^2_{\boldsymbol{L}^2(\mathcal{D})} > \|\mathbf{u}^n\|^2_{\boldsymbol{L}^2(\mathcal{D})}. \end{aligned}$$

□

Note that $\|\boldsymbol{u}^{n+1}_{\text{prel}}\|_{L^2(\mathcal{D})} \geq 1$ implies $E(\mathbf{u}^{n+1}) \leq E(\boldsymbol{u}^{n+1}_{\text{prel}})$. Hence, it is sufficient to show $E(\boldsymbol{u}^{n+1}_{\text{prel}}) \leq E(\mathbf{u}^n)$ to conclude energy dissipation $E(\mathbf{u}^{n+1}) \leq E(\mathbf{u}^n)$. This is established in the following lemma with a quantitative lower bound.

Lemma 4 *For* $\tau_n \leq \frac{1}{2}$,

$$\begin{aligned} &E(\mathbf{u}^n) - E(\boldsymbol{u}^{n+1}_{\text{prel}}) \\ &\geq \left(\tfrac{1}{\tau_n} - \tfrac{1}{2}\right)|||\boldsymbol{u}^{n+1}_{\text{prel}} - \mathbf{u}^n|||^2 - \int_{\mathcal{D}} \sum_{j=1}^{2} \tfrac{3\beta_{jj}}{4}|u^{n+1}_{\text{prel},j} - u^n_j|^4 + \tfrac{3\beta_{12}}{2}|u^{n+1}_{\text{prel},1} - u^n_1|^2|u^{n+1}_{\text{prel},2} - u^n_2|^2\, dx. \end{aligned}$$

Proof. For $\mathbf{v} = (v_1, v_2) \in \mathbf{H}^1_0(\mathcal{D})$ we express the energy E in (34) through the operator $\mathcal{L}_{\mathbf{v}}$ as

$$E(\mathbf{v}) = \tfrac{1}{2}\langle \mathcal{L}_{\mathbf{v}}\mathbf{v}, \mathbf{v}\rangle - \int_{\mathcal{D}} \tfrac{\beta_{11}}{4}|v_1|^4 + \tfrac{\beta_{22}}{4}|v_2|^4 + \tfrac{\beta_{12}}{2}|v_1|^2|v_2|^2\, \mathrm{d}x. \tag{45}$$

This yields

$$E(\mathbf{u}^n) - E(\boldsymbol{u}^{n+1}_{\text{prel}}) = \tfrac{1}{2}\Big(\langle \mathcal{L}_{\mathbf{u}^n} \mathbf{u}^n, \mathbf{u}^n \rangle - \langle \mathcal{L}_{\boldsymbol{u}^n_{\text{prel}}} \boldsymbol{u}^n_{\text{prel}}, \boldsymbol{u}^n_{\text{prel}} \rangle \Big) \qquad (46)$$

$$+ \int_{\mathcal{D}} \sum_{j=1}^{2} \tfrac{\beta_{jj}}{4} \left(|u^{n+1}_{\text{prel},j}|^4 - |u^n_j|^4 \right) + \tfrac{\beta_{12}}{2} \left(|u^{n+1}_{\text{prel},1}|^2 |u^{n+1}_{\text{prel},2}|^2 - |u^n_1|^2 |u^n_2|^2 \right) \, \mathrm{d}x.$$

To rewrite the first term in this identity, we first note that

$$\tfrac{1}{\tau_n} \langle \mathcal{L}_{\mathbf{u}^n} (\boldsymbol{u}^{n+1}_{\text{prel}} - \mathbf{u}^n), \boldsymbol{u}^{n+1}_{\text{prel}} - \mathbf{u}^n \rangle \overset{(44)}{=} -\langle \mathcal{L}_{\mathbf{u}^n} \mathbf{u}^n, \boldsymbol{u}^{n+1}_{\text{prel}} - \mathbf{u}^n \rangle. \qquad (47)$$

Together with the self-adjointness of $\mathcal{L}_{\mathbf{u}^n}$, this implies

$$\langle \mathcal{L}_{\mathbf{u}^n} (\boldsymbol{u}^{n+1}_{\text{prel}} - \mathbf{u}^n), \boldsymbol{u}^{n+1}_{\text{prel}} - \mathbf{u}^n \rangle = \langle \mathcal{L}_{\mathbf{u}^n} \boldsymbol{u}^{n+1}_{\text{prel}}, \boldsymbol{u}^{n+1}_{\text{prel}} \rangle + \langle \mathcal{L}_{\mathbf{u}^n} \mathbf{u}^n, \mathbf{u}^n \rangle - 2 \langle \mathcal{L}_{\mathbf{u}^n} \mathbf{u}^n, \boldsymbol{u}^{n+1}_{\text{prel}} \rangle$$

$$= \langle \mathcal{L}_{\mathbf{u}^n} \boldsymbol{u}^{n+1}_{\text{prel}}, \boldsymbol{u}^{n+1}_{\text{prel}} \rangle - \langle \mathcal{L}_{\mathbf{u}^n} \mathbf{u}^n, \mathbf{u}^n \rangle + 2 \langle \mathcal{L}_{\mathbf{u}^n} \mathbf{u}^n, \mathbf{u}^n \rangle - 2 \langle \mathcal{L}_{\mathbf{u}^n} \mathbf{u}^n, \boldsymbol{u}^{n+1}_{\text{prel}} \rangle \rangle$$

$$= \langle \mathcal{L}_{\mathbf{u}^n} \boldsymbol{u}^{n+1}_{\text{prel}}, \boldsymbol{u}^{n+1}_{\text{prel}} \rangle - \langle \mathcal{L}_{\mathbf{u}^n} \mathbf{u}^n, \mathbf{u}^n \rangle + \tfrac{2}{\tau_n} \langle \mathcal{L}_{\mathbf{u}^n} (\boldsymbol{u}^{n+1}_{\text{prel}} - \mathbf{u}^n), \boldsymbol{u}^{n+1}_{\text{prel}} - \mathbf{u}^n \rangle.$$

We obtain

$$\langle \mathcal{L}_{\mathbf{u}^n} \mathbf{u}^n, \mathbf{u}^n \rangle = \langle \mathcal{L}_{\mathbf{u}^n} \boldsymbol{u}^{n+1}_{\text{prel}}, \boldsymbol{u}^{n+1}_{\text{prel}} \rangle + \left(\tfrac{2}{\tau_n} - 1\right) \langle \mathcal{L}_{\mathbf{u}^n} (\boldsymbol{u}^{n+1}_{\text{prel}} - \mathbf{u}^n), \boldsymbol{u}^{n+1}_{\text{prel}} - \mathbf{u}^n \rangle.$$

We can now plug this result into equation (46). Noting that the terms $\langle \mathcal{L}_{\mathbf{u}^n} \boldsymbol{u}^{n+1}_{\text{prel}}, \boldsymbol{u}^{n+1}_{\text{prel}} \rangle$ and $\langle \mathcal{L}_{|\boldsymbol{u}^{n+1}_{\text{prel}}|} \boldsymbol{u}^{n+1}_{\text{prel}}, \boldsymbol{u}^{n+1}_{\text{prel}} \rangle$ only differ in their non-linear parts and that the linear parts cancel each other out, this yields

$$E(\mathbf{u}^n) - E(\boldsymbol{u}^{n+1}_{\text{prel}}) = \left(\tfrac{1}{\tau_n} - \tfrac{1}{2}\right) \langle \mathcal{L}_{\mathbf{u}^n} (\boldsymbol{u}^{n+1}_{\text{prel}} - \mathbf{u}^n), \boldsymbol{u}^{n+1}_{\text{prel}} - \mathbf{u}^n \rangle$$

$$- \sum_{j=1}^{2} \tfrac{\beta_{jj}}{4} \int_{\mathcal{D}} (|u^{n+1}_{\text{prel},j}|^2 - |u^n_j|^2)^2 \, \mathrm{d}x$$

$$- \tfrac{\beta_{12}}{2} \int_{\mathcal{D}} \left(|u^n_1|^2 - |u^{n+1}_{\text{prel},1}|^2\right) \left(|u^n_2|^2 - |u^{n+1}_{\text{prel},2}|^2\right) \mathrm{d}x.$$

Recalling the notation $|||\mathbf{v}|||^2 = \langle \mathcal{L}_0 \mathbf{v}, \mathbf{v} \rangle$ it follows

$$E(\mathbf{u}^n) - E(\boldsymbol{u}^{n+1}_{\text{prel}}) = \left(\tfrac{1}{\tau_n} - \tfrac{1}{2}\right) |||\boldsymbol{u}^{n+1}_{\text{prel}} - u^n|||^2$$

$$+ \left(\tfrac{1}{\tau_n} - \tfrac{1}{2}\right) \Big(\sum_{j=1}^{2} \beta_{jj} \int_{\mathcal{D}} |u^n_j|^2 |u^{n+1}_{\text{prel},j} - u^n_j|^2 \, \mathrm{d}x + \beta_{12} \int_{\mathcal{D}} |u^n_1|^2 |u^{n+1}_{\text{prel},2} - u^n_2|^2 + |u^n_2|^2 |u^{n+1}_{\text{prel},1} - u^n_1|^2 \, \mathrm{d}x \Big)$$

$$- \int_{\mathcal{D}} \left(\sum_{j=1}^{2} \tfrac{\beta_{jj}}{4} (|u^{n+1}_{\text{prel},j}|^2 - |u^n_j|^2)^2 + \tfrac{\beta_{12}}{2} \left(|u^n_1|^2 - |u^{n+1}_{\text{prel},1}|^2\right) \left(|u^n_2|^2 - |u^{n+1}_{\text{prel},2}|^2\right) \right) \mathrm{d}x.$$

A simple application of the triangle inequality and Young's inequality helps in the estimation of the last two terms, where we use

$$(|u^{n+1}_{\text{prel},j}|^2 - |u^n_j|^2)^2 \le 3 |u^{n+1}_{\text{prel},j} - u^n_j|^4 + 6 |u^{n+1}_{\text{prel},j} - u^n_j|^2 |u^n_j|^2 \qquad \text{for } j = 1, 2$$

and

$$\begin{aligned}
&(|u_1^n|^2 - |u_{\mathrm{prel},1}^{n+1}|^2)(|u_2^n|^2 - |u_{\mathrm{prel},2}^{n+1}|^2)\\
&\leq 3\left(|u_{\mathrm{prel},1}^{n+1} - u_1^n|^2 |u_{\mathrm{prel},2}^{n+1} - u_2^n|^2 + |u_{\mathrm{prel},1}^{n+1} - u_1^n|^2 |u_2^n|^2 + |u_{\mathrm{prel},2}^{n+1} - u_2^n|^2 |u_1^n|^2\right).
\end{aligned}$$

We conclude that

$$\begin{aligned}
&E(\mathbf{u}^n) - E(\boldsymbol{u}_{\mathrm{prel}}^{n+1})\\
&\geq \left(\tfrac{1}{\tau_n} - \tfrac{1}{2}\right)|||\boldsymbol{u}_{\mathrm{prel}}^{n+1} - u^n|||^2 - \tfrac{3}{2}\beta_{12}\int_{\mathcal{D}} |u_{\mathrm{prel},1}^{n+1} - u_1^n|^2 |u_{\mathrm{prel},2}^{n+1} - u_2^n|^2 \,\mathrm{d}x - \tfrac{3}{4}\sum_{j=1}^{2}\beta_{jj}\int_{\mathcal{D}} |u_{\mathrm{prel},j}^{n+1} - u_j^n|^4 \,\mathrm{d}x\\
&+ \left(\tfrac{1}{\tau_n} - 2\right)\Big(\sum_{j=1}^{2}\beta_{jj}\int_{\mathcal{D}} |u_j^n|^2 |u_{\mathrm{prel},j}^{n+1} - u_j^n|^2 \,\mathrm{d}x + \beta_{12}\int_{\mathcal{D}} |u_1^n|^2 |u_{\mathrm{prel},2}^{n+1} - u_2^n|^2 + |u_2^n|^2 |u_{\mathrm{prel},1}^{n+1} - u_1^n|^2 \,\mathrm{d}x\Big).
\end{aligned}$$

If we pick the step size τ_n such that $\tau_n \leq \frac{1}{2}$, this implies $\frac{1}{\tau_n} - 2 \geq 0$ and hence the result follows. □

With the previous lemma, we obtain the following lower bound on the energy reduction per iteration.

Lemma 5 *There exists some $\tau_{max} < 2$, depending only on the non-linear interaction parameters, $\mathcal{D}$ and the initial energy $E(\mathbf{u}^0)$, such that for all $\tau_n \in (0, \tau_{max})$,*

$$E(\mathbf{u}^n) - E(\boldsymbol{u}_{\mathrm{prel}}^{n+1}) \;\geq\; C_{\tau_n}\, |||\mathbf{u}^n - \boldsymbol{u}_{\mathrm{prel}}^{n+1}|||^2. \tag{48}$$

The constant fulfills $C_{\tau_n} \geq C_{\tau_{max}} > 0$ for all n. Furthermore,

$$\lim_{n\to\infty} \|\boldsymbol{u}_{\mathrm{prel}}^n\|_{L^2(\mathcal{D})} = 1. \tag{49}$$

Proof. First, note that

$$2\,E(\mathbf{v}) \overset{(34)}{\geq} E(\mathbf{v}) + \int_{\mathcal{D}} \tfrac{\beta_{11}}{4}|v_1|^4 + \tfrac{\beta_{22}}{4}|v_2|^4 + \tfrac{\beta_{12}}{2}|v_1|^2|v_2|^2 \,\mathrm{d}x \overset{(45)}{=} \tfrac{1}{2}\langle \mathcal{L}_{\mathbf{v}}\mathbf{v}, \mathbf{v}\rangle. \tag{50}$$

We obtain

$$\begin{aligned}
&|||\boldsymbol{u}_{\mathrm{prel}}^{n+1} - \mathbf{u}^n|||^2 = \langle \mathcal{L}_0(\boldsymbol{u}_{\mathrm{prel}}^{n+1} - \mathbf{u}^n), \boldsymbol{u}_{\mathrm{prel}}^{n+1} - \mathbf{u}^n\rangle \leq \langle \mathcal{L}_{\mathbf{u}^n}(\boldsymbol{u}_{\mathrm{prel}}^{n+1} - \mathbf{u}^n), \boldsymbol{u}_{\mathrm{prel}}^{n+1} - \mathbf{u}^n\rangle\\
&\overset{(47)}{=} -\tau_n\langle \mathcal{L}_{\mathbf{u}^n}\mathbf{u}^n, \boldsymbol{u}_{\mathrm{prel}}^{n+1} - \mathbf{u}^n\rangle \;=\; -\tau_n\langle \mathcal{L}_{\mathbf{u}^n}\mathbf{u}^n, -\tau_n\mathbf{u}^n + \tau_n\gamma_{\mathbf{u}^n}\mathcal{L}_{\mathbf{u}^n}^{-1}\mathbf{u}^n\rangle\\
&= \tau_n^2\,\langle \mathcal{L}_{\mathbf{u}^n}\mathbf{u}^n, \mathbf{u}^n\rangle - \tau_n^2\,\gamma_{\mathbf{u}^n}\,\langle \mathcal{L}_{\mathbf{u}^n}\mathbf{u}^n, \mathcal{L}_{\mathbf{u}^n}^{-1}\mathbf{u}^n\rangle \overset{\gamma_{\mathbf{u}^n}\geq 0}{\leq} \tau_n^2\,\langle \mathcal{L}_{\mathbf{u}^n}\mathbf{u}^n, \mathbf{u}^n\rangle\\
&\overset{(50)}{\leq} 4\,\tau_n^2\,E(\mathbf{u}^n).
\end{aligned}$$

Hence, for all $n \geq 0$

$$|||\boldsymbol{u}_{\mathrm{prel}}^{n+1} - \mathbf{u}^n|||^2 \;\leq\; 4\,\tau_n^2\,E(\mathbf{u}^n). \tag{51}$$

The statement (48) will now follow by induction: Let $n = 0$, then inequality (51) implies

$$|||\boldsymbol{u}^1_{\text{prel}} - \mathbf{u}^0|||^2 \;\leq\; 4\,\tau_0^2\, E(\mathbf{u}^0) < 1, \tag{52}$$

if τ_0 is chosen such that $\tau_0^2 \;<\; (4\,E(\mathbf{u}^0))^{-1}$. From Lemma 4 and Young's inequality it follows

$$\begin{aligned} E(\mathbf{u}^0) - E(\boldsymbol{u}^1_{\text{prel}}) &\geq \left(\tfrac{1}{\tau_0} - \tfrac{1}{2}\right)|||\boldsymbol{u}^1_{\text{prel}} - \mathbf{u}^0|||^2 - \tfrac{3}{4}\sum_{j=1}^{2}\beta_{jj}\int_{\mathcal{D}} |u^1_{\text{prel},j} - u^0_j|^4\,\mathrm{d}x \\ &\quad - \tfrac{3}{2}\beta_{12}\int_{\mathcal{D}} |u^1_{\text{prel},1} - u^0_1|^2|u^1_{\text{prel},2} - u^0_2|^2\,\mathrm{d}x \\ &\geq \left(\tfrac{1}{\tau_0} - \tfrac{1}{2}\right)|||\boldsymbol{u}^1_{\text{prel}} - \mathbf{u}^0|||^2 - \tfrac{3}{4}\,C_\beta\,\|\boldsymbol{u}^1_{\text{prel}} - \mathbf{u}^0\|^4_{\boldsymbol{L}^4(\mathcal{D})} \end{aligned}$$

for some constant $C_\beta > 0$ that depends on β_{11}, β_{12} and β_{22}. Due to the norm equivalence of $\|\cdot\|_{\boldsymbol{H}^1(\mathcal{D})}$ and $|||\cdot|||$ (guaranteed by Lemma 2) as well as the Sobolev embedding theorem, there exists a constant $C_{\text{Sob}} > 0$ such that

$$\|\boldsymbol{u}^1_{\text{prel}} - \mathbf{u}^0\|^4_{L^4(\mathcal{D})^2} \leq C_{\text{Sob}}\|\boldsymbol{u}^1_{\text{prel}} - \mathbf{u}^0\|^4_{\boldsymbol{H}^1(\mathcal{D})} \overset{(40)}{\leq} 16\,C_{\text{Sob}}|||\boldsymbol{u}^1_{\text{prel}} - \mathbf{u}^0|||^4 \overset{(52)}{\leq} 16\,C_{\text{Sob}}|||\boldsymbol{u}^1_{\text{prel}} - \mathbf{u}^0|||^2.$$

Denote $\tilde{C}_\beta := 12C_{\text{Sob}}C_\beta$, then a combination of the previous two estimates results in

$$E(\mathbf{u}^0) - E(\boldsymbol{u}^1_{\text{prel}}) \;\geq\; \left(\tfrac{1}{\tau_0} - \tfrac{1}{2} - \tilde{C}_\beta\right)|||\boldsymbol{u}^1_{\text{prel}} - \mathbf{u}^0|||^2 \;=:\; C_{\tau_0}\,|||\boldsymbol{u}^1_{\text{prel}} - \mathbf{u}^0|||^2.$$

The constant C_{τ_0} is strictly positive as long as $\tau_0 \leq \tau_{\max}$ for a sufficiently small $\tau_{\max}$. By Lemma 3, we also have $E(\mathbf{u}^1) \leq E(\mathbf{u}^0)$.

For the induction step, assume $E(\mathbf{u}^n) \leq E(\mathbf{u}^0)$ and let $\tau_{\max} > 0$ be sufficiently small for step 1. Then for all $\tau_n \leq \tau_{\max}$ and with equation (51) we can still verify $|||\boldsymbol{u}^{n+1}_{\text{prel}} - \mathbf{u}^n|||^2 \leq 4\tau_n^2 E(\mathbf{u}^n) \leq 4\tau_n^2 E(\mathbf{u}^0) < 1$. Consequently, we can argue as before with the same constants to verify for $C_{\tau_n} = \frac{1}{\tau_n} - \frac{1}{2} - \tilde{C}_\beta$ that $E(\mathbf{u}^n) - E(\boldsymbol{u}^{n+1}_{\text{prel}}) \geq C_{\tau_n}\,|||\boldsymbol{u}^{n+1}_{\text{prel}} - \mathbf{u}^n|||^2$.

For property (49), recall $\|\boldsymbol{u}^{n+1}_{\text{prel}}\|_{L^2(\mathcal{D})} \geq 1$ and $E(\mathbf{u}^{n+1}) \leq E(\boldsymbol{u}^{n+1}_{\text{prel}})$. The Poincaré inequality implies $\|\boldsymbol{u}^{n+1}_{\text{prel}}\|_{L^2(\mathcal{D})} - 1 \leq \|\boldsymbol{u}^{n+1}_{\text{prel}} - \mathbf{u}^n\|_{L^2(\mathcal{D})} \lesssim |||\boldsymbol{u}^{n+1}_{\text{prel}} - \mathbf{u}^n|||$. Combining this with (48) and exploiting that $\lim_{n\to\infty} E(\mathbf{u}^n) - E(\boldsymbol{u}^{n+1}_{\text{prel}}) \leq \lim_{n\to\infty} E(\mathbf{u}^n) - E(\mathbf{u}^{n+1}) = 0$ (due to $E(\mathbf{u}^n)$ being a monotonically decreasing positive sequence), statement (49) follows. □

The lemmas above can be used to prove Theorem 7.(i).

Corollary 1 (Theorem 7.(i).) *There exists $\tau_{max} < 2$ depending on $E(\mathbf{u}^0)$, $\mathcal{D}$, β_{11}, β_{22} and β_{12} such that for all $\tau_n \in (0, \tau_{max})$ and a constant $C_{\tau_n} \geq C_{\tau_{max}} > 0$*

$$E(\mathbf{u}^n) - E(\mathbf{u}^{n+1}) \;\geq\; C_{\tau_n}\,\|\mathbf{u}^{n+1} - \mathbf{u}^n\|^2_{\boldsymbol{H}^1(\mathcal{D})}.$$

Proof. The result follows from Lemmas 2, 3 and 5 if we can show $\|\mathbf{u}^{n+1} - \mathbf{u}^n\|_{\boldsymbol{H}^1(\mathcal{D})} \leq C\|\boldsymbol{u}^{n+1}_{\text{prel}} - \mathbf{u}^n\|_{\boldsymbol{H}^1(\mathcal{D})}$ for some constant $C > 0$. As a first step, we show that $\|u^n\|_{\boldsymbol{H}^1(\mathcal{D})}$

can be bounded in terms of the initial energy, where we get

$$\tfrac{1}{4}\|\mathbf{u}^n\|^2_{\boldsymbol{H}^1(\mathcal{D})} \overset{(40)}{\leq} |||\mathbf{u}^n|||^2 \;\leq\; \langle \mathcal{L}_{\mathbf{u}^n}\mathbf{u}^n, \mathbf{u}^n\rangle \overset{(50)}{\leq} \;4\,E(\mathbf{u}^n) \;\leq\; 4\,E(\mathbf{u}^0). \tag{53}$$

Hence, $\|\mathbf{u}^n\|_{\boldsymbol{H}^1(\mathcal{D})} \leq \sqrt{8}\sqrt{E(\mathbf{u}^0)} =: C_{\mathbf{u}^0}$. The corollary now follows from the fact that $\|\boldsymbol{u}^{n+1}_{\text{prel}}\|_{L^2(\mathcal{D})} \geq 1$ (as per Lemma 3):

$$\begin{aligned}
\|\mathbf{u}^{n+1} - \mathbf{u}^n\|_{\boldsymbol{H}^1(\mathcal{D})} &\leq \|\boldsymbol{u}^{n+1}_{\text{prel}}\|_{\boldsymbol{L}^2(\mathcal{D})}\|\mathbf{u}^{n+1} - \mathbf{u}^n\|_{\boldsymbol{H}^1(\mathcal{D})} \\
&= \|\boldsymbol{u}^{n+1}_{\text{prel}} - \mathbf{u}^n\|_{\boldsymbol{H}^1(\mathcal{D})} + |1 - \boldsymbol{u}^{n+1}_{\text{prel}}\|_{\boldsymbol{L}^2(\mathcal{D})}|\|\mathbf{u}^n\|_{H^1(\mathcal{D})} \\
&= \|\boldsymbol{u}^{n+1}_{\text{prel}} - \mathbf{u}^n\|_{\boldsymbol{H}^1(\mathcal{D})} + \Big(\|\boldsymbol{u}^{n+1}_{\text{prel}}\|_{\boldsymbol{L}^2(\mathcal{D})} - \|\mathbf{u}^n\|_{\boldsymbol{L}^2(\mathcal{D})}\|\Big)\|\mathbf{u}^n\|_{\boldsymbol{H}^1(\mathcal{D})} \\
&\leq \|\boldsymbol{u}^{n+1}_{\text{prel}} - \mathbf{u}^n\|_{\boldsymbol{H}^1(\mathcal{D})} + \|\boldsymbol{u}^{n+1}_{\text{prel}} - \mathbf{u}^n\|_{\boldsymbol{H}^1(\mathcal{D})}\|\mathbf{u}^n\|_{\boldsymbol{H}^1(\mathcal{D})} \\
&\leq \big(1 + C_{\mathbf{u}^0}\big)\,\|\boldsymbol{u}^{n+1}_{\text{prel}} - \mathbf{u}^n\|_{\boldsymbol{H}^1(\mathcal{D})} \overset{(40)}{\leq} 2\,\big(1 + C_{\mathbf{u}^0}\big)\,|||\boldsymbol{u}^{n+1}_{\text{prel}} - \mathbf{u}^n|||.
\end{aligned}$$

Since $\|\boldsymbol{u}^{n+1}_{\text{prel}}\|_{L^2(\mathcal{D})} \geq 1$ yields $E(\mathbf{u}^{n+1}) \leq E(\boldsymbol{u}^{n+1}_{\text{prel}})$, Lemma 5 finishes the proof. □

We are now ready to prove the remaining properties stated in Theorem 7.

Proof of Theorem 7. With property (i) already established, the remaining properties follow with classical compactness arguments.
(ii): Since the sequence $\{E(\mathbf{u}^n)\}_{n\in\mathbb{N}}$ is monotonically decreasing and bounded from below, we have existence of a limit energy $\lim_{n\to\infty} E(\mathbf{u}^n) = E_0$ and property (ii) follows.
(iii): By (53) we know that $\{\mathbf{u}^n\}_{n\in\mathbb{N}}$ is uniformly bounded in $\mathbf{H}^1_0(\mathcal{D})$. Due to the compact embedding of $H^1(\mathcal{D})$ into $L^4(\mathcal{D})$ (for $d \leq 3$), we may extract a subsequence $\{\mathbf{u}^{n_k}\}_{k\in\mathbb{N}}$ that converges to some $\mathbf{u} \in \mathbb{S}$ weakly in $\boldsymbol{H}^1(\mathcal{D})$ and strongly in $\boldsymbol{L}^4(\mathcal{D})$. Using (49), we also have

$$1 = \lim_{n\to\infty} \|\boldsymbol{u}^{n+1}_{\text{prel}}\|_{\boldsymbol{L}^2(\mathcal{D})} = \lim_{n\to\infty} \|(1 - \tau_n)\mathbf{u}^n + \tau_n\gamma_{\mathbf{u}^n}\mathcal{L}^{-1}_{\mathbf{u}^n}\mathbf{u}^n\|_{\boldsymbol{L}^2(\mathcal{D})}.$$

By the coercivity of $\langle \mathcal{L}_{\mathbf{u}}\,\cdot, \cdot\rangle$ in Conclusion 1 we also have

$$\|\mathcal{L}^{-1}_{\mathbf{u}}(\mathbf{u}^{n_k} - \mathbf{u})\|_{\boldsymbol{H}^1(\mathcal{D})} \;\lesssim\; \|\mathbf{u}^{n_k} - \mathbf{u}\|_{\boldsymbol{L}^2(\mathcal{D})} \;\longrightarrow\; 0, \tag{54}$$

i.e., $\mathcal{L}^{-1}_{\mathbf{u}}\mathbf{u}^{n_j} \longrightarrow \mathcal{L}^{-1}_{\mathbf{u}}\mathbf{u}$ in $\boldsymbol{H}^1(\mathcal{D})$. To verify convergence of $\mathcal{L}^{-1}_{|\mathbf{u}^{n_j}|}\mathbf{u}^{n_j}$ to $\mathcal{L}^{-1}_{\mathbf{u}}\mathbf{u}$ (weakly in $\boldsymbol{H}^1(\mathcal{D})$ and strongly in $\boldsymbol{L}^2(\mathcal{D})$), recall that $\mathcal{L}_{\mathbf{u}}$ and $\mathcal{L}_{\mathbf{u}^{n_j}}$ only differ from each other in their non-linear parts, which for arbitrary $\mathbf{w} \in \boldsymbol{H}^1(\mathcal{D})$ yields

$$\begin{aligned}
&|\langle \mathcal{L}_{\mathbf{u}}(\mathcal{L}^{-1}_{\mathbf{u}}\mathbf{u}^{n_k} - \mathcal{L}^{-1}_{\mathbf{u}^{n_k}}\mathbf{u}^{n_k}), \mathbf{w}\rangle| = |\langle \mathcal{L}_{\mathbf{u}^{n_k}}\mathcal{L}^{-1}_{\mathbf{u}^{n_k}}\mathbf{u}^{n_k} - \mathcal{L}_{\mathbf{u}}\mathcal{L}^{-1}_{\mathbf{u}^{n_k}}\mathbf{u}^{n_k}, \mathbf{w}\rangle| \\
&\overset{(39)}{=} \Bigg|\mathrm{Re}\int_{\mathcal{D}} \sum_{j=1}^{2}\Big(\beta_{jj}(\mathcal{L}^{-1}_{\mathbf{u}^{n_k}}\mathbf{u}^{n_k})_j(|u_j^{n_k}|^2 - |u_j|^2)\overline{w}_j\Big) + \beta_{12}(\mathcal{L}^{-1}_{\mathbf{u}^{n_k}}\mathbf{u}^{n_k})_1(|u_2^{n_k}|^2 - |u_2|^2)\overline{w}_1 \\
&\qquad + \mathrm{Re}\int_{\mathcal{D}} \beta_{12}(\mathcal{L}^{-1}_{\mathbf{u}^{n_k}}\mathbf{u}^{n_k})_2(|u_1^{n_k}|^2 - |u_1|^2)\overline{w}_2 \,\mathrm{d}x\Bigg|
\end{aligned}$$

$$\begin{aligned}
&\lesssim \| \, |\mathbf{u}^{n_k}|^2 - |\mathbf{u}|^2 \|_{L^2(\mathcal{D})} \, \| \mathcal{L}_{\mathbf{u}^{n_k}}^{-1} \mathbf{u}^{n_k} \|_{L^4(\mathcal{D})} \| \mathbf{w} \|_{L^4(\mathcal{D})} \\
&\lesssim \| \, |\mathbf{u}^{n_k}|^2 - |\mathbf{u}|^2 \|_{L^2(\mathcal{D})} \, \| \mathbf{u}^{n_k} \|_{L^2(\mathcal{D})} \| \mathbf{w} \|_{H^1(\mathcal{D})} \\
&= \| \, |\mathbf{u}^{n_k}|^2 - |\mathbf{u}|^2 \|_{L^2(\mathcal{D})} \, \| \mathbf{w} \|_{H^1(\mathcal{D})} \\
&\lesssim \| \mathbf{u}^{n_k} - \mathbf{u} \|_{\mathbf{L}^4(\mathcal{D})} \, \| \mathbf{w} \|_{H^1(\mathcal{D})} \;\; \rightarrow \;\; 0 \qquad \text{for } n_k \rightarrow \infty,
\end{aligned}$$

where we used in the last two steps that $\mathbf{u}^{n_k}$ is uniformly bounded in $\mathbf{H}^1(\mathcal{D})$ (and hence also in $\mathbf{L}^4(\mathcal{D})$) such that $\mathbf{u}^{n_k}$ converges to $\mathbf{u}$ strongly in $\mathbf{L}^4(\mathcal{D})$. With the test function $\mathbf{w} = \mathcal{L}_{\mathbf{u}}^{-1}\mathbf{u}^{n_k} - \mathcal{L}_{\mathbf{u}^{n_k}}^{-1}\mathbf{u}^{n_k}$ and the coercivity of $\mathcal{L}_{\mathbf{u}}$ we therefore obtain

$$\| \mathcal{L}_{\mathbf{u}}^{-1}\mathbf{u}^{n_k} - \mathcal{L}_{\mathbf{u}^{n_k}}^{-1}\mathbf{u}^{n_k} \|_{\mathbf{H}^1(\mathcal{D})} \;\lesssim\; \| \mathbf{u}^{n_k} - \mathbf{u} \|_{\mathbf{L}^4(\mathcal{D})} \;\; \rightarrow \;\; 0 \qquad \text{for } n_k \rightarrow \infty.$$

Together with (54), we conclude that

$$\mathcal{L}_{\mathbf{u}^{n_k}}^{-1}\mathbf{u}^{n_k} \longrightarrow \mathcal{L}_{\mathbf{u}}^{-1}\mathbf{u} \quad \text{ in } H^1(\mathcal{D})$$

and hence also

$$\gamma_{\mathbf{u}^{n_k}} \;=\; (\mathbf{u}^{n_k}, \mathcal{L}_{\mathbf{u}^{n_k}}^{-1}\mathbf{u}^{n_k})_{L^2(\mathcal{D})}^{-1} \;\longrightarrow\; (\mathbf{u}, \mathcal{L}_{\mathbf{u}}^{-1}\mathbf{u})_{L^2(\mathcal{D})}^{-1} \;=:\; \lambda.$$

Since $\|\boldsymbol{u}_{\text{prel}}^{(n_k+1)} - \mathbf{u}^{n_k}\|_{H^1(\mathcal{D})} \rightarrow 0$ (as a consequence of Lemma 5) and since τ_n is bounded from below and from above, we can pass to the limit in (43) to see that

$$\mathbf{u}^{n_k} = -\frac{1}{\tau_{n_k}}(\boldsymbol{u}_{\text{prel}}^{n_k+1} - \mathbf{u}^{n_k}) \,+\, \gamma_{\mathbf{u}^{n_k}} \, \mathcal{L}_{\mathbf{u}^{n_k}}^{-1}\mathbf{u}^{n_k} \;\longrightarrow\; \lambda \, \mathcal{L}_{\mathbf{u}}^{-1}\mathbf{u} \qquad \text{strongly in } \mathbf{H}^1(\mathcal{D})$$

for $n_k \rightarrow \infty$. From the equation above we know $\mathbf{u}^{n_k}$ must converge strongly in $\mathbf{H}^1(\mathcal{D})$ and as we already know that the weak limit is $\mathbf{u}$, we conclude that $\mathbf{u}^{n_k}$ does in fact converge strongly to $\mathbf{u}$ and that it fulfills the equation $\mathbf{u} = \lambda \, \mathcal{L}_{\mathbf{u}}^{-1}\mathbf{u}$. Applying $\mathcal{L}_{\mathbf{u}}$ to the equation yields $E'(\mathbf{u}) = \mathcal{L}_{\mathbf{u}}\mathbf{u} = \lambda(\mathbf{u}, \cdot)_{L^2(\mathcal{D})}$, i.e., $\mathbf{u}$ is a critical point of E. This proves (iii).

(iv): If one of the limits is a locally quasi-isolated ground state, then the corresponding ground state density $|\mathbf{u}|^2$ is locally unique. In particular, there is no path of constant energy that connects $\mathbf{u}$ with another ground state (aside from phase shifts). Due to the strict energy decay of the iterates $\mathbf{u}^n$, it is hence impossible for the iterates to leave a small neighborhood of the ground state for all sufficiently large n, as this would require an intermediate increase of the energy. Consequently, all density iterates $|\mathbf{u}^n|^2$ must converge to the same limit. The strong convergence in $\mathbf{L}^2$ is again due to the embedding of $H^1(\mathcal{D})$ into $L^4(\mathcal{D})$ as used before. For a mathematical formalization of the proof we refer to the arguments elaborated in [40]. □

4.4 Implementation Aspects and Computational Cost

We briefly comment on the practical realization of the metric-driven Riemannian gradient method (MDRGM) and the associated computational cost. Recall that MDRGM, as defined in Definition 4, is an abstract iterative scheme and requires a spatial discretization for its implementation.

From an algorithmic point of view, one iteration of MDRGM consists of five main steps:

(i) The assembly of system matrices to represent the discrete version of the elliptic differential operator $\mathcal{L}_{\mathbf{u}^n}$. Note that this requires nonlinear updates in each iteration to account for the influence of the current iterate $\mathbf{u}^n$.
(ii) The computation of $\mathcal{L}_{\mathbf{u}^n}^{-1}\mathbf{u}^n$ (i.e., the application of an inverse operator), which requires the solution of one linear system per iteration with the system matrix computed in (i) and the source term given by $\mathbf{u}^n$.
(iii) A linear update $u_{\text{prel}}^{n+1}(\tau) := (1-\tau)\,\mathbf{u}^n + \tau\,(\mathcal{L}_{\mathbf{u}^n}^{-1}\mathbf{u}^n, \mathbf{u}^n)_{\mathrm{L}^2(\mathcal{D})}^{-1}\,\mathcal{L}_{\mathbf{u}^n}^{-1}\mathbf{u}^n$ which only involves matrix-vector operations once $\mathcal{L}_{\mathbf{u}^n}^{-1}\mathbf{u}^n$ is available. For example, $(\mathcal{L}_{\mathbf{u}^n}^{-1}\mathbf{u}^n, \mathbf{u}^n)_{\mathrm{L}^2(\mathcal{D})}$ can be easily assembled using the mass matrix.
(iv) Normalization $u^{n+1}(\tau) := u_{\text{prel}}^{n+1}(\tau)/\|u_{\text{prel}}^{n+1}(\tau)\|_{\mathrm{L}^2(\mathcal{D})}$.
(v) Computation of the optimal step size $\tau_n = \underset{0<\tau\leq 2}{\arg\min}\; E\left(u^{n+1}(\tau)\right)$ which can be realized at negligible cost by a line search, noting that $E\left(u^{n+1}(\tau)\right)$ is a rational function in τ with explicitly computable coefficients. Finally setting $u^{n+1} := u^{n+1}(\tau^n)$.

The efficient assembly of the system matrices and the realization of the line search without relevant computational overhead are discussed in detail in [40]. In typical implementations, the linear solve (ii) constitutes the dominant part of the computational cost, while the remaining steps are minor in comparison.

Hence, up to discretization-specific constants, each iteration of MDRGM is essentially equivalent to one assembly of the discrete operator and one linear solve in the chosen approximation space. The computational complexity per iteration is therefore determined by the dimension of this space together with the cost of assembly and the efficiency of the underlying linear solver. In particular, for a discretization with N degrees of freedom, the cost per iteration scales like the solution of a linear system of size N, up to additional costs arising from the assembly of the discrete operator.

The total computational effort is thus given by the product of the number of iterations and the cost per iteration. As discussed above and observed in the numerical experiments, the number of iterations mainly depends on spectral properties of the operator and is largely independent of the discretization parameters.

In this work, MDRGM is in particular combined with metric-driven approximation spaces that admit efficient realizations via localized basis functions. This allows to keep the dimension of the discrete problem small while still capturing the characteristic features of the solution, thereby reducing the cost of each iteration without increasing the number of iterations.

4.5 Numerical Experiments

We will now demonstrate the convergence of the algorithm to a ground state of the energy functional and we will also compare the approximation properties obtained in a standard $\mathbb{P}^1$-Lagrange finite element space with the approximation properties in a metric-driven approximation space in the spirit of Section 3.2.

To fix the setting, we consider the domain $\mathcal{D} = [-1, 1]^2$ and the parameters

$$\beta_{11} = 10,\ \beta_{12} = \beta_{22} = 9,\ k_0 = 10,\ \Omega = 50 \text{ and } \delta = 0.$$

To ensure that assumption (A2) is fulfilled we shift the energy level by a constant, which is achieved by selecting the trapping potentials as $V_j(x) = \frac{\Omega+\delta+2k_0^2}{2}$ for $j = 1, 2$.

4.5.1 Metric-Driven Steepest Descent

To avoid an influence from the spatial discretization and to isolate the error from the iterative solver, we fix a sufficiently accurate approximation space for the experiments in this subsection. To be precise, the exact solution space $\mathbf{H}_0^1(\mathcal{D})$ is discretized using a (standard) $\mathbb{P}^2$-Lagrange finite element on a uniform mesh with $(2^8 - 1)^2$ degrees of freedom. Metric-driven approximation spaces are used for the experiments in the next subsection.

In the following experiments, we compare three different iterative schemes. First, we consider the metric-driven Riemannian gradient method (41) as stated in Definition 4, abbreviated here by MDRGM. The optimal step size τ_n for the MDRGM is computed with the golden-section line search. Second, we consider the "inverse iteration", abbreviated by InvIter, which is obtained by fixing $\tau = 1$ in (41). As a third method, we also include the popular GFDN method which is obtained by a modified backward Euler discretization of an L^2-gradient flow with discrete normalization (GFDN), cf. [10, 11, 30]. The iterations are formally given by

$$\mathbf{u}^{n+1} := \frac{(\mathrm{id} + \tau_n \mathcal{L}_{\mathbf{u}^n})^{-1}\mathbf{u}^n}{\|(\mathrm{id} + \tau_n \mathcal{L}_{\mathbf{u}^n})^{-1}\mathbf{u}^n\|_{L^2(\mathcal{D})}} \qquad \text{(GFDN)},$$

where we select the step size uniformly as $\tau_n = 1$ (since an adaptive computation of the optimal τ-values is computationally not feasible for the GFDN). All iterative methods are initialized by interpolating the following functions in the finite element space:

$$u_1^0(x_1, x_2) = \frac{1}{2}(x_1 - 1)^2(x_2 - 1)^2 \exp(-\frac{x_1^2 + x_2^2}{2}\mathrm{i}),$$

$$u_2^0(x_1, x_2) = (x_1 - 1)^2(x_2 - 1)^2 \exp(-\frac{x_1^2 + x_2^2}{2}\mathrm{i}).$$

In this case, the initial energy is given by $E(\mathbf{u}^0) = 74.97448979636732$. The iterations were stopped as soon as the energy difference of two subsequent iterations fell below the threshold $\varepsilon = 10^{-11}$ and our calculations were carried out using `julia`.

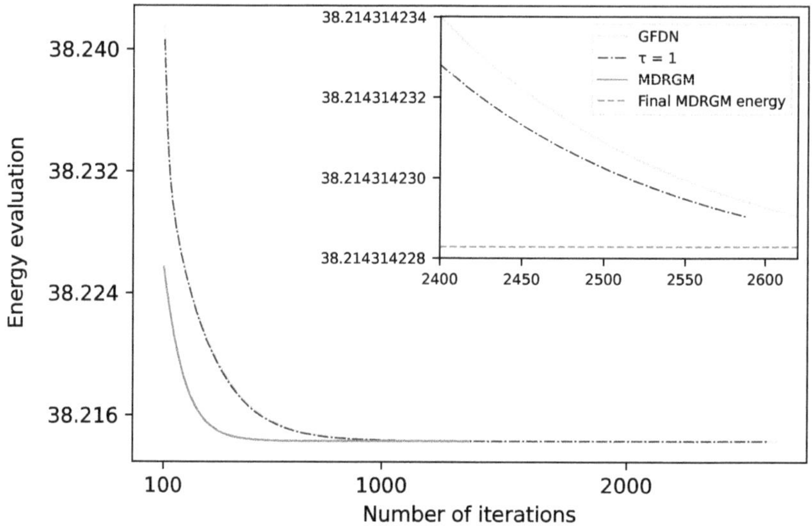

Figure 1 Comparison of the energy per iteration for the MDRGM, the inverse iteration and the GFDN.

In Figure 1 we can see the energy evolution with respect to the number of iterations for the three different schemes. We observe that the metric-driven Riemannian gradient method (MDRGM) shows the best performance. The inverse iteration and the GFDN took around 2600 iterations to reach the tolerance, while the MDRGM only required 1354 iterations. Here we note that the costs per time step are essentially the same for all three methods and that the computation of the optimal τ for the MDRGM does not cause any significant overhead. Table 1 shows the energy and eigenvalue approximation for the three methods after the final iteration. Evidently, the MDRGM is not only faster, but also yields a slightly smaller energy compared to the inverse iteration, which in turn also outputs a slightly smaller final energy compared to the GFDN.

Method	No. of iterations	Final energy	Final eigenvalue
GFDN	2620	38.214314229059	78.376223785843
InvIter	2589	38.214314229034	78.376223776962
MDRGM	1354	38.214314228286	78.376223454563

Table 1 Comparison of final results

To verify whether a found state **u** is a (local) minimizer, we recall from Section 4.1.2 that we have to check the first and the second order condition for minimizers. For that we compute the residuals in the first order condition $E'(\mathbf{u}) = \lambda(\mathbf{u}, \cdot)_{L^2(\mathcal{D})}$ and find that they range between $O(10^{-8})$ and $O(10^{-10})$ in the maximum norm for the various methods. Hence, the first order condition is fulfilled up to the numerical precision of the methods. For the second order condition, we have to verify that λ is the unique smallest eigenvalue of the operator $E''(\mathbf{u})|_{T_\mathbf{u}\mathbb{S}}$. The results are depicted in Table 2. We can confirm that λ is indeed simple and at the bottom of the spectrum. Furthermore, we observe that the smallest eigenvalues of $E''(\mathbf{u})|_{T_\mathbf{u}\mathbb{S}}$ lie very close together, which often indicates the presence of interesting physical phenomena, as demonstrated by the density plots.

i	1	2	3
GFDN	78.376223779431	78.636660348051	78.651177504685
InvIter	78.376223770582	78.636660410597	78.651177520292
MDRGM	78.376223448183	78.636662675963	78.651178085638

Table 2 The three smallest eigenvalues of $E''(\mathbf{u})|_{T_\mathbf{u}\mathbb{S}}$ sorted in ascending order. Here, **u** denotes the state after the final iteration obtained with GFDN, inverse iteration and MDRGM respectively.

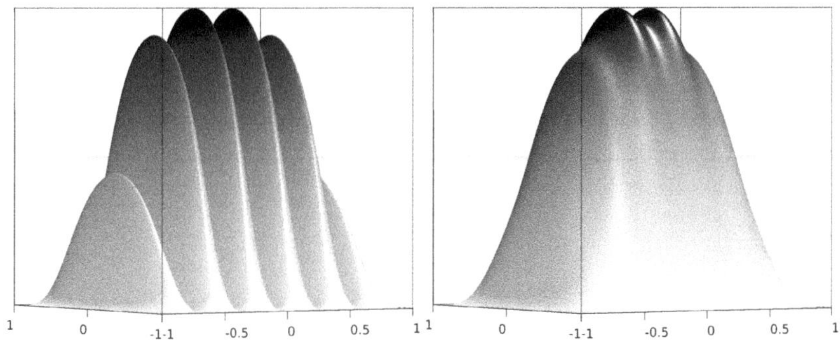

Figure 2 Converged ground state density $|\mathbf{u}|^2$. The left picture shows the density of the first component and the right picture the density of the second component.

The density plots in Figure 2 reveal the formation of disk-like structures in the first component, characterized by alternating regions of high and low density. This pattern is indicative of a supersolid phase, meaning that the state of matter exhibits properties of solids and superfluids at the same time.

4.5.2 Metric-Driven Approximation Spaces

We next investigate the convergence of the metric-driven approximation spaces (LOD) towards the fully resolved solution of (31). To this end, we consider the spaces $V_H^{\mathcal{L}_0}$ defined in (30) with respect to a sequence of coarse triangulations with mesh widths $H = 2^{-1}, \ldots, 2^{-5}$. Recall that this construction of the metric-driven space neglects nonlinear contributions and does not require updates.

In order to make the method computationally feasible, we employ two simplifications. First, we localize the correctors, i.e., we compute localized basis functions on finite element patches of width $|\log_2(H)|$, which ensures that the sparsity pattern is preserved. This follows standard practice in LOD theory [51, 37, 53, 4]. Second, we omit the coupling term in the construction of the spaces. As a consequence, the approximation spaces decouple, and the same precomputed basis can be used for both components.

Even with these simplifications, which improve the efficiency of the method, the convergence rates predicted by Theorem 5 for the simplified setting are still observed in the present, more challenging case of a spin-orbit coupled BEC; see Figure 3. Already at refinement level 4, the practical metric-driven space captures the characteristic disk-like structure of the ground state; see Figure 4.

For comparison, Figure 3 also reports the errors obtained with standard $\mathbb{P}^1$ finite elements on the same meshes. As expected, this problem-agnostic method converges, but at a significantly slower rate when measured against the number of degrees of freedom. The logarithmic communication overhead inherent to the metric-driven method improves the approximation quality substantially.

Given that the iterative solvers for this setup require order 10^3 iterations, the additional cost of precomputing the metric-driven basis is justified. Overall, these results provide a promising proof of concept for the LOD approach in this context. A more systematic and rigorous investigation will be the subject of future research.

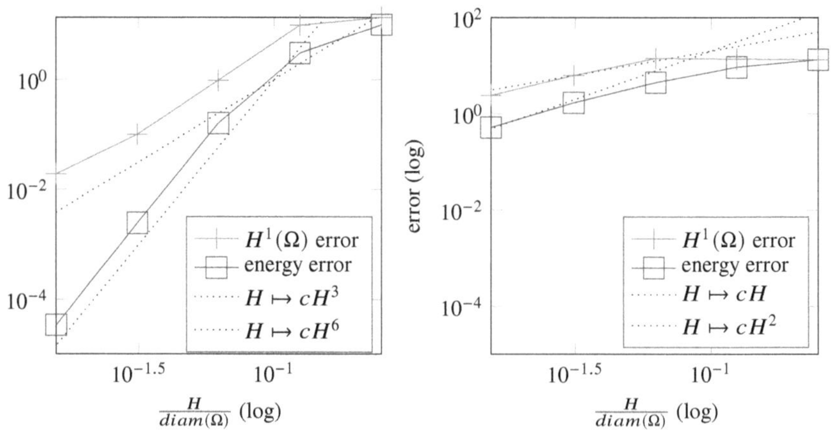

Figure 3 Convergence of the metric-driven approximation (LOD, left) compared with standard $\mathbb{P}^1$-FEM (right). Shown are the H^1 and energy errors with respect to a sequence of uniformly refined triangular meshes of width H.

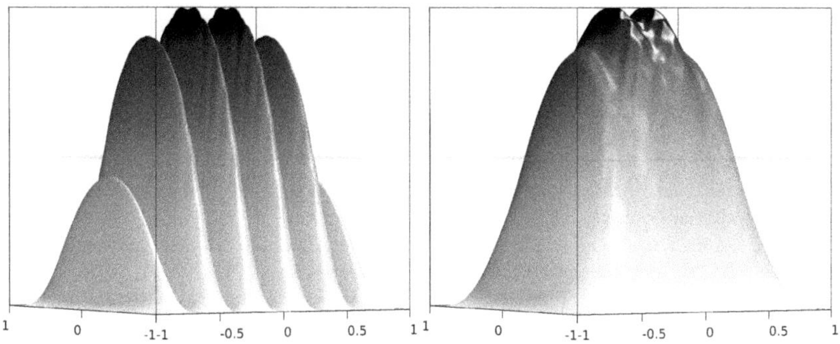

Figure 4 Ground state density $|\mathbf{u}|^2$ obtained with the LOD method on refinement level 4. Left: first component. Right: second component. The main qualitative features of the disk-like structure are already captured at this resolution.

Acknowledgements The authors acknowledge the support by the Deutsche Forschungsgemeinschaft (DFG, German Research Foundation) where P. Henning and L. Huynh are supported through the project grant 551527112 and D. Peterseim through the project grant 564828373.

References

1. A. Abdulle, W. E, B. Engquist, and E. Vanden-Eijnden. The heterogeneous multiscale method. *Acta Numer.*, 21:1–87, 2012.
2. Y. Ai, P. Henning, M. Yadav, and S. Yuan. Riemannian conjugate Sobolev gradients and their application to compute ground states of BECs. *J. Comput. Appl. Math.*, 473:Paper No. 116866, 17, 2026.
3. R. Altmann, P. Henning, and D. Peterseim. The J-method for the Gross–Pitaevskii eigenvalue problem. *Numer. Math.*, 148(3):575–610, 2021.
4. R. Altmann, P. Henning, and D. Peterseim. Numerical homogenization beyond scale separation. *Acta Numer.*, 30:1–86, 2021.
5. R. Altmann, M. Hermann, D. Peterseim, and T. Stykel. Riemannian optimization methods for ground states of multicomponent Bose–Einstein condensates. *IMA J. Numer. Anal.*, 2026+.
6. R. Altmann, D. Peterseim, and T. Stykel. Energy-adaptive Riemannian optimization on the Stiefel manifold. *ESAIM Math. Model. Numer. Anal.*, 56(5):1629–1653, 2022.
7. R. Altmann, D. Peterseim, and T. Stykel. Riemannian Newton methods for energy minimization problems of Kohn-Sham type. *J. Sci. Comput.*, 101(1):Paper No. 6, 25, 2024.
8. I. Babuska and R. Lipton. Optimal local approximation spaces for generalized finite element methods with application to multiscale problems. *Multiscale Model. Simul.*, 9(1):373–406, 2011.
9. I. Babuška, R. Lipton, P. Sinz, and M. Stuebner. Multiscale-spectral GFEM and optimal oversampling. *Comput. Methods Appl. Mech. Engrg.*, 364:112960, 28, 2020.
10. W. Bao and Y. Cai. Ground states and dynamics of spin-orbit-coupled Bose–Einstein condensates. *SIAM J. Appl. Math.*, 75(2):492–517, 2015.
11. W. Bao and Y. Cai. Mathematical models and numerical methods for spinor Bose–Einstein condensates. *Commun. Comput. Phys.*, 24(4):899–965, 2018.
12. W. Bao and Q. Du. Computing the ground state solution of Bose–Einstein condensates by a normalized gradient flow. *SIAM J. Sci. Comput.*, 25(5):1674–1697, 2004.
13. S. N. Bose. Plancks Gesetz und Lichtquantenhypothese. *Zeitschrift für Physik*, 26(1):178–181, 1924.

14. E. Cancès, R. Chakir, and Y. Maday. Numerical analysis of nonlinear eigenvalue problems. *J. Sci. Comput.*, 45(1-3):90–117, 2010.
15. E. Cancès and C. Le Bris. Can we outperform the DIIS approach for electronic structure calculations? *Int. J. Quantum Chem.*, 79(2):82–90, 2000.
16. E. Cancès and C. Le Bris. On the convergence of SCF algorithms for the Hartree–Fock equations. *M2AN Math. Model. Numer. Anal.*, 34(4):749–774, 2000.
17. C. Carstensen. Quasi-interpolation and a posteriori error analysis in finite element methods. *M2AN Math. Model. Numer. Anal.*, 33(6):1187–1202, 1999.
18. Z. Chen, J. Lu, Y. Lu, and X. Zhang. On the convergence of Sobolev gradient flow for the Gross–Pitaevskii eigenvalue problem. *SIAM J. Numer. Anal.*, 62(2):667–691, 2024.
19. E. Chung, Y. Efendiev, and T. Y. Hou. *Multiscale model reduction—multiscale finite element methods and their generalizations*, volume 212 of *Applied Mathematical Sciences*. Springer, Cham, 2023.
20. I. Danaila and P. Kazemi. A new Sobolev gradient method for direct minimization of the Gross–Pitaevskii energy with rotation. *SIAM J. Sci. Comput.*, 32(5):2447–2467, 2010.
21. I. Danaila and B. Protas. Computation of ground states of the Gross–Pitaevskii functional via Riemannian optimization. *SIAM J. Sci. Comput.*, 39(6):B1102–B1129, 2017.
22. C. Döding, P. Henning, and J. Wärnegård. A two level approach for simulating Bose-Einstein condensates by localized orthogonal decomposition. *ESAIM Math. Model. Numer. Anal.*, 58(6):2317–2349, 2024.
23. W. E and B. Engquist. The heterogeneous multiscale methods. *Commun. Math. Sci.*, 1(1):87–132, 2003.
24. Y. Efendiev, J. Galvis, and T. Y. Hou. Generalized multiscale finite element methods (GMsFEM). *J. Comput. Phys.*, 251:116–135, 2013.
25. Y. Efendiev and T. Y. Hou. *Multiscale finite element methods: Theory and applications*, volume 4 of *Surveys and Tutorials in the Applied Mathematical Sciences*. Springer, New York, 2009.
26. A. Einstein. *Quantentheorie des einatomigen idealen Gases*, pages 261–267. Sitzber. Kgl. Preuss. Akad. Wiss., 1924.
27. C. Engwer, P. Henning, A. Målqvist, and D. Peterseim. Efficient implementation of the localized orthogonal decomposition method. *Comput. Methods Appl. Mech. Engrg.*, 350:123–153, 2019.
28. Z. Feng, Q. Shu, and Q. Tang. Mass-preserving and energy-diminishing semi-implicit schemes for computing ground states of rotating Bose-Einstein condensates. *J. Sci. Comput.*, 106(1):Paper No. 30, 28, 2026.
29. Z. Feng and Q. Tang. On preconditioned Riemannian gradient methods for minimizing the Gross–Pitaevskii energy functional: algorithms, global convergence and optimal local convergence rate. ArXiv e-print 2510.13516, 2025.
30. Z. Feng, Q. Tang, and C. Wang. On the discrete normalized gradient flow for computing ground states of rotating Bose–Einstein condensates: energy dissipation and global convergence. *IMA J. Numer. Anal.*, 11 2026+.
31. M. Hauck and D. Peterseim. Super-localization of elliptic multiscale problems. *Math. Comp.*, 92(341):981–1003, 2023.
32. P. Henning. The dependency of spectral gaps on the convergence of the inverse iteration for a nonlinear eigenvector problem. *Math. Models Methods Appl. Sci.*, 33(07):1517–1544, 2023.
33. P. Henning and E. Jarlebring. The Gross–Pitaevskii equation and eigenvector nonlinearities: numerical methods and algorithms. *SIAM Rev.*, 67(2):256–317, 2025.
34. P. Henning, A. Målqvist, and D. Peterseim. A localized orthogonal decomposition method for semi-linear elliptic problems. *ESAIM Math. Model. Numer. Anal. (M2AN)*, 48(5):1331–1349, 2014.
35. P. Henning, A. Målqvist, and D. Peterseim. Two-level discretization techniques for ground state computations of Bose–Einstein condensates. *SIAM J. Numer. Anal.*, 52(4):1525–1550, 2014.
36. P. Henning and A. Persson. On optimal convergence rates for discrete minimizers of the Gross–Pitaevskii energy in localized orthogonal decomposition spaces. *Multiscale Model. Simul.*, 21(3):993–1011, 2023.
37. P. Henning and D. Peterseim. Oversampling for the Multiscale Finite Element Method. *SIAM Multiscale Model. Simul.*, 11(4):1149–1175, 2013.

38. P. Henning and D. Peterseim. Sobolev gradient flow for the Gross–Pitaevskii eigenvalue problem: global convergence and computational efficiency. *SIAM J. Numer. Anal.*, 58(3):1744–1772, 2020.
39. P. Henning and J. Wärnegård. Superconvergence of time invariants for the Gross–Pitaevskii equation. *Math. Comp.*, 91(334):509–555, 2022.
40. P. Henning and M. Yadav. Convergence of a Riemannian gradient method for the Gross–Pitaevskii energy functional in a rotating frame. *ESAIM Math. Model. Numer. Anal.*, 59(2):1145–1175, 2025.
41. P. Henning and M. Yadav. On discrete ground states of rotating Bose–Einstein condensates. *Math. Comp.*, 94(351):1–32, 2025.
42. T. Y. Hou and X.-H. Wu. A multiscale finite element method for elliptic problems in composite materials and porous media. *J. Comput. Phys.*, 134(1):169–189, 1997.
43. T. J. R. Hughes, G. R. Feijóo, L. Mazzei, and J.-B. Quincy. The variational multiscale method—a paradigm for computational mechanics. *Comput. Methods Appl. Mech. Engrg.*, 166(1-2):3–24, 1998.
44. T. J. R. Hughes and G. Sangalli. Variational multiscale analysis: the fine-scale Green's function, projection, optimization, localization, and stabilized methods. *SIAM J. Numer. Anal.*, 45(2):539–557, 2007.
45. E. Jarlebring, S. Kvaal, and W. Michiels. An inverse iteration method for eigenvalue problems with eigenvector nonlinearities. *SIAM J. Sci. Comput.*, 36(4):A1978–A2001, 2014.
46. P. Kazemi and M. Eckart. Minimizing the Gross–Pitaevskii energy functional with the Sobolev gradient – analytical and numerical results. *Int. J. Comput. Methods*, 7(3):453–475, 2010.
47. M. G. Larson and A. Målqvist. Adaptive variational multiscale methods based on a posteriori error estimation: energy norm estimates for elliptic problems. *Comput. Methods Appl. Mech. Engrg.*, 196(21-24):2313–2324, 2007.
48. J.-R. Li, J. Lee, W. Huang, S. Burchesky, B. Shteynas, F. C. Top, A. O. Jamison, and W. Ketterle. A stripe phase with supersolid properties in spin–orbit-coupled Bose–Einstein condensates. *Nature*, 543(7643):91–94, 2017.
49. C. Ma and R. Scheichl. Error estimates for discrete generalized FEMs with locally optimal spectral approximations. *Math. Comp.*, 91(338):2539–2569, 2022.
50. C. Ma, R. Scheichl, and T. Dodwell. Novel design and analysis of generalized finite element methods based on locally optimal spectral approximations. *SIAM J. Numer. Anal.*, 60(1):244–273, 2022.
51. A. Målqvist and D. Peterseim. Localization of elliptic multiscale problems. *Math. Comp.*, 83(290):2583–2603, 2014.
52. A. Målqvist and D. Peterseim. Computation of eigenvalues by numerical upscaling. *Numer. Math.*, 130(2):337–361, 2015.
53. A. Målqvist and D. Peterseim. *Numerical homogenization by localized orthogonal decomposition*, volume 5 of *SIAM Spotlights*. Society for Industrial and Applied Mathematics (SIAM), Philadelphia, PA, 2021.
54. J. W. Neuberger. *Sobolev gradients and differential equations*, volume 1670 of *Lecture Notes in Mathematics*. Springer-Verlag, Berlin, 1997.
55. H. Owhadi. Multigrid with rough coefficients and multiresolution operator decomposition from hierarchical information games. *SIAM Rev.*, 59(1):99–149, 2017.
56. H. Owhadi and C. Scovel. *Operator-adapted wavelets, fast solvers, and numerical homogenization – From a game theoretic approach to numerical approximation and algorithm design*, volume 35 of *Cambridge Monographs on Applied and Computational Mathematics*. Cambridge University Press, Cambridge, 2019.
57. H. Owhadi and L. Zhang. Metric-based upscaling. *Comm. Pure Appl. Math.*, 60(5):675–723, 2007.
58. D. Peterseim, J. Wärnegård, and C. Zimmer. Super-localised wave function approximation of Bose-Einstein condensates. *J. Comput. Phys.*, 510:Paper No. 113097, 23, 2024.
59. L. Pitaevskii and S. Stringari. *Bose—Einstein condensation*, volume 116 of *International Series of Monographs on Physics*. The Clarendon Press, Oxford University Press, Oxford, 2003.
60. S. Shrestha, J. Dekker, M. Gerritsma, S. Hulshoff, and I. Akkerman. Construction and application of an algebraic dual basis and the fine-scale Green's function for computing projections and reconstructing unresolved scales. *Comput. Methods Appl. Mech. Engrg.*, 422:Paper No. 116833, 24, 2024.

61. P. Upadhyaya, E. Jarlebring, and E. H. Rubensson. A density matrix approach to the convergence of the self-consistent field iteration. *Numer. Algebra Control Optim.*, 11(1):99–115, 2021.
62. X. Wu, Z. Wen, and W. Bao. A regularized Newton method for computing ground states of Bose–Einstein condensates. *J. Sci. Comput.*, 73(1):303–329, 2017.
63. Y. Wu, C. Liu, and Y. Cai. Normalized flows based on Sobolev gradients for computing ground states of spinor Bose-Einstein condensates. *J. Comput. Phys.*, 538:Paper No. 114153, 2025.
64. F. Xu, H. Xie, M. Xie, and M. Yue. A multigrid method for the ground state solution of Bose–Einstein condensates based on Newton iteration. *BIT*, 61(2):645–663, 2021.
65. X. Zhang and H. Wang. Second-order Sobolev gradient flows for computing ground state of ultracold Fermi gases. *J. Comput. Appl. Math.*, 451:Paper No. 116096, 15, 2024.
66. Z. Zhang. Exponential convergence of Sobolev gradient descent for a class of nonlinear eigenproblems. *Commun. Math. Sci.*, 20(2):377–403, 2022.

Concentrated Real-Pole Uniform-in-Time Approximation of the Matrix Exponential

Stefan Güttel and Shuai Shao

Abstract We propose an asymptotically optimal choice of shared concentrated real poles of a family of rational approximants of time-dependent exponential functions $\exp(-tz)$ for $z \geq 0$ and t in a positive time interval T. Our result extends a classical result by J.-E. Andersson [*J. Approx. Theory*, 32(2):85–95, 1981] on the asymptotic best rational approximation of $\exp(-z)$ with real poles. Numerical experiments demonstrate the near-optimality of our choice for various time ranges and for both small and large approximation degrees. An application of the uniform-in-time rational approximation using our proposed concentrated real poles to a linear constant-coefficient initial-value problem is also discussed.

1 Introduction

The numerical approximation of the action of a parametric matrix exponential function on a vector, $\exp(-tA)\boldsymbol{b}$, with a symmetric positive semidefinite matrix $A \in \mathbb{R}^{N\times N}$, a vector $\boldsymbol{b} \in \mathbb{R}^N$, for all $t \in T$ in a positive time interval T, is an important problem in science and engineering. For example, the exact solution of the fundamental dynamical system $\boldsymbol{f}'(t) + A\boldsymbol{f}(t) = \boldsymbol{0}$ can be written as $\boldsymbol{f}(t) = \exp(-tA)\boldsymbol{b}$, where $\boldsymbol{b} = \boldsymbol{f}(0)$ is the initial vector. Such systems arise when solving evolution problems in geophysics [9, 26, 10], chemical reactions [24, 21], epidemiology [2], network analysis [13, 5], and exponential integration [20, 16, 17]. For a comprehensive introduction to matrix functions we refer to [19].

Stefan Güttel
Department of Mathematics, The University of Manchester, United Kingdom, stefan.guettel@manchester.ac.uk

Shuai Shao
Department of Mathematics, The University of Manchester, United Kingdom, shuai.shao-2@manchester.ac.uk

B. Koren and D. Roose (eds.), *Numerical Analysis and Scientific Computing: an Anthology*, Lecture Notes in Computational Science and Engineering 156,
https://doi.org/10.1007/978-3-032-33428-2_2

The explicit evaluation of a matrix function $\exp(A)$ is usually only possible for small to medium-sized matrices. For example, MATLAB's `expm` function implements a scaling and squaring approach to compute $\exp(A)$, which first computes a Padé approximation of $\exp(A/2^s)$ and then performs s matrix multiplications with s being carefully chosen [18, 1]. This method is very reliable but the cost of dense matrix operations is $O(sN^3)$ flops, which becomes prohibitively expensive for large N. Even more problematic is the fact that the algorithm has to be rerun from scratch for each $\exp(-tA)$ when the evaluation time point t changes, unless in special cases of, e.g., constant time steps. Therefore, alternative approaches that avoid forming the matrix exponential $\exp(-tA)$ entirely and do not strongly depend on t are desirable.

A very effective approach is to construct a family of *shared-pole* rational approximants

$$r_{t,n}(z) = \alpha_0(t) + \frac{\alpha_1(t)}{z - \sigma_1} + \cdots + \frac{\alpha_n(t)}{z - \sigma_n}, \tag{1}$$

with poles σ_i and scalar residues $\alpha_i(t)$ such that

$$r_{t,n}(A)\boldsymbol{b} \approx \exp(-tA)\boldsymbol{b}$$

uniformly for all $t \in T$. We refer to n as the degree of approximation. It is important that the poles σ_i of $r_{t,n}$ do *not* depend on the time parameter t. Therefore, the computation of $r_{t,n}(A)\boldsymbol{b}$ involves solutions of only n shifted linear systems $(A - \sigma_i I)\boldsymbol{x}_i = \boldsymbol{b}$, independent of the number of time points to evaluate for. Since these linear systems are completely decoupled, they can be solved in parallel using well-developed algorithms such as sparse Cholesky factorization [15, Chaper 11] or iterative methods [27].

The shared-pole restriction may be less relevant if the number of evaluation time points is very small, allowing for a lower degree $\widehat{n}$ for the same target tolerance `tol`[1] by finding a best approximant for each time point independently. If we allow for poles to be placed anywhere in the complex plane, it is known that the set of optimal poles for $\exp(-z)$ lies on a parabolic curve [31]. However, in many applications, the degree ratio $n/\widehat{n}$ is much smaller than the number of discrete time points of interest, for various choices of `tol` and T in consideration; see, e.g., [10, Section 3.3]. In such situations, the constraint on shared poles yields a substantially smaller total number of shifted linear systems to be solved for a fixed target tolerance. Our discussion hereafter will therefore focus on shared-pole approximants only.

Reliable algorithms for computing a family of rational approximants with shared poles $\sigma_i \in \mathbb{C}$ are readily available. These include rational Krylov fitting (RKFIT) [8] and the set-valued adaptive Antoulas–Anderson (svAAA) method [23]. The RKFIT method optimizes the shared poles iteratively, using orthogonal transformations on rational Krylov spaces in order to approximately minimize a relative misfit given by the sum of squares of weighted error of each rational approximant. It then finds all the best approximants in a fixed shared-pole rational Krylov subspace via least squares projection. The svAAA method optimizes the shared interpolation nodes in the barycentric

[1] The tolerance `tol` is set for the time-uniform error, which will be introduced later in (4). More precisely, n and $\widehat{n}$ are the minimal degrees needed for the time-uniform error to fall below `tol` with and without the shared-pole condition, respectively.

representation of rational approximants in a greedy fashion, with the poles being implicitly encoded in the barycentric form. The two algorithms are extensions of [6] and [25], respectively, from a single rational approximant to a family of rational approximants with shared poles. Since in both algorithms the shared poles are computed from a generalized eigenvalue problem involving a pair of real-valued matrices, the poles appear as conjugate pairs in the complex plane.

In this paper we focus on the case where $\sigma_i < 0$ is restricted to be real. This can lead to a significant reduction in the overall computational costs for matrix-valued problems as the resulting shifted linear systems are real (i.e., no complex arithmetic required) and symmetric positive definite. Real-pole time-uniform approximations have been researched in the literature, for example in [12]. In this work, the poles are placed in the negatively reflected spectral interval of A, relating the pole optimization problem to solving a classical Zolotarev problem on the real line. However, this approach requires knowledge of A's spectral interval, and it deteriorates as the spectral interval increases. This approach hence does not typically lead to mesh-independent integrators with matrices arising from discretizations of differential operators. On the other hand, this approach leads to approximants which are good for the *infinite* time interval $T = [0, +\infty)$, which may be useful in some circumstances (e.g., if the final time point of interest is not known in advance). In contrast, our paper proposes an approach to real-pole optimization that applies to *any* symmetric positive semidefinite matrix A but requires a *finite* time interval T.

Before we proceed, let us briefly review some classical results on the rational approximation of the exponential function. For a fixed time point ($t = 1$), Kaufman and Taylor [22] conjectured that the best uniform real-pole approximant to $\exp(-z)$ has all its poles *concentrated* at a single point; that is, $\sigma_1 = \sigma_2 = \cdots = \sigma_n = \sigma$ where $\sigma = -cn$ for some positive constant c, and this conjecture was later proved by Borwein [11]. Saff *et al.* [28] showed that for $c = 1$ the associated best uniform error decreases at most $O(2^{-n})$ and they suggested that $c = 1$ is also the optimal choice [29]. In a later work by Andersson [3], it was proved that the optimal choice is $c = 1/\sqrt{2}$, and the associated best uniform error decreases like $O((\sqrt{2} + 1)^{-n})$.

In this paper, we will extend Andersson's result from [3] to a time interval T, considering the case where the poles of all rational approximants are concentrated at a point $\sigma := \sigma(n, T)$, which depends on the degree of approximation n and the time interval T. Now, the partial fraction form (1) must be changed to

$$r_{t,n}(z) = \alpha_0(t) + \frac{\alpha_1(t)}{z - \sigma} + \frac{\alpha_2(t)}{(z - \sigma)^2} + \cdots + \frac{\alpha_n(t)}{(z - \sigma)^n}. \tag{2}$$

Although the evaluation of $r_{t,n}(A)\boldsymbol{b}$ with $r_{t,n}$ in (2) involves solving dependent shifted linear systems which cannot be decoupled and computed in parallel, a benefit is that all these systems involve the same matrix $A - \sigma I$. Hence, only one Cholesky factorization needs to be computed and can be reused for all solves.

In order to assess the quality of approximation, let us assume without loss of generality that $\|\boldsymbol{b}\|_2 = 1$, and note that for all $t \in T$,

$$\begin{aligned}\|r_{t,n}(A)\boldsymbol{b} - \exp(-tA)\boldsymbol{b}\|_2 &\leq \max_{z\in\Lambda(A)} |r_{t,n}(z) - \exp(-tz)| \\ &\leq \max_{z\in[0,+\infty)} |r_{t,n}(z) - \exp(-tz)| \\ &=: \rho(n,t).\end{aligned} \tag{3}$$

Here, $\Lambda(A)$ denotes the set of A's eigenvalues. We aim to construct shared concentrated real-pole rational approximants $r_{t,n}$ such that $\rho(n,t)$ is uniformly small for all $t \in T$. In other words, we consider the *time-uniform error* $\rho(n,T)$ defined as

$$\rho(n,T) := \sup_{t\in T} \rho(n,t). \tag{4}$$

This paper is organized as follows. In Section 2 we propose a method to determine an asymptotically optimal location of a shared concentrated real pole $\sigma(n,T)$. We exploit a classical result on the asymptotically optimal concentrated real pole of a rational approximant for $\exp(-z)$ (i.e., a single time point $t = 1$) by Andersson [3]. In Section 3 we discuss efficient algorithms for the numerical evaluation of the proposed approximants. In Section 4 we provide numerical experiments which confirm the near-optimality of our choice of poles, and we also compute matrix exponential functions acting on a vector arising from an initial-value problem. In Section 5 we conclude and discuss potential future work.

2 Optimal Shared Concentrated Real Poles

For any $q > 0$, let $\mathcal{R}_q^n$ be the set of degree n real rational functions with poles concentrated at $-nq$,

$$\mathcal{R}_q^n = \left\{ \frac{p_n(z)}{(z+nq)^n} : p_n \in \mathcal{P}_n \right\}.$$

Here and in the sequel, $\mathcal{P}_n$ denotes the class of polynomials of degree at most n. Furthermore, for each time point t, let

$$\rho_n^t(q) = \inf \left\{ \sup_{z\geq 0} |r_{t,n}(z) - \exp(-tz)| : r_{t,n} \in \mathcal{R}_q^n \right\}.$$

The following theorem is one of the main results of [3].

Theorem 2.1 *For any $q > 0$,*

$$\lim_{n\to+\infty} \rho_n^1(q)^{1/n} = \tilde{H}(q),$$

where

$$\tilde{H}(q) = \exp\left(\log\left|\frac{\tilde{q}-1}{\tilde{q}+1}\right| + q\Re(\tilde{q}^2)\right), \tag{5}$$

and $\tilde{q}$ *is the root in* $\Im(z) \geq 0$ *of the equation*

$$q(z^3 - z) + 1 = 0 \tag{6}$$

that has the smallest positive real part. The minimum of $\tilde{H}$ *is* $\sqrt{2}-1$*, which is attained at* $1/\sqrt{2}$*.*

Proof. We refer the reader to [3]. □

Theorem 2.1 suggests that $-n/\sqrt{2}$ is an asymptotically optimal location for a concentrated real pole when approximating $\exp(-z)$ on the non-negative real axis, and its associated uniform error decreases like $(\sqrt{2}+1)^{-n}$ as n gets large.

We now extend the result of Theorem 2.1 to rational approximants of a family of exponential functions $\exp(-tz)$ for $t \in T := [t_{\min}, t_{\max}]$ with $0 < t_{\min} < t_{\max}$. Let

$$\rho_n^T(q) = \sup_{t\in T} \rho_n^t(q)$$

be the time-uniform error corresponding to a family of best approximants with shared poles concentrated at $-nq$, over the time interval T. We propose an asymptotically optimal choice of q that minimizes ρ_n^T for large n, which is derived from the above function $\tilde{H}$. Let us summarize the result in the following theorem.

Theorem 2.2 *For any* $q > 0$*, let* $G(q) = \lim_{n\to+\infty}\left(\rho_n^T(q)\right)^{1/n}$ *be the asymptotic rate of convergence associated with a family of best rational approximants with the concentrated real pole* $-nq$*, over the time interval* $T := [t_{\min}, t_{\max}]$ *with* $0 < t_{\min} < t_{\max}$*. Then*

$$\arg\min_{q>0} G(q) = \arg\min_{q>0} \sup_{z\in[qt_{\min}, qt_{\max}]} \tilde{H}(z), \tag{7}$$

where $\tilde{H}$ *is the function defined in* (5)*.*

Proof. For any $t \in T$, by using a change of variable $w = tz$, we have

$$\begin{aligned}
\rho_n^t(q) &= \inf\left\{\sup_{z\geq 0} |r_{t,n}(z) - \exp(-tz)| : r_{t,n} \in \mathcal{R}_q^n\right\} \\
&= \inf\left\{\sup_{w\geq 0} |r_{1,n}(w) - \exp(-w)| : r_{1,n} \in \mathcal{R}_{qt}^n\right\} \\
&= \rho_n^1(qt).
\end{aligned}$$

Therefore, we have

$$\rho_n^T(q) = \sup_{t\in T} \rho_n^t(q) = \sup_{t\in T} \rho_n^1(qt),$$

and

$$G(q) = \sup_{t\in T}\left(\lim_{n\to+\infty} \rho_n^1(qt)\right)^{1/n} = \sup_{z\in[qt_{\min}, qt_{\max}]} \tilde{H}(z), \tag{8}$$

where the second equality in (8) follows from Theorem 2.1. The proof is completed by taking arg min on the left and right-hand sides of (8). □

Theorem 2.2 allows us to determine an asymptotically optimal location of the shared concentrated poles by searching for an optimal interval with fixed endpoint ratio $\tau := t_{\max}/t_{\min}$ such that the supremum of $\tilde{H}$ over that interval is minimized. The following theorem on the monotonicity of $\tilde{H}$ guarantees that this optimization problem has a unique solution.

Theorem 2.3 *The function $\tilde{H}$ defined in Theorem 2.1 is monotonically decreasing when $q \in (0, 1/\sqrt{2})$ and increasing when $q \in [1/\sqrt{2}, +\infty)$. Moreover, $\tilde{H}$ has the following asymptotic behavior*

$$\lim_{q\to+0} \tilde{H}(q) = \lim_{q\to+\infty} \tilde{H}(q) = 1.$$

Proof. By Cardano's formula, the discriminant of the cubic equation (6) can be written as $\Delta = 1/4q^2 - 1/27$ for all $q > 0$, which has the critical value $q^* = 3\sqrt{3}/2$.

When $0 < q < 3\sqrt{3}/2$, we have $\Delta > 0$ and the three roots are given by

$$\begin{aligned} z_1 &= a + b, \\ z_2 &= a\boldsymbol{w} + b\boldsymbol{w}^2, \\ z_3 &= b\boldsymbol{w} + a\boldsymbol{w}^2, \end{aligned}$$

where $a = (-1/2q + \sqrt{\Delta})^{1/3}$ and $b = (-1/2q - \sqrt{\Delta})^{1/3}$ satisfy $b < a < 0$, and $\boldsymbol{w} = e^{2\pi i/3}$. In this case, z_1 is a negative real number, z_2 and z_3 are complex conjugate numbers on the right-half plane with $\Im(z_2) > 0$ and $\Im(z_3) < 0$. Therefore, we have

$$\begin{aligned} \tilde{H}(q) &= \left|\frac{z_2 - 1}{z_2 + 1}\right| \exp(q\Re(z_2^2)) \\ &= \frac{\left|(-\frac{1}{2}a - \frac{1}{2}b - 1) + (\frac{\sqrt{3}}{2}a - \frac{\sqrt{3}}{2}b)i\right|}{\left|(-\frac{1}{2}a - \frac{1}{2}b + 1) + (\frac{\sqrt{3}}{2}a - \frac{\sqrt{3}}{2}b)i\right|} \exp(q\Re((a\boldsymbol{w} + b\boldsymbol{w}^2)^2)) \\ &= \sqrt{\frac{(a+b)+1}{(a+b)-1}} \exp(-\frac{1}{2}q((a+b)^2 - 2)) \\ &:= \sqrt{\frac{m+1}{m-1}} \exp(-\frac{1}{2}q(m^2 - 2)). \end{aligned} \tag{9}$$

The third equality follows from the relation $ab = 1/3$ and $\boldsymbol{w}^3 = 1$, and we used a change of variable $m := a + b$ for the fourth equality. Moreover, by noticing that

$$-1 = q(a^3 + b^3) = q(a+b)((a+b)^2 - 1) = qm(m^2 - 1), \tag{10}$$

we can now represent the expression (9) merely in terms of m,

$$\tilde{H}(q) = \tilde{h}(m) = \sqrt{\frac{m+1}{m-1}} \exp\left(\frac{m^2 - 2}{2m(m^2 - 1)}\right).$$

Taking the implicit derivative with respect to q in equation (10) yields:

$$q(3m^2 - 1)m'(q) + m(m^2 - 1) = 0,$$

and hence

$$m'(q) = \frac{1}{q^2(3m^2 - 1)}.$$

Since

$$m^2 = (a + b)^2 \geq 4ab = 4/3,$$

we conclude that $m'(q) > 0$ and $m(q)$ is an increasing function for $0 < q < 3\sqrt{3}/2$. As $m(q) = -\infty$ when $q \to 0^+$ and $m(q) = -2/\sqrt{3}$ when $q \to 3\sqrt{3}/2^-$, it suffices to analyze the monotonicity of $\tilde{h}$ on the domain $(-\infty, -2/\sqrt{3})$. Note that the derivative of $\tilde{h}$ is given by

$$\tilde{h}'(m) = -\frac{1}{2}\left(\frac{m+1}{m-1}\right)^{-3/2} m^{-2}(m-1)^{-4}(m^2 - 2)(3m^2 - 1)\exp\left(\frac{m^2 - 2}{2m(m^2 - 1)}\right),$$

which is negative when $m \in (-\infty, -\sqrt{2})$ and positive when $m \in (-\sqrt{2}, -2/\sqrt{3})$. Since $m(1/\sqrt{2}) = -\sqrt{2}$, we conclude that $\tilde{H}(q)$ is monotonically decreasing when $q \in (0, 1/\sqrt{2})$ and increasing when $q \in (1/\sqrt{2}, 3\sqrt{3}/2)$. Moreover, we can obtain the asymptotic behavior

$$\lim_{q \to 0^+} \tilde{H}(q) = \lim_{m \to -\infty} \tilde{h}(m) = 1.$$

When $q > 3\sqrt{3}/2$, we have $\Delta < 0$ and all three roots of equation (6) are real, which can be obtained as the intersection of the real-valued function $g_1(x) = x^3 - x$ and the constant function $g_2(x) = -1/q$. Therefore, $\tilde{q}$ is the smallest positive real root of (6). Since $g_1'(x) < 0$ for $0 < x < 1/\sqrt{3}$ and $g_1'(x) > 0$ for $1/\sqrt{3} < x < 1$, and $g_1(1/\sqrt{3}) < -1/q < 0 = g_1(0) = g_1(1)$, we have $\tilde{q} \in (0, 1/\sqrt{3})$. By representing q in terms of $\tilde{q}$, $\tilde{H}$ can also be written as a function of $\tilde{q}$:

$$\tilde{H}(q) = \phi(\tilde{q}) = \left(\frac{1-\tilde{q}}{1+\tilde{q}}\right)\exp\left(\frac{\tilde{q}}{1-\tilde{q}^2}\right). \tag{11}$$

Taking the derivative yields

$$\phi'(\tilde{q}) = \frac{1 - 3\tilde{q}^2}{(\tilde{q} - 1)(\tilde{q} + 1)^3}\exp\left(\frac{\tilde{q}}{1-\tilde{q}^2}\right),$$

which is negative for all $\tilde{q} \in (0, 1/\sqrt{3})$, and hence implying ϕ is a monotonically decreasing function with $\tilde{q}$. When q increases, its associated $\tilde{q}$ decreases, which implies ϕ increases and hence $\tilde{H}$ increases. Moreover, we can obtain another asymptotic behavior

$$\lim_{q \to +\infty} \tilde{H}(q) = \lim_{\tilde{q} \to 0^+} \phi(\tilde{q}) = 1.$$

The continuity of $\tilde{H}$ at the critical value $3\sqrt{3}/2$ can be ensured by noticing

$$\lim_{q\to 3\sqrt{3}/2^-} \tilde{H}(q) = \lim_{m\to -2/\sqrt{3}^-} \tilde{h}(m) = (2-\sqrt{3})\exp(\sqrt{3}/2),$$

and

$$\lim_{q\to 3\sqrt{3}/2^+} \tilde{H}(q) = \lim_{\tilde{q}\to 1/\sqrt{3}^-} \phi(\tilde{q}) = (2-\sqrt{3})\exp(\sqrt{3}/2).$$

This completes the proof. □

Figure 1 is a numerical plot of $\tilde{H}(q)$ using MATLAB for a million logarithmically spaced points q in $[10^{-6}, 10^{6}]$, with the minimal point $(1/\sqrt{2}, \sqrt{2}-1)$ marked by a red asterisk. The plot confirms our results in Theorem 2.3. Based on the simple structure of $\tilde{H}$, we can numerically determine the right-hand side of (7) as the unique $\widehat{q} > 0$ such that $\tilde{H}(\widehat{q}t_{\min}) = \tilde{H}(\widehat{q}t_{\max})$. Any other choice corresponds to a shifted interval with the same endpoint ratio τ, which would lead to a larger supremum.

Table 1 collects values of $\widehat{q}$ and $G(\widehat{q})$ for 5 different time ranges $T = [t_{\min}, t_{\max}] = 1, [10^{-1}, 1], [10^{-2}, 1], [10^{-3}, 1], [10^{-4}, 1]$, respectively. Correspondingly, the asymptotically optimal locations of the shared concentrated pole for rational approximants of degree n in these time ranges are around $\sigma(n, T_0) = -0.71n$, $\sigma(n, T_1) = -1.70n$, $\sigma(n, T_2) = -2.67n$, $\sigma(n, T_3) = -4.31n$, $\sigma(n, T_4) = -7.47n$, respectively.

While Table 1 records optimal pole parameters for the case $t_{\max} = 1$, a simple rescaling of time as used in Theorem 2.2 can be used to get the optimal pole for any time interval $[t_{\min}, t_{\max}]$. The convergence rate only depends on the ratio $\tau = t_{\max}/t_{\min}$.

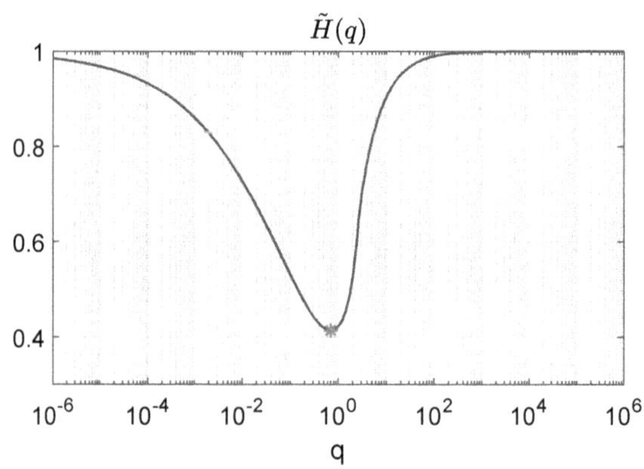

Figure 1 The plot of $\tilde{H}$ as a function of q.

Table 1 Asymptotically optimal parameters for approximating $\exp(-tz)$ uniformly for $z \geq 0$ and over a time interval $T = [t_{\min}, t_{\max}]$. The optimal pole is at $-n\widehat{q}$ and the expected convergence rate is $G(\widehat{q})$.

T	$t_{\min}$	$t_{\max}$	$\widehat{q}$	$G(\widehat{q})$
T_0	1	1	0.71	0.41
T_1	10^{-1}	1	1.70	0.49
T_2	10^{-2}	1	2.67	0.65
T_3	10^{-3}	1	4.31	0.79
T_4	10^{-4}	1	7.47	0.87

3 Numerical Evaluation of the Approximants

Before verifying the near-optimality of our proposed concentrated real poles, we briefly discuss several numerical methods for efficiently computing $\{r_{t,n}(A)\boldsymbol{b}\}_{t\in T}$, the action of a family of rational matrix functions on a vector, where $r_{t,n}$ is of the form (2) with the choice of σ as suggested in Section 2.

3.1 Rational Chebyshev Interpolation

The first method is to construct $r_{t,n}(z)$ as the rational Chebyshev interpolant of $\exp(-tz)$ for $z \in [0, +\infty)$. By considering the Möbius transformation $\widehat{z} = (z - \sigma)/(z + \sigma)$, this is equivalent to finding the polynomial Chebyshev interpolant $p_{t,n}(\widehat{z})$ to the function $s_{t,\sigma}(\widehat{z}) = \exp(-t\sigma(1 + \widehat{z})/(1 - \widehat{z}))$ for $\widehat{z} \in [-1, 1]$. If we represent $p_{t,n}(\widehat{z})$ as

$$p_{t,n}(\widehat{z}) = \sum_{k=0}^{n} c_{t,k} T_k(\widehat{z}),$$

where $T_k(\widehat{z}) = \cos(k \arccos(\widehat{z}))$ is the degree-k Chebyshev polynomial of the first kind, then the coefficients $\{c_{t,k}\}$ can be computed from samples of $s_{t,\sigma}$ at Chebyshev points using fast Fourier Transform (FFT) [30, Chaper 4]. Now we have

$$r_{t,n}(A)\boldsymbol{b} = p_{t,n}(\widehat{A})\boldsymbol{b} = \sum_{k=0}^{n} c_{t,k} T_k(\widehat{A})\boldsymbol{b} \tag{12}$$

where

$$\widehat{A} = (A - \sigma I)(A + \sigma I)^{-1}.$$

The vector-valued Chebyshev approximants $T_k(\widehat{A})\boldsymbol{b}$ in (12) can be computed efficiently by running the three-term recurrence

$$T_{k+1}(\widehat{A})\boldsymbol{b} = 2\widehat{A}T_k(\widehat{A})\boldsymbol{b} - T_{k-1}(\widehat{A})\boldsymbol{b}, \quad T_0(\widehat{A})\boldsymbol{b} = \boldsymbol{b}, \quad T_1(\widehat{A})\boldsymbol{b} = \widehat{A}\boldsymbol{b}.$$

Note that these $n + 1$ vectors need to be generated only once and could be summed with different coefficients $\{c_{t,k}\}$ if the evaluation at various time points t is required. Overall, this amounts to solving n shifted linear systems independent of the number of time points t to be evaluated for. Furthermore, this Chebyshev interpolation approach guarantees near-best approximation quality, since the supremum approximation error of the Chebyshev interpolant is at most a factor $2 + 2\log(n + 1)/\pi$ larger than that of the best approximation [30, Theorem 16.1].

3.2 Shift-and-Invert Arnoldi Method

As an alternative we consider a shift-and-invert Arnoldi approach as in [32] with our optimized poles. Let a symmetric positive semidefinite matrix A, a vector $\boldsymbol{b}$, the pole $\sigma < 0$ and an integer n be given. We generate a Krylov basis matrix $V_n \in \mathbb{R}^{N\times n}$ with orthonormal columns that satisfies the Arnoldi relation

$$(A - \sigma I)^{-1} V_n = V_n H_n + h_{n+1,n} \boldsymbol{v}_{n+1} \boldsymbol{e}_n^T,$$

where $H_n \in \mathbb{R}^{n\times n}$ is symmetric tridiagonal, $\boldsymbol{v}_{n+1} \perp \operatorname{colspan}(V_n)$, and $\boldsymbol{e}_j$ denotes the j-th canonical unit vector in $\mathbb{R}^n$. The computation of this decomposition requires the solution of n shifted linear systems. The shift-and-invert Arnoldi approximation to $\exp(-tA)\boldsymbol{b}$ is then given by

$$f_{t,n} = V_n \exp(-t[H_n^{-1} + \sigma I])(\|\boldsymbol{b}\|_2 \boldsymbol{e}_1).$$

This formula involves the evaluation of an $n\times n$ matrix function, typically much smaller than the original problem size $N \gg n$.

As opposed to the rational Chebyshev interpolant, the rational function $r_{t,n}$ underlying the shift-and-invert Arnoldi approximant $f_{t,n} = r_{t,n}(A)\boldsymbol{b}$ has the potential to adapt to the discrete spectrum of A, and to the spectral components in $\boldsymbol{b}$. This property is known as *spectral adaptivity* and it can lead to significant improvements in approximation accuracy; see, e.g., [4].

4 Numerical Experiments

This section is dedicated to numerical experiments. In Section 4.1 we demonstrate that our proposed choice of poles derived using the approach in Section 2 is almost indistinguishable from using best approximants produced by the Remez algorithm. An application to a linear constant-coefficient initial-value problem is presented in Section 4.2.

4.1 Verifying Near-Optimality

Let us provide numerical evidence for the near-optimality of our choice of poles. Figure 2 illustrates the time-uniform error $\rho_n^T(q)$ for various values of q, under degree $n = 20$, $T = T_1, T_2, T_3, T_4$, respectively. Here, we discretize each T into 41 logarithmically spaced time points, and use the scaled reciprocal $\gamma := (nq)^{-1} = -\sigma^{-1}$ as the horizontal axis for a clearer visualization of our plots. The numerical method we use to compute $\rho_n^t(q)$ for all $t \in T$ is to solve an associated best polynomial approximation problem on $[-1, 1]$ via the Chebfun `'minimax'` command [14] by exploiting the following equivalence transformation (13) using a change of variable $\widehat{z} = (\gamma z + 1)/(\gamma z - 1)$,

$$\begin{aligned}\rho_n^t(q) &= \inf\left\{\sup_{z\in[0,+\infty)} |r_{t,n}(z) - \exp(-tz)| : r_{t,n} \in \mathcal{R}_q^n\right\}\\ &= \inf\left\{\sup_{\widehat{z}\in[-1,1]} |p_n(\widehat{z}) - \exp(t\gamma^{-1}(1+\widehat{z})/(1-\widehat{z}))| : p_n \in \mathcal{P}_n\right\}.\end{aligned} \tag{13}$$

(While the function $\exp(t\gamma^{-1}(1+\widehat{z})/(1-\widehat{z}))$ is not defined at $\widehat{z} = 1$, this is a removable singularity when setting the function value to be zero at that point.) Hence finding the optimal $r_{t,n}^* \in \mathcal{R}_q^n$ is equivalent to a best polynomial approximation problem. We can use this relation to numerically search for the optimal $q^* > 0$ that minimizes $\rho_n^t(q)$.

In Figure 2, the black circles mark the corresponding time-uniform error $\rho_{20}^T(\widehat{q})$ computed using our choice of concentrated poles $\sigma(20, T_1), \sigma(20, T_2), \sigma(20, T_3), \sigma(20, T_4)$, and they are visually close to the time-uniform errors $\rho_{20}^T(q^*)$, associated with the truly optimal computed concentrated pole parameter q^*, which are marked by red circles.

Table 2 collects the values of $\rho_{20}^T(\widehat{q})$ and $\rho_{20}^T(q^*)$ for $T = T_1, T_2, T_3, T_4$, and one can see that our choices of concentrated poles are near-best in all these cases.

Table 2 Time-uniform approximation errors $\rho_{20}^T(\widehat{q})$ and $\rho_{20}^T(q^*)$, achieved with our asymptotically optimal poles versus the true optimal poles, respectively, for different time intervals T_1, T_2, T_3, T_4.

T	$\rho_{20}^T(\widehat{q})$	$\rho_{20}^T(q^*)$
T_1	7.75×10^{-8}	7.15×10^{-8}
T_2	2.44×10^{-5}	2.41×10^{-5}
T_3	1.40×10^{-3}	1.30×10^{-3}
T_4	2.02×10^{-2}	1.59×10^{-2}

Figure 3 shows the time-uniform error using our choices of poles, $\rho_n^T(\widehat{q})$, and the computed truly optimal poles, $\rho_n^T(q^*)$, for degrees n ranging from 1 to 40, and time ranges $T = T_0, T_1, T_2, T_3, T_4$. It can be seen from the figure that our choice of poles is near-optimal for all degrees and time ranges in consideration. The time-uniform error under the single time point T_0 stagnates for degree $n \geq 36$ as the level of machine precision is reached.

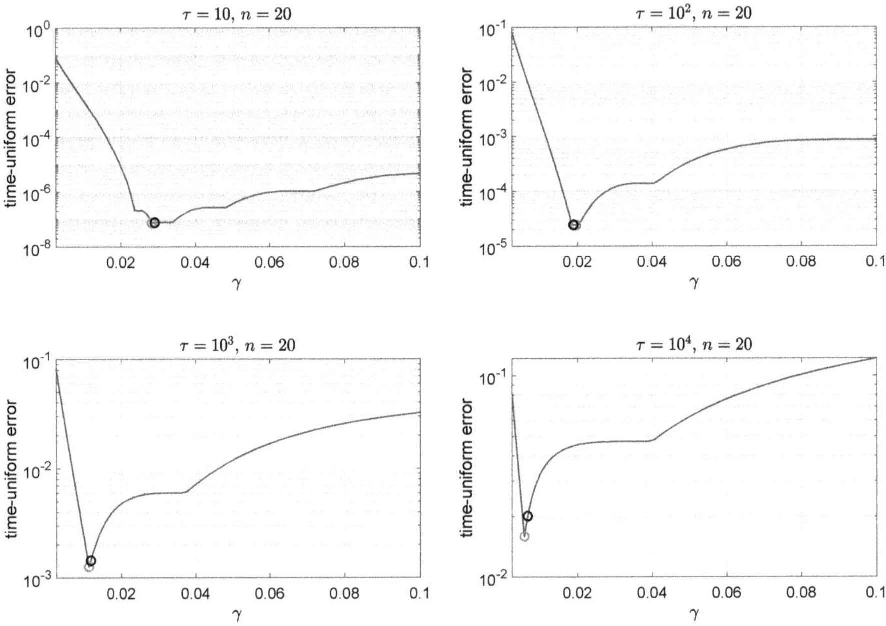

Figure 2 Time-uniform error when using our choice of concentrated poles (marked by black circles) and numerically computed optimal choice of poles (marked by red circles), for different time ratios $\tau = t_{\max}/t_{\min}$. The degree is $n = 20$ in all cases.

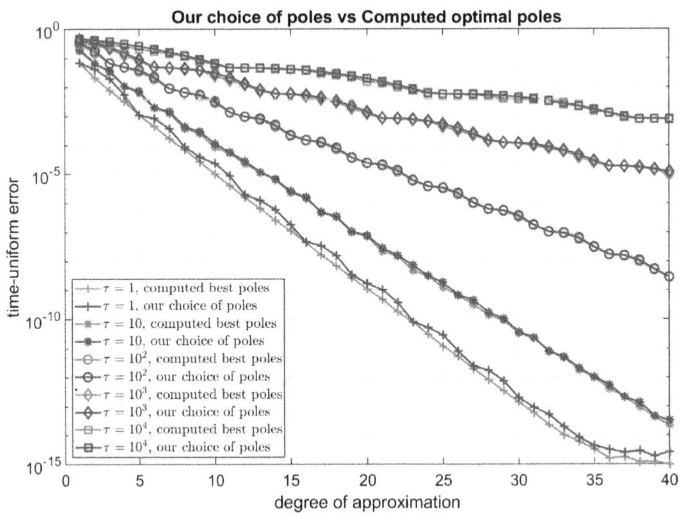

Figure 3 Time-uniform error using our choice of poles and the numerically computed optimal poles for the ranges T_0, T_1, T_2, T_3, T_4, for degree n from 1 to 40.

4.2 An Application to Matrix Exponential Function

We are now concerned with the approximation of the action of matrix exponential functions on a vector using our time-uniform approximants. Let us consider the linear constant-coefficient initial value problem

$$\boldsymbol{u}'(t) + L\boldsymbol{u}(t) = \boldsymbol{0}, \qquad \boldsymbol{u}(0) = \boldsymbol{u}_0. \tag{14}$$

Here, $L \in \mathbb{R}^{4761\times4761}$ is a sparse symmetric matrix arising from a finite-difference discretization of the scaled 2D Laplace operator -0.2Δ on the domain $\Omega = [-1, 1]^2$ with homogeneous Dirichlet boundary conditions, and $\boldsymbol{u}_0$ corresponds to the discretization of the initial function $u_0(x, y) = (1-x^2)(1-y^2)e^x$ on Ω. We are interested in obtaining the solution $\boldsymbol{u}(t)$ of (14) at several time points $T = \{t_j\}_{j=1}^{s}$. A similar example has been considered in [31] and [8, Section 6.2] (see also [7] for the corresponding MATLAB code).

Figure 4 shows the 2-norm error of the rational approximation $\widehat{\boldsymbol{u}}(t) := r_{t,n}(L)\boldsymbol{u}_0$ to the exact solution $\boldsymbol{u}(t) := \exp(-tL)\boldsymbol{u}_0$ for 41 log-spaced time points in $[10^{-3}, 1]$ using the form (2), and concentrated poles σ selected according to Theorem 2.2, along with its associated error bound $\rho(n, t)$ in (3), for degrees $n = 8, 14, 20, 26$. The approaches we use to compute $\widehat{\boldsymbol{u}}(t)$ are the rational Chebyshev interpolation and the shift-and-invert Arnoldi method, which are described in Section 3. We also compute an "approximate error bound" by running the rational Chebyshev method for a diagonal matrix having "dense" spectrum on $[0, +\infty)$ and with a vector of all ones. This allows us to estimate the scalar approximation error of the rational interpolants. We find that the error bounds at the two time endpoints attain almost the same level, which is the maximum throughout the time range. This aligns with our choice of concentrated poles, which is constructed by minimizing the right-hand side of the equation (7), balancing the approximation over the whole time interval. We also see that the shift-and-invert Arnoldi method delivers significantly better approximants than the bound suggests. This is due to the spectral adaptation of the underlying rational functions. These effects would reduce if we chose a finer discretization of the differential operator or a rougher right-hand side vector (e.g., a random $\boldsymbol{u}_0$).

5 Conclusions

We have proposed an asymptotically optimal choice of shared concentrated real poles for the rational approximation of a family of time-dependent exponential functions. The result is a generalization of [3] to any time interval.

In future work, we plan to further explore the possibilities of finding an optimal set of non-concentrated real poles supported on the negative real axis. We also intend to investigate extensions to the shared-pole approximation of functions such as the φ_j-functions, which are closely related to the exponential function and often appear in the context of exponential integrators [20]. More work could be devoted to a detailed

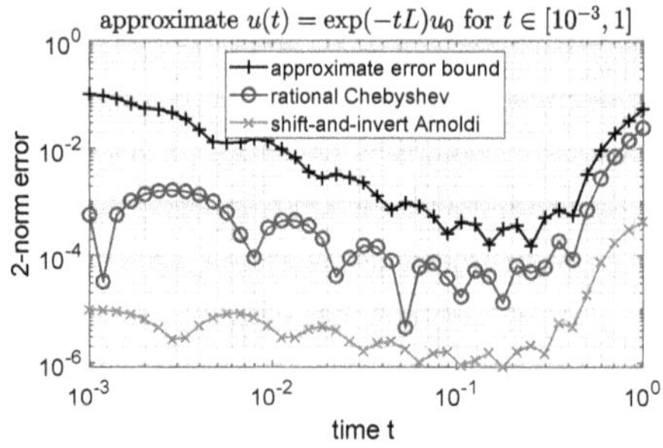

(a) $n = 8$, $\tau = 10^3$.

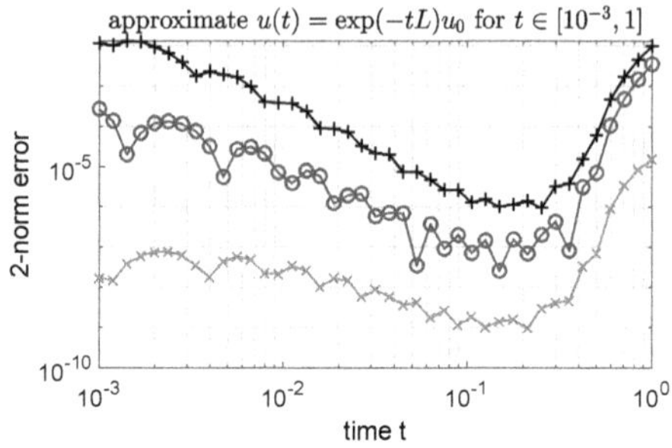

(b) $n = 14$, $\tau = 10^3$.

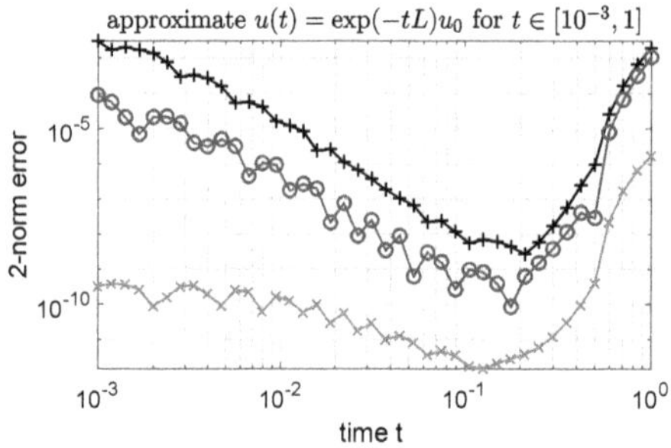

(c) $n = 20$, $\tau = 10^3$.

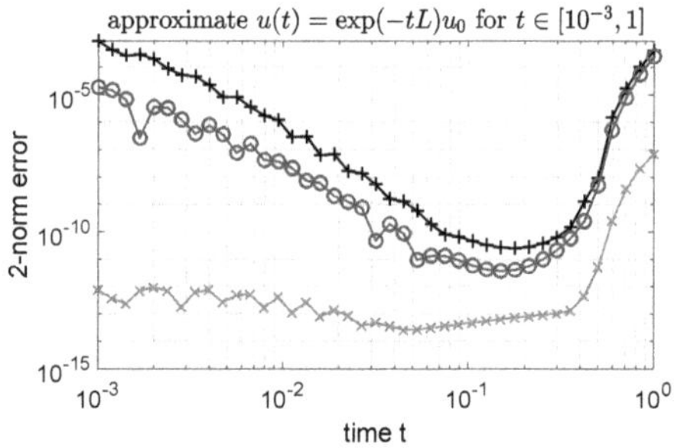

(d) $n = 26$, $\tau = 10^3$.

Figure 4 Plot of $\|\widehat{\boldsymbol{u}}(t) - \boldsymbol{u}(t)\|_2$ using rational Chebyshev interpolation (blue), shift-and-invert Arnoldi (red), and their approximate error bound (black), for the time points T = `logspace(-3,0,41)` and degrees $n = 8, 14, 20, 26$.

comparison of the performance of the different shared-pole rational approximants (with free complex poles, real poles, or concentrated real poles) in the context of large-scale computations as they arise, e.g., in Geophysics forward modeling and in inverse problems.

Acknowledgments

We are grateful to Yoshihito Kazashi, Nick Trefethen, and Marcus Webb for providing useful suggestions. S. G. acknowledges funding from the UK's Engineering and Physical Sciences Research Council (EPSRC grant EP/Z533786/1) and the Royal Society (RS Industry Fellowship IF/R1/231032). S. S. acknowledges a Dean's Doctoral Scholarship from the University of Manchester.

References

1. A. H. Al-Mohy and N. J. Higham. A new scaling and squaring algorithm for the matrix exponential. *SIAM Journal on Matrix Analysis and Applications*, 31(3):970–989, 2010.
2. L. J. Allen. An introduction to stochastic epidemic models. In *Mathematical Epidemiology*, pages 81–130. Springer, 2008.
3. J.-E. Andersson. Approximation of $\exp(-x)$ by rational functions with concentrated negative poles. *Journal of Approximation Theory*, 32(2):85–95, 1981.
4. B. Beckermann and S. Güttel. Superlinear convergence of the rational Arnoldi method for the approximation of matrix functions. *Numerische Mathematik*, 121(2):205–236, 2012.
5. M. Benzi, E. Estrada, and C. Klymko. Ranking hubs and authorities using matrix functions. *Linear Algebra and its Applications*, 438(5):2447–2474, 2013.
6. M. Berljafa and S. Güttel. Generalized rational Krylov decompositions with an application to rational approximation. *SIAM Journal on Matrix Analysis and Applications*, 36(2):894–916, 2015.
7. M. Berljafa and S. Güttel. Pole optimization for exponential integration. 2015. `https://guettel.com/rktoolbox/examples/html/example_expint.html`.
8. M. Berljafa and S. Güttel. The RKFIT algorithm for nonlinear rational approximation. *SIAM Journal on Scientific Computing*, 39(5):A2049–A2071, 2017.
9. R.-U. Börner, O. G. Ernst, and S. Güttel. Three-dimensional transient electromagnetic modelling using rational Krylov methods. *Geophysical Journal International*, 202(3):2025–2043, 2015.
10. R.-U. Börner and S. Güttel. Fast parallel transient electromagnetic modeling using a uniform-in-time approximation to the exponential. *Geophysical Journal International*, 243(1):ggaf319, 2025.
11. P. B. Borwein. Rational approximations with real poles to $\exp(-x)$ and x^n. *Journal of Approximation Theory*, 38(3):279–283, 1983.
12. V. Druskin, L. Knizhnerman, and M. Zaslavsky. Solution of large scale evolutionary problems using rational Krylov subspaces with optimized shifts. *SIAM Journal on Scientific Computing*, 31(5):3760–3780, 2009.
13. E. Estrada, N. Hatano, and M. Benzi. The physics of communicability in complex networks. *Physics Reports*, 514(3):89–119, 2012.
14. S.-I. Filip, Y. Nakatsukasa, L. N. Trefethen, and B. Beckermann. Rational minimax approximation via adaptive barycentric representations. *SIAM Journal on Scientific Computing*, 40(4):A2427–A2455, 2018.
15. G. H. Golub and C. F. Van Loan. *Matrix Computations, 4th edition.* JHU Press, 2013.
16. S. Güttel. *Rational Krylov methods for operator functions*. PhD thesis, Technische Universität Bergakademie Freiberg, 2010.
17. S. Güttel. Rational Krylov approximation of matrix functions: Numerical methods and optimal pole selection. *GAMM-Mitteilungen*, 36(1):8–31, 2013.
18. N. J. Higham. The scaling and squaring method for the matrix exponential revisited. *SIAM Journal on Matrix Analysis and Applications*, 26(4):1179–1193, 2005.
19. N. J. Higham. *Functions of Matrices: Theory and Computation*. SIAM, 2008.
20. M. Hochbruck and A. Ostermann. Exponential integrators. *Acta Numerica*, 19:209–286, 2010.
21. T. Jahnke and W. Huisinga. Solving the chemical master equation for monomolecular reaction systems analytically. *Journal of Mathematical Biology*, 54(1):1–26, 2007.
22. E. Kaufman Jr and G. Taylor. Uniform approximation with rational functions having negative poles. *Journal of Approximation Theory*, 23(4):364–378, 1978.
23. P. Lietaert, K. Meerbergen, J. Pérez, and B. Vandereycken. Automatic rational approximation and linearization of nonlinear eigenvalue problems. *IMA Journal of Numerical Analysis*, 42(2):1087–1115, 2022.
24. B. Munsky and M. Khammash. The finite state projection algorithm for the solution of the chemical master equation. *The Journal of Chemical Physics*, 124(4), 2006.
25. Y. Nakatsukasa, O. Sète, and L. N. Trefethen. The AAA algorithm for rational approximation. *SIAM Journal on Scientific Computing*, 40(3):A1494–A1522, 2018.

26. C. Qiu, S. Güttel, X. Ren, C. Yin, Y. Liu, B. Zhang, and G. Egbert. A block rational Krylov method for 3-D time-domain marine controlled-source electromagnetic modelling. *Geophysical Journal International*, 218(1):100–114, 2019.
27. Y. Saad. *Iterative Methods for Sparse Linear Systems, 2nd edition*. SIAM, Philadelphia, PA, 2003.
28. E. Saff, A. Schönhage, and R. Varga. Geometric convergence to e^{-z} by rational functions with real poles. *Numerische Mathematik*, 25(3):307–322, 1975.
29. E. Saff and R. Varga. Some open problems concerning polynomials and rational functions. In *Padé and Rational Approximation (Proceedings of Conference on Rational Approximation with Emphasis on Applications of Padé Approximants, Tampa, Florida, 1976)*, pages 483–488, 1977.
30. L. N. Trefethen. *Approximation Theory and Approximation Practice, extended edition*. SIAM, 2019.
31. L. N. Trefethen, J. A. C. Weideman, and T. Schmelzer. Talbot quadratures and rational approximations. *BIT Numerical Mathematics*, 46(3):653–670, 2006.
32. J. van den Eshof and M. Hochbruck. Preconditioning Lanczos approximations to the matrix exponential. *SIAM Journal on Scientific Computing*, 27(4):1438–1457, 2006.

Adaptive Robust Interior Penalty Discontinuous Galerkin Methods

Zhaonan Dong, Emmanuil H. Georgoulis and Natalia Kopteva

Abstract This work is concerned with the development of an adaptive algorithm for the so-called robust interior penalty discontinuous Galerkin (RIPDG) method for an elliptic model problem, based on new rigorous *a posteriori* error bounds. The RIPDG methodology offers a combined robust behavior with respect to sharp local variation in mesh-size, polynomial degree and diffusion contrast simultaneously. Thus, special attention is given to the dependence of the resulting *a posteriori* error bounds in these quantities. The proof employs a special H^1- conforming reconstruction of the numerical solution, along with a Helmholtz decomposition argument. A series of numerical experiments showcasing the reliability of the proven *a posteriori* error bounds are included.

1 Introduction

Discontinuous Galerkin (DG) methods are now established as a general-purpose methodology for the numerical approximation of partial differential equation (PDE) problems of various types. A significantly attractive attribute of DG methods is the

Zhaonan Dong
Inria, 48 Rue Barrault, 75647 Paris, France AND CERMICS, ENPC, Institut Polytechnique de Paris, CNRS, 6 & 8 avenue B. Pascal, 77455 Marne-la-Vallée, France.
zhaonan.dong@inria.fr

Emmanuil H. Georgoulis
The Maxwell Institute for Mathematical Sciences & Department of Mathematics, School of Mathematical and Computer Sciences, Heriot-Watt University, Edinburgh EH14 4AS, United Kingdom, AND Department of Mathematics, School of Applied Mathematical and Physical Sciences, National Technical University of Athens, Zografou 15780, Greece, AND IACM-FORTH, Crete, Greece.
e.georgoulis@hw.ac.uk

Natalia Kopteva
Department of Mathematics and Statistics University of Limerick Limerick, V94 T9PX, Ireland.
Natalia.Kopteva@ul.ie

B. Koren and D. Roose (eds.), *Numerical Analysis and Scientific Computing: an Anthology*, Lecture Notes in Computational Science and Engineering 156,
https://doi.org/10.1007/978-3-032-33428-2_3

extreme flexibility in the mesh design they admit (see, e.g., [14, 9, 10, 13, 11] and the references therein). This is particularly pertinent in the context of adaptive algorithms.

At the same time, the proof of *a posteriori* error bounds that drive mathematically justifiable adaptive methodologies is typically considerably more technical for discontinuous Galerkin methods than the corresponding conforming FEMs. The derivation of various *a posteriori* error bounds and the convergence of adaptive algorithms for discontinuous Galerkin methods for elliptic problems has received considerable interest during the last 20 years or so; see [29, 30, 27, 1, 2, 22, 6, 12, 34] and other works. A key technical tool for the proof of such bounds is the reconstruction/recovery of a piecewise discontinuous numerical solution into the natural (conforming) function space of the exact PDE problem solution. In the celebrated work of Karakashian & Pascal [29], the DG solution is recovered into a globally continuous, element-wise polynomial via nodal averaging. Later works involve the use of Helmholtz decomposition [1, 2] or, more recently, 'orthogonal' H^1-reconstructions [20, 12, 17] and $\boldsymbol{H}(\mathbf{curl})$-reconstructions [18].

With the exception of [22, 34], the above developments concern the classical interior penalty DG (IPDG) method for elliptic problems of Arnold [4]. Ern & Stephansen [22] and Kopteva & Rankin [34] consider IPDG methods involving weighted averages in the respective numerical fluxes to address the issues of high contrast diffusion and high local variation in meshsize, respectively. The, so-called, *robust interior penalty discontinuous Galerkin* (RIPDG) method for elliptic problems, introduced in [19] is a variant of the classical IPDG admitting robust behavior with respect to sharp local variation in mesh-size, polynomial degree, and diffusion contrast simultaneously, by a careful selection of weighted averages in the definition of the respective numerical fluxes. RIPDG unifies and extends previous works on the development of IPDG methods with weighted averages [23, 3, 25], which have been, in turn, motivated by developments in domain decomposition [21, 8]; see also the recent developments in [34].

This work is concerned with the proof of residual-type *a posteriori* error bounds in the energy norm for the hp-version *robust* IPDG method. The analysis is based on a variant of the 'orthogonal' H^1-reconstruction from [20, 12, 17], along with a new localised estimation of the diffusion tensor. The analysis tracks carefully the local variations in meshsize, polynomial degree and diffusion contrast. Moreover, in an important deviation from available *a posteriori* error analyses for IPDG methods, we also discuss the case of 'rough', i.e., essentially bounded from above and below, diffusion tensors. Indeed, to the best of our knowledge, the combined treatment of possibly highly locally variable rough diffusion coefficients, rendering essentially the employment of the robust version of IPDG, has not been addressed in the literature.

For simplicity, we focus on the basic second order elliptic boundary value problem on an open bounded polyhedral Lipschitz domain $\Omega \subset \mathbb{R}^d$, $d \in \mathbb{N}$: find $u \in H^1(\Omega)$ such that

$$\begin{aligned} -\nabla \cdot (a\nabla u) + cu &= f \quad \text{in } \Omega; \\ u &= g \quad \text{on } \partial\Omega, \end{aligned} \tag{1}$$

with $f \in L_2(\Omega)$, $g \in H^{\frac{1}{2}}(\partial\Omega)$ and a positive definite diffusion tensor $a \in [L_\infty(\Omega)]^{d\times d}$, and a non-negative $c \in L_\infty(\Omega)$. The problem (1) is understood in the weak sense. Considerable extensions to the above problem are possible, but we refrain from exhausting full generality in the interest of clarity of the presentation.

The remainder of this work is structured as follows. The numerical methods are presented in Section 2, along with proofs of coercivity and continuity. Section 3 is concerned with the proof of *a posteriori* error bounds for the reconstructed/'conforming' part of the error. The analysis is completed in Section 4 with the proof of computable upper bound for the non-conforming remainder of the error. Finally, in Section 5 we present some numerical investigations.

2 Interior Penalty Discontinuous Galerkin Methods

We consider a family of shape-regular simplicial meshes. Each mesh $\mathcal{T}$ consists of mutually disjoint open simplicial elements $K \in \mathcal{T}$, so that $\cup_{K\in\mathcal{T}}\bar{K} = \bar{\Omega}$. Let also $h_K :=$ diam(K), the diameter of $K \in \mathcal{T}$, and define the mesh-function $\mathbf{h} : \cup_{K\in\mathcal{T}}K \to \mathbb{R}_+$ by $\mathbf{h}|_K = h_K$, $K \in \mathcal{T}$. We let $\omega_1(K)$ denote the collection of elements $\hat{K} \in \mathcal{T}$ sharing at least one vertex with the element K. Similarly, $\omega_k(K)$ denotes the collection of elements $\hat{K} \in \mathcal{T}$ sharing at least one vertex with $\omega_{k-1}(K)$.

The set of mesh vertices is denoted by $\mathcal{V}_h$ and is decomposed into the subset of interior vertices, $\mathcal{V}_h^{\mathrm{i}}$, and the subset of boundary vertices, $\mathcal{V}_h^{\mathrm{b}}$. For any $\boldsymbol{a} \in \mathcal{V}_h$, $\mathcal{T}_{\boldsymbol{a}}$ denotes the collection of mesh cells which share $\boldsymbol{a}$ and $\omega_{\boldsymbol{a}}$ in the corresponding open subdomain.

Also, let $\Gamma := \cup_{K\in\mathcal{T}}\partial K$ denote the mesh skeleton and let $\Gamma_{\mathrm{int}} := \Gamma\backslash\partial\Omega$. The mesh skeleton Γ is decomposed into $(d-1)$–dimensional simplicial *faces* F, shared by at most two elements. Let also ∇_h denote the *broken gradient* defined as $\nabla_h v|_K := \nabla(v|_K)$, $K \in \mathcal{T}$. The set of mesh faces is denoted by $\mathcal{F}_h$. In addition, we define $\mathcal{F}_{\boldsymbol{a}}$ as the collection of faces in $\mathcal{F}_h$ which share $\boldsymbol{a}$.

The *hp-discontinuous Galerkin space* $S_{\mathcal{T}}^{\mathbf{p}}$ with respect to $\mathcal{T}$ is defined as

$$S_{\mathcal{T}}^{\mathbf{p}} := \{v \in L_2(\Omega) : v|_K \in \mathcal{P}_{p_K}(K),\ K \in \mathcal{T}\},$$

where $\mathcal{P}_{p_K}(K)$ denotes the space of d-variate polynomials of total degree up to p_K on K; we collect all polynomial degrees in a vector $\mathbf{p} := \{p_K : K \in \mathcal{T}\}$.

Let K_+ and K_- be two adjacent elements of $\mathcal{T}$ sharing a face $F \subset \partial K_+ \cap \partial K_- \subset \Gamma_{\mathrm{int}}$. For v and $\mathbf{q}$ element-wise continuous scalar- and vector-valued functions, respectively, we define the *weighted average* across F by

$$\{\!\{v\}\!\}_{w_F}|_F := w_+v_+|_F + w_-v_-|_F, \quad \{\!\{\mathbf{q}\}\!\}_{w_F}|_F := w_+\mathbf{q}_+|_F + w_-\mathbf{q}_-|_F,$$

with $w_F := (w_+, w_-) \in \mathbb{R}_+^2$ such that $w_+ + w_- = 1$, with $v_\pm|_F$ respectively denoting the trace of v from the element $K_\pm$ on F, and correspondingly for $\mathbf{q}$. Also, we define the *jump* across F by

$$[\![v]\!]|_F := v_+ - v_-, \quad [\![\mathbf{q}]\!]|_F := \mathbf{q}_+ - \mathbf{q}_-,$$

using the convention that the global element numbering of v_+ is greater than that of v_- to determine the sign. In addition, $\mathbf{n}_F$ is denoting the unit outward normal vector on F pointing from v_+ to v_-. We collect all weight pairs in a *weight function* $\mathbf{w} : \Gamma_{\mathrm{int}} \to \mathbb{R}_+^2$, with $\mathbf{w}|_F = w_F$, for each face $F \subset \Gamma_{\text{int}}$. On a boundary face $F \subset \partial\Omega \cap \partial K$, we set $\{\!\{v\}\!\} := v$, $\{\!\{\mathbf{q}\}\!\} := \mathbf{q}$, $[[v]] := v$, and $[[\mathbf{q}]] := \mathbf{q}$, respectively. In view of the latter, we extend $\mathbf{w}$ to Γ by setting $\mathbf{w}|_{\partial\Omega} = (1, 0)$ with $K_- = \emptyset$.

Assuming $a \in [W^{1,\infty}(\Omega, \mathcal{T})]^{d\times d}$, with $W^{1,\infty}(\Omega, \mathcal{T})$ denoting the 'broken' space, containing functions which are element-wise in $W^{1,\infty}(K)$, $K \in \mathcal{T}$, the (weighted) interior penalty discontinuous Galerkin (IPDG) method reads: find $u_h \in S^{\mathbf{p}}_{\mathcal{T}}$, such that

$$B(u_h, v_h) = \ell(v_h), \qquad \text{for all } v_h \in S^{\mathbf{p}}_{\mathcal{T}}, \tag{2}$$

with

$$B(u_h, v_h) := \int_\Omega \left(a\nabla_h u_h \cdot \nabla_h v_h + c u_h v_h\right) \mathrm{d}x + \int_\Gamma \sigma [[u_h]]\,[[v_h]]\,\mathrm{d}s - \int_\Gamma \left(\{\!\{(a\nabla u_h)\cdot \mathbf{n}\}\!\}_{\mathbf{w}}[[v_h]] + \theta\{\!\{(a\nabla v_h)\cdot\mathbf{n}\}\!\}_{\mathbf{w}}[[u_h]]\right) \mathrm{d}s,$$

and

$$\ell(v_h) := \int_\Omega f v_h \,\mathrm{d}x + \int_{\partial\Omega} g(\sigma v_h - \theta(a\nabla v_h)\cdot \mathbf{n})\,\mathrm{d}s,$$

with $\theta \in [-1, 1]$ and $\sigma : \Gamma \to \mathbb{R}$ the, so-called, *discontinuity-penalization* or *penalty* function, of which the precise definition, along with that of the weights $\mathbf{w}$ is a central concern in this work.

The choice $\theta = 1$ in (2) yields the *symmetric* version, while the choice $\theta = -1$ the non-symmetric version of the weighted IPDG method.

The selection $w_F = (1/2, 1/2)$, for all $F \subset \Gamma_{\text{int}}$, corresponds to the classical IPDG method of Wheeler [38] and of Arnold [4]. The case $w_F = (1, 0)$ or $w_F = (0, 1)$ for all $F \subset \Gamma_{\text{int}}$ gives the original IPDG method of Baker [5]. Also, in [21, 8], the weighted average choice $w_F = (\delta_+, \delta_-)$ with

$$\delta_\pm := \frac{\alpha_\mp}{\alpha_+ + \alpha_-},$$

with $\alpha_\pm := a|_{K_\pm}$, was proposed to ensure robustness with respect to variation of element-wise constant scalar diffusion coefficients, while in [23] the selection $\alpha_\pm := \mathbf{n}_F^T a|_{K_\pm}\mathbf{n}_F$ for element-wise constant diffusion tensors was provided, upon combining the selections in [21, 8] and [24]. Also, weighted averages involving local mesh-sizes have been proposed for treating interface problems [3, 25] on non-matching grids on either side of the interface; see also [34]. Finally, in [19], the so-called *robust* interior penalty DG method is presented admitting a choice of σ yielding robust dependence not only with respect to the local variation of the diffusion, but also with respect to the local *direction-wise* variations in mesh-size and polynomial degree.

Remark 1 The above IPDG method (2) is well defined for $a \in [W^{1,\infty}(\Omega, \mathcal{T})]^{d\times d}$. The definition of the method is unclear if $a \in [L_\infty(\Omega)]^{d\times d}$, as the face integrals involving the normal fluxes, $\{\!\{a\nabla w_h\}\!\}_{\mathbf{w}}$, are not necessarily integrable on the skeleton Γ. Typical remedies involve the local projection of the diffusion tensor to element-wise constant functions or, polynomial, in general functions at the cost of introducing an additional consistency error.

To ensure the bilinear form $B : S^{\mathbf{p}}_{\mathcal{T}} \times S^{\mathbf{p}}_{\mathcal{T}} \to \mathbb{R}$ accepts functions in $H^1(\Omega)$ as arguments, we define its extension, still denoted by $B : \mathcal{S}\times\mathcal{S} \to \mathbb{R}$, for $\mathcal{S} := H^1(\Omega)+S^{\mathbf{p}}_{\mathcal{T}}$, given by

$$B(w, v) := \int_\Omega (a\nabla_h w \cdot \nabla_h v + cwv)\,\mathrm{d}x + \int_\Gamma \sigma [\![w]\!] [\![v]\!]\,\mathrm{d}s$$
$$- \int_\Gamma \left(\{\!\{a\Pi_{\mathbf{p-1}}(\nabla w) \cdot \mathbf{n}\}\!\}_{\mathbf{w}} [\![v]\!] + \theta \{\!\{a\Pi_{\mathbf{p-1}}(\nabla v) \cdot \mathbf{n}\}\!\}_{\mathbf{w}} [\![w]\!]\right) \mathrm{d}s,$$

for $w, v \in \mathcal{S}$, with $\Pi_{\mathbf{p-1}} : L_2(\Omega) \to S^{\mathbf{p-1}}_{\mathcal{T}}$ denoting the component-wise orthogonal L_2-projection onto $S^{\mathbf{p-1}}_{\mathcal{T}}$, i.e., the space of one degree less elementwise than the original approximation space. Note that B collapses to the original bilinear form in (2) when restricted to $S^{\mathbf{p}}_{\mathcal{T}} \times S^{\mathbf{p}}_{\mathcal{T}}$, i.e., it is an extension. For completeness, we also set

$$\ell(v) := \int_\Omega f v\,\mathrm{d}x + \int_{\partial\Omega} g(\sigma v - \theta a \Pi_{\mathbf{p-1}}(\nabla v) \cdot \mathbf{n})\,\mathrm{d}s,$$

noting that $B(u, v) = \ell(v)$, for all $v \in H^1(\Omega)$, with $u \in H^1(\Omega)$ denoting the exact solution to the problem (1).

We now discuss the coercivity and continuity of the bilinear form $B(\cdot, \cdot)$ with respect to the norm

$$|||w||| := \left(\|\sqrt{a}\nabla_h w\|^2_\Omega + \|\sqrt{c}w\|^2_\Omega + \|\sqrt{\sigma}[\![w]\!]\|^2_\Gamma\right)^{\frac{1}{2}}.$$

We have

$$B(v, v) = \|\sqrt{a}\nabla_h v\|^2_\Omega + \|\sqrt{c}v\|^2_\Omega + \|\sqrt{\sigma}[\![v]\!]\|^2_\Gamma$$
$$- (1 + \theta) \int_\Gamma \{\!\{a\Pi_{\mathbf{p-1}}(\nabla v) \cdot \mathbf{n}\}\!\}_{\mathbf{w}} [\![v]\!]\,\mathrm{d}s.$$

Now, for a generic face F shared by two elements K_+ and K_-, we have

$$- \int_\Gamma \{\!\{a\Pi_{\mathbf{p-1}}(\nabla v) \cdot \mathbf{n}\}\!\}_{\mathbf{w}} [\![v]\!]\,\mathrm{d}s$$
$$\geq - \sum_{F\subset\Gamma} \Big(\sum_{*\in\{+,-\}} w_* \|a|_{K_*}\mathbf{n}\|_{L_\infty(F)} \|\Pi_{p_{K_*}-1}(\nabla v)|_{K_*}\|_F \Big) \|[\![v]\!]\|_F$$
$$\geq - \sum_{F\subset\Gamma} \Big(\sum_{*\in\{+,-\}} w_* \|a|_{K_*}\mathbf{n}\|_{L_\infty(F)} \sqrt{\frac{p_{K_*}(p_{K_*} + d - 1)|F|}{d|K_*|}} \|\nabla v\|_{K_*} \Big) \|[\![v]\!]\|_F,$$

using the trace inverse estimate $\|v\|^2_F \leq (p + 1)(p + d)|F|/(d|K|)\|v\|^2_K$ for any $v \in \mathcal{P}_p(K)$ [37]. Setting

$$\zeta_* := \left((d + 1)\|a|_{K_*}\mathbf{n}\|_{L_\infty(F)} \|a^{-1/2}\|_{L_\infty(K_*)} \sqrt{\frac{p_{K_*}(p_{K_*} + d - 1)|F|}{d|K_*|}} \right)^{-1},$$

with $* \in \{+,-\}$, for brevity, the above gives

$$
\begin{aligned}
&\left|\int_\Gamma \{\!\{a\Pi_{\mathbf{p-1}}(\nabla v)\cdot \mathbf{n}\}\!\}_{\mathbf{w}}[\![v]\!]\,\mathrm{d}s\right| \\
&\le \sum_{F\subset\Gamma}\Big(\sum_{*\in\{+,-\}}\frac{w_*}{(d+1)\zeta_*}\|\sqrt{a}\nabla v\|_{K_*}\Big)\|[\![v]\!]\|_F \\
&\le \sum_{F\subset\Gamma}\Big(\sum_{*\in\{+,-\}}\frac{1}{d+1}\|\sqrt{a}\nabla v\|_{K_*}\Big)\max_{*\in\{+,-\}}\frac{w_*}{\zeta_*}\|[\![v]\!]\|_F \\
&\le \bigg(\sum_{F\subset\Gamma}\Big(\sum_{*\in\{+,-\}}\frac{1}{d+1}\|\sqrt{a}\nabla v\|_{K_*}\Big)^2\bigg)^{\frac12}\bigg(\sum_{F\subset\Gamma}\max_{*\in\{+,-\}}\frac{w_*^2}{\zeta_*^2}\|[\![v]\!]\|_F^2\bigg)^{\frac12} \\
&\le \bigg(\sum_{F\subset\Gamma}\sum_{*\in\{+,-\}}\frac{2}{(d+1)^2}\|\sqrt{a}\nabla v\|_{K_*}^2\bigg)^{\frac12}\bigg(\sum_{F\subset\Gamma}\max_{*\in\{+,-\}}\frac{w_*^2}{\zeta_*^2}\|[\![v]\!]\|_F^2\bigg)^{\frac12} \\
&\le 2\|\sqrt{a}\nabla v\|_\Omega\bigg(\sum_{F\subset\Gamma}\max_{*\in\{+,-\}}\frac{w_*^2}{\zeta_*^2}\|[\![v]\!]\|_F^2\bigg)^{\frac12} \\
&\le \frac{1}{3+\theta}\|\sqrt{a}\nabla v\|_\Omega^2+(3+\theta)\bigg(\sum_{F\subset\Gamma}\max_{*\in\{+,-\}}\frac{w_*^2}{\zeta_*^2}\|[\![v]\!]\|_F^2\bigg),
\end{aligned}
\tag{3}
$$

since each element has exactly $d+1$ $(d-1)$-dimensional faces.

Selecting

$$\tau|_F = (1+\theta)(3+\theta)\max_{*\in\{+,-\}}\frac{w_*^2}{\zeta_*^2},$$

on every face $F\subset\Gamma$, we deduce

$$B(v,v)\ge \frac{2}{3+\theta}\|\sqrt{a}\nabla_h v\|_\Omega^2+\|\sqrt{c}v\|_\Omega^2+\|\sqrt{\sigma-\tau}[\![v]\!]\|_\Gamma^2.$$

A good choice for σ, as we shall see below, satisfies the equation $(1+\theta)\sigma=(3+\theta)\tau$, thereby dictating the selection

$$\sigma|_F := (3+\theta)^2\max_{*\in\{+,-\}}\frac{w_*^2}{\zeta_*^2},$$

for each $F\subset\Gamma$, whenever $\theta\in(-1,1]$ and σ can be selected as an arbitrary non-negative number $\sigma\ge 0$, whenever $\theta=-1$. With the above choice of σ, we deduce the coercivity of the bilinear form:

$$B(v,v)\ge\frac{2}{3+\theta}\Big(\|\sqrt{a}\nabla_h v\|_\Omega^2+\|\sqrt{\sigma}[\![v]\!]\|_\Gamma^2\Big)+\|\sqrt{c}v\|_\Omega^2\ge\frac{2}{3+\theta}|\!|\!|v|\!|\!|^2. \tag{4}$$

The continuity of the bilinear form $B(\cdot,\cdot)$ follows using similar arguments. To this end, working as in (3), we have, for $\theta \in (-1,1]$,

$$\begin{aligned} B(w,v) &\leq |||w||| |||v||| + \frac{2}{3+\theta}\Big(\|\sqrt{a}\nabla w\|_{\Omega}\|\sqrt{\sigma}[[v]]\|_{\Gamma} + \|\sqrt{a}\nabla v\|_{\Omega}\|\sqrt{\sigma}[[w]]\|_{\Gamma}\Big) \\ &\leq \frac{5+\theta}{3+\theta}|||w||| |||v|||. \end{aligned} \tag{5}$$

For $\theta = -1$, we have, instead

$$B(w,v) \leq (1 + C_{\mathrm{NIP}})|||w||| |||v|||, \tag{6}$$

with $C_{\mathrm{NIP}} := 2\| \max_{*\in\{+,-\}} w_*(\zeta_*\sigma)^{-1}\|_{L_\infty(\Gamma)}$, for any $\sigma : \Gamma \to \mathbb{R}$ non-negative function.

We now comment on particular cases. For the classical IP(DG) for $\theta \in (-1,1]$ [38, 4, 28], we have $w_* = \frac{1}{2}$, thereby we deduce

$$\sigma_{\mathrm{IP}} = \left(\frac{3+\theta}{4}\right)^2 \max_{*\in\{+,-\}} \zeta_*^{-2},$$

while for the Robust IP(DG) [19], we select $w_* = \frac{\zeta_*}{\zeta_+ + \zeta_-}$, thereby, deducing

$$\sigma_{\mathrm{RIP}} = (3+\theta)^2(\zeta_+ + \zeta_-)^{-2}.$$

The terminology 'robust' stems from the fact that σ_{RIP} remains bounded under extreme local variations of the diffusion coefficient, mesh-size, polynomial degree; we refer to [19] for a detailed discussion. We note that the above analysis extends slightly (and corrects a missing factor of 4 in) the respective developments in [19].

We now return to the question of the definition of an IPDG method for elliptic problems for which the diffusion tensor is just assumed to be essentially bounded, viz., $a \in [L_\infty(\Omega)]^{d\times d}$, and positive definite. Instead of projecting a into a piece-wise constant or polynomial function, possibly subordinate to the mesh, we consider the weighted IPDG variant reading: find $u_h \in S_{\mathcal{T}}^{\mathbf{p}}$, such that

$$\mathcal{B}(u_h, v_h) = \mathrm{l}(v_h), \qquad \text{for all } v_h \in S_{\mathcal{T}}^{\mathbf{p}}, \tag{7}$$

with

$$\begin{aligned} \mathcal{B}(u_h, v_h) :=& \int_\Omega \left(a\nabla_h u_h \cdot \nabla_h v_h + c u_h v_h\right) \mathrm{d}x + \int_\Gamma \sigma [[u_h]]\,[[v_h]] \,\mathrm{d}s \\ &- \int_\Gamma \left(\{\!\{\Pi_{\mathbf{p-1}}(a\nabla u_h)\cdot \mathbf{n}\}\!\}_{\mathbf{w}}[[v_h]] + \theta\{\!\{\Pi_{\mathbf{p-1}}(a\nabla v_h)\cdot \mathbf{n}\}\!\}_{\mathbf{w}}[[u_h]]\right) \mathrm{d}s, \end{aligned}$$

and

$$\mathrm{l}(v_h) := \int_\Omega f v_h \,\mathrm{d}x + \int_{\partial\Omega} g(\sigma v_h - \theta\Pi_{\mathbf{p-1}}(a\nabla v_h)\cdot \mathbf{n}) \,\mathrm{d}s.$$

The key difference between (7) and the (R)IPDG method (2) is the application of orthogonal L_2-projection onto the finite element space for the numerical fluxes. In practice, this requires to compute the orthogonal L_2 projection of $a\nabla\phi_i$ for each local basis function ϕ_i. This is a local operation, made easier when using elemental L_2-orthogonal bases, affecting only the assembly step, which is fully parallelisable, e.g., on a GPU chip. We note that the complexity of performing the projections is entirely comparable to first projecting a and then applying the (R)IPDG method (2).

To show coercivity and continuity properties, we work similarly to the above. In particular, we have

$$\mathcal{B}(v,v) = \|\sqrt{a}\nabla_h v\|_\Omega^2 + \|\sqrt{c}v\|_\Omega^2 + \|\sqrt{\sigma}[\![v]\!]\|_\Gamma^2 - (1+\theta)\int_\Gamma \{\!\{\Pi_{\mathbf{p}-\mathbf{1}}(a\nabla v)\cdot\mathbf{n}\}\!\}_{\mathbf{w}}[\![v]\!]\,\mathrm{d}s,$$

so that, for a generic face F shared by two elements K_+ and K_-, we get

$$\begin{aligned}
&-\int_\Gamma \{\!\{\Pi_{\mathbf{p}-\mathbf{1}}(a\nabla v)\cdot\mathbf{n}\}\!\}_{\mathbf{w}}[\![v]\!]\,\mathrm{d}s\\
&\geq -\sum_{F\subset\Gamma}\Big(\sum_{*\in\{+,-\}} w_*\|\Pi_{p_{K_*}-1}(a\nabla v)|_{K_*}\|_F\Big)\|[\![v]\!]\|_F\\
&\geq -\sum_{F\subset\Gamma}\Big(\sum_{*\in\{+,-\}} w_*\sqrt{\frac{p_{K_*}(p_{K_*}+d-1)|F|}{d|K_*|}}\|a\nabla v\|_{K_*}\Big)\|[\![v]\!]\|_F,
\end{aligned}$$

using the trace inverse estimate and $\|\mathbf{n}\|_{L_\infty(F)}\leq 1$. Setting, now,

$$\bar{\zeta}_* := \left((d+1)\|\sqrt{a}\|_{L_\infty(K_*)}\sqrt{\frac{p_{K_*}(p_{K_*}+d-1)|F|}{d|K_*|}}\right)^{-1},$$

for $*\in\{+,-\}$, the above gives

$$\begin{aligned}
&\Big|\int_\Gamma \{\!\{\Pi_{\mathbf{p}-\mathbf{1}}(a\nabla v)\cdot\mathbf{n}\}\!\}_{\mathbf{w}}[\![v]\!]\,\mathrm{d}s\Big|\\
&\leq \sum_{F\subset\Gamma}\Big(\sum_{*\in\{+,-\}}\frac{w_*}{(d+1)\bar{\zeta}_*}\|\sqrt{a}\nabla v\|_{K_*}\Big)\|[\![v]\!]\|_F\\
&\leq \frac{1}{3+\theta}\|\sqrt{a}\nabla v\|_\Omega^2 + (3+\theta)\Big(\sum_{F\subset\Gamma}\max_{*\in\{+,-\}}\frac{w_*^2}{\bar{\zeta}_*^2}\|[\![v]\!]\|_F^2\Big),
\end{aligned} \tag{8}$$

as before. Selecting, now, $\bar{\tau}|_F = (1+\theta)(3+\theta)\max_{*\in\{+,-\}} w_*^2\bar{\zeta}_*^{-2}$, on every face $F\subset\Gamma$, we deduce

$$B(v,v)\geq \frac{2}{3+\theta}\|\sqrt{a}\nabla_h v\|_\Omega^2 + \|\sqrt{c}v\|_\Omega^2 + \|\sqrt{\sigma-\bar{\tau}}[\![v]\!]\|_\Gamma^2.$$

We select σ satisfying $(1+\theta)\sigma = (3+\theta)\bar{\tau}$, thereby

$$\sigma|_F := (3+\theta)^2 \max_{*\in\{+,-\}} \frac{w_*^2}{\bar{\zeta}_*^2},$$

for each $F \subset \Gamma$, for $\theta \in (-1,1]$, arbitrary $\sigma \geq 0$, for $\theta = -1$. With the above choice of σ, we deduce both the coercivity (4) and continuity (5) of the bilinear form, the latter for $\theta \in (-1,1]$, while for $\theta = -1$, we have instead (6) with the constant C_{NIP} replaced by $\bar{C}_{\mathrm{NIP}} := 2\| \max_{*\in\{+,-\}} w_* (\bar{\zeta}_* \sigma)^{-1} \|_{L_\infty(\Gamma)}$, for any $\sigma : \Gamma \to \mathbb{R}$ non-negative function.

Remark 2 The variant (7) can also be defined when the diffusion tensor vanishes locally upon adopting the convention that $\bar{\zeta}_*^{-1} = 0$ in regions where $a = 0$ also. For an alternative IPDG method that is robust with respect to diffusion degeneration we refer to [24] and its 'robust' extension in [19, Section 6].

3 *A Posteriori* Error Bounds for a Conforming Reconstruction

A key ingredient of a typical proof of *a posteriori* error bounds in the energy norm for non-conforming methods is a recovery of the approximate solution u_h into the solution space. Indeed, employing the stability of local nodal averaging operators, Karakashian & Pascal proved the first rigorous *a posteriori* error bounds in the energy norm for the IPDG method discretising elliptic problems [31]. An alternative recovery/reconstruction employing the PDE problem has been recently used in [12] for the proof of *a posteriori* error bounds in the energy norm for IPDG methods on polygonal/polyhedral meshes on simply connected domains. More recently, it has been extended to hybrid high-order methods on simplicial meshes on the polytopic Lipschitz domain [17].

It is useful to define some contrast factors related to the diffusion coefficient a. For all $K \in \mathcal{T}$ and all $\boldsymbol{a} \in \mathcal{V}_h$, we set

$$\mu_{\omega_k(K)}(a) := a_K^\sharp (a^\flat_{\omega_k(K)})^{-1}, \tag{9}$$

with $a_K^\sharp := \sup_{\mathbf{x}\in K} |a(\mathbf{x})|$, $a_K^\flat := \inf_{\mathbf{x}\in K} |a(\mathbf{x})|$ and $a^\flat_{\omega_k(K)} := \inf_{\mathbf{x}\in\omega_k(K)} |a(\mathbf{x})|$, where $\omega_k(K)$ denotes the patch containing k-th layer elements. In addition,

$$\mu_{\boldsymbol{a}}(a) := a_{\boldsymbol{a}}^\sharp (a_{\boldsymbol{a}}^\flat)^{-1}, \tag{10}$$

with $a_{\boldsymbol{a}}^\sharp := \sup_{\mathbf{x}\in\omega_{\boldsymbol{a}}} |a(\mathbf{x})|$ and $a_{\boldsymbol{a}}^\flat := \inf_{\mathbf{x}\in\omega_{\boldsymbol{a}}} |a(\mathbf{x})|$. Moreover, it is convenient to set

$$a_F^\flat := \min(a_{K_+}^\flat, a_{K_-}^\flat), \quad \forall F \subset \partial K_+ \cap \partial K_-, \tag{11a}$$

$$a_F^\flat := a_K^\flat, \quad \forall F \subset \partial K \subset \partial\Omega. \tag{11b}$$

Following [17], we define the potential reconstruction $u_c \in H_g^1(\Omega)$ constructed from the DG solution $u_h \in S_{\mathcal{T}}^{\mathbf{p}}$, with $H_g^1(\Omega) := \{v \in H^1(\Omega) \mid v|_{\partial\Omega} = g\}$.

For the *a posteriori* error analysis, we begin by decomposing the error as

$$e := u - u_h = (u - u_c) + (u_c - u_h) =: e_c + e_d. \tag{12}$$

Observe the identity $B(e_d, v) = 0$ for any $v \in H^1(\Omega)$. Then, we have

$$\begin{aligned}\frac{2}{3+\theta}|||e_c|||^2 \le B(e_c, e_c) &= B(e, e_c) - B(e_d, e_c) = \ell(e_c) - B(u_h, e_c) - B(e_d, e_c)\\ &= (\ell(e_c - v_h) - B(u_h, e_c - v_h)) - B(e_d, e_c) = T_1 + T_2,\end{aligned} \tag{13}$$

for any $v_h \in S_{\mathcal{T}}^{\mathbf{p}}$. Setting $\eta := e_c - \Pi_{\mathbf{p}} e_c$ for brevity, we perform element-wise integration by parts to deduce

$$\begin{aligned}T_1 = &\sum_{K\in\mathcal{T}} \int_K \Big((f + \nabla \cdot (a\nabla u_h) - cu_h)\eta \,\mathrm{d}x - \int_{\partial K} a\nabla u_h \cdot \mathbf{n}\eta \,\mathrm{d}s \Big)\\ &+ \int_\Gamma \big(\{\!\{(a\nabla u_h)\cdot \mathbf{n}\}\!\}_{\mathbf{w}} [\![\eta]\!] + \theta \{\!\{a\Pi_{\mathbf{p-1}}(\nabla\eta)\cdot\mathbf{n}\}\!\}_{\mathbf{w}} [\![u_h]\!] - \sigma [\![u_h]\!][\![\eta]\!] \big) \,\mathrm{d}s\\ &+ \int_{\partial\Omega} g(\sigma\eta - \theta a \Pi_{\mathbf{p-1}}(\nabla\eta)\cdot\mathbf{n}) \,\mathrm{d}s\\ = &\sum_{K\in\mathcal{T}} \int_K (f + \nabla\cdot(a\nabla u_h) - cu_h)\eta \,\mathrm{d}x - \int_{\Gamma_{\text{int}}} [\![(a\nabla u_h)\cdot\mathbf{n}]\!] \{\!\{\eta\}\!\}_{\mathbf{1-w}} \,\mathrm{d}s\\ &+ \int_\Gamma \big(\theta\{\!\{a\Pi_{\mathbf{p-1}}(\nabla\eta)\cdot\mathbf{n}\}\!\}_{\mathbf{w}}[\![u_h]\!] - \sigma[\![u_h]\!][\![\eta]\!]\big)\,\mathrm{d}s\\ &+ \int_{\partial\Omega} g(\sigma\eta - \theta a\Pi_{\mathbf{p-1}}(\nabla\eta)\cdot\mathbf{n})\,\mathrm{d}s,\end{aligned} \tag{14}$$

since $\Pi_{\mathbf{p-1}}(\nabla u_h) = \nabla u_h$ and, upon employing the identity [34, eq. (3.15)]:

$$\begin{aligned}&\sum_{K\in\mathcal{T}} \int_{\partial K} a\nabla u_h \cdot \mathbf{n}\eta \,\mathrm{d}s\\ &= \int_\Gamma \{\!\{(a\nabla u_h)\cdot\mathbf{n}\}\!\}_{\mathbf{w}}[\![\eta]\!]\,\mathrm{d}s + \int_{\Gamma_{\text{int}}} [\![(a\nabla u_h)\cdot\mathbf{n}]\!]\{\!\{\eta\}\!\}_{\mathbf{1-w}}\,\mathrm{d}s,\end{aligned}$$

respectively, with $(\mathbf{1} - \mathbf{w})|_F := 1 - w_F$, for each $F \subset \Gamma_{\text{int}}$.

We still need to select suitable $v_h \in S_{\mathcal{T}}^{\mathbf{p}}$ in (21). Selecting $v_h := \Pi_{\mathbf{p}} e_c$, we have

$$\|\eta\|_K \le C_{L_2}^{\text{app}} h_K p_K^{-1} \|\nabla e_c\|_K, \tag{15}$$

and

$$\|\eta\|_F \le C_{L_2}^{\text{app}} \sqrt{h_K p_K^{-1}} \|\nabla e_c\|_K, \tag{16}$$

for a suitable constant $C_{L_2}^{\text{app}} > 0$, independent of h_K, p_K and of e_c, but maybe dependent on the element shape. A computable upper bound to $C_{L_2}^{\text{app}}$ can be found in [16, 35]. Moreover, we have

$$\|\nabla\eta\|_K \le C_{L_2,K}^{\text{st}} p_K^{\frac{1}{2}} \|\nabla e_c\|_K. \tag{17}$$

with $C^{\mathrm{st}}_{L_2,K} := Cp_K^{\frac{1}{2}}$, see [15, Appendix C].

Defining now the residuals $\mathcal{R}_{\mathrm{el}} : \Omega\backslash\Gamma \to \mathbb{R}$, $\mathcal{R}_{\mathrm{fa}} : \Gamma_{\mathrm{int}} \to \mathbb{R}$, and $\mathcal{R}_{\mathrm{pe}} : \Gamma \to \mathbb{R}$, with

$$\mathcal{R}_{\mathrm{el}}|_K := (a_K^\flat)^{-\frac{1}{2}} h_K p_K^{-1} r_{\mathrm{el},K},$$

whereby $r_{\mathrm{el},K} := (I - \Pi_{p_K})\big(f + \nabla\cdot(a\nabla u_h) - cu_h\big)|_K$, and

$$\mathcal{R}_{\mathrm{fa}}|_F := (a_F^\flat)^{-\frac{1}{2}} \max_{*\in\{+,-\}} (1-w_*)\sqrt{2(d+1)h_{K_*}p_{K_*}^{-1}}\,[\![(a\nabla u_h)\cdot\mathbf{n}]\!],$$

$$\mathcal{R}_{\mathrm{pe}}|_F := \begin{cases} \sqrt{2(d+1)\max_{*\in\{+,-\}}(a_{K_*}^\flat)^{-1}h_{K_*}p_{K_*}^{-1}}\,\sigma[\![u_h]\!], & \text{if } F\subset\Gamma_{\mathrm{int}};\\ \sqrt{2(d+1)(a_K^\flat)^{-1}h_K p_K^{-1}}\,\sigma(u_h-g), & \text{if } F\subset\partial K\cap\partial\Omega. \end{cases}$$

Returning to (14), we further estimate each term on the right-hand side: for the first term we have

$$\begin{aligned}\sum_{K\in\mathcal{T}}\int_K r_{\mathrm{el},K}\eta\,\mathrm{d}x &\le \sum_{K\in\mathcal{T}} C^{\mathrm{app}}_{L_2}\|(h_K p_K^{-1})r_{\mathrm{el},K}\|_K\|\nabla e_c\|_K\\ &\le \sum_{K\in\mathcal{T}} C^{\mathrm{app}}_{L_2}\|\mathcal{R}_{\mathrm{el}}\|_K\|\sqrt{a}\nabla e_c\|_K \le C^{\mathrm{app}}_{L_2}\|\mathcal{R}_{\mathrm{el}}\|_\Omega\|\sqrt{a}\nabla e_c\|_\Omega.\end{aligned}$$

For the second term in the right-hand side of (14), we have

$$\begin{aligned}&\int_{\Gamma_{\mathrm{int}}}[\![(a\nabla u_h)\cdot\mathbf{n}]\!]\{\!\{\eta\}\!\}_{\mathbf{1-w}}\,\mathrm{d}s\\ &\le C^{\mathrm{app}}_{L_2}\sum_{F\subset\Gamma_{\mathrm{int}}}\|[\![(a\nabla u_h)\cdot\mathbf{n}]\!]\|_F\sum_{*\in\{+,-\}}(1-w_*)\sqrt{h_{K_*}p_{K_*}^{-1}}\|\nabla e_c\|_{K_*}\\ &\le C^{\mathrm{app}}_{L_2}\sum_{F\subset\Gamma_{\mathrm{int}}}\max_{*\in\{+,-\}}(1-w_*)\sqrt{(a_{K_*}^\flat)^{-1}h_{K_*}p_{K_*}^{-1}}\|[\![(a\nabla u_h)\cdot\mathbf{n}]\!]\|_F\sum_{*\in\{+,-\}}\|\sqrt{a}\nabla e_c\|_{K_*}\\ &\le C^{\mathrm{app}}_{L_2}\Big(\sum_{F\subset\Gamma_{\mathrm{int}}}\frac{1}{2(d+1)}\|\mathcal{R}_{\mathrm{fa}}\|_F^2\Big)^{\frac{1}{2}}\Big(\sum_{F\subset\Gamma_{\mathrm{int}}}\Big(\sum_{*\in\{+,-\}}\|\sqrt{a}\nabla e_c\|_{K_*}\Big)^2\Big)^{\frac{1}{2}}\\ &\le C^{\mathrm{app}}_{L_2}\Big(\sum_{F\subset\Gamma_{\mathrm{int}}}(d+1)^{-1}\|\mathcal{R}_{\mathrm{fa}}\|_F^2\Big)^{\frac{1}{2}}\Big(\sum_{F\subset\Gamma_{\mathrm{int}}}\sum_{*\in\{+,-\}}\|\sqrt{a}\nabla e_c\|_{K_*}^2\Big)^{\frac{1}{2}}\\ &\le C^{\mathrm{app}}_{L_2}\|\mathcal{R}_{\mathrm{fa}}\|_{\Gamma_{\mathrm{int}}}\|\sqrt{a}\nabla e_c\|_\Omega.\end{aligned}$$

To estimate the third term in the right-hand side of (14), we set

$$\begin{aligned}\xi_* &:= \zeta_*\|a^{-1/2}\|_{L_\infty(K_*)}\\ &= \left((d+1)\|a|_{K_*}\mathbf{n}\|_{L_\infty(F)}\sqrt{\frac{p_{K_*}(p_{K_*}+d-1)|F|}{d|K_*|}}\right)^{-1}.\end{aligned} \tag{18}$$

For brevity, we split the interior skeleton Γ_{int} and the boundary, we work as for (8) and, additionally, we use (17), getting, respectively,

$$
\begin{aligned}
&\Big|\int_{\Gamma_{\text{int}}} \{\!\!\{a\Pi_{\mathbf{p-1}}(\nabla\eta)\cdot\mathbf{n}\}\!\!\}_{\mathbf{w}}[\![u_h]\!]\,\mathrm{d}s\Big| \\
&\le \sum_{F\subset\Gamma_{\text{int}}}\Big(\sum_{*\in\{+,-\}}\frac{w_*}{(d+1)\xi_*}\|\nabla\eta\|_{K_*}\Big)\|[\![u_h]\!]\|_F \\
&\le \sum_{F\subset\Gamma_{\text{int}}}\Big(\sum_{*\in\{+,-\}}\frac{w_*C^{\text{st}}_{L_2,K^*}}{(d+1)\xi_*}\|\nabla e_c\|_{K_*}\Big)\|[\![u_h]\!]\|_F \\
&\le \sum_{F\subset\Gamma_{\text{int}}}\Big(\sum_{*\in\{+,-\}}\frac{w_*C^{\text{st}}_{L_2,K^*}}{(d+1)\xi_*(a^{\flat}_{K_*})^{\frac12}}\|\sqrt{a}\nabla e_c\|_{K_*}\Big)\|[\![u_h]\!]\|_F \\
&\le \sum_{F\subset\Gamma_{\text{int}}}\Big(\sum_{*\in\{+,-\}}(d+1)^{-1}\|\sqrt{a}\nabla e_c\|_{K_*}\Big) \\
&\qquad\times\max_{*\in\{+,-\}}\frac{w_*}{\xi_*}C^{\text{st}}_{L_2,K^*}(a^{\flat}_{K_*})^{-\frac12}\|[\![u_h]\!]\|_F \\
&\le\Big(\sum_{F\subset\Gamma_{\text{int}}}\sum_{*\in\{+,-\}}\frac{2}{(d+1)^2}\|\sqrt{a}\nabla e_c\|^2_{K_*}\Big)^{\frac12} \\
&\qquad\times\Big(\sum_{F\subset\Gamma_{\text{int}}}\max_{*\in\{+,-\}}\Big(\frac{w_*}{\xi_*}\frac{C^{\text{st}}_{L_2,K^*}}{(a^{\flat}_{K_*})^{\frac12}}\Big)^2\|[\![u_h]\!]\|^2_F\Big)^{\frac12} \\
&\le\sqrt{\frac{2}{d+1}}\|\sqrt{a}\nabla e_c\|_{\Omega}\Big(\sum_{F\subset\Gamma_{\text{int}}}\max_{*\in\{+,-\}}\Big(\frac{w_*}{\xi_*}\frac{C^{\text{st}}_{L_2,K^*}}{(a^{\flat}_{K_*})^{\frac12}}\Big)^2\|[\![u_h]\!]\|^2_F\Big)^{\frac12} \\
&= \|\chi[\![u_h]\!]\|_{\Gamma_{\text{int}}}\|\sqrt{a}\nabla e_c\|_{\Omega},
\end{aligned}
\tag{19}
$$

for $\theta\in[-1,1]$, with $\chi:\Gamma\to\mathbb{R}$, defined face-wise by

$$
\chi|_F := \sqrt{2}(d+1)^{-\frac12}\max_{*\in\{+,-\}}\frac{w_*}{\xi_*}\frac{C^{\text{st}}_{L_2,K^*}}{(a^{\flat}_{K_*})^{\frac12}},
$$

for $F\subset\Gamma_{\text{int}}$ and $\chi|_F := \sqrt{2}(d+1)^{-\frac12}\xi_+^{-1}C^{\text{st}}_{L_2,K^+}(a^{\flat}_{K_+})^{-\frac12}$ for $F\subset\partial\Omega$ (the latter to be required below). Also, we have

$$
\begin{aligned}
&\int_{\Gamma_{\text{int}}}\sigma[\![u_h]\!][\![\eta]\!]\,\mathrm{d}s \\
&\le\sum_{F\subset\Gamma_{\text{int}}}\|\sigma[\![u_h]\!]\|_F\|[\![\eta]\!]\|_F \\
&\le C^{\text{app}}_{L_2}\sum_{F\subset\Gamma_{\text{int}}}\|\sigma[\![u_h]\!]\|_F\Big(\sum_{*\in\{+,-\}}\sqrt{h_{K_*}p^{-1}_{K_*}}\|\nabla e_c\|_{K_*}\Big)
\end{aligned}
$$

$$\le C_{L_2}^{\rm app} \sum_{F\subset\Gamma_{\rm int}} \max_{*\in\{+,-\}} \sqrt{(a_{K_*}^{\flat})^{-1} h_{K_*} p_{K_*}^{-1}} \|\sigma [\![u_h]\!]\|_F \Big(\sum_{*\in\{+,-\}} \|\sqrt{a}\nabla e_c\|_{K_*}\Big)$$
$$\le \sqrt{2} C_{L_2}^{\rm app} \Big(\sum_{F\subset\Gamma_{\rm int}} \max_{*\in\{+,-\}} \frac{h_{K_*}}{a_{K_*}^{\flat} p_{K_*}} \|\sigma [\![u_h]\!]\|_F^2 \Big)^{\frac12} \Big(\sum_{F\subset\Gamma_{\rm int}} \sum_{*\in\{+,-\}} \|\sqrt{a}\nabla e_c\|_{K_*}^2 \Big)^{\frac12}$$
$$\le C_{L_2}^{\rm app} \Big(2(d+1) \sum_{F\subset\Gamma_{\rm int}} \max_{*\in\{+,-\}} \frac{h_{K_*}}{a_{K_*}^{\flat} p_{K_*}} \|\sigma [\![u_h]\!]\|_F^2 \Big)^{\frac12} \|\sqrt{a}\nabla e_c\|_\Omega$$
$$= C_{L_2}^{\rm app} \|\mathcal{R}_{\rm pe}\|_{\Gamma_{\rm int}} \|\sqrt{a}\nabla e_c\|_\Omega.$$

For the respective boundary terms, working analogously to the respective interior terms, we deduce

$$\int_{\partial\Omega} (u_h - g)(\theta a \Pi_{\mathbf{p-1}}(\nabla\eta)\cdot\mathbf{n} u_h - \sigma u_h \eta)\,\mathrm{d}s$$
$$\le (\|\chi(u_h-g)\|_{\partial\Omega} + C_{L_2}^{\rm app} \|\mathcal{R}_{\rm pe}^{L_2}\|_{\partial\Omega}) \|\sqrt{a}\nabla e_c\|_\Omega.$$

Combining the above, and performing straightforward further estimation, we obtain the following result.

Theorem 1 *Assume $a \in [W^{1,\infty}(\Omega,\mathcal{T})]^{d\times d}$, let $u \in H^1(\Omega)$ be the solution to (1) and let $u_h \in S_{\mathcal{T}}^{\mathbf{p}}$ be its approximation satisfying (2). Then for any $u_c \in H_g^1(\Omega)$, and e_c defined in (12), we have the error bound*

$$\begin{aligned} |\!|\!| e_c |\!|\!| \le C_1 \Big(\|\mathcal{R}_{\rm el}\|_\Omega + \|\mathcal{R}_{\rm fa}\|_{\Gamma_{\rm int}} + \|\mathcal{R}_{\rm pe}\|_{\Gamma_{\rm int}} + \|\mathcal{R}_{\rm pe}\|_{\partial\Omega} \Big) \\ + C_2 \Big(\|\chi [\![u_h]\!]\|_{\Gamma_{\rm int}} + \|\chi(u_h-g)\|_{\partial\Omega} \Big) + C_3 |\!|\!| e_d |\!|\!|, \end{aligned} \tag{20}$$

whereby $C_2 := \frac{3+\theta}{2}$, $C_1 := C_{L_2}^{\rm app} C_2$ and $C_3 := \frac{5+\theta}{2}$, for $\theta \in [-1,1]$.

Proof. Combining the above estimates, we get

$$\begin{aligned} \frac{2}{3+\theta} |\!|\!| e_c |\!|\!|^2 \le C_{L_2}^{\rm app} \Big(\|\mathcal{R}_{\rm el}\|_\Omega + \|\mathcal{R}_{\rm fa}\|_{\Gamma_{\rm int}} + \|\mathcal{R}_{\rm pe}\|_{\Gamma_{\rm int}} + \|\mathcal{R}_{\rm pe}\|_{\partial\Omega} \Big) \|\sqrt{a}\nabla e_c\|_\Omega \\ + \Big(\|\chi [\![u_h]\!]\|_{\Gamma_{\rm int}} + \|\chi(u_h-g)\|_{\partial\Omega} \Big) \|\sqrt{a}\nabla e_c\|_\Omega + \frac{5+\theta}{3+\theta} |\!|\!| e_d |\!|\!| |\!|\!| e_c |\!|\!|. \end{aligned}$$

for $\theta \in [-1,1]$. Since $\|\sqrt{a}\nabla e_c\|_\Omega \le |\!|\!| e_c |\!|\!|$, elementary calculations give the result. □

We note that (20) is *not* an *a posteriori* error bound, as it contains e_d on the right-hand side. A computable bound will be obtained below by showing an *a posteriori* error bound for $|\!|\!| e_d |\!|\!|$.

Next, we also show a bound for the variant of the classical weighted IPDG method (7), which is valid for $a \in [L_\infty(\Omega)]^{d\times d}$ as well. In this case, noting that $\Pi_{\mathbf{p-1}}(a\nabla u_h)|_K \in L_2(K)$, for all $K \in \mathcal{T}$, we add and subtract the term

$$\int_\Omega \Pi_{\mathbf{p-1}}(a\nabla_h u_h)\cdot\nabla_h\eta\,\mathrm{d}x$$

on the right-hand side of the error equation, to deduce

$$\begin{aligned}
\mathfrak{l}(\eta) - \mathcal{B}(u_h,\eta) = &\sum_{K\in\mathcal{T}}\int_K (f + \nabla\cdot\Pi_{\mathbf{p-1}}(a\nabla u_h) - cu_h)\hat{\eta}\,\mathrm{d}x\\
&-\int_{\Gamma_{\rm int}} [[\Pi_{\mathbf{p-1}}(a\nabla u_h)\cdot\mathbf{n}]]\{\!\{\eta\}\!\}_{\mathbf{1-w}}\,\mathrm{d}s\\
&-\int_{\Gamma_{\rm int}} (\theta\{\!\{\Pi_{\mathbf{p-1}}(a\nabla\eta)\cdot\mathbf{n}\}\!\}_{\mathbf{w}}[[u_h]] - \sigma[[u_h]][[\eta]])\,\mathrm{d}s\\
&+\int_{\partial\Omega}(g-u_h)(\sigma\eta - \theta\Pi_{\mathbf{p-1}}(a\nabla\eta)\cdot\mathbf{n})\,\mathrm{d}s\\
&+\int_{\Omega}(I-\Pi_{\mathbf{p-1}})(a\nabla_h u_h)\cdot\nabla_h\eta\,\mathrm{d}x,
\end{aligned}\tag{21}$$

for $\eta = e_c - v_h$, $v_h \in S_{\mathcal{T}}^{\mathbf{p}}$.

We select $v_h := \Pi_{\mathbf{p}} e_c$ and define the residuals $\mathfrak{R}_{\rm el} : \Omega\backslash\Gamma \to \mathbb{R}$, and $\mathfrak{R}_{\rm fa} : \Gamma_{\rm int} \to \mathbb{R}$, with

$$\begin{aligned}
\mathfrak{R}_{\rm el}|_K &:= (a_K^\flat)^{-\frac{1}{2}} h_K p_K^{-1}(I-\Pi_{p_K})(f-cu_h),\\
\mathcal{R}_{\rm fa}|_F &:= (a_F^\flat)^{-\frac{1}{2}}\max_{*\in\{+,-\}}(1-w_*)\sqrt{s(d+1)h_K p_K^{-1}}[[\Pi_{\mathbf{p-1}}(a\nabla u_h)\cdot\mathbf{n}]],
\end{aligned}$$

as well as the non-standard "oscillation" term $\mathfrak{R}_{\rm a} : \Omega\backslash\Gamma \to \mathbb{R}$, given by

$$\mathfrak{R}_{\rm a}|_K := (a_K^\flat)^{-\frac{1}{2}} C_{L_2,K}^{\rm st}(I-\Pi_{\mathbf{p-1}})(a\nabla u_h)|_K$$

for all $K \in \mathcal{T}$.

Returning to (21), we estimate each term on the right-hand side: for the first term we have

$$\begin{aligned}
\sum_{K\in\mathcal{T}}\int_K (f + \nabla\cdot\Pi_{\mathbf{p-1}}(a\nabla u_h) - cu_h)\eta\,\mathrm{d}x &= \int_\Omega (f-cu_h)\eta\,\mathrm{d}x\\
&\le C_{L_2}^{\rm app}\|\mathfrak{R}_{\rm el}\|_\Omega\|\sqrt{a}\nabla e_c\|_\Omega.
\end{aligned}$$

For the third term, we let

$$\bar{\xi}_* := \left((d+1)\|a\|_{L_\infty(K_*)}\sqrt{\frac{p_{K_*}(p_{K_*}+d-1)|F|}{d|K_*|}}\right)^{-1}\tag{22}$$

and, following similar steps as in (19), we deduce

$$\left|\int_{\Gamma_{\rm int}}\{\!\{\Pi_{\mathbf{p-1}}(a\nabla\eta)\cdot\mathbf{n}\}\!\}_{\mathbf{w}}[[u_h]]\,\mathrm{d}s\right| \le \|\bar{\chi}[[u_h]]\|_{\Gamma_{\rm int}}\|\sqrt{a}\nabla e_c\|_\Omega,$$

for $\bar{\chi} : \Gamma \to \mathbb{R}$, with

$$\bar{\chi}|_F := \sqrt{2}(d+1)^{-\frac{1}{2}} \max_{*\in\{+,-\}} (w_* \bar{\xi}_*^{-1}) C^{\mathrm{st}}_{L_2,K^*} (a^\flat_{K_*})^{-\frac{1}{2}},$$

for every face $F \subset \Gamma_{\mathrm{int}}$ and $\bar{\chi}|_F := \sqrt{2}(d+1)^{-\frac{1}{2}} \bar{\xi}_+ C^{\mathrm{st}}_{L_2,K^+} (a^\flat_{K_+})^{-\frac{1}{2}}$ for $F \subset \partial\Omega$. The bound for the discontinuity-penalisation component from (19) is still valid.

For the boundary terms, we have

$$\begin{aligned}
&\int_{\partial\Omega} (u_h - g)\big(\theta \Pi_{\mathbf{p}-1}(a\nabla\eta)\cdot \mathbf{n} u_h - \sigma u_h \eta\big)\,\mathrm{d}s \\
&\le \big(\|\bar{\chi}(u_h - g)\|_{\partial\Omega} + C^{\mathrm{app}}_{L_2} \|\mathcal{R}_{\mathrm{pe}}\|_{\partial\Omega}\big)\|\sqrt{a}\nabla e_c\|_{\Omega}.
\end{aligned}$$

Finally, for the last term on the right-hand side of (21), we have

$$\int_{\Omega} (I - \Pi_{\mathbf{p}-1})(a\nabla_h u_h)\cdot \nabla_h \eta \,\mathrm{d}x \le \|\mathfrak{R}_a\|_{\Omega} \|\sqrt{a}\nabla e_c\|_{\Omega}.$$

Combining the above, along with straightforward arguments, we arrive at the following result.

Theorem 2 *For $a \in [L_\infty(\Omega)]^{d\times d}$, let $u \in H^1(\Omega)$ be the solution to* (1) *and let $u_h \in S^{\mathbf{p}}_{\mathcal{T}}$ be its approximation satisfying* (7)*. Then, we have the bound*

$$\begin{aligned}
|||e_c||| \le{}& C_1\Big(\|\mathfrak{R}_{\mathrm{el}}\|_{\Omega} + \|\mathfrak{R}_{\mathrm{fa}}\|_{\Gamma_{\mathrm{int}}} + \|\mathcal{R}_{\mathrm{pe}}\|_{\Gamma_{\mathrm{int}}} + \|\mathcal{R}_{\mathrm{pe}}\|_{\partial\Omega}\Big) \\
&+ C_2\Big(\|\bar{\chi}[\![u_h]\!]\|_{\Gamma_{\mathrm{int}}} + \|\bar{\chi}(u_h - g)\|_{\partial\Omega} + \|\mathfrak{R}_a\|_{\Omega}\Big) + C_3|||e_d|||,
\end{aligned} \tag{23}$$

whereby $C_2 := \frac{3+\theta}{2}\|a^{-\frac{1}{2}}\|_{L_\infty(\Omega)}$ and $C_1 := C^{\mathrm{app}}_{L_2} C_2$, and $C_3 := \frac{5+\theta}{2}$, for $\theta \in [-1, 1]$.

4 Completing the *a Posteriori* Error Bounds

To arrive at a completely computable upper bound, we need to further estimate the non-conforming part of the error $|||e_d|||$.

We start by defining the function $u_c \in H^1_g(\Omega)$ introduced in (12). We solve local minimization problems in $H^1(\omega_{\boldsymbol{a}})$ with suitable boundary conditions for every mesh vertex $\boldsymbol{a} \in \mathcal{V}_h$ (recall that $\omega_{\boldsymbol{a}}$ denotes the star associated with the vertex $\boldsymbol{a}$). Let $\psi_{\boldsymbol{a}}$ be the hat basis function equal to 1 at $\boldsymbol{a}$ and having a support in the vertex star $\omega_{\boldsymbol{a}}$. Recall that the hat basis functions satisfy the following partition-of-unity property:

$$\sum_{\boldsymbol{a}\in\mathcal{V}_h} \psi_{\boldsymbol{a}} = 1. \tag{24}$$

Before introducing the potential reconstruction, we introduce an auxiliary SIPDG bilinear form $A : H^1(\Omega, \mathcal{T}) \times H^1(\Omega, \mathcal{T}) \to \mathbb{R}$, given by

$$\begin{aligned} A(w, v) := & \int_\Omega a\nabla_h w \cdot \nabla_h v \,\mathrm{d}x + \int_\Gamma \sigma [[w]]\,[[v]] \,\mathrm{d}s \\ & - \int_\Gamma \left(\{\!\{a\Pi_{\mathbf{p-1}}(\nabla w)\cdot \mathbf{n}\}\!\}_{\mathbf{w}}[[v]] + \{\!\{a\Pi_{\mathbf{p-1}}(\nabla v)\cdot \mathbf{n}\}\!\}_{\mathbf{w}}[[w]]\right) \mathrm{d}s, \end{aligned} \tag{25}$$

for $w, v \in H^1(\Omega, \mathcal{T})$. As $A(\cdot,\cdot)$ is a particular case of the bilinear form $B(\cdot,\cdot)$, considered in Section 2, following the analysis in the latter, we have

$$A(w, w) \geq \frac{1}{2}|||w|||_d^2 \quad \text{and} \quad A(v, w) \leq 2|||v|||_d|||w|||_d, \tag{26}$$

with $|||v|||_d := \left(\|\sqrt{a}\nabla_h w\|_\Omega^2 + \|\sqrt{\sigma}[[w]]\|_\Gamma^2\right)^{\frac{1}{2}}$.

Definition 1 (Patchwise and global potential reconstruction) For all $\boldsymbol{a} \in \mathcal{V}_h$, let $u_c^{\boldsymbol{a}} \in H_g^1(\omega_{\boldsymbol{a}})$ be the solution of the following well-posed problem:

$$A(u_c^{\boldsymbol{a}}, v_{\boldsymbol{a}}) = A(\psi_{\boldsymbol{a}} u_h, v_{\boldsymbol{a}}), \qquad \forall v_{\boldsymbol{a}} \in H_0^1(\omega_{\boldsymbol{a}}), \tag{27}$$

with

$$H_g^1(\omega_{\boldsymbol{a}}) := \{v \in H^1(\omega_{\boldsymbol{a}}) \mid v|_{\partial\omega_{\boldsymbol{a}}\cap\partial\Omega} = \psi_{\boldsymbol{a}} g \text{ and } v|_{\partial\omega_{\boldsymbol{a}}\cap\Omega} = 0\}. \tag{28}$$

An equivalent definition is

$$u_c^{\boldsymbol{a}} := \arg \min_{\rho_{\boldsymbol{a}} \in H_g^1(\omega_{\boldsymbol{a}})} A(\rho_{\boldsymbol{a}} - (\psi_{\boldsymbol{a}} u_h), \rho_{\boldsymbol{a}} - (\psi_{\boldsymbol{a}} u_h)). \tag{29}$$

Then, extending $u_c^{\boldsymbol{a}}$ by zero to Ω, we set

$$u_c := \sum_{\boldsymbol{a}\in\mathcal{V}_h} u_c^{\boldsymbol{a}}. \tag{30}$$

Notice that we indeed have $u_c \in H_g^1(\Omega)$ as required in (12); this follows from the partition-of-unity property (24) and the definition (28) of $H_g^1(\omega_{\boldsymbol{a}})$.

Lemma 1 (Local nonconforming error) *Let $u_c^{\boldsymbol{a}}$ solve (27) and set $e_d^{\boldsymbol{a}} := u_c^{\boldsymbol{a}} - \psi_{\boldsymbol{a}} u_h$ for all $\boldsymbol{a} \in \mathcal{V}_h$. The following holds:*

$$\|\sqrt{a}\nabla_h e_d^{\boldsymbol{a}}\|_{\omega_{\boldsymbol{a}}} \leq C_{\mathrm{d}}^{\boldsymbol{a}}\left(\|\tilde{\chi}[[u_h]]\|_{\mathcal{F}_{\boldsymbol{a}}\cap\Gamma_{\mathrm{int}}}^2 + \|\tilde{\chi}(u_h - g)\|_{\mathcal{F}_{\boldsymbol{a}}\cap\partial\Omega}^2 + \|\mathfrak{R}_{\mathrm{ta}}^{\mathrm{QI}}\|_{\mathcal{F}_{\boldsymbol{a}}}^2\right)^{\frac{1}{2}}, \tag{31}$$

with

$$\mathcal{R}_{\mathrm{ta}}^{\mathrm{QI}}|_F := \begin{cases} \sqrt{(a_F^\flat)^{-1}\max_{*\in\{+,-\}} h_{K_*} p_{K_*}^{-1}}\,[[\nabla u_h \times \mathbf{n}]], & \textit{if } F \subset \mathcal{F}_{\boldsymbol{a}} \cap \Gamma_{\mathrm{int}}; \\ \sqrt{(a_F^\flat)^{-1} h_{K_+} p_{K_+}^{-1}}\,(\nabla(u_h - g_{\mathrm{D}}) \times \mathbf{n}_\Omega), & \textit{if } F \subset \mathcal{F}_{\boldsymbol{a}} \cap \partial\Omega, \end{cases} \tag{32}$$

where the constant $C_{\mathrm{d}}^{\boldsymbol{a}}$ depends on the shape-regularity of the mesh, on γ and on $\mu_{\boldsymbol{a}}(A)^{\frac{1}{2}}$, but is independent of h and p.

Proof. The proof primarily follows the approach detailed in [12, Section 3.1], applied to each local vertex patch. The key ingredients include the Helmholtz decomposition for simply connected domains [26, Theorem 3] and the hp-optimal quasi-interpolation operator introduced in [32]. The proof can be derived following techniques in the proof of [17, Lemma 5.4] for the H^1-reconstruction. □

Next, we will derive the L^2-norm error bound for $e_d^{\boldsymbol{a}}$.

Lemma 2 (Local nonconforming L_2–norm error) *Let $u_c^{\boldsymbol{a}}$ solve (27) and set $e_d^{\boldsymbol{a}} := u_c^{\boldsymbol{a}} - \psi_{\boldsymbol{a}} u_h$ for all $\boldsymbol{a} \in \mathcal{V}_h$. The following holds:*

$$\|\sqrt{c} e_d^{\boldsymbol{a}}\|_{\omega_{\boldsymbol{a}}} \leq \tilde{C}_{\mathrm{d}}^{\boldsymbol{a}} h_{\omega_{\boldsymbol{a}}} \left(\|\tilde{\chi}_* [\![u_h]\!]\|^2_{\mathcal{F}_{\boldsymbol{a}} \cap \Gamma_{\mathrm{int}}} + \|\tilde{\chi}_* (u_h - g)\|^2_{\mathcal{F}_{\boldsymbol{a}} \cap \partial\Omega} + \|\mathfrak{R}^{\mathrm{QI}}_{\mathrm{ta},*}\|^2_{\mathcal{F}_{\boldsymbol{a}}} \right)^{\frac{1}{2}}, \quad (33)$$

with

$$\tilde{\chi}_*|_F := \begin{cases} \sqrt{2 c_{\boldsymbol{a}}^{\sharp}} (d+1)^{-\frac{1}{2}} C^{\mathrm{st}}_{\mathrm{QI}} \max_{* \in \{+,-\}} \frac{w_*}{\xi_*}, & \text{if } F \subset \mathcal{F}_{\boldsymbol{a}} \cap \Gamma_{\mathrm{int}}; \\ \sqrt{2 c_{\boldsymbol{a}}^{\sharp}} (d+1)^{-\frac{1}{2}} C^{\mathrm{st}}_{\mathrm{QI}} \frac{w_+}{\xi_+}, & \text{if } F \subset \mathcal{F}_{\boldsymbol{a}} \cap \partial\Omega, \end{cases} \quad (34)$$

and

$$\mathcal{R}^{\mathrm{QI}}_{\mathrm{ta},*}|_F := \begin{cases} \sqrt{c_{\boldsymbol{a}}^{\sharp}} \max_{* \in \{+,-\}} h_{K_*} p_{K_*}^{-1} [\![\nabla u_h \times \mathbf{n}]\!], & \text{if } F \subset \mathcal{F}_{\boldsymbol{a}} \cap \Gamma_{\mathrm{int}}; \\ \sqrt{c_{\boldsymbol{a}}^{\sharp}} h_{K_+} p_{K_+}^{-1} (\nabla(u_h - g_{\mathrm{D}}) \times \mathbf{n}_{\Omega}), & \text{if } F \subset \mathcal{F}_{\boldsymbol{a}} \cap \partial\Omega, \end{cases} \quad (35)$$

where the constant $\tilde{C}_{\mathrm{d}}^{\boldsymbol{a}}$ depends on the shape-regularity of the mesh and on γ, but is independent of h and p.

Proof. First, recalling the definition of $u_c^{\boldsymbol{a}}$, we have $[\![e_d^{\boldsymbol{a}}]\!]|_{\mathcal{F}_{\boldsymbol{a}} \cap \Gamma_{\mathrm{int}}} = -[\![\psi_{\boldsymbol{a}} u_h]\!]|_{\mathcal{F}_{\boldsymbol{a}} \cap \Gamma_{\mathrm{int}}}$ and $e_d^{\boldsymbol{a}}|_{\partial\omega_{\boldsymbol{a}} \setminus \mathcal{F}_{\boldsymbol{a}}} = 0$. Next, applying the broken Poincaré inequality from [7], we have

$$\|e_d^{\boldsymbol{a}}\|^2_{\omega_{\boldsymbol{a}}} \leq C^2_{\mathrm{bPF}} h^2_{\omega_{\boldsymbol{a}}} \Bigg(\|\nabla_h e_d^{\boldsymbol{a}}\|^2_{\omega_{\boldsymbol{a}}} + \|h^{-\frac{1}{2}} \Pi_0^F [\![\psi_{\boldsymbol{a}} u_h]\!]\|^2_{\mathcal{F}_{\boldsymbol{a}} \cap \Gamma_{\mathrm{int}}} + \|h^{-\frac{1}{2}} \Pi_0^F (\psi_{\boldsymbol{a}} (u_h - g))\|^2_{\mathcal{F}_{\boldsymbol{a}} \cap \partial\Omega} \Bigg),$$

where $C_{\mathrm{bPF}} > 0$ only depends on the shape-regularity of the mesh. Invoking the L^2-stability of the L^2-orthogonal projection, $\|\psi_{\boldsymbol{a}}\|_{L_\infty(\omega_{\boldsymbol{a}})} = 1$, mesh shape regularity and the result (31) by choosing the diffusion tensor a to be the identity matrix, we infer

$$\begin{aligned} \|\sqrt{c} e_d^{\boldsymbol{a}}\|^2_{\omega_{\boldsymbol{a}}} &\leq c_{\boldsymbol{a}}^{\sharp} (C_{\mathrm{bPF}} h_{\omega_{\boldsymbol{a}}})^2 \Bigg(\|\nabla_h e_d^{\boldsymbol{a}}\|^2_{\omega_{\boldsymbol{a}}} + \|h^{-\frac{1}{2}} [\![u_h]\!]\|^2_{\mathcal{F}_{\boldsymbol{a}} \cap \Gamma_{\mathrm{int}}} + \|h^{-\frac{1}{2}} (u_h - g)\|^2_{\mathcal{F}_{\boldsymbol{a}} \cap \partial\Omega} \Bigg) \\ &\leq (\tilde{C}_{\mathrm{d}}^{\boldsymbol{a}} h_{\omega_{\boldsymbol{a}}})^2 \Bigg(\|\tilde{\chi}_* [\![u_h]\!]\|^2_{\mathcal{F}_{\boldsymbol{a}} \cap \Gamma_{\mathrm{int}}} + \|\tilde{\chi}_* (u_h - g)\|^2_{\mathcal{F}_{\boldsymbol{a}} \cap \partial\Omega} + \|\mathfrak{R}^{\mathrm{QI}}_{\mathrm{ta},*}\|^2_{\mathcal{F}_{\boldsymbol{a}}} \Bigg). \end{aligned}$$

□

Remark 3 (L_2–norm bound) The above L_2–norm error bound of $e_d^{\boldsymbol{a}}$ is h-optimal but p-suboptimal by one order in the DG norm. Moreover, as we do not assume the reaction coefficient c to be strictly positive, the above L_2–norm error bound is independent of the contrast of c.

Theorem 3 (Global bound on nonconforming error) *The following holds:*

$$\begin{aligned}
&\|\sqrt{a}\nabla_h e_d\|_\Omega^2 + \|\sqrt{c}e_d\|_\Omega^2 \\
&\leq C_{\mathrm{d}}^2 \sum_{K\in\mathcal{T}} \Big(\|\tilde{\chi}[\![u_h]\!]\|_{\partial K\cap\Gamma_{\mathrm{int}}}^2 + \|\tilde{\chi}(u_h-g)\|_{\partial K\cap\partial\Omega}^2 \\
&\qquad + \|\mathfrak{R}_{\mathrm{ta}}^{\mathrm{QI}}\|_{\partial K}^2 + h_K^2\|\tilde{\chi}_*[\![u_h]\!]\|_{\partial K\cap\Gamma_{\mathrm{int}}}^2 \\
&\qquad + h_K^2\|\tilde{\chi}_*(u_h-g)\|_{\partial K\cap\partial\Omega}^2 + h_K^2\|\mathfrak{R}_{\mathrm{ta},*}^{\mathrm{QI}}\|_{\partial K}^2 \Big),
\end{aligned} \tag{36}$$

where the constant C_{d} depends on the shape-regularity of the mesh, γ, $\max_{\boldsymbol{a}\in\mathcal{V}_h}\chi_{\boldsymbol{a}}(A)^{\frac{1}{2}}$ and $\max_{\boldsymbol{a}\in\mathcal{V}_h}(c_{\boldsymbol{a}}^\sharp)^{\frac{1}{2}}$, but is independent of h and k.

Proof. Using that $u_c|_K = \sum_{\boldsymbol{a}\in\mathcal{V}_K} u_c^{\boldsymbol{a}}|_K$ for all $K\in\mathcal{T}$, where $\mathcal{V}_K$ is the set of vertices of K, the partition-of-unity property (24), and the triangle inequality, we infer that

$$\begin{aligned}
\|\sqrt{a}\nabla_h e_d\|_\Omega^2 &= \sum_{K\in\mathcal{T}} \|\sqrt{a}\nabla(u_c - u_h|_K)\|_K^2 \\
&= \sum_{K\in\mathcal{T}} \Big\| \sum_{\boldsymbol{a}\in\mathcal{V}_K} \sqrt{a}\nabla(u_c^{\boldsymbol{a}} - \psi_{\boldsymbol{a}} u_h|_K)\Big\|_K^2 \\
&\leq \sum_{K\in\mathcal{T}}\sum_{\boldsymbol{a}\in\mathcal{V}_K} (d+1)\|\sqrt{a}\nabla(u_c^{\boldsymbol{a}} - \psi_{\boldsymbol{a}} u_h|_K)\|_K^2 \\
&= \sum_{\boldsymbol{a}\in\mathcal{V}_h} (d+1)\|\sqrt{a}\nabla_h e_d^{\boldsymbol{a}}\|_{\omega_{\boldsymbol{a}}}^2.
\end{aligned}$$

Similarly, we infer

$$\begin{aligned}
\|\sqrt{c}e_d\|_\Omega^2 &= \sum_{K\in\mathcal{T}} \|\sqrt{c}(u_c - u_h|_K)\|_K^2 \\
&= \sum_{K\in\mathcal{T}} \Big\| \sum_{\boldsymbol{a}\in\mathcal{V}_K} \sqrt{c}(u_c^{\boldsymbol{a}} - \psi_{\boldsymbol{a}} u_h|_K)\Big\|_K^2 \\
&\leq \sum_{K\in\mathcal{T}}\sum_{\boldsymbol{a}\in\mathcal{V}_K} (d+1)\|\sqrt{c}(u_c^{\boldsymbol{a}} - \psi_{\boldsymbol{a}} u_h|_K)\|_K^2 \\
&= \sum_{\boldsymbol{a}\in\mathcal{V}_h} (d+1)\|\sqrt{c}e_d^{\boldsymbol{a}}\|_{\omega_{\boldsymbol{a}}}^2.
\end{aligned}$$

Invoking Lemma 1, Lemma 2 and using that $h_{\omega_{\boldsymbol{a}}} \approx h_K$ for all $K\in\omega_{\boldsymbol{a}}$ and all $\boldsymbol{a}\in\mathcal{V}_h$ completes the proof. □

Upon completing the upper bound, some comments on lower bounds are also in order. The individual *a posteriori* error estimators of the previous theorem are similar to known bounds in the literature [29, 12]. There are, however, key differences. For instance, the use of robust weighted averages in conjunction with the localized treatment of the diffusion tensor, that can lead to smaller constants in certain scenarios of localized high contrast in the coefficients. Also, tracking the local polynomial degree in the constants is also of interest for high-order finite element spaces. Nevertheless, the proof of respective lower bounds follows completely analogous steps to the cases of the pure Poisson problem. Thus, we prefer not to include these in the discussion here in the interest of brevity, as there are no essential improvements in this context, compared to the results in the literature.

5 Numerical Examples

In this section, we present numerical examples to illustrate our theoretical results.

5.1 Example 1: Adaptive Algorithm for Singular Solution

The first example we consider is a diffusion problem on the L-shaped domain $\Omega := (-1,1)^2 \setminus \{(0,1) \times (-1,0)\}$, with the exact solution in polar coordinates:

$$u = r^{\frac{2}{3}} \sin(2\theta/3).$$

We test an adaptive algorithm driven by the a posteriori error estimator from Section 4. The adaptive algorithm starts from a coarse mesh and uses the estimator to mark mesh cells for refinement through a bulk-chasing criterion (also known as Dörfler's marking). The adaptive algorithm can be classically described as

$$\text{SOLVE} \longrightarrow \text{ESTIMATE} \longrightarrow \text{MARK} \longrightarrow \text{REFINE}.$$

We first test the convergence rate of the above adaptive algorithm with $p \in \{1, 2, 3, 4\}$ and a bulk-chasing criterion set to 30%. The energy error and the a posteriori error estimator are reported in Figures 1 and 2 with convergence rates computed as a function of numerical degrees of freedom (DoFs). We observe that the energy error and the a posteriori error estimator converge at the optimal rate $O(\text{DoFs}^{-\frac{p}{2}})$. Moreover, we observe in Figure 3 that the effectivity index remains well behaved as a function of DoFs and slightly increases with the polynomial degree $p \geq 1$, with values between 1.5 and 2.8 asymptotically. In addition, we notice that there are pre-asymptotic regimes in which the effectivity index grows to 3.5. We point out that although the solution exhibits a corner singularity, it appears that the two smooth functions require more degrees of freedom to resolve than the singularity. This behavior can be observed in the sequence of meshes produced by the adaptive algorithm with $p = 4$, shown in Figure 4. One possible explanation is that the exponential functions induce a larger error in the DG norm than the corner singularity in the pre-asymptotic regime.

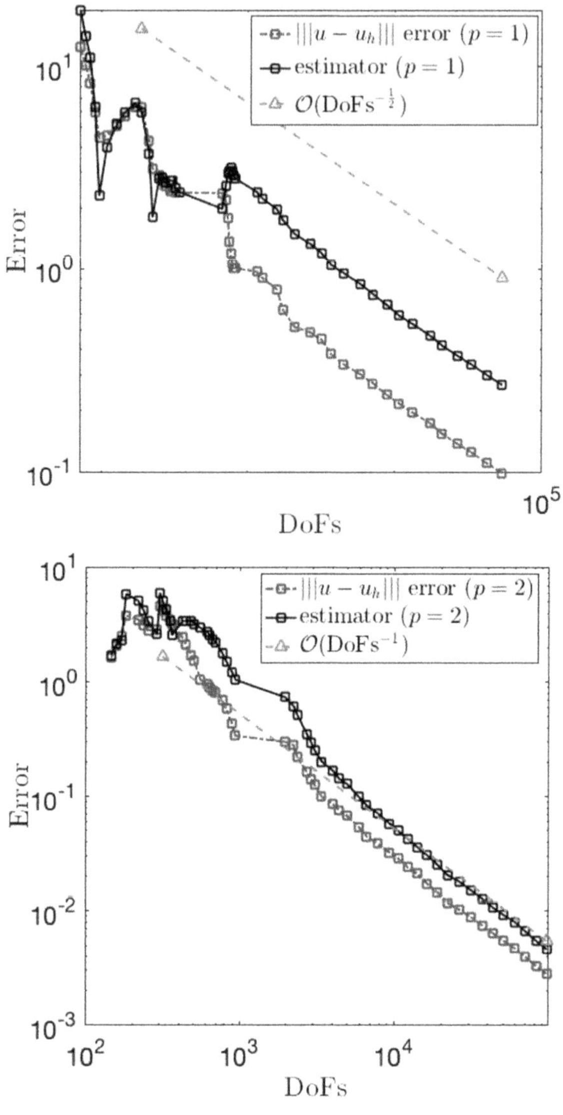

Figure 1 Example 1. DG-norm error and a posteriori error estimator as a function of DoFs for $p \in \{1, 2\}$.

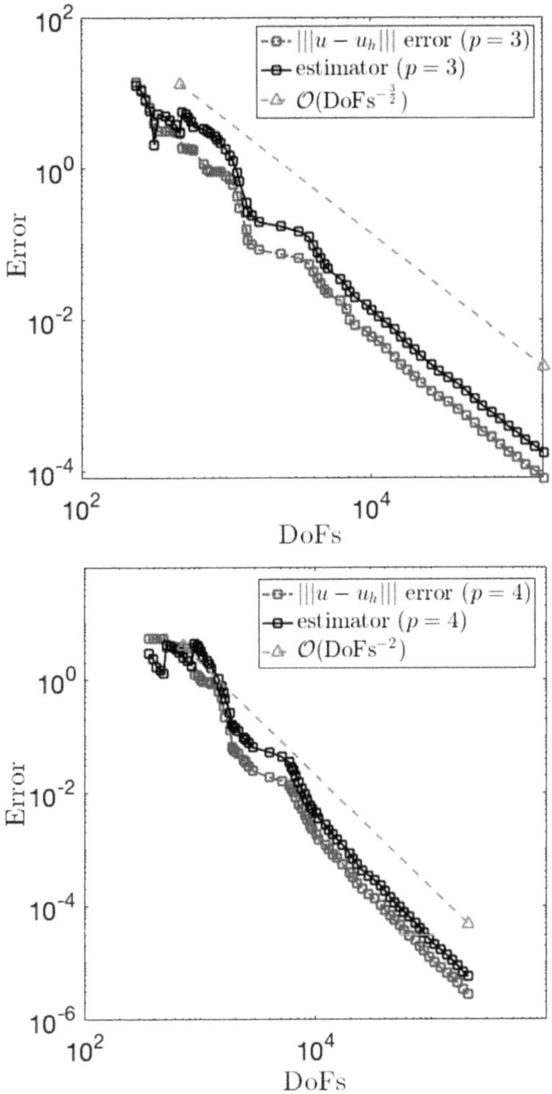

Figure 2 Example 1. DG-norm error and a posteriori error estimator as a function of DoFs for $p \in \{3, 4\}$.

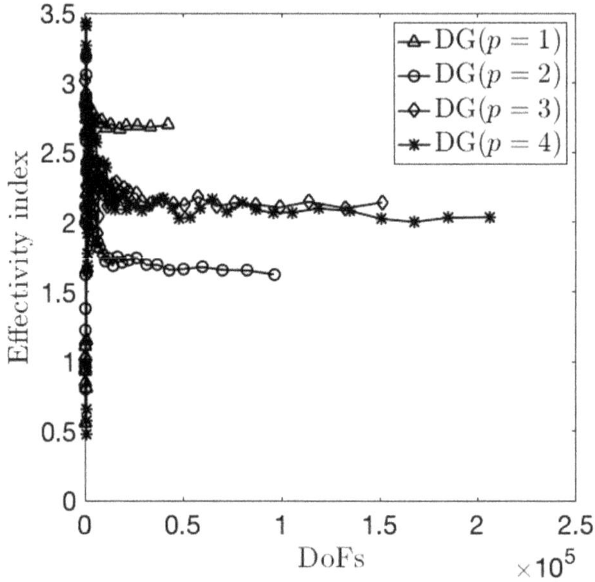

Figure 3 Example 1. Effectivity index as a function of DoFs for $p \in \{1, 2, 3, 4\}$.

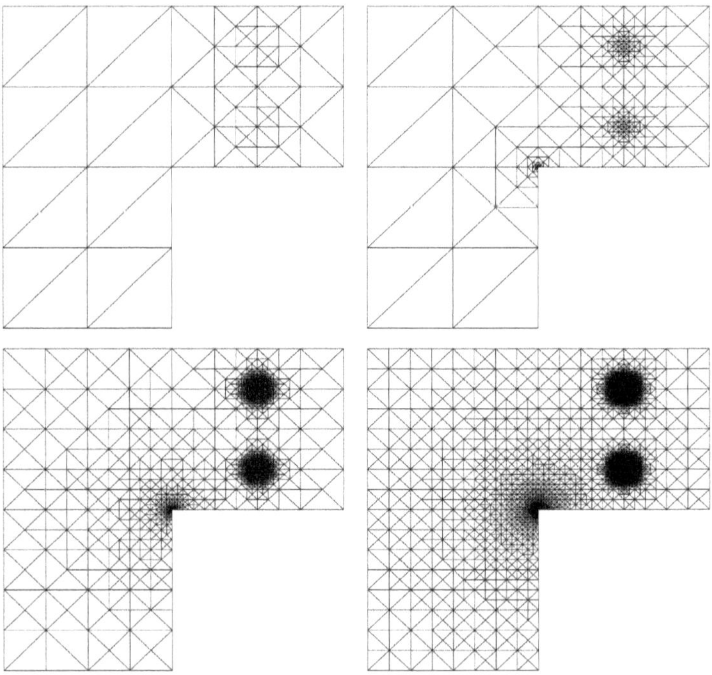

Figure 4 Example 1. The sequence of meshes generated by the adaptive DG algorithm with $p = 4$.

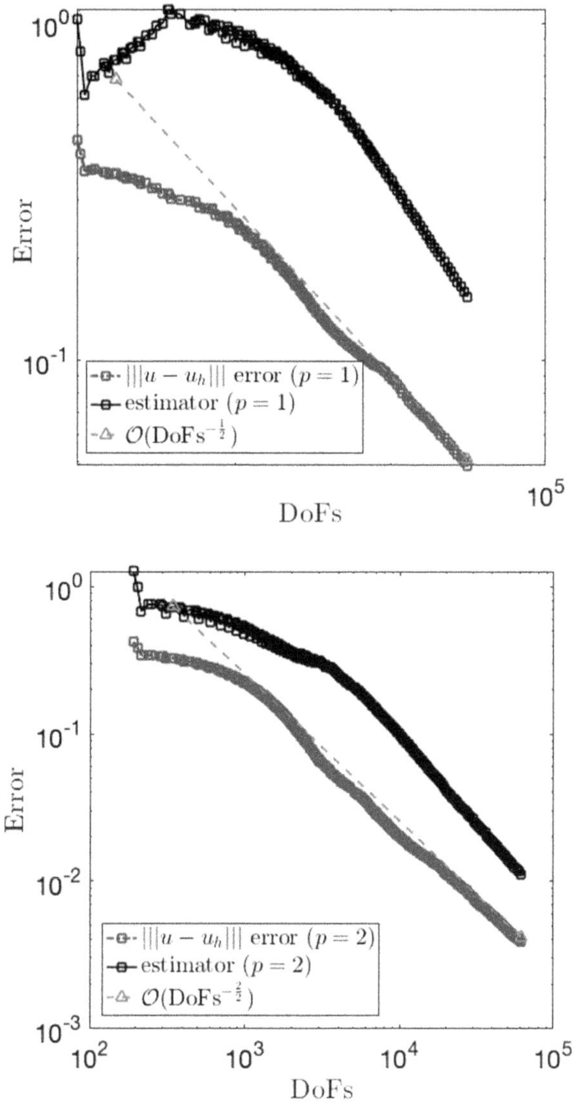

Figure 5 Example 2. DG-norm error and a posteriori error estimator as a function of DoFs for $p \in \{1, 2\}$.

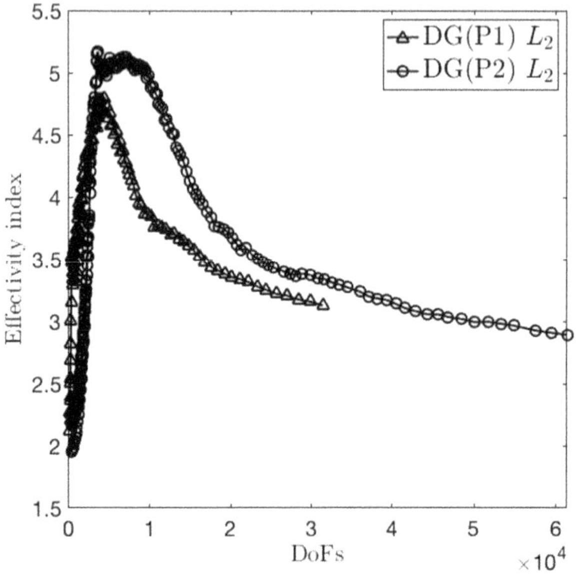

Figure 6 Example 2. Effectivity index as a function of DoFs for $p \in \{1, 2\}$.

5.2 Example 2: Adaptive Algorithm for Kellogg's Test Case

Our last example is Kellogg's test case [33], i.e., a diffusion problem on the square domain $\Omega := (-1, 1)^2$ with a checkerboard pattern for the diffusion tensor $a := b\mathbb{I}_{2\times 2}$ with $\mathbb{I}_{2\times 2}$, the 2×2 identity matrix. We select $b := 1$ for $xy < 0$, and $b = 161.4476387975881$ otherwise. The exact solution (for zero right-hand side and suitable Dirichlet boundary conditions) takes the form $u := r^{\alpha}\phi(\theta)$ in polar coordinates, where the explicit expression for the function ϕ can be found in [36, Section 5.3]. In particular, we have $\alpha = 0.1$, so that $u \in H^{1.1-\varepsilon}(\Omega)$ with $\varepsilon > 0$ arbitrarily small.

We test the convergence rate of the above adaptive algorithm with $p \in \{1, 2\}$ and a bulk-chasing criterion set to 10%. The energy error and the a posteriori error estimator are reported in Figure 5 with convergence rates computed as a function of the numerical DoFs. We observe that the energy error and the a posteriori error estimator converge at the optimal rate $O(\text{DoFs}^{-\frac{p}{2}})$. Moreover, we observe in Figure 6 that the effectivity index remains well-behaved as a function of DoFs and slightly increases with the polynomial degree $p \geq 1$, with values between 3 and 3.2 asymptotically. In addition, we notice that there are pre-asymptotic regimes in which the effectivity index grows to 5.5. We observe that there is a strong corner singularity at the origin. This behavior can be observed by the adaptive meshes generated by the adaptive algorithm with $p = 2$, shown in Figure 7.

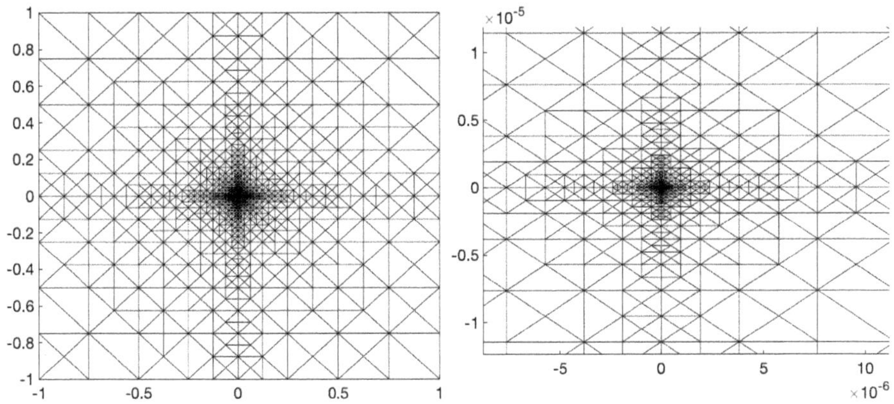

Figure 7 Example 2. 10228 meshes generated by the adaptive DG algorithm with $p = 2$ in the domain (left) and in the subdomain $(-1e - 5, 1e - 5)^2$ (right).

References

1. M. Ainsworth. A synthesis of a posteriori error estimation techniques for conforming, non-conforming and discontinuous Galerkin finite element methods. In *Recent advances in adaptive computation*, volume 383 of *Contemp. Math.*, pages 1–14. Amer. Math. Soc., Providence, RI, (2005).
2. M. Ainsworth and R. Rankin. Fully computable bounds for the error in nonconforming finite element approximations of arbitrary order on triangular elements. *SIAM J. Numer. Anal.*, 46(6):3207–3232, 2008.
3. C. Annavarapu, M. Hautefeuille, and J. E. Dolbow. A robust Nitsche's formulation for interface problems. *Comput. Methods Appl. Mech. Engrg.*, 225/228:44–54, 2012.
4. D. Arnold. An interior penalty finite element method with discontinuous elements. *SIAM J. Numer. Anal.*, 19(4):742–760, 1982.
5. G. Baker. Finite element methods for elliptic equations using nonconforming elements. *Math. Comp.*, 31(137):45–59, 1977.
6. A. Bonito and R. H. Nochetto. Quasi-optimal convergence rate of an adaptive discontinuous Galerkin method. *SIAM J. Numer. Anal.*, 48(2):734–771, 2010.
7. S. C. Brenner. Poincaré-Friedrichs inequalities for piecewise H^1 functions. *SIAM J. Numer. Anal.*, 41(1):306–324, 2003.
8. E. Burman and P. Zunino. A domain decomposition method based on weighted interior penalties for advection-diffusion-reaction problems. *SIAM J. Numer. Anal.*, 44(4):1612–1638, 2006.
9. A. Cangiani, Z. Dong, E. Georgoulis, and P. Houston. hp–version discontinuous Galerkin methods for advection–diffusion–reaction problems on polytopic meshes. *ESAIM: M2AN*, 50(3):699–725, 2016.
10. A. Cangiani, Z. Dong, and E. H. Georgoulis. hp-version space-time discontinuous Galerkin methods for parabolic problems on prismatic meshes. *SIAM J. Sci. Comput.*, 39(4):A1251–A1279, 2017.
11. A. Cangiani, Z. Dong, and E. H. Georgoulis. hp-version discontinuous Galerkin methods on essentially arbitrarily-shaped elements. *Math. Comp.*, 91(333):1–35, 2021.
12. A. Cangiani, Z. Dong, and E. H. Georgoulis. A posteriori error estimates for discontinuous Galerkin methods on polygonal and polyhedral meshes. *SIAM J. Numer. Anal.*, 61(5):2352–2380, 2023.

13. A. Cangiani, Z. Dong, E. H. Georgoulis, and P. Houston. *hp-version discontinuous Galerkin methods on polygonal and polyhedral meshes.* SpringerBriefs in Mathematics. Springer, Cham, (2017).
14. A. Cangiani, E. Georgoulis, and P. Houston. hp–Version discontinuous Galerkin methods on polygonal and polyhedral meshes. *Math. Models Methods Appl. Sci.*, 24(10):2009–2041, 2014.
15. C. Carstensen, B. Gräßle, and N. T. Tran. Adaptive hybrid high-order method for guaranteed lower eigenvalue bounds. *Numer. Math.*, 156(3):813–851, 2024.
16. A. Chernov. Optimal convergence estimates for the trace of the polynomial L^2-projection operator on a simplex. *Math. Comp.*, 81(278):765–787, 2012.
17. Z. Dong and A. Ern. hp-error analysis of mixed-order hybrid high-order methods for elliptic problems on simplicial meshes. working paper or preprint, Oct. (2024).
18. Z. Dong and A. Ern. $\boldsymbol{H}(\mathbf{curl})$-reconstruction of piecewise polynomial fields with application to hp-a posteriori nonconforming error analysis for Maxwell's equations. *accepted in SIAM Journal on Numerical Analysis*, 2026.
19. Z. Dong and E. H. Georgoulis. Robust interior penalty discontinuous Galerkin methods. *J. Sci. Comput.*, 92(2):Paper No. 57, 23, 2022.
20. Z. Dong, L. Mascotto, and O. J. Sutton. Residual-based a posteriori error estimates for hp-discontinuous Galerkin discretizations of the biharmonic problem. *SIAM J. Numer. Anal.*, 59(3):1273–1298, 2021.
21. M. Dryja. On discontinuous Galerkin methods for elliptic problems with discontinuous coefficients. *Comput. Methods Appl. Math.*, 3(1):76–85, 2003.
22. A. Ern and A. Stephansen. A posteriori energy-norm error estimates for advection-diffusion equations approximated by weighted interior penalty methods. *J. Comp. Math.*, 26(4):488–510, 2008.
23. A. Ern, A. F. Stephansen, and P. Zunino. A discontinuous Galerkin method with weighted averages for advection-diffusion equations with locally small and anisotropic diffusivity. *IMA J. Numer. Anal.*, 29(2):235–256, 2009.
24. E. Georgoulis and A. Lasis. A note on the design of hp-version interior penalty discontinuous Galerkin finite element methods for degenerate problems. *IMA J. Numer. Anal.*, 26(2):381–390, 2006.
25. T. Gustafsson, R. Stenberg, and J. Videman. On Nitsche's method for elastic contact problems. *SIAM J. Sci. Comput.*, 42(2):B425–B446, 2020.
26. R. Hiptmair and C. Pechstein. Regular decompositions of vector fields - continuous, discrete, and structure-preserving. Technical Report 2019-18, Seminar for Applied Mathematics, ETH Zürich, Switzerland, (2019).
27. P. Houston, D. Schötzau, and T. P. Wihler. Energy norm a posteriori error estimation of hp-adaptive discontinuous Galerkin methods for elliptic problems. *Math. Models Methods Appl. Sci.*, 17(1):33–62, 2007.
28. P. Houston, C. Schwab, and E. Süli. Discontinuous hp-finite element methods for advection-diffusion-reaction problems. *SIAM J. Numer. Anal.*, 39(6):2133–2163, 2002.
29. O. Karakashian and F. Pascal. A posteriori error estimates for a discontinuous Galerkin approximation of second-order elliptic problems. *SIAM J. Numer. Anal.*, 41(6):2374–2399, 2003.
30. O. Karakashian and F. Pascal. Convergence of adaptive discontinuous Galerkin approximations of second-order elliptic problems. *SIAM J. Numer. Anal.*, 45(2):641–665, 2007.
31. O. A. Karakashian and F. Pascal. A posteriori error estimates for a discontinuous Galerkin approximation of second-order elliptic problems. *SIAM J. Numer. Anal.*, 41(6):2374–2399, 2003.
32. M. Karkulik and J. M. Melenk. Local high-order regularization and applications to hp-methods. *Comput. Math. Appl.*, 70(7):1606–1639, 2015.
33. B. R. Kellogg. On the Poisson equation with intersecting interfaces. *Appl. Anal.*, 4(2):101–129, 1974.
34. N. Kopteva and R. Rankin. Pointwise a posteriori error estimates for discontinuous Galerkin methods for singularly perturbed reaction-diffusion equations. *SIAM J. Numer. Anal.*, 61(4):1938–1961, 2023.
35. J. Melenk and T. Wurzer. On the stability of the boundary trace of the polynomial L_2-projection on triangles and tetrahedra. *Comput. Math. Appl.*, 67(4):944–965, 2014.

36. P. Morin, R. H. Nochetto, and K. G. Siebert. Convergence of adaptive finite element methods. *SIAM Rev.*, 44(4):631–658, 2002.
37. T. Warburton and J. S. Hesthaven. On the constants in hp-finite element trace inverse inequalities. *Comput. Methods Appl. Mech. Engrg.*, 192(25):2765–2773, 2003.
38. M. Wheeler. An elliptic collocation-finite element method with interior penalties. *SIAM J. Numer. Anal.*, 15(1):152–161, 1978.

Complexity Theory Meets Ordinary Differential Equations

Adalbert Fono, Noah Wedlich, Holger Boche and Gitta Kutyniok

Abstract This contribution investigates the computational complexity of simulating linear ordinary differential equations (ODEs) on digital computers. We provide an exact characterization of the complexity blowup for a class of ODEs of arbitrary order based on their algebraic properties, extending the characterization of first order ODEs shown in [6]. Complexity blowup indeed arises in most ODEs (except for certain degenerate cases) and means that there exists a low complexity input signal, which can be generated on a Turing machine in polynomial time, leading to a corresponding high complexity output signal of the system in the sense that the computation time for determining an approximation up to n significant digits grows faster than any polynomial in n. Similarly, we derive an analogous blowup criterion for a subclass of first-order systems of linear ODEs. Finally, we discuss the implications for the simulation of analog systems governed by ODEs and exemplarily apply our framework to a simple model of neuronal dynamics—the leaky integrate-and-fire neuron—heavily employed in neuroscience.

Adalbert Fono
Ludwig-Maximilians-Universität München, fono@math.lmu.de

Noah Wedlich
Ludwig-Maximilians-Universität München, N.Wedlich@campus.lmu.de

Holger Boche
Technical University of Munich, boche@tum.de
Munich Center for Quantum Science and Technology (MCQST)
Munich Quantum Valley (MQV), BMBF Research Hub 6G-life

Gitta Kutyniok
Ludwig-Maximilians-Universität München, kutyniok@math.lmu.de
University of Tromsø, Munich Center for Machine Learning (MCML), DLR-German Aerospace Center

B. Koren and D. Roose (eds.), *Numerical Analysis and Scientific Computing: an Anthology*, Lecture Notes in Computational Science and Engineering 156,
https://doi.org/10.1007/978-3-032-33428-2_4

1 Introduction

The study of differential equations is one of the most important areas of mathematics, with applications ranging from physics and engineering to biology and Artificial Intelligence. It allows us to model and analyze dynamical systems, describing how quantities change over time or space. The mathematical theory of differential equations, concerned with finding (explicit) solutions, proving their existence and uniqueness under certain conditions, or deducing their qualitative properties, is a well-studied (though not completed) field.

However, with the advent of modern computational methods, the numerical solution, emulation, and simulation of differential equations have become increasingly important. For instance, the design and development of analog systems, as well as optimization testing, strongly depend on numerical simulations [3, 21, 44, 14, 40, 7, 1, 9]. Similarly, emulation of such systems as in neuroscience, e.g., the behaviour of neural networks as well as single neurons and synapses, is a key step in advancing knowledge and exploiting the properties of (biological) brains [19, 10, 30]. Understanding the feasibility and computational complexity of these tasks is therefore practically relevant.

One obstacle in this analysis is the fact that analog systems are inherently dependent on continuous (physical) quantities but are almost always simulated on digital computers. Digital processing relies on manipulating a discrete set of values in a step-by-step manner, i.e., digital computers employ rational numbers to approximate the exact continuous quantities. Although discretizing the inputs and parameters of an analog system to obtain a rational description is typically not a severe restriction, the true output of the rational analog system is nevertheless often irrational. For instance, even for (linear) ordinary differential equations with rational coefficients, the corresponding solution may take irrational values when evaluated at rational time points. Hence, the output of the (digital) simulation and the analog system operate in different domains.

This dichotomy is an often-neglected aspect in the simulation of analog systems on digital hardware and the subsequent mathematical analysis assessing the correctness and accuracy of the simulation. However, there are certain restrictions under which analog systems can actually be simulated on digital machines with error control. Here, not only the aspired simulation is carried out, but additionally, the correctness of the simulation up to an arbitrary but fixed error is guaranteed. The described property illustrates the problem of the *computability* of a task. In fact, this condition cannot be satisfied, among others, for the solution of the wave equation [35], the Fourier transform [33, 4], and spectral factorization [5]. This means that for a specific problem, such as the wave equation with rational coefficients, there exist computable initial conditions such that the corresponding solution does not take computable values for certain rational inputs.

Going beyond the computability of a task, a similarly important question is the one of *complexity*, i.e., assuming that a system is computable in the above sense, what is the computational complexity for achieving a simulation with a given maximum approximation error? Complexity is understood here as the necessary number of computational steps to achieve the desired accuracy (which, by the computability assumption, is feasible in principle). Of particular interest is the occurrence of a *complexity blowup* describing the situation where a low-complexity input to the simulation of an analog system entails a high-complexity computation to obtain the associated output.

More formally, assume that we want to simulate a given system $S : u \mapsto y$ mapping an input signal u onto an output signal y on a digital computer. For a specific input u_s and associated output $y_s = S(u_s)$, the number of computational steps to generate an approximation up to an error of 2^{-n} of u_s and y_s represents the complexity/computation time of the computation of the input and output, respectively. Classically, in computer science, low complexity refers to tasks where the computation time grows at most polynomially in the relevant parameter, with the understanding that polynomial-time problems are feasible to solve in practice, whereas super-polynomial problems are typically infeasible. Hence, computational complexity is crucial when assessing the practical feasibility of the implementation of a specific problem. Here, we will consider systems S with a fixed set of parameters/coefficients, each polynomial-time computable, so that a potential complexity blowup traces back to the inherent complexity of S and not to other external factors.

1.1 Contribution

In this work, we show that any non-trivial linear ordinary differential equation (ODE) of the form

$$\sum_{i=0}^{m} a_i y^{(i)} = \sum_{j=0}^{n} b_j u^{(j)}, \quad m, n \in \mathbb{N}, a_i, b_j \in \mathbb{C}, u \in C^n([0,1], \mathbb{C}), y \in C^m([0,1], \mathbb{C}),$$

exhibits complexity blowup on digital machines modeled by Turing machines. Here, $y^{(i)}$ and $u^{(j)}$ denote the i-th and j-th derivative of the univariate functions y and u, respectively. Note that we implicitly assume for simplicity that $y(t)$ is defined for all $t \in [0, 1]$, i.e., the solution of the ODE does not grow arbitrarily large at some τ satisfying $0 < \tau < 1$. However, we believe that this assumption is not necessary and can be removed by refining our proof strategy. Indeed, for linear ODEs with constant coefficients and smooth input signal u—a key scenario for our considerations—, the solution y is a continuous function that is smooth on $(0, 1]$ [15]. Finally, the ODE is non-trivial if y cannot be written as $c_n u^{(n)} + \cdots + c_0 u$ for some $c_0, \ldots, c_n \in \mathbb{C}$.

In this setting, we establish a link between the algebraic structure of ODEs and their computational complexity measured in terms of accuracy. In particular, we show that there exists a smooth input signal u and coefficients a_i, b_j, $i = 1, \ldots, m$, $j = 1, \ldots, n$, which can be computed efficiently, i.e., in polynomial time, such that the corresponding solution of the ODE is $\#P$-complete. Under the assumption $P \neq NP$, it follows that the solution is not efficiently, i.e., in polynomial time, computable. Furthermore, we show that this is an upper bound on the complexity of the solution: if the input signal can be computed in polynomial time, then the solution is at most in $\#P$. This extends the analysis of the first-order case ($m = n = 1$) in [6] to higher-order equations ($m, n \in \mathbb{N}$). In both settings, the proof strategies build on foundational work on the computational complexity of real functions and operators [22, 27]. The key ingredient is Friedman's seminal result about the complexity of integration that shows the existence of smooth polynomial-time computable functions on real intervals whose integral/accumulation functions are $\#P$-complete [17].

Beyond theoretical interest, our result forms a foundation on which further discussions about the complexity of emulating analog (electrical) circuits can be based. In particular, non-trivial ODEs basically comprise all relevant applications in electrical engineering and physics based on linear systems such as any linear time-invariant (LTI) system of arbitrary order, including, for instance, simple LR low-pass and RC low-pass electrical circuits [6]. Hence, the simulation of all these dynamics on digital hardware is prone to a complexity blowup.

Moreover, the most basic description of neuronal dynamics is based on electrical circuits governed by linear ODE systems such as the (leaky) integrate and fire (LIF) model [19, 18]. To account for these potentially more complex behaviours, we also begin to extend our observations to systems of linear ODEs, i.e., equations of the form

$$\sum_{i=0}^{m} A_i \cdot y^{(i)} = \sum_{j=0}^{n} B_j \cdot u^{(j)}, \quad A_i, B_j \in \mathbb{C}^{d \times d}, u \in C^n([0,1], \mathbb{C}^d), y \in C^m([0,1], \mathbb{C}^d)$$

with $m, n, d \in \mathbb{N}$. Due to the different underlying algebraic structures of the domain (compared to single ODEs), we cannot directly adapt the previous techniques to derive an analogous result. However, for systems with order $m = 1$ and under further technical assumptions, which are, for instance, met by the LIF model, we can still prove the existence of complexity blowup in the above sense. Obtaining a full description of the complexity blowup phenomenon for general systems, while its occurrence is still expected, is left for future work. Similarly, it is an interesting question whether and to what degree complexity blowup arises in more general dynamics such as linear time-invariant differential-algebraic equations. Here, the increased intrinsic complexity of the dynamics suggests that complexity blowup may still naturally occur although certain properties of linear ODEs do not hold any more [8, 31, 37, 39, 38].

1.2 Related Work

There is a rich body of literature on the computability and complexity of differential equations; a good introduction and summary are provided in [20]. We briefly review the most relevant results for our work here.

With regard to ODEs, it has been shown that the solutions of computable ODEs, i.e. equations of the form $y' = f(t, y)$ with computable initial condition $y(0)$ and sufficiently regular (i.e. Lipschitz) right-hand side f, are computable [11]. Thus, we will implicitly assume the computability of the solution throughout this work on linear ODEs. When considering complexity, it was shown that solutions can always be computed in PSPACE [20, Theorem 3.3.2.], i.e., solved by a Turing machine using a polynomial amount (in the input size) of space or equivalently polynomial many tapes. Furthermore, there exist polynomial-time computable ODEs (i.e., instead of mere computability, one requires polynomial-time computability of the initial condition and right-hand side) whose solution is PSPACE-complete [20, Theorem 3.3.3.]. This indicates that, in general, the solution of ODEs can be highly complex. In contrast, we show that by restricting to linear ODEs, the complexity of the solution is reduced to $\#P$-completeness.

Regarding PDEs, several works have analyzed the complexity of specific equations. These include the wave equation [35], the inhomogeneous Schrödinger equation [46], and Navier–Stokes equations [41]. It has also been shown that the solutions of certain linear PDEs can be PSPACE-computable [25], again showing that by generalizing our setting (now from ordinary to partial differential equations but maintaining the linearity), the complexity of the solution can increase significantly.

1.3 Outline

We begin by introducing the necessary concepts in complexity theory and linear ODEs in Section 2. Subsequently, in Section 3, we state our results, starting with determining the exact complexity of first-order linear ODEs. Next, we extend this to higher orders by induction on the order, exploiting the algebraic structure of linear ODEs. Finally, we state a similar result for a subclass of first-order systems of linear ODEs. We conclude by discussing the practical relevance of these results, highlighting a potential application in neuronal dynamics, in Section 4 followed by the proofs of our results in Section 5.

1.4 Notation

We write $\{0,1\}^*$ for the set of all finite binary strings over $\{0,1\}$ and denote by $|x|$ the length of a string $x \in \{0,1\}^*$. Similarly, $|A|$ denotes the cardinality of a set A.

We use $f : \mathcal{X} \rightharpoonup \mathcal{Y}$ to indicate that f is a partial function from $\mathcal{X}$ to $\mathcal{Y}$. Moreover, for $\mathcal{X} \subset \mathbb{R}$ and $\mathcal{Y} \subset \mathbb{C}^d$, $C^n(\mathcal{X}, \mathcal{Y})$ denotes the class of n-times continuously differentiable functions from $\mathcal{X}$ to $\mathcal{Y}$.

For a ring $\mathcal{R}$, $\mathcal{R}[X]$ denotes the ring of polynomials in the indeterminate variable X with coefficients in $\mathcal{R}$. Additionally, for $P(X), Q(X) \in \mathcal{R}[X]$, we write $Q(X) \mid P(X)$ if $P(X) = Q(X) \cdot S(X)$ for some $S(X) \in \mathcal{R}[X]$, i.e., $P(X)$ divides $Q(X)$, and $Q(X) \nmid P(X)$ otherwise. Furthermore, $\mathcal{Z}(P)$ denotes the multiset of roots of $P(X)$.

2 Preliminaries

This section presents the necessary background from complexity theory and the theory of ODEs.

2.1 Computable Analysis and Complexity Theory

The Theory of Computability presented here is based on the computational model of the Turing machine [43]. This is an abstract machine that represents a natural and intuitive model of computation [12, 2]. In fact, by the Church–Turing Thesis [13], a problem can be solved by an intuitively effective method if and only if it can be solved by a Turing machine. Thus, by reasoning about Turing machines, we can obtain results on the intrinsic complexity of a given problem.

2.1.1 Computable Analysis

Classical computability theory is based on Turing machines, which operate in an inherently discrete domain [43, 12, 2]. Turing machines operate on strings on a tape over an alphabet such as $\{0, 1\}$. Hence, they intrinsically compute partial (word) functions $\{0, 1\}^* \rightharpoonup \{0, 1\}^*$, where $\{0, 1\}^*$ denotes the set of all finite binary strings. Thus, Turing machines operate with inputs and outputs in $\{0, 1\}^*$. By introducing (partial) surjective functions $\{0, 1\}^* \rightharpoonup \mathcal{X}$, one can consider Turing machines operating on any countable set $\mathcal{X}$, e.g., using binary numbers to encode natural numbers. Similarly, Turing machines computing binary functions $\{0, 1\}^* \times \{0, 1\}^* \rightharpoonup \{0, 1\}^*$, i.e., Turing machines with two (or any finite number of) binary input strings, can be reduced to Turing machines operating on $\{0, 1\}^* \rightharpoonup \{0, 1\}^*$ by, for instance, introducing a separator symbol in the alphabet. Thus, we often assume Turing machines directly operating on $\mathbb{N}$ and/or $\mathbb{Q}$ for (notational) simplicity.

Crucially, a Turing machine cannot directly operate on infinite objects such as real numbers or real functions. However, we can extend classical computability theory to infinite objects by means of finite approximations [34, 45, 26]. In particular, we consider Turing machines with a single tape and count the number of necessary computation steps to achieve a prescribed approximation accuracy given as input. Since only one computation step is possible per base time unit, this is the computation time. As already mentioned in the context of PSPACE, the Turing machine could also have multiple parallel tapes, which would reflect the necessary computation space.

Definition 1 (Computable Real number) A real number $x \in \mathbb{R}$ is *computable* if there exists a Turing machine $M : \mathbb{N} \to \mathbb{Q}$ such that

$$|x - M(n)| \leq 2^{-n} \quad \forall n \in \mathbb{N}. \tag{1}$$

In this case, we say that M converges effectively to x, written as $M \xrightarrow{\text{eff}} x$.

Remark 1 We write $\mathbb{R}_c \subsetneq \mathbb{R}$ for the (dense) set of computable real numbers. A complex number $z = a + bi \in \mathbb{C}$ is computable if both $a, b \in \mathbb{R}_c$ are computable. Similarly, a matrix $A \in \mathbb{C}^{m \times n}$ is computable if all its entries are computable. We write $\mathbb{C}_c$ and $\mathbb{C}_c^{m \times n}$ for the set of computable complex numbers and computable complex matrices, respectively.

To quantify the complexity of computing real numbers, the number of iterations to reach the approximation rate prescribed in (1) can be considered.

Definition 2 (Polynomial-Time Computable Numbers) A real number $x \in \mathbb{R}$ is *polynomial-time computable* if it is computable by a Turing machine M and if there is a polynomial $P(X) \in \mathbb{N}[X]$ such that (1) holds after at most $P(n)$ steps of M.

A complex number or matrix is polynomial-time computable if its components are polynomial-time computable. We write $\mathbb{R}_p$, $\mathbb{C}_p$, and $\mathbb{C}_p^{m\times n}$ for the set of polynomial-time computable real numbers, complex numbers, and complex matrices, respectively.

To define computable functions, we employ Oracle Turing machines [26]; an equivalent characterization can be achieved via Weihrauch's type-two effectivity approach [45]. An Oracle Turing machine $M.$ is simply an ordinary Turing machine equipped with the feature to query oracle tapes. Conceptually, one can think of an oracle γ as a function from $\mathbb{N}$ to $\mathbb{Q}$, which is able to calculate the function value $\gamma(\cdot)$ and print it on its tape in a single operation. Hereby, an oracle may process potentially impossible computations by digital means, e.g., an oracle γ could represent the effective convergence to a non-computable real number $x \in \mathbb{R}$. Thus, informally speaking, the function value $f(x)$ of a computable real function f might be computed (in a certain sense) by an associated Oracle Turing machine $M.$ equipped with the oracle γ denoted as M_γ. However, this is not feasible when considering only Turing machines representing digital computations. But the crucial point is that the actual machine $M.$ is solely responsible for the computation of the function f, and the oracle γ is only used to provide approximations to the input value x. Consequently, this allows us to separate the computation of the inputs from the computation of the function evaluation. Thus, Oracle Turing machines provide a tool to clearly assess the intrinsic computability and in a second step the complexity of functions.

Definition 3 (Computable Function) A function $f : \mathbb{R} \to \mathbb{R}$ is *computable* if there exists an Oracle Turing machine $M. : \mathbb{N} \to \mathbb{Q}$ such that for all $t \in \mathbb{R}$ and all oracles $\gamma : \mathbb{N} \to \mathbb{Q}$ with $\gamma \xrightarrow{\text{eff}} t$ we have

$$\left|f(t) - M_\gamma(n)\right| \le 2^{-n} \quad \forall n \in \mathbb{N}. \tag{2}$$

In this case, we write $M. \xrightarrow{\text{eff}} f$.

Analogously to polynomial-time computable numbers, we can define polynomial-time computable functions. However, there is a subtle but important issue not yet discussed. Simply assuming that an oracle effectively converges to a given input is not sufficient since the convergence can be slow which potentially artificially introduces a higher complexity. For instance, an oracle could produce a rational approximation $\frac{p}{q}$ of some $x \in [0, 1]$ with accuracy 2^{-n}, where p, q have both binary/decimal length 2^n. Then, the associated Turing machine would already require exponential time to just read the result of the oracle call. To resolve this issue one would need to measure (polynomial) complexity both relative to the accuracy of the output and the size of the oracle encoding the input. For our needs, it suffices to employ dyadic rational numbers instead of all rational numbers as a basis for computable real numbers, i.e., to substitute

$M(n)$ by $M(n)/2^n$ in (1) and require that $M : \mathbb{N} \to \mathbb{Z}$ instead of $M : \mathbb{N} \to \mathbb{Q}$ [23]. This leads to the following definition.

Definition 4 (Polynomial-Time Computable Function) A function $f : [0,1] \to \mathbb{R}$ is *polynomial-time computable* if there exists an Oracle Turing machine $M. : \mathbb{N} \to \mathbb{Q}$ such that for all $t \in [0,1]$ and all oracles $\gamma : \mathbb{N} \to \mathbb{Z}$ with $\frac{\gamma}{2^n} \xrightarrow{\text{eff}} t$ we have

$$\left|f(t) - M_{\gamma/2^n}(n)\right| \leq 2^{-n} \quad \forall n \in \mathbb{N} \tag{3}$$

and there exists a polynomial $P(X) \in \mathbb{N}[X]$ such that (3) holds after at most $P(n)$ steps of $M.$.

Remark 2 A complex-valued function $f : \mathbb{R} \to \mathbb{C}$ is *computable* if both its imaginary and its real parts are computable, i.e., there exists an Oracle Turing machine $M. : \mathbb{N} \to \mathbb{Q}[i] := \{p + iq \mid p, q \in \mathbb{Q}\}$ such that (2) holds. Analogously, f is polynomial-time computable if both its imaginary and real parts are polynomial-time computable.

Moreover, a function $f : [0,1] \to \mathbb{C}^d$ with $d \in \mathbb{N}$ is called computable (resp. polynomial-time computable) if all its components $f_i : [0,1] \to \mathbb{C}$ for $1 \leq i \leq d$ are computable (resp. polynomial-time computable). We write $\mathrm{Comp}([0,1], \mathbb{C}^d)$ and $\mathrm{Pol}([0,1], \mathbb{C}^d)$ for computable and polynomial-time computable functions $f : [0,1] \to \mathbb{C}^d$, respectively. If $d = 1$, we write $\mathrm{Comp}([0,1])$ and $\mathrm{Pol}([0,1])$. This and the following definitions and theorems can be straightforwardly extended to functions on any compact interval $[a,b]$, however, for simplicity, we only consider functions and ODEs on the interval $[0,1]$.

Remark 3 The class of polynomial-time computable functions is denoted by *FP*.

Equipped with the notion of (polynomial-time) computable functions, we can now formally introduce the concept of a complexity blowup.

Definition 5 (Complexity Blowup) Consider $\mathcal{F} := \{f \mid f : [0,1] \to \mathbb{C}^d\}$ and an operator $T : \mathcal{F} \to \mathcal{F}$. We say that T *preserves computability* or *polynomial-time computability* if $T\left(\mathrm{Comp}([0,1]; \mathbb{C}^d)\right) \subseteq \mathrm{Comp}([0,1]; \mathbb{C}^d)$ or $T\left(\mathrm{Pol}([0,1]; \mathbb{C}^d)\right) \subseteq \mathrm{Pol}([0,1]; \mathbb{C}^d)$, respectively. A computability preserving operator T exhibits *complexity blowup* if $T\left(\mathrm{Pol}([0,1]; \mathbb{C}^d)\right) \not\subseteq \mathrm{Pol}([0,1]; \mathbb{C}^d)$.

Remark 4 One needs to distinguish between an operator being (polynomial-time) computable and preserving (polynomial-time) computability. Although the latter follows from the former, the converse is false [17, 24].

It turns out that the complexity of functions and the occurrence of a complexity blowup are linked to the classical complexity classes on binary strings, which we review next.

2.1.2 Complexity Classes

The most well-known complexity classes P (solvable in polynomial time) and NP (verifiable in polynomial time) are associated with decision problems, i.e., problems expressed by partial functions mapping from $\{0,1\}^*$ to $\{0,1\}$. Since our analysis treats function evaluation problems, considering decision problems, i.e., deciding membership, is no longer sufficient. Instead, a different approach is generally needed to assess the complexity of function evaluation, and a suitable tool is provided by counting problems expressed by partial function mappings $\{0,1\}^* \rightharpoonup \{0,1\}^*$.

Definition 6 (Function Problem) A *function problem* is defined by a partial function $F : \{0,1\}^* \rightharpoonup \{0,1\}^*$. We say that a Turing machine M *computes* F, if for all $x \in \{0,1\}^*$ we have $M(x) = F(x)$. Moreover, F is in

- the class FP, if it can be computed by a (deterministic) polynomial-time Turing machine.
- the class $\#P$, if there exist a (deterministic) polynomial-time Turing machine M and a polynomial $P \in \mathbb{N}[X]$ such that for all $x \in \{0,1\}^*$ we have

$$\nu(F(x)) = \left|\left\{y \in \{0,1\}^{P(|x|)} \mid M(x,y) = 1\right\}\right|,$$

  where $\nu : \{0,1\}^* \to \mathbb{N}$ is a bijection encoding strings into natural numbers.

In terms of Oracle Turing machines on real numbers the classes FP and $\#P$, which (to a certain degree) can be thought of as analogs of P and NP for function evaluation problems, can be characterized in the following way: FP (as already described in Definition 4) denotes the class of functions that can be computed by a Function-oracle Turing machine in polynomial time, whereas $\#P$ denotes the class of functions that enumerate the number of computations that end in an accepting state of polynomial-time Function-oracle Turing machines. Hence, a problem in $\#P$ asks for the number of polynomial-time certificates for an input; in comparison, NP asks whether such a certificate exists. Note that $FP \subseteq \#P$ holds, but it is unknown whether $FP = \#P$, i.e., whether all problems in $\#P$ can be solved in polynomial time. Since $FP = \#P$ implies $P = NP$ (or equivalently, $P \neq NP$ implies $FP \neq \#P$), it is commonly assumed that this is not the case. Moreover, the importance of the class $\#P$ is demonstrated by the polynomial hierarchy and Toda's theorem [42, 2]. The polynomial hierarchy describes a hierarchy of complexity classes based on the following iteration:

- $\mathcal{S}_0 := P$ and $\mathcal{V}_0 := P$
- For all $k \in \mathbb{N}$ we have $\mathcal{S}_{k+1} := P^{\mathcal{V}_k}$ and $\mathcal{V}_{k+1} := NP^{\mathcal{V}_k}$, where $P^{\mathcal{V}_k}$ and $NP^{\mathcal{V}_k}$ denote the class of problems in P and NP when using an oracle for $\mathcal{V}_k$ problems, respectively.

Thus, for example, we have $\mathcal{V}_1 = NP$ and $\mathcal{S}_2 = P^{NP}$, i.e., the class of problems that can be solved in polynomial time using an oracle for NP problems. Crucially, Toda's theorem states that the entire polynomial hierarchy is contained in $P^{\#P}$ [42]. Thus, if there were a polynomial-time algorithm for any $\#P$-complete problem, it would allow us to solve any problem in the polynomial hierarchy in polynomial time, which is commonly deemed highly unlikely.

2.2 (Systems of) Linear Ordinary Differential Equations

Next, we formally introduce linear ODEs, focusing on their algebraic structure.

Definition 7 (Linear Ordinary Differential Equation) A *linear ordinary differential equation* of order $(m, n) \in \mathbb{N}^2$ is of the form

$$\sum_{i=0}^{m} a_i y^{(i)} = \sum_{k=0}^{n} b_k u^{(k)},$$

where $a_i, b_j \in \mathbb{C}$, $a_m, b_n \neq 0$ are constants, $u \in C^{n+1}([0,1], \mathbb{C})$ is the inhomogeneous generator, and $y \in C^{m+1}([0,1], \mathbb{C})$ is the unknown function. Since $a_m \neq 0$, we may always assume without loss of generality that $a_m = 1$. Moreover, we write $\mathrm{SOL}_{m,n}((a_i)_{i=0}^m, (b_k)_{k=0}^n) : u \mapsto y$ for the solution operator of the ODE.

We will express our main result about the complexity blowup of linear ODEs in terms of the properties of the characteristic polynomials, which we introduce next.

Definition 8 (Characteristic Polynomial) Consider a linear ODE of order (m, n), i.e. of the form

$$\sum_{i=0}^{m} a_i y^{(i)} = \sum_{k=0}^{n} b_k u^{(k)}.$$

We define the *characteristic polynomials* $P_y(X)$ and $P_u(X)$ of the ODE as

$$P_y(X) = \sum_{i=0}^{m} a_i X^i \in \mathbb{C}[X] \quad \text{and} \quad P_u(X) = \sum_{k=0}^{n} b_k X^k \in \mathbb{C}[X].$$

Next, we briefly extend our framework to systems of linear ODEs that essentially couple several scalar ODEs.

Definition 9 (System of linear ODE and its characteristic polynomial) A *d-system of linear ODEs* of order $(m, n) \in \mathbb{N}^2$ is of the form

$$\sum_{i=0}^{m} A_i \cdot y^{(i)} = \sum_{j=0}^{n} B_j \cdot u^{(j)}$$

where $A_i, B_j \in \mathbb{C}^{d \times d}$ are constant, $u \in C^{n+1}([0,1], \mathbb{C}^d)$ is the inhomogeneous generator, and $y \in C^{m+1}([0,1], \mathbb{C}^d)$ is the unknown function. We write $P_y(X)$, $P_u(X)$ for its *characteristic polynomials* given by

$$P_y(X) = \sum_{i=0}^{m} A_i X^i \in \mathbb{C}^{d \times d}[X], \quad P_u(X) = \sum_{j=0}^{n} B_j X^j \in \mathbb{C}^{d \times d}[X]$$

and $\mathrm{SOL}^d_{m,n}((A_i)_{i=0}^m, (B_j)_{j=0}^n) : u \mapsto y$ for the solution operator.

Finally, we describe the general setting we are interested in studying the complexity blowup phenomenon. In particular, we are interested in cases where the inputs and coefficients of an ODE are regular/of low complexity a priori, so that a potential increase in the complexity of the output is indeed generated by the intrinsic complexity of the problem.

Definition 10 (Polynomial-Time Computable (System of) ODEs) A system of ODEs of the form

$$\sum_{i=0}^{m} A_i \cdot y^{(i)} = \sum_{j=0}^{n} B_j \cdot u^{(j)}$$

with $A_i, B_j \in \mathbb{C}^{d\times d}$ and $u \in C^{n+1}([0,1], \mathbb{C}^d)$ is (polynomial-time) computable, if all coefficients A_i, B_j, all initial conditions $y^{(i)}(0)$ with $0 \le i \le m-1$, and the inhomogeneous generator u are (polynomial-time) computable.

3 Complexity Blowup in Linear ODEs

This section presents our results, providing an exact characterization of the complexity of linear ODEs. We begin by examining first-order ODEs, building on the work in [6]. Next, we extend this to higher-order ODEs by induction. We conclude by proving an analogous result for certain first-order systems of ODEs.

Subsequently, we will always assume that our ODEs are well-behaved in the sense of Definition 10, i.e., the parameter and the input of the ODE are polynomial-time computable. The question we ask is whether, in this well-behaved setting, one can also expect the corresponding solution to be well-behaved in the sense of low complexity, i.e., polynomial-time computability. In other words, does the solution operator preserve polynomial-time computability? Since, by construction, the potential complexity of the solution cannot arise due to external factors (high-complexity inputs or parameters), but only by the implicit properties of the ODE, we can assess whether an actual increase in complexity from input to output signal occurs.

The proofs of the subsequent results can be found in Section 5.

3.1 Linear ODEs

Our main result shows that the answer to the above question, namely, whether the solution to a polynomial-time computable ODE is also polynomial-time computable in general, is false, assuming that $FP \neq \#P$. We actually obtain the stronger result that for any linear ODE of order (m, n) with $P_y(X) \nmid P_u(X)$, there is a smooth and polynomial-time computable inhomogeneous generator such that the solution is $\#P$-complete. Thus, an algorithm that could solve such a polynomial-time computable linear ODE in polynomial time could be exploited to solve every problem in the polynomial hierarchy (as described in Subsection 2.1.2) in polynomial time (by virtue of the $\#P$-complete instance), including, for instance, NP-complete problems.

Theorem 1 *If $FP \neq \#P$, a polynomial-time computable linear ODE of order (m,n) exhibits complexity blowup if and only if $P_y(X) \nmid P_u(X)$.*

The condition $P_y(X) \nmid P_u(X)$ describes a property of the algebraic structure of the ODE. If it is violated, then the ODE is trivial in a certain sense. Informally—a rigorous treatment is provided in Subsection 5.1—, a polynomial $P(X) \in \mathbb{C}[X]$ of the form

$$P(X) = \sum_{i=0}^{n} a_i X^i$$

can be linked to the polynomial differential operator $P(\frac{d}{dt})$ as

$$P\left(\frac{d}{dt}\right) := \sum_{i=0}^{n} a_i \frac{d^i}{dt^i}.$$

Then an ODE of order (m,n) can be stated succinctly as

$$P_y\left(\frac{d}{dt}\right) y(t) = P_u\left(\frac{d}{dt}\right) u(t) \quad \text{for } t \in [0,1].$$

Thus, assuming $P_y(X) \mid P_u(X)$, we have $P_u(X) = P_y(X) \cdot Q(X)$ for some $Q(X) \in \mathbb{C}[X]$ and the ODE can now be written as

$$P_y\left(\frac{d}{dt}\right) y = P_y\left(\frac{d}{dt}\right) Q\left(\frac{d}{dt}\right) u.$$

Hence, it is clear that $y := Q\left(\frac{d}{dt}\right) u$ is a solution to the ODE, which is polynomial-time computable. Our result shows that for all feasible inhomogeneous generators, this is actually the only possibility for the solution to be polynomial-time computable.

Next, we present more details of the proof of Theorem 1. It is essentially divided into two steps. First, the complexity of the solution of linear ODEs of order $(1,n)$ is explicitly analyzed.

Proposition 1 *For any (1,n)-order polynomial-time computable ODE of the form*

$$y' + a_0 y = \sum_{j=0}^{n} b_j u^{(j)}$$

the solution y is in $\#P$. Furthermore, there exists $u \in C^\infty([0,1],\mathbb{C}) \cap \mathrm{Pol}([0,1])$ such that $\mathrm{SOL}_{1,n}(a_0,(b_j)_{j=0}^n)(u)$ is $\#P$-complete if and only if $P_u(-a_0) \neq 0$.

Note that the condition $P_u(-a_0) \neq 0$ is equivalent to $P_y(X) \nmid P_u(X)$ since $P_y(X) \mid P_u(X)$ holds if and only if $\mathcal{Z}(P_y) \subseteq \mathcal{Z}(P_u)$ (c.f. Lemma 1) and $-a_0$ is the only root of P_y.

The proof strategy for linear ODEs of order (1,1), treated in [6], also extends to linear ODEs of order (1,n) with some necessary modifications. The main idea is to rely on the solution formula, which we will exploit to explicitly encode a high-complexity problem into the solution of a linear ODE.

Proposition 2 *The solution of a linear ODE of order* $(1, n)$*, i.e. of the form*

$$y' + a_0 y = \sum_{k=0}^{n} b_k u^{(k)},$$

is given by

$$y(t) = e^{-a_0 t} y(0) + \sum_{j=0}^{n} \int_0^t b_j e^{-a_0(t-\tau)} u^{(j)}(\tau) d\tau.$$

The encoding of the high-complexity problem, in fact, a $\#P$-complete problem, is built on the following well-known result about the complexity of integration [17].

Theorem 2 *There exists a smooth polynomial-time computable function* $f : [0, 1] \to \mathbb{C}$ *such that the function* g *given by*

$$g(t) = \int_0^t f(\tau) d\tau$$

is $\#P$*-complete.*

Remark 5 Under the stronger assumption that f is analytic, the statement no longer holds, i.e., in this case g is indeed polynomial-time computable as well. Also note that this is not a statement about polynomial-time computability of an operator on continuous functions. It is a point-wise statement in the sense that even if $FP = \#P$ holds true there might be no general efficient way (in the sense of polynomial-time computability) to obtain a procedure to compute the integral from a procedure to compute the function; indeed, this was shown in [24].

Putting these two observations together, we obtain the characterization of the complexity of linear ODEs of order $(1, n)$ presented in Proposition 1.

The second step in the proof of Theorem 1 is to extend the first-order result to arbitrary linear ODEs by (polynomial-time) reduction techniques exploiting connections between the roots of the characteristic polynomials and certain polynomial reductions.

Proposition 3 *For any* (m, n)*-order polynomial-time computable ODE of the form*

$$\sum_{i=0}^{m} a_i y^{(i)} = \sum_{j=0}^{n} b_j u^{(j)},$$

the solution y *is in* $\#P$*. Furthermore, there exists* $u \in C^\infty([0, 1], \mathbb{C}) \cap \mathrm{Pol}([0, 1])$ *such that* $\mathrm{SOL}_{m,n}((a_i)_{i=0}^m, (b_j)_{j=0}^n)(x)$ *is* $\#P$*-complete if and only if* $\mathcal{Z}(P_y) \not\subseteq \mathcal{Z}(P_u)$.

Finally, Theorem 1 is now a direct consequence of Proposition 3 together with the next observation linking polynomial divisibility with roots (see, e.g., [32]).

Lemma 1 *Let F be an algebraically closed field, and let $P(X), Q(X) \in F[X]$ be polynomials. Then we have*

$$P(X) \mid Q(X) \iff \mathcal{Z}(P) \subseteq \mathcal{Z}(Q).$$

3.2 Systems of Linear ODEs

We now extend our findings to systems of linear ODEs. This is hindered by the fact that $\mathbb{C}^{d\times d}[X]$, respectively $\mathbb{C}_p^{d\times d}[X]$, is neither algebraically closed nor commutative. Thus, we cannot directly adapt the techniques used for single ODEs. However, in certain simplified settings that more closely mirror the properties of single ODEs, we can still obtain similar results.

Theorem 3 *If $FP \neq \#P$, a polynomial-time computable d-system of linear ODEs of order $(1, n)$*

$$A_1 y' + A_0 y = \sum_{j=0}^{n} B_j u^{(j)}$$

with A_1 invertible, such that for all j the commutativity condition

$$A_1^{-1} A_0 \cdot A_1^{-1} B_j = A_1^{-1} B_j \cdot A_1^{-1} A_0$$

holds, exhibits complexity blowup if and only if $P_y(-A_1^{-1} A_0) \neq 0$.

Remark 6 To prove Theorem 3, we employ the same proof strategy as in Proposition 1, i.e., use the explicit solution formula and encode a high-complexity problem. In particular, the solution formula contains the terms

$$e^{-A_1^{-1} A_0 (t-\tau)} B_j u^{(j)}(t).$$

In one dimension, the exponential and B_j commute, which allows us to eliminate the exponential in all terms simultaneously. However, in the general case, this no longer holds, which means that we cannot simplify the terms, and the proof strategy no longer applies. However, the additional assumptions in Theorem 3 (artificially) create a setting where the exponential and the coefficients commute.

Under the given assumptions, the proof of Theorem 3 is now completely analogous to the single case treated in Proposition 1 since we can rely on an explicit solution formula by a straightforward extension of Proposition 2.

Proposition 4 *The solution of a d-system of linear ODEs of order $(1, n)$*

$$A_1 y' + A_0 y = \sum_{j=0}^{n} B_j u^{(j)}$$

with A_1 invertible is given by

$$y(t) = e^{-A_1^{-1}A_0 t}y(0) + \sum_{j=0}^{n} \int_0^t e^{-A_1^{-1}A_0(t-\tau)} A_1^{-1} B_j u^{(j)}(\tau) d\tau,$$

where integration is understood component-wise.

Crucially, we cannot generalize this result to higher-order systems of ODEs by the proof technique used in Proposition 3 since it again relies on the fact that the base field is algebraically closed.

Although we did not yet fully prove it, we believe that (a variant of) the general result in Theorem 1 also extends to higher-dimensional systems. In particular, it seems unlikely that the increased complexity induced by the coupling of multiple ODEs would result in a solution of lower complexity in general. Essentially, it suffices to obtain an analogous result to Proposition 1, i.e., an exact characterization of the occurrence of complexity blowup in first-order systems of ODEs, since higher-order systems can be reduced to first-order systems.

4 Discussion

Now, we discuss the details and general implications of our findings. Subsequently, we exemplarily apply our framework to the modeling of neuronal dynamics, which are an important tool in understanding/studying biological neural networks and to a certain degree inspired modern deep learning architectures/dynamics.

4.1 Interpretation and Contextualization of Results

The results in Theorems 1 and 3 show that complexity blowup generally occurs in a wide range of simple models, in particular, (systems of) linear ODEs. Thus, it also seems likely that similar findings extend to more complex models, so that complexity blowup may be considered a rather general phenomenon. Although this appears to be a severe limitation in simulating/solving certain problems, there are also reasons to believe that the practical impact is less significant. In fact, deriving mathematical guarantees for low-complexity preserving operators in fairly general settings—e.g., simple linear ODES with polynomial-time computable coefficients—is infeasible due to our worst-case analysis, resulting in complexity blowup. However, we wish to mention the following points.

1. The results are not constructive, but only the existence of complexity blowup is shown, since the main auxiliary result in Theorem 2 (complexity blowup in integration) is also only an existence result. Therefore, strengthening the results qualitatively or quantitatively is a hard problem, since characterizing the admissible input signals that lead to complexity blowup is difficult.

2. In our framework, complexity blowup is measured with respect to the accuracy of a computation. More exactly, the computation time is measured relative to the desired output accuracy. Thus, even in the presence of a (theoretical) complexity blowup, it may be negligible if low accuracies are sufficient or the blowup kicks in only at high accuracies.
3. Finally, the theory describes the worst-case complexity, i.e., the potential for a complexity blowup based on the existence of certain polynomial-time computable input signals. Hence, obtaining mathematical guarantees for maintaining low complexity via the solution operator is in a sufficiently general setting not feasible since the complexity blowup inducing signal(s) may be admissible input(s). However, transitioning from a worst-case to a (more practicable) average-case analysis is difficult because it requires (beyond the characterization of complexity blowup inducing inputs) a suitable probability distribution over $\mathrm{Pol}([0,1];\mathbb{C}^d)$.

An interesting open problem arising from this discussion is whether there exist universal (to a certain degree) input signals that trigger complexity blowup simultaneously in distinct ODEs. In other words, are the complexity blowup inducing input signals specifically adapted to a specific ODE (as we construct them in our proofs) or does a common input signal exist for certain classes of ODEs?

4.2 Case Study: Complexity Blowup in Neuronal Dynamics

Modeling and analyzing neural dynamics (as displayed in, for instance, human brains) is a key element of neuroscience. On the one hand, the simulation of said dynamics enables their study in controlled environments. On the other hand, the power and efficiency of biological neural networks make them a viable computational framework. Indeed, unlocking the benefits of brain-like properties in computational models has not only led to current AI systems powered by deep learning and artificial neural networks (ANNs) but also potentially offers further advantages. Going beyond the loose analogy of artificial neurons and biological neurons in ANNs, spiking neural networks (SNNs) are envisioned to be more closely aligned with the structure of computations in biological neural networks, thereby exploiting their intrinsic benefits, foremost their energy efficiency [28, 30, 36, 16].

The adoption of spikes—asynchronous, point-like electrical pulses in the biological context—as a means of communication between neurons represents the key innovation of SNNs [19, 18]. The dynamic of spikes is governed (in the most basic model, still retaining biological plausibility) by a system of (linear) ODEs. Hence, a crucial question is what the computational cost of simulating the dynamics or turning the framework into a computational model on digital hardware actually is. Is it plausible to expect the advantages of the biological model in the digital computing paradigm? We approach an answer by examining the algorithmic complexity of approximating the neuronal dynamics on digital hardware via the introduced complexity analysis of linear ODEs.

4.2.1 Leaky Integrate and Fire Dynamics

We consider a simple model class of neuronal dynamics still retaining biological plausibility, the *Integrate-and-Fire* (IF) models [18]. They generally consist of two components: (i) the time evolution of the *potential* $V(t)$ of a neuron given a *current* $I(t)$ and (ii) the spike generation mechanism through thresholding operations. In the *Leaky IF* (LIF) model, a time-continuous dynamical system describes the evolution of the potential via a system of linear differential equations,

$$\begin{cases} \tau_m \frac{\mathrm{d}V}{\mathrm{d}t}(t) = -(V(t) - V_{\text{rest}}) + I(t), \\ \tau_s \frac{\mathrm{d}I}{\mathrm{d}t}(t) = -I(t) + I_e(t), \end{cases} \tag{4}$$

where $\tau_m, \tau_s > 0$ are the *membrane* and *synaptic time constants*, respectively, and $I_e(t)$ denotes an *external input current*. When V reaches a certain value, the *firing threshold* $\vartheta > 0$, the neuron emits a spike, and subsequently V is reset to the *resting potential* V_{rest} while the spike triggers postsynaptic currents in downstream neurons. The temporal dynamics in a layered network of spiking neurons are then typically (mathematically) abstracted by conceiving spikes as point processes localized in time expressed as a sum of Dirac delta distributions, the so-called *spike train* $S(t) = \sum_s \delta(t-s)$, where s iterates over the spike times of a neuron.

4.2.2 Complexity Blowup in LIF Dynamics

Our objective is to understand the complexity of emulating the dynamics in (4) on digital hardware by examining the potential occurrence of a complexity blowup in this context. Note that (4) constitutes a system of first-order linear ODEs with constant coefficients, driven by the input signal $I_e(t)$ and producing the output signal $(V(t), I(t))$. Recall that these dynamics describe the behaviour of LIF neurons only up to the point where the potential reaches the threshold. Therefore, it is necessary to simulate the dynamics/compute the threshold crossing to obtain the first spike time of a LIF neuron. The actual firing and reset mechanisms, which affect subsequent dynamics, are not even considered in this simplified formulation. Nevertheless, it turns out that complexity blowup can already occur in this reduced setting.

Theorem 4 *If FP* $\neq$ *#P, the LIF model in* (4) *on* $[0, 1]$ *with polynomial-time computable parameters, i.e., coefficients and initial conditions, exhibits complexity blowup.*

Proof. Observe that we can rewrite the LIF model from (4) as

$$A_1 y' + A_0 y = B_0 u,$$

with

$$u = \begin{pmatrix} 0 \\ I_e \end{pmatrix}, \quad y = \begin{pmatrix} u \\ I \end{pmatrix}, \quad A_1 = \begin{pmatrix} 1 & 0 \\ 0 & 1 \end{pmatrix}, \quad A_0 = \begin{pmatrix} \tau_m^{-1} & -\tau_m^{-1} \\ 0 & \tau_s^{-1} \end{pmatrix}, \quad B_0 = \begin{pmatrix} \tau_s^{-1} & 0 \\ 0 & \tau_s^{-1}, \end{pmatrix},$$

where we neglected constants terms (i.e., V_{rest}) without loss of generality. Now, the statement immediately follows from verifying the conditions in Theorem 3. □

This result demonstrates that simulating the dynamics of a LIF neuron on digital hardware can be computationally expensive (or even practically infeasible) when an accurate determination of spike times is required. However, since the above theorem relies on a non-constructive existence proof for the specific input signal, we can neither meaningfully characterize its properties nor identify the subset of inputs with analogous effects on the solution operator. Additionally, for a given configuration of a LIF neuron the complexity blowup in (4) may only occur at a time τ after the potential $V(\tau)$ already crossed the threshold. This implies that the complexity blowup has no actual impact on the computations in the LIF model since only the dynamics up to the threshold crossing are relevant for the spike times.

Moreover, in practice, the admissible input signals to networks of LIF neurons are typically more constrained. Specifically, the input signals are simplified to (weighted) spike trains, i.e., weighted Dirac delta distributions. Although this represents an unrealistic theoretical abstraction of the variety of neuro-physiological inputs in biological neurons, it nevertheless preserves essential features of their dynamics. Within our complexity framework, we conclude that—in this restricted setting—two opposite but equally restrictive situations may arise. First, the input signals themselves may already possess high complexity, which renders the question of output complexity less meaningful. Conversely, by constraining to a subset of polynomial-time computable input signals (which are intended to approximate the Dirac delta in a practical scenario) or if the complexity blowup occurs after the threshold crossing, the operator may in fact be polynomial-time computable on this subset or restricted domain. Deepening our understanding and providing a clearer characterization of the complexity blowup phenomenon in this setting is an interesting open problem.

5 Proofs

In this section, we present the proofs of the results stated in Section 3.

5.1 General Observations and Preliminaries

The set $\mathbb{C}_p$ is an algebraically closed field and $\mathbb{R}_p \subset \mathbb{C}_p$ is a subfield [29]. Based on this structure, we collect some useful properties about the complexity of functions and operations employed in the subsequent analysis.

Proposition 5

1. Let $\mathbb{A} \in \{\mathbb{R}, \mathbb{C}, \mathbb{C}^{d\times d}\}$. Then we have

$$\left\{x + y \mid x \in \mathbb{A}_p, y \in \mathbb{A}_p^c\right\} \cup \left\{x \cdot y \mid x \in \mathbb{A}_p^*, y \in \mathbb{A}_p^c\right\} \subseteq \mathbb{A}_p^c,$$

where $\mathbb{A}_p^*$ denotes the set of elements with a (multiplicative) inverse in $\mathbb{A}_p$.

2. *The set* $\mathrm{Pol}([0, 1], \mathbb{C}^d)$ *is a* $\mathbb{C}_p$*-algebra.*
3. *Let* $f \in \mathrm{Pol}([0, 1]) \cap C^{n+1}([0, 1], \mathbb{C})$*. Then* $f^{(i)} \in \mathrm{Pol}([0, 1])$ *for all* $i = 1, \ldots, n$*.*
4. *Let* $z \in \mathbb{C}_p$*. Then* $f : [0, 1] \to \mathbb{R}$ *given by* $f(x) = e^{zx}$ *is in* $\mathrm{Pol}([0, 1])$*.*
5. *Let* $f : [0, 1] \to \mathbb{C}$ *be #P-complete, let* $g \in \mathrm{Pol}([0, 1], \mathbb{C})$ *and* $0 \neq z \in \mathbb{C}_p$ *be polynomial-time computable. Then the function*

$$h(t) := zf(t) + g(t)$$

is also #P-complete.
6. *If* $f : [0, 1] \to \mathbb{C}$ *is in #P, then so is* $\int_0^t f(t)dt$*.*
7. *If* $f' : [0, 1] \to \mathbb{C}$ *is #P-complete, then* f *is also #P-complete.*

Proof.
1. Assume $x + y =: z$ is in $\mathbb{A}_p$ with $x \in \mathbb{A}_p$ and $y \in \mathbb{A}_p^c$. Then $y = z - x$ is also in $\mathbb{A}_p$, which is a contradiction. Multiplication follows analogously.
2. Addition was shown in [6], multiplication follows analogously, and scalar multiplication is a special case of multiplication.
3. This follows immediately from Lemma 4.1 in [27]
4. By Euler's formula $e^{a+ib} = e^a(\cos(b) + i\sin(b))$, it suffices to show that the real exponential, sine, and cosine functions are polynomial-time computable. The real exponential case was shown in [6], and the other cases are analogous.
5. This quickly follows from the transitivity of Turing reductions.
6. This was shown in [22].
7. We can apply the algorithm used in the proof of Theorem 5.2 in [27], to calculate f' from f in polynomial time. Thus, f' reduces to f in polynomial time, and f is #P-hard. By the previous proof, f must also be in #P, and hence #P-complete. □

To characterize linear ODEs, we have already introduced their characteristic polynomials in Definition 8. To motivate the appearance of these characteristic polynomials more formally, consider linear differential operators of order n with constant coefficients

$$D : \{f : [0, 1] \to \mathbb{C} \mid f \text{ is } n\text{-times differentiable}\} \to \{f : [0, 1] \to \mathbb{C}\},$$
$$f \mapsto \sum_{k=0}^{n} a_k f^{(k)}(t), \quad a_k \in \mathbb{C}, n \in \mathbb{N}.$$

We write $\mathcal{P}([0, 1])$ and $\mathcal{P}_n([0, 1])$ for the set of all differential operators with constant coefficients on $[0, 1]$ and those of order n, respectively. Now we can associate with each $D \in \mathcal{P}_n(I)$ a specific polynomial $P_D(X) \in \mathbb{C}[X]$:

$$D = \sum_{k=0}^{n} a_k \frac{d^k}{dt^k} \implies P_D(X) = \sum_{k=0}^{n} a_k X^k.$$

By abuse of notation, we will often write $D = P_D(\frac{d}{dt})$ for simplicity so that a linear ODE as introduced in Definition 7 can be expressed via its characteristic polynomials as

$$P_y\left(\frac{d}{dt}\right)y = P_u\left(\frac{d}{dt}\right)u.$$

Moreover, one easily verifies the following equalities, which we will repeatedly apply in the remainder: For $P(X), Q(X) \in \mathbb{C}[X]$ we have

$$\begin{aligned}(P(X)+Q(X))\left(\frac{d}{dt}\right) &= P\left(\frac{d}{dt}\right)+Q\left(\frac{d}{dt}\right) \quad \text{and}\\ (P(X)\cdot Q(X))\left(\frac{d}{dt}\right) &= P\left(\frac{d}{dt}\right)\cdot Q\left(\frac{d}{dt}\right).\end{aligned}$$

Another tool we need for our analysis is the reduction of the degree of polynomials. For $y \in R$ and a polynomial $P(X) \in R[X]$ over a ring R with

$$P(X) = \sum_{i=0}^{m} a_i X^i, \quad a_i \in R,$$

we set

$$\begin{aligned}\mathcal{R}_z[P](X) &:= \sum_{i=0}^{m-1} b_i(z)X^i \in R[X], \quad \text{where}\\ b_i(z) &:= \begin{cases} a_m, & \text{if } i = m-1,\\ a_{i+1} + z \cdot b_{i+1}(z), & \text{otherwise.}\end{cases}\end{aligned} \tag{5}$$

In particular, $\mathcal{R}_z[P](X)$ is the polynomial obtained through the polynomial division of $P(X)$ by $(X-z)$ with remainder term $b_{-1}(y) := a_0 + z \cdot b_0(z)$, i.e.,

$$P(X) = (X-z)\cdot \mathcal{R}_z[P](X) + b_{-1}(z).$$

This can be easily verified by direct calculation. In this setting, we immediately obtain the following characterizations.

Proposition 6 *Let F be a field, and let $P(X) \in F[X]$ be a polynomial.*

- *With the notation introduced above, for $\sigma \in F$ the following statements are equivalent: (i) $\sigma \in \mathcal{Z}(P)$, (ii) $b_{-1}(\sigma) = 0$, (iii) $\sigma b_0(\sigma) = -a_0$.*
- *For $\sigma \in \mathcal{Z}(P)$ we have $\mathcal{Z}(\mathcal{R}_\sigma[P]) = \mathcal{Z}(P) \setminus \{\sigma\}$.*

5.2 Proof of Complexity Blowup in Linear ODEs

Proof of Proposition 1. According to Theorem 2, there exists $\tilde{f} \in \mathrm{Pol}([0,1]) \cap C^\infty([0,1],\mathbb{C})$ such that

$$\tilde{g}(t) := \int_0^t \tilde{f}(\tau)d\tau$$

is #P-complete. Set $u(t) := e^{-a_0 t}\tilde{f}(t)$ and observe that $u \in \mathrm{Pol}([0,1]) \cap C^\infty([0,1], \mathbb{C})$ as well via Proposition 5.2 and 5.4. Since for $i \in \{1, \dots, n\}$ we have

$$u^{(i)}(t) = \sum_{j=0}^{i} \binom{i}{j} (-a_0)^{i-j} e^{-a_0 t} \tilde{f}^{(j)}(t) = e^{-a_0 t} \sum_{j=0}^{n} \binom{i}{j} (-a_0)^{i-j} \tilde{f}^{(j)}(t),$$

it follows that

$$\begin{aligned}
\sum_{i=0}^{n} \int_0^t b_i e^{-a_0(t-\tau)} u^{(i)}(\tau)\, d\tau &= \sum_{i=0}^{n} \int_0^t b_i e^{-a_0 t} \sum_{j=0}^{n} \binom{i}{j} (-a_0)^{i-j} \tilde{f}^{(j)}(\tau)\, d\tau \\
&= e^{-a_0 t} \sum_{j=0}^{n} \left(\sum_{i=0}^{n} \binom{i}{j} (-a_0)^{i-j} b_i \right) \int_0^t \tilde{f}^{(j)}(\tau)\, d\tau \\
&= e^{-a_0 t} \sum_{j=0}^{n} \lambda(j) \int_0^t \tilde{f}^{(j)}(\tau)\, d\tau \\
&= e^{-a_0 t} \left(\sum_{j=1}^{n} \lambda(j) f^{(j-1)}(t) + \lambda(0) \tilde{g}(t) \right) \\
&= e^{-a_0 t} \sum_{j=0}^{n-1} \lambda(j+1) f^{(j)}(t) + e^{-a_0 t} \lambda(0) \tilde{g}(t)
\end{aligned}$$

with $\lambda : \mathbb{N} \to \mathbb{C}_p$ given by

$$\lambda(j) := \sum_{i=0}^{n} \binom{i}{j} (-a_0)^{i-j} b_i.$$

Thus, by the solution formula in Proposition 2 we obtain that

$$\begin{aligned}
\mathrm{SOL}_{1,n}(a_0, (b_j)_{j=0}^n)(u) &= e^{-a_0 t} y(0) + \sum_{i=0}^{n} \int_0^t b_i e^{-a_0(t-\tau)} u^{(i)}(\tau)\, d\tau \\
&= e^{-a_0 t} y(0) + e^{-a_0 t} \sum_{j=0}^{n-1} \lambda(j+1) f^{(j)}(t) + e^{-a_0 t} \lambda(0) \tilde{g}(t).
\end{aligned}$$

Note that Proposition 5.2, 5.3, and 5.4 imply that

$$e^{-a_0 t} \lambda(0),\ e^{-a_0 t} \sum_{j=0}^{n-1} \lambda(j+1) f^{(j)}(t), \text{ and } e^{-a_0 t} y(0) \in \mathrm{Pol}([0,1]).$$

Hence, $\mathrm{SOL}_{1,n}(a_0, (b_j)_{j=0}^n)(u)$ is #P-complete if

$$0 \neq \lambda(0) = \sum_{i=0}^{n} \binom{i}{0} (-a_0)^{i-0} b_i = \sum_{i=0}^{n} b_i (-a_0)^i = P_u(-a_0)$$

follows from Proposition 5.5. Thus, it remains to show that if $P_u(-a_0) = 0$, then the solution is always polynomial-time computable. Observe that by definition, $P_y(X)$ is a polynomial of degree one with a single root $-a_0$. Therefore, assuming $P_u(-a_0) = 0$ implies that $\{-a_0\} = \mathcal{Z}(P_y) \subset \mathcal{Z}(P_u)$, which by Lemma 1 is equivalent to $P_y(X) \mid P_u(X)$. Thus, there exists $Q(X) \in \mathbb{C}[X]$ with $P_u(X) = P_y(X) \cdot Q(X)$ so that the ODE can be rewritten as

$$P_y\left(\frac{d}{dt}\right) y = P_u\left(\frac{d}{dt}\right) u = (P_y(X) \cdot Q(X)) \left(\frac{d}{dt}\right) u = P_y\left(\frac{d}{dt}\right)\left(Q\left(\frac{d}{dt}\right) u\right). \tag{6}$$

Comparing the left-hand and right-hand side shows that $y_p := Q\left(\frac{d}{dt}\right) u$ is a particular solution of the ODE, which is polynomial-time computable by the structural properties of $\mathrm{Pol}([0,1])$ given in Proposition 5.2 and 5.3 for $u \in \mathrm{Pol}([0,1]) \cap C^{n+1}([0,1], \mathbb{C})$. Since the general solution of the ODE can be decomposed into the homogeneous solution y_h and a particular solution y_p we get

$$\mathrm{SOL}_{1,n}(a_0, (b_j)_{j=0}^n)(u) = y_h + y_p. \tag{7}$$

Note that by Proposition 2 the homogeneous solution is given by $y_h(t) = e^{-a_0 t} y(0)$, i.e., y_h is polynomial-time computable via Proposition 5.2 and 5.4, and thereby $y_h + y_p \in \mathrm{Pol}([0,1])$.

Finally, note that by applying Proposition 5.6 to the solution formula from Proposition 2 it follows immediately that y is at most in #P, independent of P_y. □

Proof of Proposition 2. We consider the first-order ODE

$$y' + a_0 y = \sum_{j=0}^{n} b_j u^{(j)}$$

where $a_0, b_j \in \mathbb{C}_p$ and $u \in \mathrm{Pol}([0,1])$ is in $C^{n+1}([0,1], \mathbb{C})$. Multiplying by $e^{a_0 t}$, we get

$$\begin{aligned} e^{a_0 t} y'(t) + a_0 e^{a_0 t} y(t) &= \frac{d}{dt}(e^{a_0 t} y(t)) \\ &= \sum_{j=0}^{n} b_j e^{a_0 t} u^{(j)}(t). \end{aligned}$$

Taking the integral, we get:

$$e^{a_0 t} y(t) - y(0) = \sum_{j=0}^{n} \int_0^t b_j e^{a_0 \tau} u^{(j)}(\tau) d\tau.$$

By rearranging, we obtain the desired formula. □

Proof of Proposition 3. First, observe that for $u \in \mathrm{Pol}([0,1]) \cap C^n([0,1],\mathbb{C})$, if $P_y(X) \mid P_u(X)$, then the corresponding output of the ODE is polynomial-time computable by the same argument as in the proof of Proposition 1, in particular (6) and (7). Hence, by Lemma 1 we know that $\mathcal{Z}(P_y) \subset \mathcal{Z}(P_u)$ implies for $u \in \mathrm{Pol}([0,1]) \cap C^n([0,1],\mathbb{C})$ that the corresponding solution $y = \mathrm{SOL}_{m,n}((a_i)_{i=0}^m, (b_j)_{j=0}^n)(u) \in \mathrm{Pol}([0,1])$ is polynomial-time computable.

Thus, assume that $\mathcal{Z}(P_y) \not\subseteq \mathcal{Z}(P_u)$, i.e., there exists $\tilde{\sigma} \in \mathcal{Z}(P_y) \setminus \mathcal{Z}(P_u)$. We show that there exists $u \in C^\infty([0,1],\mathbb{C}) \cap \mathrm{Pol}([0,1])$ such that $\mathrm{SOL}_{m,n}((a_i)_{i=0}^m, (b_j)_{j=0}^n)(u)$ is #P-complete by induction on m. The base-case $(1,n)$ was proven in Proposition 1. Therefore, assume that the claim holds for all polynomial-time computable ODEs of order (k,n) with $k < m$ and consider a polynomial-time computable ODE of order (m,n)

$$\sum_{i=0}^{m} a_i y^{(i)} = \sum_{j=0}^{n} c_j u^{(j)}, \quad a_i, c_i \in \mathbb{C}_p.$$

We now make the following observation: If there exist $b_1, \ldots, b_{m-1}, \sigma \in \mathbb{C}_p$ such that

$$\sum_{i=0}^{m} a_i y^{(i)} = \sum_{i=0}^{m-1} b_i (y' - \sigma y)^{(i)}, \tag{8}$$

then $\tilde{y} := y' - \sigma y$ is a solution of the polynomial-time computable ODE

$$\sum_{i=0}^{m-1} b_i \tilde{y}^{(i)} = \sum_{j=0}^{n} c_j u^{(j)}. \tag{9}$$

Hence, by the induction hypothesis we find $u \in \mathrm{Pol}([0,1]) \cap C^\infty([0,1],\mathbb{C})$ such that $\tilde{y} = \mathrm{SOL}_{m-1,n}((b_i)_{i=0}^{m-1}, (c_j)_{j=0}^n)(u)$ is #P-complete provided that $\mathcal{Z}(\tilde{P}_y) \not\subseteq \mathcal{Z}(P_u)$ with $\tilde{P}_y(X) := \sum_{i=0}^{m-1} b_i X^i$ being the characteristic polynomial of the left-hand side corresponding to the modified ODE (9). Then, by Proposition 5.7, y is also #P-complete. Note that, independently of the algebraic structure, we can also apply this construction to immediately show that the solution of the ODE is at most #P.

It is left to prove the existence of $b_1, \ldots, b_{m-1}, \sigma$ with the given properties and the fact that $\mathcal{Z}(\tilde{P}_y) \not\subseteq \mathcal{Z}(P_u)$ (or equivalently $\tilde{P}_y(X) \nmid P_u(X)$) holds. Unfolding the condition in (8) gives

$$a_m y^{(m)} + \sum_{i=1}^{m-1} a_i y^{(i)} + a_0 y = b_{m-1} y^{(m)} + \sum_{i=1}^{m-1} (b_{i-1} - \sigma b_i) y^{(i)} - \sigma b_0 y.$$

Hence, comparing the coefficients yields

$$a_m = b_{m-1}, \quad b_{i-1} = a_i + \sigma b_i \text{ for } i = 1, \dots, m-1, \quad \text{and} \quad \sigma b_0 = -a_0.$$

Recalling the definition of polynomial reduction in (5) and its properties in Proposition 6, we notice that the condition is satisfied for $\tilde{P}_y(X) = \mathcal{R}_\sigma[P_y](X)$ where $\sigma \in \mathcal{Z}(P_y)$ and additionally $b_i, \sigma \in \mathbb{C}_p$ since $\in \mathbb{C}_p$ is an algebraically closed field (note that σ is in $\mathbb{C}_p$, since it is a root of a polynomial in $\mathbb{C}_p[X]$). Since $|\mathcal{Z}(P_y)| = \deg(P_y) = m \geq 2$ we can in particular choose $\sigma \neq \tilde{\sigma}$ so that again by Proposition 6

$$\mathcal{Z}(\tilde{P}_y) = \mathcal{Z}(\mathcal{R}_\sigma[P_y]) = \mathcal{Z}(P_y) \setminus \{\sigma\},$$

i.e., $\tilde{\sigma} \in \mathcal{Z}(\tilde{P}_y) \setminus \mathcal{Z}(P_u)$ or in other words $\mathcal{Z}(\tilde{P}_y) \not\subseteq \mathcal{Z}(P_u)$. □

Proof of Theorem 1. Proposition 3 (together with Lemma 1) characterizes the complexity of the solution of a polynomial-time computable linear ODE. In particular, there exists $u \in \text{Pol}([0,1]) \cap C^\infty([0,1], \mathbb{C})$ such that the corresponding solution is #P-complete if and only if the characteristic polynomials of the ODE satisfy $P_y(X) \nmid P_u(X)$. Thus, assuming $FP \neq \#P$, the solution is not polynomial-time computable, i.e., the ODE exhibits complexity blowup. □

5.3 Proof of Complexity Blowup in Systems of linear ODEs

We can immediately extend our proof technique from Proposition 1 to systems of linear ODEs, assuming that the coefficients (now in the form of matrices) of the ODE in question commute.

Proof of Theorem 3. By the commutativity conditions, we can rewrite the solution formula from Proposition 4 as

$$y(t) = e^{-A_1^{-1}A_0 t} y(0) + \sum_{j=0}^{n} \int_0^t A_1^{-1} B_j e^{-A_1^{-1}A_0(t-\tau)} u^{(j)}(\tau) d\tau.$$

We can then apply an analogous construction as in the proof of Proposition 1, using the matrix exponential instead of the scalar exponential. Note that the matrix exponential (considered as a function) with a polynomial-time computable coefficient matrix—which is the case under the given assumptions—is a polynomial-time computable function. This follows from a direct extension of the one-dimensional case in Proposition 5.4. □

Acknowledgments

A. Fono and N. Wedlich were supported by the project "Next Generation AI Computing (gAIn)", funded by the Bavarian Ministry of Science and the Arts and the Saxon Ministry for Science, Culture, and Tourism.

This work of H. Boche was supported in part by the Federal Ministry for Research, Technology and Space of Germany (BMFTR) in the programme of "Souverän.Digital.Vernetzt", joint project 6G-life, project identification number 16KISK002. H. Boche was also partially supported by the project "Next Generation AI Computing (gAIn)", funded by the Bavarian Ministry of Science and the Arts and the Saxon Ministry for Science, Culture, and Tourism.

G. Kutyniok was supported in part by the Munich Center for Machine Learning (BMFTR) as well as the German Research Foundation under Grants DFG-SPP-2298, KU 1446/31-1 and KU 1446/32-1. She also acknowledges support by the Konrad Zuse School of Excellence in Reliable AI (DAAD) and the project "Next Generation AI Computing (gAIn)", which is funded by the Bavarian Ministry of Science and the Arts and the Saxon Ministry for Science, Culture and Tourism.

References

1. E. Albuquerque and M. Silva. A comparison by simulation and by measurement of the substrate noise generated by CMOS, CSL, and CBL digital circuits. *IEEE Transactions on Circuits and Systems I: Regular Papers*, 52(4):734–741, 2005.
2. S. Arora and B. Barak. *Computational Complexity: A Modern Approach*. Cambridge University Press, USA, 1st edition, 2009.
3. P. Bauer and L.-J. Leclerc. A computer-aided test for the absence of limit cycles in fixed-point digital filters. *IEEE Transactions on Signal Processing*, 39(11):2400–2410, 1991.
4. H. Boche and U. J. Mönich. Turing computability of Fourier transforms of bandlimited and discrete signals. *IEEE Transactions on Signal Processing*, 68:532–547, 2020.
5. H. Boche and V. Pohl. On the algorithmic solvability of the spectral factorization and the calculation of the Wiener filter on Turing machines. In *2019 IEEE International Symposium on Information Theory (ISIT)*, pages 2459–2463, 2019.
6. H. Boche and V. Pohl. Complexity blowup in simulating analog linear time-invariant systems on digital computers. *IEEE Transactions on Signal Processing*, 69:5005–5020, 2021.
7. A. Brambilla and D. D'Amore. Method for steady-state simulation of strongly nonlinear circuits in the time domain. *IEEE Transactions on Circuits and Systems I: Fundamental Theory and Applications*, 48:885 – 889, 08 2001.
8. K. E. Brenan, S. L. Campbell, and L. R. Petzold. *Numerical Solution of Initial-Value Problems in Differential-Algebraic Equations*. Society for Industrial and Applied Mathematics, 1995.
9. W. Chen, X. Han, C.-W. Chang, G. Liu, and R. Domer. Out-of-order parallel discrete event simulation for transaction level models. *IEEE Transactions on Computer-Aided Design of Integrated Circuits and Systems*, 33:1859–1872, 12 2014.
10. D. V. Christensen, R. Dittmann, B. Linares-Barranco, A. Sebastian, M. Le Gallo, A. Redaelli, S. Slesazeck, T. Mikolajick, S. Spiga, S. Menzel, I. Valov, G. Milano, C. Ricciardi, S.-J. Liang, F. Miao, M. Lanza, T. J. Quill, S. T. Keene, A. Salleo, J. Grollier, D. Markovic, A. Mizrahi, P. Yao, J. J. Yang, G. Indiveri, J. P. Strachan, S. Datta, E. Vianello, A. Valentian, J. Feldmann, X. Li, W. H. Pernice, H. Bhaskaran, S. Furber, E. Neftci, F. Scherr, W. Maass, S. Ramaswamy, J. Tapson, P. Panda, Y. Kim, G. Tanaka, S. Thorpe, C. Bartolozzi, T. A. Cleland, C. Posch,

S.-C. Liu, G. Panuccio, M. Mahmud, A. N. Mazumder, M. Hosseini, T. Mohsenin, E. Donati, S. Tolu, R. Galeazzi, M. E. Christensen, S. Holm, D. Ielmini, and N. Pryds. 2022 Roadmap on neuromorphic computing and engineering. *Neuromorph. Comput. Eng.*, 2(2), 2022. 022501.
11. P. Collins and D. S. Graça. Effective Computability of Solutions of Ordinary Differential Equations The Thousand Monkeys Approach. *Electronic Notes in Theoretical Computer Science*, 221:103–114, Dec. 2008.
12. S. Cooper. *Computability Theory*. Chapman and Hall/CRC, New York, USA, 1st edition, 2004.
13. B. J. Copeland. The Church-Turing Thesis. In E. N. Zalta, editor, *The Stanford Encyclopedia of Philosophy*. Metaphysics Research Lab, Stanford University, Summer 2020 edition, 2020.
14. L. B. de Sá and A. Mesquita. Evolutionary synthesis of low-sensitivity equalizers using adjacency matrix representation. In *Proceedings of the 10th Annual Conference on Genetic and Evolutionary Computation*, GECCO '08, page 1283–1290, New York, NY, USA, 2008. Association for Computing Machinery.
15. G. Doetsch. *Handbuch der Laplace-Transformation: Band 3: Anwendungen der Laplace-Transformation*. Lehrbücher und Monographien aus dem Gebiete der exakten Wissenschaften. Springer, Basel, 2014.
16. A. Fono, M. Singh, E. Araya, P. C. Petersen, H. Boche, and G. Kutyniok. Sustainable AI: Mathematical foundations of spiking neural networks. *arXiv:2503.02013*, 2025.
17. H. Friedman. The computational complexity of maximization and integration. *Advances in Mathematics*, 53(1):80–98, 1984.
18. W. Gerstner and W. M. Kistler. *Spiking Neuron Models: Single Neurons, Populations, Plasticity*. Cambridge University Press, 2002.
19. W. Gerstner, W. M. Kistler, R. Naud, and L. Paninski. *Neuronal Dynamics: From Single Neurons to Networks and Models of Cognition*. Cambridge University Press, 2014.
20. D. S. GraÇa and N. Zhong. Computability of Differential Equations. In V. Brattka and P. Hertling, editors, *Handbook of Computability and Complexity in Analysis*, pages 71–99. Springer International Publishing, Cham, 2021. Series Title: Theory and Applications of Computability.
21. R. Jain, P. Yang, and T. Yoshino. FIRGEN: a computer-aided design system for high performance FIR filter integrated circuits. *IEEE Transactions on Signal Processing*, 39(7):1655–1668, 1991.
22. A. Kawamura. *Computational Complexity in Analysis and Geometry*. PhD thesis, University of Toronto, 2011.
23. A. Kawamura and S. Cook. Complexity theory for operators in analysis. *ACM Trans. Comput. Theory*, 4(2), 2012.
24. A. Kawamura, N. Müller, C. Rösnick, and M. Ziegler. Computational benefit of smoothness: Parameterized bit-complexity of numerical operators on analytic functions and Gevrey's hierarchy. *Journal of Complexity*, 31(5):689–714, 2015.
25. A. Kawamura, F. Steinberg, and M. Ziegler. On the computational complexity of the Dirichlet problem for Poisson's equation. *Mathematical Structures in Computer Science*, 27(8):1437–1465, 2017.
26. K.-I. Ko. *Complexity Theory of Real Functions*. Birkhäuser Boston Inc., USA, 1991.
27. K.-I. Ko and H. Friedman. Computational complexity of real functions. *Theoretical Computer Science*, 20(3):323–352, 1982.
28. W. Maass. Networks of spiking neurons: the third generation of neural network models. *Neural networks*, 10(9):1659–1671, 1997.
29. T. Matsui. On Polynomial Time Computable Numbers. *arXiv:cs/0608067*, 2006.
30. A. Mehonic, D. Ielmini, K. Roy, O. Mutlu, S. Kvatinsky, T. Serrano-Gotarredona, B. Linares-Barranco, S. Spiga, S. Savel'ev, A. G. Balanov, N. Chawla, G. Desoli, G. Malavena, C. Monzio Compagnoni, Z. Wang, J. J. Yang, S. G. Sarwat, A. Sebastian, T. Mikolajick, S. Slesazeck, B. Noheda, B. Dieny, T.-H. A. Hou, A. Varri, F. Brückerhoff-Plückelmann, W. Pernice, X. Zhang, S. Pazos, M. Lanza, S. Wiefels, R. Dittmann, W. H. Ng, M. Buckwell, H. R. J. Cox, D. J. Mannion, A. J. Kenyon, Y. Lu, Y. Yang, D. Querlioz, L. Hutin, E. Vianello, S. S. Chowdhury, P. Mannocci, Y. Cai, Z. Sun, G. Pedretti, J. P. Strachan, D. Strukov, M. Le Gallo, S. Ambrogio, I. Valov, and R. Waser. Roadmap to neuromorphic computing with emerging technologies. *APL Materials*, 12(10):109201, 10 2024.
31. L. Petzold. Differential/algebraic equations are not ODE's. *SIAM Journal on Scientific and Statistical Computing*, 3(3):367–384, 1982.

32. C. C. Pinter. *A book of abstract algebra: 2nd edition.* Dover Books on Mathematics. Courier Corporation, 2010.
33. M. B. Pour-El and I. Richards. Noncomputability in analysis and physics: A complete determination of the class of noncomputable linear operators. *Advances in Mathematics*, 48(1):44–74, 1983.
34. M. B. Pour-El and J. I. Richards. *Computability in Analysis and Physics.* Perspectives in Logic. Cambridge University Press, 2017.
35. M. B. Pour-El and N. Zhong. The wave equation with computable initial data whose unique solution is nowhere computable. *Mathematical Logic Quarterly*, 43(4):499–509, Nov. 2006.
36. N. Rathi, I. Chakraborty, A. Kosta, A. Sengupta, A. Ankit, P. Panda, and K. Roy. Exploring neuromorphic computing based on spiking neural networks: Algorithms to hardware. *ACM Comput. Surv.*, 55(12), 2023.
37. G. Reissig. Differential-algebraic equations and impasse points. *IEEE Transactions on Circuits and Systems I: Fundamental Theory and Applications*, 43(2):122–133, 1996.
38. G. Reissig and H. Boche. On singularities of autonomous implicit ordinary differential equations. *IEEE Transactions on Circuits and Systems I: Fundamental Theory and Applications*, 50(7):922–931, 2003.
39. G. Reißig, H. Boche, and P. I. Barton. On inconsistent initial conditions for linear time-invariant differential-algebraic equations. *IEEE Trans. Circuits Systems I Fund. Theory Appl.*, 49(11):1646–1648, 2002.
40. S. Salcedo-Sanz, F. Cruz-Roldan, C. Heneghan, and X. Yao. Evolutionary design of digital filters with application to subband coding and data transmission. *IEEE Transactions on Signal Processing*, 55(4):1193 – 1203, 2007.
41. S. Sun, N. Zhong, and M. Ziegler. Computability of the solutions to Navier-Stokes equations via effective approximation. In D.-Z. Du and J. Wang, editors, *Complexity and Approximation - In Memory of Ker-I Ko*, Lecture Notes in Computer Science (including subseries Lecture Notes in Artificial Intelligence and Lecture Notes in Bioinformatics), pages 80–112. Springer, 2020.
42. S. Toda. PP is as hard as the polynomial-time hierarchy. *SIAM Journal on Computing*, 20(5):865–877, 1991.
43. A. M. Turing. On computable numbers, with an application to the Entscheidungs-problem. *Proc. Lond. Math. Soc.*, s2-42(1):230–265, 1936.
44. K. Uesaka and M. Kawamata. Synthesis of low-sensitivity second-order digital filters using genetic programming with automatically defined functions. *Signal Processing Letters, IEEE*, 7:83 – 85, 05 2000.
45. K. Weihrauch. *Computable Analysis: An Introduction.* Springer-Verlag, Berlin, Heidelberg, 2000.
46. K. Weihrauch and N. Zhong. Is the linear Schrödinger propagator Turing computable? In J. Blanck, V. Brattka, and P. Hertling, editors, *Computability and Complexity in Analysis*, pages 369–377, Berlin, Heidelberg, 2001. Springer Berlin Heidelberg.

Some Recent Advances in Time Integration Methods for Kinetic Equations with Time-Scale Separation

Emil Løvbak, Vince Maes and Giovanni Samaey

Abstract Time integration remains a bottleneck in multiscale simulation of stiff ODEs and PDEs. This chapter surveys recent developments in time integration from a particle-based perspective. The discussed topics include deterministic approaches, stochastic Monte Carlo approaches, asymptotic-preserving schemes, multilevel time integrators, and adjoint optimization techniques. As a concrete example, we focus on kinetic-diffusion Monte Carlo (KDMC) algorithms and their hybridizations, emphasizing their role in bridging deterministic and stochastic formulations.

1 Introduction

In many chemical, physical, and biological systems, macroscopic, coherent behavior arises from interactions between microscopic entities (molecules, cells, individuals in a population) with each other and their environment. In many cases, a macroscopic model (such as the Navier–Stokes equations for fluid flow or a reaction-diffusion equation) has been formally derived to quantitatively describe behavior at this level. However, one regularly encounters systems that can only be modeled with sufficient accuracy at a more microscopic, fine scale than the scale at which one wants to observe or analyse the system. In such cases, while one observes the emergence of coarse-scale, macroscopic behavior in practice, directly modeling it may be impossible or impractical without simplifying assumptions that are hard to justify.

Emil Løvbak
Scientific Computing Center, Karlsruhe Institute of Technology, emil.loevbak@kit.edu

Vince Maes
Department of Computer Science, KU Leuven, vince.maes@kuleuven.be

Giovanni Samaey
Department of Computer Science, KU Leuven, giovanni.samaey@kuleuven.be

B. Koren and D. Roose (eds.), *Numerical Analysis and Scientific Computing: an Anthology*, Lecture Notes in Computational Science and Engineering 156,
https://doi.org/10.1007/978-3-032-33428-2_5

This review chapter focuses on recent advancements in time integration methods for a specific class of models characterized by multiscale dynamics, namely kinetic equations. Kinetic equations describe a gas as a collection of particles undergoing instantaneous collisions interspersed with ballistic motion. These equations have applications in various scientific and engineering fields, including astrophysics, aerospace and nuclear engineering, semiconductors, fusion processes in plasmas, biology, finance and social sciences. The common structure of kinetic equations involves a combination of a linear transport term and one or more interaction terms, which collectively dictate the time evolution of the particle distribution in position-velocity phase space.

From a numerical perspective, solving kinetic equations presents significant challenges due to the inherent computational cost [36]. Beyond the curse of dimensionality, kinetic equations also face unique difficulties. A first challenge relates to the computational cost of collision operator evaluation: the evaluation of the Boltzmann collision operator involves computing multidimensional integrals at each point in the physical space. This computational burden is significant and necessitates efficient algorithms [42]. In this review, we ignore this challenge by considering only the linear Boltzmann–Batnagar–Gross–Krook (Boltzmann-BGK) collision operator. Second, collision rates in practical applications can vary over multiple time scales throughout the spatial domain. Hence, computational methods must be able to tackle these different regimes, some of which contain stiff dynamics [35, 65, 64, 29]. It is this second challenge that is the main focus of the present review.

Historically, two types of approaches have been pursued to tackle kinetic equations numerically: deterministic methods, such as finite volume, semi-Lagrangian and spectral schemes, see [36] for an overview, and probabilistic methods, such as Direct Simulation Monte Carlo (DSMC) schemes [8]. Both approaches have strengths and weaknesses. Deterministic methods can normally reach high orders of accuracy. On the other hand, stochastic methods are often faster, especially for steady-state problems, but typically exhibit lower convergence rates and challenges in describing non-stationary and slow-motion flows. This review gives an overview of the efforts over the past decades to obtain efficient time integrators for kinetic equations, with an emphasis on the work performed in the research team of the authors in the context of plasma edge simulations in fusion energy reactors. We start by setting the mathematical context and introducing the model problems exhibiting time-scale separation in Section 2, while also providing a brief introduction to stochastic Monte Carlo methods. We then give an overview of deterministic asymptotic-preserving deterministic methods that circumvent the time-scale separation issue in Section 3. Next, we turn to hybrid kinetic-fluid methods (Section 4), and to asymptotic-preserving particle methods (Section 5). To eliminate the bias introduced by these methods, we combine them with the multilevel Monte Carlo method in Section 6. The improved time integration methods for kinetic equations allow for an enormous cost reduction for simulation intensive tasks. As an example, Section 7, discusses some recent developments on PDE-constrained optimization, using Monte Carlo simulation. Section 8 contains a conclusion and outlook.

2 Kinetic Equations

Kinetic equations describe the behaviour of particles at the mesoscopic level. A kinetic equation is typically of the form

$$\partial_t f(\mathbf{x},\mathbf{v},t) + \mathbf{v}\cdot\nabla_{\mathbf{x}} f(\mathbf{x},\mathbf{v},t) + \frac{\mathbf{F}(\mathbf{x},\mathbf{v},t)}{m}\cdot\nabla_{\mathbf{v}} f(\mathbf{x},\mathbf{v},t) = Q[f](\mathbf{x},\mathbf{v},t), \tag{1}$$

where $f(\mathbf{x},\mathbf{v},t)$ represents the particle distribution function in phase space, $\mathbf{F}(\mathbf{x},\mathbf{v},t)$ represents a force field that accelerates the particles, m is the mass of a particle, and $Q[f](\mathbf{x},\mathbf{v},t)$ represents a collision operator. In Section 2.1, we first simplify the collision operator in the Boltzmann equation to the linear Boltzmann-BGK operator, such that we can model collisions as random velocity changes. We then discuss the time-scale separation issues that lead to stiffness in Section 2.2. Finally, we discuss the particle view and corresponding stochastic simulation methods in Section 2.4.

2.1 From the Boltzmann Equation to Linear Boltzmann-BGK

2.1.1 The Boltzmann Equation and the Boltzmann-BGK Equation

The Boltzmann equation [53, 17, 128] describes how particles behave in a rarefied gas, and is of the form (1), with $\mathbf{F}(\mathbf{x},\mathbf{v},t)=0$. Applications of this equation range from micro-electromechanical systems (MEMS), to re-entry problems of space vehicles [12]. The collision operator of the Boltzmann equation is given by

$$Q[f](\mathbf{x},\mathbf{v},t) = \int_{\mathbb{R}^3}\int_{\mathbb{S}^2} B(|\mathbf{v}-\mathbf{v}_*|,\sigma)\Big(f(\mathbf{v}')f(\mathbf{v}'_*) - f(\mathbf{v})f(\mathbf{v}_*)\Big)d\mathbf{v}_* d\sigma,$$

with $B(|\mathbf{v}-\mathbf{v}_*|,\sigma)$ the collision kernel and

$$\mathbf{v}' = \frac{\mathbf{v}+\mathbf{v}_*}{2} + \frac{1}{2}|\mathbf{v}-\mathbf{v}_*|\sigma, \qquad \mathbf{v}'_* = \frac{\mathbf{v}+\mathbf{v}_*}{2} - \frac{1}{2}|\mathbf{v}-\mathbf{v}_*|\sigma.$$

Due to the complexity of the Boltzmann collision operator, it is often approximated using the so-called BGK-approximation [7]:

$$Q[f](\mathbf{x},\mathbf{v},t) = R(x,t)\left(\mathcal{V}(\mathbf{x},\mathbf{v},t)\rho(\mathbf{x},t) - f(\mathbf{x},\mathbf{v},t)\right),$$

where $R(x,t)$ represents the collision rate, and $\mathcal{V}(\mathbf{x},\mathbf{v},t)$ represents the local equilibrium distribution, which in the case of the Boltzmann equation is a local Maxwellian:

$$\mathcal{V}(\mathbf{x},\mathbf{v},t) = \frac{1}{(2\pi\sigma^2(\mathbf{x},t))^{3/2}}\exp\left(-\frac{1}{2}\frac{(\mathbf{v}-\mathbf{u}(\mathbf{x},t))^2}{\sigma^2(\mathbf{x},t)}\right). \tag{2}$$

The local number density $\rho(\mathbf{x},t)$, the local mean velocity $\mathbf{u}(\mathbf{x},t)$, and the local variance on the velocity $\sigma^2(\mathbf{x},t)$ all depend on the local particle distribution function $f(\mathbf{x},\mathbf{v},t)$.

Introducing the notation

$$\langle\cdot\rangle = \int_{\mathbb{R}^3} \cdot d\mathbf{v}$$

to indicate averaging over velocity space, we write

$$\rho(\mathbf{x},t) = \langle f(\mathbf{x},\mathbf{v},t)\rangle, \tag{3}$$

$$\rho(\mathbf{x},t)\mathbf{u}(\mathbf{x},t) = \langle \mathbf{v} f(\mathbf{x},\mathbf{v},t)\rangle, \tag{4}$$

$$\rho(\mathbf{x},t)\sigma^2(\mathbf{x},t) = \left\langle (\mathbf{v}-\mathbf{u}(\mathbf{x},t))^2 f(\mathbf{x},\mathbf{v},t)\right\rangle. \tag{5}$$

The Boltzmann-BGK equation is then given by

$$\partial_t f(\mathbf{x},\mathbf{v},t) + \mathbf{v}\cdot\nabla_{\mathbf{x}} f(\mathbf{x},\mathbf{v},t) = R(x,t)\left(\mathcal{V}(\mathbf{x},\mathbf{v},t)\rho(\mathbf{x},t) - f(\mathbf{x},\mathbf{v},t)\right), \tag{6}$$

which is a non-linear equation due to the dependence of the mean velocity $\mathbf{u}(\mathbf{x},t)$ and the variance on the velocity $\sigma^2(\mathbf{x},t)$ on the local particle distribution function $f(\mathbf{x},\mathbf{v},t)$.

2.1.2 Nonlinear Versus Linear Boltzmann-BGK Equations

The nonlinearity of equation (6) with Maxwellian (2) and moments defined as (3)-(5) is a mathematical consequence of the physical principle that each collision involves two particles of the particle distribution function $f(\mathbf{x},\mathbf{v},t)$. The nonlinear Boltzmann-BGK collision operator therefore has three collisional invariants: $\{1,\mathbf{v},|\mathbf{v}|^2\}$, corresponding to the conservation of mass, momentum, and energy.

However, there are also many kinetic equations that describe collisions of a population of particles with a background medium. In such situations, collisions amount to a random change in velocity, which does not need to preserve momentum and energy. A particle just gets a new velocity randomly drawn from some Maxwellian distribution (2), but the mean velocity $\mathbf{u}(\mathbf{x},t)$ and the velocity variance $\sigma^2(\mathbf{x},t)$ *no longer satisfy* (4)-(5). Applications in which particles collide with a background medium include neutron transport [16], radiative transfer [19], neutral particle transport in fusion energy [125], semiconductors [99], trace gases in chemistry [112], and chemotaxis [114].

As an example of a linear kinetic equation, we take a linear Boltzmann-BGK equation, which is again of the form (6) with a Maxwellian equilibrium distribution (2), but this time with the values of $\mathbf{u}(\mathbf{x},t)$ and $\sigma^2(\mathbf{x},t)$ given by the background medium, rendering the Maxwellian and therefore the kinetic equation linear. For clarity, we now explicitly write the linear Boltzmann-BGK equation as

$$\partial_t f(\mathbf{x},\mathbf{v},t) + \mathbf{v}\cdot\nabla_{\mathbf{x}} f(\mathbf{x},\mathbf{v},t) = R(x,t)\left(\mathcal{M}(\mathbf{x},\mathbf{v},t)\rho(\mathbf{x},t) - f(\mathbf{x},\mathbf{v},t)\right),$$

with

$$\mathcal{M}(\mathbf{x},\mathbf{v},t) = \frac{1}{(2\pi\sigma_p^2(\mathbf{x},t))^{3/2}} \exp\left(-\frac{1}{2}\frac{(\mathbf{v}-\mathbf{u}_p(\mathbf{x},t))^2}{\sigma_p^2(\mathbf{x},t)}\right). \tag{7}$$

This way, the definition of the moments (3)-(5) is maintained and the origin of linearity of the equation, i.e., the fact that velocities are drawn from a Maxwellian of a different species, is marked by the subscript p. (The subscript p refers to the context of plasma edge simulation in fusion energy, where the distribution f represents the population of neutral particles, and collisions are with the surrounding plasma.) The linear Boltzmann-BGK collision operator only has one collisional invariant: $\{1\}$, and therefore only conserves mass. This also makes sense, as no mass is generated or lost during collisions with the background medium, but there is an exchange of momentum and energy between the particle population under consideration and the background medium.

2.2 Scalings, Stiffness, and Macroscopic Behaviour

Kinetic equations, such as the linear Boltzmann-BGK equation (6) can become stiff under relevant physical conditions. Among these conditions are the high-collisional regimes in which the collision rate $R(x,t)$ becomes extremely large in some part of the domain. Mathematically, we can introduce such assumptions on the order of magnitude of the different physical parameters using a scaling parameter $0 \leq \varepsilon \ll 1$ and relating all quantities in the equation to the scaling parameter ε. A set of assumptions on the order of magnitude of physical parameters is called a scaling. The mathematical idealization of a scaling can be achieved by taking the asymptotic limit $\varepsilon \longrightarrow 0$. In such an asymptotic limit, the terms of the kinetic equation with different orders in ε decouple. When also introducing a Hilbert expansion or Chapman–Enskog expansion [17] for the particle distribution function $f(x,v,t)$, solving the decoupled equations leads to a set of macroscopic fluid models corresponding to the kinetic equation that is valid under the introduced physical assumptions.

2.2.1 A Physical View on Scale Separation

The easiest scaling is to assume that all quantities are $O(1)$, except for the collision rate, which we normalize as $R(\mathbf{x},t) = \varepsilon^{-1}\tilde{R}(\mathbf{x},t)$, meaning intuitively that we assume that the collision rate is an order $1/\varepsilon$ bigger than the other quantities in the kinetic equation. The scaled linear Boltzmann-BGK equation then becomes

$$\partial_t f(\mathbf{x},\mathbf{v},t) + \mathbf{v}\cdot\nabla_{\mathbf{x}} f(\mathbf{x},\mathbf{v},t) = \varepsilon^{-1}\tilde{R}(x,t)\left(\mathcal{M}(\mathbf{x},\mathbf{v},t)\rho(\mathbf{x},t) - f(\mathbf{x},\mathbf{v},t)\right).$$

This scaling is often referred to as the *hydrodynamic scaling*.

Another scaling that is frequently encountered in the literature is the *diffusive scaling*. In the diffusive scaling, the assumption is that the collision rate and the variance on the velocity are large and can be normalized as $R(\mathbf{x},t) = \varepsilon^{-2}\tilde{R}(\mathbf{x},t)$ and $\sigma^2(\mathbf{x},t) = \varepsilon^{-2}\tilde{\sigma}^2(\mathbf{x},t)$. As the variance scales as the velocity squared, the velocity should be normalized as $\mathbf{v} = \varepsilon^{-1}\tilde{\mathbf{v}}$. All other quantities are again assumed to be $O(1)$. The diffusive scaling linear Boltzmann-BGK equation is given by

$$\partial_t f(\mathbf{x},\tilde{\mathbf{v}},t)+\varepsilon^{-1}\tilde{\mathbf{v}}\cdot\nabla_{\mathbf{x}} f(\mathbf{x},\tilde{\mathbf{v}},t)=\varepsilon^{-2}\tilde{R}(\mathbf{x},t)\left(\mathcal{M}(\mathbf{x},\tilde{\mathbf{v}},t)\rho(\mathbf{x},t)-f(\mathbf{x},\tilde{\mathbf{v}},t)\right).$$

The Maxwellian distribution also contains a scaling parameter due to the scaling of the velocity and the variance on the velocity:

$$\mathcal{M}(\mathbf{x},\tilde{\mathbf{v}},t)=\frac{1}{(2\pi\varepsilon^{-2}\tilde{\sigma}_p^2(\mathbf{x},t))^{3/2}}\exp\left(-\frac{1}{2}\frac{(\varepsilon^{-1}\tilde{\mathbf{v}}-\mathbf{u}_p(\mathbf{x},t))^2}{\varepsilon^{-2}\tilde{\sigma}_p^2(\mathbf{x},t)}\right).$$

In both the hydrodynamic and the diffusive scaling, decreasing values of ε mean that collisions become increasingly frequent, introducing stiffness into the equation.

2.2.2 A Mathematical View on Scale Separation

An alternative view on the involved space and time-scale separations arises by rescaling space and time. To make the timescale separation explicit, we introduce again the small parameter ε, which we now use to rescale space and time as

$$\tilde{\mathbf{x}}=\varepsilon\mathbf{x},\qquad \tilde{t}=\varepsilon^{1+\eta}t,$$

with the parameter η taking values in the range $[0,1]$. Here, $\tilde{\mathbf{x}}$ and $\tilde{t}$ correspond with the macroscopic scale. Rescaling space in this way, we can interpret ε as the ratio between two length scales, the mean free path between collisions, and the domain size. Hence, when working with the rescaled variables, we have "zoomed out" sufficiently that we only see the "slower" dynamics of the macroscopic variables (3)–(5), while $\mathbf{x}$ and t represent a scale where we see particle collision dynamics. The parameter η allows distinguishing between the hydrodynamic and diffusive scaling introduced before.

2.3 Macroscopic Limits in Hydrodynamic and Diffusive Scaling

We consider the rescaled kinetic equation

$$\partial_t f(\mathbf{x},\mathbf{v},t)+\frac{1}{\varepsilon^{\eta}}\mathbf{v}\cdot\nabla_{\mathbf{x}} f(\mathbf{x},\mathbf{v},t)=\frac{R(\mathbf{x},t)}{\varepsilon^{1+\eta}}\left(\mathcal{M}(\mathbf{x},\mathbf{v},t)\rho(\mathbf{x},t)-f(\mathbf{x},\mathbf{v},t)\right),\tag{8}$$

where $\varepsilon>0$ is a dimensionless parameter, $\eta\geq 0$ controls the relative scaling between transport and collisions, and all tildes have been removed for notational simplicity.

Let us assume in the present derivation that the Maxwellian $\mathcal{M}(\mathbf{x},\mathbf{v},t)$ is independent of t and satisfies

$$\langle\mathcal{M}(\mathbf{x},\mathbf{v})\rangle=1,\qquad \langle\mathbf{v}\,\mathcal{M}(\mathbf{x},\mathbf{v})\rangle=\varepsilon^{\eta}\mathbf{u}(\mathbf{x}),\tag{9}$$

where the last product needs to be understood component-wise. We thus assume the mean velocity to be scaled as ε^{η}.

To identify the macroscopic behaviour as $\varepsilon \to 0$, we perform a Chapman–Enskog expansion of f around the local equilibrium $\mathcal{M}\rho$,

$$f(\mathbf{x},\mathbf{v},t) = \mathcal{M}(\mathbf{x},\mathbf{v})\rho(\mathbf{x},t) + \varepsilon g_\varepsilon(\mathbf{x},\mathbf{v},t), \tag{10}$$

subject to the solvability condition $\langle g_\varepsilon \rangle = 0$. Substituting (10) into (8) yields

$$\partial_t\left(\mathcal{M}\rho + \varepsilon g_\varepsilon\right) + \frac{1}{\varepsilon^\eta}\mathbf{v}\cdot\nabla_\mathbf{x}\left(\mathcal{M}\rho + \varepsilon g_\varepsilon\right) = -\frac{R}{\varepsilon^\eta} g_\varepsilon.$$

Then, taking the mean over velocity space and using (9), we obtain

$$\partial_t\rho + \nabla\cdot(\mathbf{u}\rho) + \varepsilon^{1-\eta}\nabla\cdot\langle \mathbf{v} g_\varepsilon\rangle = 0.$$

Considering the terms of order $\mathcal{O}(\varepsilon^{-\eta})$, we obtain

$$g_\varepsilon = -\frac{1}{R}\mathbf{v}\cdot\nabla\left(\mathcal{M}(\mathbf{x},\mathbf{v})\rho(\mathbf{x},t)\right),$$

giving rise to

$$\partial_t\rho + \nabla\cdot(\mathbf{u}\rho) = \varepsilon^{1-\eta}\,\nabla\cdot\left(\frac{1}{R}\left\langle \mathbf{v}\otimes\mathbf{v}\nabla(\mathcal{M}\rho)\right\rangle\right),$$

or, assuming $\mathcal{M}$ to be independent of $\mathbf{x}$ and defining the diffusion matrix $\mathbf{D} = \langle \mathbf{v}\otimes\mathbf{v}\mathcal{M}\rangle$,

$$\partial_t\rho + \nabla\cdot(\mathbf{u}\rho) = \varepsilon^{1-\eta}\,\nabla\cdot\left(\frac{1}{R}\mathbf{D}\nabla\rho\right). \tag{11}$$

The macroscopic limit (11) depends on the scaling parameter η. In the hydrodynamic scaling ($\eta = 0$), the diffusion term disappears. Hence, the leading-order behaviour is governed by advection. In the diffusive scaling ($\eta = 1$), transport and collisions balance in the limit of vanishing ε and the limit $\varepsilon \to 0$ yields a drift–diffusion equation. In the remainder of this text, we focus on the diffusive scaling $\eta = 1$ and the associated drift–diffusion limit. For further discussion and derivations of such macroscopic limits we refer to texts such as [3, 4, 18, 122].

2.4 Stochastic Particle Discretizations and Basic Monte Carlo Methods

The aim of this section is to use particle schemes to simulate the diffusively scaled linear Boltzmann-BGK equation

$$\partial_t f(\mathbf{x},\mathbf{v},t) + \frac{\mathbf{v}}{\varepsilon}\cdot\nabla_\mathbf{x} f(\mathbf{x},\mathbf{v},t) = \frac{1}{\varepsilon^2}\left(\mathcal{M}(\mathbf{v})\rho(\mathbf{x},t) - f(\mathbf{x},\mathbf{v},t)\right), \tag{12}$$

i.e., (8) with $\eta = 1$. For convenience, we here assume $R(\mathbf{x},t) \equiv 1$. We consider two approaches, which both fall under the category of direct simulation Monte Carlo methods [8]: (i) an event driven approach in Section 2.4.1, where we mimic actual physics of the particles; and (ii) a fixed-size time step approach in Section 2.4.2, which introduces a simulation bias but reduces complexity. Both approaches suffer from time step restrictions that lead to unacceptably high simulation costs in the limit $\varepsilon \to 0$, despite the limit itself being well defined [37].

2.4.1 Event Driven Monte Carlo Simulation

Each particle trajectory is fully defined at each moment in time t by its position $X(t)$ and velocity $V(t)$. We represent the distribution of particles by an ensemble of P particles, with indices $p \in \{1,\ldots,P\}$ as

$$\left\{(X_p(t), V_p(t))\right\}_{p=1}^{P}.$$

In the absence of external forces, we can model an individual particle's trajectory as a velocity-jump process, i.e., a particle with index p undergoes a series of collisions at moments in time t_p^n, $n = 0\ldots N$, and has a fixed velocity V_p^n for $t_p^n \leq t < t_p^{n+1}$ which determines its direction and speed of travel [83]. The time between collisions is exponentially distributed with expected time between collisions ε^2, i.e.,

$$t_p^{n+1} - t_p^n = \tau_p^n \sim \mathcal{E}\left(\frac{1}{\varepsilon^2}\right). \tag{13}$$

We therefore have the following velocity-jump stochastic process:

1. **Free flight.** Given the particle's position X_p^n and velocity V_p^n, compute the particle's position at the next collision as

$$X_p^{n+1} = X_p^n + \tau_p^n \frac{V_p^n}{\varepsilon}, \quad \tau_p^n \sim \mathcal{E}\left(\frac{1}{\varepsilon^2}\right).$$

2. **Collision.** Sample a new velocity $V_p^{n+1} \sim \mathcal{M}\left(\mathbf{v}; X_p^{n+1}\right)$.

With this scheme, the collisions are distributed exactly as the particle dynamics prescribes. As a consequence, this approach is bias-free. The main drawback is the fact that the time step size cannot be selected, and is different in each time step. Given that the scheme resolves all collisions, the computational cost of this method increases drastically when ε tends to 0.

2.4.2 Fixed-Size Time Step Monte Carlo Simulation

We consider the simulation of (12) with particle schemes using a fixed finite time step of size Δt. At each time step n we thus have an ensemble of P particles

$$\left\{\left(X_{p,\Delta t}^n, V_{p,\Delta t}^n\right)\right\}_{p=1}^{P}, \tag{14}$$

where we use the subscript Δt to denote the fixed time step discretization. We now take an operator splitting approach, which is first order in the time step Δt [119], to simulate the ensemble (14). For (12), operator splitting results in two actions for each time step:

1. **Transport step.** Each particle's position is updated based on its velocity

$$X^{n+1}_{p,\Delta t} = X^n_{p,\Delta t} + \Delta t \frac{V^n_{p,\Delta t}}{\varepsilon}. \tag{15}$$

2. **Collision step.** Between kinetic steps, each particle's velocity is either left unchanged (no collision) or re-sampled from $\mathcal{M}\left(\mathbf{v}; X^{n+1}_p\right)$ (collision), i.e., given a $V^{n,\prime}_{p,\Delta t} \sim \mathcal{M}\left(\mathbf{v}; X^{n+1}_p\right)$,

$$V^{n+1}_{p,\Delta t} = \begin{cases} V^{n,\prime}_{p,\Delta t}, & \text{with probability } p_{c,\Delta t} = 1 - \exp\left(-\frac{\Delta t}{\varepsilon^2}\right), \\ V^n_{p,\Delta t}, & \text{otherwise.} \end{cases} \tag{16}$$

This scheme is able to perform simulations with a fixed time step size, but introduces a simulation bias that scales linearly with Δt for sufficiently small $\Delta t \ll \varepsilon^2$. However, we are still forced to take small time steps for two reasons: particle velocities scale as ε^{-1}, and the probability of collision increases quickly when ε decreases.

3 Deterministic Time Integration and Multiscale Methods

In this review, we are interested in the time discretization of linear kinetic equations of the form (8) in the limit of infinite collisionality, when ε tends to zero and a limiting macroscopic advection-diffusion equation emerges of the form (11). In this section, we overview two approaches to design computationally efficient deterministic time integration methods. In Section 3.1, we discuss asymptotic-preserving methods, algorithms that are stable in ε and converge to a scheme for the limiting macroscopic equation when ε tends to 0. In Section 3.2, we consider approaches that aim at simulating directly the time evolution of the macroscopic state variables that appear in the limiting equation, using localized bursts of simulation with the kinetic equation.

3.1 Deterministic Asymptotic-Preserving Schemes

There is currently a large research effort in the design of algorithms for (8) that are stable in ε and converge to a scheme for the limiting equation (11) when ε tends to 0; such schemes are called asymptotic-preserving in the sense of Jin [64].

In [64, 70], separating the distribution function f into its odd and even parts in the velocity variable results in a coupled system of transport equations where the stiffness appears only in the source term, allowing to use a time-splitting technique with implicit treatment of the source term; see also related work in [64, 77, 76]. Since then, many similar schemes have also been developed, with various goals such as being well-balanced in the rarefied regime, i.e., $\varepsilon \gg 0$ [52] and capturing stiffness in the relaxation term of the limiting advection-diffusion equation [68]. This approach has also been combined with more complex numerical methods such as Runge–Kutta schemes [11], discontinuous Galerkin [120, 121], polynomial chaos [67] and collision operator decomposition [70]. In addition, similar schemes have also been developed in the hydrodynamic limit [5, 34, 35], including a recent scheme for mixtures [9].

Another approach to developing asymptotic-preserving schemes is to apply operator splitting. In [69], a splitting approach was applied to a diffusively scaled kinetic equation. An alternative approach is to first decompose the kinetic equation into a system of equations on a pair $\{f_0(\mathbf{x},\mathbf{v},t), f_1(\mathbf{x},\mathbf{v},t)\}$, derived from a truncated expansion

$$f(\mathbf{x},\mathbf{v},t) = f_0(\mathbf{x},\mathbf{v},t) + \varepsilon f_1(\mathbf{x},\mathbf{v},t) + O\left(\varepsilon^2\right),$$

which can then be split into slow and fast components. This approach has been taken in both the diffusive limit [75, 77] and the Navier Stokes limit [76]. In [89], a similar splitting approach was applied to the diffusive limit of a kinetic equation under the micro-macro decomposition (10). Schemes in both diffusive and Euler limits have been derived for the Vlasov equations in [27].

Recent work has also considered the macroscopic limit as a low-dimensional subspace in the space of possible solutions of the kinetic equation. Based on this view, one can then show that the solution of a kinetic equation close to the macroscopic limit can be captured by a low-rank approximation [41, 40]. Another recent line of work, the unified gas kinetic schemes [134, 135], consider finite volume solvers with scale-adaptive fluxes based on the analytical solution of the kinetic equation along characteristic lines. For a more extensive introduction to asymptotic-preserving schemes for kinetic equations, we refer to the review articles [65, 36]. Recently, another review article was published as [66].

3.2 Projective Integration Methods

A robust and fully explicit method, which allows for time integration of (two-scale) stiff systems with arbitrary order of accuracy in time, is projective integration. Projective integration was proposed in [46] for stiff systems of ordinary differential equations with a clear gap in their eigenvalue spectrum. In such stiff problems, the fast modes, corresponding to the Jacobian eigenvalues with large negative real parts, decay quickly, whereas the slow modes correspond to eigenvalues of smaller magnitude and are the solution components of practical interest. Projective integration allows a stable yet explicit integration of such problems by first taking a few small (inner) steps using a step size δt with a simple, explicit method, until the transients corresponding to the fast modes have died out, and subsequently projecting (extrapolating) the solution forward in time over a large (outer) time step of size $\Delta t \gg \delta t$. In [81], projective integration was analyzed for kinetic equations with a diffusive scaling. An arbitrary order version, based on Runge-Kutta methods, has been proposed recently in [79], where it was also analyzed for kinetic equations with an advection-diffusion limit. In [80], the scheme was used to construct an explicit, flexible, arbitrary order method for general nonlinear hyperbolic conservation laws, based on relaxation to a kinetic equation. Alternative approaches to obtain a higher-order projective integration scheme have been proposed in [88, 126]. These methods fit within research efforts on generic numerical methods for multiscale simulation [39, 73].

For problems exhibiting more than a single fast time scale, telescopic projective integration was proposed [47]. In these methods, the projective integration idea is applied recursively. Starting from an inner integrator at the fastest time scale, a projective integration method is constructed with a time step that corresponds to the second-fastest time scale. This projective integration method is then considered as the inner integrator of a projective integration method on yet a coarser level. By repeating this idea, telescopic projective integration methods construct a hierarchy of projective levels in which each outer integrator step on a certain level serves as an inner integrator step one level higher. The idea was studied and tested for linear kinetic equations in [101]. These methods turn out to have a computational cost that is essentially independent of the stiffness of the collision operator. A generalization of projective and telescopic projective integration for nonlinear Boltzmann-BGK and Boltzmann kinetic equations was presented in [100]. Those methods are of arbitrary order in time, fully explicit, and general (they do not exploit any particular form of the collision operator).

We do not call projective integration methods asymptotic-preserving as such, because we cannot explicitly evaluate the scheme for $\varepsilon = 0$ to obtain a classical numerical scheme for the limiting equation. Nevertheless, projective and telescopic projective integration methods share important features with asymptotic-preserving methods. In particular, their computational cost does (in many cases) not depend on the stiffness of the problem. To be specific, it was shown in [101], for linear kinetic equations, that the number of inner time steps at each level of the telescopic hierarchy is independent of the small-scale parameter ε, as is the step size of the outermost integrator. The only parameter in the method that may depend on ε is the *number* of levels in the telescopic hierarchy. For systems in which the spectrum of the collision operator falls apart into a set of clearly separated clusters (each corresponding to a specific time scale), the number of levels equals the number of spectral clusters. In this situation, the computational cost is completely independent of the scaling parameter ε. When the collision operator represents a continuum of time scales, the number of projective integration levels increases logarithmically with ε [46].

4 Hybrid Fluid–Kinetic Schemes

An alternative class of methods are hybrid methods. Such methods are particularly useful when the collision rate varies heavily in space and time. The collision rate might be low in some portions of the domain, making Monte Carlo simulation a cheap and reliable choice for performing the simulations. In other parts of the domain, however, the collision rate might become extremely high, making Monte Carlo computations computationally prohibitive. At the same time, with increasing collision rate, the macroscopic limiting equation becomes more accurate. The goal of hybrid fluid-kinetic schemes is to obtain a method that dynamically switches to a particle-based Monte Carlo solver in low-collisional regions, where a detailed kinetic description is required, and falls back on a fluid-based solver, e.g., a finite volume or finite element code, in high-collisional regions, where the macroscopic fluid equations are accurate. Below, we briefly introduce three hybrid fluid-kinetic schemes that aim to achieve this: spatially hybrid models (in Section 4.1), δf or low-variance methods (in Section 4.2), and micro-macro decomposition methods (in Section 4.3). We also mention that the UGKS methods discussed in Section 3.1 have recently been extended to hybrid methods such as the unified gas-kinetic wave-particle (UGKWP) method [92, 135] and the simplified unified wave-particle (SUWP) method [93].

4.1 Spatially Hybrid (Domain-Decomposition) Models

4.1.1 Subdivision Strategies

Spatially hybrid methods decompose the domain in subdomains where an expensive kinetic treatment is required and subdomains where a reduced-order macroscopic model is valid. This decomposition is typically in the position $\mathbf{x}$, but there are also works about domain decomposition in the velocity variable $\mathbf{v}$ [25]. The idea of a domain decomposition method is to subdivide the domain and to solve the problem in each subdomain with the cheapest model that has an acceptable accuracy. The method's performance therefore strongly depends on the position of the interfaces between the subdomain, which might move in transient simulations. There are three main approaches.

First, we have the domain decomposition methods with sharp interfaces [62, 32, 25, 2, 12, 130, 74, 10, 86, 20] where the interfaces can be interpreted as internal boundary conditions that couple the dynamics in the different subdomains. Such internal boundary conditions, or interface conditions, typically require that the state variables (the fluid moments) or the fluxes are matched.

The second approach uses a buffer region instead of a sharp interface [30, 28, 132]. The buffer zone allows for a smooth transition between different subdomains. In a buffer zone between two subdomains, a transition function interpolates between the solutions of the two different models used in the subdomains, i.e., both solution representations exist in the buffer zone and are combined as a weighted average. The advantage over sharp interface methods is that the transition is smoother and does not require boundary (matching) conditions, avoiding jumps in the solution at the interface.

The third approach is a cell-based domain decomposition [43, 45, 82, 90, 38, 115, 110, 31], which is a special case of the two previously discussed approaches. The idea is to define a grid on the simulation domain and decide for each grid cell if it belongs to the kinetic or the fluid subdomain. This decision is made based on some so-called regime detection criterion. If there is an overlapping region, then the cell-based domain decomposition is a special case of a domain decomposition method with a buffer region.

If there is no overlapping region, then the sharp interface is defined by the grid cell edges between cells that belong to a different subdomain.

Some work on implementations that take care of the coupling between the different subdomains can be found in [109, 108, 63]. Desired properties of such implementations include a decoupling of the solvers of the different subdomains, modularization of the coupling, efficiency and parallelizability of the coupled code, and flexibility to operate with different subdomain solvers and coupling strategies such as flux-based or state variable-based couplings.

4.1.2 Solver Choice in Each Subdomain

The different subdomains have different types of solvers. The main division that has to be made is the division between particle-particle, deterministic-deterministic and particle-deterministic. In particle-particle, the subdomains are all treated with a Monte Carlo code, but the equations governing the particle behaviour change when crossing an interface [32]. Because particles reside at both sides of the interface, the coupling is typically straightforward. In deterministic-deterministic, one typically has a high-dimensional grid at one side of the interface and a low-dimensional grid at the other side of the interface [43, 25, 30, 28, 2, 74, 82, 86, 90]. The solution at the interface then has to be transferred between the two grids by lifting and interpolation operations in order to exchange information over the interface.

The most difficult coupling follows from particle-deterministic interfaces, where there are particles at one side of the interface and a grid-based solution at the other side of the interface such as in [20], the AMAR method [45], the MPC method [12, 130], the method of Valentinuzzi where the transition in the buffer zone occurs using artificially introduced evaporation and condensation processes [132], the method of Blommaert which uses a sharp interface with flux matching coupling conditions [10], and the overlapping region based methods [109, 108, 63, 38, 115, 110, 31]. One then has to discard particles when they move from the particle to the deterministic subdomain and one has to sample particles for fluxes going from the deterministic to the particle subdomain. An extensive overview of this kind of domain decomposition methods is given by Boyd and Deschenes [12]. Note that for this kind of domain decomposition methods, noise from the particle subdomain gets transported to the deterministic subdomain, potentially leading to stability issues.

4.2 The δf Method and Low-Variance Particle Methods

The δf method [1, 33, 13] and low-variance methods aim at reducing the variance of Monte Carlo methods, such that we can simulate a lower number of particles without losing accuracy. The class of δf methods, sometimes also called low-variance deviational simulation Monte Carlo (LVDSMC) methods [123, 124] do this by splitting the distribution in an equilibrium part and a deviational part

$$f(\mathbf{x}, \mathbf{v}, t) = f_{eq}(\mathbf{x}, \mathbf{v}) + f_d(\mathbf{x}, \mathbf{v}, t),$$

where the equilibrium part is chosen stationary. These methods are the simulation equivalent of what is called a control variate in Monte Carlo integration [1]. To get a significant variance reduction, the distribution $f_{eq}(\mathbf{x}, \mathbf{v})$ should be a good approximation of the actual solution $f(\mathbf{x}, \mathbf{v}, t)$.

For kinetic equations in high-collisional regimes, the equilibrium with the background medium provides a good choice for $f_{eq}(\mathbf{x}, \mathbf{v})$. Alternatively, a $f_{eq}(\mathbf{x}, \mathbf{v})$ can be obtained by running a cheap simulation (possibly with a reduced-order model) and using that solution. For transient simulations, the equilibrium part $f_{eq}(\mathbf{x}, \mathbf{v})$ can be a good approximation at some times, but a bad approximation at other times. Therefore, some methods allow to update the equilibrium part from time to time using an update rule [124]. The deviational part $f_d(\mathbf{x}, \mathbf{v}, t)$ is simulated using a Monte Carlo solver. As only the deviational part is simulated, there can only be noise on that part of the solution, so provided that the stationary part (the control variate) is a good approximation of the solution such that the deviational part is small, the variance should be reduced.

4.3 Micro–Macro Decompositions and Moment Closures

Micro-Macro decompositions [23, 26, 24, 21, 22, 57, 58, 59] are an extension of the δf-method, where the equilibrium part is also allowed to be time dependent and follows from the solution of a macroscopic equation. The macroscopic evolution equations are chosen such that they govern the K lowest-order moments of $f(\mathbf{x}, \mathbf{v}, t)$, such that we have

$$\int \mathbf{v}^k f_{eq}(\mathbf{x}, \mathbf{v}, t) d\mathbf{v} = \int \mathbf{v}^k f(\mathbf{x}, \mathbf{v}, t) d\mathbf{v}, \quad k = 0..K-1$$
$$\int \mathbf{v}^k f_d(\mathbf{x}, \mathbf{v}, t) d\mathbf{v} = 0, \quad k = 0..K-1.$$

The deviational part is ideally simulated using a Monte Carlo solver, as for the δf methods; however in some works a grid-based solver is used for the deviational part.

Note that different hybrid fluid-kinetic schemes can be combined. For example, in [28, 56, 82] micro-macro decomposition is combined with domain decomposition.

5 Asymptotic-Preserving Monte Carlo Schemes

In previous sections, we introduced the concept of deterministic asymptotic-preserving schemes for kinetic equations, as well as hybrid methods that combine deterministic solvers for a limiting model with a stochastic solver for a kinetic equation. In this section, we discuss an approach of performing hybrid simulations, fully at the Monte Carlo level using stochastic asymptotic-preserving schemes.

The majority of stochastic asymptotic-preserving schemes in the literature for kinetic equations have considered fluid models as a limiting equation. A prototypical example is so-called time-relaxed Monte Carlo methods, first proposed in [44], generalized in [117, 118] and applied to a 2D fluid simulation in [116]. However, we focus on an alternative asymptotic-preserving Monte Carlo method that, instead of using a fluid limit, uses the diffusive scaling and considers asymptotic convergence at the level of individual trajectories, as sketched in Figure 1.

At this level, the convergence in the diffusive scaling typically takes the form of showing that a velocity-jump process, modelling individual collision events of a given kinetic trajectory, converges to a Brownian motion. A formal derivation of this fact can be found in, e.g., [83]. Intuitively, one can see this from the fact that when the collision rate becomes infinitely large, one sums an infinite number of independently sampled velocities when a particle moves for a finite amount of time. Suitably scaled, this sum will then converge to a normally distributed increment through the law of large numbers, corresponding to a discretized Brownian motion.

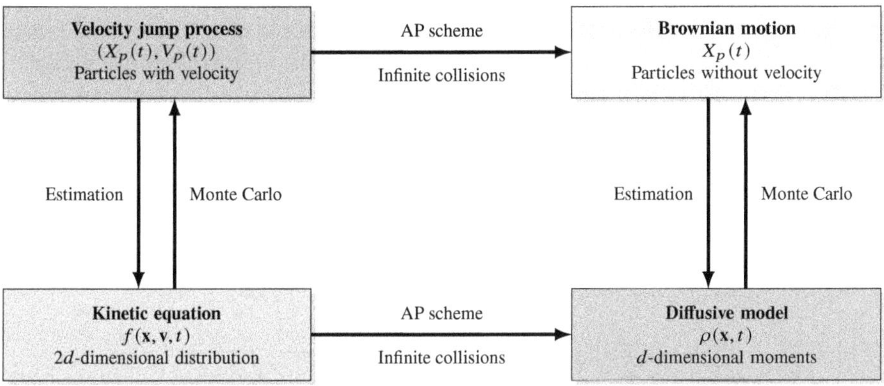

Figure 1 Diagram showing the relationship between Monte Carlo discretizations and asymptotic-preserving schemes in the diffusive limit. Asymptotic-preserving schemes allow us to capture the dynamics of the models on the right of the diagram as the infinite collision rate limit of the models on the left of the diagram.

As model problem, we start from equation (12), in which we further assume isotropic diffusion such that the diffusion tensor $\mathbf{D}$ can be replaced by a scalar D. In this section, we discuss two approaches to construct an asymptotic-preserving particle method in the diffusive limit. In Section 5.1, we first derive an asymptotic-preserving Monte Carlo scheme as a Monte Carlo discretization of the modified equation

$$\partial_t f + \frac{\varepsilon \mathbf{v}}{\varepsilon^2 + \Delta t} \cdot \nabla_{\mathbf{x}} f = \frac{D \Delta t}{\varepsilon^2 + \Delta t} \cdot \Delta_{\mathbf{x}} f + \frac{1}{\varepsilon^2 + \Delta t} (\mathcal{M}(\mathbf{v})\rho - f), \tag{17}$$

which can be obtained by constructing an IMEX scheme and replacing the finite difference in time with the time derivative while maintaining all other occurrences of Δt, see [37]. As a consequence, the coefficients of (17) explicitly contain the simulation time step Δt. Next, in Section 5.2, we present an improved scheme called kinetic-diffusion Monte Carlo [103] that follows directly from an interpretation of the modified equation at the Monte Carlo level. We further note that a hybrid scheme was developed using a similar approach in [24].

5.1 A First Asymptotic-Preserving Particle Method for the Diffusive Limit

Equation (17) has the following properties [37]:

1. For $\varepsilon^2/\Delta t \to 0$, (17) converges to the diffusion equation (11) with rate $O(\varepsilon^2/\Delta t)$.
2. For $\Delta t/\varepsilon^2 \to 0$, (17) converges to the kinetic equation (12) with rate $O(\Delta t/\varepsilon^2)$.

We can interpret the modified equation as a combination of the diffusion equation (11) and the original kinetic equation (12), with each contribution weighted in function of Δt. For large time steps, diffusion dominates over kinetic effects, whereas for small time steps, kinetic effects are dominant. When fixing Δt, the first property leads to the modified kinetic equation having the same diffusive limit in ε as the original kinetic equation. When fixing ε and decreasing Δt, we recover the original kinetic equation.

At the particle level, we can make the same observations, since diffusion corresponds to a Brownian motion and the normally distributed random increments arise naturally from the sum of independent free path increments using the central limit theorem.

Particle trajectories corresponding with (17) are generated as follows:

1. **Transport-diffusion step.** Update the particle given its velocity $V^n_{p,\Delta t}$ and a Brownian increment with diffusion coefficient $D_{\Delta t} = \dfrac{D\Delta t}{\varepsilon^2 + \Delta t}$, i.e.,

$$X^{n+1}_{p,\Delta t} = X^n_{p,\Delta t} + \Delta t \frac{\varepsilon}{\varepsilon^2 + \Delta t} V^n_{p,\Delta t} + \sqrt{2\Delta t}\sqrt{D_{\Delta t}}\,\xi^n_p, \tag{18}$$

   in which we generate $\xi^n_p \sim \mathcal{N}(0, I)$.

2. **Collision step.** Perform a collision with probability $p_{c,\Delta t} = \dfrac{\Delta t}{\varepsilon^2 + \Delta t}$ by sampling $\mathcal{M}_{\Delta t}(v)$, i.e.,

$$V^{n+1}_{p,\Delta t} = \begin{cases} V^{n,\prime}_{p,\Delta t}, & \text{with probability } p_{c,\Delta t}, \\ V^n_{p,\Delta t}, & \text{otherwise (with probability } p_{nc,\Delta t} = 1 - p_{c,\Delta t}), \end{cases} \tag{19}$$

   where $V^{n,\prime}_{p,\Delta t} \sim \mathcal{M}(\mathbf{v}; X^{n+1}_p)$.

We present the scheme (18)–(19) as Algorithm 1. This algorithm generates stable trajectories for $\Delta t \gg \varepsilon^2$. However, these trajectories have a bias that scales linearly in Δt, when compared to those generated by (15)–(16), see [37].

Algorithm 1: Simulating a particle until time t' with IMEX APMC. Note that we introduce the notation $p_{nc,\Delta t} = 1 - p_{c,\Delta t}$.

$t \leftarrow 0, \quad \left\{X^0_{p,\Delta t}, V^0_{p,\Delta t}\right\} \sim f(\mathbf{x}, \mathbf{v}, 0)$
while $t < t'$ **do**
    $\xi^n_p \sim \mathcal{N}(0,1)$
    $X^{n+1}_{p,\Delta t} \leftarrow$ (18) {Transport-diffusion step}
    $\chi^n_p \sim \mathcal{U}([0,1])$
    if $\chi^n_p \geq p_{nc,\Delta t}$ **then** {Collision}
        $V^{n,\prime}_{p,\Delta t} \sim \mathcal{M}_{\Delta t}(\mathbf{v})$
        $V^{n+1}_{p,\Delta t} \leftarrow V^{n,\prime}_{p,\Delta t}$
    else {No collision}
        $V^{n+1}_{p,\Delta t} \leftarrow V^{n}_{p,\Delta t}$
    end if
    $n \leftarrow n+1, \quad t \leftarrow t + \Delta t$
end while

5.2 The Kinetic-Diffusion Monte Carlo (KDMC) Scheme

5.2.1 Definition of the KDMC Algorithm

In [105], the scheme presented in Algorithm 1 was adapted to an event-driven Monte Carlo method using free flight times between collisions as produced by (13). The general idea behind this scheme is to sample the exponentially distributed time τ^n_p until the next collision and then to fill up the remainder of a given time Δt with diffusive behavior to produce an asymptotic-preserving scheme.

The simulation strategy works as follows. First, an exponentially distributed free-flight time τ^n_p is sampled for particle p at time step n. After comparing τ^n_p with Δt one of two cases applies:

1. **Kinetic step.** If the remaining time τ^n_p is larger than Δt, a fully kinetic step is performed, i.e.,
$$X^{n+1}_{p,\Delta t} = X^n_{p,\Delta t} + \Delta t \frac{V^n_{p,\Delta t}}{\varepsilon},$$
and the velocity is kept for the next time step, i.e.,
$$V^{n+1}_{p,\Delta t} = V^n_{p,\Delta t}.$$
2. **Kinetic-diffusion step.** If the time to the next collision $\tau^n_p < \Delta t$, a time step with combined kinetic and diffusive behavior is performed, i.e.,
$$X^{n+1}_{p,\Delta t} = X^n_{p,\Delta t} + \tau^n_p \frac{V^n_{p,\Delta t}}{\varepsilon} + \Delta W,$$
where ΔW is a normally distributed increment for time $\Delta t_W = \Delta t - \tau^n_p$, conditioned on the next time step's velocity $V^{n+1}_{p,\Delta t} = \frac{1}{\varepsilon} V^{n,\prime}_{p,\Delta t}$, where $V^{n,\prime}_{p,\Delta t} \sim \mathcal{M}(\mathbf{v}; X^{n+1}_p)$. This

correlation is achieved by performing the diffusive step with mean

$$\mathbb{E}\left[\Delta W\right] = \mathbf{u}\Delta t_W + \left(V_{p,\Delta t}^{n+1} - \mathbf{u}\right)\varepsilon^2\left(1 - e^{-\Delta t_W/\varepsilon^2}\right), \tag{20}$$

and variance [85]

$$\begin{aligned}\mathbb{V}\left[\Delta W\right] = 2\tilde{\nu}\varepsilon^2&\left(2e^{-\Delta t_W/\varepsilon^2} + \frac{\Delta t_W}{\varepsilon^2} + \frac{\Delta t_W}{\varepsilon^2}e^{-\Delta t_W/\varepsilon^2} - 2\right)I\\ &+ \varepsilon^4\left(1 - \frac{2}{\varepsilon^2}\Delta t_W e^{-\Delta t_W/\varepsilon^2} - e^{-2\Delta t_W/\varepsilon^2}\right)\left(V_{p,\Delta t}^{n+1} - \mathbf{u}\right)^{\otimes 2},\end{aligned} \tag{21}$$

where the exponent $\otimes 2$ is used to denote the outer product of a vector with itself. Here, $\mathbf{u}$ is the mean flow velocity of the Maxwellian $\mathcal{M}(\mathbf{v})$, defined in (7).

This scheme is presented as Algorithm 2. As opposed to Algorithm 1, the KD scheme does not have a time-step dependent post-collisional distribution $\mathcal{M}(\mathbf{v})$. Note also that a new τ_p^n can be sampled in each time step as we assume that the time between subsequent collisions is exponentially distributed.

Algorithm 2: Simulating a particle until time t' with KD APMC.

$t \leftarrow 0,\quad \left\{X_{p,\Delta t}^0, V_{p,\Delta t}^0\right\} \sim f(\mathbf{x},\mathbf{v},0)$
while $t < t'$ **do**
  $\tau \sim \mathcal{E}\left(\frac{1}{\varepsilon^2}\right)$
  $\tau \leftarrow \min(\tau, \Delta t)$
  $X_{p,\Delta t}^{n,\prime} \leftarrow X_{p,\Delta t}^n + \tau\frac{1}{\varepsilon}V_{p,\Delta t}^n$
  if $\tau = \Delta t$ **then** {Kinetic step}
    $V_{p,\Delta t}^{n+1} \leftarrow V_{p,\Delta t}^n$
    $X_{p,\Delta t}^{n+1} \leftarrow X_{p,\Delta t}^{n,\prime}$
  else {Kinetic-diffusive step}
    $V_{p,\Delta t}^{n,\prime} \sim \mathcal{M}_{\Delta t}(\mathbf{v})$
    $V_{p,\Delta t}^{n+1} \leftarrow V_{p,\Delta t}^{n,\prime}$
    $\Delta t_W \leftarrow \Delta t - (\tau \bmod \Delta t)$
    $\Delta W \sim \mathcal{N}((20),(21))$
    $X_{p,\Delta t}^{n+1} \leftarrow X_{p,\Delta t}^{n,\prime} + \Delta W$
  end if
  $n \leftarrow n+1,\quad t \leftarrow t + \Delta t$
end while

5.2.2 Comparing KDMC with the First Asymptotic-Preserving Scheme

The key difference between the two schemes is as follows. The IMEX scheme simultaneously simulates kinetic and diffusive behavior in each time step, combining both contributions in a weighted fashion. The KD scheme alternates between kinetic and diffusive behavior, only simulating one of the two phenomena at a given moment in time. Although its implementation is more complex than the IMEX scheme, the KD scheme has two key advantages:

1. The KD scheme avoids performing an extra diffusive increment in each time step, when the time step size is smaller than the time to the next collision. This means that this scheme is more efficient when simulating with very small time steps, at least in terms of the number of floating point operations.
2. The KD scheme also achieves a better convergence rate by half an order in terms of decreasing Δt, due to the fact that collisions are no longer restricted to times which are a multiple of Δt. This improved rate was both proven and demonstrated numerically in [105].

As of yet, a full comparison of these schemes has not been done to assess whether these advantages result in more efficient computations. Note that the KD scheme has been derived for a more general set of velocity distributions as it does not require the distribution $\mathcal{M}(\mathbf{v})$ to be symmetric. While there is no reason to believe a non-symmetric variant of the IMEX scheme cannot be derived, the technical work of performing the derivation is yet to be done as the derivation in [37] does not generalize trivially.

5.2.3 Error and Convergence Analysis Across Kinetic–Diffusive Regimes

Based on the properties of (17), we expect the KDMC scheme to converge to a Monte Carlo scheme for the original kinetic equation (6) in the limit $\Delta t/\varepsilon^2 \to 0$, and to a diffusion equation (11) in the limit $\Delta t/\varepsilon^2 \to \infty$. In [104], these limits were studied in the 1D-setting with homogeneous parameters by considering the Wasserstein-1 distance between the particle distribution produced by a KDMC simulation and the corresponding kinetic simulation at a fixed simulation end time. The derived bounds in this work are, respectively, $O(\Delta t^{1.5})$ for $\Delta t \to 0$ [104], and $O(\varepsilon^3)$ for $\varepsilon \to 0$.

Recently, a more elaborate error analysis of KDMC was performed in the L2-norm for the moments of the particle distribution such as density, momentum and energy [131]. Here, both cases with homogeneous and with space-dependent model parameters were considered, theoretically and numerically, showing that one can expect respective convergence rates of $O(\Delta t)$ (in the diffusive limit) and $O(\varepsilon^2)$ (in the kinetic limit).

In [85], an initial numerical study was conducted of a two-dimensional implementation of KDMC in the Eiron code [84]. Here a simple test case was considered, where particles were simulated up to a fixed end time and a histogram was constructed. In the limit $\varepsilon \to 0$, the same convergence rate of $O(\varepsilon^2)$ was observed in the L_2-norm. In the limit $\Delta t \to 0$, a much faster rate $O(\Delta t^3)$ was observed. This faster rate is likely due to symmetries in the considered test-problem.

6 Multilevel and Multiscale Monte Carlo Time Integration

The asymptotic-preserving schemes introduced in the previous section perform well in the two limits $\varepsilon^2 \gg \Delta t$ and $\Delta t \gg \varepsilon^2$. However, when $\varepsilon^2 \approx \Delta t$, these schemes produce a noticeable bias without a corresponding speed-up. In this section, we discuss how asymptotic-preserving Monte Carlo schemes have been combined with the multilevel Monte Carlo method to eliminate this bias.

The section is structured as follows. In Section 6.1, we introduce multilevel Monte Carlo. Next, in Section 6.2, we sketch out the idea behind asymptotic-preserving multilevel Monte Carlo methods. These methods require the generation of correlated pairs of trajectories with differing discretization time step sizes Δt. In Section 6.3, we outline how such pairs can be generated for the two schemes introduced in Section 5.

6.1 The Multilevel Monte Carlo Method

We assume to be interested in a (time-dependent) quantity of interest

$$y(t) = \int_{\mathbb{X}} \int_{\mathbb{V}} F(\mathbf{x}, \mathbf{v}) f(\mathbf{x}, \mathbf{v}, t) d\mathbf{x} d\mathbf{v},$$

which is computed as the average of a scoring function $F(\mathbf{x}, \mathbf{v})$ over the particle distribution. Note that, in many applications including fusion energy, we are interested in space-dependent quantities (such as mass, momentum and energy sources). These can be included in the presented setup by incorporating the characteristic function of the elements of a finite element discretization in scoring function.

Estimating $y(t)$ using a standard Monte Carlo estimator is done as

$$\widehat{Y}(t^n) = \frac{1}{P} \sum_{p=1}^{P} F\left(X^n_{p,\Delta t}, V^n_{p,\Delta t}, n\Delta t\right), \quad t \approx n\Delta t \quad n \in \{0, \ldots, N\}. \tag{22}$$

The core idea behind the multilevel Monte Carlo method (MLMC) [49, 55] is to avoid the bias-variance trade-off, implied by the fixed cost budget $N \times P$ (number of time steps times number of particles), by combining estimators based on trajectories with different time step sizes. This is done by starting with a coarse time step size Δt_0. At this coarse level, we can cheaply simulate a large number of trajectories P_0, as the number of required time steps N_0 is small. This estimator has a large bias, but a low variance, and is given by

$$\widehat{Y}_0(t^n) = \frac{1}{P_0} \sum_{p=1}^{P_0} F\left(X^n_{p,\Delta t_0}, V^n_{p,\Delta t_0}, n\Delta t_0\right), \quad t \approx n\Delta t_0 \quad n \in \{0, \ldots, N_0\}. \tag{23}$$

The estimate $\widehat{Y}_0(t^n)$ is then refined using a sequence of L difference estimators at levels $\ell = 1, \ldots, L$. Each difference estimator uses an ensemble of P_ℓ particle pairs. Each particle pair consists of two coupled particles: a particle with a fine time step Δt_ℓ and a particle with a coarse time step $\Delta t_{\ell-1} = M\Delta t_\ell$, with the refinement factor M a positive integer. The difference estimators take the form

$$\widehat{Y}_\ell(t) = \frac{1}{P_\ell}\sum_{p=1}^{P_\ell}\left(F\left(X_{p,\Delta t_\ell}^{nM}, V_{p,\Delta t_\ell}^{nM}, nM\Delta t_\ell\right) - F\left(X_{p,\Delta t_{\ell-1}}^{n}, V_{p,\Delta t_{\ell-1}}^{n}, n\Delta t_{\ell-1}\right)\right),$$

$$t \approx n\Delta t_{\ell-1} = nM\Delta t_\ell, \quad n \in \{0, \ldots, N_{\ell-1}\}. \tag{24}$$

Remark 1 (Notation) Due to the bias of the time discretization method, the samples $F\left(X_{p,\Delta t_\ell}^{n}, V_{p,\Delta t_\ell}^{n}, n\Delta t_\ell\right)$ at each level are generated by a level-dependent random variable, which we denote as F_ℓ. The corresponding estimator is then given by

$$\widehat{F}_\ell(t) = \frac{1}{P_\ell}\sum_{p=1}^{P_\ell} F\left(X_{p,\Delta t_\ell}^{nM}, V_{p,\Delta t_\ell}^{nM}, nM\Delta t_\ell\right), \qquad t \approx n\Delta t_{\ell-1} = nM\Delta t_\ell,$$

such that we can write $\widehat{Y}_\ell(t) = \widehat{F}_\ell(t) - \widehat{F}_{\ell-1}(t)$.

The variance of the random variables in the estimator (24) can be decomposed as

$$\mathbb{V}\left[F_\ell - F_{\ell-1}\right] = \mathbb{V}\left[F_\ell\right] + \mathbb{V}\left[F_{\ell-1}\right] - 2\,\mathrm{Cov}\left(F_\ell, F_{\ell-1}\right).$$

Therefore, if we can ensure that particle pairs in the difference estimators are sufficiently correlated, then we expect that

$$\mathbb{V}\left[\widehat{F}_\ell - \widehat{F}_{\ell-1}\right] \leq \mathbb{V}\left[\widehat{F}_\ell\right].$$

As such, fewer particle simulations will be needed for a given variance when using the multilevel estimator (24), compared to a single level estimator using a time step size Δt_ℓ. We achieve correlation in the sampled quantity of interest by correlating simulation of each particle pair, which intuitively can be understood as an attempt to let two particles follow approximately the same trajectory for two different simulation accuracies.

One can interpret the difference estimator as using the fine simulation to estimate the bias in the coarse simulation. Given a sequence of levels $\ell \in \{0, \ldots, L\}$, with decreasing step sizes, and the corresponding estimators given by (23)–(24), the multilevel Monte Carlo estimator for the quantity of interest $Y(\mathbf{x}, t)$ is computed by the telescoping sum

$$\widehat{Y}(\mathbf{x}, t) = \sum_{\ell=0}^{L}\widehat{Y}_\ell(\mathbf{x}, t). \tag{25}$$

Due to linearity of the expectation, the expectation of (25) is the same as that of (22), with the finest time step $\Delta t = \Delta t_L$. Given a sufficiently quick reduction in the number of simulated (pairs of) trajectories P_ℓ as ℓ increases, it is possible to show that the multilevel Monte Carlo estimator is able to achieve the same mean square error as the classical Monte Carlo estimator at a lower computational cost.

For a detailed overview of the multilevel Monte Carlo method and its properties, we refer the reader to the survey paper [50] and the PhD thesis [127]. For an overview of recent developments in multilevel Monte Carlo methods specifically tailored to hyperbolic systems, we refer to [60].

6.2 Asymptotic-Preserving Multilevel Coupling (AP-MLMC)

We aim to correlate simulated particle trajectories whose time steps at levels ℓ and $\ell-1$ are related through $\Delta t_{\ell-1} = M\Delta t_\ell$. To this end, we define a sub-step index $m \in \{0,\ldots,M-1\}$ at the fine level ℓ, i.e., we write $X^{n,m}_{p,\Delta t_\ell} \approx X_p(n\Delta t_{\ell-1} + m\Delta t_\ell) \equiv X_p((nM+m)\Delta t_\ell)$. We now consider a coupled pair of simulations spanning a time step with size $\Delta t_{\ell-1}$: (i) a simulation at level $\ell-1$, using a single time step of size $\Delta t_{\ell-1}$ and (ii) a simulation at level ℓ, using M time steps of size Δt_ℓ:

$$\begin{cases} X^{n+1}_{p,\Delta t_{\ell-1}} = X^n_{p,\Delta t_{\ell-1}} + \mathcal{S}_{\Delta t_{\ell-1}}(X^n_{p,\Delta t_{\ell-1}}, V^n_{p,\Delta t_{\ell-1}}, \zeta^n_{p,\ell-1}). \\ X^{n+1,0}_{p,\Delta t_\ell} = X^{n,0}_{p,\Delta t_\ell} + \sum_{m=0}^{M-1} \mathcal{S}_{\Delta t_\ell}(X^{n,m}_{p,\Delta t_\ell}, V^{n,m}_{p,\Delta t_\ell}, \zeta^{n,m}_{p,\ell}). \end{cases}$$

Here, $\mathcal{S}_{\Delta t_{\ell-1}}$ and $\mathcal{S}_{\Delta t_\ell}$ represent the taking of a single forward time step with the given discrete numerical scheme, using a set of random numbers $\zeta^n_{p,\ell-1}$, respectively $\zeta^{n,m}_{p,\ell}$. It is these time steppers that use the random numbers we wish to correlate, such as post-collisional velocities and Brownian increments.

We aim to correlate $X^{N_\ell}_{p,\Delta t_\ell} = X^{N_{\ell-1},0}_{p,\Delta t_\ell}$ and $X^{N_{\ell-1}}_{p,\Delta t_{\ell-1}}$. To this end, we will not draw independent samples for values $\zeta^n_{p,\ell-1}$ at level $\ell-1$, but instead compute these as a function of the values $\zeta^{n,m}_{p,\ell}$ for $m = 0,\ldots,M-1$ in the fine simulation, i.e.,

$$\zeta^n_{p,\ell-1} = \varphi(\zeta^{n,0}_{p,\ell},\ldots,\zeta^{n,M-1}_{p,\ell}),$$

with a function φ that ensures that the $\zeta^n_{p,\ell-1}$ still follow the correct statistical distribution. Hence, to produce correlated trajectories, one first performs the M independent fine simulation steps and then computes the value $\zeta^n_{p,\ell-1}$ to produce the corresponding coarse simulation step.

6.3 Variance Reduction Through Correlated Time Steps

We are now ready to describe the concrete coupling strategies for the two asymptotic-preserving particle schemes of Section 5. In Section 6.3.1, we show how this can be achieved for the scheme described in Section 5.1. In Section 6.3.2, we discuss correlating paths of the kinetic-diffusion Monte Carlo scheme of Section 5.2.

6.3.1 Coupling for the First Asymptotic-Preserving Scheme

In each time step using the asymptotic-preserving particle scheme in Algorithm 1, there are two sources of stochastic behavior. First, a new Brownian increment ξ_p^n is generated for each particle in each transport-diffusion step (18). Second, in each collision step (19), a fraction of particles randomly gets a new velocity $V_p^{n,\prime}$ for use in the following time step. Particle trajectories can be coupled by correlating the random numbers used for the individual particles in the transport-diffusion and collision phase of each time step.

For the scheme in Algorithm 1, the correlation function φ consists of two parts.

1. **Course diffusive increments.** The diffusive increments in the coarse simulation are constructed as
$$\xi_{p,\ell-1}^n = \sqrt{\omega_\ell}\,\xi_{p,\ell-1,W}^n + \sqrt{1-\omega_\ell}\,\xi_{p,\ell-1,T}^n, \tag{26}$$
with $\omega_\ell \in [0,1]$ a level dependent weighting parameter,
$$\xi_{p,\ell-1,W}^n = \frac{1}{\sqrt{M}} \sum_{m=0}^{M-1} \xi_{p,\ell}^{n,m} \quad \text{and} \quad \xi_{p,\ell-1,T}^n = \left(\mathbb{V}\left[\sum_{m=0}^{M-1} \bar{V}_{p,\ell}^{n,m} \right] \right)^{-\frac{1}{2}} \sum_{m=0}^{M-1} \bar{V}_{p,\ell}^{n,m},$$
where the notation $\bar{V}$ denotes velocities that have been rescaled to have unit variance and the variance term is given by
$$\mathbb{V}\left[\sum_{m=0}^{M-1} \bar{V}_{p,\ell}^{n,m} \right] = M + 2\frac{p_{nc,\Delta t_\ell}\left(\left(p_{nc,\Delta t_\ell}\right)^M + M p_{c,\Delta t_\ell} - 1 \right)}{\left(p_{c,\Delta t_\ell}\right)^2}. \tag{27}$$
Originally, a coupling scheme was proposed in [98] in which $\omega_\ell = 1$, i.e., the particle velocities at the finer level did not influence the Brownian increments at the coarser level. However, in [97], it was later shown that the optimal value of ω_ℓ satisfies $0 < \omega_\ell < 1$. Intuitively, this can be understood by the fact that the coefficients in (17) depend on Δt, making the diffusive part of the increment (18) have relatively more effect when using coarser time steps. This increased effect corresponds to the fact that the IMEX scheme resolves fewer individual collision events for larger Δt, with more collisions being replaced by diffusion.
When the velocities $\bar{V}_{p,\ell}^{n,m}$ are not sampled from a Gaussian distribution, further corrections are needed to ensure consistency in the telescoping sum. For further details, we refer to [97].
2. **Coarse kinetic increments.** In the simulation at level ℓ, each of the M time steps of size Δt_ℓ contains a collision with probability $p_{c,\Delta t_\ell} = \Delta t_\ell/\varepsilon^2 + \Delta t_\ell$. The choice of whether to simulate a collision is made by sampling a random number $\chi_{p,\ell}^{n,m} \sim \mathcal{U}([0,1])$. A collision takes place if
$$\chi_{p,\ell}^{n,m} \geq p_{nc,\Delta t_\ell} = 1 - p_{c,\Delta t_\ell} = \frac{\varepsilon^2}{\varepsilon^2 + \Delta t_\ell},$$
i.e., a new velocity is randomly drawn from $\mathcal{M}_{\Delta t_\ell}$ at the end of that time step. The approach proposed in [98] applies a similar comparison

$$\left(\chi_{p,\ell}^{n,\max}\right)^M \geq p_{nc,\Delta t_{\ell-1}} = \frac{\varepsilon^2}{\varepsilon^2 + \Delta t_{\ell-1}},$$

in the correlated simulation at level $\ell - 1$ with

$$\chi_{p,\ell}^{n,\max} = \max_m \chi_{p,\ell}^{n,m}, \qquad m \in \{0,\dots,M-1\}.$$

If a collision takes place in the coarse simulation then the new velocity is chosen by setting $V_{p,\ell-1}^{n,\prime} = V_{p,\ell}^{n,M-1,\prime}$.

6.3.2 Kinetic-Diffusion Monte Carlo

We now outline the correlation strategy for the KDMC scheme in Algorithm 2 in the case of homogeneous collision rates.

1. **Free flight times.** Following the scheme's dynamics, all time steps in both the coarse and fine trajectories start with a kinetic free flight. The approach for correlating the free flights starting at time $n\Delta t_{\ell-1}$ is, in the general case, to select the free flight from the fine trajectory starting from, or overlapping this moment in time. Given the memorylessness property of the exponential distribution, the end time of this free flight will be an exponentially distributed amount of time away from $n\Delta t_{\ell-1}$ and hence this difference can be used as a free flight time in the coarse simulation. One then ensures that the free flights occur with the same velocity by reusing the fine velocity in the coarse trajectory.
2. **Coarse diffusion.** The coarse diffusion in the KDMC scheme is generated similarly to (26), namely by summing the diffusive increments in the corresponding time span of the fine trajectory and adding this sum to the sum of kinetic-free flights not included in the correlation of the free flights. These different contributions are then weighted according to the amount of time they occupied. The main difference with the coupling for Algorithm 1 is that the expression (27) is significantly simplified due to the fact that all velocities in the KDMC scheme are independent.

For the non-homogeneous setting, the potentially different collision rates in coarse and fine correlated trajectories mean that care must be taken in which free-flight events from the fine simulation to use when sampling free-flight times in the coarse simulation. For details, we refer to [106].

7 Adjoint and Optimization Aspects

Improved simulation methods for kinetic equations have led to an increasing interest in design and parameter identification problems—often called inverse problems—in particular in the context of engineering applications. In this section, we discuss some recent developments in tackling inverse problems, in particular when using the stochastic time-stepping methods discussed previously in this text. The most straightforward approach to solving such problems is through PDE-constrained optimization. Here, one seeks to minimize an objective functional, typically depending on the residual, i.e., the difference between a solution to the kinetic equation and some reference data. This functional implicitly depends on the parameter or design being considered through the numerical solution of the equation. As an alternative to the PDE-constrained optimization approach, we note that one can solve inverse problems through a Bayesian framework [129]. Combining algorithms for Bayesian inversion with Monte Carlo simulation is still an open area of research. Hence, we do not elaborate on this topic here, other than to mention a recent result [96], studying likelihood ratios as they appear in the Metropolis–Hastings algorithm [102, 54].

In the PDE-constrained optimization setting, we write

$$\min_{\boldsymbol{\Omega}} \mathcal{J}(\mathbf{q},\boldsymbol{\Omega}), \quad \text{subject to} \quad \mathcal{B}(\mathbf{q};\boldsymbol{\Omega}) = 0. \tag{28}$$

Here, $\mathcal{J}(\mathbf{q},\boldsymbol{\Omega})$ is an objective functional depending on a design parameter $\boldsymbol{\Omega} \in \mathbb{R}^{\Omega}$ and PDE state $\mathbf{q} \in \mathbb{R}^{q}$ and $\mathcal{B}(\mathbf{q};\boldsymbol{\Omega})$ is a system of equations, representing a spatially and temporally discretized PDE. Although we directly assume the discrete setting, similar strategies can be worked out where $\mathcal{B}(\mathbf{q};\boldsymbol{\Omega})$ is an undiscretized PDE. For a more generic derivation and a discussion of these two approaches, we refer to [94, Chapter 5] as well as [51, 72].

We assume the existence of a solver for $\mathbf{q}$ given $\boldsymbol{\Omega}$, but that it is not feasible to write an explicit function $\mathbf{q}(\boldsymbol{\Omega})$. Hence, the Jacobian $\mathrm{d}\mathbf{q}/\mathrm{d}\boldsymbol{\Omega}$ can only be approximated numerically and it becomes infeasible to apply the chain rule

$$\frac{\mathrm{d}\mathcal{J}}{\mathrm{d}\boldsymbol{\Omega}}(\mathbf{q}(\boldsymbol{\Omega}),\boldsymbol{\Omega}) = \frac{\partial \mathcal{J}}{\partial \boldsymbol{\Omega}}(\mathbf{q},\boldsymbol{\Omega}) + \frac{\partial \mathcal{J}}{\partial \mathbf{q}}(\mathbf{q},\boldsymbol{\Omega})\frac{\mathrm{d}\mathbf{q}}{\mathrm{d}\boldsymbol{\Omega}}(\boldsymbol{\Omega})$$

to compute the gradient $\frac{\mathrm{d}\mathcal{J}}{\mathrm{d}\boldsymbol{\Omega}}$ when $\mathbf{q}$ is high-dimensional.

Finite differences could be a reasonable strategy to compute $\mathrm{d}\mathcal{J}/\mathrm{d}\boldsymbol{\Omega}$, but have the drawback of requiring a number of PDE solves equal to the gradient length plus one. They also typically only achieve a relative precision of approximately $\sqrt{\varepsilon_{\mathrm{mach}}}$, with $\varepsilon_{\mathrm{mach}}$ the machine precision, due to subtraction-induced cancellation errors [111, Ch. 8]. Instead, we use the adjoint method, avoiding both this exploding computational cost and reduced accuracy.

7.1 Adjoint Approaches in Deterministic Integrators

We now substitute the problem (28) with that of finding fixed points of the Lagrangian

$$\mathcal{L}(\mathbf{q},\mathbf{q}^*,\boldsymbol{\Omega}) = \mathcal{J}(\mathbf{q},\boldsymbol{\Omega}) + (\mathbf{q}^*,\mathcal{B}(\mathbf{q};\boldsymbol{\Omega})), \tag{29}$$

where $\mathbf{q}^* \in \mathbb{R}^{q^*}$ is a vector of adjoint variables or Lagrange multipliers. Note that, in the Lagrangian formulation, we treat the variables $\boldsymbol{\Omega}, \mathbf{q}$ and $\mathbf{q}^*$ as independent. Also observe that if the constraints $\mathcal{B}(\mathbf{q};\boldsymbol{\Omega}) = 0$ hold, (29) reduces to (28) and, as a consequence, the gradients of (28) and (29) are equivalent.

By setting the derivatives of (29) in $\mathbf{q}^*$, $\mathbf{q}$ and $\boldsymbol{\Omega}$ to zero one can show that the fixed points of (29) correspond to a solution of the KKT system [71, 78]:

$$\frac{\partial \mathcal{L}}{\partial \mathbf{q}^*} = \mathcal{B}(\mathbf{q},\boldsymbol{\Omega}) = 0, \qquad \text{State equation} \tag{30}$$

$$\frac{\partial \mathcal{L}}{\partial \mathbf{q}}^{\top} = \frac{\partial \mathcal{J}}{\partial \mathbf{q}}^{\top}(\mathbf{q},\boldsymbol{\Omega}) + \frac{\partial \mathcal{B}}{\partial \mathbf{q}}^{\top}(\mathbf{q},\boldsymbol{\Omega})\mathbf{q}^* = 0, \qquad \text{Adjoint equation} \tag{31}$$

$$\frac{\partial \mathcal{L}}{\partial \boldsymbol{\Omega}}^{\top} = \frac{\partial \mathcal{J}}{\partial \boldsymbol{\Omega}}^{\top}(\mathbf{q},\boldsymbol{\Omega}) + \frac{\partial \mathcal{B}}{\partial \boldsymbol{\Omega}}^{\top}(\mathbf{q},\boldsymbol{\Omega})\mathbf{q}^* = 0. \qquad \text{Design equation} \tag{32}$$

One can iteratively solve the system (30)–(32) to find a solution as follows:

1. Given a proposed $\boldsymbol{\Omega}^i$, solve (30) to obtain $\mathbf{q}^i$.
2. Given $\mathbf{q}^i$ and $\boldsymbol{\Omega}^i$, solve (31) to obtain $\mathbf{q}^{*,i}$.
3. Given $\mathbf{q}^{*,i}$, $\mathbf{q}^i$ and $\boldsymbol{\Omega}^i$, compute a descent direction

$$\Delta\boldsymbol{\Omega} = -\frac{\partial \mathcal{J}}{\partial \boldsymbol{\Omega}}^{\top}(\mathbf{q},\boldsymbol{\Omega}) - \frac{\partial \mathcal{B}}{\partial \boldsymbol{\Omega}}^{\top}(\mathbf{q},\boldsymbol{\Omega})\mathbf{q}^*$$

from (32) and update $\boldsymbol{\Omega}^i \mapsto \boldsymbol{\Omega}^{i+1}$.

Note that after linearization, (30) and (31) both require the solution of a linear system with the matrix $\partial\mathcal{B}/\partial\mathbf{q}$ or its transpose, but (32) involves at most a matrix-vector product. This means that, conceptually, for a single iteration over the KKT system, the majority of the work is expected to lie in solving the equations (30) and (31).

7.2 Adjoint Monte Carlo Time Integration

In the Monte Carlo setting, the state vector $\mathbf{q}$ consists of a vector Q, containing all information of individual steps taken in each particle trajectory, potentially supplemented with additional quantities such as summary statistics of the particle distribution. That is, we assume an ensemble

$$\{Q_{p,n}\}_{p=1}^{P},$$

describing the states of particles $p \in 1,\ldots,P$ at time steps $n = 0,\ldots N_p$, with N_p potentially depending on p. Each $Q_{p,n}$ consists of the necessary information to fully describe a particle's state at the given moment in time, e.g., the position $X_{p,n}$, velocity $V_{p,n}$ and/or weight $W_{p,n}$. The aggregate Monte Carlo state vector is given by

$$Q = \left[\left[Q_{1,0}{}^{\top} \cdots Q_{1,N_1}{}^{\top}\right] \cdots \left[Q_{P,0}{}^{\top} \cdots Q_{P,N_P}{}^{\top}\right]\right]^{\top}.$$

The portion of the constraint $\mathcal{B}(\mathbf{q};\boldsymbol{\Omega}) = 0$ pertaining to Q then consists of all particle update equations. Assuming independent particle trajectories and Markovian particle updates, one can then explicitly write out individual equations in the constraint as

$$\mathcal{B}_{p,n+1}(Q_{p,n+1}, Q_{p,n}; \mathbf{q}, \mathbf{\Omega}) = 0, \quad n = 0, \ldots, N_p - 1,$$

supplemented by an initial condition $\widehat{\mathcal{B}}_{p,0}(Q_{p,0}; \widehat{\mathbf{q}}, \widehat{\mathbf{\Omega}}) = 0$, for each particle $p = 1, \ldots, P$.

Ignoring initial conditions, the Jacobian $\partial \widehat{\mathcal{B}}(Q, \widehat{K}_{\mathcal{T}})/\partial Q$ is block-diagonal with P blocks of the form

$$\frac{\partial \widehat{\mathcal{B}}_p}{\partial Q}(Q, \widehat{K}_{\mathcal{T}}) = \begin{bmatrix} \frac{\partial \mathcal{B}_{p,1}}{\partial Q_{p,1}} & & & \\ \frac{\partial \mathcal{B}_{p,2}}{\partial Q_{p,1}} & \frac{\partial \mathcal{B}_{p,2}}{\partial Q_{p,2}} & & \\ & \ddots & \ddots & \\ & & \frac{\partial \mathcal{B}_{p,N}}{\partial Q_{p,N-1}} & \frac{\partial \mathcal{B}_{p,N}}{\partial Q_{p,N}} \end{bmatrix}. \tag{33}$$

Note that (33) is blockwise lower-triangular, with a single off-diagonal of blocks. Conceptionally, one can thus interpret time-stepping over a particle trajectory with linear dynamics as solving a matrix-vector system (33), with a suitably chosen right-hand side, using forward-substitution. In the adjoint equation (31), we retrieve a similar system of equations, based on the transposed Jacobian which becomes blockwise upper-triangular. Backward-substitution on this system can then be rewritten as a Monte Carlo simulation running backwards in time over the same trajectories. For further reading on adjoint Monte Carlo simulation, we refer to [14, 15, 95] for similar discrete approaches and [91] for an approach based on the continuous adjoint.

The need for simulating identical trajectories in time-reversed fashion leads to a major bottleneck in terms of memory requirements for practical simulations, as storing all trajectories can easily require in the order of tens of GB. Ultimately, we are not interested in computing a full vector Q, hence, we aim to avoid storing full trajectory data by recomputing it on the fly.

7.3 Checkpointing and Time-Reversal Strategies

For most practical Monte Carlo simulations, it is impossible to store full trajectories of all simulated particles in memory. A similar problem occurs in algorithmic differentiation when computing gradients through a back-trace of computer programs. In that setting, a standard workaround for this problem has been developed, called checkpointing, see, e.g., [107, Sec. 5]. In this approach, intermediate computation states are stored so that the relevant portions of the forward simulation can be recomputed when their states are needed in the gradient simulation. This recomputation causes an additional runtime of at least that of the original forward simulation to the gradient computation. The checkpointing itself often also introduces a non-negligible overhead [48].

We aim to do better than the checkpointing approach by computing Monte Carlo trajectories in a fully reversed fashion. To do so, we use a pseudorandom number generator that can be reversed in time. The idea here is that, if the random numbers used to generate the trajectories can be generated in reverse order, one can recompute the corresponding time-reversed trajectory using a modified time-stepping strategy. Rather than use the default C++ generator, we use the reversible PCG generator [113] for this purpose. In [95], a first reversed implementation of this generator was presented, together with its application in the context of adjoint Monte Carlo simulation for the one-dimensional diffusion equation.

PCGs consist of an internal linear congruential generator that generates pseudorandom integer values χ_z in the range $[0, Z^2 - 1]$ through a recurrence

$$\chi_{z+1} = a\chi_z + c \mod Z^2,$$

with a and c suitably chosen integers modulo Z^2, with Z the number of possible generator output values. Even for well chosen values a, c and Z, LCGs have poor statistical properties, as demonstrated by benchmark tests such as TestU01 [87]. PCGs improve on the poor statistics of the underlying LCG by applying a 1-way parameterized permutation to its state χ_z to generate an output value. The internal state update is fully reversible by the Hull-Dobell theorem [61]. Given a reversed sequence of internal states, one can apply the output permutation to this sequence to get the corresponding reversed sequence of output values.We also refer to [6] for a description on the reversal of several commonly used random number generators, including an independently developed reversible PCG.

7.4 Gradient Computation in PDE-Constrained Optimization

In [95], the adjoint method with reversible pseudorandom number generators was applied to a PDE-constrained optimization problem of cooling a 1D rod of which the temperature profile $\theta(x,t)$ depends on space $x \in [0, L]$ and time $t \in \mathbb{R}_{\geq 0}$. The rod is further cooled by a fluid at temperature zero with a rate $u(x)$. Formally, this problem is written as

$$\begin{aligned} \min_{u(x)} \mathcal{J}(\theta(x,t), u(x,t)) &= \int_0^\infty \int_0^L \frac{1}{2}\theta(x,t)^2 \mathrm{d}x\, \mathrm{d}t + \nu \int_0^L \frac{1}{2} u(x)^2 \mathrm{d}x, \\ \text{subject to } \mathcal{B}(\theta(x,t); u(x,t)) &= \frac{\partial}{\partial t}\theta(x,t) - \frac{\partial^2}{\partial x^2}\theta(x,t) + u(x)\theta(x,t) = 0, \\ \theta(x,0) = \theta_0(x), &\quad \theta(0,t) = \theta(L,t), \end{aligned}$$

with $\nu \in \mathbb{R}_{\geq 0}$ a regularization parameter.

In Figure 2, we show the resulting runtime and memory usage, against the number of simulated particles, of both the forward simulation of the constraint equation and the backward simulation of the adjoint equation. The reversible approach is compared to a non-reversible implementation, where all paths are stored in memory between the two simulations. From the figure it is clear that once the memory requirement of storing all paths reaches approximately 8 GB, or half of the available RAM on the system, the

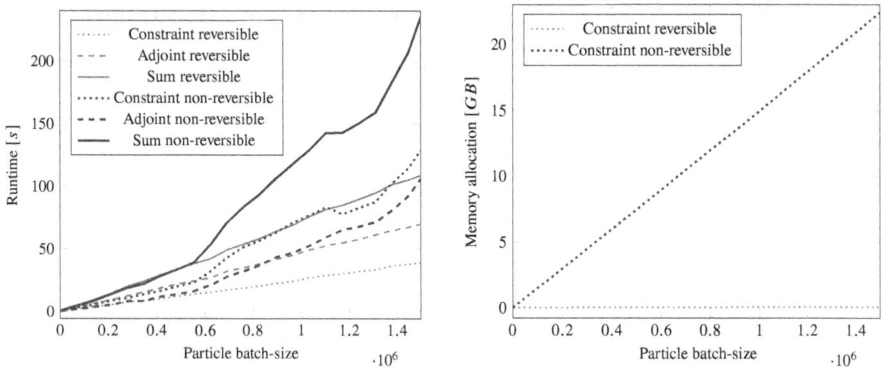

(a) Batch simulation runtime **(b)** Memory allocation during simulation

Figure 2 The increase in runtime and memory usage of a reversible and non-reversible constraint and corresponding adjoint simulations as the number of particles increases. As memory requirements approach the available RAM on the system (16GB), the non-reversible simulation's runtime increases drastically, while the reversible implementation maintains its linear scaling.

non-reversible simulation's runtime grows significantly faster than that of the reversible simulation signifying the gains possible when using this approach.

In [94, Ch. 6], the reversible generator approach was applied to a one-dimensional plasma-edge model inspired by tokamak scrap-off layer simulations such as performed with the SOLPS-ITER software package [133]. This edge-plasma simulation is performed with a finite-volume method, but depends on sources for plasma mass and momentum based on interactions with neutral particles which have not reached an equilibrium velocity distribution. Hence, these neutral particles are modelled with a kinetic equation, solved using a Monte Carlo method. It was shown that one can produce gradient computations with a similar computational cost to the forward simulation, using the reversible adjoint method.

8 Conclusions and Outlook

Kinetic equations, describing a gas as a collection of particles subject to ballistic motion interrupted by instantaneous collision events, play a key role in a variety of physical phenomena. In this review, we have given an overview of some recent developments in time integration methods for simulating kinetic equations. We have mentioned some deterministic simulation methods, but the main focus has been on stochastic methods. We also covered techniques for solving inverse problems based on stochastic simulations.

The work presented here, is only a subset of recent developments in the simulation of kinetic equations. As simulation techniques improve, practitioners can tackle increasingly challenging real-world problems. For example, within fusion energy research, there is a growing interest in tackling design problems through optimization, requiring a large number of simulations due to iteration and, while Bayesian inversion is still out of reach in many cases, it will likely become more feasible in the coming years. Hence, we expect the different topics discussed in this work to continue as active research areas going forward.

Acknowledgements

Emil Løvbak was funded by the Deutsche Forschungsgemeinschaft (DFG, German Research Foundation) – Project-ID 563450842. Vince Maes is an SB PhD fellow of the Research Foundation Flanders (FWO), funded by grant 1S64723N. Part of this research was funded by the Research Foundation Flanders (FWO) under grant 3E211335. Parts of this work have been carried out within the framework of the EUROfusion Consortium, funded by the European Union via the Euratom Research and Training Programme (Grant Agreement No 101052200 — EUROfusion). Views and opinions expressed are, however, those of the author(s) only and do not necessarily reflect those of the European Union or the European Commission. Neither the European Union nor the European Commission can be held responsible for them.

References

1. A. Y. Aydemir. A unified Monte Carlo interpretation of particle simulations and applications to non-neutral plasmas. *Physics of Plasmas*, 1(4):822–831, Apr. 1994.
2. G. Bal and Y. Maday. Coupling of transport and diffusion models in linear transport theory. *ESAIM: Mathematical Modelling and Numerical Analysis*, 36(1):69–86, Jan. 2002.
3. C. Bardos, F. Golse, and D. Levermore. Fluid Dynamic Limits of Kinetic Equations. I. Formal Derivations. *Journal of Statistical Physics*, 63:323–344, Apr. 1991.
4. C. Bardos, F. Golse, and D. Levermore. Fluid Dynamic Limits of Kinetic Equations. II. Convergence Proofs for the Boltzmann Equation. *Communications on Pure and Applied Mathematics*, 46(5):667–753, June 1993.
5. M. Bennoune, M. Lemou, and L. Mieussens. Uniformly stable numerical schemes for the Boltzmann equation preserving the compressible Navier–Stokes asymptotics. *Journal of Computational Physics*, 227(8):3781–3803, Apr. 2008.
6. T. Bewley. Reversible random number generation. Technical report, May 2024.
7. P. L. Bhatnagar, E. P. Gross, and M. Krook. A Model for Collision Processes in Gases. I. Small Amplitude Processes in Charged and Neutral One-Component Systems. *Physical Review*, 94(3):511–525, May 1954.
8. G. A. Bird. *The DSMC Method*. G. A. Bird, S.l., 1.2 edition, 2013.
9. M. Bisi, W. Boscheri, G. Dimarco, M. Groppi, and G. Martalò. A new mixed Boltzmann-BGK model for mixtures solved with an IMEX finite volume scheme on unstructured meshes. *Applied Mathematics and Computation*, 433:127416, Nov. 2022.
10. M. Blommaert, N. Horsten, P. Börner, and W. Dekeyser. A spatially hybrid fluid-kinetic neutral model for SOLPS-ITER plasma edge simulations. *Nuclear Materials and Energy*, 19:28–33, May 2019.
11. S. Boscarino, L. Pareschi, and G. Russo. Implicit-Explicit Runge–Kutta Schemes for Hyperbolic Systems and Kinetic Equations in the Diffusion Limit. *SIAM Journal on Scientific Computing*, 35(1):A22–A51, Jan. 2013.

12. I. D. Boyd and T. R. Deschenes. Hybrid Particle-Continuum Numerical Methods for Aerospace Applications. DTIC Number ADA588168, Michigan University, Ann Arbor, Department of Aerospace Engineering, 2011.
13. S. Brunner, E. Valeo, and J. A. Krommes. Collisional delta-f scheme with evolving background for transport time scale simulations. *Phys. Plasmas*, 6(12):19, 1999.
14. R. Caflisch, D. Silantyev, and Y. Yang. Adjoint DSMC for nonlinear Boltzmann equation constrained optimization. *Journal of Computational Physics*, 439:110404, Aug. 2021.
15. R. Caflisch and Y. Yang. Adjoint Monte Carlo Method. In J. A. Carrillo and E. Tadmor, editors, *Active Particles, Volume 4*, pages 461–505. Springer Nature Switzerland, Cham, Switzerland, 2024.
16. K. M. Case and P. F. Zweifel. *Linear Transport Theory*. Addison-Wesley Educational Publishers Inc, Jan. 1967.
17. C. Cercignani. *Mathematical Methods in Kinetic Theory*. Springer US, Boston, MA, 1969.
18. C. Cercignani, R. Illner, and M. Pulvirenti. *The Mathematical Theory of Dilute Gases*, volume 106 of *Applied Mathematical Sciences*. Springer, New York, NY, 1994.
19. S. Chandrasekhar. *Radiative Transfer*. The International Series of Monographs on Physics. Oxford: Clarendon, 1950.
20. S. Chou and D. Baganoff. Kinetic Flux–Vector Splitting for the Navier–Stokes Equations. *Journal of Computational Physics*, 130(2):217–230, Jan. 1997.
21. A. Crestetto, N. Crouseilles, G. Dimarco, and M. Lemou. Asymptotically complexity diminishing schemes (ACDS) for kinetic equations in the diffusive scaling. *Journal of Computational Physics*, 394:243–262, Oct. 2019.
22. A. Crestetto, N. Crouseilles, G. Dimarco, and M. Lemou. A new deviational asymptotic preserving Monte Carlo method for the homogeneous Boltzmann equation. *Communications in Mathematical Sciences*, 18(8):2305–2339, 2020.
23. A. Crestetto, N. Crouseilles, and M. Lemou. Kinetic/fluid micro-macro numerical schemes for Vlasov-Poisson-BGK equation using particles. *Kinetic & Related Models*, 5(4):787–816, 2012.
24. A. Crestetto, N. Crouseilles, and M. Lemou. A particle micro-macro decomposition based numerical scheme for collisional kinetic equations in the diffusion scaling. *Communications in Mathematical Sciences*, 16(4):887–911, Oct. 2018.
25. N. Crouseilles, P. Degond, and M. Lemou. A hybrid kinetic/fluid model for solving the gas dynamics Boltzmann–BGK equation. *Journal of Computational Physics*, 199(2):776–808, Sept. 2004.
26. N. Crouseilles, G. Dimarco, and M. Lemou. Asymptotic preserving and time diminishing schemes for rarefied gas dynamic. *Kinetic & Related Models*, 10(3):643–668, 2017.
27. N. Crouseilles and M. Lemou. An asymptotic preserving scheme based on a micro-macro decomposition for Collisional Vlasov equations: Diffusion and high-field scaling limits. *Kinetic and Related Models*, 4(2):441–477, May 2011.
28. P. Degond, G. Dimarco, and L. Mieussens. A Multiscale Kinetic-Fluid Solver with Dynamic Localization of Kinetic Effects. *Journal of Computational Physics*, 229(13):4907–4933, July 2010.
29. P. Degond, G. Dimarco, and L. Pareschi. The moment-guided Monte Carlo method. *International Journal for Numerical Methods in Fluids*, 67(2):189–213, Sept. 2011.
30. P. Degond, S. Jin, and L. Mieussens. A smooth transition model between kinetic and hydrodynamic equations. *Journal of Computational Physics*, 209(2):665–694, Nov. 2005.
31. R. Delgado-Buscalioni and P. V. Coveney. Continuum-particle hybrid coupling for mass, momentum, and energy transfers in unsteady fluid flow. *Physical Review E*, 67(4):046704, Apr. 2003.
32. J. D. Densmore, T. J. Urbatsch, T. M. Evans, and M. W. Buksas. A hybrid transport-diffusion method for Monte Carlo radiative-transfer simulations. *Journal of Computational Physics*, 222(2):485–503, Mar. 2007.
33. R. E. Denton and M. Kotschenreuther. δf Algorithm. *Journal of Computational Physics*, 119(2):283–294, July 1995.
34. G. Dimarco and L. Pareschi. Exponential Runge–Kutta Methods for Stiff Kinetic Equations. *SIAM Journal on Numerical Analysis*, 49(5):2057–2077, Jan. 2011.

35. G. Dimarco and L. Pareschi. High order asymptotic-preserving schemes for the Boltzmann equation. *Comptes Rendus Mathématique*, 350(9-10):481–486, May 2012.
36. G. Dimarco and L. Pareschi. Numerical methods for kinetic equations. *Acta Numerica*, 23:369–520, May 2014.
37. G. Dimarco, L. Pareschi, and G. Samaey. Asymptotic-Preserving Monte Carlo Methods for Transport Equations in the Diffusive Limit. *SIAM Journal on Scientific Computing*, 40(1):A504–A528, Jan. 2018.
38. A. Dupuis, E. M. Kotsalis, and P. Koumoutsakos. Coupling lattice Boltzmann and molecular dynamics models for dense fluids. *Physical Review E*, 75(4):046704, Apr. 2007.
39. W. E, B. Engquist, X. Li, W. Ren, and E. Vanden-Eijnden. Heterogeneous multiscale methods: a review. *Communications in Computational Physics*, 2(3):367–450, 2007.
40. L. Einkemmer, J. Hu, and Y. Wang. An asymptotic-preserving dynamical low-rank method for the multi-scale multi-dimensional linear transport equation. *Journal of Computational Physics*, 439:110353, Aug. 2021.
41. L. Einkemmer, J. Hu, and L. Ying. An Efficient Dynamical Low-Rank Algorithm for the Boltzmann-BGK Equation Close to the Compressible Viscous Flow Regime. *SIAM Journal on Scientific Computing*, 43(5):B1057–B1080, Jan. 2021.
42. F. Filbet, C. Mouhot, and L. Pareschi. Solving the Boltzmann equation in n log2 n. *SIAM Journal on Scientific Computing*, 28(3):1029–1053, 2006.
43. F. Filbet and T. Rey. A Hierarchy of Hybrid Numerical Methods for Multiscale Kinetic Equations. *SIAM Journal on Scientific Computing*, 37(3):A1218–A1247, Jan. 2015.
44. E. Gabetta, L. Pareschi, and G. Toscani. Relaxation Schemes for Nonlinear Kinetic Equations. *SIAM Journal on Numerical Analysis*, 34(6):2168–2194, Dec. 1997.
45. A. L. Garcia, J. B. Bell, W. Y. Crutchfield, and B. J. Alder. Adaptive Mesh and Algorithm Refinement Using Direct Simulation Monte Carlo. *Journal of Computational Physics*, 154(1):134–155, Sept. 1999.
46. C. Gear and I. Kevrekidis. Projective methods for stiff differential equations: Problems with gaps in their eigenvalue spectrum. *SIAM Journal on Scientific Computing*, 24(4):1091–1106, 2003.
47. C. Gear and I. G. Kevrekidis. Telescopic projective methods for parabolic differential equations. *Journal of Computational Physics*, 187(1):95–109, 2003.
48. A. H. Gebremedhin and A. Walther. An introduction to algorithmic differentiation. *WIREs Data Mining and Knowledge Discovery*, 10(1):e1334, Jan. 2020.
49. M. B. Giles. Multilevel Monte Carlo Path Simulation. *Operations Research*, 56(3):607–617, June 2008.
50. M. B. Giles. Multilevel Monte Carlo methods. *Acta Numerica*, 24:259–328, May 2015.
51. M. B. Giles and N. A. Pierce. An Introduction to the Adjoint Approach to Design. *Flow, Turbulence and Combustion*, 65:393–415, Dec. 2000.
52. L. Gosse and G. Toscani. An asymptotic-preserving well-balanced scheme for the hyperbolic heat equations. *Comptes Rendus Mathématique*, 334(4):337–342, Feb. 2002.
53. H. Grad. Principles of the Kinetic Theory of Gases. In S. Flügge, editor, *Thermodynamik der Gase / Thermodynamics of Gases*, volume 3 / 12, pages 205–294. Springer Berlin Heidelberg, Berlin, Heidelberg, 1958.
54. W. K. Hastings. Monte Carlo sampling methods using Markov chains and their applications. *Biometrika*, 57(1):97–109, Apr. 1970.
55. S. Heinrich. Multilevel Monte Carlo Methods. In S. Margenov, J. Waśniewski, and P. Yalamov, editors, *Large-Scale Scientific Computing*, volume 2179 of *Lecture Notes in Scientific Computing*, pages 58–67. Springer Berlin Heidelberg, Berlin, Germany, 2001.
56. N. Horsten, M. Groth, W. Dekeyser, W. Van Uytven, S. Carli, and JET Contributors. Combination of micro–macro and spatially hybrid fluid-kinetic approach for hydrogenic plasma edge neutrals. *Contributions to Plasma Physics*, 62(5-6), June 2022.
57. N. Horsten, G. Samaey, and M. Baelmans. Hybrid fluid-kinetic neutral model based on a kinetic correction on a fluid solution. *Contributions to Plasma Physics*, 58(6-8):703–709, July 2018.
58. N. Horsten, G. Samaey, and M. Baelmans. Hybrid fluid-kinetic model for neutral particles in the plasma edge. *Nuclear Materials and Energy*, 18:201–207, Jan. 2019.

59. N. Horsten, G. Samaey, and M. Baelmans. A hybrid fluid-kinetic model for hydrogenic atoms in the plasma edge of tokamaks based on a micro-macro decomposition of the kinetic equation. *Journal of Computational Physics*, 409:109308, May 2020.
60. J. Hu, S. Jin, J. Li, and L. Zhang. On multilevel Monte Carlo methods for deterministic and uncertain hyperbolic systems. *Journal of Computational Physics*, 475:111847, Feb. 2023.
61. T. E. Hull and A. R. Dobell. Random Number Generators. *SIAM Review*, 4(3):230–254, July 1962.
62. R. Illner and H. Neunzert. Domain decomposition: Linking kinetic and aerodynamic descriptions, 1993. Berichte der Arbeitsgruppe Technomathematik (AGTM Report) 90, Technische Universität Kaiserslautern.
63. P. Jarmatz and P. Neumann. MaMiCo: Parallel Noise Reduction for Multi-instance Molecular-Continuum Flow Simulation. In J. M. F. Rodrigues, P. J. S. Cardoso, J. Monteiro, R. Lam, V. V. Krzhizhanovskaya, M. H. Lees, J. J. Dongarra, and P. M. Sloot, editors, *Computational Science – ICCS 2019*, volume 11539, pages 451–464. Springer International Publishing, Cham, 2019.
64. S. Jin. Efficient Asymptotic-Preserving (AP) Schemes For Some Multiscale Kinetic Equations. *SIAM Journal on Scientific Computing*, 21(2):441–454, Jan. 1999.
65. S. Jin. Asymptotic preserving (AP) schemes for multiscale kinetic and hyperbolic equations: A review. *Rivista di Matematica della Università di Parma*, 3:177–216, 2012.
66. S. Jin. Asymptotic-Preserving Schemes for Multiscale Physical Problems. *Acta Numerica*, 31:415–489, June 2022.
67. S. Jin, H. Lu, and L. Pareschi. Efficient Stochastic Asymptotic-Preserving Implicit-Explicit Methods for Transport Equations with Diffusive Scalings and Random Inputs. *SIAM Journal on Scientific Computing*, 40(2):A671–A696, Jan. 2018.
68. S. Jin and L. Pareschi. Discretization of the Multiscale Semiconductor Boltzmann Equation by Diffusive Relaxation Schemes. *Journal of Computational Physics*, 161(1):312–330, June 2000.
69. S. Jin, L. Pareschi, and G. Toscani. Diffusive Relaxation Schemes for Multiscale Discrete-Velocity Kinetic Equations. *SIAM Journal on Numerical Analysis*, 35(6):2405–2439, Jan. 1998.
70. S. Jin, L. Pareschi, and G. Toscani. Uniformly Accurate Diffusive Relaxation Schemes for Multiscale Transport Equations. *SIAM Journal on Numerical Analysis*, 38(3):913–936, Jan. 2000.
71. W. Karush. Minima of Functions of Several Variables with Inequalities as Side Conditions. Master's thesis, University of Chicago, Dec. 1939.
72. S. M. Kast. An Introduction to Adjoints and Output Error Estimation in Computational Fluid Dynamics. *arXiv:1712.00693*, Dec. 2017.
73. I. G. Kevrekidis, C. W. Gear, J. M. Hyman, P. G. Kevrekidid, O. Runborg, and C. Theodoropoulos. Equation-free, coarse-grained multiscale computation: Enabling mocroscopic simulators to perform system-level analysis. *Communications in Mathematical Sciences*, 1(4):715–762, 2003.
74. A. Klar. Domain Decomposition for Kinetic Problems with Nonequilibrium States. *European Journal of Mechanics - B/Fluids*, 15:203–216, 1995.
75. A. Klar. An Asymptotic-Induced Scheme for Nonstationary Transport Equations in the Diffusive Limit. *SIAM Journal on Numerical Analysis*, 35(3):1073–1094, Jan. 1998.
76. A. Klar. An Asymptotic Preserving Numerical Scheme for Kinetic Equations in the Low Mach Number Limit. *SIAM Journal on Numerical Analysis*, 36(5):1507–1527, Jan. 1999.
77. A. Klar. A Numerical Method for Kinetic Semiconductor Equations in the Drift-Diffusion Limit. *SIAM Journal on Scientific Computing*, 20(5):1696–1712, Jan. 1999.
78. H. Kuhn and A. Tucker. Nonlinear Programming. In *Proceedings of the Second Berkeley Symposium on Mathematical Statistics and Probability*, volume 2, pages 481–492, Berkeley, California, 1951. University of California Press.
79. P. Lafitte, A. Lejon, and G. Samaey. A high-order asymptotic-preserving scheme for kinetic equations using projective integration. *SIAM Journal on Numerical Analysis*, 54(1):1–33, 2016.
80. P. Lafitte, W. Melis, and G. Samaey. A high-order relaxation method with projective integration for solving nonlinear systems of hyperbolic conservation laws. *Journal of Computational Physics*, 340:1–25, 2017.
81. P. Lafitte and G. Samaey. Asymptotic-preserving projective integration schemes for kinetic equations in the diffusion limit. *SIAM Journal on Scientific Computing*, 34(2):A579–A602, 2012.

82. T. Laidin. Hybrid kinetic/fluid numerical method for the Vlasov-BGK equation in the diffusive scaling. *Kinetic and Related Models*, 16(6):913–947, 2023.
83. B. Lapeyre, É. Pardoux, and R. Sentis. *Introduction to Monte Carlo Methods for Transport and Diffusion Equations*, volume 6 of *Oxford Texts in Applied and Engineering Mathematics*. Oxford University Press, Oxford, UK, 2003.
84. O. Lappi, H. Leggate, Y. Marandet, J. Åström, K. Heljanko, and D. V. Borodin. Scalable Domain-decomposed Monte Carlo Neutral Transport for Nuclear Fusion. *arXiv:2511.04489*, 2025.
85. O. Lappi, E. Løvbak, T. Steel, and G. Samaey. 2D implementation of Kinetic-diffusion Monte Carlo in Eiron. *arXiv:2509.19140*, Sept. 2025.
86. S. Lautenbach and R. Grauer. Multiphysics Simulations of Collisionless Plasmas. *Frontiers in Physics*, 6:113, Oct. 2018.
87. P. L'Ecuyer and R. Simard. TestU01: A C library for empirical testing of random number generators. *ACM Transactions on Mathematical Software*, 33(4):1–40, Aug. 2007.
88. S. L. Lee and C. W. Gear. Second-order accurate projective integrators for multiscale problems. *Journal of Computational and Applied Mathematics*, 201(1):258–274, 2007.
89. M. Lemou and L. Mieussens. A New Asymptotic Preserving Scheme Based on Micro-Macro Formulation for Linear Kinetic Equations in the Diffusion Limit. *SIAM Journal on Scientific Computing*, 31(1):334–368, Jan. 2008.
90. Q. Li, J. Lu, and W. Sun. Diffusion approximations and domain decomposition method of linear transport equations: asymptotics and numerics. *Journal of Computational Physics*, 292:141–167, July 2015.
91. Q. Li, L. Wang, and Y. Yang. Monte Carlo Gradient in Optimization Constrained by Radiative Transport Equation. *SIAM Journal on Numerical Analysis*, 61(6):2744–2774, 2023.
92. C. Liu, Y. Zhu, and K. Xu. Unified gas-kinetic wave-particle methods I: Continuum and rarefied gas flow. *Journal of Computational Physics*, 401:108977, Jan. 2020.
93. S. Liu, C. Zhong, and M. Fang. Simplified unified wave-particle method with quantified model-competition mechanism for numerical calculation of multiscale flows. *Physical Review E*, 102(1):013304, July 2020.
94. E. Løvbak. *Multilevel and Adjoint Monte Carlo Methods for Plasma Edge Neutral Particle Models*. PhD thesis, KU Leuven, Leuven, Sept. 2023.
95. E. Løvbak, F. Blondeel, A. Lee, L. Vanroye, A. Van Barel, and G. Samaey. Reversible Random Number Generation for Adjoint Monte Carlo Simulation of the Heat Equation. In A. Hinrichs, P. Kritzer, and F. Pillichshammer, editors, *Monte Carlo and Quasi-Monte Carlo Methods*, volume 460, pages 451–468. Springer, Cham, Switzerland, 2024.
96. E. Løvbak and S. Krumscheid. An Investigation into the Distribution of Ratios of Particle Solver-based Likelihoods. *arXiv:2508.05303*, Aug. 2025.
97. E. Løvbak and G. Samaey. Accelerated simulation of Boltzmann-BGK equations near the diffusive limit with asymptotic-preserving multilevel Monte Carlo. *SIAM Journal on Scientific Computing*, 45(4):A1862–A1889, Aug. 2023.
98. E. Løvbak, G. Samaey, and S. Vandewalle. A multilevel Monte Carlo method for asymptotic-preserving particle schemes in the diffusive limit. *Numerische Mathematik*, 148(1):141–186, May 2021.
99. P. A. Markowich, C. A. Ringhofer, and C. Schmeiser. *Semiconductor Equations*. Springer-Verlag, Vienna, Austria, 1990.
100. W. Melis, T. Rey, and G. Samaey. Projective and telescopic projective integration for the nonlinear BGK and Boltzmann equations. *The SMAI Journal of Computational Mathematics*, 5:53–88, 2019.
101. W. Melis and G. Samaey. Telescopic projective integration for linear kinetic equations with multiple relaxation times. *Journal of Scientific Computing*, 76(2):697–726, 2018.
102. N. Metropolis, A. W. Rosenbluth, M. N. Rosenbluth, A. H. Teller, and E. Teller. Equation of State Calculations by Fast Computing Machines. *The Journal of Chemical Physics*, 21(6):1087–1092, June 1953.
103. B. Mortier, M. Baelmans, and G. Samaey. Kinetic-diffusion asymptotic-preserving Monte Carlo algorithms for plasma edge neutral simulation. *Contributions to Plasma Physics*, page e201900134, Dec. 2019.

104. B. Mortier, M. Baelmans, and G. Samaey. A Kinetic-Diffusion Asymptotic-Preserving Monte Carlo Algorithm for the Boltzmann-BGK Model in the Diffusive Scaling. *SIAM Journal on Scientific Computing*, 44(2):A720–A744, Apr. 2022.
105. B. Mortier, V. Maes, and G. Samaey. Estimation as a post-processing step for random walk approximations of the Boltzmann-BGK model. *Contributions to Plasma Physics*, 62(5-6):e202100197, July 2022.
106. B. Mortier, P. Robbe, M. Baelmans, and G. Samaey. Multilevel asymptotic-preserving Monte Carlo for kinetic-diffusive particle simulations of the Boltzmann-BGK equation. *Journal of Computational Physics*, 450:110736, Feb. 2022.
107. U. Naumann and J. du Toit. Adjoint algorithmic differentiation tool support for typical numerical patterns in computational finance. *Journal of Computational Finance*, 21(4):23–57, Feb. 2018.
108. P. Neumann and X. Bian. MaMiCo: Transient multi-instance molecular-continuum flow simulation on supercomputers. *Computer Physics Communications*, 220:390–402, Nov. 2017.
109. P. Neumann, H. Flohr, R. Arora, P. Jarmatz, N. Tchipev, and H.-J. Bungartz. MaMiCo: Software design for parallel molecular-continuum flow simulations. *Computer Physics Communications*, 200:324–335, Mar. 2016.
110. X. B. Nie, S. Y. Chen, W. N. E, and M. O. Robbins. A continuum and molecular dynamics hybrid method for micro- and nano-fluid flow. *Journal of Fluid Mechanics*, 500:55–64, Jan. 2004.
111. J. Nocedal and S. J. Wright. *Numerical Optimization*. Springer Series in Operations Research. Springer, New York, NY, 2nd edition, 2006.
112. B. Nowakowski and J. Popielawski. The kinetic theory of the effect of chemical reaction on diffusion of a trace gas. *The Journal of Chemical Physics*, 100(10):7602–7609, May 1994.
113. M. E. O'Neill. PCG: A Family of Simple Fast Space-Efficient Statistically Good Algorithms for Random Number Generation. Technical Report HMC-CS-2014-0905, Harvey Mudd College, Claremont, CA, Sept. 2014.
114. H. G. Othmer and T. Hillen. The Diffusion Limit of Transport Equations II: Chemotaxis Equations. *SIAM Journal on Applied Mathematics*, 62(4):1222–1250, Jan. 2002.
115. S. T. O'Connell and P. A. Thompson. Molecular dynamics–continuum hybrid computations: A tool for studying complex fluid flows. *Physical Review E*, 52(6):R5792–R5795, Dec. 1995.
116. L. Pareschi. Hybrid Multiscale Methods for Hyperbolic and Kinetic Problems. In T. Goudon, E. Sonnendrucker, and D. Talay, editors, *ESAIM: Proceedings*, volume 15, pages 87–120, Nov. 2005.
117. L. Pareschi and R. E. Caflisch. An Implicit Monte Carlo Method for Rarefied Gas Dynamics. I. The Space Homogeneous Case. *Journal of Computational Physics*, 154(1):90–116, Sept. 1999.
118. L. Pareschi and G. Russo. Time Relaxed Monte Carlo Methods for the Boltzmann Equation. *SIAM Journal on Scientific Computing*, 23(4):1253–1273, Jan. 2001.
119. L. Pareschi and S. Trazzi. Numerical solution of the Boltzmann equation by time relaxed Monte Carlo (TRMC) methods. *International Journal for Numerical Methods in Fluids*, 48(9):947–983, July 2005.
120. Z. Peng, Y. Cheng, J.-M. Qiu, and F. Li. Stability-enhanced AP IMEX-LDG schemes for linear kinetic transport equations under a diffusive scaling. *Journal of Computational Physics*, 415:109485, Aug. 2020.
121. Z. Peng and F. Li. Asymptotic Preserving IMEX-DG-S Schemes for Linear Kinetic Transport Equations Based on Schur Complement. *SIAM Journal on Scientific Computing*, 43(2):A1194–A1220, Jan. 2021.
122. F. Poupaud. Diffusion approximation of the linear semiconductor Boltzmann equation: Analysis of boundary layers. *Asymptotic Analysis*, 4(4):293–317, 1991.
123. G. A. Radtke and N. G. Hadjiconstantinou. Variance-reduced particle simulation of the Boltzmann transport equation in the relaxation-time approximation. *Physical Review E*, 79(5):056711, May 2009.
124. G. A. Radtke, J.-P. M. Péraud, and N. G. Hadjiconstantinou. On efficient simulations of multiscale kinetic transport. *Philosophical Transactions of the Royal Society A: Mathematical, Physical and Engineering Sciences*, 371(1982):20120182, Jan. 2013.
125. D. Reiter, C. May, D. Coster, and R. Schneider. Time dependent neutral gas transport in tokamak edge plasmas. *Journal of Nuclear Materials*, 220-222:987–992, Apr. 1995.

126. R. Rico-Martinez, C. W. Gear, and I. G. Kevrekidis. Coarse projective kMC integration: forward/reverse initial and boundary value problems. *Journal of Computational Physics*, 196(2):474–489, 2004.
127. P. Robbe. *Multilevel Uncertainty Quantification Methods for Robust Design of Industrial Applications*. PhD thesis, KU Leuven, Leuven, Belgium, Dec. 2019.
128. L. Saint-Raymond. *Hydrodynamic Limits of the Boltzmann Equation*, volume 1971 of *Lecture Notes in Mathematics*. Springer Berlin Heidelberg, Berlin, Heidelberg, 2009.
129. D. Sanz-Alonso, A. M. Stuart, and A. Taeb. *Inverse Problems and Data Assimilation*. London Mathematical Society Student Texts. Cambridge University Press, Cambridge, UK, 2023.
130. T. E. Schwartzentruber and I. D. Boyd. Progress and future prospects for particle-based simulation of hypersonic flow. *Progress in Aerospace Sciences*, 72:66–79, Jan. 2015.
131. Z. Tang, J. Koellermeier, E. Løvbak, and G. Samaey. Numerical analysis of fluid estimation for source terms in neutral particles simulation. *arXiv:2509.11883*, 2025.
132. M. Valentinuzzi, Y. Marandet, H. Bufferand, G. Ciraolo, and P. Tamain. Two-phases hybrid model for neutrals. *Nuclear Materials and Energy*, 18:41–45, Jan. 2019.
133. S. Wiesen, D. Reiter, V. Kotov, M. Baelmans, W. Dekeyser, A. S. Kukushkin, S. W. Lisgo, R. A. Pitts, V. Rozhansky, G. Saibene, I. Veselova, and S. Voskoboynikov. The new SOLPS-ITER code package. *Journal of Nuclear Materials*, 463:480–484, Aug. 2015.
134. K. Xu and J.-C. Huang. A unified gas-kinetic scheme for continuum and rarefied flows. *Journal of Computational Physics*, 229(20):7747–7764, Oct. 2010.
135. Y. Zhu and K. Xu. The first decade of unified gas kinetic scheme. *arXiv:2102.01261*, Feb. 2021.

Randomized Orthogonalization and Krylov Subspace Methods: Principles and Algorithms

Jean-Guillaume de Damas, Laura Grigori, Igor Simunec and Edouard Timsit

Abstract We present an overview of randomized orthogonalization techniques that construct a well-conditioned basis whose sketch is orthonormal. Randomized orthogonalization has recently emerged as a powerful paradigm for reducing the computational and communication cost of state-of-the-art orthogonalization procedures on parallel architectures, while preserving, and in some cases improving, their numerical stability. This approach can be employed within Krylov subspace methods to mitigate the cost of orthogonalization, yielding a randomized Arnoldi relation. We review the main variants of the randomized Gram–Schmidt and Householder QR algorithms, and discuss their application to Krylov methods for the solution of large-scale linear algebra problems, such as linear systems of equations, eigenvalue problems, the evaluation of matrix functions, and matrix equations.

1 Introduction

Krylov subspace methods are among the most powerful and widely used techniques for solving large-scale numerical linear algebra problems, such as linear systems of equations, eigenvalue problems, matrix equations, and matrix function evaluations.

Jean-Guillaume de Damas
Sorbonne University, Paris, France. Work done while the author was at INRIA, Paris, France. jean-guillaume.de-damas@inria.fr

Laura Grigori
Institute of Mathematics, EPFL, Lausanne, Switzerland, and PSI Center for Scientific Computing, Theory and Data, Villigen, Switzerland. laura.grigori@epfl.ch

Igor Simunec
Institute of Mathematics, EPFL, Lausanne, Switzerland. igor.simunec@epfl.ch

Edouard Timsit
Éducation Nationale, Académie de Versailles, France. Part of this work was done while the author was at INRIA, Paris, France. edouard-gaston.timsit@ac-versailles.fr

B. Koren and D. Roose (eds.), *Numerical Analysis and Scientific Computing: an Anthology*, Lecture Notes in Computational Science and Engineering 156,
https://doi.org/10.1007/978-3-032-33428-2_6

Given a matrix $A \in \mathbb{R}^{n\times n}$ and a vector $\mathbf{b} \in \mathbb{R}^n$, these methods iteratively construct the sequence of Krylov subspaces

$$\mathcal{K}_m(A, \mathbf{b}) = \operatorname{span}\{\mathbf{b}, A\mathbf{b}, \dots, A^{m-1}\mathbf{b}\},$$

which are used as search spaces to find an approximate solution to a given problem. An orthonormal basis of $\mathcal{K}_m(A, \mathbf{b})$ is typically constructed with the Arnoldi process, which produces a decomposition $AQ_m = Q_{m+1}\underline{H}_m$, where the columns of Q_m are the orthonormal basis vectors and $\underline{H}_m$ is an upper Hessenberg matrix that contains the orthogonalization coefficients representing the projection of A onto $\mathcal{K}_m(A, \mathbf{b})$. Each step of the Arnoldi process involves a matrix-vector product with A followed by orthogonalization of the new vector against all previous basis vectors through a Gram–Schmidt process, for a total computational cost of $O(m\,\mathtt{mv}(A) + nm^2)$ for m iterations, where $\mathtt{mv}(A)$ denotes the cost of a matrix-vector product with A. Consequently, if we assume that a matrix-vector product with A has a cost of $O(n)$, which holds for instance when A is a sparse matrix, the orthogonalization step becomes the dominant cost when m is moderately large, or when the computation is performed on a parallel computer, and this often limits the practical efficiency of Krylov subspace methods.

Several strategies have been proposed in the literature to address this computational bottleneck. Incomplete or truncated orthogonalization schemes reduce the number of inner products at the expense of a loss of numerical stability. Restarting techniques limit the dimension of the Krylov subspace and periodically restart the iteration to control memory usage and computational cost, but they may incur convergence delays or stagnations [67].

Randomization has emerged as a powerful technique for solving large-scale problems by enabling dimensionality reduction through random projections and subspace embeddings [1, 48, 70]. It has been applied successfully to different linear algebra problems, including solving least squares problems [29, 65] and computing low-rank matrix approximations (see, e.g., [56, 60, 79] for details). More recently, randomized techniques have also been introduced in the context of Krylov subspace methods. Randomized or sketched Krylov subspace methods replace exact orthogonalization in the Gram–Schmidt process with operations performed on vectors that belong to a low-dimensional sketched space, obtained by applying a random sketching matrix $\Omega \in \mathbb{R}^{\ell\times n}$ to the basis vectors, which acts as an oblivious subspace embedding. In addition to having the embedding property, the matrix Ω should be cheap to apply to vectors, so a variety of sparse and structured matrices have been considered in the literature, such as the subsampled randomized Hadamard transform and s-hashing sketching matrices. This approach has the potential to reduce the computational and communication costs of building the Krylov basis [5,61], and instead of an orthonormal basis Q_m, it produces a sketch-orthonormal basis V_m such that its sketch ΩV_m contains orthonormal columns. It allows exploiting mixed-precision arithmetic and optimized computational kernels, while providing numerical guarantees with high probability.

Variants of this randomized approach have been explored for the solution of linear systems [5,6,42,61,75], eigenvalue problems [6,26,27,61], for the evaluation of matrix functions [22,41,63] and the solution of matrix equations [64]. This work provides a general introduction to the use of randomized orthogonalization within Krylov subspace methods and its application for the solution of different linear algebra problems. Section 2 describes the oblivious subspace embedding property and different sketching matrices that satisfy this property. Section 3 reviews randomized orthogonalization techniques, including randomized Gram–Schmidt and randomized Householder QR. Randomized Krylov subspace methods are discussed in Section 4, while their usage to solve linear systems and eigenvalue problems is presented in Sections 5 and 6, respectively. Randomized approaches for matrix functions and matrix equations are introduced in Sections 7 and 8, respectively.

1.1 Notation

We introduce here some general notation that we use throughout this manuscript. We denote matrices with uppercase letters, and vectors with bold lowercase letters. We denote by I_n the identity matrix of dimension n, and omit the subscript when it can easily be inferred from the context. The columns of the identity matrix of dimension n are denoted by $\mathbf{e}_1, \ldots, \mathbf{e}_n$. We denote by 0_n the vector of zeros of length n, and by $0_{n\times m}$ a zero-matrix of size $n \times m$; occasionally, the subscripts may be omitted if there is no ambiguity on the dimensions. We use calligraphic letters to denote a vector subspace $\mathcal{W} \subset \mathbb{R}^n$, and we denote by $A\mathcal{W}$ the image of $\mathcal{W}$ under the action of the matrix A. We denote by $\|\mathbf{x}\|$ the Euclidean norm of a vector $\mathbf{x} \in \mathbb{R}^n$, and by $\|A\|_2$ and $\|A\|_F$ the spectral and Frobenius norms of a matrix $A \in \mathbb{R}^{n\times n}$, respectively. The singular values of A in nonincreasing order are denoted by $\sigma_1(A), \ldots, \sigma_n(A)$.

2 Sketching and Embeddings

Randomized algorithms rely on sketching, a dimensionality reduction technique that allows to embed high-dimensional subspaces into lower-dimensional ones while approximately preserving their geometry, such as inner products between vectors in the subspace [70]. These random linear maps (see early references [1, 20, 24, 48]) preserve enough information to enable, with high probability, solving a wide range of linear algebra problems accurately. This is demonstrated, for example, in early work on overdetermined least squares problems [29, 65]. We begin by introducing the definition of the ϵ-embedding property [70, 79].

Definition 1 Let $\mathcal{W} \subset \mathbb{R}^n$ be an m-dimensional vector subspace and $\epsilon \in \,]0,1[$. We say that $\Omega \in \mathbb{R}^{\ell\times n}$, $m \le \ell$, is an ϵ-embedding of $\mathcal{W}$ if and only if

$$\forall \mathbf{w} \in \mathcal{W}, \quad (1-\epsilon)\|\mathbf{w}\|^2 \le \|\Omega\mathbf{w}\|^2 \le (1+\epsilon)\|\mathbf{w}\|^2. \qquad (\epsilon\text{-embedding property}) \tag{1}$$

The vector $\Omega\mathbf{w} \in \Omega\mathcal{W} \subset \mathbb{R}^\ell$ is called the *sketch* of $\mathbf{w} \in \mathcal{W} \subset \mathbb{R}^n$. The ϵ-embedding property can be interpreted as the restriction of Ω to $\mathcal{W}$ being nearly isometric, that is, it maps $\mathcal{W}$ to a new vector subspace $\Omega\mathcal{W} \subset \mathbb{R}^\ell$ with very little distortion. It

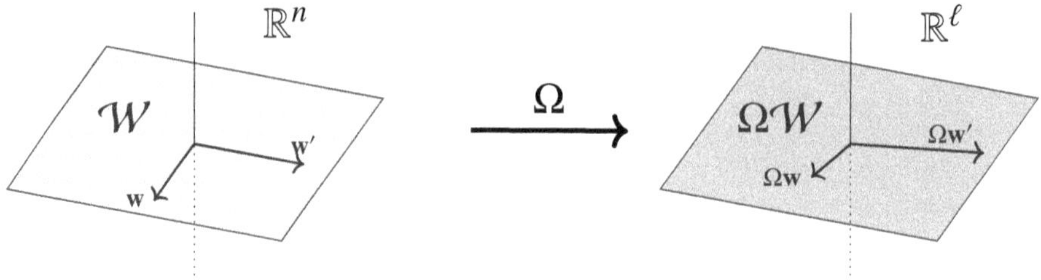

Figure 1 ϵ-embedding of a vector subspace $\mathcal{W} \subset \mathbb{R}^n$, with minor distortion of norms and angles, and conservation of the dimension of $\mathcal{W}$.

follows from (1) that $\mathrm{Ker}(\Omega) \cap \mathcal{W} = \{0\}$, and hence $\dim \mathcal{W} = \dim \Omega\mathcal{W}$; however, the latter lies in a vector space of much smaller dimension. We illustrate this property in Figure 1, where the norms of $\mathbf{w}, \mathbf{w}' \in \mathcal{W}$ and $\Omega\mathbf{w}, \Omega\mathbf{w}'$ are slightly different, the angles between $\mathbf{w}, \mathbf{w}'$ and $\Omega\mathbf{w}, \Omega\mathbf{w}'$ are slightly different, but the dimensions of $\mathcal{W}$, $\Omega\mathcal{W}$ are the same. Although Ω does not induce a proper inner product, as $\Omega^T\Omega$ is only positive semi-definite, it has been shown that it defines a proper norm when restricted to the embedded space $\mathcal{W}$ with $\|\mathbf{w}\|^2_{\Omega^T\Omega} = \mathbf{w}^T\Omega^T\Omega\mathbf{w}$ for $\mathbf{w} \in \mathcal{W}$; see, e.g., [7].

When considering a matrix $W \in \mathbb{R}^{n\times m}$ such that $\mathcal{W} = \mathrm{Range}(W)$, the ϵ-embedding property allows to establish spectral relations between W and its sketch ΩW. For example, it is derived in [35] that

$$\sqrt{1-\epsilon} \leq \frac{\sigma_j(\Omega W)}{\sigma_j(W)} \leq \sqrt{1+\epsilon} \qquad \text{for } j = 1, \ldots, m,$$

which in turn implies that

$$\mathrm{Cond}(W) \leq \sqrt{\frac{1+\epsilon}{1-\epsilon}} \cdot \mathrm{Cond}(\Omega W), \tag{2}$$

where $\mathrm{Cond}(M)$ denotes the spectral condition number of the matrix M.

Such ϵ-embedding sketching matrices can be efficiently obtained by drawing from simple random distributions $\mathcal{D}$ over $\mathbb{R}^{\ell\times n}$ while satisfying (1) with high probability, without knowledge of $\mathcal{W}$.

Definition 2 [79, Definition 2.2] Let $\mathcal{D}$ be a distribution over matrices of $\mathbb{R}^{\ell\times n}$ with $m \leq \ell$. We say that $\mathcal{D}$ is an (ϵ, δ, m)-oblivious subspace embedding (OSE) if and only if for any $\Omega \in \mathbb{R}^{\ell\times n}$ drawn from $\mathcal{D}$ and any given m-dimensional subspace $\mathcal{W} \subset \mathbb{R}^n$, Ω is an ϵ-embedding of $\mathcal{W}$ with probability at least $1-\delta$.

Depending on the distribution $\mathcal{D}$, it is possible to construct an OSE for which the order of magnitude of ℓ may be $O(\epsilon^{-2}m)$ or $O(\epsilon^{-2}m\log(m))$, the calculation $\Omega\mathbf{x}$ may be as cheap as $n\log(n)$ flops, and the storage cost of Ω may be as small as that of $\ell + n$ integers. Historically, such distributions $\mathcal{D}$ have been first described as yielding embeddings Ω of a finite set E_d of d vectors (and not of the subspace they generate). Indeed, for $\delta \in]0,1[$, and for an integer ℓ greater than a modest multiple of $\epsilon^{-2}\log(d/\delta)$, the celebrated Johnson–Lindenstrauss lemma [48] shows that there exist distributions $\mathcal{D}$ over $\mathbb{R}^{\ell \times n}$ such that, for any given set E_d of d vectors, the following event occurs with probability at least $1 - \delta$:

$$\forall\, \mathbf{x}_i \neq \mathbf{x}_j \in E_d, \quad (1-\epsilon)\|\mathbf{x}_i - \mathbf{x}_j\|^2 \leq \|\Omega \mathbf{x}_i - \Omega \mathbf{x}_j\|^2 \leq (1+\epsilon)\|\mathbf{x}_i - \mathbf{x}_j\|^2.$$

Let us now outline some concrete OSEs. We first outline the Gaussian sketching distribution. To draw from this distribution, we simply draw each entry of the matrix $\Omega \in \mathbb{R}^{\ell \times n}$ independently from $\mathcal{N}(0, 1)$, and scale the resulting matrix by $\ell^{-1/2}$. Provided that the sampling size ℓ is set to

$$\ell = O\left(\frac{1}{\epsilon^2}(m + \log(1/\delta))\right), \tag{3}$$

this distribution is a (ϵ, δ, m)-OSE [79, Theorem 2.3]. This requirement on the sampling size coincides with that of a Johnson–Lindenstrauss transform for an exponential number of arbitrary points [79, Theorem 2.1]. In that sense, it is optimal [53]. In practice, setting the sampling size $\ell = O(m)$, we get an embedding matrix Ω with a parameter $\epsilon \approx 1/2$ with high probability [56]. Despite its favorable theoretical properties, the main downside of a Gaussian sketching matrix is that it is dense and unstructured and thus costly to store and apply.

We next outline the s-hashing distribution, also known as sparse sign distribution. To draw $\Omega \in \mathbb{R}^{\ell \times n}$ from this distribution, we randomly choose s entries in each column of Ω, randomly set them to $\{-1/s^{1/2},\ +1/s^{1/2}\}$, and set all other entries to zero, resulting in sparse columns of unit norm. This results in a sparse matrix Ω with exactly ns nonzero entries. The more balanced the rows of $\mathcal{W}$, the lower we can set s and ℓ, as shown in [18]. From [21], the s-hashing ensemble is an (ϵ, δ, m)-OSE provided that the sampling size ℓ and the parameter s satisfy

$$\ell = O\left(\frac{1}{\epsilon^2}(m + \log(m/\delta))\right) \quad \text{and} \quad s \geq O\left(\frac{1}{\epsilon}\log(m/\delta)^{5/2} + \log(m/\delta)^4\right), \tag{4}$$

which highlights a trade-off between the sampling size ℓ and the number s of nonzero entries in each column. However, it has been experimentally observed that a constant parameter $s = 8$ and a sampling size $\ell = O(m\log(m))$ produce an embedding matrix Ω with parameter $\epsilon \approx 1/2$ with high probability for a wide variety of applications [77].

We finally outline the subsampled randomized Hadamard transform (SRHT), or SRHT distribution. Assuming that $n = 2^p$, to draw $\Omega \in \mathbb{R}^{\ell \times n}$ from the SRHT distribution, we first draw a diagonal matrix $D \in \mathbb{R}^{n \times n}$ whose diagonal entries are signs ± 1 drawn uniformly at random. We then apply the Walsh–Hadamard transform $H \in \mathbb{R}^{n \times n}$, defined by

$$H_1 := \begin{bmatrix} 1 & 1 \\ 1 & -1 \end{bmatrix} \in \mathbb{R}^{2\times 2}, \quad H_j := \begin{bmatrix} H_{j-1} & H_{j-1} \\ H_{j-1} & -H_{j-1} \end{bmatrix} \in \mathbb{R}^{2^j \times 2^j} \quad j \geq 2, \tag{5}$$

$$H := \frac{1}{\sqrt{n}} H_p \in \mathbb{R}^{n \times n}. \tag{6}$$

When applied to a vector, this transformation uniformly distributes its mass across all its entries, with high probability [76]. We then draw ℓ rows of the identity matrix, uniformly at random and without replacement, to form $P \in \mathbb{R}^{\ell \times n}$ (applying P to a vector is equivalent to sampling entries of this vector uniformly at random without replacement). A final scaling is required to compensate for the sampling:

$$\Omega = \sqrt{\frac{n}{\ell}} PHD \in \mathbb{R}^{\ell \times n}.$$

The matrix $H_p \in \mathbb{R}^{2^p \times 2^p}$ is a structured matrix, built recursively. For this reason, it can be applied without being formed by means of a fast recursive routine, such as the Fast-Walsh–Hadamard transform, which requires only $O(n \log(n))$ flops. In the frequent case where $2^p < n < 2^{p+1}$, the input matrix can simply be padded with a block of zeros, so that it fits the application of H_{p+1}. The SRHT distribution is an (ϵ, δ, m)-OSE for a sampling size ℓ such that [5, 7, 79]

$$\ell = O\left(\frac{1}{\epsilon^2}(\sqrt{m} + \sqrt{\log(n/\delta)})^2 \log(m/\delta)\right). \tag{7}$$

Assuming that $m \gg \log(n)$, it is shown in [76] that the sampling size ℓ can be set to $O(m \log(m))$, producing an embedding matrix Ω with parameter $\epsilon \approx 1/2$ with high probability for a wide variety of applications. The $\log(m)$ factor in the sampling size ℓ is necessary in the worst case: see, for instance, [43, Remark 11.2].

3 Computation of a Well-Conditioned Basis through Randomization

In this section, we outline the theoretical principles underlying the randomized orthogonalization framework. We then present three approaches for computing the randomized QR decomposition of a tall-and-skinny matrix W, namely the randomized Cholesky QR algorithm, the randomized Gram–Schmidt process, and the randomized Householder QR factorization.

3.1 General Discussion

Let $\mathcal{W} \subset \mathbb{R}^n$ be an m-dimensional subspace, and let $W \in \mathbb{R}^{n \times m}$ be a full-rank matrix such that $\text{Range}(W) = \mathcal{W}$. In many applications, the construction of an orthonormal basis of $\mathcal{W}$ is a key algorithmic component. For example, an orthonormal basis Q of $\mathcal{W}$ is constructed when solving an overdetermined least squares problem with coefficient matrix W, or when $\mathcal{W}$ is a Krylov subspace employed in the solution of a linear system or eigenvalue problem. An orthonormal basis of $\mathcal{W}$ can be constructed via a Householder QR factorization or a Gram–Schmidt process. Both algorithms construct

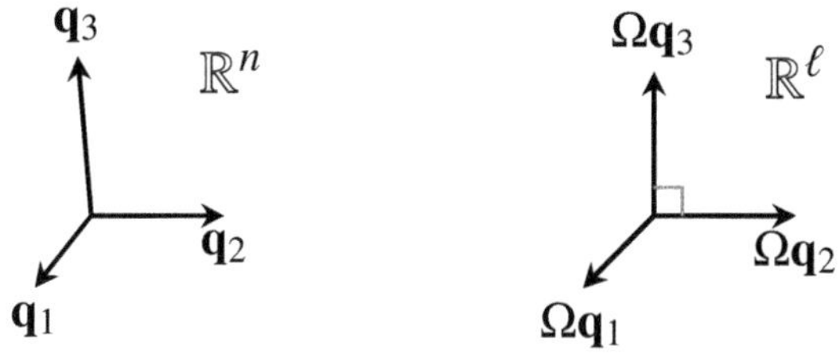

Figure 2 Sketch-orthogonal basis (left) and its orthogonal sketch (right).

the factorization $W = QR$, where $Q \in \mathbb{R}^{n\times m}$ has orthonormal columns and $R \in \mathbb{R}^{m\times m}$ is upper triangular and both have a computational cost of $O(nm^2)$.

Let us denote by $\mathcal{P}_{\mathcal{W}}$ the orthogonal projector to $\mathcal{W}$. We recall that this projector satisfies the following properties:

- for any $\mathbf{x} \in \mathbb{R}^n$ we have $\mathcal{P}_{\mathcal{W}}\mathbf{x} = \operatorname{argmin}_{\mathbf{w}\in\mathcal{W}}\|\mathbf{x} - \mathbf{w}\|$,
- we have $\mathcal{P}_{\mathcal{W}} = QQ^T$, where Q is an orthonormal basis of $\mathcal{W}$,
- we have $\mathcal{P}_{\mathcal{W}} = ZZ^+$ for an arbitrary basis Z of $\mathcal{W}$, where $Z^+ = (Z^TZ)^{-1}Z^T$ denotes the Moore–Penrose pseudoinverse of Z.

In this section, we introduce the concept of randomized QR factorization and present efficient algorithms for its computation. Assume that $\Omega \in \mathbb{R}^{\ell\times n}$ is an ϵ-embedding for $\mathcal{W}$, and consider the decomposition:

$$W = QR, \qquad \Omega W = \Omega Q \cdot R = SR, \quad (\Omega Q)^T\Omega Q = S^TS = I_m, \quad R \text{ upper triangular.} \tag{8}$$

We refer to this decomposition as a *randomized QR factorization* of W. Note that the columns of $Q = [\mathbf{q}_1 \cdots \mathbf{q}_m]$ are not orthogonal in general, but their sketches $\{\Omega\mathbf{q}_1, \ldots, \Omega\mathbf{q}_m\}$ form an orthonormal basis of $\operatorname{Range}(\Omega W)$, as illustrated in Figure 2. Although Q is not orthonormal, it is guaranteed to be extremely well-conditioned due to the ϵ-embedding property of Ω. Indeed, (2) implies that we have

$$\operatorname{Cond}(Q) \le \sqrt{\frac{1+\epsilon}{1-\epsilon}} \cdot \operatorname{Cond}(\Omega Q) = \sqrt{\frac{1+\epsilon}{1-\epsilon}}.$$

This property is fundamental to successfully applying the decomposition (8) to the solution of a variety of linear algebra problems, as we discuss in the following sections.

Before we present the algorithms to compute a randomized QR factorization, we introduce the *sketch-orthogonal projector* $\mathcal{P}^{\Omega}_{\mathcal{W}}$ to $\mathcal{W}$. Given an embedding Ω of $\mathcal{W} = \operatorname{Range}(W)$, given a randomized QR factorization (8), we define:

$$\forall \mathbf{x} \in \mathbb{R}^n, \quad \mathcal{P}^{\Omega}_{\mathcal{W}}\mathbf{x} = Q(\Omega Q)^T\Omega\mathbf{x}.$$

Using the sketch-orthogonality relation $(\Omega Q)^T\Omega Q = I$, we verify that $\mathcal{P}^{\Omega}_{\mathcal{W}}$ is indeed a projector, that is, it satisfies $(\mathcal{P}^{\Omega}_{\mathcal{W}})^2 = \mathcal{P}^{\Omega}_{\mathcal{W}}$. More precisely, it is an *oblique* (non-orthogonal) projector. We remark that it satisfies the identity

$$\Omega \cdot \mathcal{P}_{\mathcal{W}}^{\Omega} = \mathcal{P}_{\Omega\mathcal{W}} \cdot \Omega, \tag{9}$$

where $\mathcal{P}_{\Omega\mathcal{W}} = \Omega Q(\Omega Q)^T$ is the orthogonal projector onto $\Omega\mathcal{W} \subset \mathbb{R}^\ell$. The sketch-orthogonal projector satisfies the following properties:

- for any $\mathbf{x} \in \mathbb{R}^n$ we have $\mathcal{P}_{\mathcal{W}}^{\Omega}\mathbf{x} = \operatorname{argmin}_{\mathbf{w}\in\mathcal{W}} \|\Omega(\mathbf{x} - \mathbf{w})\|$,
- for any $\mathbf{x} \in \mathbb{R}^n$ we have $\mathbf{x} - \mathcal{P}_{\mathcal{W}}^{\Omega}\mathbf{x} \perp^{\Omega} \mathcal{W}$, where we use $\perp^{\Omega}$ to denote the sketch-orthogonality condition $\Omega(\mathbf{x} - \mathcal{P}_{\mathcal{W}}^{\Omega}\mathbf{x}) \perp \Omega\mathcal{W}$,
- given an arbitrary basis Z of $\mathcal{W}$, the sketch-orthogonal projector can be equivalently written as $\mathcal{P}_{\mathcal{W}}^{\Omega} = Z(\Omega Z)^+\Omega$.

The sketch-orthogonal projector can be used as an approximation of the standard orthogonal projector [5, 6]. Let $\mathbf{b} \in \mathbb{R}^n$, and assume that Ω is an ϵ-embedding for $\mathcal{W} + \mathrm{Range}(\mathbf{b})$. Recalling the ε-embedding property and (9), we have

$$\begin{aligned}\|\mathbf{b} - \mathcal{P}_{\mathcal{W}}^{\Omega}\mathbf{b}\| \le \frac{1}{\sqrt{1-\epsilon}}\|\Omega\mathbf{b} - \Omega\mathcal{P}_{\mathcal{W}}^{\Omega}\mathbf{b}\| &= \frac{1}{\sqrt{1-\epsilon}}\|\Omega\mathbf{b} - \mathcal{P}_{\Omega\mathcal{W}}\Omega\mathbf{b}\| \\ &= \frac{1}{\sqrt{1-\epsilon}} \min_{\mathbf{z}\in\Omega\mathcal{W}} \|\Omega\mathbf{b} - \mathbf{z}\|,\end{aligned}$$

where for the last equality, we used the optimality property of the orthogonal projector $\mathcal{P}_{\Omega\mathcal{W}}$. For any fixed $\mathbf{w} \in \mathcal{W}$, we have $\mathbf{z} = \Omega\mathbf{w} \in \Omega\mathcal{W}$, so again using the ε-embedding property, we get

$$\|\mathbf{b} - \mathcal{P}_{\mathcal{W}}^{\Omega}\mathbf{b}\| \le \frac{1}{\sqrt{1-\epsilon}}\|\Omega(\mathbf{b} - \mathbf{w})\| \le \sqrt{\frac{1+\epsilon}{1-\epsilon}}\|\mathbf{b} - \mathbf{w}\|.$$

By taking the minimum over $\mathbf{w} \in \mathcal{W}$, we finally obtain

$$\|\mathbf{b} - \mathcal{P}_{\mathcal{W}}^{\Omega}\mathbf{b}\| \le \sqrt{\frac{1+\epsilon}{1-\epsilon}} \cdot \min_{\mathbf{w}\in\mathcal{W}} \|\mathbf{b} - \mathbf{w}\|. \tag{10}$$

This shows that the sketch-orthogonal projection of $\mathbf{b}$ onto $\mathcal{W}$ is a *quasi-optimal minimizer* of the distance between $\mathbf{b}$ and $\mathcal{W}$, so the sketch-orthogonal projector $\mathcal{P}_{\mathcal{W}}^{\Omega}$ acts as an approximation of the orthogonal projector $\mathcal{P}_{\mathcal{W}}$. This property is illustrated in Figure 3.

This property has a straightforward implication for the solution of a least squares problem. Indeed, let us denote by $\mathbf{x}^\star$ the solution of the sketched least squares problem

$$\mathbf{x}^\star = \operatorname*{argmin}_{\mathbf{x}\in\mathbb{R}^m} \|\Omega W\mathbf{x} - \Omega\mathbf{b}\|.$$

Then we have $W\mathbf{x}^\star = \mathcal{P}_{\mathcal{W}}^{\Omega}\mathbf{b}$, and from (10) it follows that $\mathbf{x}^\star$ satisfies

$$\|W\mathbf{x}^\star - \mathbf{b}\| \le \sqrt{\frac{1+\epsilon}{1-\epsilon}} \cdot \min_{\mathbf{x}\in\mathbb{R}^m} \|W\mathbf{x} - \mathbf{b}\|,$$

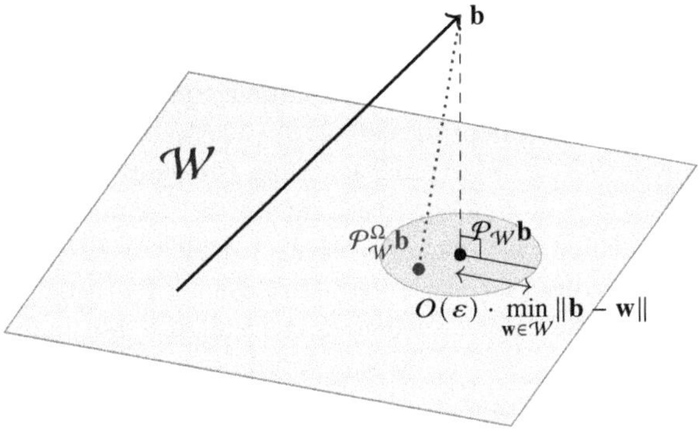

Figure 3 Quasi-optimality of the sketched projection $\mathcal{P}^{\Omega}_{\mathcal{W}}\mathbf{b}$.

i.e., it is an approximate solution of the corresponding non-sketched least squares problem (see, e.g., [70] and [61, Section 2.2]). Furthermore, the randomized QR factorization in (8) yields a closed-form formula for the computation of $\mathbf{x}^\star$:

$$W\mathbf{x}^\star = \mathcal{P}^{\Omega}_{\mathcal{W}}\mathbf{b} \quad \Longrightarrow \quad QR\mathbf{x}^\star = Q(\Omega Q)^T\Omega\mathbf{b} \quad \Longrightarrow \quad \mathbf{x}^\star = R^{-1}(\Omega Q)^T\Omega\mathbf{b}.$$

We conclude this introductory section on the randomized QR factorization by outlining a framework for computing it, which underpins the randomized QR processes that we present in the following sections. Given a full-rank matrix $W \in \mathbb{R}^{n\times m}$ and an ϵ-embedding $\Omega \in \mathbb{R}^{\ell\times n}$ for $\mathcal{W} = \mathrm{Range}(W)$, a randomized QR factorization of W can be computed using the following simple procedure:

1. Compute the sketch $\Omega W \in \mathbb{R}^{\ell\times m}$.
2. Compute a QR factorization $\Omega W = SR$.
3. Set $Q = WR^{-1}$.

Then it follows that $\Omega Q = S$ and Q, S and R satisfy (8). The idea for this randomized orthogonalization procedure originates from [65], where the R factor is used as a preconditioner for the solution of a least squares problem, and is presented in [6, Algorithm 2.3]. This framework is sometimes simply called *randomized QR* [45, 65], or sometimes *randomized Cholesky QR* [4, 6], or *randomized preconditioning* or *sketch-and-precondition* [34, 57], and it is often used as a preconditioner for subsequent deterministic algorithms. For instance, in [65], where the authors propose to solve a least squares problem, the obtained factor R and the solution $\mathbf{x}^\star$ to the sketched least squares are used, respectively, as a preconditioner and a starting point for conjugate gradient iterations.

In the foundational work [65], and in many sketch-and-precondition papers [34, 57, 58], the authors obtain the triangular factor R from ΩW by pivoted (strong) rank-revealing factorization, rather than a simple QR factorization. Moreover, the randomized QR algorithm can be followed by an efficient algorithm for the computation of a QR factorization, such as standard deterministic CholeskyQR, to obtain an orthogonal factor Q [6]. For example, in [34, 57, 58], the matrix W is preconditioned with the truncated, pivoted factor R and this is followed by a standard deterministic CholeskyQR factorization. All of these algorithms are efficiently parallelizable, since they require a constant number of synchronizations. In addition, randomized QR + CholeskyQR hybrids perform most of their flops through BLAS3 kernels.

Instead of applying R^{-1} to W in order to obtain Q explicitly, one can also keep the basis W and compute $Q\mathbf{x}$ as $W(R^{-1}\mathbf{x})$, i.e., applying R^{-1} to the input vectors instead. This approach, often called *whitening of the basis*, is widely used in randomized Krylov subspace methods [41,61,63,64]. We refer to Section 4.2 for further details.

To close this section, we emphasize that, as in standard orthogonalization processes, there are multiple methodologies for the sketch orthogonalization of W. Although they are all equivalent in exact arithmetic, they accumulate rounding errors in different ways when performed in floating point arithmetic, and they suffer from these errors in various ways.

3.2 Randomized Gram–Schmidt

In this section we present the randomized Gram–Schmidt process [5,6] for computing the randomized QR decomposition (8) of a tall-and-skinny matrix $W \in \mathbb{R}^{n\times m}$. This randomized process is inspired by the deterministic Gram–Schmidt process, which we briefly recall here.

Let us denote by $\mathbf{w}_1, \dots \mathbf{w}_m$ the columns of W, and by $\mathcal{W}_j = \mathrm{Range}([\mathbf{w}_1, \dots, \mathbf{w}_j])$. The Gram–Schmidt process constructs an orthonormal basis $Q = [\mathbf{q}_1, \dots, \mathbf{q}_m]$ of $\mathrm{Range}(W)$ by iteratively subtracting from $\mathbf{w}_{j+1}$ its projection onto $\mathcal{W}_j$. More precisely, the algorithm sets $\mathbf{q}_1 = \mathbf{w}_1/\|\mathbf{w}_1\|$, and then for $j = 1, \dots, m-1$ we set

$$\widetilde{\mathbf{q}}_{j+1} = (I - \mathcal{P}_{\mathcal{W}_j})\mathbf{w}_{j+1}, \qquad \mathbf{q}_{j+1} = \widetilde{\mathbf{q}}_{j+1}/\|\widetilde{\mathbf{q}}_{j+1}\|. \tag{11}$$

The practical implementation of the Gram–Schmidt process depends on the specific implementation of the projector $\mathcal{P}_{\mathcal{W}_j}$. Letting $Q_j = [\mathbf{q}_1, \dots, \mathbf{q}_j]$, we have $\mathcal{P}_{\mathcal{W}_j} = Q_j Q_j^T$ and thus we can implement (11) as

$$\widetilde{\mathbf{q}}_{j+1} = (I - Q_j Q_j^T)\mathbf{w}_{j+1}, \qquad \mathbf{q}_{j+1} = \widetilde{\mathbf{q}}_{j+1}/\|\widetilde{\mathbf{q}}_{j+1}\|,$$

which corresponds to the *classical Gram–Schmidt* (CGS) algorithm. The main advantage of the CGS implementation is that it performs the inner products $Q_j \mathbf{w}_{j+1}$ by exploiting matrix-vector BLAS2 routines. Using the orthogonality of the columns of Q_j, we have $I - Q_j Q_j^T = \prod_{k=1}^{j}(I - \mathbf{q}_k\mathbf{q}_k^T)$, so we can also write (11) as

$$\widetilde{\mathbf{q}}_{j+1} = \prod_{k=1}^{j}(I - \mathbf{q}_k\mathbf{q}_k^T)\mathbf{w}_{j+1}, \qquad \mathbf{q}_{j+1} = \widetilde{\mathbf{q}}_{j+1}/\|\widetilde{\mathbf{q}}_{j+1}\|,$$

which corresponds to the *modified Gram–Schmidt* (MGS) algorithm. This algorithm computes the inner products with the columns of Q_j sequentially, so it is slower than CGS on modern computational architectures, but it has better numerical stability. In general, the stability of CGS and MGS can be improved by applying the projector $I - \mathcal{P}_{\mathcal{W}_j}$ twice, leading to the CGS2 and MGS2 algorithms. The numerical stability of different implementations of the Gram–Schmidt process and its relation with the condition number of W has been extensively studied in the literature, see, e.g., [19].

The randomized Gram–Schmidt process essentially replaces the orthogonal projector $\mathcal{P}_{\mathcal{W}_j}$ in (11) with the oblique projector $\mathcal{P}^{\Omega}_{\mathcal{W}_j}$ to construct a basis Q that is now sketch-orthogonal. The first column of Q is set as $\mathbf{q}_1 = \mathbf{w}_1/\|\mathbf{w}_1\|$, and for $j = 1, \ldots, m-1$ we compute

$$\widetilde{\mathbf{q}}_{j+1} = (I - \mathcal{P}^{\Omega}_{\mathcal{W}_j})\mathbf{w}_{j+1}, \qquad \mathbf{q}_{j+1} = \widetilde{\mathbf{q}}_{j+1}/\|\Omega\widetilde{\mathbf{q}}_{j+1}\|. \tag{12}$$

It immediately follows from the properties of the sketch-orthogonal projector $\mathcal{P}^{\Omega}_{\mathcal{W}_j}$ that $\mathbf{q}_{j+1} \perp^{\Omega} \mathcal{W}_j$, which in turn implies that $\Omega Q = [\Omega\mathbf{q}_1, \ldots, \Omega\mathbf{q}_m]$ is an orthonormal basis of Range(ΩW). Recall that since $Q_j = [\mathbf{q}_1, \ldots, \mathbf{q}_j]$ is a sketch-orthogonal basis of $\mathcal{W}_j$, we have $\mathcal{P}^{\Omega}_{\mathcal{W}_j} = Q_j(\Omega Q_j)^T\Omega = Q_j(\Omega Q_j)^+\Omega$ and we can more explicitly write

$$\widetilde{\mathbf{q}}_{j+1} = \mathbf{w}_{j+1} - Q_j\mathbf{h}_j, \qquad \mathbf{h}_j = (\Omega Q_j)^+\Omega\mathbf{w}_{j+1} = \underset{h\in\mathbb{R}^j}{\operatorname{argmin}}\|\Omega Q_j\mathbf{h} - \Omega\mathbf{w}_{j+1}\|. \tag{13}$$

From (13) we see that the main difference between the standard and randomized Gram–Schmidt processes is that the latter replaces inner products of vectors of length n with inner products of much shorter sketched vectors of length ℓ or the solution of an $\ell \times j$ least squares problem, drastically reducing the cost of this operation. Since the product $Q_j\mathbf{h}_j$ still needs to be performed, the overall computational cost is roughly half of that of the deterministic Gram–Schmidt process.

The solution of the least squares problem for the computation of $\mathbf{h}_j$ in (13) is crucial for the implementation of the randomized Gram–Schmidt process. By exploiting the fact that $\Omega\mathbf{Q}_j$ is sketch-orthogonal, we can simply compute $\mathbf{h}_j = (\Omega Q_j)^T\Omega\mathbf{w}_{j+1}$ and obtain the *randomized classical Gram–Schmidt* algorithm. However, since ΩQ_j is a small $\ell \times j$ matrix, we can afford to solve the least squares problem with a more expensive method to achieve better numerical stability, without assuming that ΩQ_j has orthonormal columns, which in general is not true in finite-precision arithmetic. We refer to [5, Section 2.3] for further details. We also mention that this process can be combined with deterministic reorthogonalization to obtain a basis with orthonormal columns; see for instance [47, Section 3.1], where an orthogonal projector is obtained as a combination of the randomized Gram–Schmidt projector and either CGS or MGS.

The randomized Gram–Schmidt process (RGS) is presented in Algorithm 1. As detailed in [5], its flop cost is dominated by $nm^2 + 2mt$ flops, where t is the flop cost of sketching one vector. With SRHT, we thus get $nm^2 + 2nm\log(n)$ flops, namely half of the number of flops of Gram–Schmidt processes. This algorithm can be executed with two synchronizations per step: one to sketch $\mathbf{w}_j$ in line 5, and a second to sketch the updated vector $\mathbf{w}$ in line 8. If $\mathbf{w}_{j+1}, \mathbf{w}_{j+2} \ldots$ are already available at step j (e.g., in s-step Krylov methods or block Krylov methods), they can be sketched in line 5 already at step j, saving one synchronization for those next steps. If the whole matrix $W \in \mathbb{R}^{n\times m}$ is available, it can be sketched in a single synchronization, allowing to remove line 5 completely, which allows RGS to be executed with a single synchronization per step.

Algorithm 1 Randomized Gram–Schmidt process

Input: $W \in \mathbb{R}^{n\times m}$ full-rank, $\Omega \in \mathbb{R}^{\ell\times n}$ that is an ϵ-embedding for Range(W)
Output: $Q \in \mathbb{R}^{n\times m}$, $S \in \mathbb{R}^{\ell\times m}$ and $R \in \mathbb{R}^{m\times m}$ such that $W = QR$, $S = \Omega Q$, $S^T S = I_m$, R upper triangular

1: **function** Randomized-Gram–Schmidt(W, Ω)
2: $\quad \mathbf{z} \leftarrow \Omega \mathbf{w}_1$
3: $\quad R_1 \leftarrow [\|\mathbf{z}\|]$, $Q_1 \leftarrow [\mathbf{w}_1/\|\mathbf{z}\|]$, $S_1 \leftarrow [\mathbf{z}/\|\mathbf{z}\|]$
4: $\quad$ **for** $j = 2 : m$ **do**
5: $\quad\quad \mathbf{z} \leftarrow \Omega \mathbf{w}_j$
6: $\quad\quad \mathbf{r} \leftarrow S_{j-1}^{+}\mathbf{z}$ # use a stable method to solve the least squares problem
7: $\quad\quad \mathbf{w} \leftarrow \mathbf{w}_j - Q_{j-1}\mathbf{r}$
8: $\quad\quad \mathbf{z} \leftarrow \Omega \mathbf{w}$
9: $\quad\quad R_j \leftarrow \left[\begin{array}{c|c} R_{j-1} & \mathbf{r} \\ \hline 0_{1\times(j-1)} & \|\mathbf{z}\| \end{array}\right], \quad Q_j \leftarrow [Q_{j-1} \mid \mathbf{w}/\|\mathbf{z}\|], \quad S_j \leftarrow [S_{j-1} \mid \mathbf{z}/\|\mathbf{z}\|]$
10: $\quad$ **end for**
11: $\quad$ **return** $Q = Q_m$, $S = S_m$, $R = R_m$
12: **end function**

This communication cost is identical to that of CGS. As in CGS, most of the flops between synchronizations in RGS can be carried out by BLAS2 routines.

We emphasize that the sketch in line 8 is crucial for numerical stability, as shown in the finite-precision analysis in [5]. An algorithm that replaces this sketch with a formula inferring $\Omega\mathbf{w}$ from $\mathbf{z} - S_{j-1}\mathbf{r} \in \mathbb{R}^{\ell}$ would not qualify as an implementation of RGS, but rather as an implementation of the *sketch-and-precondition* framework. We also emphasize that line 6 is specified by which orthogonalization method is chosen by the user to orthogonalize ΩW. It is crucial to select one that is stable enough to handle the successful orthogonalization of ΩW.

We illustrate in Figure 4 the numerical efficiency of RGS compared to CGS on a medium-difficulty example. The input matrix $W \in \mathbb{R}^{n\times m}$ is initialized in double precision through an SVD formula, with singular values decreasing exponentially from 10^2 to 10^{-2}. CGS is tested in single precision. RGS is tested in single precision, and in mixed precision (with low-dimensional operations done in double precision and remaining operations in single precision), with the results cast into single precision. The number of rows of the input matrix W is $n = 100\,000$. The sketching matrix is drawn from an 8-hash distribution, with sampling size set to $\ell = 2m = 600$. We see in Figure 4a that the basis obtained with CGS loses orthogonality slowly in the first steps, and then much more quickly, near the point where $W^T W$ becomes numerically singular. At the same time, the sketch of the basis generated by RGS is numerically orthogonal, which in turn explains the small condition number of the resulting basis. As illustrated in Figure 4b, all factorization errors remain very small, with CGS performing marginally better. Employing mixed precision for RGS further enhances the factorization accuracy.

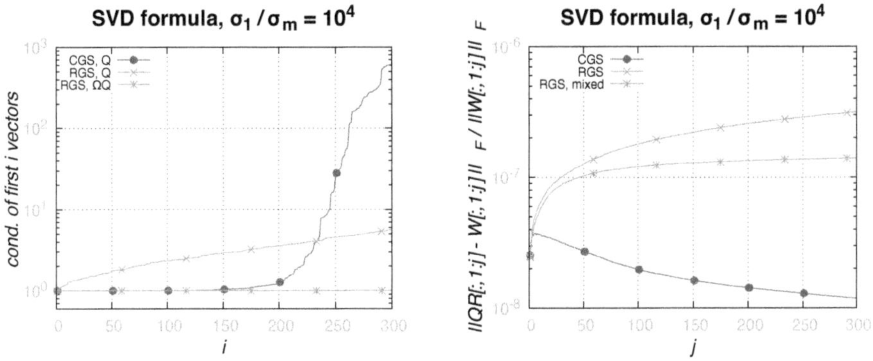

(a) Condition number of basis Q and *a posteriori* sketch ΩQ built by RGS, compared with CGS.

(b) Relative Frobenius factorization errors of RGS in both single and mixed precision, compared with CGS.

Figure 4 Comparison of CGS and RGS on a medium-difficulty example.

3.3 Randomized Householder QR

We now introduce a randomized version of the celebrated Householder QR factorization. We only outline here the elements that are directly used in this factorization. More general properties can be found in [39]. We give first a brief summary of the standard Householder QR.

The Householder QR is an orthogonalization process alternative to the Gram–Schmidt process. The central operator of the process is the *Householder reflector*:

$$P = I_n - \beta \mathbf{u}\mathbf{u}^T, \quad \mathbf{u} \in \mathbb{R}^n \setminus \{0\}, \quad \beta = 2/\|\mathbf{u}\|^2.$$

It is an orthogonal reflector, i.e., it satisfies $P^T P = P^2 = PP^T = I_n$. The Householder process is derived from the ability to easily generate a Householder reflector P that annihilates all the entries in a given vector $\mathbf{w}$ below a given index. Indeed, for some $\mathbf{w}_j \in \mathbb{R}^n$, denoting by $\mathbf{w}'$ the vector formed by $j-1$ zeros followed by the last $n-j+1$ entries in $\mathbf{w}_j$, we may define

$$\rho_j := \|\mathbf{w}'\|, \quad \sigma_j := \operatorname{sign}(\mathbf{e}_j^T \mathbf{w}'), \quad \mathbf{u}_j := \mathbf{w}' + \sigma_j \|\mathbf{w}'\| \mathbf{e}_j, \quad \beta_j := 2/\|\mathbf{u}_j\|^2 \tag{14}$$

and verify that $P_j = I_n - \beta_j \mathbf{u}_j \mathbf{u}_j^T$ annihilates all entries of $\mathbf{w}_j$ strictly below the j-th index, while not modifying the first $j-1$ entries of any vector. We can thus triangularize an arbitrary matrix $W \in \mathbb{R}^{n \times m}$ with Householder reflectors: we generate P_1 that annihilates the first column $\mathbf{w}_1$ below the first index, and apply it to the whole matrix; then we generate P_2 that annihilates the updated second column $P_1 \mathbf{w}_2$ below the second index and does not modify the first row of $P_1 W$, and apply it to the whole matrix, and continue similarly on the following columns, as illustrated in Figure 5. We obtain:

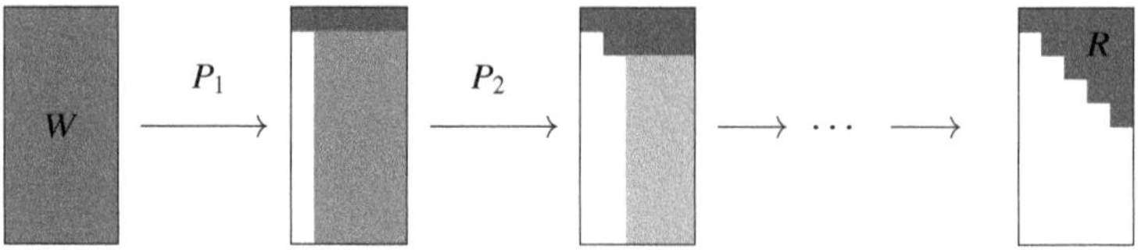

Figure 5 Triangularizing W through Householder reflections.

$$P_m P_{m-1} \cdots P_1 W = \begin{bmatrix} R \\ 0_{(n-m)\times m} \end{bmatrix} \implies W = P_1 \cdots P_m \begin{bmatrix} R \\ 0_{(n-m)\times m} \end{bmatrix}.$$

As shown in [71], the reflectors can be aggregated as follows:

$$W = (I_n - UTU^T) \begin{bmatrix} R \\ 0_{(n-m)\times m} \end{bmatrix},$$
$$U = [\mathbf{u}_1 \ \cdots \ \mathbf{u}_m] \in \mathbb{R}^{n\times m}, \quad T \in \mathbb{R}^{m\times m} \text{ upper triangular.}$$

Let us now outline how this process can be randomized in order to obtain a randomized QR factorization equivalent to (8). We use the following wrapper Ψ of any sketching matrix Ω:

$$\Psi = \left[\begin{array}{c|c} I_m & \\ \hline & \Omega \end{array}\right] \in \mathbb{R}^{(\ell+m)\times n}, \quad \Omega \in \mathbb{R}^{\ell\times(n-m)}. \tag{15}$$

Computing the sketch of a vector $\mathbf{x} \in \mathbb{R}^n$ with Ψ thus consists of sketching the last $n-m$ coordinates of $\mathbf{x}$ with $\Omega \in \mathbb{R}^{\ell\times(n-m)}$, and concatenating the result to the first m coordinates of $\mathbf{x}$. For simplicity, we denote $\ell' = \ell + m$ throughout this section. We then define the randomized Householder reflector associated with Ψ as any matrix of the form

$$P = I_n - \beta \mathbf{u} (\Psi\mathbf{u})^T \Psi \in \mathbb{R}^{n\times n}, \quad \mathbf{u} \in \mathbb{R}^n \setminus \mathrm{Ker}(\Psi), \quad \beta = 2/\|\Psi\mathbf{u}\|^2.$$

We can verify that $P \in \mathbb{R}^{n\times n}$ defined in this way satisfies $P^2 = I_n \neq P^T P$, i.e., it is an oblique reflection, with respect to the hyperplane $\{\mathbf{x} \in \mathbb{R}^n : (\Psi\mathbf{u})^T\Psi\mathbf{x} = 0\}$, and with $\mathbf{u}$ being mapped to $-\mathbf{u}$. We also verify that $\Psi P = P'\Psi$ for some standard Householder reflector $P' \in \mathbb{R}^{\ell'\times\ell'}$. We illustrate the two related reflectors in Figure 6.

For any vector $\mathbf{w}_j \in \mathbb{R}^n$, $j \leq m$, using the same notation for $\mathbf{w}'$, we can straightforwardly *randomize* the formulas in (14):

$$\rho_j := \|\Psi\mathbf{w}'\|, \quad \sigma_j := \mathrm{sign}(\mathbf{e}_j^T\mathbf{w}'), \quad \mathbf{u}_j := \mathbf{w}' + \sigma_j\|\Psi\mathbf{w}'\|\mathbf{e}_j, \quad \beta_j := 2/\|\Psi\mathbf{u}_j\|^2. \tag{16}$$

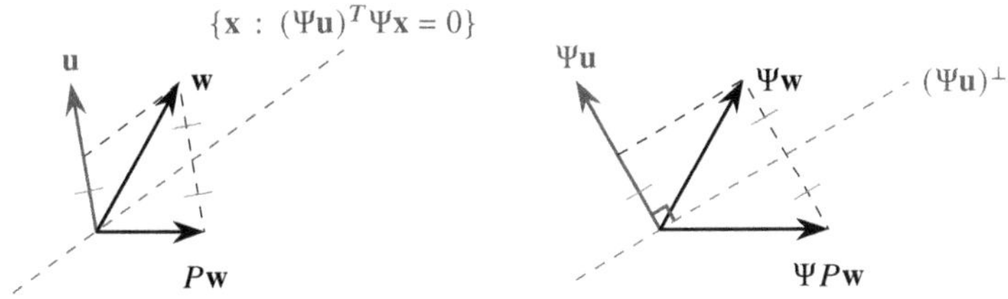

Figure 6 Randomized Householder reflector $P \in \mathbb{R}^{n\times n}$ (left) and induced Householder reflector $P' \in \mathbb{R}^{(\ell+m)\times(\ell+m)}$ on the sketched space (right).

The associated randomized Householder reflector $P_j = I_n - \beta_j \mathbf{u}_j(\Psi\mathbf{u}_j)^T\Psi$ also annihilates the last $n-j$ coordinates of $\mathbf{w}$, while not modifying the first $j-1$ entries of any vector. We can thus proceed as in the standard Householder process and generate randomized Householder reflectors $P_1 \dots P_m \in \mathbb{R}^{n\times n}$ satisfying

$$W = P_1 \cdots P_m \begin{bmatrix} R \\ 0_{(n-m)\times m} \end{bmatrix} = \left(I_n - UT(\Psi U)^T\Psi\right)\begin{bmatrix} R \\ 0_{(n-m)\times m} \end{bmatrix} \in \mathbb{R}^{n\times n},$$

$$\Psi W = \underbrace{\Psi\left(I_n - UT(\Psi U)^T\Psi\right)\begin{bmatrix} R \\ 0_{(n-m)\times m} \end{bmatrix}}_{\text{RHQR fact. of W}} = \underbrace{\left(I_{\ell'} - \Psi U\,T(\Psi U)^T\right)\begin{bmatrix} R \\ 0_{\ell\times m} \end{bmatrix}}_{\text{HQR fact. of } \Psi W} \in \mathbb{R}^{\ell'\times m}.$$

All these elements yield Algorithm 2 (RHQR). As detailed in [39], this algorithm has the same computational and communication cost as RGS, while also leveraging mainly BLAS2 routines between sketches. Compared with the original Householder process, the randomized paradigm allows to aggregate the reflectors without synchronizations and for a negligible arithmetic cost. RHQR with SRHT is thus twice as cheap as non-aggregated Householder QR, and thrice as cheap as aggregated Householder QR.

We illustrate the numerical stability of RHQR in Figure 7 for a difficult example. The input matrix $W \in \mathbb{R}^{n\times m}$ is initialized in double precision with an SVD formula, with its singular values decreasing exponentially from 10^4 to 10^{-4}. It is then cast in single precision. Householder QR is tested in single precision. RGS and RHQR are tested in single and in mixed precisions (with low-dimensional operations done in double precision). The number of rows of the input matrix W is $n = 100\,000$. The sketching matrix is drawn from an 8-hash distribution, with sampling size set to $\ell = 2m = 600$. We see in Figure 7a that the sketch of the basis computed by RGS loses orthogonality. When using mixed precision, this phenomenon occurs later. The loss of orthogonality in RGS, and the growth of the condition number of the basis, are very well mitigated by the use of mixed precision. Meanwhile, RHQR maintains the orthogonality of the sketch of the basis, which explains the small condition number of the basis itself. The use of mixed precision in RHQR does not substantially improve the condition number of the basis, as numerical sketched orthogonality is already achieved with single precision. The basis obtained with Householder QR also achieves numerical orthogonality. We see in Figure 7b that all factorization errors of randomized algorithms are small, with a noticeable advantage when compared to Householder QR. In single precision, the

Algorithm 2 Randomized Householder QR (left-looking)

Input: Matrix $W = [\mathbf{w}_1 \mid \cdots \mid \mathbf{w}_m] \in \mathbb{R}^{n\times m}, \Omega \in \mathbb{R}^{\ell\times(n-m)}, m < \ell \ll n - m$
Output: $U \in \mathbb{R}^{n\times m}, S \in \mathbb{R}^{(\ell+m)\times m}$, $T, R \in \mathbb{R}^{m\times m}$ such that $S = \Psi U$ and $W = (I - UT(\Psi U)^T\Psi)\cdot \left[R;\ 0_{(n-m)\times m}\right]$

function RHQR(W, Ω)
  $\mathbf{z} \leftarrow \Psi \mathbf{w}_1$
  Define $\rho_1, \sigma_1, \mathbf{u}_1,\ \mathbf{s}_1 = \Psi\mathbf{u}_1, \beta_1$ as in equations (16)
  $U_1 \leftarrow [\mathbf{u}_1],\ S_1 \leftarrow [\mathbf{s}_1],\ T_1 \leftarrow [\beta_1],\ R_1 \leftarrow [-\sigma_1\rho_1]$
  for $j = 2 : m$ **do**
    $\mathbf{z} \leftarrow \Psi \mathbf{w}_j$
    $\mathbf{w} \leftarrow \mathbf{w}_j - U_{j-1}T_{j-1}^T S_{j-1}^T \mathbf{z}$
    $\mathbf{z} \leftarrow \Psi \mathbf{w}$
    Define $\rho_j, \sigma_j, \mathbf{u}_j,\ \mathbf{s}_j = \Psi\mathbf{u}_j, \beta_j$ as in equations (16)
    $U_j \leftarrow \left[U_{j-1} \mid \mathbf{u}_j\right],\ S_j \leftarrow \left[S_{j-1} \mid \mathbf{s}_j\right],\ R_j \leftarrow \begin{bmatrix} R_{j-1} & (\mathbf{z})_{1:j-1} \\ 0_{1\times(j-1)} & -\sigma_j\rho_j \end{bmatrix},$
    $T_j \leftarrow \begin{bmatrix} T_{j-1} & -\beta_j T_{j-1} S_{j-1}^T \mathbf{s}_j \\ 0_{1\times(j-1)} & \beta_j \end{bmatrix}$
  end for
  return $R_m,\ U_m,\ S_m,\ T_m$
end function

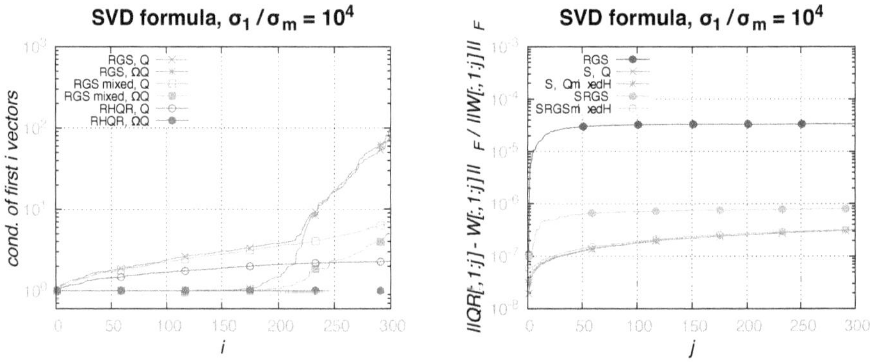

(a) Condition number of basis Q and *a posteriori* sketch ΩQ built by RHQR, compared with RGS.

(b) Relative Frobenius factorization errors of RHQR in both single and mixed precisions, compared with RGS and Householder QR.

Figure 7 Comparison of RHQR and RGS on a difficult example.

factorization error of RGS is slightly better than that of RHQR. The use of mixed precision allows RHQR to attain the same factorization error as RGS in both precision settings.

3.4 Block Sketch-Orthogonalization

When computing multiple matrix–vector products, substantial speedups can be achieved by replacing successive BLAS-2 operations $\mathbf{c}_1 = A\mathbf{b}_1, \mathbf{c}_2 = A\mathbf{b}_2, \ldots, \mathbf{c}_b = A\mathbf{b}_b$ with a single BLAS-3 matrix–matrix multiplication $C = AB$, where $B = [\mathbf{b}_1, \ldots, \mathbf{b}_b]$. Indeed, this allows to reduce data movement between different levels of the memory hierarchy. BLAS-3 kernels can be exploited by using block algorithms that partition the input matrix $W \in \mathbb{R}^{n\times bm}$ into blocks $W_1, \ldots, W_m \in \mathbb{R}^{n\times b}$. Such a block strategy is used in LAPACK's `xgeqrf` routines for computing the QR decomposition of dense matrices. Rather than performing a single iterative loop with bm orthogonalization steps for $W \in \mathbb{R}^{n\times bm}$, the computation is organized into two nested loops: an outer loop over the m blocks and an inner loop over the b vectors within each block. At step $jb + 1$ (first vector of the $(j + 1)$-th block), having built the matrix $\mathbf{Q}_j = [\mathbf{q}_1\ \mathbf{q}_2\ \cdots\ \mathbf{q}_{bj}] \in \mathbb{R}^{n\times bj}$ and its sketch $\Omega\mathbf{Q}_j \in \mathbb{R}^{\ell\times bj}$, assuming that the whole block $W_{j+1} \in \mathbb{R}^{n\times m}$ is available, we orthogonalize all vectors $W_{j+1}\mathbf{e}_1, \ldots, W_{j+1}\mathbf{e}_m$ with matrix-matrix operations only (BLAS3):

1. Sketch $Z_{j+1} \leftarrow \Omega W_{j+1}$.
2. Orthogonalize against current basis $W'_{j+1} \leftarrow W_{j+1} - \mathbf{Q}_j(\Omega\mathbf{Q}_j)^T Z_{j+1}$.
3. Orthogonalize W'_{j+1} with a single loop, BLAS2 algorithm.

This results in the randomized block Gram–Schmidt process described in [6]. The RHQR process, thanks to its cost-free aggregation of reflectors, can be easily expressed as a block algorithm [39]. We emphasize that these block algorithms are mathematically equivalent to single-loop BLAS2 algorithms and they perform the same number of flops. The only difference is that more flops are performed through BLAS3 kernels.

The sketch Z_{j+1} is necessary in all randomized algorithms (the input matrix W_{j+1} must be sketched at least once). Since Z_{j+1} is a matrix of small dimensions that approximately preserves the condition number of W_{j+1}, this condition number can thus be efficiently estimated. If it is small, less stable but faster algorithms can be used for the orthogonalization of the block, leading to even more significant speedups.

3.5 Bi-Orthogonalization

In [38], a randomized two-sided Gram–Schmidt algorithm is introduced to compute sketch-biorthogonal bases associated with two subspaces $\mathcal{X}$ and $\mathcal{Y}$ of the same dimension. This algorithm computes two bases P and Q such that Range$(Q) = \mathcal{X}$ and Range$(P) = \mathcal{Y}$, satisfying the sketch-biorthogonality condition $(\Omega P)^T\Omega Q = I$. This approach is computationally cheaper and numerically more stable than the two-sided Gram–Schmidt deterministic process, and it often constructs bases that are better conditioned than those obtained by deterministic algorithms which impose the biorthogonality condition $P^T Q = I$. We refer to [38] for further details.

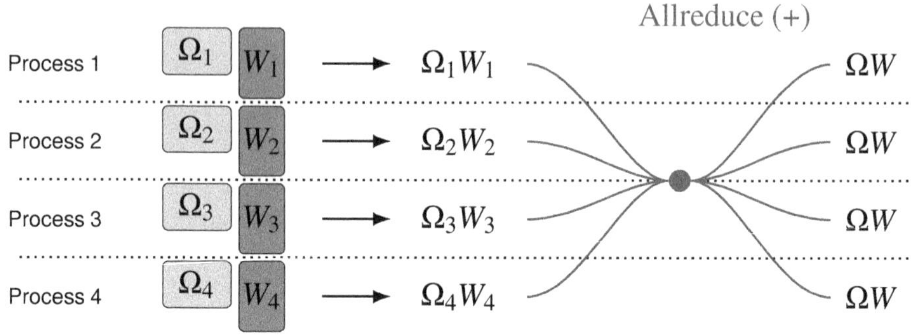

Figure 8 Sketching a matrix W partitioned into 4 blocks of rows over 4 processes. The sketching matrix Ω (Gaussian or s-hashing) is partitioned into 4 blocks of columns $\Omega_1, \Omega_2, \Omega_3, \Omega_4$.

3.6 Computation in Mixed Precision and on Parallel Computers

Randomized algorithms benefit not only from optimized kernels, but also from mixed precision and reduced communication on parallel architectures. On a parallel computer, the matrix $W \in \mathbb{R}^{n\times m}$ is typically distributed over processes by using a block row distribution, as in [28]. Sketching W can be performed efficiently in parallel. Consider for example Ω to be a dense Gaussian or an s-hashing sketching. As displayed in Figure 8, the sketching matrix is partitioned into blocks of columns and can be generated locally on each process with no communication. Once each process computes a local sketch $\Omega_i W_i$, an Allreduce communication among processes is required to sum the local sketches and compute $\Omega W = \sum_{i=1}^{p} \Omega_i W_i$, where p is the number of processes. The RCholeskyQR process can be performed very efficiently using a single synchronization (required by the sketch), as outlined in Figure 9. Once ΩW is computed, each process computes its QR factorization, and then the computed R factor is used locally to compute a block of the orthogonal factor as $Q_i = W_i R^{-1}$.

For each vector, the Gram–Schmidt process involves computing the projection coefficients onto the current basis, and then updating the vector by removing these projected components. In a parallel setting, the goal is to compute the projection coefficients and have them available on every process so that the second stage of the algorithm can be performed independently, without further communication. For randomized methods, this second stage has the same asymptotic cost as deterministic algorithms, namely, nm^2 floating-point operations. Their main advantage typically lies in how efficiently they build and replicate the small matrix ΩQ from which these projection coefficients are computed.

We remark that both RGS in Algorithm 1 and RHQR in Algorithm 2 require between one and two synchronizations per step, depending on how the sketches can be grouped. This is m or $2m$ synchronizations overall, which is the same computational cost as Classical Gram–Schmidt. As mentioned before, the computational cost of the two algorithms is dominated by the substitution process (nm^2 flops), and the total sketching cost, namely $2mt$ flops, where t is the cost of sketching one vector. Thus, with SRHT, the

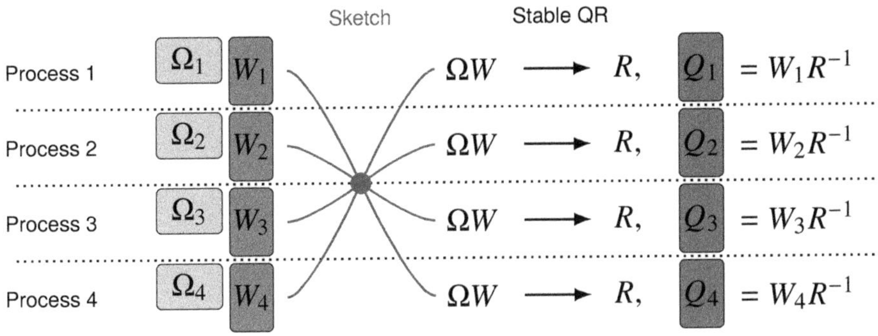

Figure 9 RCholeskyQR on 4 processes.

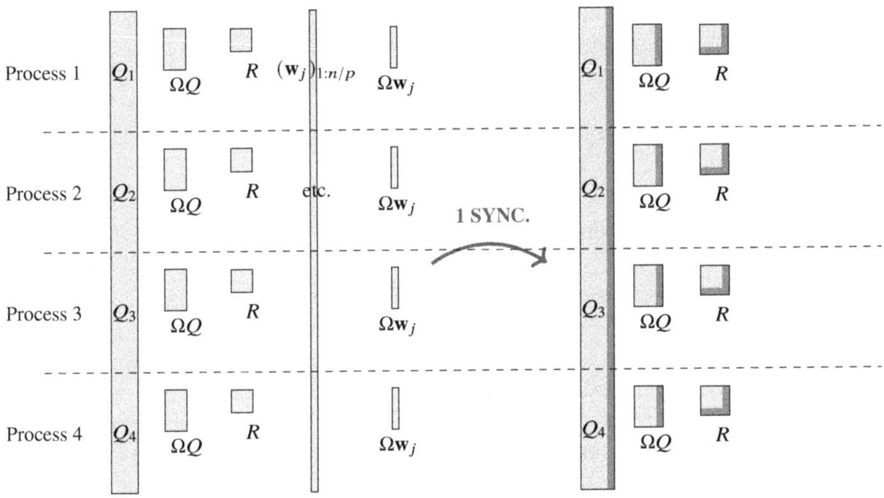

Figure 10 Scattering of data in one step of RGS

total cost of both algorithms is $nm^2 + 2nm\log(m)$ flops asymptotically, which is half of the number of flops of standard Gram–Schmidt processes. We illustrate the distribution of data in one step of RGS, when performed on a parallel computer, in Figure 10. The matrix $Q \in \mathbb{R}^{n\times m}$ is distributed on each node by blocks of contiguous rows, and so is the next incoming input basis vector $\mathbf{w}_j \in \mathbb{R}^n$. The sketches ΩQ and $\boldsymbol{\Omega}\mathbf{w}$ are available on each processor, and so is $R \in \mathbb{R}^{m\times m}$. Each processor can compute locally $(\Omega Q_{1:j-1})^T \boldsymbol{\Omega}\mathbf{w}_j$ and thus update the first $j-1$ entries of $R\mathbf{e}_j$. It also allows each processor to perform its part of the vector refresh $\mathbf{w}_j \leftarrow \mathbf{w}_j - Q_{1:j-1}(\Omega Q_{1:j-1})^T \boldsymbol{\Omega}\mathbf{w}_j$. The result is then sketched, which allows every processor to compute the j-th entry of $R\mathbf{e}_j$, scale its part of the refreshed $\mathbf{w}_j$, and thus obtain the new basis vector $\mathbf{q}_j \in \mathbb{R}^n$. All the matrices are updated with the vectors computed in this step.

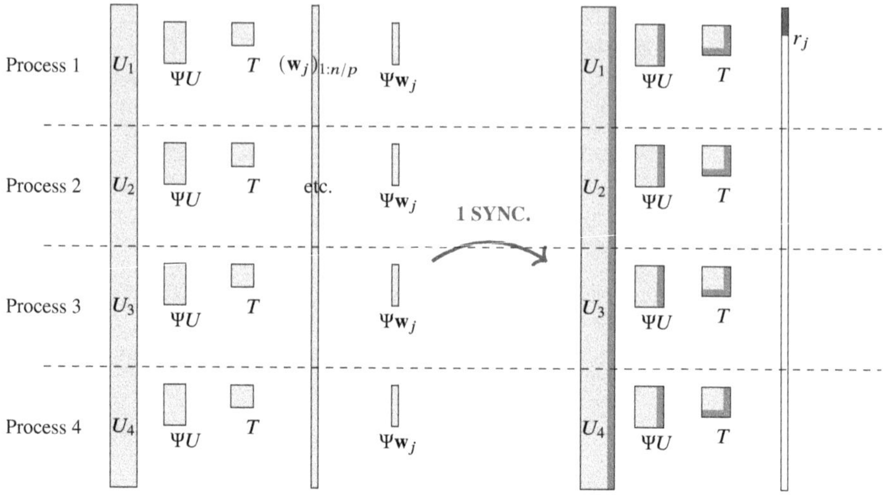

Figure 11 Scattering of data in one step of RHQR

We illustrate the distribution of data in one step of RHQR, when performed on a parallel computer, in Figure 11. The tall-and-skinny matrix $U \in \mathbb{R}^{n\times m}$ is distributed by blocks of contiguous rows, as the input (it can be written in its place), and so is the next basis vector $\mathbf{w} \in \mathbb{R}^n$. The sketches $\Psi U \in \mathbb{R}^{(\ell+m)\times m}$ and $\Psi\mathbf{w} \in \mathbb{R}^{\ell+m}$ are available on each processor, and so is $T \in \mathbb{R}^{m\times m}$. All processors can thus compute locally $T^T(\Psi U)^T\Psi\mathbf{w}$, which allows each processor to perform its part of the vector refresh $\mathbf{w} \leftarrow \mathbf{w} - UT^T(\Psi U)^T\Psi\mathbf{w}$. The resulting, refreshed vector is then sketched, which allows every processor to compute its share of the associated randomized Householder vector. All the matrices are then updated with the vectors computed in this step.

It is very natural to use a mixed-precision setting in the case of randomized orthogonalization. A common approach is to store the high-dimensional matrix in a *coarse* floating point format (typically 32 bits, or even 16 bits when the CPU supports it), while casting and storing low-dimensional matrices (mainly the sketches and the triangular factors) in a *fine* floating point format (typically 64, or even 128 bits when the CPU supports it). In the case of RCholeskyQR in Figure 9 for instance, the matrix W is stored in coarse precision. Then Ω (often not stored explicitly) is applied, using either coarse or fine arithmetic operations. The result ΩW is stored in fine precision, and the R factor is computed from ΩW in fine precision. The computed R factor is finally cast to coarse precision, and the substitution for Q is performed in coarse precision.

A critical aspect of finite-precision analysis of randomized algorithms is the *forward accuracy* of the sketching step, namely the bounding of the magnitude of $\|\Omega\mathbf{x} - \mathbf{fl}[\Omega\mathbf{x}]\|/\|\Omega\mathbf{x}\|$, where $\mathbf{fl}[\Omega\mathbf{x}]$ denotes the finite-precision result of the routine applying Ω to $\mathbf{x}$. If this forward error is sufficiently small, the stability of the process can potentially reduce to the accuracy of the orthogonalization of the sketch [5, 39, 45]. The authors of [5] describe the accuracy of the sketching process, performed in fine precision $\mathbf{u}$, for any sketching matrix $\Omega \in \mathbb{R}^{\ell\times n}$, basing their analysis on the following backward error [46]

$$\mathbf{fl}[\Omega\mathbf{x}] = (\Omega + \Delta\Omega) \cdot \mathbf{x}, \quad |\Delta\Omega| \leq O(n^{1/2})|\Omega|\mathbf{u} \quad \text{(w.h.p.)}.$$

By restricting Ω to some specific OSE distributions, the accuracy of the sketching step can be even more favorable. The authors of [39] base their analysis on the observation that the application of the SRHT, in fine precision $\mathbf{u}$, using the Fast Walsh–Hadamard Transform, is *backward stable*, with fine constants even in the worst-case:

$$\forall \mathbf{x} \in \mathbb{R}^n, \quad \mathbf{fl}[\Omega\mathbf{x}] = \Omega(\mathbf{x} + \Delta\mathbf{x}), \quad \|\Delta\mathbf{x}\| \leq O(\log(n))\|\mathbf{x}\|\mathbf{u} \quad \text{(SRHT)}.$$

The authors of [5] consider a mixed-precision model for the analysis of Randomized Gram–Schmidt. Denoting a fine unit roundoff $\mathbf{u}$ and a coarse unit roundoff $\mathbf{f}$, $0 < \mathbf{u} \ll \mathbf{f}$, considering the ϵ-embedding property for a vector subspace of dimension $O(m)$, they show that, if $O(m^2)\mathrm{Cond}(W)\mathbf{f} < 1$ and $O(m^{1/2}n^{3/2} + m^{3/2}\ell)\mathbf{u} < \mathbf{f}$, then the output of RGS in Algorithm 1 satisfies

$$\mathrm{Cond}(\mathbf{fl}[Q]) \leq (1 + O(\epsilon))\left(1 + O(m^2)\mathrm{Cond}(W)\mathbf{u}\right). \tag{17}$$

Finally, using SRHT sketching and assuming that $O(\log(n) + \ell + m)\mathbf{u} \leq \mathbf{f}$, the authors of [39] show that the output of the RHQR process in Algorithm 2 satisfies

$$\mathrm{Cond}(\mathbf{fl}[Q]) \leq (1 + O(\epsilon))(1 + O(\ell m^{3/2})\mathbf{f}),$$
$$\|(\mathbf{fl}[QR] - W)\mathbf{e}_j\| \leq (1 + O(\epsilon))O(\ell m^{3/2})\|W\mathbf{e}_j\|\mathbf{f}.$$

It is often observed that the factorization error of randomized algorithms is smaller in practice than that of standard orthogonalization. Indeed, with fast and stable sketching processes, the coefficients of the matrix R driving the substitution are potentially obtained with much fewer flops and better numerical stability than those of the deterministic R. This phenomenon has very concrete consequences, for instance, in the case of randomized GMRES, as we outline in the next sections.

4 Krylov Subspace Methods

Krylov subspace methods are a valuable tool for the solution of various problems in numerical linear algebra. We review here the fundamental definitions and properties associated with Krylov subspaces, and we introduce the corresponding randomized versions.

Given a matrix $A \in \mathbb{R}^{n \times n}$ and a vector $\mathbf{b} \in \mathbb{R}^n$, the associated Krylov subspace of dimension m is defined as $\mathcal{K}_m(A, \mathbf{b}) = \mathrm{span}\{\mathbf{b}, A\mathbf{b}, \ldots, A^{m-1}\mathbf{b}\}$. Defining $\mathbf{q}_1 = \mathbf{b}/\|\mathbf{b}\|_2$, the Arnoldi process [67, Algorithm 6.2] can be used to generate an orthonormal basis $Q_m = [\mathbf{q}_1 \cdots \mathbf{q}_m]$ for $\mathcal{K}_m(A, \mathbf{b})$. This basis satisfies the Arnoldi relation

$$AQ_m = Q_{m+1}\underline{H}_m = Q_m H_m + h_{m+1,m}\mathbf{q}_{m+1}\mathbf{e}_m^T, \qquad \text{with} \quad \underline{H}_m = \begin{bmatrix} H_m \\ h_{m+1,m}\mathbf{e}_m^T \end{bmatrix}, \quad (18)$$

where $\mathbf{q}_{m+1} \perp Q_m$ and $H_m \in \mathbb{R}^{m\times m}$ is an upper Hessenberg matrix containing the orthogonalization coefficients. Since Q_{m+1} has orthonormal columns, it follows from (18) that $Q_m^T A Q_m = H_m$.

4.1 Randomized Arnoldi and Krylov Factorizations

Let W_m be any basis of $\mathcal{K}_m(A,\mathbf{b})$ where each new basis vector $\mathbf{w}_{j+1}$ is generated iteratively as a linear combination of $A\mathbf{w}_j$ and the columns of the current basis W_j. Such a basis satisfies the Arnoldi-like relation

$$AW_m = W_{m+1}\underline{L}_m = W_m L_m + \mathbf{w}_{m+1}\boldsymbol{\ell}_m^T, \qquad \boldsymbol{\ell}_m := \ell_{m+1,m}\mathbf{e}_m, \quad (19)$$

where $\underline{L}_m \in \mathbb{R}^{(m+1)\times m}$ is upper Hessenberg but W_m does not necessarily have orthonormal columns. In this case, we have the identity

$$W_m^+ A W_m = L_m + W_m^+ \mathbf{w}_{m+1}\boldsymbol{\ell}_m^T. \quad (20)$$

The randomized Arnoldi process [5] given in Algorithm 3 constructs a basis of $\mathcal{K}_m(A,\mathbf{b})$ employing the randomized Gram–Schmidt algorithm to generate a sketch-orthonormal basis V_m and an associated Hessenberg matrix $\underline{G}_m$, which also satisfy the randomized Arnoldi relation

$$AV_m = V_{m+1}\underline{G}_m = V_m G_m + \mathbf{v}_{m+1}\mathbf{g}_m^T, \qquad \mathbf{g}_m := g_{m+1,m}\mathbf{e}_m. \quad (21)$$

In this case, by multiplying (21) from the left by $(\Omega V_m)^T\Omega$ and using the fact that V_m satisfies $(\Omega V_{m+1})^T\Omega V_{m+1} = I$, we obtain the following

$$(\Omega V_m)^T \Omega A V_m = G_m.$$

This identity allows us to use the Hessenberg matrix G_m generated by the randomized Arnoldi process to efficiently construct approximate solutions for different problems, ranging from the solution of linear systems and matrix equations to the evaluation of matrix functions and the computation of eigenvalues.

A major distinction of the randomized Arnoldi algorithm compared to deterministic Arnoldi is that the method does not reduce to a short recurrence when the matrix A is symmetric. In the deterministic case $H_m = Q_m^T A Q_m$ from (18) is symmetric and upper Hessenberg, hence it is necessarily tridiagonal. Since its entries are inner products between successive Krylov basis vectors, it yields the three-term Lanczos recurrence and requires orthogonalization only against the two most recent basis vectors at each step; see [52]. By contrast, in the randomized Arnoldi process the Hessenberg matrix $G_m = (\Omega V_m)^T \Omega A V_m$ from (21) arising from the sketched oblique projection may be non-symmetric, and in general it is not similar to the tridiagonal matrix H_m generated

Algorithm 3 Randomized Arnoldi process [5]

Input: $A \in \mathbb{R}^{n\times n}$, a starting vector $\mathbf{b} \in \mathbb{R}^n$, a Krylov dimension m, a matrix $\Omega \in \mathbb{R}^{\ell\times n}$ such that $\mathrm{Ker}(\Omega) \cap \mathcal{K}_m(A, \mathbf{b}) = \emptyset$.

Output: A randomized Arnoldi factorization $AV_m = V_{m+1}\underline{G}_m = V_m G_m + \mathbf{v}_{m+1}\mathbf{g}_m^T$ as in (21) with $S_{m+1} = \Omega V_{m+1}$ explicitly computed.

function RANDOMIZED ARNOLDI($A, \mathbf{b}, m, \Omega$)
  $\mathbf{z}_1 \leftarrow \Omega\mathbf{b}$
  $V_1 \leftarrow [\mathbf{b}/\|\mathbf{z}\|]$, $S_1 \leftarrow [\mathbf{z}/\|\mathbf{z}\|]$
  $\underline{G}_0 = 0_{1\times 0}$
  for $j = 1 : m$ **do**
    $\mathbf{v}_{j+1} \leftarrow A\mathbf{v}_j$
    $\mathbf{z} \leftarrow \Omega\mathbf{v}_{j+1}$
    $\mathbf{g}_j \leftarrow S_j^+\mathbf{z} \in \mathbb{R}^j$ # use a method stable enough to handle S_j^+
    $\mathbf{v}_{j+1} \leftarrow \mathbf{v}_{j+1} - V_j\mathbf{g}_j$
    $\mathbf{z} \leftarrow \Omega\mathbf{v}_{j+1}$
    $g_{j+1,j} \leftarrow \|\mathbf{z}\|$
    $\underline{G}_j \leftarrow \left[\begin{array}{c|c} \underline{G}_{j-1} & \mathbf{g}_j \\ \hline 0_{1\times(j-1)} & g_{j+1,j} \end{array}\right]$
    $V_{j+1} \leftarrow [V_j \mid \mathbf{v}_{j+1}/g_{j+1,j}]$, $S_{j+1} \leftarrow [S_j \mid \mathbf{z}/g_{j+1,j}]$
  end for
  return $V_{m+1}, S_{m+1}, \underline{G}_m$
end function

by the Lanczos method. Consequently, no short recurrence is obtained in general, and one must exercise care when using G_m to approximate eigenvalues of A as later developed in Section 6.2, since spurious or complex eigenvalues may appear. Restoring the similarity between the Hessenberg matrices H_m and G_m at a small additional cost, and thus recovering real approximate eigenvalues for symmetric matrices in randomized Arnoldi, has been explored in [37].

4.2 Whitening

In several recent works [41, 61, 63, 64], it has been proposed to construct the sketch-orthonormal basis V_m implicitly. First, a non-orthogonal basis W_m is generated with a cheap procedure, such as a k-truncated Arnoldi process (see, e.g., [67, Chapter 6.4.2]), in which the orthogonalization of a new vector is only performed against the last k basis vectors, and then a QR factorization of the sketched basis $\Omega W_m = S_m R_m$ is computed. A sketch-orthonormal basis of $\mathcal{K}_m(A, \mathbf{b})$ can then be obtained as $V_m = W_m R_m^{-1}$. This process is often called *whitening* in the literature and coincides with an implicit application of the randomized Cholesky QR framework described at the end of Section 3.1.

The main advantage of this approach is that it allows for a cheaper, implicit computation of the sketch-orthonormal basis V_m. Indeed, note that explicitly forming $V_m = W_m R_m^{-1}$ requires $O(nm^2)$ operations, which coincides with the computational cost of a direct sketch-orthogonalization via randomized Gram–Schmidt. However, the main advantage of the whitening strategy is that, in many cases, there is no need to form the basis V_m explicitly. For example, if the desired solution to a certain problem has the form $V_m \mathbf{y}_m$ for some $\mathbf{y}_m \in \mathbb{R}^m$, it can be computed as $W_m(R_m^{-1}\mathbf{y}_m)$. This operation only costs $O(mn + m^2)$, since R_m^{-1} is applied to a vector of length m, thus reducing the asymptotic computational complexity of the method, provided that the basis W_m is computed using a cheap procedure. When W_m is constructed using the k-truncated Arnoldi process, the orthogonalization cost in the computation of W_m drops to $O(kmn)$, and implicit whitening of the basis (that is, the computation of the QR factorization $\Omega W_m = S_m R_m \in \mathbb{R}^{\ell \times m}$, where $\ell = O(m)$) requires $O(m^3)$ operations, making this approach asymptotically cheaper than directly performing the randomized Gram–Schmidt process. In exact arithmetic, the (implicitly computed) whitened basis V_m coincides with the basis obtained by randomized Gram–Schmidt, ensuring that the two approaches produce the same approximations for any task that employs a sketch-orthonormal basis of $\mathcal{K}_m(A, \mathbf{b})$. An ulterior advantage of the k-truncated Arnoldi process is that only the $k + 1$ vectors need to be kept in memory during the generation of the basis W_m, making the whitening approach viable in a low-memory setting. If the full basis W_m is required to form the approximate solution, it can be generated again with a two-pass approach, without the need to store it in full: see, for instance, [41, Section 4.2] and [16, 33].

The main limitation of whitening is that the non-orthogonal basis W_m typically becomes severely ill-conditioned even for moderate m, and as a consequence both the QR factorization $\Omega W_m = S_m R_m$ and the application of R_m^{-1} may suffer from numerical instability, eventually yielding approximations that in finite precision may diverge significantly from those obtained with a sketch-orthonormal basis V_m constructed explicitly through a randomized Gram–Schmidt process. Although the numerical behavior of whitening within Krylov subspace methods is still not completely understood theoretically, encouraging numerical results have been observed in several applications, often even better than what one would expect from the growth of $\kappa(W_m)$ [22, 63]. Various selective orthogonalization strategies have been explored in [42] as alternatives to k-truncated Arnoldi, aiming to mitigate the ill-conditioning of the non-orthogonal basis W_m before applying whitening.

When using whitening to sketch-orthogonalize the basis W_m, the following relations may be useful. Similarly to (20), we have

$$(\Omega W_m)^+ \Omega A W_m = L_m + \mathbf{z}_m \boldsymbol{\ell}_m^T, \qquad \mathbf{z}_m := (\Omega W_m)^+ \Omega \mathbf{w}_{m+1},$$

which gives us the following alternative expression for the Arnoldi relation (19),

$$A W_m = W_m (L_m + \mathbf{z}_m \boldsymbol{\ell}_m^T) + \widehat{\mathbf{v}}_{m+1} \boldsymbol{\ell}_m^T, \qquad \widehat{\mathbf{v}}_{m+1} := \mathbf{w}_{m+1} - W_m \mathbf{z}_m, \tag{22}$$

where $\widehat{\mathbf{v}}_{m+1} \perp^{\Omega} W_m$. In other words, we can add a rank-one perturbation to L_m in order to obtain an Arnoldi relation in which the last basis vector $\widehat{\mathbf{v}}_{m+1}$ is sketch-orthogonal to the basis W_m.

The vector $\mathbf{z}_m$ can be alternatively written in terms of a QR factorization of the sketched basis ΩW_{m+1}. Indeed, assuming that we have the QR factorization

$$\Omega W_{m+1} = \begin{bmatrix} S_m & \mathbf{s}_{m+1} \end{bmatrix} \begin{bmatrix} R_m & \mathbf{r}_m \\ & \rho_{m+1} \end{bmatrix},$$

we have

$$\mathbf{z}_m = (\Omega W_m)^+ \Omega \mathbf{w}_{m+1} = R_m^{-1} S_m^T (S_m \mathbf{r}_m + \rho_{m+1} \mathbf{s}_{m+1}) = R_m^{-1} \mathbf{r}_m. \tag{23}$$

A variant of the identity (22) with the expression for $\mathbf{z}_m$ in (23) is given in [63, Proposition 3.1], where it is referred to with the name *sketched Arnoldi relation*.

Multiplying (22) by R_m^{-1} on the right, we obtain

$$A W_m R_m^{-1} = W_m R_m^{-1} (R_m L_m R_m^{-1} + \mathbf{z}_m \boldsymbol{\ell}_m^T R_m^{-1}) + \widehat{\mathbf{v}}_{m+1} \boldsymbol{\ell}_m^T R_m^{-1}.$$

Using the fact that R_m is upper triangular, it is easy to see that $\mathbf{e}_m^T R_m^{-1} = \rho_m^{-1} \mathbf{e}_m^T$, where $\rho_m = [R_m]_{m,m}$. Recalling that $\boldsymbol{\ell}_m = \ell_{m+1,m} \mathbf{e}_m$ and $V_m = W_m R_m^{-1}$, we get

$$A V_m = V_m (\widehat{L}_m + \rho_m^{-1} \mathbf{z}_m \boldsymbol{\ell}_m^T) + \rho_m^{-1} \widehat{\mathbf{v}}_{m+1} \boldsymbol{\ell}_m^T, \qquad \widehat{L}_m := R_m^{-1} L_m R_m. \tag{24}$$

Multiplying (24) by $(\Omega V_m)^T \Omega$ from the left and using $(\Omega V_m)^T \Omega V_m = I$ and $(\Omega V_m)^T \Omega \widehat{\mathbf{v}}_{m+1} = 0$, we obtain

$$G_m = (\Omega V_m)^T \Omega A V_m = \widehat{L}_m + \rho_m^{-1} \mathbf{z}_m \boldsymbol{\ell}_m^T. \tag{25}$$

The identity (24) is called *whitened-sketched Arnoldi relation* in [63], and (25) provides an explicit expression for the Hessenberg matrix associated with the whitened basis V_m, which can be evaluated cheaply by only using the $\underline{L}_m$ and the upper triangular factor from the QR factorization of ΩW_{m+1}.

5 Solution of Linear Systems

Let us consider the linear system $A\mathbf{x} = \mathbf{b}$, with $A \in \mathbb{R}^{n \times n}$ and $\mathbf{b} \in \mathbb{R}^n$. Subspace projection methods [67] are a very effective tool for solving large-scale linear systems. These methods seek an approximate solution $\mathbf{x}_m$ which satisfies the two following conditions:

- $\mathbf{x}_m$ is contained in the Krylov subspace $\mathcal{K}_m(A, \mathbf{b})$,
- the residual $\mathbf{b} - A\mathbf{x}_m$ satisfies the *Petrov–Galerkin condition* $\mathbf{b} - A\mathbf{x}_m \perp \mathcal{L}_m$, where $\mathcal{L}_m \subset \mathbb{R}^n$ is an m-dimensional subspace.

When $\mathcal{L}_m = \mathcal{K}_m(A, \mathbf{b})$, the resulting method is usually known as an orthogonal projection method, while with a more general $\mathcal{L}_m$ we obtain an oblique projection method.

5.1 Krylov Methods for Linear Systems

We start by briefly reviewing the well-known GMRES and FOM methods for the solution of $A\mathbf{x} = \mathbf{b}$.

GMRES [69] takes $\mathcal{L}_m = A\mathcal{K}_m(A, \mathbf{b})$, so it seeks an approximate solution $\widetilde{\mathbf{x}}_m^{\mathrm{G}} \in \mathcal{K}_m(A, \mathbf{b})$ whose residual satisfies

$$\mathbf{b} - A\widetilde{\mathbf{x}}_m^{\mathrm{G}} \perp A\mathcal{K}_m(A, \mathbf{b}).$$

Recalling the Arnoldi relation (18) and writing $\widetilde{\mathbf{x}}_m^{\mathrm{G}} = Q_m \widetilde{\mathbf{y}}_m^{\mathrm{G}}$ with $\widetilde{\mathbf{y}}_m^{\mathrm{G}} \in \mathbb{R}^m$, we can equivalently rewrite this condition as

$$\begin{aligned} 0 = (AQ_m)^T(\mathbf{b} - AQ_m\widetilde{\mathbf{y}}_m^{\mathrm{G}}) &= (Q_{m+1}\underline{H}_m)^T(\mathbf{b} - Q_{m+1}\underline{H}_m\widetilde{\mathbf{y}}_m^{\mathrm{G}}) \\ &= \underline{H}_m^T\widetilde{\beta}\mathbf{e}_1 - \underline{H}_m^T\underline{H}_m\widetilde{\mathbf{y}}_m^{\mathrm{G}}, \end{aligned}$$

where $\widetilde{\beta} \in \mathbb{R}$ is defined from the identity $\mathbf{b} = \widetilde{\beta}Q_m\mathbf{e}_1$. Observe that these are the normal equations associated with a least squares problem, so we can write the GMRES approximate solution as

$$\widetilde{\mathbf{x}}_m^{\mathrm{G}} = Q_m\widetilde{\mathbf{y}}_m^{\mathrm{G}} \qquad \text{where} \qquad \widetilde{\mathbf{y}}_m^{\mathrm{G}} = \underset{\mathbf{y}\in\mathbb{R}^m}{\operatorname{argmin}} \|\underline{H}_m\mathbf{y} - \widetilde{\beta}\mathbf{e}_1\|. \tag{26}$$

From the condition $(AQ_m)^T(\mathbf{b} - AQ_m\widetilde{\mathbf{y}}_m^{\mathrm{G}}) = 0$, we also obtain that

$$\widetilde{\mathbf{x}}_m^{\mathrm{G}} = \underset{\mathbf{x}\in\mathcal{K}_m(A,\mathbf{b})}{\operatorname{argmin}} \|A\mathbf{x} - \mathbf{b}\|,$$

i.e., the GMRES solution $\widetilde{\mathbf{x}}_m^{\mathrm{G}}$ minimizes the residual over the Krylov subspace $\mathcal{K}_m(A, \mathbf{b})$.

On the other hand, the Full Orthogonalization Method (FOM) [2, 67] employs $\mathcal{L}_m = \mathcal{K}_m(A, \mathbf{b})$, thus requiring that the residual is orthogonal to the Krylov subspace $\mathcal{K}_m(A, \mathbf{b})$, a condition that is also known as *Galerkin condition*. Let us denote by $\widetilde{\mathbf{x}}_m^{\mathrm{F}}$ the approximate solution after m iterations of FOM. Recalling the Arnoldi relation (18) and writing $\widetilde{\mathbf{x}}_m^{\mathrm{F}} = Q_m\widetilde{\mathbf{y}}_m^{\mathrm{F}}$ with $\widetilde{\mathbf{y}}_m^{\mathrm{F}} \in \mathbb{R}^m$, we can rewrite the Galerkin condition as

$$\begin{aligned} 0 = Q_m^T(\mathbf{b} - AQ_m\widetilde{\mathbf{y}}_m^{\mathrm{F}}) &= Q_m^T(\mathbf{b} - (Q_mH_m + h_{m+1,m}\mathbf{q}_{m+1}\mathbf{e}_m^T)\widetilde{\mathbf{y}}_m^{\mathrm{F}}) \\ &= \widetilde{\beta}\mathbf{e}_1 - H_m\widetilde{\mathbf{y}}_m^{\mathrm{F}}, \end{aligned}$$

where we used the orthogonality of the columns of Q_{m+1} in the last equality. This yields the expression for the FOM approximate solution

$$\widetilde{\mathbf{x}}_m^{\mathrm{F}} = Q_m\widetilde{\mathbf{y}}_m^{\mathrm{F}}, \qquad \text{where} \qquad H_m\widetilde{\mathbf{y}}_m^{\mathrm{F}} = \widetilde{\beta}\mathbf{e}_1. \tag{27}$$

When A is symmetric positive definite, FOM simplifies to the conjugate gradient (CG) method [44]. In this case, the approximate solution $\widetilde{\mathbf{x}}_m^{\mathrm{F}}$ can be iteratively updated using short recurrences and it additionally satisfies the optimality property

$$\widetilde{\mathbf{x}}_m^{\mathrm{F}} = \underset{\widehat{\mathbf{x}} \in \mathcal{K}_m(A,\mathbf{b})}{\operatorname{argmin}} \|\mathbf{x} - \widehat{\mathbf{x}}\|_A, \qquad \text{where} \qquad \|\mathbf{z}\|_A = (\mathbf{z}^T A \mathbf{z})^{1/2}.$$

5.2 Randomized Krylov Methods for Linear Systems

In this section, we present algorithms that employ a sketch-orthonormal basis of the Krylov subspace $\mathcal{K}_m(A, \mathbf{b})$ to solve the linear system $A\mathbf{x} = \mathbf{b}$, following the presentation in [61] and [75]. These algorithms replace the Petrov–Galerkin condition imposed by standard Krylov methods with a similar condition on the sketched residual. Specifically, they seek an approximate solution $\mathbf{x}_m$ that satisfies:

- $\mathbf{x}_m$ belongs to the Krylov subspace $\mathcal{K}_m(A, \mathbf{b})$,
- the residual $\mathbf{b} - A\mathbf{x}_m$ satisfies the *sketched Petrov–Galerkin condition* $\mathbf{b} - A\mathbf{x}_m \perp^{\Omega} \mathcal{L}_m$, where $\mathcal{L}_m \subset \mathbb{R}^n$ is an m-dimensional subspace and $\Omega \in \mathbb{R}^{\ell \times n}$ is an ϵ-embedding for $\mathcal{K}_m(A, \mathbf{b})$.

In analogy with the deterministic Krylov methods, choosing $\mathcal{L}_m = \mathcal{K}_m(A, \mathbf{b})$ or $\mathcal{L}_m = A\mathcal{K}_m(A, \mathbf{b})$ leads, respectively, to the *randomized FOM* and *randomized GMRES* algorithms (also known as sketched FOM and sketched GMRES in the literature). In the following, we assume that we have a sketch-orthonormal basis V_m of $\mathcal{K}_m(A, \mathbf{b})$ and an associated Hessenberg matrix $\underline{G}_m$ which satisfy the Arnoldi-like relation (21). We define $\beta \in \mathbb{R}$ according to the identity $\mathbf{b} = \beta V_m \mathbf{e}_1$. We mention that the sketched Galerkin and sketched Petrov–Galerkin conditions have been used in the context of model order reduction in [7,8].

In simulations using deterministic Krylov methods, m is usually a maximum number of iterations set by the user, for instance by considering time or memory limitations. Assuming that $m \ll n$, since randomized methods yield quasi-optimal results (see below), the same m then indicates a sampling size ℓ for the sketching matrix $\Omega \in \mathbb{R}^{\ell \times n}$ (see equations 4and 7).

5.2.1 Randomized GMRES

Randomized GMRES (rGMRES) seeks an approximate solution $\mathbf{x}_m^{\mathrm{G}} \in \mathcal{K}_m(A, \mathbf{b})$ such that its associated residual satisfies the *sketched Petrov–Galerkin condition*

$$\Omega(\mathbf{b} - A\mathbf{x}_m^{\mathrm{G}}) \perp \Omega A \mathcal{K}_m(A, \mathbf{b}).$$

Using (21) and writing $\mathbf{x}_m^{\mathrm{G}} = V_m \mathbf{y}_m^{\mathrm{G}}$, we have

$$\begin{aligned} 0 = (\Omega A V_m)^T \Omega(\mathbf{b} - A V_m \mathbf{y}_m^{\mathrm{G}}) &= (\Omega V_{m+1} \underline{G}_m)^T (\Omega \mathbf{b} - \Omega V_{m+1} \underline{G}_m \mathbf{y}_m^{\mathrm{G}}) \\ &= \underline{G}_m^T \beta \mathbf{e}_1 - \underline{G}_m^T \underline{G}_m \mathbf{y}_m^{\mathrm{G}}. \end{aligned}$$

Similarly to the above derivation of GMRES, these are the normal equations associated with the least squares problem

$$\mathbf{y}_m^{\mathrm{G}} = \underset{\mathbf{y}\in\mathbb{R}^m}{\operatorname{argmin}} \|\underline{G}_m\mathbf{y} - \beta\mathbf{e}_1\|. \tag{28}$$

From the condition $(\Omega A V_m)^T(\Omega\mathbf{b} - \Omega A V_m \mathbf{y}_m^{\mathrm{G}}) = 0$, we also see that $\mathbf{x}_m^{\mathrm{G}}$ solves the least squares problem

$$\mathbf{x}_m^{\mathrm{G}} = \underset{\mathbf{x}\in\mathcal{K}_m(A,\mathbf{b})}{\operatorname{argmin}} \|\Omega(\mathbf{b} - A\mathbf{x})\|.$$

In other words, the rGMRES solution $\mathbf{x}_m^{\mathrm{G}}$ minimizes the sketched residual $\Omega(\mathbf{b} - A\mathbf{x})$ among all $\mathbf{x}$ in the Krylov subspace $\mathcal{K}_m(A,\mathbf{b})$. If Ω is an ϵ-embedding of $\mathcal{K}_{m+1}(A,\mathbf{b})$, we get

$$\|\mathbf{b} - A\mathbf{x}_m^{\mathrm{G}}\| \leq \sqrt{\frac{1+\epsilon}{1-\epsilon}} \underset{\widehat{\mathbf{x}}\in\mathcal{K}_m(A,\mathbf{b})}{\operatorname{argmin}} \|\mathbf{b} - A\widehat{\mathbf{x}}\|,$$

so rGMRES achieves a quasi-optimal residual, up to the multiplicative factor $\sqrt{(1+\epsilon)/(1-\epsilon)}$.

We emphasize that, even though the sequence of sketched residuals of rGMRES is decreasing, it is not true in general that the sequence of residuals of rGMRES is also decreasing, especially in consecutive iterations where the sketched residual stagnates.

As in GMRES, we remark that the residual norm can be evaluated with cheap formulas in the Krylov basis:

$$\begin{aligned}\|\mathbf{b} - A\mathbf{x}_m^{\mathrm{G}}\| \leq \frac{1}{\sqrt{1-\epsilon}}\|\Omega(\mathbf{b} - A\mathbf{x}_m^{\mathrm{G}})\| &= \frac{1}{\sqrt{1-\epsilon}}\|\Omega V_{m+1}(\beta\mathbf{e}_1 - \underline{G}_m\mathbf{y}_m^{\mathrm{G}})\| \\ &= \frac{1}{\sqrt{1-\epsilon}}\|\beta\mathbf{e}_1 - \underline{G}_m\mathbf{y}_m^{\mathrm{G}}\|.\end{aligned}$$

Of course, in finite-precision arithmetic, these last two equalities are only as good as the factorization error of the orthogonalization process used, and the true condition number of the computed basis. As to the first aspect, the factorization error of randomized QR processes is often better than that of deterministic processes (since it requires fewer flops). As to the second aspect, randomization often allows users to choose stabler methods. For these reasons, we often observe experimentally that the true residual is slightly smaller with rGMRES than with GMRES.

5.2.2 Randomized FOM

Randomized FOM (rFOM) searches for an approximate solution $\mathbf{x}_m^{\mathrm{F}} \in \mathcal{K}_m(A,\mathbf{b})$ such that its residual satisfies the *sketched Galerkin condition*

$$\Omega(\mathbf{b} - A\mathbf{x}_m^{\mathrm{F}}) \perp \Omega\mathcal{K}_m(A,\mathbf{b}).$$

Using (21) and writing $\mathbf{x}_m^{\mathrm{F}} = V_m \mathbf{y}_m^{\mathrm{F}}$, we can rewrite this condition as

$$0 = (\Omega V_m)^T \Omega(\mathbf{b} - A V_m \mathbf{y}_m^{\mathrm{F}}) = (\Omega V_m)^T (\Omega \mathbf{b} - (\Omega V_m G_m + g_{m+1,m} \Omega \mathbf{v}_{m+1} \mathbf{e}_m^T) \mathbf{y}_m^{\mathrm{F}})$$
$$= \beta \mathbf{e}_1 - G_m \mathbf{y}_m^{\mathrm{F}},$$

where for the last equality we exploited the orthogonality of the columns of ΩV_{m+1}. This yields the expression for the randomized FOM approximate solution

$$\mathbf{x}_m^{\mathrm{F}} = V_m \mathbf{y}_m^{\mathrm{F}}, \qquad \text{where} \qquad G_m \mathbf{y}_m^{\mathrm{F}} = \beta \mathbf{e}_1. \tag{29}$$

Note that when A is symmetric positive definite, the randomized FOM approximation (29) cannot in general be obtained with short-term recurrences, in contrast with the deterministic case, where FOM reduces to the CG method. Indeed, as mentioned already in Section 4.1 for the Lanczos process, the core reason behind the short recurrence in CG is the fact that the Hessenberg matrix H_m in the Arnoldi relation (18) is tridiagonal when A is symmetric. On the other hand, when we generate a sketch-orthogonal basis V_m through the randomized Arnoldi process, we have the relation (21), which implies $G_m = (\Omega V_m)^T \Omega A V_m = V_m^T \Omega^T \Omega A V_m$, hence this matrix is not symmetric in general, unless $\Omega^T \Omega$ commutes with A, which is usually not the case. We refer to [75, Section 4] for a more in-depth discussion.

It is shown in [23] that the sequence of FOM approximants $\widetilde{\mathbf{x}}_1^{\mathrm{F}}, \widetilde{\mathbf{x}}_2^{\mathrm{F}} \ldots$, for an arbitrary operator $A \in \mathbb{R}^{n \times n}$, yields a sequence of quasi-optimal residual norms $\|\mathbf{b} - A\widetilde{\mathbf{x}}_1^{\mathrm{F}}\|_2$, $\|\mathbf{b} - A\widetilde{\mathbf{x}}_2^{\mathrm{F}}\|_2, \ldots$, with occasional spikes when the minimum residual sequence (produced by GMRES) stagnates. This property is founded on [23, Theorem 1], which only uses the Hessenberg matrix. As long as the randomized Arnoldi process does not break, this property extends to the sequence of residuals associated with randomized FOM approximations:

$$\forall\, 2 \le k \le m, \quad \|\Omega \mathbf{r}_{k-1}^{\mathrm{G}}\| < \|\Omega \mathbf{r}_k^{\mathrm{G}}\|,$$

$$\|\Omega \mathbf{r}_k^{\mathrm{F}}\| \le \frac{\|\Omega \mathbf{r}_k^{G}\|}{\sqrt{1 - (\|\Omega \mathbf{r}_k^{\mathrm{G}}\| / \|\Omega \mathbf{r}_{k-1}^{\mathrm{G}}\|)^2}} =: \alpha_k \|\Omega \mathbf{r}_k^{\mathrm{G}}\|, \quad 2 \le k \le m.$$

This bound is valid for both the symmetric and non-symmetric cases $A \in \mathbb{R}^{n \times n}$.

In the case where $A \in \mathbb{R}^{n \times n}$ is symmetric and positive definite, FOM (which is equivalent to CG) produces a sequence that minimizes the A-norm of the error, $\|\mathbf{x} - \widetilde{\mathbf{x}}_k^{\mathrm{F}}\|_A$. On the other hand, the behavior of $\|\mathbf{x} - \mathbf{x}_k^{\mathrm{F}}\|_A$ is a more complex topic. One of the main difficulties surrounding this question is that the standard sketching framework does not yield a simple concept of *sketched energy norm*, as $(\mathbf{x}, \mathbf{y}) \mapsto (\Omega A \mathbf{x})^T \Omega \mathbf{y}$ is not even symmetric in general.

For inputs of medium difficulty $A \in \mathbb{R}^{n \times n}$ where A is symmetric or non-symmetric, the cost of rRFOM might be higher than that of CG or BiCG. However, it is well known that these algorithms, based on short recurrences, lose orthogonality and hence for very ill-conditioned symmetric inputs FOM is often preferred to CG. In these cases, where we may require the Householder-Arnoldi iteration, randomized FOM based on RGS2-Arnoldi or RHQR-Arnoldi performs well, achieving the same stability as MGS2-

Arnoldi and Householder–Arnoldi, for half of the number of flops (or even a third) and much fewer synchronizations of a parallel computer.

For both rGMRES and rFOM, the solution at the m-th iteration step is given in the form $\mathbf{x}_m = V_m \mathbf{y}_m$, where $\mathbf{y}_m \in \mathbb{R}^m$. This implies that, if the sketch-orthogonal basis V_m is obtained implicitly via a whitening procedure, we can compute the approximate solution as $\mathbf{x}_m = W_m(R_m^{-1}\mathbf{y}_m)$, where W_m is a basis constructed, for instance, with the k-truncated Arnoldi process and R_m is the upper triangular factor of the QR factorization of ΩW_m. The Hessenberg matrix associated with V_m, required for the solution of (28) and (29), can be obtained from the whitened-sketched Arnoldi relation (24).

6 Solution of Eigenvalue Problems

Finding the eigenvalues and eigenvectors of a matrix $A \in \mathbb{R}^{n\times n}$ is a fundamental task in numerical linear algebra, with applications ranging from structural engineering, where eigenvectors represent natural modes of vibration whose frequencies are given by the eigenvalues, to quantum chemistry, where the lowest eigenvector corresponds to the ground state of a molecule, and to machine learning, where principal component analysis reduces dimensionality by focusing on dominant eigendirections [49, 62].

The standard eigenvalue problem consists of finding pairs $(\lambda_i, \mathbf{x}_i)$ indexed by $i \in \mathcal{I}$ of scalars $\lambda_i \in \mathbb{C}$ and unit-norm eigenvectors $\mathbf{x}_i \in \mathbb{C}^n$ such that

$$A\mathbf{x}_i = \lambda_i \mathbf{x}_i. \tag{30}$$

Since eigenvalues are the roots of the characteristic polynomial of A, the Abel–Ruffini theorem implies that there is no general algebraic formula in radicals for polynomials of degree greater than four. Consequently, for matrices of dimension $n \geq 5$ the eigenvalue problem is solved in practice by iterative numerical methods that compute approximate eigenpairs [50, 68].

The set $\mathcal{I}$ of eigenpairs that can be computed in practice depends on the size of the matrix and the structure of the problem. For moderate n, the state-of-the-art method to compute all eigenpairs $\mathcal{I} = \{1, \ldots, n\}$ is the shifted QR algorithm. This algorithm produces a sequence of matrices A_k similar to A that converges (up to ordering) to the real Schur form of A; in particular, one obtains an upper quasi-triangular matrix with eigenvalues on the diagonal (and 2×2 blocks for complex conjugate pairs):

$$\lim_{k\to\infty} A_k = \begin{bmatrix} \lambda_1 & & \star \\ & \ddots & \\ 0 & & \lambda_n \end{bmatrix}. \tag{31}$$

In step k, approximate eigenvalues are read from the diagonal (or block eigenvalues) of A_k, and eigenvectors can be recovered from accumulated similarity transformations; convergence is typically monitored by the magnitudes of the off-diagonal entries. Each step of the QR algorithm consists of computing the QR factorization of A_k, resulting in an arithmetic cost of $O(n^3)$ for the method. The scheme has been improved in terms of both the number of operations and stability by first reducing A to upper Hessenberg form using $O(n^3)$ operations, and then resorting to the Francis implicit QR step with bulge-chasing at each step k for $O(n^2)$ operations. See [50, 59, 78] for details on the shifted QR algorithm and [36] for the real Schur form and the Francis implicit QR. For large-scale eigenvalue problems with $n \geq 10^4$, however, this is still impractical as at most a linear cost in n is affordable.

6.1 Rayleigh–Ritz

For large eigenvalue problems, a small subset $I \subset \{1, \ldots, n\}$ of m eigenpairs is generally sought, with m small relative to n. The Rayleigh–Ritz method projects A onto an m-dimensional subspace and solves the reduced eigenvalue problem to obtain approximations to the desired eigenpairs; its convergence and accuracy depend on the choice of the projection subspace and on the spectral properties of A. More precisely, given an m-dimensional subspace $\mathcal{K}_m$, the Rayleigh–Ritz method seeks an approximate eigenvector $\widetilde{\mathbf{x}}$ and an approximate eigenvalue $\widetilde{\lambda}$ by imposing the following two constraints:

1. The approximate eigenvector (Ritz vector) $\widetilde{\mathbf{x}}$ belongs to $\mathcal{K}_m$.
2. The residual vector $A\widetilde{\mathbf{x}} - \widetilde{\lambda}\widetilde{\mathbf{x}}$ is orthogonal to $\mathcal{K}_m$.

The orthogonality of the residual to $\mathcal{K}_m$ fixes the m degrees of freedom that arise when seeking $\widetilde{\mathbf{x}}$ in $\mathcal{K}_m$ and is known as the Galerkin condition. Let $Q_m \in \mathbb{R}^{n\times m}$ be an orthogonal basis for $\mathcal{K}_m$ and write $\widetilde{\mathbf{x}} = Q_m\mathbf{y} \in \mathcal{K}_m$. The Galerkin condition can be written as

$$Q_m^T(AQ_m\mathbf{y} - \widetilde{\lambda}Q_m\mathbf{y}) = 0, \tag{32}$$

$$Q_m^T AQ_m\mathbf{y} = \widetilde{\lambda}\mathbf{y}. \tag{33}$$

Hence, the pair $(\widetilde{\lambda}, \mathbf{y})$ is an exact eigenpair of the small operator $H_m = Q_m^T AQ_m \in \mathbb{R}^{m\times m}$, which represents the orthogonal projection of A onto $\mathcal{K}_m$. Multiplying (33) by Q_m and using $\mathbf{y} = Q_m^T Q_m\mathbf{y}$ gives

$$\mathcal{P}_{\mathcal{K}_m} A\mathcal{P}_{\mathcal{K}_m}\widetilde{\mathbf{x}} = \widetilde{\lambda}\widetilde{\mathbf{x}}, \tag{34}$$

where $\mathcal{P}_{\mathcal{K}_m}$ is the orthogonal projector onto $\mathcal{K}_m$ represented by $Q_mQ_m^T$. This reinforces the characterization of the Rayleigh–Ritz method as an orthogonal projection approach that solves a small eigenvalue problem for H_m to obtain approximate eigenpairs of A.

The quality of a Ritz pair, namely the Ritz eigenvalue $\widetilde{\lambda}$ and the Ritz vector $\widetilde{\mathbf{x}}$, as an approximation of an eigenpair of A depends strongly on the subspace $\mathcal{K}_m$, in particular on the distance between an exact eigenvector $\mathbf{x}$ of A and $\mathcal{K}_m$. In practice, Krylov subspaces are a natural and effective choice for $\mathcal{K}_m$. The Arnoldi procedure introduced in Section 4 constructs, from a starting vector $\mathbf{b}$, an orthogonal basis Q_m for the Krylov subspace $\mathcal{K}_m(A, \mathbf{b})$ and simultaneously computes the small projected Hessenberg matrix H_m, as given by (18).

Since H_m is an $m \times m$ matrix, it is possible to compute its eigenpairs $(\widetilde{\lambda}_i, \mathbf{y}_i)$ (for instance, with a shifted QR algorithm) and form Ritz vectors $\widetilde{\mathbf{x}}_i = Q_m \mathbf{y}_i$, for $i = 1, \dots m$. Multiplying (18) by $\mathbf{y}_i$ yields

$$A\widetilde{\mathbf{x}}_i - \widetilde{\lambda}_i \widetilde{\mathbf{x}}_i = h_{m+1,m} \mathbf{q}_{m+1} \mathbf{e}_m^T \mathbf{y}_i. \tag{35}$$

Consequently, the residual norm satisfies $\|A\widetilde{\mathbf{x}}_i - \widetilde{\lambda}_i \widetilde{\mathbf{x}}_i\| = \|h_{m+1,m} \mathbf{q}_{m+1} \mathbf{e}_m^T \mathbf{y}_i\|$. Using $\|\mathbf{q}_{m+1}\| = 1$, one obtains a simple and inexpensive expression to monitor the quality of the approximate i-th eigenpair at the end of the Arnoldi method:

$$\|A\widetilde{\mathbf{x}}_i - \widetilde{\lambda}_i \widetilde{\mathbf{x}}_i\| = |h_{m+1,m}| \cdot |\mathbf{e}_m^T \mathbf{y}_i| \quad \text{for } i = 1, \dots, m. \tag{36}$$

The Arnoldi method [2], originally developed as an extension of the Lanczos method for non-symmetric matrices [52], underpins many restarted schemes with filtering, such as the implicitly restarted Arnoldi algorithm and the Krylov-Schur method. The convergence results and the bounds on the distance between the eigenvectors of A and the Krylov subspace $\mathcal{K}_m(A, \mathbf{b})$ are given in particular in [9, 10, 68].

6.2 Randomized Rayleigh–Ritz

In the Arnoldi algorithm, each step consists in applying A to the last vector of the Krylov basis and orthogonalizing the result against the existing basis. Provided that A is structured so that the cost of a matrix-vector product scales linearly with n, as is the case for a sparse A, the dominant arithmetic cost of the method is the orthogonalization step. Reducing this cost with randomized orthogonalization procedures has been the subject of recent work, including [6, 26, 42, 61]. The approach in [6] uses a randomized (block) Gram–Schmidt process [5] within the Arnoldi procedure to obtain a sketch-orthonormal basis for the Krylov subspace, as described in Section 4.1 with Algorithm 3.

Let V_m and G_m denote the sketch-orthogonal basis and the associated Hessenberg matrix generated by the randomized Arnoldi process Algorithm 3. The structure of the resulting randomized Arnoldi decomposition (21) mirrors the deterministic Arnoldi factorization (18), and therefore a randomized Rayleigh–Ritz procedure follows naturally. The theory relies on the oblique projector $\mathcal{P}_{\mathcal{K}_m}^{\Omega}$ defined in [6, 26] by

$$\mathcal{P}_{\mathcal{K}_m}^{\Omega} \mathbf{x} = \underset{\mathbf{v} \in \mathcal{K}_m}{\operatorname{argmin}} \|\Omega(\mathbf{x} - \mathbf{v})\| \tag{37}$$

for $\mathbf{x} \in \mathbb{R}^n$, which we introduced in Section 3.1. If $(\widetilde{\lambda}, \mathbf{y})$ is an eigenpair of G_m, multiplying (21) by $\mathbf{y}$ gives

$$A(V_m \mathbf{y}) - \widetilde{\lambda}(V_m \mathbf{y}) = g_{m+1,m} \mathbf{v}_{m+1} \mathbf{e}_m^T \mathbf{y}. \tag{38}$$

Defining the Ritz vector $\widetilde{\mathbf{x}} = V_m \mathbf{y} \in \mathcal{K}_m$, the residual $A\widetilde{\mathbf{x}} - \widetilde{\lambda}\widetilde{\mathbf{x}}$ is sketch-orthogonal to $\mathcal{K}_m(A, \mathbf{b})$ because it is proportional to $\mathbf{v}_{m+1}$:

$$\Omega(A\widetilde{\mathbf{x}} - \widetilde{\lambda}\widetilde{\mathbf{x}}) \perp \Omega\mathcal{K}_m. \tag{39}$$

Equation (39) is the sketched Galerkin condition. Since $\widetilde{\mathbf{x}} \in \mathcal{K}_m(A, \mathbf{b})$, this framework is naturally called a sketched or randomized Rayleigh–Ritz approach. It is shown in [6] that a pair satisfying (39) is an exact eigenpair of $\mathcal{P}^{\Omega}_{\mathcal{K}_m} A \mathcal{P}^{\Omega}_{\mathcal{K}_m}$:

$$\mathcal{P}^{\Omega}_{\mathcal{K}_m} A \mathcal{P}^{\Omega}_{\mathcal{K}_m} \widetilde{\mathbf{x}} = \widetilde{\lambda}\widetilde{\mathbf{x}}. \tag{40}$$

Since $G_m = (\Omega V_m)^T \Omega A V_m$, [26] further shows that G_m is the representation of $\mathcal{P}^{\Omega}_{\mathcal{K}_m} A \mathcal{P}^{\Omega}_{\mathcal{K}_m}$ in the basis V_m, i.e. $\mathcal{P}^{\Omega}_{\mathcal{K}_m} A \mathcal{P}^{\Omega}_{\mathcal{K}_m} \mathbf{w} = V_m G_m \mathbf{z}$ when $\mathbf{w} = V_m \mathbf{z}$. In particular, the characteristic polynomial p_{G_m} of G_m satisfies

$$p_{G_m} = \underset{p \in \mathcal{PM}_m}{\operatorname{argmin}} \|\Omega p(A)\mathbf{b}\|, \tag{41}$$

where $\mathcal{PM}_m$ is the set of monic polynomials of degree m. If p_{H_m} is the characteristic polynomial of the Hessenberg matrix H_m that originates from an orthogonal projection, then

$$\|p_{G_m}(A)\mathbf{b}\| \le \sqrt{\frac{1+\epsilon}{1-\epsilon}} \|p_{H_m}(A)\mathbf{b}\| = \sqrt{\frac{1+\epsilon}{1-\epsilon}} \min_{p \in \mathcal{PM}_m} \|p(A)\mathbf{b}\|, \tag{42}$$

which follows from the ϵ-embedding property (1). As in the deterministic setting, the quality of a Ritz pair $(\widetilde{\lambda}, \widetilde{\mathbf{x}})$ depends on the spectral properties of A and on the Krylov subspace $\mathcal{K}_m(A, \mathbf{b})$. Bounds on the residual $\|(A - \widetilde{\lambda} I)\widetilde{\mathbf{x}}\|$ and on the distance of an exact eigenvector $\mathbf{x}$ from $\mathcal{K}_m$ are derived in [6, 26]; these results are outside the scope of this review but are useful to characterize the convergence of randomized Arnoldi. The main conclusion is that the randomized Arnoldi method is an oblique projection technique on a Krylov subspace, delivering accuracy comparable to deterministic Arnoldi while reducing cost via sketch-orthogonalization.

We conclude this section by describing practical convergence monitoring. Given an eigenpair $(\widetilde{\lambda}_i, \mathbf{y}_i)$ of G_m, (38) implies

$$\|A\widetilde{\mathbf{x}}_i - \widetilde{\lambda}_i \widetilde{\mathbf{x}}_i\| = |g_{m+1,m}| \cdot \|\mathbf{v}_{m+1}\| \cdot |\mathbf{e}_m^T \mathbf{y}_i|. \tag{43}$$

Unlike the deterministic case, $\|\mathbf{v}_{m+1}\|$ need not be equal to one. Instead, [61] provides the bound, for $i = 1, \ldots, m$,

$$\sqrt{\frac{1-\epsilon}{1+\epsilon}} \|\Omega(A\widetilde{\mathbf{x}}_i - \widetilde{\lambda}_i \widetilde{\mathbf{x}}_i)\| \le \frac{\|A\widetilde{\mathbf{x}}_i - \widetilde{\lambda}_i \widetilde{\mathbf{x}}_i\|}{\|\widetilde{\mathbf{x}}_i\|} \le \sqrt{\frac{1+\epsilon}{1-\epsilon}} \|\Omega(A\widetilde{\mathbf{x}}_i - \widetilde{\lambda}_i \widetilde{\mathbf{x}}_i)\|. \tag{44}$$

Since $\|\Omega(A\widetilde{\mathbf{x}}_i - \widetilde{\lambda}_i \widetilde{\mathbf{x}}_i)\| = |g_{m+1,m}| \cdot |\mathbf{e}_m^T \mathbf{y}_i|$, this quantity is readily available during the randomized Arnoldi iteration and provides a good approximation of the relative residual $\|A\widetilde{\mathbf{x}}_i - \widetilde{\lambda}_i \widetilde{\mathbf{x}}_i\| / \|\widetilde{\mathbf{x}}_i\|$ in view of (44).

6.3 Restarting Strategies for Krylov Subspace Methods

A typical issue arising from Krylov subspace methods such as the Arnoldi procedure is the quadratic growth, with the number of steps m, of the arithmetic cost of orthogonalization together with the accompanying memory required to store a growing Krylov basis V_m of vectors in $\mathbb{R}^n$. To mitigate these problems, several restarting strategies have been proposed to produce a new Arnoldi factorization from a new starting vector $\mathbf{b}^+$ that retains much of the information from a previous length-m factorization. Popular methods include the implicitly restarted Arnoldi (IRA) and the Krylov–Schur (KS) approaches [54, 73, 74]. These approaches have been extended to the randomized Arnoldi setting in [26, 27].

Assume we have a randomized Arnoldi decomposition (21) with Krylov basis $\mathcal{K}_m(A, \mathbf{b})$ and we wish to restart it. The randomized implicitly restarted Arnoldi method (rIRA) [26] is based on polynomial filtering; see [68, Chapter 7]. If the starting vector $\mathbf{b} \in \mathbb{R}^n$ admits the expansion

$$\mathbf{b} = \sum_{i=1}^{n} \alpha_i \mathbf{x}_i, \tag{45}$$

with $\mathbf{x}_i$ the eigenvectors of A, then for any polynomial p one has

$$p(A)\mathbf{b} = \sum_{i=1}^{n} \alpha_i p(\lambda_i) \mathbf{x}_i. \tag{46}$$

If the goal is to approximate the k dominant eigenvectors $\mathbf{x}_1, \ldots, \mathbf{x}_k$ of A (with $k \leq m$), one may use $\mathbf{b}^+ = p(A)\mathbf{b}$ as a new starting vector, where p is chosen to be large on $\lambda_1, \ldots, \lambda_k$ and small on the remaining eigenvalues. Then $\mathcal{K}_m(A, \mathbf{b}^+)$ is likely to contain high-quality Ritz vectors for the desired eigenpairs. The key advantage of IRA is that $p(A)\mathbf{b}$ can be applied implicitly by performing shifted QR steps on the small Hessenberg matrix G_m obtained from the previous Arnoldi factorization; this property carries over to rIRA. In practice the polynomial p has degree q, equal to the number of shifted QR steps performed, and its roots are the shifts. There exist different strategies to choose these shifts (for example Chebyshev polynomials [73, Chapter 4]), but a common and successful approach is to pick the $q = m - k$ unwanted Ritz values of G_m as the shifts. When aiming for the k largest eigenvalues of A, one typically designates the q Ritz values of smallest modulus as unwanted. This promotes the Ritz vectors associated with the largest Ritz values in the subsequent Arnoldi iteration and yields practical convergence to the desired eigenvalues. Convergence is proven in [26] in a more restrictive setting of fixed shifts over the iteration. It is also shown how sketch-orthonormalization preserves IRA's beneficial properties while reducing computational cost.

We summarize the main steps of the randomized Implicitly Restarted Arnoldi algorithm below. The method is illustrated in Figure 12 and further details are in [26].

1. Compute the eigenvalues $\mu_1, \ldots, \mu_m$ of G_m and select k wanted eigenvalues among them (e.g., those of largest or smallest modulus).

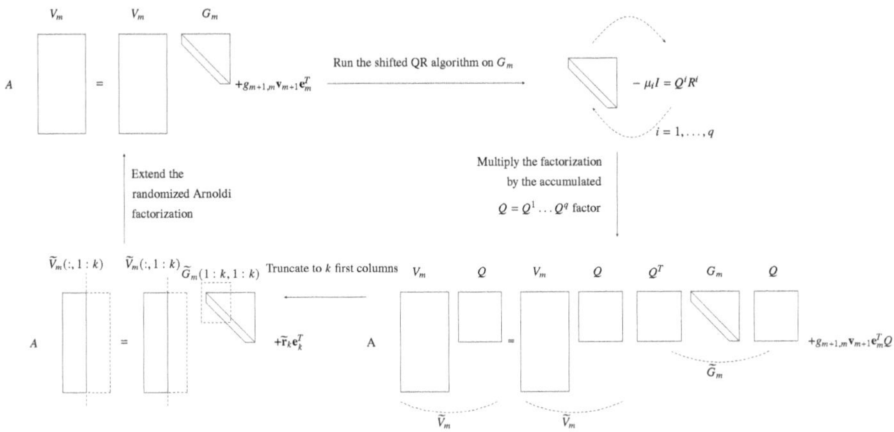

Figure 12 Visual representation of a cycle of the rIRA method.

2. Apply $q = m - k$ steps of the shifted QR algorithm on G_m using the q unwanted eigenvalues as shifts.
3. Recover a new length-k randomized Arnoldi factorization by multiplying the previous factorization by the accumulated orthogonal transformations from the shifted QR steps and truncating to length k.
4. Extend this factorization to length m using the randomized Arnoldi method and repeat from step 1.

The IRA method has been implemented in the ARPACK library [55] and has seen wide adoption. However, [54, 74] observed that the shifted QR algorithm can suffer from loss of forward stability, which motivated the development of the Krylov–Schur method [74].

The Krylov–Schur method is based on a generalization of the Arnoldi decomposition, referred to as the Krylov decomposition:

$$AW_m = W_m B_m + \mathbf{w}_{m+1} \mathbf{z}_m^T. \tag{47}$$

The distinction from an Arnoldi decomposition (18) is that the columns of $[W_m \ \mathbf{w}_{m+1}] \in \mathbb{R}^{n \times (m+1)}$ are only required to be linearly independent rather than orthonormal, B_m must only be full-rank and $\mathbf{z}_m$ may be arbitrary. This relaxation removes structural constraints present in Arnoldi factorizations; indeed, both Arnoldi factorization (18) and randomized Arnoldi factorization (21) are special cases of Krylov decompositions. Working with Krylov decompositions permits replacing the shifted QR steps used in IRA by numerically stable Schur form re-orderings; see [51] for details. This idea is the basis for Stewart's Krylov–Schur algorithm.

In [27] it is shown that a randomized Arnoldi decomposition can be obtained from a Krylov decomposition and that a sketch-orthonormal Krylov basis integrates naturally into the Krylov–Schur framework. Sketch-orthonormalizing W_m and translating $\mathbf{w}_{m+1}$ so that it is sketch-orthogonal to the basis yields a randomized Arnoldi factorization. This observation leads to a randomized Krylov–Schur (rKS) algorithm that combines the stability of Schur reordering with the efficiency and scalability of sketch-orthonormalization.

The Krylov–Schur method also incorporates a simple deflation procedure [50, 74], which was extended in rKS in [27]. When eigenpairs have converged, they can be removed from the active subspace by sketch-orthogonalizing the remaining Krylov vectors against the converged vectors, producing a partial sketch-orthonormal Schur factorization for A. That is, $AQ_m = Q_m T_m$ where Q_m is sketch-orthonormal and T_m is block upper-triangular. If among the k sought eigenpairs, there are q converged eigenpairs whose residual errors are smaller than a value η, it is shown in [27] that the sketch-orthonormal deflation procedure is equivalent to continuing the rKS method by seeking $k - q$ eigenpairs of a slightly perturbed matrix $A + E$ with

$$\|E\|_F \leq \sqrt{q}\sqrt{\frac{1+\epsilon}{1-\epsilon}}\eta. \tag{48}$$

We conclude this section with numerical experiments for the rKS method on its efficiency and accuracy, that are similar to those in [27]. A number of $k = 40$ eigenvalues of tri-diagonal synthetic matrices are sought, with two types of spectra, harmonic and geometric. This means that their diagonal entries are $1 + 1/i^2$ and 0.99^i for $i = 1, \dots, n$, respectively. The entries on their subdiagonal and superdiagonal are Gaussian noise:

$$A_{i+1,i} = \frac{g_{i+}}{100}, \; A_{i-1,i} = \frac{g_{i-}}{100}, \tag{49}$$

where $g_{i\pm}$ are drawn independently from $\mathcal{N}(0, 1)$. The eigenvalues are sought with a precision of 10^{-10} on the sketch residuals, that is $\|\Omega(A\widetilde{\mathbf{x}}_i - \widetilde{\lambda}\widetilde{\mathbf{x}}_i)\| \leq 10^{-10}$ from (44). The Ritz vectors are sought in Krylov subspaces of dimension $m = 2k = 80$. For the harmonic spectrum, the 40 eigenvalues of smallest modulus (SM) are sought, whereas for the geometric spectrum it is the ones of largest modulus (LM). Experiments are conducted with the Julia programming language in its version 1.10 [15], and for the sketching matrix $\Omega \in \mathbb{R}^{\ell \times n}$ we use an s-hashing matrix with sampling size $\ell = 2m$ and sparsity parameter $s = 8$.

In Figure 13 the input dimension n of $A \in \mathbb{R}^{n \times n}$ increases from $n_{min} = 10^5$ to $n_{max} = 5 \times 10^6$ and the execution times of IRA, KS and rKS are compared. It can be observed that the rKS method runs faster than IRA and KS, with a speed-up between 2× and 3× thanks to sketch-orthonormalization. In Figure 14, the quality of the obtained Ritz eigenpairs is compared. The left panel shows the evolution of the maximum over $i = 1, \dots, 40$ of $\|\Omega(A\widetilde{\mathbf{x}}_i - \widetilde{\lambda}\widetilde{\mathbf{x}}_i)\|$ at the end of each restart for KS and rKS, demonstrating that both methods converge in roughly the same number of iterations with similar convergence behavior. The right panel displays the real parts of the obtained eigenvalues, using the IRA method (via Julia's `eigs` function from ARPACK) as a reference. All three methods find the same approximate eigenvalues, demonstrating that the faster randomized approach rKS delivers accurate solutions for these problems.

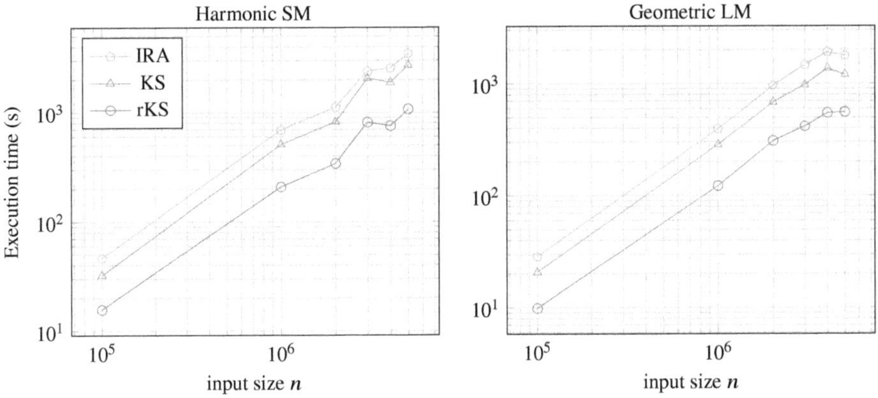

Figure 13 Execution time of KS and rKS for an increasing input dimension n. A total of 40 eigenvalues are sought, those of smallest modulus for the harmonic spectrum (left) and those of largest modulus for the geometric spectrum (right). The time label is in logarithmic scale, meaning that rKS being constantly below KS and IRA represents here a speed-up of around 3 times faster.

7 Evaluation of Matrix Functions

Given a matrix $A \in \mathbb{R}^{n\times n}$ and a function f that is analytic on and inside a contour $\Gamma \subset \mathbb{C}$ which encloses the spectrum of A, the matrix function $f(A)$ can be defined as

$$f(A) = \frac{1}{2\pi i}\int_\Gamma f(t)(tI - A)^{-1}\,dt.$$

We refer to [46] for other equivalent definitions for $f(A)$. The computation of matrix functions arises in many areas, such as the solution of partial differential equations [17], network analysis [12, 32], and electronic structure computations [13]. In these applications, one is often interested in the computation of $f(A)\mathbf{b}$ for a given vector $\mathbf{b} \in \mathbb{R}^n$, rather than the full matrix function $f(A)$. When A is large and sparse, the computation of $f(A)$ through methods such as Schur–Parlett [25] is usually infeasible as it has a computational cost of $O(n^3)$, especially if the (generally dense) matrix $f(A)$ is too large to store explicitly. In this setting, Krylov subspace methods are the most popular methods for the approximation of $f(A)\mathbf{b}$ [30, 66]. In this section, we discuss the computation of $f(A)\mathbf{b}$ with randomized Krylov methods, which has been recently investigated in [22, 41, 63]. We are going to present the different approximations that have been proposed in the literature, emphasizing the links and equivalences between the different approaches.

A well-established way to approximate $f(A)\mathbf{b}$ is to use a projection onto the Krylov subspace $\mathcal{K}_m(A, \mathbf{b})$, which yields the Arnoldi approximation [30, 66]

$$\mathbf{f}_m := \widetilde{\beta} Q_m f(H_m)\mathbf{e}_1, \tag{50}$$

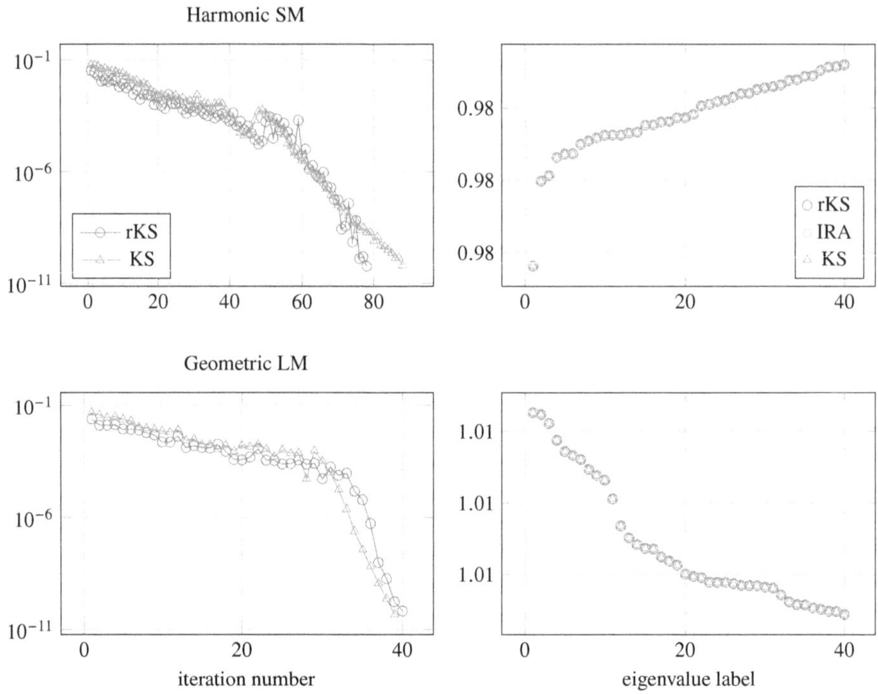

Figure 14 Left: maximum over i of the residual errors $\|\Omega(A\tilde{\mathbf{x}}_i - \tilde{\lambda}\tilde{\mathbf{x}}_i)\|$, against the number of iterations of KS and rKS. Convergence is declared when this maximum reaches 10^{-10}. Right: obtained Ritz eigenvalues for KS and rKS compared to the reference IRA.

where Q_m and H_m satisfy the (orthogonal) Arnoldi relation (18) and $\mathbf{b} = \widetilde{\beta} Q_m \mathbf{e}_1$. This approximation is exact when f is a polynomial of degree up to $m-1$, and it is equivalent to computing $p_{m-1}(A)\mathbf{b}$, where p_{m-1} is a polynomial which interpolates f at the eigenvalues of H_m, see for instance [66, Theorem 3.3].

Given an arbitrary basis W_m of $\mathcal{K}_m(A, \mathbf{b})$ and the associated Hessenberg matrix $\underline{L}_m$ satisfying the Arnoldi-like relation (19), the approximation (50) satisfies

$$\mathbf{f}_m = W_m f(W_m^+ A W_m) W_m^+ \mathbf{b} = \widetilde{\beta} W_m f(L_m + W_m^+ \mathbf{w}_{m+1} \boldsymbol{\ell}_m^T)\mathbf{e}_1, \tag{51}$$

where we used (20) and $W_m^+\mathbf{b} = W_m^+(\widetilde{\beta}\mathbf{w}_1) = \widetilde{\beta}\mathbf{e}_1$ to rewrite the identity. See, for instance, [22, Lemma 3.1] for a proof of the equivalence between (50) and (51). When the basis W_m is not orthonormal, the downside of the approximation (51) is that computing $W_m^+\mathbf{w}_{m+1}$ requires the solution of the least squares problem

$$W_m^+\mathbf{w}_{m+1} = \underset{\mathbf{h}\in\mathbb{R}^m}{\operatorname{argmin}} \|W_m\mathbf{h} - \mathbf{w}_{m+1}\|.$$

When W_m is sketch-orthonormal it follows from (2) that $\mathrm{Cond}(W_m) \le \sqrt{(1+\varepsilon)/(1-\varepsilon)}$, so W_m is well-conditioned and $W_m^+ \mathbf{w}_{m+1}$ can be computed by solving a least squares problem with the LSQR algorithm. This approach is proposed in [22, Algorithm 3.1]. For a general basis W_m, in [22, Algorithm 3.2] it is proposed to solve the least squares problem for $W_m^+ \mathbf{w}_{m+1}$ by using Blendenpik [3].

A further alternative that has been explored in [22, Algorithm 3.3] is to completely ignore the rank-one update in (51) and approximate $f(A)\mathbf{b}$ with

$$\widetilde{\mathbf{f}}_m = \widetilde{\beta} W_m f(L_m) \mathbf{e}_1,$$

but for a general basis W_m, the matrix $f(L_m)$ can be quite far from $f(L_m + W_m^+ \mathbf{w}_{m+1} \boldsymbol{\ell}_m^T)$, so $\widetilde{\mathbf{f}}_m$ may be an inaccurate approximation of $f(A)\mathbf{b}$. However, with a sketch-orthonormal basis of $\mathcal{K}_m(A, \mathbf{b})$, we show below that an approximation of this form has a natural link with (50). Assume that we have a sketch-orthonormal basis V_m and a corresponding Hessenberg matrix $\underline{G}_m$ which satisfy the randomized Arnoldi relation (21), and let $\mathbf{b} = \beta V_m \mathbf{e}_1$. Then we have

$$\mathbf{f}_m^{\Omega} := \beta V_m f(G_m) \mathbf{e}_1 = V_m f\left((\Omega V_m)^T \Omega A V_m\right) (\Omega V_m)^T \Omega \mathbf{b}, \tag{52}$$

where for the last equality we used the sketch-orthonormality of V_m and the identity $G_m = (\Omega V_m)^T \Omega A V_m$. This approximation is introduced in [41] with the name *sketched FOM* (sFOM). To describe its connection with (50), we follow the presentation in [41, Section 2] and consider a function that admits an integral representation of the form

$$f(z) = \int_\Gamma (t+z)^{-1} \, d\mu(t),$$

which includes as special cases both Stieltjes functions [14] and the Cauchy integral representation for analytic functions. The integral expression for f translates to the matrix function representation

$$f(A)\mathbf{b} = \int_\Gamma (tI + A)^{-1} \mathbf{b} \, d\mu(t) = \int_\Gamma \mathbf{x}(t) \, d\mu(t), \qquad \text{where} \quad (tI + A)\mathbf{x}(t) = \mathbf{b}. \tag{53}$$

For this class of functions, the Arnoldi approximation (50) can be interpreted as follows. For each $t \in \Gamma$, let $\mathbf{x}_m(t) := \widetilde{\beta} Q_m (tI + H_m)^{-1} \mathbf{e}_1$ be the approximate solution to the shifted linear system $(tI + A)\mathbf{x}(t) = \mathbf{b}$ after m iterations of FOM. Then, we have

$$\mathbf{f}_m = \widetilde{\beta} Q_m f(H_m) \mathbf{e}_1 = \int_\Gamma \widetilde{\beta} Q_m (tI + H_m)^{-1} \mathbf{e}_1 \, d\mu(t) = \int_\Gamma \mathbf{x}_m(t) \, d\mu(t).$$

The residuals $\mathbf{r}_m(t) = \mathbf{b} - (tI + A)\mathbf{x}_m(t)$ are orthogonal to $\mathcal{K}_m(A, \mathbf{b}) = \mathrm{Range}(Q_m)$.

The sFOM approximation (52) stems from the following observation: instead of imposing the orthogonality condition on $\mathbf{r}_m(t)$ exactly, we can alternatively solve the shifted linear systems using randomized FOM, i.e., impose that the sketched residuals are orthogonal to the sketch of the Krylov subspace. In other words, for each $t \in \Gamma$ we look for an approximate solution $\mathbf{x}_m^{\Omega}(t) \in \mathcal{K}_m(A, \mathbf{b})$ such that the residual $\mathbf{r}_m^{\Omega}(t) =$

$\mathbf{b} - (tI + A)\mathbf{x}_m^\Omega(t)$ satisfies the sketched Galerkin condition $\Omega\mathbf{r}_m^\Omega(t) \perp \Omega\mathcal{K}_m(A, \mathbf{b})$. This implies that $\mathbf{x}_m^\Omega(t) = V_m\mathbf{y}_m^\Omega(t)$ with

$$(\Omega V_m)^T(\Omega\mathbf{b} - \Omega(tI + A)V_m\mathbf{y}_m^\Omega(t)) = \mathbf{0},$$

and hence, similarly to the derivation of randomized FOM in Section 5.2.2, we have

$$\mathbf{x}_m^\Omega(t) = \beta V_m(tI + (\Omega V_m)^T\Omega A V_m)^{-1}\mathbf{e}_1 = \beta V_m(tI + G_m)^{-1}\mathbf{e}_1, \tag{54}$$

where we used the identities $\mathbf{b} = \beta\mathbf{v}_1$, $(\Omega V_m)^T\Omega V_m = I$ and $(\Omega V_m)^T\Omega A V_m = G_m$. It then follows that

$$\int_\Gamma \mathbf{x}_m^\Omega(t)d\mu(t) = \int_\Gamma \beta V_m(tI + G_m)^{-1}\mathbf{e}_1 = \beta V_m f(G_m)\mathbf{e}_1 = \mathbf{f}_m^\Omega,$$

showing that the approximation $\mathbf{f}_m^\Omega$ can be obtained by imposing a sketched Galerkin condition on the residuals of the shifted linear systems in the integral expression (53) for $f(A)\mathbf{b}$. The relation between (52) and (50) then mimics the one between the approximate solutions for linear systems obtained with randomized FOM (29) and FOM (27). We refer to [41, Section 2] for further details on the sFOM approximation for $f(A)\mathbf{b}$.

The authors of [41] also consider a sketched GMRES approximation, in which the shifted linear systems $(tI + A)\mathbf{x}(t) = \mathbf{b}$ are solved by using randomized GMRES instead of randomized FOM. However, this approximation to $f(A)\mathbf{b}$ has no simple closed-form expression, so it requires using a quadrature rule to evaluate the integral expression of f. We refer to [41, Section 3] for additional details.

Both in [22] and in [41], it is proposed to compute the approximation (52) by using a sketch-orthonormal basis V_m obtained implicitly through the whitening approach described in Section 4.2. Specifically, a non-orthonormal basis W_m for $\mathcal{K}_m(A, \mathbf{b})$ is constructed using the k-truncated Arnoldi process, and then a sketch-orthonormal basis $V_m = W_m R_m^{-1}$ is obtained, where $\Omega W_m = S_m R_m$ is a QR factorization. In this setting, we can compute the approximation (52) without ever forming the basis V_m explicitly. Indeed, we have [41, Section 2.1]

$$\mathbf{f}_m^\Omega = V_m f(G_m)\beta\mathbf{e}_1 = W_m R_m^{-1} f\left(S_m^T\Omega A W_m R_m^{-1}\right) S_m^T\Omega\mathbf{b},$$

and in this expression we only need to apply R_m^{-1} to the right of $S_m^T\Omega A V_m \in \mathbb{R}^{m\times m}$ and to the left of the vector $f(S_m^T\Omega A V_m R_m^{-1})S_m^T\Omega\mathbf{b} \in \mathbb{R}^m$, for a cost of $O(m^3)$. Recall that, on the other hand, explicitly computing $V_m R_m^{-1}$ would cost $O(nm^2)$, which would eliminate the computational advantage of the whitening approach.

The approximation (52) is also studied in [63], where the authors also employ the k-truncated Arnoldi process combined with whitening in order to implicitly construct the sketch-orthonormal basis V_m. In particular, they derive the identities (24) and (25) and use them to rewrite (52) as

$$\mathbf{f}_m^\Omega = W_m R_m^{-1} f(\widehat{L}_m + \rho_m^{-1}\mathbf{z}_m\boldsymbol{\ell}_m^T)\beta\mathbf{e}_1, \tag{55}$$

see [63, Algorithm 1] for the implementation details. in [63, Section 7], the authors argue that the approximation (55) is quite robust in finite-precision arithmetic despite the potentially ill-conditioned matrix R_m.

We also mention that in the recent preprint [40], sketch-orthogonalization is proposed as a means to reduce the orthogonalization costs of restarted Krylov methods for matrix functions [31]. Although the authors do not provide a rigorous convergence analysis, the method they present exhibits competitive performance in their numerical tests, occasionally even converging in fewer iterations compared to the deterministic restarted method; see, e.g., [40, Figure 3].

8 Solution of Matrix Equations

Matrix Sylvester equations appear in numerous applications, for instance in model order reduction and in the discretization of certain partial differential equations; we refer to [11, 72] for additional details. In several applications, the right-hand side is low-rank and the matrix Sylvester equation can be written in the form

$$AX + XB = C_1 C_2^T, \tag{56}$$

with $A, B \in \mathbb{R}^{n\times n}$ and $C_1, C_2 \in \mathbb{R}^{n\times r}$, with $r \ll n$. In this setting, efficient approaches for the solution of (56) are often based on projection on the polynomial block Krylov subspaces $\mathcal{K}_m(A, C_1)$ and $\mathcal{K}_m(B^T, C_2)$, or on extended and rational block Krylov subspaces; we refer to the review paper [72] for further details and references.

In [64], it was proposed to use randomized sketching to reduce the cost of orthogonalization within Krylov methods for the solution of (56). In this section, we briefly describe the approach presented in [64]. Assume that we generate two non-orthonormal bases W_m^A and $W_m^B \in \mathbb{R}^{n\times mr}$ of $\mathcal{K}_m(A, C_1)$ and $\mathcal{K}_m(B^T, C_2)$, using a truncated block Arnoldi procedure, where orthogonalization is only performed against the previous k blocks, which leads to the Arnoldi relations

$$AW_m^A = W_{m+1}^A \underline{H}_m^A \qquad \text{and} \qquad B^T W_m^B = W_{m+1}^B \underline{H}_m^B,$$

where $\underline{H}_m^A$ and $\underline{H}_m^B \in \mathbb{R}^{(m+1)r\times mr}$ are block upper Hessenberg matrices with upper bandwidth kr. Assume that we have two ε-subspace embeddings Ω_A and Ω_B for $\mathcal{K}_{m+1}(A, C_1)$ and $\mathcal{K}_{m+1}(B^T, C_2)$, respectively. Then, we can use the basis whitening approach to cheaply compute sketch-orthonormal bases of the two block Krylov subspaces: given the QR factorizations $\Omega^A W_m^A = Q_m^A T_m^A$ and $\Omega^B W_m^B = Q_m^B T_m^B$, the whitened bases

$$V_m^A = W_m^A (T_m^A)^{-1} \qquad \text{and} \qquad V_m^B = W_m^B (T_m^B)^{-1}$$

are sketch-orthogonal due to the ϵ-embedding property. Associated to the bases V_m^A and V_m^B are the modified block upper Hessenberg matrices $G_m^A = T_m^A H_m^A (T_m^A)^{-1}$ and $G_m^B = T_m^B H_m^B (T_m^B)^{-1}$. For further details we refer to the whitened-sketched Arnoldi relations presented in [64, Section 2.3], which generalize (24) to the block case.

Projection methods for the Sylvester matrix equation (56) usually look for an approximate solution of the form $X_m = Q_m^A Y_m (Q_m^B)^T$, with corresponding residual $R_m = AX_m + X_m B - C_1 C_2^T$, where Q_m^A and Q_m^B are orthonormal bases of the block Krylov subspaces $\mathcal{K}_m(A, C_1)$ and $\mathcal{K}_m(B^T, C_2)$, and $Y_m \in \mathbb{R}^{mr \times mr}$ is computed by solving a smaller projected problem. The authors of [64] propose a sketched-and-truncated method that searches for a solution of the form $X_m = V_m^A Y_m (V_m^B)^T$, and imposes on the associated residual matrix R_m the following *sketched Galerkin condition*

$$(\Omega^A V_m^A)^T (\Omega^A R_m (\Omega^B)^T)(\Omega^B V_m^B) = 0, \tag{57}$$

which can be satisfied by taking as Y_m the solution of the projected equation

$$(G_m^A + \widehat{G}^A E_m^T) Y_m + Y_m (G_m^B + \widehat{G}^B E_m^T)^T = E_1 \beta_1 \beta_2^T E_1^T, \tag{58}$$

where $\widehat{G}^A$ and $\widehat{G}^B \in \mathbb{R}^{n \times r}$ are suitable block vectors which can be obtained when computing the whitened bases V_m^A and V_m^B, and the block scalars $\beta_1, \beta_2 \in \mathbb{R}^{r \times r}$ can be determined from the expressions $C_1 = W_m^A E_1 \beta_1$, $C_2 = W_m^B E_1 \beta_2$, where $E_1 \in \mathbb{R}^{mr \times r}$ contains the first r columns of an $mr \times mr$ identity matrix. We refer to [64, Section 3] for a detailed description of their algorithm and explicit expressions for $\widehat{G}^A$ and $\widehat{G}^B$, and to [64, Algorithm 1] for a pseudocode.

9 Conclusions and Outlook

The randomized orthogonalization framework provides efficient ways to construct sketch-orthogonal bases that are very well-conditioned. The randomized Gram–Schmidt and Householder QR algorithms have excellent numerical stability properties and significantly lower communication costs on parallel architectures. These techniques can be employed within Krylov subspace methods to lower the cost of orthogonalization, and the resulting randomized Arnoldi relation can be used to construct approximate solutions to linear systems of equations, eigenvalue problems, to evaluate matrix functions, and to solve matrix equations.

Two main approaches can be identified in this context. On one hand, we can construct a sketch-orthogonal basis explicitly, for instance via a randomized Gram–Schmidt process, which requires roughly half the number of flops with respect to the deterministic Gram–Schmidt process, has communication costs which are comparable to those of CGS, and numerical stability which is comparable to that of MGS. On the other hand, we can construct a non-orthogonal basis with a cheap procedure, such as the k-truncated Arnoldi process, and then obtain a sketch-orthogonal basis implicitly through whitening, by computing a QR factorization of the sketched basis. This second approach is asymptotically cheaper than randomized Gram–Schmidt, as the sketch-orthogonal basis is never formed explicitly, but its main issue is that the basis constructed with k-truncated Arnoldi quickly becomes ill-conditioned, and this may have a negative impact on the convergence of the approximate solutions extracted from the Krylov subspace. Nevertheless, although less robust than the first, the approach employing whitening is computationally efficient, and it very often performs well in practice, although this behavior is still not completely understood theoretically.

Several research directions are still open. For instance, it is not known whether it is possible to obtain a sketch-orthogonal basis of a Krylov subspace with an algorithm that has the same numerical stability as randomized Gram–Schmidt or randomized Householder QR, and the same computational efficiency as whitening. Also, in the case where A is symmetric, we still lack a framework in which the small projected problem would also be symmetric, which is not the case in the current randomized Arnoldi processes. This phenomenon is undesirable especially when computing eigenvalues of a symmetric matrix, since the eigenvalues of the projected matrix are not guaranteed to be real. Finally, for simulations that can only afford the storage of a few high-dimensional vectors, we still lack a randomized short-recurrence framework, that could for instance be inspired by Bi-Lanczos, BiCG or CG. Although these algorithms are already very efficient, they perform only a few vector-vector operations between synchronizations, which are often the limiting factor preventing their scalability. A randomized version of these algorithms which would save some synchronizations in exchange for a few additional flops could result in noticeable speedups.

In addition, the development of a standard library for randomized orthogonalization routines would enable their use in real applications. This would make it easier to benchmark these algorithms and gain feedback on their numerical behavior in large-scale applications.

Acknowledgment

The first, second, and fourth authors acknowledge funding from the European Research Council (ERC) under the European Union's Horizon 2020 research and innovation program, grant agreement No 810367.

References

1. N. Ailon and B. Chazelle. Approximate nearest neighbors and the fast Johnson–Lindenstrauss transform. In *Proceedings of the Thirty-Eighth Annual ACM Symposium on Theory of Computing (STOC)*, pages 557–563, New York, NY, USA, 2006. Association for Computing Machinery.
2. W. E. Arnoldi. The principle of minimized iterations in the solution of the matrix eigenvalue problem. *Quarterly of Applied Mathematics*, 9(1):17–29, 1951.
3. H. Avron, P. Maymounkov, and S. Toledo. Blendenpik: supercharging Lapack's least-squares solver. *SIAM J. Sci. Comput.*, 32(3):1217–1236, 2010.
4. O. Balabanov. Randomized Cholesky QR factorizations. *arXiv preprint arXiv:2210.09953*, 2022.
5. O. Balabanov and L. Grigori. Randomized Gram–Schmidt process with application to GMRES. *SIAM J. Sci. Comput.*, 44(3):A1450–A1474, 2022.
6. O. Balabanov and L. Grigori. Randomized block Gram-Schmidt process for the solution of linear systems and eigenvalue problems. *SIAM J. Sci. Comput.*, 47(1):A553–A585, 2025.
7. O. Balabanov and A. Nouy. Randomized linear algebra for model reduction. Part I: Galerkin methods and error estimation. *Adv. Comput. Math.*, 45(5-6):2969–3019, 2019.

8. O. Balabanov and A. Nouy. Randomized linear algebra for model reduction—part II: minimal residual methods and dictionary-based approximation. *Adv. Comput. Math.*, 47(2):Paper No. 26, 54, 2021.
9. M. Bellalij, G. Meurant, and H. Sadok. The distance of an eigenvector to a Krylov subspace and the convergence of the Arnoldi method for eigenvalue problems. *Linear Algebra and its Applications*, 504:387–405, sep 2016.
10. M. Bellalij, Y. Saad, and H. Sadok. Further analysis of the Arnoldi process for eigenvalue problems. *SIAM Journal on Numerical Analysis*, 48(2):393–407, jan 2010.
11. P. Benner and J. Saak. Numerical solution of large and sparse continuous time algebraic matrix Riccati and Lyapunov equations: a state of the art survey. *GAMM-Mitt.*, 36(1):32–52, 2013.
12. M. Benzi and P. Boito. Matrix functions in network analysis. *GAMM-Mitt.*, 43(3):e202000012, 36, 2020.
13. M. Benzi, P. Boito, and N. Razouk. Decay properties of spectral projectors with applications to electronic structure. *SIAM Rev.*, 55(1):3–64, 2013.
14. C. Berg. Stieltjes-Pick-Bernstein-Schoenberg and their connection to complete monotonicity. In J. Mateu and E. Porcu, editors, *Positive Definite Functions: From Schoenberg to Space-Time Challenges*, pages 15–45, 2008. Preprint available at `https://web.math.ku.dk/~berg/manus/castellon.pdf`.
15. J. Bezanson, A. Edelman, S. Karpinski, and V. B. Shah. Julia: A fresh approach to numerical computing. *SIAM Review*, 59(1):65–98, 2017.
16. A. Boriçi. Fast methods for computing the Neuberger operator. In A. Frommer, T. Lippert, B. Medeke, and K. Schilling, editors, *Numerical Challenges in Lattice Quantum Chromodynamics*, pages 40–47, Berlin, Heidelberg, 2000. Springer Berlin Heidelberg.
17. M. A. Botchev and L. A. Knizhnerman. ART: adaptive residual-time restarting for Krylov subspace matrix exponential evaluations. *J. Comput. Appl. Math.*, 364:112311, 14, 2020.
18. J. Bourgain, S. Dirksen, and J. Nelson. Toward a unified theory of sparse dimensionality reduction in Euclidean space. In *Proceedings of the forty-seventh annual ACM symposium on Theory of Computing*, pages 499–508, 2015.
19. E. Carson, K. Lund, M. Rozložník, and S. Thomas. Block Gram-Schmidt algorithms and their stability properties. *Linear Algebra Appl.*, 638:150–195, 2022.
20. M. Charikar, K. Chen, and M. Farach-Colton. Finding frequent items in data streams. In P. Widmayer, S. Eidenbenz, F. Triguero, R. Morales, R. Conejo, and M. Hennessy, editors, *Automata, Languages and Programming*, pages 693–703, Berlin, Heidelberg, 2002. Springer Berlin Heidelberg.
21. S. Chenakkod, M. Dereziński, and X. Dong. Optimal subspace embeddings: Resolving Nelson-Nguyen conjecture up to sub-polylogarithmic factors. *arXiv preprint arXiv:2508.14234*, 2025.
22. A. Cortinovis, D. Kressner, and Y. Nakatsukasa. Speeding up Krylov subspace methods for computing $f(A)b$ via randomization. *SIAM J. Matrix Anal. Appl.*, 45(1):619–633, 2024.
23. J. Cullum and A. Greenbaum. Relations between Galerkin and norm-minimizing iterative methods for solving linear systems. *SIAM Journal on Matrix Analysis and Applications*, 17(2):223–247, 1996.
24. S. Dasgupta and A. Gupta. An elementary proof of a theorem of Johnson and Lindenstrauss. *Random Structures & Algorithms*, 22(1):60–65, 2003.
25. P. I. Davies and N. J. Higham. A Schur–Parlett algorithm for computing matrix functions. *SIAM J. Matrix Anal. Appl.*, 25(2):464–485, 2003.
26. J.-G. de Damas and L. Grigori. Randomized implicitly restarted Arnoldi method for the nonsymmetric eigenvalue problem. *SIAM Journal on Matrix Analysis and Applications*, 46(4):2395–2422, 2025.
27. J.-G. de Damas and L. Grigori. Randomized Krylov–Schur eigensolver with deflation. *arXiv preprint arXiv:2508.05400*, 2025.
28. J. Demmel, L. Grigori, M. Hoemmen, and J. Langou. Communication-optimal parallel and sequential QR and LU factorizations. *SIAM Journal on Scientific Computing*, 34(1):A206–A239, 2012.
29. P. Drineas, M. W. Mahoney, and S. Muthukrishnan. Sampling algorithms for ℓ_2 regression and applications. In *Proceedings of the 17th Annual ACM–SIAM Symposium on Discrete Algorithms (SODA)*, pages 1127–1136, 2006.

30. V. L. Druskin and L. A. Knizhnerman. Two polynomial methods for calculating functions of symmetric matrices. *Zh. Vychisl. Mat. i Mat. Fiz.*, 29(12):1763–1775, 1989.
31. M. Eiermann and O. G. Ernst. A restarted Krylov subspace method for the evaluation of matrix functions. *SIAM J. Numer. Anal.*, 44(6):2481–2504, 2006.
32. E. Estrada and D. J. Higham. Network properties revealed through matrix functions. *SIAM Rev.*, 52(4):696–714, 2010.
33. A. Frommer and V. Simoncini. Matrix functions. In *Model order reduction: theory, research aspects and applications*, volume 13 of *Math. Ind.*, pages 275–303. Springer, Berlin, 2008.
34. J. E. Garrison and I. C. Ipsen. A randomized preconditioned Cholesky-QR algorithm. *arXiv preprint arXiv:2406.11751*, 2024.
35. A. C. Gilbert, J. Y. Park, and M. B. Wakin. Sketched SVD: Recovering spectral features from compressive measurements. *arXiv preprint arXiv:1211.0361*, 2012.
36. G. H. Golub and C. F. Van Loan. *Matrix Computations*. Johns Hopkins Studies in the Mathematical Sciences. Johns Hopkins University Press, Baltimore, MD, fourth edition, 2013.
37. L. Grigori, D. Kressner, N. Shao, and I. Simunec. Restoring similarity in randomized Krylov methods with applications to eigenvalue problems and matrix functions. *arXiv preprint arXiv:2601.10248*, 2026.
38. L. Grigori, L. Piccinini, and I. Simunec. Randomized biorthogonalization through a two-sided Gram–Schmidt process. *arXiv preprint arXiv:2509.04386*, 2025.
39. L. Grigori and E. Timsit. Randomized Householder QR. *arXiv preprint arXiv:2405.10923*, 2024.
40. N. L. Guidotti, P.-G. Martinsson, J. A. Acebrón, and J. Monteiro. Accelerating a restarted Krylov method for matrix functions with randomization. *arXiv preprint arXiv:2503.22631*, 2025.
41. S. Güttel and M. Schweitzer. Randomized sketching for Krylov approximations of large-scale matrix functions. *SIAM J. Matrix Anal. Appl.*, 44(3):1073–1095, 2023.
42. S. Güttel and I. Simunec. A sketch-and-select Arnoldi process. *SIAM J. Sci. Comput.*, 46(4):A2774–A2797, 2024.
43. N. Halko, P. G. Martinsson, and J. A. Tropp. Finding structure with randomness: probabilistic algorithms for constructing approximate matrix decompositions. *SIAM Rev.*, 53(2):217–288, 2011.
44. M. R. Hestenes and E. Stiefel. Methods of conjugate gradients for solving linear systems. *J. Research Nat. Bur. Standards*, 49:409–436, 1952.
45. A. J. Higgins, D. B. Szyld, E. G. Boman, and I. Yamazaki. Analysis of randomized Householder-Cholesky QR factorization with multisketching. *Numer. Math.*, 157(5):1695–1737, 2025.
46. N. J. Higham. *Functions of Matrices: Theory and Computation*. Society for Industrial and Applied Mathematics (SIAM), Philadelphia, PA, 2008.
47. Y. Jang and L. Grigori. Randomized orthogonalization process with reorthogonalization. *Numer. Linear Algebra Appl.*, 32(4):Paper No. e70029, 14, 2025.
48. W. B. Johnson and J. Lindenstrauss. Extensions of Lipschitz mappings into a Hilbert space. In *Conference in modern analysis and probability (New Haven, Conn., 1982)*, volume 26 of *Contemp. Math.*, pages 189–206. Amer. Math. Soc., Providence, RI, 1984.
49. I. T. Jolliffe and J. Cadima. Principal component analysis: a review and recent developments. *Philosophical Transactions of the Royal Society A: Mathematical, Physical and Engineering Sciences*, 374(2065):20150202, Apr. 2016.
50. D. Kressner. *Numerical Methods for General and Structured Eigenvalue Problems*. Springer, 2005.
51. D. Kressner. Block algorithms for reordering standard and generalized Schur forms. *ACM Transactions on Mathematical Software*, 32(4):521–532, Dec. 2006.
52. C. Lanczos. An iteration method for the solution of the eigenvalue problem of linear differential and integral operators. *Journal of research of the National Bureau of Standards*, 45(4):255–282, 1950.
53. K. G. Larsen and J. Nelson. The Johnson–Lindenstrauss lemma is optimal for linear dimensionality reduction. *arXiv preprint arXiv:1411.2404*, 2014.
54. R. Lehoucq and D. Sorensen. Deflation techniques for an implicitly restarted Arnoldi iteration. *SIAM Journal on Matrix Analysis and Applications*, 17(4):789–821, 1995.
55. R. B. Lehoucq, D. C. Sorensen, and C. Yang. *ARPACK users' guide: solution of large-scale eigenvalue problems with implicitly restarted Arnoldi methods*. SIAM, 1998.

56. P.-G. Martinsson and J. A. Tropp. Randomized numerical linear algebra: Foundations and algorithms. *Acta Numerica*, 29:403–572, 2020.
57. M. Meier, Y. Nakatsukasa, A. Townsend, and M. Webb. Are sketch-and-precondition least squares solvers numerically stable? *SIAM Journal on Matrix Analysis and Applications*, 45(2):905–929, 2024.
58. M. Melnichenko, O. Balabanov, R. Murray, J. Demmel, M. W. Mahoney, and P. Luszczek. CholeskyQR with randomization and pivoting for tall matrices (CQRRPT). *SIAM J. Matrix Anal. Appl.*, 46(3):1701–1734, 2025.
59. G. S. Miminis and C. C. Paige. Implicit shifting in the QR and related algorithms. *SIAM Journal on Matrix Analysis and Applications*, 12(2):385–400, apr 1991.
60. R. Murray, J. Demmel, M. W. Mahoney, N. B. Erichson, M. Melnichenko, O. A. Malik, L. Grigori, P. Luszczek, M. Dereziński, M. E. Lopes, T. Liang, H. Luo, and J. Dongarra. Randomized numerical linear algebra : A perspective on the field with an eye to software. *arXiv preprint arXiv:2302.11474*, 2023.
61. Y. Nakatsukasa and J. A. Tropp. Fast and accurate randomized algorithms for linear systems and eigenvalue problems. *SIAM Journal on Matrix Analysis and Applications*, 45(2):1183–1214, 2024.
62. H. J. Pain. *The Physics of Vibrations and Waves*. Wiley, Apr. 2005.
63. D. Palitta, M. Schweitzer, and V. Simoncini. Sketched and truncated polynomial Krylov methods: evaluation of matrix functions. *Numer. Linear Algebra Appl.*, 32(1):Paper No. e2596, 16, 2025.
64. D. Palitta, M. Schweitzer, and V. Simoncini. Sketched and truncated polynomial Krylov subspace methods: matrix Sylvester equations. *Math. Comp.*, 94(354):1761–1792, 2025.
65. V. Rokhlin and M. Tygert. A fast randomized algorithm for overdetermined linear least-squares regression. *Proceedings of the National Academy of Sciences*, 105(36):13212–13217, 2008.
66. Y. Saad. Analysis of some Krylov subspace approximations to the matrix exponential operator. *SIAM J. Numer. Anal.*, 29(1):209–228, 1992.
67. Y. Saad. *Iterative methods for sparse linear systems*. Society for Industrial and Applied Mathematics, Philadelphia, PA, second edition, 2003.
68. Y. Saad. *Numerical methods for large eigenvalue problems*, volume 66 of *Classics in Applied Mathematics*. Society for Industrial and Applied Mathematics (SIAM), Philadelphia, PA, revised edition, 2011.
69. Y. Saad and M. H. Schultz. GMRES: a generalized minimal residual algorithm for solving nonsymmetric linear systems. *SIAM J. Sci. Statist. Comput.*, 7(3):856–869, 1986.
70. T. Sarlós. Improved approximation algorithms for large matrices via random projections. In *Proceedings of the 47th Annual IEEE Symposium on Foundations of Computer Science (FOCS)*, pages 143–152, USA, 2006. IEEE Computer Society.
71. R. Schreiber and C. Van Loan. A storage-efficient WY representation for products of Householder transformations. *SIAM Journal on Scientific and Statistical Computing*, 10(1):53–57, 1989.
72. V. Simoncini. Computational methods for linear matrix equations. *SIAM Rev.*, 58(3):377–441, 2016.
73. D. C. Sorensen. Implicit application of polynomial filters in a k-step Arnoldi method. *SIAM Journal on Matrix Analysis and Applications*, 13(1):357–385, jan 1992.
74. G. W. Stewart. A Krylov–Schur algorithm for large eigenproblems. *SIAM Journal on Matrix Analysis and Applications*, 23(3):601–614, jan 2002.
75. E. Timsit, L. Grigori, and O. Balabanov. Randomized orthogonal projection methods for Krylov subspace solvers. *arXiv preprint arXiv:2302.07466*, 2023.
76. J. A. Tropp. Improved analysis of the subsampled randomized Hadamard transform. *Advances in Adaptive Data Analysis*, 3(01n02):115–126, 2011.
77. J. A. Tropp, A. Yurtsever, M. Udell, and V. Cevher. Streaming low-rank matrix approximation with an application to scientific simulation. *SIAM Journal on Scientific Computing*, 41(4):A2430–A2463, 2019.
78. D. S. Watkins. Understanding the QR algorithm. *SIAM Review*, 24(4):427–440, Oct. 1982.
79. D. P. Woodruff et al. Sketching as a tool for numerical linear algebra. *Foundations and Trends in Theoretical Computer Science*, 10(1–2):1–157, 2014.

Randomized Algorithms for Low-Rank Matrix and Tensor Decompositions

Katherine J. Pearce and Per-Gunnar Martinsson

Abstract This survey concerns randomized algorithms in numerical linear algebra for low-rank decompositions of matrices and tensors. The survey begins with a review of classical matrix algorithms that can be accelerated by randomized dimension reduction, such as the singular value decomposition (SVD) or interpolative (ID) and CUR decompositions. Recent advances in randomized dimensionality reduction are discussed, including new methods for fast matrix sketching and sampling techniques. Randomized dimension reduction maps are incorporated into classical matrix algorithms for fast low-rank matrix approximations. The extension of randomized matrix algorithms to tensors is then explored for several low-rank tensor decompositions in the CP and Tucker formats, including the higher-order SVD, ID, and CUR decomposition.

1 Introduction

Numerical linear algebra (NLA) is a cornerstone of applied mathematics, scientific computing, and data science. Throughout its long history, NLA algorithmic development has been driven by two competing imperatives, accuracy and efficiency, with the extant challenge to strike the right balance between them.

Randomization has long played a role in NLA, for instance in providing starting points for iterative algorithms or as a means to avoid communication-intensive pivoted factorizations when solving linear systems [172]. However, randomized methods were rather niche within the mainstream literature. Concerns about reproducibility likely played a role, and perhaps also concerns that randomized methods could not be executed in high-accuracy mode, since that is a characteristic of Monte Carlo methods based on

Katherine J. Pearce
Oden Institute & Dept. of Mathematics, University of Texas at Austin, katherine.pearce@austin.utexas.edu

Per-Gunnar Martinsson
Oden Institute & Dept. of Mathematics, University of Texas at Austin, pgm@oden.utexas.edu

B. Koren and D. Roose (eds.), *Numerical Analysis and Scientific Computing: an Anthology*, Lecture Notes in Computational Science and Engineering 156,
https://doi.org/10.1007/978-3-032-33428-2_7

the central limit theorem. (For additional discussion of the history of randomization in NLA, see [119, 139, 149].) The techniques considered in this survey, which we refer to as randomized numerical linear algebra (RNLA), emerged in the early 2000s and have since then attracted significant interest.

RNLA in the 21st century: By the early 2000s, randomization had led to major developments in ground-breaking applications, such as the notable Google PageRank algorithm [26, 106, 165]. Randomized sampling was also being used to compute low-rank matrix decompositions [2, 89, 169].

Over the past two decades, RNLA has developed into a firmly established research area, garnering widespread attention from both practitioners and theorists. In 2006, a three-part series of papers laid the theoretical groundwork for randomized sampling in approximate matrix multiplication [73], low-rank matrix decomposition (QR and SVD) [74], and CUR decomposition [75]. Around the same time, [66, 67] investigated adaptive sampling for low-rank approximation; [187] proposed a relative error SVD algorithm that utilized randomized subspace embeddings; and [147] introduced what would be later referred to as the "randomized SVD (RSVD)".

Numerical analysts and theoretical computer scientists also provided compelling empirical evidence of the improved performance of RNLA algorithms for practical applications [67, 87, 89, 137, 146, 184]. These results inspired subsequent investigations of randomized algorithms expressly for linear algebra tasks, such as preconditioning or rangefinding, paired with strong probabilistic guarantees [10, 98, 99, 148, 183]. Twenty years later, the scope of research in RNLA now encompasses nearly every aspect of linear algebra, including eigenvalue problems [25, 128, 201]; general linear solvers [13, 47, 159]; orthogonalization [12, 90, 114]; matrix function approximation [8, 55, 177]; trace estimation [100, 176, 209]; and many, many more.

Why the R in RNLA? The rise in popularity of randomized algorithms is due in large part to the commensurate progress made in computing technology. In June 2006, the performance of the world's top supercomputer BlueGene/L was measured at 280.6 Tera-flops/s (floating point operations per second) on the High Performance Linpack (HPL) benchmark, a program that solves a large dense system of linear equations by LU with partial pivoting. As of November 2025, that rate has grown to 1.7 Exa-flops/s, achieved by today's top supercomputer El Capitan.

Modern computing architectures have made it possible to solve increasingly larger-scale problems; however, as evidenced by both theory and practice, the performance of classical algorithms is limited by runtime scaling and the cost of data movement. As a point of reference, the amount of data collected by the Event Horizon Telescope Collaboration to generate the first image of a black hole in 2019 was over 5 Petabytes, which is almost the full storage capacity of El Capitan. In other words, classical matrix algorithms with cubic cost in the input size, such as orthogonalization, are untenable given the large-scale problems now considered in scientific investigations. For the best practical performance, it is necessary to develop algorithms and implementations that can fully leverage computing resources.

Perhaps the greatest benefit of randomization is that it gives us the ability to restructure algorithms to attenuate computational demands on whichever computing resources most hinder performance, such as flops, matrix access, or data movement. For example, randomization admits algorithms that require just a single pass over the input data [135, 203, 204], which is especially advantageous for large-scale problems or streaming data. Randomization can also significantly reduce the number of arithmetic operations required by classical algorithms; e.g., for dense $m \times n$ overdetermined least squares problems with $m \gg n$, randomization yields solutions in roughly $O(mn + n^3)$ operations vs. $O(mn^2)$ with classical approaches [81, 133, 164, 184]. In short, randomized algorithms produce accurate solutions with high probability while reducing working storage and communication costs, offering improved asymptotic complexity, admitting straightforward parallelization and GPU acceleration, and enabling practical investigations that would be otherwise impossible. For an expanded discussion, see [149].

RNLA for tensors: The successful incorporation of randomization into classical matrix algorithms inspired similar work to exploit randomization to accelerate tensor computations. Originally proposed in 1927 as a means of working with multi-way physical and chemical measurements [107, 108], tensors are now ubiquitous in scientific computing and data science. Applications that involve tensorial data include 3D image reconstruction [18, 92, 221]; signal processing [45, 154, 190]; quantum chemistry and physics [23, 174, 179]; machine learning [6, 115, 168]; high-dimensional partial differential equations (PDEs) [24, 59, 120]; among many others. For detailed surveys of tensors and their applications, we refer the reader to [15, 88, 96, 126, 170].

Unfortunately, tensor practitioners suffer harsh consequences of the curse of dimensionality. For a d-way array of uniform size n, the storage costs scale as $O(n^d)$, and most algorithms for analyzing the data scale significantly worse. The key contribution of randomization in this context has been to reveal new algorithms that involve minimal data movement whose complexities scale linearly (or near-linearly) with the amount of input data; these modern randomized algorithms for tensors build upon the foundational work of [30, 56, 77, 140] from the mid-to-late 2000s. Two decades later, tensor practitioners increasingly rely on randomization for efficient tensor computations, backed by the strong accuracy guarantees of well-established randomized matrix algorithms. Randomized algorithms have been developed for nearly every type of low-rank tensor decomposition in the literature, including the CP format [138, 212, 216], the Tucker format [5, 44, 127], the Tensor Train format [43, 112, 136], as well as other specialized representations [14, 16, 36, 182, 195, 215, 224].

Scope: Our intent with this survey is to bring attention to recent work extending the ideas of RNLA to the world of tensors. We have chosen to include a thorough review of the matrix case in order to make the survey self-contained, and to introduce notation and ideas that will be brought to bear in the tensor regime. (This also provided us the opportunity to review some recent advances on topics such as "fast" random embeddings.) To keep the survey focused, we had to omit several important and interesting topics. For matrices, we do not consider the vast body of work on randomized algorithms for eigenvalues and eigenvalue problems (see, e.g., [105, 159, 188]) or optimization (see, e.g., [60, 158]). We also suppress the details of most theoretical results in favor of more background material (see, e.g., [33, 119, 149]). We refer the reader to [156] for a software-focused survey of RNLA, and to [63] for a survey of RNLA from the perspective of machine learning. For tensors, we do not consider the tensor train format or other more specialized tensor decompositions and refer the reader to [15, 96].

Outline: Sect. 2–4 focus on the matrix setting, beginning with a review of requisite linear algebra material and classical matrix decomposition algorithms. We next survey existing methods of randomized dimension reduction in addition to promising new developments in randomized sketching and sampling. We end the first part of our chapter by discussing recent work on randomized algorithms for low-rank matrix decompositions, which are the foundation of our randomized algorithms for tensors. Sect. 5 contains the necessary tensor prerequisites, as well as overviews of the two major tensor formats that we consider in our work: CP and Tucker. We then summarize key developments in randomized algorithms to compute low-rank tensor decompositions in these formats in Sect. 6. We close with a few final thoughts on RNLA in Sect. 7.

2 Matrix Preliminaries

We begin by summarizing notation and key concepts in linear algebra that are frequently referenced in our work. We introduce our notational conventions in Sect. 2.1 and summarize several fundamental matrix decompositions, as well as algorithms to compute them, in Sect. 2.2-2.5.

2.1 Notation

Throughout this chapter, we work over the real numbers $\mathbb{R}$ or complex numbers $\mathbb{C}$ (using $\mathbb{F}$ to signify either), and we let $[m]$ denote the integers $1, 2, \ldots, m$.

Vectors are denoted by bold lowercase Roman or Greek letters (e.g., $\mathbf{x}, \boldsymbol{\omega}$), whereas matrices are denoted by bold uppercase letters (e.g., $\mathbf{X}, \boldsymbol{\Omega}$). We use $\mathbf{0}$ to denote the zero matrix and $\mathbf{I}$ for the identity matrix, with their dimensions made explicit by subscripts when needed. We generally reserve Greek letters for random vectors and matrices.

To refer to coordinates of vectors, we use parentheses with subscripts, i.e. $(\mathbf{x})_i$ refers to the ith coordinate of $\mathbf{x} \in \mathbb{F}^n$. We also adopt the notation of Golub and Van Loan [95] to reference submatrices. Namely, if $\mathbf{A} \in \mathbb{F}^{m \times n}$, and $I = [i_1, i_2, \ldots, i_k] \subseteq [m]$ and $J = [j_1, j_2, \ldots, j_l] \subseteq [n]$ are (row and column, resp.) index sets, then $\mathbf{A}(I, J)$ denotes

$$\mathbf{A}(I, J) = \begin{bmatrix} \mathbf{A}(i_1, j_1) & \mathbf{A}(i_1, j_2) & \ldots & \mathbf{A}(i_1, j_l) \\ \vdots & \vdots & & \vdots \\ \mathbf{A}(i_k, j_1) & \mathbf{A}(i_k, j_2) & \ldots & \mathbf{A}(i_k, j_l) \end{bmatrix} \in \mathbb{F}^{k \times \ell}.$$

The abbreviation $\mathbf{A}(I, :)$ refers to the submatrix $\mathbf{A}(I, [n])$, analogously for $\mathbf{A}(:, J)$.

A vector $\mathbf{x} \in \mathbb{F}^n$ is measured in the Euclidean or ℓ_2-norm $\|\mathbf{x}\|_2 = \left(\sum_i |(\mathbf{x})_i|^2\right)^{1/2}$. A matrix $\mathbf{A} \in \mathbb{F}^{m \times n}$ may be equipped with the corresponding operator norm, $\|\mathbf{A}\|_2 = \sup_{\|\mathbf{x}\|=1} \|\mathbf{A}\mathbf{x}\|_2$, or with the Frobenius norm $\|\mathbf{A}\|_{\mathrm{F}} = (\sum_{i,j} |\mathbf{A}(i,j)|^2)^{1/2}$.

The (Hermitian) transpose of $\mathbf{A} \in \mathbb{F}^{m \times n}$ is denoted by $\mathbf{A}^*$, and the Moore–Penrose pseudoinverse of $\mathbf{A}$ is denoted by $\mathbf{A}^\dagger$. A matrix $\mathbf{U}$ is said to be *orthonormal* if its columns are orthonormal, i.e. $\mathbf{U}^*\mathbf{U} = \mathbf{I}$.

The trace of a square matrix $\mathbf{A} \in \mathbb{F}^{n \times n}$ is the sum of its diagonal elements $\sum_{i=1}^n \mathbf{A}(i,i)$. For matrices $\mathbf{A}, \mathbf{B} \in \mathbb{F}^{m \times n}$, we can define the inner product $\langle \mathbf{A}, \mathbf{B} \rangle = \mathrm{trace}(\mathbf{A}^*\mathbf{B})$, so $\|\mathbf{A}\|_{\mathrm{F}}^2 = \langle \mathbf{A}, \mathbf{A} \rangle = \mathrm{trace}(\mathbf{A}^*\mathbf{A})$.

The abbreviation *i.i.d.* stands for "independent and identically distributed" to describe random variables. The probability of a random event is denoted by $\mathbb{P}[\cdot]$, shortened to p if clear in context. The expectation of a random variable is denoted by $\mathbb{E}[\cdot]$.

A random vector $\mathbf{x}$ is isotropic if $\mathbb{E}[\mathbf{x}\mathbf{x}^*] = \mathbf{I}$. If $\mathbb{F} = \mathbb{R}$, the Rademacher distribution is the uniform distribution on the set $\{\pm 1\}$, or Uniform$\{\pm 1\}$, and if $\mathbb{F} = \mathbb{C}$, a complex Rademacher random variable has the form $(\rho_1 + i\rho_2)/\sqrt{2}$, where ρ_1 and ρ_2 are *i.i.d* Uniform$\{\pm 1\}$. The standard normal or Gaussian distribution with mean 0 and variance 1 is denoted by $\mathcal{N}(0, 1)$.

2.2 The Singular Value Decomposition

Every matrix $\mathbf{A} \in \mathbb{F}^{m \times n}$ admits a *singular value decomposition (SVD)*, given by

$$\underset{m\times n}{\mathbf{A}} = \underset{m\times r}{\mathbf{U}} \; \underset{r\times r}{\Sigma} \; \underset{r\times n}{\mathbf{V}^*}, \tag{1}$$

where $r = \min(m, n)$, $\mathbf{U}$ and $\mathbf{V}$ are orthonormal, and Σ is diagonal. The columns $\{\mathbf{u}_i\}_{i=1}^r$ and $\{\mathbf{v}_i\}_{i=1}^r$ of $\mathbf{U}$ and $\mathbf{V}$, respectively, are called the left and right singular vectors of $\mathbf{A}$. The diagonal elements $\{\sigma_i\}_{i=1}^r$ of Σ are the singular values of $\mathbf{A}$, ordered so that $\sigma_1 \geq \sigma_2 \geq \cdots \geq \sigma_r \geq 0$. The rank of $\mathbf{A}$ is the number of nonzero singular values.

To obtain a rank-k approximation of $\mathbf{A}$, we can truncate its SVD after the first k terms: $\mathbf{A}_k = \sum_{i=1}^k \sigma_i \mathbf{u}_i \mathbf{v}_i^*$. By the Eckart–Young theorem [79], $\mathbf{A}_k$ is the best possible rank-k approximation of $\mathbf{A}$, with approximation error given by the singular values:

$$\|\mathbf{A} - \mathbf{A}_k\|_2 = \sigma_{k+1} \quad \text{and} \quad \|\mathbf{A} - \mathbf{A}_k\|_{\mathrm{F}} = \left(\sum_{j=k+1}^{r} \sigma_j^2\right)^{1/2}. \tag{2}$$

Frequently, the Golub–Reinsch algorithm [94] is used to compute an SVD, as in the built-in MATLAB function

$$\left[\mathbf{U}, \Sigma, \mathbf{V}\right] = \mathtt{svd}(\mathbf{A}).$$

We adopt this notation when referring to the computation of the SVD, as well as $\mathtt{svd}(\mathbf{A}, k)$ to represent the function outputting the truncation $\mathbf{A}_k$.

2.3 The QR Decomposition

Every matrix $\mathbf{A} \in \mathbb{F}^{m\times n}$ admits a *QR decomposition* $\mathbf{A} = \mathbf{QR}$ where $\mathbf{Q}$ is an orthonormal matrix of size $m\times r$, and where $\mathbf{R}$ is an $r\times n$ upper triangular matrix, where $r = \min(m, n)$. In what follows, we are particularly interested in *column pivoted* QR factorizations, which take the form

$$\underset{m\times n}{\mathbf{A}}\ \underset{n\times n}{\mathbf{P}} = \underset{m\times r}{\mathbf{Q}}\ \underset{r\times n}{\mathbf{R}}, \tag{3}$$

for some permutation matrix $\mathbf{P}$. The point here is that it is often possible to pick a permutation matrix $\mathbf{P}$ in such a way that the truncation of (3) to the dominant terms forms a good low rank approximation to $\mathbf{A}$. Common algorithms such as Golub–Businger with standard column pivoting perform well in practice, and more advanced pivoting techniques come with provable guarantees [97], cf. Sect 2.5.1. These methods are greedy, and can be halted after a fixed number of steps k. We use the notation

$$\left[\mathbf{Q}, \mathbf{R}, \mathbf{P}\right] = \mathtt{qr}(\mathbf{A}) \quad \text{and} \quad \left[\mathbf{Q}_k, \mathbf{R}_k, \mathbf{P}\right] = \mathtt{qr}(\mathbf{A}, k), \tag{4}$$

to refer to functions for computing either a full or a partial QR factorization. It is sometimes convenient to store the permutation using an index vector J such that $\mathbf{AP} = \mathbf{A}(:\,, J)$. If we need only the factor $\mathbf{Q}$, of which the columns form an orthonormal basis for the range of $\mathbf{A}$, we use the notation

$$\mathbf{Q} = \mathtt{col}(\mathbf{A}) \quad \text{and} \quad \mathbf{Q}_k = \mathtt{col}(\mathbf{A}, k). \tag{5}$$

2.4 The LU Decomposition

Every matrix $\mathbf{A} \in \mathbb{F}^{m\times n}$ admits an *LU decomposition* given by

$$\underset{m\times m}{\mathbf{P}}\ \underset{m\times n}{\mathbf{A}} = \underset{m\times r}{\mathbf{L}}\ \underset{r\times n}{\mathbf{U}}, \tag{6}$$

where $r = \min(m, n)$, $\mathbf{P}$ is a permutation matrix, $\mathbf{L}$ is lower triangular and $\mathbf{U}$ is upper triangular. We can halt the factorization after computation of the first k terms to obtain a "partial" LU factorization. We let

$$\left[\mathbf{L}, \mathbf{U}, I\right] = \mathtt{lu}(\mathbf{A}) \quad \text{and} \quad \left[\mathbf{L}_k, \mathbf{U}_k, I\right] = \mathtt{lu}(\mathbf{A}, k) \tag{7}$$

denote functions that yield either the full or partial LU factorization, respectively, where $\mathbf{P}$ is stored as an index vector I such that $\mathbf{PA} = \mathbf{A}(I, :\,)$.

2.5 The Interpolative and CUR Decompositions

Every matrix $\mathbf{A} \in \mathbb{F}^{m\times n}$ of rank k admits *interpolative decompositions* (ID) and a *CUR decomposition* of the form

$$\begin{aligned} \mathbf{A} &= \underset{m\times k}{\mathbf{AR}^\dagger}\underset{k\times n}{\mathbf{R}}, && \textit{Row ID} \\ \mathbf{A} &= \underset{m\times k}{\mathbf{C}}\underset{k\times n}{\mathbf{C}^\dagger\mathbf{A}}, && \textit{Column ID} \\ \mathbf{A} &= \underset{m\times k}{\mathbf{C}}\underset{k\times k}{\mathbf{U}}\underset{k\times n}{\mathbf{R}}, && \textit{CUR} \end{aligned} \tag{8}$$

where $\mathbf{R} = \mathbf{A}(I, :)$ for k linearly independent rows indexed by I; $\mathbf{C} = \mathbf{A}(:, J)$ for k linearly independent columns indexed by J; and $\mathbf{U} = \mathbf{C}^\dagger\mathbf{AR}^\dagger$. The columns of $\mathbf{C}$ or rows of $\mathbf{R}$ are known as skeletons. If used for rank-k approximation, the CUR factorization $\mathbf{CUR}$ as defined in (8), with $\mathbf{U} = \mathbf{C}^\dagger\mathbf{AR}^\dagger$, is known as the CUR of "best approximation," or CURBA [171], as it minimizes the Frobenius norm error for that choice of $\mathbf{C}$ and $\mathbf{R}$.

In practice, however, it is prohibitively expensive to access the full matrix $\mathbf{A}$ to compute $\mathbf{U} = \mathbf{C}^\dagger\mathbf{AR}^\dagger$, which costs $O(mnk)$ operations for a rank-k approximation of a dense $m \times n$ matrix. Instead, the CUR with "cross approximation" (CURCA) [171], where $\mathbf{U} = \mathbf{A}(I, J)^{-1}$, or more generally $\mathbf{U} = \mathbf{A}(I, J)^\dagger$, is often used for sublinear-time factorizations, as no additional matrix access is needed beyond the skeletons. However, $\mathbf{A}(I, J)^\dagger$ can be highly ill-conditioned, so it is recommended to use the stable implementation proposed in [171]. Namely, the ε-pseudoinverse of $\mathbf{A}(I, J)$ is computed to approximate $\mathbf{A}(I, J)^\dagger$, where the singular values of $\mathbf{A}(I, J)$ less than ε are truncated prior to computing the pseudoinverse. Additionally, oversampling rows and columns by an amount proportional to the target rank is also recommended for greater accuracy and stability; see [171] for implementation details. Crucially, though, the skeletons I and J should always be computed conditionally upon one another in the following sense: without loss of generality, if column skeletons $\mathbf{C} = \mathbf{A}(:, J)$ are computed first, then the row skeletons $\mathbf{R} = \mathbf{A}(I, :)$ should be computed using $\mathbf{C}$, not the full matrix $\mathbf{A}$.

Of course, the accuracy of a rank-k ID or CUR approximation also depends on the choice of skeletons. It was shown in [66] that it is possible to construct an ID satisfying

$$\|\mathbf{A} - \mathbf{CC}^\dagger\mathbf{A}\|_{\mathrm{F}} \le \sqrt{k+1}\,\|\mathbf{A} - \mathbf{A}_k\|_{\mathrm{F}}, \tag{9}$$

where the coefficient $\sqrt{k+1}$ is *worst-case* optimal (typically a loose bound, which cannot be improved). In practice, greedy pivoting algorithms, based on QR, LU, or the SVD, are frequently used to choose skeletons for ID and CUR decompositions. We discuss several in the following sections.

2.5.1 ID/CUR from QR with Column Pivoting (CPQR)

For an ID or CUR decomposition using CPQR, column pivots are selected from the active submatrix of $\mathbf{A}$, which is updated after each pivot selection. Assuming $m \geq n$, let $\mathbf{A}^{(t)}$ represent the active submatrix, or the resulting matrix after t steps of pivoting and updating for $t = 0, \ldots, n-2$, with $\mathbf{A}^{(0)} = \mathbf{A}$. At the $(t+1)$-th step, CPQR chooses the column pivot j_{t+1} as the column of maximal Euclidean norm in the active submatrix,

$$j_{t+1} = \arg \max_{t+1 \leq j \leq n} \|\mathbf{A}^{(t)}([t+1:m], j)\|_2.$$

The selected k (column) skeletons are then $[j_1, \ldots, j_k]$. (Row skeletons could analogously be computed via CPQR on the transpose.) Once a pivot entry is selected, column j_{t+1} is swapped with column $t+1$, the normalized Householder reflection vector $\mathbf{v}$ is formed, and the lower right block of the active submatrix, starting from the $(t+1)$-st main diagonal entry, is updated as $\mathbf{A}^{(t+1)} = \mathbf{A}^{(t)} - 2\mathbf{v}\mathbf{v}^*\mathbf{A}^{(t)}$.

In practice, CPQR chooses pivots in an order that reveals the rank of $\mathbf{A}$, so that the chosen skeletons form a good basis for its row or column space. While there are adversarial cases for which CPQR can fail to be rank-revealing (e.g., Kahan-type matrices [117]), these are rarely encountered; cf. [198]. There are variations of CPQR with better theoretical guarantees, such as rank-revealing QR [37, 38] or strong rank-revealing QR [97], but they are more costly to implement.

2.5.2 ID/CUR from LU with Partial Pivoting (LUPP)

For an ID or CUR decomposition using LUPP, row pivots are selected from the active submatrix of $\mathbf{A}$, which is updated after each pivot selection via Schur complements (e.g., [95, Algorithm 3.2.1]). Let $\mathbf{A}^{(t)}$ be the active submatrix, the resulting matrix after t steps of pivoting and updating for $t = 0, 1, \ldots, n-2$, with $\mathbf{A}^{(0)} = \mathbf{A}$. At the $(t+1)$-st step of LUPP, the largest-magnitude column entry of the active submatrix is selected as the next pivot entry:

$$i_{t+1} = \arg \max_{t+1 \leq i \leq m} |\mathbf{A}^{(t)}(i, t+1)|.$$

The active submatrix is then updated: defining vectors $\mathbf{r} = [t+2:m]$ and $\mathbf{c} = [t+2:n]$,

$$\mathbf{A}^{(t+1)}(\mathbf{r}, \mathbf{c}) = \mathbf{A}^{(t)}(\mathbf{r}, \mathbf{c}) - \frac{\mathbf{A}^{(t)}(\mathbf{r}, t+1)\mathbf{A}^{(t)}(t+1, \mathbf{c})}{\mathbf{A}^{(t)}(t+1, t+1)}, \tag{10}$$

where $\mathbf{A}^{(t)}(t+1, t+1)$ has been updated to $\mathbf{A}^{(t)}(i_{t+1}, t+1)$ following row permutation. If expressed as the output of the function $[\mathbf{L}, \mathbf{U}, I] = \texttt{lu}(\mathbf{A})$, the row skeletons at the $(t+1)$-th step are given by $I(1:t+1)$.

This algorithm leads to an exponential upper bound on the entries of $\mathbf{U}$, which is tight for certain matrices (e.g., Kahan-type matrices [117]). In the general case, LU is not rank-revealing unless *full* rather than partial pivoting is used, cf. [167]. We will later see that in combination with randomized mixing in one direction, LUPP provides an algorithm for pivot selection that is both very fast and very reliable.

2.5.3 ID/CUR from One Truncated SVD

Previously, the most efficient algorithms to compute column skeletons satisfying (9) required $O(kmn\min(m,n))$ operations [53, 65]. Recently, in [163], an algorithm is presented that constructs a column ID satisfying (9) in $O(mnk)$ operations, plus the cost of one rank-k truncated SVD; it is now colloquially known as Osinsky's method. Osinsky's method works with any arbitrary orthonormal basis for the row space, but to satisfy (9), we use the first k right singular vectors.

Let $\mathbf{A}_k =: \mathbf{Z} = \mathbf{U\Sigma V}^*$ (suppressing subscripts) be a rank-k truncated SVD of $\mathbf{A}$. Consider

$$\widetilde{\mathbf{A}} = \mathbf{A} - \mathbf{AVV}^*,$$

the norm of which is the orthogonal projection error of the rows of $\mathbf{A}$ onto the span of $\mathbf{V}$. We will iteratively build an oblique projector out of the previously selected columns, and then select the column that leads to the smallest increase in the oblique projection error. Define $\widetilde{\mathbf{C}} = \mathbf{C} - (\mathbf{AVV}^*)(:, J)$. Then, if $\mathbf{W} = [\mathbf{V}(J, :)]^{-*}\mathbf{V}^*$,

$$\begin{aligned} \|\mathbf{A} - \mathbf{CW}\|_{\mathrm{F}} &= \|\widetilde{\mathbf{A}} + \mathbf{AVV}^* - \widetilde{\mathbf{C}}\mathbf{W} - \mathbf{AVV}(:, J)^*\mathbf{W}\|_{\mathrm{F}} \qquad (11)\\ &= \|\widetilde{\mathbf{A}} - \widetilde{\mathbf{C}}\mathbf{W}\|_{\mathrm{F}}. \qquad (12) \end{aligned}$$

Since $\widetilde{\mathbf{A}}$ and $\mathbf{W}$ are not affected by replacing $\mathbf{V}^*$ with any unitary transformation $\mathbf{QV}^*$, we will also assume Householder reflectors are successively applied to $\mathbf{V}^*$ so that $\mathbf{V}(J, :)^*$ is upper-triangular.

Now, assume we have already selected $i-1$ columns; without loss of generality, choose the first $i-1$ columns of $\mathbf{A}$. Let

$$\begin{aligned} \mathbf{C}^{(i-1)} &= \mathbf{A}(:, [i-1]),\\ \widetilde{\mathbf{C}}^{(i-1)} &= \mathbf{C}^{(i-1)} - (\mathbf{AVV}^*)(:, [i-1])\\ \mathbf{W}^{(i-1)} &= \mathbf{V}^*([i-1], [i-1])^{-1}\mathbf{V}^*([i-1], :)\\ \widetilde{\mathbf{A}}^{(i-1)} &= \widetilde{\mathbf{A}} - \widetilde{\mathbf{C}}^{(i-1)}\mathbf{W}^{(i-1)} \end{aligned}$$

Then we would like to express the error at the ith step

$$\widetilde{\mathbf{A}}^{(i)} := \widetilde{\mathbf{A}} - \widetilde{\mathbf{C}}^{(i)}\mathbf{W}^{(i)}, \qquad (13)$$

in terms of $\widetilde{\mathbf{A}}^{(i-1)}$, which will inform how we should choose the next pivot; suppose that the next pivot will be i for notational convenience. We also assume that $\mathbf{V}^*([i-1], [i-1])$ is already upper triangular. At the ith step, let

$$\widetilde{\mathbf{C}}^{(i)} = \left[\widetilde{\mathbf{C}}^{(i-1)}\ \widetilde{\mathbf{c}}_i\right] := \left[\widetilde{\mathbf{C}}^{(i-1)}\ \widetilde{\mathbf{A}}(:,i)\right], \tag{14}$$

and partition $\mathbf{W}^{(i)}$ as

$$\begin{bmatrix} [\mathbf{V}^*([i-1],[i-1])]^{-1} & -\mathbf{V}^*([i-1],[i-1])^{-1}\mathbf{V}^*([i-1],i)/\mathbf{V}^*(i,i) \\ \mathbf{0} & 1/\mathbf{V}^*(i,i) \end{bmatrix} \begin{bmatrix} \mathbf{V}^*([i-1],:) \\ \mathbf{V}^*(i,:) \end{bmatrix}. \tag{15}$$

Multiplying (14) with (15) and plugging into (13) yields

$$\widetilde{\mathbf{A}}^{(i)} = \widetilde{\mathbf{A}}^{(i-1)} - \widetilde{\mathbf{A}}^{(i-1)}(:,i)[\mathbf{V}^*(i:k,i)]^{\dagger}\mathbf{V}^*(i:k,:). \tag{16}$$

To minimize the Frobenius norm error at the ith step, choose the column j such that

$$j = \arg\min_{j \geq i} \|\widetilde{\mathbf{A}}^{(i-1)}(:,j)\|_2 / \|\mathbf{V}^*(i:k,j)\|_2. \tag{17}$$

Again, without loss of generality, suppose $j = i$ and apply a Householder reflector so that $\mathbf{V}^*([i],[i])$ is upper triangular. Then

$$\sum_{\ell=1}^{n} \|\widetilde{\mathbf{A}}^{(i-1)}(:,\ell)\|_2^2 = \|\widetilde{\mathbf{A}}^{(i-1)}\|_{\mathrm{F}}^2, \tag{18}$$

$$\sum_{\ell=1}^{n} \|\mathbf{V}^*(i:k,\ell)\|_2^2 = \|\mathbf{V}^*(i:k,:)\|_{\mathrm{F}}^2 = k - i + 1. \tag{19}$$

Using the fact that $\sum_\ell \alpha_\ell / \sum_\ell \beta_\ell \geq \min_\ell \alpha_\ell/\beta_\ell$ for non-negative α_ℓ and positive $\sum_\ell \beta_\ell$, we can upper bound the ratio of (17) by the ratio of (18) and (19), so that

$$\begin{aligned} \|\widetilde{\mathbf{A}}^{(i-1)}(:,i)\mathbf{V}^*(i:k,i)^{\dagger}\mathbf{V}^*(i:k,:)\|_2^2 &= \|\widetilde{\mathbf{A}}^{(i-1)}(:,i)\mathbf{V}^*(i:k,i)^{\dagger}\|_2^2 \\ &= \frac{\|\widetilde{\mathbf{A}}^{(i-1)}(:,i)\|_2^2}{\|\mathbf{V}^*(i:k,i)\|_2^2} \leq \frac{\|\widetilde{\mathbf{A}}^{(i-1)}\|_{\mathrm{F}}^2}{k-i+1}. \end{aligned}$$

The rows of $\widetilde{\mathbf{A}}^{(i-1)}$ are orthogonal to $\mathbf{V}^*(i:k,:)$, and the rows of $\widetilde{\mathbf{A}}$ are orthogonal to $\mathbf{V}^*$ by construction, which allows us to estimate the error at the ith step as

$$\begin{aligned} \|\widetilde{\mathbf{A}}^{(i)}\|_{\mathrm{F}}^2 &= \|\widetilde{\mathbf{A}}^{(i-1)}\|_{\mathrm{F}}^2 + \|\widetilde{\mathbf{A}}^{(i-1)}(:,i)\mathbf{V}^*(i:k,i)^{\dagger}\mathbf{V}^*(i:k,:)\|_2^2 \\ &\leq \|\widetilde{\mathbf{A}}^{(i-1)}\|_{\mathrm{F}}^2 + \frac{\|\widetilde{\mathbf{A}}^{(i-1)}\|_{\mathrm{F}}^2}{k-i+1}. \end{aligned}$$

Thus, over all $i-1$ previous steps, we have

$$\|\widetilde{\mathbf{A}}^{(i)}\|_{\mathrm{F}}^2 \leq \|\widetilde{\mathbf{A}}\|_{\mathrm{F}}^2 + \frac{i}{k-i+1}\|\widetilde{\mathbf{A}}\|_{\mathrm{F}}^2, \tag{20}$$

and setting $i = k$ in (20), we achieve the desired bound in (9) in $O(mnk)$ operations, plus the cost of the truncated SVD.

3 Randomized Dimensionality Reduction: Sketching and Sampling

In this section, we discuss ways that randomization can be integrated into classical algorithms in numerical linear algebra to compute low-rank decompositions more efficiently. Namely, we will often rely on special linear maps that map the row or column space of an input matrix to a smaller-dimensional space in such a way that the original geometry is roughly preserved, called *dimensionality reduction maps* (DRMs). If a linear map is drawn from a random matrix distribution that achieves geometry-preserving dimensionality reduction with high probability, it is known as a randomized DRM.

Randomized DRMs are often associated with randomized subspace embeddings, which are linear maps drawn from random distributions that satisfy the following condition with high probability. Let $\mathbf{A} \in \mathbb{F}^{m\times n}$ be a matrix of rank $k \leq \min(m, n)$ and $\varepsilon \in (0, 1)$ a distortion parameter. A linear map $\mathbf{\Gamma} : \mathbb{F}^m \to \mathbb{F}^d$ is a subspace embedding of $\mathbf{A}$ with distortion ε if

$$(1 - \varepsilon)\|\mathbf{A}\mathbf{x}\| \leq \|\mathbf{\Gamma}\mathbf{A}\mathbf{x}\| \leq (1 + \varepsilon)\|\mathbf{A}\mathbf{x}\| \quad \forall\, \mathbf{x} \in \mathbb{F}^n. \tag{21}$$

Usually, $k \leq d \ll m$. The minimal value of d required for $\mathbf{\Gamma}$ to satisfy (21) depends on the type of embedding $\mathbf{\Gamma}$, the rank k of $\mathbf{A}$, and the distortion parameter ϵ.

More generally, if $\mathcal{S}$ is any probability distribution over linear maps $\mathbb{F}^m \to \mathbb{F}^d$, then $\mathbf{\Gamma} \sim \mathcal{S}$ is a randomized subspace embedding if (21) holds over $\mathcal{S}$ for all $\mathbf{A} \in \mathbb{F}^{m\times n}$ with at least constant probability. If a randomized embedding can be constructed to map an unknown subspace into a lower-dimensional space (i.e. without knowing anything about the input space except its dimension), then it is said to be an oblivious subspace embedding (OSE) [187]. Many times in practice, randomized DRMs are treated as embeddings, but we will see in Sect. 3.1 that satisfying (21) is sufficient but not necessary for the smaller-dimensional space to be a good proxy for the input.

In Sect. 3.1, we review randomized sketching, a procedure that produces random linear combinations of input matrix coordinates, which we say is "coordinate-mixing." We then focus on randomized sampling of matrix coordinates in Sect. 3.2. We end with an important application of randomized DRMs that will be relevant for our discussion of tensor decompositions: solving overdetermined least squares problems. For detailed treatments, the reader is referred to [33, 99, 145, 149, 219].

3.1 Randomized Matrix Sketching

The application of a randomized DRM $\mathbf{\Gamma}$ to $\mathbf{A}$ is known as randomized sketching. We refer to $\mathbf{\Gamma}\mathbf{A}$ as a *row sketch* of $\mathbf{A}$, with a *column sketch* $\mathbf{A}\mathbf{\Omega}$ defined analogously. However, we generally reserve the term "sketching" for randomized DRMs that are coordinate-mixing, as opposed to methods that sample coordinates, which we will simply call sampling, cf. Sect. 3.2.

Ideally, the random matrix used for sketching should roughly preserve the geometry of the row (or column) space of $\mathbf{A}$. Commonly used randomized DRMs $\mathbf{\Gamma}$ include:

- **Gaussian matrices** ([99, Section 4.1], [149, Section 8.3], [219, Theorem 2.3]): $\mathbf{\Gamma} \in \mathbb{F}^{d\times m}$ with entries $\mathbf{\Gamma}(i,j) \sim \mathcal{N}(0, 1/d)$.
- **Subsampled randomized trigonometric transforms (SRTT)** ([99, Section 4.6], [149, Section 9.3], [199], [219, Theorem 2.4]): $\mathbf{\Gamma} \in \mathbb{F}^{d\times m}$ defined as

$$\mathbf{\Gamma} = \sqrt{m/d}\,\mathbf{\Pi}_{m\to d}\mathbf{F}\mathbf{\Phi}\mathbf{\Pi}_m.$$

  Here, $\mathbf{\Pi}_{m\to d} \in \mathbb{F}^{m\times d}$ is a uniformly random selection of d out of m rows, whereas $\mathbf{\Pi}_m \in \mathbb{F}^{m\times m}$ is a random permutation of the m rows. $\mathbf{F} \in \mathbb{F}^{m\times m}$ is a discrete Fourier transform, and $\mathbf{\Phi} := \operatorname{diag}(\phi_1, \ldots, \phi_n)$ has *i.i.d.* Rademacher entries.
- **Sparse sign matrices** ([118], [149, Section 9.2], [160], [219]): $\mathbf{\Gamma} \in \mathbb{F}^{d\times m}$ defined as

$$\mathbf{\Gamma} = \sqrt{1/\zeta}\,\left[\mathbf{s}_1, \ldots, \mathbf{s}_m\right]$$

  for sparsity parameter $2 \le \zeta \le d$, where each column $\mathbf{s}_j \in \mathbb{F}^d$, $j = 1, \ldots, m$ contains ζ *i.i.d.* Rademacher random variables at uniformly random coordinates.
- **CountSketch matrices** ([39, 151, 160]): $\mathbf{\Gamma} \in \mathbb{F}^{d\times m}$ defined as a sparse sign matrix with sparsity $\zeta = 1$. We advise against using CountSketch matrices on their own with sparse matrix inputs, cf. [82, Figure B.6].
- **SparseStack matrices** ([33, 52, 118]): $\mathbf{\Gamma} \in \mathbb{F}^{d\times m}$ defined as

$$\mathbf{\Gamma} = \frac{1}{\sqrt{\zeta}}\begin{bmatrix} \rho_{1,1}\mathbf{e}_{s_{1,1}} & \cdots & \rho_{1,m}\mathbf{e}_{s_{1,m}} \\ \rho_{2,1}\mathbf{e}_{s_{2,1}} & \cdots & \rho_{2,m}\mathbf{e}_{s_{2,m}} \\ \vdots & \ddots & \vdots \\ \rho_{\zeta,1}\mathbf{e}_{s_{\zeta,1}} & \cdots & \rho_{\zeta,m}\mathbf{e}_{s_{\zeta,m}} \end{bmatrix},$$

  where $d = \zeta b$ for some block size b and sparsity parameter ζ, $\rho_{i,j}$ are *i.i.d.* Rademacher variables, $\mathbf{e}_i \in \mathbb{F}^b$ is the ith standard basis vector, and $s_{i,j} \in [b]$ is chosen uniformly. In other words, the SparseStack test matrix is comprised of ζ independent copies of a (scaled) CountSketch matrix, stacked on top of one another.
- **LESS-Uniform matrices** ([33, Definition 4.1] [62], [156]): $\mathbf{\Gamma} \in \mathbb{F}^{d\times m}$ defined by its d rows

$$\omega = \sqrt{\frac{m}{\zeta}}\sum_{i=1}^{\zeta}\rho_i\mathbf{e}_{s_i} \in \mathbb{F}^m, \tag{22}$$

  where ζ is the sparsity parameter per row, $\rho_1, \ldots \rho_\zeta$ are *i.i.d.* Rademacher random variables, the indices $s_1, \ldots, s_\zeta$ are sampled uniformly from $[m]$ without replacement, and $\mathbf{e}_{s_i}$ is the s_i-th standard basis vector. We emphasize that the sparsity parameter for LESS-Uniform matrices controls the number of nonzeros *per row*, whereas the sparsity parameter for Sparse Sign and SparseStack matrices determines the number of nonzeros *per column*. Thus, LESS-Uniform matrices are sparser than the other sparse random matrix constructions, which makes them more suitable for problems where the input matrix is incoherent, and even then, increasing the sparsity parameter ζ may be necessary.

- **SparseRTT matrices** ([33, Section 4.1], [199]): $\mathbf{\Gamma} \in \mathbb{F}^{d\times m}$ defined as

$$\mathbf{\Gamma} = \mathbf{SFD}, \tag{23}$$

where $\mathbf{D} \in \mathbb{F}^{m\times m}$ is a random diagonal matrix populated with *i.i.d.* Rademacher random entries, $\mathbf{F} \in \mathbb{F}^{m\times m}$ is a discrete Fourier transform, and $\mathbf{S} \in \mathbb{F}^{d\times m}$ is the transpose of an LESS-Uniform matrix with exactly ζ nonzero entries per column.

These distributions yield randomized DRMs that work well in practice, with theoretical guarantees summarized in Table 1. Gaussian test matrices are OSEs that enjoy a superior lower bound on the requisite embedding dimension d for (21), but the costs of storage and matrix-vector products for Gaussians are much more expensive than for "structured" embeddings like SRTTs or sparse sign matrices. Put simply, Gaussian sketching makes for straightforward analysis, but structured random matrix sketching makes for better performance. The reader is referred to [149, Sections 8,9] for a more thorough treatment.

Randomized DRM $\mathbf{\Gamma} \in \mathbb{F}^{d\times m}$	Dimension d for (21)	Storage	Mat-vec cost
Gaussian	$d = \Omega\left(k/\varepsilon^2\right)$	$O(md)$	$O(md)$
SRTT	$d = \Omega\left(k \log k/\varepsilon^2\right)$	$O(m)$	$O(m \log d)$
Sparse sign	$d = \Omega\left(k \log k/\varepsilon^2\right),\ \zeta = \Omega\left(\log k/\varepsilon\right)$	$O(\zeta m)$	$O(\zeta m)$
CountSketch	$d = \Omega\left(k^2/\varepsilon^2\right)$	$O(m \log d)$	$O(m)$
SparseStack	$d = \Omega\left(k \log k/\varepsilon^2\right),\ \zeta = \Omega\left(\log k/\varepsilon\right)$	$O(\zeta m)$	$O(\zeta m)$

Table 1 Summary of coordinate-mixing randomized DRMs commonly used for (row-)sketching rank-k matrix $\mathbf{A} \in \mathbb{F}^{m\times n}$. Theoretical lower bounds on the dimension d for $\mathbf{\Gamma}$ to satisfy the OSE property are provided, as well as the cost to store $\mathbf{\Gamma}$ and the cost to apply $\mathbf{\Gamma}$ to a vector (in floating point operations).

While the embedding dimensions in Table 1 guarantee the OSE property (21), it has been observed that smaller values of d are often sufficient for many linear algebra tasks in practice when utilizing any of the randomized DRMs of Table 1 (with the exception of CountSketch for sparse matrices); i.e., $d = k + 10$ in the event of fast singular value decay, or more generally, $d = 2k$, cf. Sect. 4.1. Recently, this observation was thoroughly investigated from the perspective of oblivious subspace *injections*, versus embeddings, which relaxes the OSE property as follows. Random matrix $\mathbf{\Gamma} : \mathbb{F}^m \to \mathbb{F}^d$ is an oblivious subspace injection (OSI) of $\mathbf{A}$ with injectivity α if it satisfies the injectivity property

$$\alpha\|\mathbf{Ax}\| \leq \|\mathbf{\Gamma Ax}\| \quad \forall\, \mathbf{x} \in \mathbb{F}^n, \tag{24}$$

and the isotropy property that $\mathbb{E}[\|\mathbf{\Gamma Ax}\|^2] = \|\mathbf{Ax}\|^2$ for all $\mathbf{x} \in \mathbb{F}^n$.

In other words, an OSI satisfies the lower bound requirement in (21), taking $\alpha = 1-\varepsilon$, so that it does not annihilate any vector in the column space of $\mathbf{A}$. Arguably, the "injectivity" parameter α is more crucial for the performance of randomized sketching than the "dilation" parameter $(1+\varepsilon)$ in the OSE definition [33]. All of the random matrices defined above satisfy the isotropy property, and if we focus instead on the weaker injectivity property of OSIs, we can obtain theoretical guarantees on the performance of randomized sketching that more closely align with what is observed in practice. In Table 2, we report the sketching dimension d required to satisfy the OSI property, compared to the OSE property.

DRM $\mathbf{\Gamma} \in \mathbb{F}^{d\times m}$	Dimension d for ε-OSE (21)	Dimension d for 1/2-OSI (24)
Gaussian	$d = \Omega\left(k/\varepsilon^2\right)$	$d = O(k)$,
SRTT	$d = \Omega\left(k \log k/\varepsilon^2\right)$	$d = O(k \log k)$
Sparse sign	$d = \Omega\left(k \log k/\varepsilon^2\right),\ \zeta = \Omega\left(\log k/\varepsilon\right)$	$d = O(k), \zeta = O(\mu(\mathbf{A}) \cdot \log k)$
CountSketch	$d = \Omega\left(k^2/\varepsilon^2\right)$	$d = O(k^2)$
SparseStack	$d = \Omega\left(k \log k/\varepsilon^2\right),\ \zeta = \Omega\left(\log k/\varepsilon\right)$	$d = O(k), \zeta = O(\log k)$
SparseRTT	$d = O(k), \zeta = O(\log^3 k)^{\dagger}$	$d = O(k), \zeta = O(\log k)$

Table 2 Summary of randomized DRMs for row-sketching rank-k $\mathbf{A} \in \mathbb{F}^{m\times n}$. Theoretical bounds on the dimension d for $\mathbf{\Gamma}$ to satisfy the OSI property (24) vs. the OSE property (21) are provided, cf. [33, 202]. For OSIs, we enforce a constant injectivity parameter, i.e. $\alpha \geq 1/2$, which is generally not attainable for OSEs due to the dilation parameter. For the OSE column of SparseRTTs, the dagger † signifies that a different sparse matrix is used for the OSE property, cf. [33, Section 4.1].

With the OSI property guiding our choice of random sketching matrix, we enjoy more favorable sketching dimensions for the random matrices considered above. In the OSI framework, Gaussian random matrices are still the gold standard for accuracy, but suffer the same performance drawbacks in terms of storage and matrix-vector product, or mat-vec, costs. Structured random matrices offer significantly better performance, and specifically, sparse random matrices exhibit the best empirical performance, as demonstrated by many practical investigations [46, 49, 83, 99, 118, 149, 160, 200, 219], and now also supported by the theoretical guarantees of [33, 202] in Table 2.

As an additional boon, the OSI values of d that are reported in Table 2 are even still too pessimistic in practice. The results of [33] provide compelling evidence that SparseStack matrices are the superior choice of DRM; it is conjectured that SparseStack matrices can achieve the OSI property with $d = O(k)$ and constant sparsity $\zeta = O(1)$. This conjecture is supported numerically, with $d = 2k$ and $\zeta = 4$ used in the experiments of [33] to demonstrate that the low-rank approximations obtained from the SparseStack sketches are of comparable accuracy to those obtained from Gaussian sketches. Thus, to sketch dense matrices in general, recent findings suggest that either a Sparse Sign or SparseStack embedding with $\zeta = 4$ and embedding dimension $d = 2k$ work well. When sparse linear algebra libraries are unavailable, sketching with fast trigonometric transforms (SRTT or SparseRTT) is a reliable alternative.

Finally, we note that for input matrices that arise from tensors (cf. Sect. 5.3.1), we can construct Khatri–Rao sketching matrices that exploit underlying structure [33, 194] [156, Section 7.2], formed as Kronecker products of isotropic random vectors. Often, these random vectors are drawn from (real) Gaussian or Rademacher distributions, but it is advised in [33] that the complex spherical distribution be used to avoid near-zero injectivity in the worst-case. We do not assume any specific isotropic distribution unless otherwise specified.

3.2 Randomized Matrix Sampling

In the previous section, we observed that a random sketch has rows (or columns) that are random linear combinations of the rows (or columns) of $\mathbf{A}$. However, we can also construct randomized DRMs that sample coordinates of $\mathbf{A}$ according to a probability distribution. We use the separate term randomized sampling for these maps.

To illustrate, suppose that we want to draw d random *i.i.d.* row indices of a matrix $\mathbf{A} \in \mathbb{F}^{m\times n}$ according to probabilities $(p_1, \ldots, p_m)$. Common sampling methods include:

- **Uniform** ([149, Section 9.6.2], [199, Lemma 3.4], [119]): $p_i = 1/m$.
- **Squared (row) norms** ([89]): $p_i = \|\mathbf{a}_i\|_2^2/\|\mathbf{A}\|_{\mathrm{F}}^2$, where $\mathbf{a}_i^*$ is the ith row of $\mathbf{A}$.
- **Row leverage scores** ([73, 78], [149, Section 9.6.3]):

$$p_i = l_i/n \ \text{ for leverage score } l_i = \|\mathbf{a}_i\|^2_{(\mathbf{A}^*\mathbf{A})^{-1}}, \tag{25}$$

  where $\mathbf{a}_i^*$ is the ith row of $\mathbf{A}$ and the norm is defined as $\|\mathbf{v}\|_{\mathbf{B}} = \sqrt{\mathbf{v}^*\mathbf{B}\mathbf{v}}$.
- **Determinantal point processes (DPP)** ([64, Section 5.2], [132]): If $S \subseteq [m]$,

$$p_S = \frac{\det\left((\mathbf{A}\mathbf{A}^*)(S, S)\right)}{\det\left(\mathbf{I} + \mathbf{A}\mathbf{A}^*\right)}, \tag{26}$$

  where index set S, $|S| = d$, is selected all-at-once. However, in practice, rows corresponding to S are selected iteratively, as in [109]. Let $\mathbf{A}\mathbf{A}^* = \sum_i \lambda_i \mathbf{u}_i \mathbf{u}_i^*$ denote its eigendecomposition with $\lambda_i \in [0, 1]$ for all i. Let $s_i \sim \text{Bernoulli}(\lambda_i)$, and let $\mathbf{U}$ be the matrix with columns $\mathbf{u}_i$. By [109, Theorem 7] and [64, Theorem 12], we can use $\sum_i s_i \mathbf{u}_i \mathbf{u}_i^*$ as a proxy for DPP sampling as follows, letting $\mathbf{v}_i^*$ represent the ith row of $\mathbf{U}$:

  1. Sample vector $\mathbf{v}_i$ with $p_i \propto \|\mathbf{v}_i\|_2^2$
  2. Project all vectors $\mathbf{v}_j$ onto the subspace orthogonal to $\mathbf{v}_i$
  3. Update the probabilities and repeat $d - 1$ times.

We can express the action of randomized sampling as a randomized DRM $\mathbf{\Gamma} \in \mathbb{F}^{d\times m}$. The lower bounds on the minimal dimension that is required for a random sampling matrix $\mathbf{\Gamma}$ to satisfy (21) are summarized in Table 3, along with the cost of computing the associated probabilities. The values of d required for a randomized sampling matrix

$\mathbf{\Gamma}$ to satisfy (21) depend not only on the rank k of $\mathbf{A}$ and distortion ε, but also, in some cases, on quantities associated with the matrix $\mathbf{A}$ that are usually very expensive to compute.

For uniform sampling, the requisite embedding dimension d depends on the coherence μ of $\mathbf{A}$, defined as $\mu(\mathbf{A}) = \max_{1\le i\le m} \frac{n}{m} l_i$, or the maximum leverage score l_i in (25). The coherence is a measure of how evenly information in the matrix is distributed; high coherence indicates that its key information is very localized and that uniform sampling may perform poorly as a result; see, e.g., [113]. In spite of this drawback, uniform sampling is a popular choice for large-scale linear algebra computations as a cost-effective OSE. However, it is unclear how to best choose the embedding dimension d in practice since the coherence is typically unavailable, cf. [149, Section 9.6.4].

Squared norm sampling and leverage score sampling also yield randomized subspace embeddings, but not oblivious ones, though we can still analyze the embedding dimensions required for them to satisfy (21). For squared norm sampling, the value of d depends on the condition number κ of $\mathbf{A}^*\mathbf{A}$, defined as the ratio of its largest to smallest eigenvalue, which is not readily available in practice. The embedding dimension for leverage score sampling only depends on k and ϵ, but the sampling probabilities are very expensive. In particular, we note that leverage scores are as expensive to directly compute as $\mathtt{svd}(\mathbf{A})$, rendering their computation infeasible for most practical purposes. However, there exist fast methods to approximate them, cf. [7, 48, 76, 157].

DPP sampling has nearly optimal theoretical guarantees, but naive implementations incur a cost of $O(m^3)$ (after forming $\mathbf{AA}^*$). Given access to $\mathbf{AA}^*$, the DPP-VFX algorithm of [61] samples exactly from a DPP distribution with pre-processing cost $O(mk^{10}+k^{15})$. In particular, DPP-VFX was the first to achieve $O(m \cdot \text{poly}(k))$ cost for the first sample and $O(\text{poly}(k))$ for subsequent samples; however, there is still a runtime bottleneck of $\Omega(m)$ due to the m marginals needed for the sampling distribution. The α-DPP algorithm of [32] improves on DPP-VFX by incorporating uniform sub-sampling before computing marginals to avoid the $\Omega(m)$ bottleneck.

Sampling distribution	Dimension d	Complexity of distribution
Uniform	$d = \Omega\left(\mu \log k/\varepsilon^2\right)$	$O(1)$
Squared norms	$d = \Omega\left(\kappa \log k/\varepsilon^2\right)$	$O(mn)$
Leverage scores	$d = \Omega\left(k \log k/\varepsilon^2\right)$	$O(mn^2)$

Table 3 Summary of randomized sampling distributions commonly used to sample rows of rank-k matrix $\mathbf{A} \in \mathbb{F}^{m\times n}$. Theoretical lower bounds on the embedding dimension d are given for (21), as well as the asymptotic complexity of computing the associated probability distribution, cf. [64, 149]. We note that with DPP sampling, $d \sim k/\varepsilon + k - 1$ is generally sufficient for low-rank approximation, though it does not guarantee (21).

3.3 Application: Randomized DRMs for Overdetermined Least Squares

As an illustrative application of randomized DRMs that will also be useful for our tensor algorithms, we consider the overdetermined least squares problem, in which we solve

$$\mathbf{x}_* = \arg\min_{\mathbf{x}\in\mathbb{R}^n} \|\mathbf{A}\mathbf{x} - \mathbf{b}\|_2, \tag{27}$$

for given $\mathbf{A} \in \mathbb{R}^{m\times n}$, $\mathbf{b} \in \mathbb{R}^m$, and $m \gg n$. Using CPQR for (27) requires $O(mn^2)$ operations, which is undesirable for large m.

A natural idea is to apply a randomized DRM to $\mathbf{A}$ and $\mathbf{b}$ in (27), a strategy known as "sketch-and-solve" [187]. Namely, we draw a randomized DRM $\mathbf{\Gamma} \in \mathbb{F}^{d\times m}$ and solve

$$\widehat{\mathbf{x}} = \arg\min_{\mathbf{x}\in\mathbb{R}^n} \|\mathbf{\Gamma}(\mathbf{A}\mathbf{x} - \mathbf{b})\|_2. \tag{28}$$

If $\mathbf{\Gamma}$ embeds the k-dimensional column space of $\begin{bmatrix}\mathbf{A} & \mathbf{b}\end{bmatrix}$ with distortion $\varepsilon \in (0, 1)$, then

$$\|\mathbf{A}\mathbf{x}_* - \mathbf{b}\|_2 \le \|\mathbf{A}\widehat{\mathbf{x}} - \mathbf{b}\|_2 \le \frac{1+\varepsilon}{1-\varepsilon}\|\mathbf{A}\mathbf{x}_* - \mathbf{b}\|_2$$

when $d \approx k/\varepsilon^2$.

Randomized DRMs are also utilized in other methods of solving (27). Sketched GMRES [159, Algorithm 1.1] includes a k-truncated Arnoldi process to quickly compute an approximate basis of the Krylov subspace. Sketch-and-precondition methods are also popular [10, 150, 152]. We discuss alternative randomized DRMs for overdetermined least squares problems that arise in low-rank tensor factorization algorithms in Sect. 6.

4 Randomized Algorithms for Low-Rank Matrix Decompositions

In this section, we illustrate several ways in which randomized DRMs are often used to compute low-rank matrix decompositions. We introduce the rangefinder problem and show how it can be solved with randomized DRMs in Sect. 4.1. We then describe how randomized DRMs can be used to estimate the error in the corresponding low-rank approximation for adaptive rangefinding. These ideas are foundational to many widely-utilized randomized algorithms for fast, robust low-rank matrix approximation: the randomized SVD (Sect. 4.2) and randomized ID/CUR (Sect. 4.3).

4.1 Randomized Rangefinding

At the heart of many linear algebraic tasks lies the *rangefinder problem*, finding a lower-dimensional space that captures the action of a given matrix. To do this, for an input matrix $\mathbf{A} \in \mathbb{F}^{m\times n}$ and target rank $k \le \min(m, n)$, we compute an orthonormal

Algorithm 1 Randomized Rangefinder

Require: $\mathbf{A} \in \mathbb{F}^{m\times n}$, target rank k
Ensure: Orthonormal $\mathbf{Q} \in \mathbb{F}^{m\times k}$ such that $\mathbf{A} \approx \mathbf{QQ}^*\mathbf{A}$
1: Draw randomized DRM $\mathbf{\Omega}^* \in \mathbb{F}^{\ell\times n}$ for $k < \ell \leq \min(m, n)$
2: Form sketch $\mathbf{Y} = \mathbf{A\Omega}$.
3: Compute $\mathbf{Q} = \texttt{col}(\mathbf{Y}, k)$.

matrix $\mathbf{Q} \in \mathbb{F}^{m\times k}$ of which columns approximately span the column space of $\mathbf{A}$. We then obtain a rank-k approximation given by $\mathbf{QQ}^*\mathbf{A}$, with approximation error

$$||\mathbf{A} - \mathbf{QQ}^*\mathbf{A}|| = ||(\mathbf{I} - \mathbf{QQ}^*)\mathbf{A}||, \tag{29}$$

measured in either the Frobenius or spectral norm in our work. The Frobenius norm is often preferable for ease of analysis but can be less informative than the spectral norm.

The rangefinder problem can be handled effectively with randomized DRMs, as in Algorithm 1, which produces an orthonormal matrix $\mathbf{Q}$ via QR factorization of an $m \times \ell$ random sketch of $\mathbf{A}$ for desired embedding dimension $\ell = O(k)$. The total cost of Algorithm 1 includes the cost of simulating the $n \times \ell$ random test matrix, $O(mn\ell)$ operations to sketch, and $O(mk^2)$ operations for the QR factorization. Sparsity or fast matrix-vector primitives with $\mathbf{A}$ can lend additional performance improvements.

We mention a few implementation details here and refer to [99, 149] for thorough treatments. For Gaussian sketching matrices with $\ell = k + p$, [99, Theorem 10.1] guarantees the expected error of the randomized rangefinder is comparable to the best rank-k approximation error from the truncated SVD, for very small p, e.g., $p = 5$, as long as the trailing singular values of $\mathbf{A}$ are small. The theory for Gaussians often furnishes performance heuristics for other randomized DRMs, which frequently behave like Gaussians despite the more pessimistic theoretical bounds, cf. Table 1. The simple addition of *a posteriori* "certificates of accuracy" provides additional assurance.

4.1.1 Error Estimation and Certificates of Accuracy

The approximation error (29) can be estimated reliably and efficiently with randomized sketching, which serves as a "certificate of accuracy" that the basis given by $\mathbf{Q}$ approximately spans the range of $\mathbf{A}$. Let $\mathbf{Q} \in \mathbb{F}^{m\times k}$ be an orthonormal matrix of which the columns potentially form a basis for the range of $\mathbf{A} \in \mathbb{F}^{m\times n}$. Let $\mathbf{\Phi} \in \mathbb{F}^{n\times s}$ be a Gaussian matrix that is independent from any random matrix used to compute $\mathbf{Q}$. In practice, s is a small fixed value, such as $s = 10$. We form a sketch, often called the auxiliary sample,

$$\mathbf{Z} = \mathbf{A\Phi}. \tag{30}$$

From $\mathbf{Z}$, we obtain an inexpensive Gaussian sketch of the approximation error (29):

$$(\mathbf{I} - \mathbf{QQ}^*)\mathbf{Z} = (\mathbf{I} - \mathbf{QQ}^*)\mathbf{A\Phi}. \tag{31}$$

Algorithm 2 Adaptive Randomized Rangefinder

Require: $\mathbf{A} \in \mathbb{F}^{m \times n}$, error tolerance $\varepsilon > 0$, block size b
Ensure: Orthonormal $\mathbf{Q}$ such that $\|\mathbf{A} - \mathbf{Q}\mathbf{Q}^*\mathbf{A}\| \leq \varepsilon$ with high probability
1: $i = 1$
2: Draw Gaussian random matrix $\boldsymbol{\Omega} \in \mathbb{F}^{n \times b}$
3: $\mathbf{Y} = \mathbf{A}\boldsymbol{\Omega}$
4: $\mathbf{Q}_i = \texttt{col}(\mathbf{Y})$
5: **while** $\texttt{norm\_est}(\mathbf{Y}) > \varepsilon$ **do** ▹ Error estimation as in (32)
6: $\quad i = i + 1$
7: $\quad$ Draw independent Gaussian $\boldsymbol{\Omega} \in \mathbb{F}^{n \times b}$
8: $\quad \mathbf{Y} = \mathbf{A}\boldsymbol{\Omega}$
9: $\quad \mathbf{Y} = \mathbf{Y} - \sum_{j=1}^{i-1} \mathbf{Q}_j(\mathbf{Q}_j^*\mathbf{Y})$
10: $\quad \mathbf{Q}_i = \texttt{col}(\mathbf{Y})$
11: **end while**
12: $\mathbf{Q} = \begin{bmatrix} \mathbf{Q}_1 & \mathbf{Q}_2 & \cdots & \mathbf{Q}_i \end{bmatrix}$

We can then estimate the Frobenius norm of (31) as

$$\|(\mathbf{I} - \mathbf{Q}\mathbf{Q}^*)\mathbf{A}\|_\mathrm{F}^2 \approx \frac{1}{s}\|(\mathbf{I} - \mathbf{Q}\mathbf{Q}^*)\mathbf{Z}\|_\mathrm{F}^2, \tag{32}$$

noting that $\operatorname{trace}(\mathbf{B}) = \omega^*\mathbf{B}\omega$ in expectation for any matrix $\mathbf{B}$ and isotropic random vector ω. Then to reduce the variance, we take the average of s random variables in (32), using the auxiliary sample $\mathbf{Z}$. The error sketch avails us of the powerful theory of Gaussian matrices, regardless of how the column space approximation was obtained. For spectral norm estimation, see [149, Sects. 5-6].

4.1.2 Rank-Adaptive Rangefinding

We have so far assumed the target approximation rank can be passed as an argument to the randomized rangefinder. Unfortunately, the target rank for an approximation is often unknown *a priori*. The method of error estimation above also enables a rank-adaptive randomized rangefinder algorithm (Algorithm 2). For ease of notation, let $\texttt{norm\_est}(\mathbf{Y})$ denote the rangefinder error estimated as in (32) using an auxiliary sample.

Let $\mathbf{A} \in \mathbb{F}^{m \times n}$ and let b be the number of columns processed at a time. In practice, $10 \leq b \leq 100$ is usually appropriate. In each iteration of Algorithm 2, after line 9,

$$\mathbf{Y} = (\mathbf{I} - \mathbf{Q}\mathbf{Q}^*)\mathbf{A}\boldsymbol{\Omega}, \tag{33}$$

where $\mathbf{Q}$ is the currently constructed basis. Critically, the random test matrix $\boldsymbol{\Omega}$ in (33) is independent of the random matrices used to compute $\mathbf{Q}$.

We note that Algorithm 2 can be susceptible to round-off error, which may lead to numerical instability. Namely, this becomes an issue if the newly-drawn sample $\mathbf{Y}$ in line 8 is nearly spanned by the range of the previously computed basis. In practice, re-orthogonalization may be necessary to avoid stability issues, cf. [93].

Algorithm 3 Randomized SVD

Require: $\mathbf{A} \in \mathbb{F}^{m\times n}$, target rank k
Ensure: Orthonormal matrices $\mathbf{U} \in \mathbb{F}^{m\times k}$, $\mathbf{V} \in \mathbb{F}^{n\times k}$, and diagonal $\mathbf{\Sigma} \in \mathbb{F}^{k\times k}$ such that $\mathbf{A} \approx \mathbf{U\Sigma V}^*$
1: Compute $\mathbf{Q}$ = `RandomizedRangefinder`($\mathbf{A}$, k) ▷ Algorithm 1
2: Form $\mathbf{B} = \mathbf{Q}^*\mathbf{A} \in \mathbb{F}^{k\times n}$
3: Compute $[\widehat{\mathbf{U}}, \mathbf{\Sigma}, \mathbf{V}] = \mathrm{svd}(\mathbf{B})$
4: Form $\mathbf{U} = \mathbf{Q}\widehat{\mathbf{U}}$

4.2 Randomized SVD

The randomized SVD of [99] (Algorithm 3) is an important application of rangefinding, with approximation error equal to the randomized rangefinder error.

Given an input matrix $\mathbf{A} \in \mathbb{F}^{m\times n}$, our goal is to compute an approximate rank-k truncated singular value decomposition $\mathbf{U\Sigma V}^*$. Suppose that $\mathbf{Q} \in \mathbb{F}^{m\times k}$ is the output of Algorithm 1, corresponding to rank-k approximation $\mathbf{QQ}^*\mathbf{A}$. Let $\mathbf{B} = \mathbf{Q}^*\mathbf{A} \in \mathbb{F}^{k\times n}$ and compute its SVD $\mathbf{B} = \widehat{\mathbf{U}}\mathbf{\Sigma V}^*$. We observe that for $\mathbf{U} := \mathbf{Q}\widehat{\mathbf{U}}$,

$$||\mathbf{A} - \mathbf{QQ}^*\mathbf{A}|| = ||\mathbf{A} - \mathbf{QB}|| = ||\mathbf{A} - \mathbf{Q}\widehat{\mathbf{U}}\mathbf{\Sigma V}^*|| = ||\mathbf{A} - \mathbf{U\Sigma V}^*||. \tag{34}$$

Thus, the approximation error of $\mathbf{U\Sigma V}^*$ is exactly the approximation error of $\mathbf{QQ}^*\mathbf{A}$. If the target rank k is unknown *a priori*, Algorithm 2 can be used instead.

The total cost of Algorithm 3 includes the cost of rangefinding, plus $O(mnk)$ floating-point operations for $\mathbf{B} = \mathbf{Q}^*\mathbf{A}$ in line 2. We note that the cost of the SVD in line 3 is only $O(k^2 n)$. The storage cost is $O((m+n)k)$.

The recent work of [84] introduces a cost-effective *a posteriori* error estimation method for the randomized SVD. Given rank-k randomized SVD $\mathbf{X} = \mathbf{U\Sigma V}^* \in \mathbb{F}^{m\times n}$ corresponding to Gaussian DRM $\mathbf{\Omega} = \begin{bmatrix}\omega_1 & \dots & \omega_k\end{bmatrix}$, the mean-square error is defined

$$\mathrm{MSE}(\mathbf{X}_{k-1}, \mathbf{A}) = \mathbb{E}||\mathbf{A} - \mathbf{X}_{k-1}||_{\mathrm{F}}^2,$$

where $\mathbf{X}_{k-1}$ is a rank-$(k-1)$ randomized SVD. By [84, Theorem 2.1], the following is an unbiased estimator for $\mathrm{MSE}(\mathbf{X}_{k-1}, \mathbf{A})$, called the leave-one-out error estimator:

$$\mathrm{MSE}(\mathbf{X}_{k-1}, \mathbf{A}) = \mathbb{E}\left[\frac{1}{k}\sum_{j=1}^{k} ||(\mathbf{A} - \mathbf{X}^{(j)})\omega_j||^2\right], \tag{35}$$

where $\mathbf{X}^{(j)} = \mathbf{Q}^{(j)}(\mathbf{Q}^{(j)})^*\mathbf{A}$ is called a replicate, formed from $\mathbf{Q}^{(j)}$ output by Algorithm 1 using $\mathbf{\Omega}$ without its j-th column ω_j. Implemented as written, the right-hand side of (35) is too expensive for practical use. However, by [84, Equation 4.2],

$$\mathbf{Q}^{(j)}(\mathbf{Q}^{(j)})^* = \mathbf{Q}(\mathbf{I} - \mathbf{t}_j\mathbf{t}_j^*)\mathbf{Q}^*, \tag{36}$$

where $\mathbf{t}_j$, $j = 1, \dots, k$, are the normalized columns of $(\mathbf{R}^*)^{-1}$ in the QR factorization defined by $\mathbf{Q}$ in step 1 of the randomized SVD (now also storing the $\mathbf{R}$ matrix). Then we

Algorithm 4 Randomized Row ID with Sketched CPQR

Require: $\mathbf{A} \in \mathbb{F}^{m \times n}$, target rank k, oversampling p
Ensure: Row indices I and row interpolation matrix $\mathbf{X} \in \mathbb{F}^{m \times k}$ such that $\mathbf{A} \approx \mathbf{X}\mathbf{A}(I, :)$
1: Draw randomized embedding $\boldsymbol{\Omega}^* \in \mathbb{F}^{(k+p) \times n}$
2: Form $\mathbf{Y} = \mathbf{A}\boldsymbol{\Omega} \in \mathbb{F}^{m \times (k+p)}$
3: Compute $[\sim, \mathbf{R}, P] = \mathtt{qr}(\mathbf{Y}^*)$
4: Set $I = P(1 : k)$
5: Set $\mathbf{X}(P, :) = \begin{bmatrix} \mathbf{I} & \mathbf{R}_{1:k,1:k}^{-1} \mathbf{R}_{1:k,k+1:m} \end{bmatrix}^*$

can obtain each replicate from (36) inexpensively as $\mathbf{X}^{(j)} = \mathbf{U}(\mathbf{I} - \widehat{\mathbf{U}}^* \mathbf{t}_j \mathbf{t}_j^* \widehat{\mathbf{U}}) \boldsymbol{\Sigma} \mathbf{V}^*$, which allows us to compute (35) without any additional rangefinding via

$$\mathrm{MSE}(\mathbf{X}_{k-1}, \mathbf{A}) \approx \frac{1}{k} \sum_{j=1}^{k} \left\| [(\mathbf{R}^*)^{-1}](:, j) \right\|^{-2}.$$

4.3 Randomized ID/CUR

In both the randomized rangefinder and the randomized SVD, orthonormal matrices are computed of which the columns form an approximate basis for the row or column space of an input matrix $\mathbf{A}$. However, for some problems (e.g., if $\mathbf{A}$ is sparse or non-negative), it may be preferable to have a "natural basis," consisting of rows or columns of $\mathbf{A}$, as in the ID/CUR of Sect. 2.5. In this section, we provide an overview of randomized methods to compute approximate ID/CUR factorizations, including recent work to improve skeleton selection and performance [71, 72, 173, 180]. We first illustrate how a random sketch can be used as a proxy for the input matrix in ID/CUR approximations.

Let $\mathbf{A} \in \mathbb{F}^{m \times n}$ be of exact rank k for notational convenience. Let $\mathbf{Y} \in \mathbb{F}^{m \times k}$ have columns that span the range of $\mathbf{A}$, not necessarily orthogonal. By definition, $\mathbf{A} = \mathbf{Y}\mathbf{F}$ for some matrix $\mathbf{F} \in \mathbb{F}^{k \times n}$. Compute a row ID of $\mathbf{Y}$, e.g., with CPQR via $[\sim, \sim, I] = \mathtt{qr}(\mathbf{Y}^*, k)$ or LUPP via $[\sim, \sim, I] = \mathtt{lu}(\mathbf{Y}, k)$, so that $\mathbf{Y} = \mathbf{X}\mathbf{Y}(I, :)$. We now observe that index set I and row interpolation matrix $\mathbf{X}$ form a row ID of $\mathbf{A}$:

$$\mathbf{A} = \mathbf{Y}\mathbf{F} = [\mathbf{X}\mathbf{Y}(I, :)]\, \mathbf{F} = \mathbf{X}\, [\mathbf{Y}(I, :)\mathbf{F}] = \mathbf{X}\mathbf{A}(I, :).$$

Thus, to compute a row ID of $\mathbf{A}$, it suffices to compute a row ID of a matrix $\mathbf{Y}$ of which the columns span the range of $\mathbf{A}$, and since we do not need $\mathbf{Y}$ to be orthogonal, we can save on computational cost using the fact that the range of $\mathbf{Y} = \mathbf{A}\boldsymbol{\Omega}$ is itself an accurate approximation to the range of $\mathbf{A}$ with high probability [99, Theorems 10.5 and 10.6]. We observe that the randomized rangefinder error $\|\mathbf{A} - \mathbf{Q}\mathbf{Q}^*\mathbf{A}\|$ using $\mathbf{Y} = \mathbf{A}\boldsymbol{\Omega}$ is equal to $\|\mathbf{A} - \mathbf{Y}\mathbf{Y}^\dagger\mathbf{A}\|$, where $\mathbf{Y}\mathbf{Y}^\dagger$ is the orthogonal projector onto the column space of $\mathbf{Y}$ (or, equivalently, the column space of $\mathbf{Q}$). Pseudocode for this randomized row ID procedure (with sketched CPQR) is provided in Algorithm 4. In the next few sections, we discuss some of the many ways that randomization has been incorporated into fast algorithms for robust skeleton selection in computing low-rank ID and CUR factorizations.

4.3.1 With Rank-Adaptive Randomized Sketching and Greedy Pivoting

The CPQR and LUPP algorithms of Sect. 2.5.1-2.5.2 both select pivots (here, skeleton indices) via greedy pivoting, which involves updating the active submatrix in each iteration. The index of the row or column of maximal norm is selected, then the active submatrix (initialized as the input matrix) is successively updated by projecting onto the orthogonal complement of the selected skeletons.

A major benefit of these "updating" greedy pivoting procedures is that the skeleton selection at each step is informed by the previous ones; however, the drawback is that greedy pivoting algorithms are inherently sequential, and the number of skeletons needed to achieve a certain level of ID/CUR accuracy can be quite suboptimal for certain adversarial inputs. The goal then is to develop randomized algorithms that still benefit from active submatrix updates, while also admitting fast, hardware-efficient implementations. Moreover, because the target rank for a low-rank approximation is often unknown *a priori* in practice, these algorithms should also be rank-adaptive, given a user-prescribed error tolerance.

Rank-adaptive randomized LUPP: If using CPQR for skeleton selection, the rank-adaptive version of the randomized rangefinder (Sect. 4.1.2) already builds an orthonormal basis incrementally from independent random sketches or samples. However, if we are using LUPP as the skeleton selection method for improved efficiency and better parallelizability, we want to estimate the approximation error *without* computing an orthonormal basis. We illustrate how this can be done for a row ID, using the rank-adaptive LUPP-based method of [173], which builds on the fixed-rank algorithm of [72].

Given $\varepsilon \in (0, 1)$ and $\mathbf{A} \in \mathbb{F}^{m\times n}$, after t iterations, the algorithm outputs a target rank k_t, row skeletons $I_s^{(t)}$ with $|I_s^{(t)}| = k_t$, and interpolation matrix $\mathbf{W}^{(t)}$ such that our certificate of accuracy (Sect. 4.1.1) for $\|\mathbf{A}-\mathbf{W}^{(t)}\mathbf{R}^{(t)}\|_\mathrm{F}$ is below ε, for $\mathbf{R}^{(t)} = \mathbf{A}(I_s^{(t)}, :)$. Fix a block size b (e.g., $b = 20$), draw a random sketching matrix $\mathbf{\Omega}^{(0)} \in \mathbb{F}^{n\times b}$, and form the sketch $\mathbf{Y}^{(0)} = \mathbf{A}\mathbf{\Omega}^{(0)} \in \mathbb{F}^{m\times b}$. Using $\mathrm{lu}(\mathbf{Y}^{(0)}, b)$, we compute

$$\mathbf{P}^{(0)}\mathbf{Y}^{(0)} = \mathbf{L}^{(0)}\mathbf{U}^{(0)},$$

so that the first b skeletons $I_s^{(0)}$ are the first b indices of $\mathbf{P}^{(0)}[1\dots m]^*$. To estimate $\|\mathbf{A} - \mathbf{W}^{(0)}\mathbf{R}^{(0)}\|_\mathrm{F}$, we draw an independent random matrix $\mathbf{\Omega}^{(1)} \in \mathbb{F}^{n\times b}$ from the same distribution as $\mathbf{\Omega}^{(0)}$ and form $\mathbf{Y}^{(1)} = \mathbf{A}\mathbf{\Omega}^{(1)}$. By [173, Equation 20],

$$\begin{aligned}\|\mathbf{A} - \mathbf{W}^{(0)}\mathbf{R}^{(0)}\|_\mathrm{F} &\approx \|\mathbf{A}\mathbf{\Omega}^{(1)} - \mathbf{W}^{(0)}\mathbf{R}^{(0)}\mathbf{\Omega}^{(1)}\|_\mathrm{F} \\ &= \|\mathbf{P}^{(0)}\mathbf{Y}^{(1)}(b+1:m,:) - \mathbf{L}_2^{(0)}(\mathbf{L}_1^{(0)})^{-1}(\mathbf{P}^{(0)}\mathbf{Y}^{(1)})(1:b,:)\|_\mathrm{F},\end{aligned}$$

where $\mathbf{L}_1^{(0)} = \mathbf{L}^{(0)}(1:b,:)$ and $\mathbf{L}_2^{(0)} = \mathbf{L}^{(0)}(b+1:m,:)$. If the given tolerance is not met, we LU-factorize $[\mathbf{Y}^{(0)}\ \mathbf{Y}^{(1)}]$ re-using the previous factorization, and repeat. The computational complexity for, e.g., a row ID is $O(mnk + mk^2 + k^3)$, where $k = tb$ is the resulting rank after t iterations.

Rank-adaptive CUR with a small sketch: The "Iterative CUR" algorithm of [180] computes a CUR factorization rank-adaptively, using a single small sketch of the residual matrix $\mathbf{S} = \mathbf{A} - \mathbf{CUR}$. (Here, $\mathbf{U}$ is computed as $\mathbf{A}(I, J)^{\dagger}$; not via the optimal but more expensive formula $\mathbf{C}^{\dagger}\mathbf{A}\mathbf{R}^{\dagger}$.) The column residual is initialized as a row sketch, $\mathbf{S}_0^{(C)} = \mathbf{GA} \in \mathbb{F}^{1.1b \times n}$, where typically $5 \leq b \leq 250$, and the row and column skeleton matrices $\mathbf{R}$ and $\mathbf{C}$ are empty. Column skeletons J are first computed from $\mathbf{S}_0^{(C)}$ (e.g., using greedy pivoting), and the row residual is formed as $\mathbf{S}^{(R)} = \mathbf{A}(:, J)$. Row skeletons I are then computed from $\mathbf{S}^{(R)}$ via greedy pivoting, and the column residual is computed: $\mathbf{S}^{(C)} = \mathbf{S}_0^{(C)} - \mathbf{GCUR}$, where now $\mathbf{C} = \mathbf{A}(:, J)$, $\mathbf{R} = \mathbf{A}(I, :)$, and $\mathbf{U} = \mathbf{A}(I, J)^{\dagger}$. (Note that $\mathbf{GC}$ can be obtained without additional mat-vecs from $\mathbf{S}_0^{(C)}$.) This procedure is repeated, with new skeletons appended to previously selected ones until the desired approximation accuracy is achieved. The computational complexity is $O(mn + (m+n)k^2 + k^3)$ for the resulting CUR approximation of rank $k = tb$ after t iterations.

4.3.2 With Rank-Adaptive Random Pivoting

There have been several recent advances in coordinate sampling-based methods for ID/CUR decompositions, in which greedy pivoting selection methods are replaced by randomized ones.

Randomly Pivoted Cholesky/QR: In [47], the randomly pivoted (RP) Cholesky algorithm is introduced for positive semi-definite matrices $\mathbf{A} \in \mathbb{F}^{m \times m}$. Given a target rank k, the algorithm outputs a set of pivot indices $S = \{s_1, \ldots, s_k\}$ akin to skeletons, and a matrix $\mathbf{F} \in \mathbb{F}^{m \times k}$ for approximate Cholesky factorization $\widehat{\mathbf{A}} = \mathbf{F}\mathbf{F}^*$, which is the factorized form of the column Nyström approximation $\widehat{\mathbf{A}} = \mathbf{A}(:, S)\mathbf{A}(S, S)^{\dagger}\mathbf{A}(S, :)$. By selecting pivots with probability proportional to the magnitudes of diagonal entries of the residual matrix in each iteration, the RP Cholesky algorithm returns the Cholesky-factorized form of the Nyström approximation in $O(km)$ matrix entry evaluations and $O(k^2m)$ additional arithmetic operations, with $O(km)$ storage requirements.

We begin by defining $\mathbf{d} = \mathrm{diag}(\mathbf{A})$ and initializing the matrix $\mathbf{F}$ as $\mathbf{0}$. After sampling each pivot $s_i \sim \mathbf{d}/\sum_{j=1}^{m}\mathbf{d}(j)$, we form $\mathbf{g} = \mathbf{A}(:, s_i)$ and remove any overlap with previously selected columns via

$$\mathbf{g} = \mathbf{g} - \mathbf{F}(:, 1 : (i-1))\mathbf{F}(s_i, 1 : (i-1))^*.$$

The matrix $\mathbf{F}$ is then updated via $\mathbf{F}(:, i) = \mathbf{g}/\sqrt{\mathbf{g}(s_i)}$, and the vector $\mathbf{d}$ of diagonal entries of the residual matrix is updated via $\mathbf{d} = \mathbf{d} - |\mathbf{F}(:, i)|^2$. To achieve an ε-approximation, $d \geq k/\varepsilon + k\log(\frac{1}{\varepsilon\eta})$ columns are required, where η is the relative trace-norm error, cf. [47, Theorem 2.3]. A blocked variant is also provided in [47, Algorithm 3]. RP Cholesky may only be used with positive semi-definite matrices, but randomly pivoted QR (RPQR, introduced in [66, 67]) is suitable for general matrices. In RPQR, the column norm maximization criterion is replaced by column norm sampling. However, in their nascent presentations, the blocked versions of these algorithms (RPCholesky and RPQR) are vulnerable to selecting pivot columns that are nearly linearly dependent.

Robust Blockwise Random Pivoting: The recent work of [71], inspired by RP Cholesky and RPQR, introduces robust blockwise random pivoting (RBRP) for matrix IDs to avoid selecting nearly linearly dependent pivots. As above, RBRP replaces greedy pivoting based on squared column norms (in CPQR) with squared-norm sampling. The "robustness" of RBRP refers to a filtering step that eliminates redundant points within each block of selected skeleton candidates. Namely, suppose that we sample b row skeleton candidates S_t for some fixed block size b. We then form

$$\mathbf{V} = \mathbf{A}(S_t, :)^* - \mathbf{Q}^{(t-1)}((\mathbf{Q}^{(t-1)})^* \mathbf{A}(S_t, :)^*) \in \mathbb{F}^{n \times b},$$

where $\mathbf{Q}^{(t-1)}$ is an orthonormal basis for the previously selected skeletons. Given some filtering tolerance τ, we use CPQR to compute the factorization

$$\mathbf{V}(:, P) = \mathbf{QR} \approx \mathbf{Q}(:, 1 : b')\mathbf{R}(1 : b', :),$$

where $b' = \max\left\{i \in [b] \mid \|\mathbf{R}(i : b, i : b)\|_F^2 \geq \tau \|\mathbf{R}\|_F^2\right\}$, so that

$$\|\mathbf{V}(:, P) - \mathbf{Q}(:, 1 : b')\mathbf{R}(1 : b', :)\|_F^2 = \|\mathbf{R}(b' + 1 : b, b' + 1 : b)\|_F^2 < \tau \|\mathbf{V}\|_F^2.$$

This leads to a time complexity of roughly $O(mnk + nk^2)$ for resulting skeleton cardinality k, as RBRP is also rank-adaptive, and the optimal associated interpolation matrix $\mathbf{W} = \mathbf{AA}(S, :)^\dagger$ can be computed efficiently in $O(mk^2)$ time.

Random pivoting with rejection sampling: The issue of sampling nearly linearly dependent columns in blocked random pivoting algorithms is also addressed in [85], a follow-up to [47] that utilizes rejection sampling.

The algorithm is similar to the blocked variant proposed in the initial work on RP Cholesky. First, a block of independent and identically distributed random pivots is selected as before. Then, a candidate pivot s is kept, or accepted, with probability

$$p(s) = \frac{\widetilde{\mathbf{A}}(s, s) - \widetilde{\mathbf{A}}(s, J)\widetilde{\mathbf{A}}(J, J)^\dagger \widetilde{\mathbf{A}}(J, s)}{\widetilde{\mathbf{A}}(s, s)},$$

where $\widetilde{\mathbf{A}}$ is the residual matrix from the previous iteration, and J is the set of the already-accepted pivots from the current block of candidates. Now, with possible rejections, the accuracy of the resulting approximation in general improves over simple RP Cholesky, and though slightly more matrix access may be needed, there is still a practical speed-up in [85] due to the blockwise computations.

Adaptive random pivoting: The adaptive random pivoting (ARP) method proposed in [54] may be thought of as a randomization of Osinsky's method (Sect. 2.5.3). Given $n \times k$ matrix $\mathbf{V}$ with orthonormal columns approximating a basis for the row space of matrix $\mathbf{A} \in \mathbb{F}^{m \times n}$ (e.g. the k truncated right singular vectors), the ARP algorithm determines the first pivot by leverage score sampling on $\mathbf{V}$. The selected row is removed by orthogonal projection, and the next index is selected via leverage scores of the updated residual matrix, for a total cost of $O(nk^2)$ operations.

5 Tensor Preliminaries

This section marks the beginning of our exploration of tensors, where we present tensors as natural higher-order generalizations of matrices, from 2-way to d-way arrays. We begin with formal notation and definitions in Sect. 5.1 and discuss important matrix and tensor operations in Sect. 5.2. While there are many different ways to represent tensors, we focus on two fundamental tensor decomposition formats: the canonical polyadic (CP) and Tucker. In Sect. 5.3, we lay out key concepts and practicalities for each of them, including algorithms to compute them. We then discuss particular low-rank tensor decompositions that can be represented in each format, such as higher-order extensions of the matrix SVD and ID/CUR. Throughout this section, we show how the classical matrix methods of Sect. 2 form the foundation of those for tensors.

5.1 Notation and Definitions

A *tensor* $\mathcal{X} \in \mathbb{F}^{N_1 \times N_2 \times \cdots \times N_d}$ is a d-way array with entries $\mathcal{X}_{i_1, i_2, \ldots, i_d}$, for $i_1 \in [N_1]$, $i_2 \in [N_2]$, ..., $i_d \in [N_d]$. A d-way tensor is said to be of *order* d, and each of its d ways is referred to as a *mode*; a matrix, for instance, is a tensor of order 2, with mode 1 comprised of its rows and mode 2 comprised of its columns. A *mode-j fiber* of a tensor is a vector obtained by fixing all indices except for the jth, e.g., a column of a matrix is a mode-1 fiber, and a row is a mode-2 fiber. The *mode-j slices* of a tensor are the sub-tensors $\mathcal{X}(:, \ldots, :, i_j, :, \ldots, :)$ obtained by fixing a single index $i_j \in [N_j]$; for example, slices of 3-mode tensors are matrices.

We define $N = \prod_{j=1}^{d} N_j =: \text{size}(\mathcal{X})$, and $N^{(-j)} = \prod_{k=1, k \neq j}^{d} N_k = N/N_j$. Through a rearrangement of entries known as *matricization*, a d-mode tensor can be "unfolded" into a matrix. The *mode-j unfolding* of a d-mode tensor $\mathcal{X}$, denoted $\mathbf{X}_{(j)} \in \mathbb{F}^{N_j \times N^{(-j)}}$, $j = 1, \ldots, d$, is a matrix of which the columns are the mode-j fibers of $\mathcal{X}$.

In our work, the (Frobenius) *norm* of a tensor $\mathcal{X} \in \mathbb{F}^{N_1 \times N_2 \times \cdots \times N_d}$, simply denoted by $\|\mathcal{X}\|$, is the square root of the sum of squares of all elements, analogous to the Frobenius norm for matrices. A tensor $\mathcal{X} \in \mathbb{F}^{N_1 \times N_2 \times \cdots \times N_d}$ is said to be *rank-one* if it can be expressed as a vector outer product (denoted by $\circ$) of d vectors, i.e. $\mathcal{X} = \mathbf{v}^{(1)} \circ \mathbf{v}^{(2)} \circ \cdots \circ \mathbf{v}^{(d)}$.

5.2 Matrix and Tensor Products

We now define operations involving matrices and tensors used throughout our work.

The *Kronecker product* of matrices $\mathbf{A} \in \mathbb{F}^{I\times J}$ and $\mathbf{B} \in \mathbb{F}^{K\times L}$, denoted $\mathbf{A} \otimes \mathbf{B}$, is the $(IK) \times (JL)$ matrix with entries

$$(\mathbf{A} \otimes \mathbf{B})(K(i-1)+k, L(j-1)+\ell) = \mathbf{A}(i,j)\mathbf{B}(k,\ell).$$

The *Khatri–Rao product* of matrices $\mathbf{A} \in \mathbb{F}^{I\times K}$ and $\mathbf{B} \in \mathbb{F}^{J\times K}$, denoted $\mathbf{A} \odot \mathbf{B}$, is the $(IJ) \times K$ matrix with entries

$$(\mathbf{A} \odot \mathbf{B})(:,k) = \mathbf{A}(:,k) \otimes \mathbf{B}(:,k)$$

for $k = 1, \ldots, K$.

The *Hadamard product* of matrices $\mathbf{A} \in \mathbb{F}^{I\times J}$ and $\mathbf{B} \in \mathbb{F}^{I\times J}$, denoted $\mathbf{A} * \mathbf{B}$, is the $I \times J$ matrix of entries

$$(\mathbf{A} * \mathbf{B})(i,j) = \mathbf{A}(i,j)\mathbf{B}(i,j).$$

The *tensor-times-matrix (TTM) product* is the tensor resulting from the mode-j product of a tensor $\mathcal{X} \in \mathbb{F}^{N_1\times\cdots\times N_j\times\cdots\times N_d}$ with a matrix $\mathbf{U} \in \mathbb{F}^{K\times N_j}$, denoted by $\mathcal{Y} = \mathcal{X} \times_j \mathbf{U} \in \mathbb{F}^{N_1\times\cdots\times N_{j-1}\times K\times N_{j+1}\times\cdots\times N_d}$, with entries

$$\mathcal{Y}(i_1,\ldots,i_{j-1},k,i_{j+1},\ldots,i_d) = \sum_{i_j=1}^{N_j} \mathcal{X}(i_1,\ldots,i_d)\mathbf{U}(k,i_j).$$

We can express this more concisely with mode-j matricization: $\mathbf{Y}_{(j)} = \mathbf{U}\mathbf{X}_{(j)}$.

The *Multi-TTM product* extends the TTM product to multiple modes. For $\mathcal{X} \in \mathbb{F}^{N_1\times\cdots\times N_j\times\cdots\times N_d}$ and matrices $\mathbf{U}_j \in \mathbb{F}^{K_j\times N_j}$ for $j \in [d]$, a Multi-TTM is given by

$$\mathcal{Y} = \mathcal{X} \times_1 \mathbf{U}_1 \times_2 \mathbf{U}_2 \cdots \times_d \mathbf{U}_d.$$

The order of multiplication does not matter, cf. [15, Proposition 3.19]. Equivalently, after mode-j matricization,

$$\mathbf{Y}_{(j)} = \mathbf{U}_j\mathbf{X}_{(j)}\left(\mathbf{U}_d \otimes \cdots \otimes \mathbf{U}_{j+1} \otimes \mathbf{U}_{j-1} \otimes \cdots \otimes \mathbf{U}_1\right)^*.$$

5.3 Tensor Decompositions

This work considers two fundamental formats of tensor decompositions: the canonical polyadic (CP) decomposition and the Tucker decomposition. The CP decomposition has been known by many different names since its introduction in 1927 in [107, 108], cf. [126, Table 3.1]. This decomposition is often viewed as a higher-order extension of the matrix SVD, as it is formed from rank-one components. The Tucker decomposition [206, 207, 208] is a representation of a tensor as a Multi-TTM product of a core tensor with a factor matrix for each mode. Like the CP decomposition, the Tucker decomposition has been known by several different names in the literature, cf. [126, Table 4.1], and the Tucker decomposition can also be viewed as an extension of the matrix SVD, in the form of higher-order principal component analysis (PCA).

5.3.1 The Canonical Polyadic (CP) Decomposition

The CP decomposition of a tensor $\mathcal{X} \in \mathbb{F}^{N_1 \times \cdots \times N_d}$ expresses it as a sum of R rank-one tensors:

$$\mathcal{X} = \sum_{r=1}^{R} \mathbf{a}_r^{(1)} \circ \mathbf{a}_r^{(2)} \circ \cdots \circ \mathbf{a}_r^{(d)}, \quad \text{where } \mathbf{a}_r^{(j)} \in \mathbb{F}^{N_j} \text{ for } j = 1, \ldots, d. \tag{37}$$

For example, the CP decomposition of a 3-mode tensor is visualized in Fig. 1.

The *factor matrices* of the CP decomposition are formed from the vectors corresponding to each of the d modes. Namely, the mode-j factor matrix is given by

$$\mathbf{A}^{(j)} = \left[\mathbf{a}_1^{(j)}\ \mathbf{a}_2^{(j)}\ \ldots\ \mathbf{a}_R^{(j)}\right] \in \mathbb{F}^{N_j \times R}.$$

We can express (37) in terms of mode-j unfoldings using the Khatri–Rao product:

$$\mathbf{X}_{(j)} = \mathbf{A}^{(j)} \left(\mathbf{A}^{(d)} \odot \cdots \odot \mathbf{A}^{(j+1)} \odot \mathbf{A}^{(j-1)} \odot \cdots \odot \mathbf{A}^{(1)}\right)^{*}.$$

The notation of [124] is frequently used as shorthand for the CP decomposition:

$$\mathcal{X} = [\![\mathbf{A}^{(1)}, \mathbf{A}^{(2)}, \ldots, \mathbf{A}^{(d)}]\!] := \sum_{r=1}^{R} \mathbf{a}_r^{(1)} \circ \cdots \circ \mathbf{a}_r^{(d)}. \tag{38}$$

If we enforce unit length, a weight vector $\lambda \in \mathbb{F}^R$ may be incorporated into (38), so that

$$\mathcal{X} = [\![\lambda; \mathbf{A}^{(1)}, \mathbf{A}^{(2)}, \ldots, \mathbf{A}^{(d)}]\!] := \sum_{r=1}^{R} \lambda_r \mathbf{a}_r^{(1)} \circ \cdots \circ \mathbf{a}_r^{(d)}. \tag{39}$$

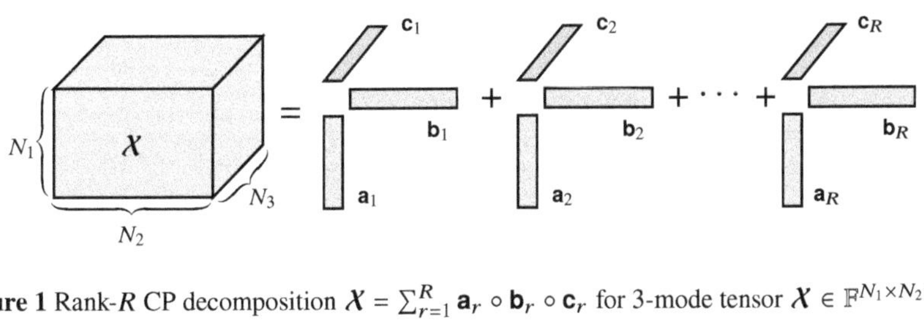

Figure 1 Rank-R CP decomposition $\mathcal{X} = \sum_{r=1}^{R} \mathbf{a}_r \circ \mathbf{b}_r \circ \mathbf{c}_r$ for 3-mode tensor $\mathcal{X} \in \mathbb{F}^{N_1 \times N_2 \times N_3}$.

Algorithm 5 Alternating Least Squares for CP Decomposition (CP-ALS)

Require: Input tensor $\mathcal{X} \in \mathbb{F}^{N_1 \times \cdots \times N_d}$, number of rank-one components R
Ensure: CP decomposition $\widehat{\mathcal{X}} = [\![\lambda; \mathbf{A}^{(1)}, \ldots, \mathbf{A}^{(d)}]\!]$
Initialize $\mathbf{A}^{(j)} \in \mathbb{F}^{N_j \times R}$, $j = 1, \ldots, d$ ▷ e.g., $\mathbf{A}^{(j)} = \mathtt{randn}(N_j, R)$
while not converged **do**
  for $j = 1, \ldots, d$ **do**
    $\mathbf{A}^{(j)} = \mathbf{X}_{(j)} \left[\left(\mathbf{A}^{(d)} \odot \cdots \odot \mathbf{A}^{(j+1)} \odot \mathbf{A}^{(j-1)} \odot \cdots \odot \mathbf{A}^{(1)} \right)^* \right]^\dagger$
    $\lambda_r = \left\| \mathbf{A}^{(j)}(:, r) \right\|_2, \ r = 1, \ldots, R$
    $\mathbf{A}^{(j)}(:, r) = \mathbf{A}^{(j)}(:, r) / \lambda_r, \ r = 1, \ldots, R$
    Update $\widehat{\mathcal{X}} = [\![\lambda; \mathbf{A}^{(1)}, \ldots, \mathbf{A}^{(d)}]\!]$
  end for
end while

The CP decomposition may be computed by the *alternating least squares* (ALS) method [34, 102] in Algorithm 5. Given a tensor $\mathcal{X} \in \mathbb{F}^{N_1 \times \cdots \times N_d}$, CP-ALS outputs $\widehat{\mathcal{X}} = [\![\lambda; \mathbf{A}^{(1)}, \ldots, \mathbf{A}^{(d)}]\!]$ with R rank-one components through a sequence of linear least squares problems, where all but one factor matrix is fixed:

$$\min_{\widehat{\mathbf{A}} \in \mathbb{F}^{N_j \times R}} \left\| \mathbf{X}_{(j)} - \widehat{\mathbf{A}} \left(\mathbf{A}^{(d)} \odot \cdots \odot \mathbf{A}^{(j+1)} \odot \mathbf{A}^{(j-1)} \odot \cdots \odot \mathbf{A}^{(1)} \right)^* \right\|_F^2. \tag{40}$$

The solution of (40) is given by

$$\mathbf{A}^{(j)} := \mathbf{X}_{(j)} \left[\left(\mathbf{A}^{(d)} \odot \cdots \odot \mathbf{A}^{(j+1)} \odot \mathbf{A}^{(j-1)} \odot \cdots \odot \mathbf{A}^{(1)} \right)^* \right]^\dagger, \tag{41}$$

where the columns of $\mathbf{A}^{(j)}$ are normalized to determine λ_r for $r = 1, \ldots, R$. This process is repeated for each mode until some convergence criterion is met; see Sect. 5.3.3.

If Algorithm 5 is naively implemented, the per-iteration cost to compute a CP decomposition is dominated by $N = \mathrm{size}(\mathcal{X})$, which is computationally undesirable. Fortunately, since its introduction in 1970, CP-ALS has benefited from many modifications for improved efficiency. One strategy that has garnered attention incorporates a preprocessing step to compress the input tensor before executing Algorithm 5, known as CANDELINC [27, 35], which may be achieved through a Tucker decomposition. Another strategy for computing CP decompositions beyond the scope of the current work is "all-at-once" optimization, cf. [1, 192, 210].

5.3.2 Tucker Decomposition

The Tucker decomposition expresses a tensor $\mathcal{X} \in \mathbb{F}^{N_1 \times \cdots \times N_d}$ as a d-way Multi-TTM

$$\mathcal{X} = \mathcal{G} \times_1 \mathbf{A}^{(1)} \times_2 \mathbf{A}^{(2)} \cdots \times_d \mathbf{A}^{(d)} =: [\![\mathcal{G}; \mathbf{A}^{(1)}, \mathbf{A}^{(2)}, \ldots, \mathbf{A}^{(d)}]\!], \tag{42}$$

Algorithm 6 Higher-order SVD for Tucker Decomposition (HOSVD)

Require: Input tensor $\mathcal{X} \in \mathbb{F}^{N_1 \times \cdots \times N_d}$, factor matrix target ranks $R_1, \ldots, R_d$
Ensure: Tucker decomposition $\widehat{\mathcal{X}} = [\![\mathcal{G}; \mathbf{A}^{(1)}, \ldots, \mathbf{A}^{(d)}]\!]$
1: **for** $j = 1, \ldots, d$ **do**
2: $[\mathbf{A}^{(j)}, \sim, \sim] = \mathsf{svd}(\mathbf{X}_{(j)}, R_j)$
3: **end for**
4: $\mathcal{G} = \mathcal{X} \times_1 \left(\mathbf{A}^{(1)}\right)^* \cdots \left(\mathbf{A}^{(d)}\right)^*$

where $\mathcal{G} \in \mathbb{F}^{R_1 \times \cdots \times R_d}$ is called the core tensor, and $\mathbf{A}^{(j)} \in \mathbb{F}^{N_j \times R_j}$ is the factor matrix for mode $j = 1, \ldots, d$. We can express (42) in terms of mode-j unfoldings by

$$\mathbf{X}_{(j)} = \mathbf{A}^{(j)} \mathbf{G}_{(j)} \left(\mathbf{A}^{(d)} \otimes \cdots \otimes \mathbf{A}^{(j+1)} \otimes \mathbf{A}^{(j-1)} \otimes \cdots \otimes \mathbf{A}^{(1)}\right)^* . \tag{43}$$

A Tucker decomposition of a 3-mode tensor is visualized in Fig. 2.

A prototypical algorithm to compute the Tucker decomposition is Tucker's "Method 1" [208], which has come to be known as the *higher-order SVD* (HOSVD) [57], summarized in Algorithm 6. For an input tensor $\mathcal{X} \in \mathbb{F}^{N_1 \times \cdots N_d}$, each factor matrix $\mathbf{A}^{(j)}$ is computed as the leading-R_j left singular vectors of the mode-j unfolding $\mathbf{X}_{(j)}$. If $R_j < \operatorname{rank}(\mathbf{X}_{(j)})$, the decomposition is said to be "truncated." When $R_j \ll N_j$, we say the tensor is compressed. The core tensor is a Multi-TTM of the original tensor with the factor matrix transposes. The total cost is $O(dn^{d+1} + \sum_{j=1}^{d} k^j n^{d-j+1})$, letting $n = N_1 = \cdots = N_d$ and $k = R_1 = \cdots = R_d$ for notational simplicity.

For improved efficiency, the *sequentially truncated HOSVD* (ST-HOSVD) is often used [211]. In the ST-HOSVD, once the mode-j factor matrix is computed, the tensor is compressed via mode-j TTM before the next factor matrix is computed; i.e., we first compute $\mathbf{A}^{(1)}$ and form $\mathcal{Y} = \mathcal{X} \times_1 \mathbf{A}^{(1)}$, then compute $\mathbf{A}^{(2)}$ from $\mathcal{Y}$, then form $\mathcal{Z} = \mathcal{Y} \times_2 \mathbf{A}^{(2)}$, and so on. ST-HOSVD successively reduces the size of the tensor used in subsequent computations, and the core tensor is available immediately after computing the factor matrix for the last mode. The total cost is $O(\sum_{j=1}^{d} k^{j-1} n^{d-j+2} + k^j n^{d-j})$.

Unlike the matrix SVD, the (ST-)HOSVD does not yield optimal approximation error. It has been shown that the relative error with (ST-)HOSVD is within $\sqrt{d}$ of an optimal Tucker decomposition of a given rank for a d-mode tensor, cf. [15, Chapter 7]. However, we can use the HOSVD output as a starting point for an ALS-style algorithm to

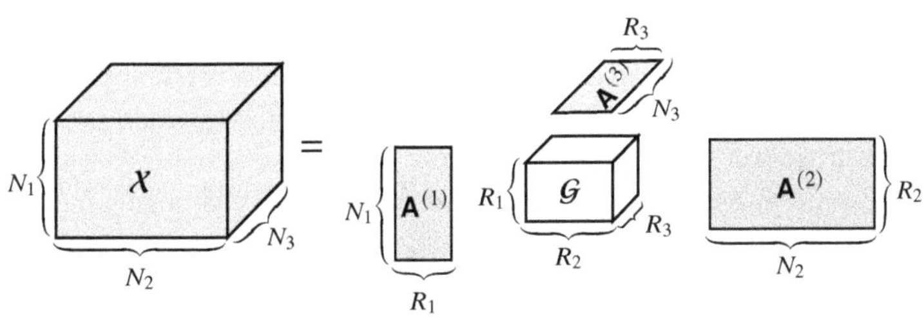

Figure 2 Rank-(R_1, R_2, R_3) Tucker decomposition $\mathcal{X} = \mathcal{G} \times_1 \mathbf{A}^{(1)} \times_2 \mathbf{A}^{(2)} \times_3 \mathbf{A}^{(3)}$ of $\mathcal{X} \in \mathbb{F}^{N_1 \times N_2 \times N_3}$.

Algorithm 7 Higher-order Orthogonal Iteration for Tucker Decomposition (HOOI)

Require: Input tensor $\mathcal{X} \in \mathbb{F}^{N_1 \times \cdots \times N_d}$, factor matrix ranks $R_1, \ldots, R_d$
Ensure: Tucker decomposition $\widehat{\mathcal{X}} = [\![\mathcal{G}; \mathbf{A}^{(1)}, \ldots, \mathbf{A}^{(d)}]\!]$
1: Initialize $\mathbf{A}^{(j)} \in \mathbb{F}^{N_j \times R_j}$, $j = 2, \ldots, d$
2: **while** not converged **do**
3: **for** $j = 1, \ldots, d$ **do**
4: $\mathbf{A}^{(j)} = \arg\min_{\mathbf{A} \in \mathbb{F}^{N_j \times R_j}} \left\| \left(\bigotimes_{i=d, i \neq j}^{1} \mathbf{A}^{(j)} \right) \mathbf{G}^*_{(j)} \mathbf{A}^* - \mathbf{X}^*_{(j)} \right\|_F^2$
5: **end for**
6: $\mathbf{g} = \arg\min_{\mathbf{z} \in \mathbb{F}^{\prod_j R_j \times 1}} \left\| \left(\bigotimes_{j=d}^{1} \mathbf{A}^{(j)} \right) \mathbf{z} - \mathbf{x} \right\|_2^2$ ▷ Column vectorization $\mathbf{x}$ of $\mathcal{X}$
7: Re-fold vectorization $\mathbf{g}$ into $\mathcal{G} \in \mathbb{F}^{R_1 \times \cdots \times R_d}$
8: **end while**

compute Tucker decompositions, known as *higher-order orthogonal iteration* (HOOI) or Tucker-ALS [129, 57]. Pseudocode for HOOI is given in Algorithm 7. We seek to minimize the approximation error $\|\mathcal{X} - \widehat{\mathcal{X}}\|$, now over $\mathcal{G}$ and $\mathbf{A}^{(1)}, \ldots, \mathbf{A}^{(d)}$, subject to the constraints that $\mathcal{G} \in \mathbb{F}^{R_1, \ldots, R_d}$ and that $\mathbf{A}^{(j)} \in \mathbb{F}^{N_j \times R_j}$ is orthogonal for $j = 1, \ldots, d$. This minimization problem can be transformed into a matrix least squares problem in each mode j, so that

$$\mathbf{A}^{(j)} = \arg \min_{\widehat{\mathbf{A}} \in \mathbb{F}^{N_j \times R_j}} \left\| \left(\mathbf{A}^{(d)} \otimes \cdots \otimes \mathbf{A}^{(j+1)} \otimes \mathbf{A}^{(j-1)} \otimes \cdots \otimes \mathbf{A}^{(1)} \right) \mathbf{G}^*_{(j)} \widehat{\mathbf{A}}^* - \mathbf{X}^*_{(j)} \right\|_F^2 .$$

The solution to this least squares problem is given by the leading left singular vectors of each unfolding $\mathbf{X}_{(j)}$ [58, 124]. The core tensor can be computed as a column vector $\mathbf{g} = \arg\min_{\mathbf{z} \in \mathbb{F}^{\prod_j R_j \times 1}} \left\| \left(\bigotimes_{j=d}^{1} \mathbf{A}^{(j)} \right) \mathbf{z} - \mathbf{x} \right\|_2^2$, where $\mathbf{x}$ is the vectorization or unfolding of tensor $\mathcal{X}$ into a large column vector of size $N \times 1$. To get $\mathcal{G}$, we "re-fold" $\mathbf{g} \in \mathbb{F}^{\prod_j R_j}$ back into a tensor of dimension $\mathbb{F}^{R_1 \times \cdots \times R_d}$. If the factor matrices are computed as the leading R_j left singular vectors of unfoldings, then the optimal core tensor is given by $\mathcal{G} = \mathcal{X} \times_1 \left(\mathbf{A}^{(1)} \right)^* \times \cdots \times_d \left(\mathbf{A}^{(d)} \right)^*$. The per-iteration cost of Algorithm 7 is $O(dkn^d + \sum_{j=1}^{d} k^j n^{d-j+1})$, letting $k = R_1 = \cdots = R_d$ and $n = N_1 = \cdots = N_d$.

While Algorithm 7 will converge to a local solution where the approximation error no longer decreases, it may not converge to the global optimum [58, 129]. Modified versions with better guarantees have been proposed, such as HOOI with Newton-Grassmann optimization [80], but these come at greater computational cost. The Tucker decomposition may also be used as a preprocessing compression step to achieve better performance in CP-ALS, known as CANDELINC or a Tucker+CP approach [27]. The HOOI algorithm can also be formulated in terms of a maximization problem, solved via power iteration; see, e.g., [111, 161].

In the next section, we offer some practical guidelines for choosing a tensor format and target rank for a given problem, beginning with a discussion of tensor rank and ending with a comparison of the presented tensor formats.

5.3.3 Rank, Uniqueness, and the Choice of Tensor Representation

The *rank* of a tensor $\mathcal{X}$ is the minimal number of rank-one tensors that are required for equality in (38):

$$\operatorname{rank}(\mathcal{X}) = \min\{R \in \mathbb{N} \mid \mathcal{X} = [\![\mathbf{A}^{(1)}, \ldots, \mathbf{A}^{(d)}]\!],\ \mathbf{A}^{(j)} \in \mathbb{F}^{N_j \times R}\ \forall j \in [d]\}. \quad (44)$$

Mode ordering and TTM products with nonsingular matrices do not affect tensor rank. However, it is NP-hard to determine [110]. In practice, the target ranks for low-rank tensor decompositions are chosen heuristically. For example, we might choose the smallest rank R for a CP decomposition that significantly reduces the relative error, in comparison to rank $R-1$, cf. [15, Section 9.4.1]. Generally, the optimization procedure in CP-ALS is repeated until the error no longer decreases significantly.

For Tucker decompositions, it is useful to consider the j-rank of a tensor, denoted by $\operatorname{rank}_j(\mathcal{X})$ and defined as the column rank of its mode-j unfolding $\mathbf{X}_{(j)}$ [131, 57]. If $R_j = \operatorname{rank}_j(\mathcal{X})$ for $j = 1, \ldots, d$, then we say that $\mathcal{X}$ is a rank-$(R_1, \ldots, R_d)$ tensor (distinct from its rank, defined above). We can compute an exact rank-$(R_1, \ldots, R_d)$ Tucker decomposition for any tensor $\mathcal{X}$ with $R_j = \operatorname{rank}_j(\mathcal{X})$ for $j = 1, \ldots, d$. However, if $R_j < \operatorname{rank}_j(\mathcal{X})$ for any j, the decomposition is inexact and more difficult to compute. In practice, we often prescribe a core j-rank for $j = 1, \ldots, d$ that meets some error tolerance or compression ratio [15, Section 4.2].

Whether it is better to represent an input tensor in the CP or Tucker format is generally problem-dependent. Under mild assumptions, the CP decomposition of a tensor is unique, up to scaling and permutation of its rank-one components [19, 103, 102, 130, 131, 189]. The same cannot be said for Tucker decompositions; the mode-j factor matrix can be multiplied by a nonsingular matrix and the core tensor updated via mode-j TTM with its inverse. As a result, the factors of a Tucker decomposition can be less interpretable in the context of the original data than those of a CP decomposition, though analyzing the Tucker factors may be viewed as multilinear PCA, cf. [51, 91, 190, 220].

Despite its utility in data compression, the Tucker decomposition still suffers from the curse of dimensionality: letting $k = R_1 = \cdots = R_d$ and $n = N_1 = \cdots = N_d$, the cost of storing a Tucker decomposition is $O(k^d + dkn)$. By contrast, the storage cost of a CP decomposition is linear in its rank, but its rank may be very large to achieve the desired accuracy. In short, the notion of rank for tensor decompositions is more nuanced than for matrices, and the choice of representation and target rank (or j-rank) should be informed by the desired trade-off between approximation accuracy and computational performance. For more detailed treatments, we refer the reader to [15, 57, 96, 126].

5.3.4 Low-Rank Tensor Factorizations

In the previous sections, we introduced two fundamental tensor representations, the CP and the Tucker decomposition, and we summarized their key properties. In this section, we survey low-rank tensor factorizations that can be expressed in each format.

Orthogonal factor matrices

Recall from Sect. 5.3.2 that the (ST-)HOSVD and HOOI methods, used to compute Tucker decompositions, return orthonormal factor matrices $\mathbf{A}^{(j)}$ for $j = 1, \ldots, d$. Alternatively, there are algorithms for Tucker decompositions where orthogonality of factor matrices is not enforced, such as algorithms for the tensor ID/CUR factorizations in which factor matrices are computed as submatrices of unfoldings. Other examples beyond the scope of the current work include diagonalization algorithms for Tucker decompositions [196, 197] or algorithms that compute non-negative or sparse Tucker factorizations [121, 155]. These algorithms may be preferable in specific application domains, such as image processing [191, 225], but in general, many tensor computations are greatly simplified when the Tucker factor matrices are orthogonal. Moreover, the (ST-)HOSVD with enforced orthogonality comes with quasi-optimality guarantees. As such, with the exception of the tensor ID/CUR, we focus on algorithms for Tucker decompositions that compute orthogonal (in fact, orthonormal) factor matrices. There is also historical precedence for this decision, as the seminal algorithms to compute Tucker decompositions enforced the orthogonality of factor matrices [57, 58, 77, 122, 207, 208, 211].

Finally, we note that the factor matrices in a CP decomposition are not orthogonal in general. Some tensors may admit CP decompositions with orthogonal factors [9, 182], called ODECO tensors; however, ODECO tensors occur with probability 0 [122, 123, 125]. This is in contrast with Tucker factor matrices, which can be transformed straightforwardly into orthogonal matrices, cf. [15, Sections 5.4, 17.1.4]. As such, in this work, we focus solely on Tucker decompositions in our discussion of randomized algorithms that compute orthogonal factor matrices in Sect. 6.

Tensor ID and CUR factorizations

As in the matrix setting (Sect. 2.5), a tensor ID or CUR factorization comprises actual entries of the input tensor. The advantages of ID/CUR factorizations for tensors are similar to those for matrices. The tensor ID/CUR factorization can be interpreted more easily in the context of the original tensorial data than factorizations involving multi-linear transformations of the data. Moreover, the ID/CUR preserves the structure of the input tensor (e.g., sparsity or non-negativity), and it can be computed efficiently with minimal storage requirements.

Analogous to the different versions of matrix ID/CUR (e.g., row, column, double-sided, CUR), there are several variants in the tensor setting that have been investigated since the ID/CUR was first extended to 3-mode tensors in [77]. In general, these variants fall into one of two categories:

1. The core tensor (in the Tucker format) is a sub-tensor of the input tensor, or
2. The factor matrices (in the CP or Tucker format) are submatrices of unfoldings.

The seminal work of [140] that introduced "tensor CUR factorizations" is most similar to tensor ID/CUR algorithms in the first category. In their algorithm, a rank-(N_1, N_2, R) core tensor is constructed by selecting R indices from one mode and keeping all indices from the other two. Core tensor skeleton selection methods for 3-mode tensor CUR were generalized to d-mode tensors in [31]. The "structure-preserving ST-HOSVD" of [153], initially developed for sparse rank-$(R_1, \ldots, R_d)$ tensors, may also be interpreted as a tensor CUR factorization of the first category. The "Chidori" and "Fiber" tensor CUR factorizations with rank-$(R_1, \ldots, R_d)$ core tensors were characterized in [29], inspired by those in [31]. The recently proposed "CoreID" in [222] is similar to the CUR factorizations of [29]. In all of these methods, the entries of the associated factor matrices are computed based on the specific form of the factorization and on the skeletons selected for the core tensor, but the factor matrices are not themselves submatrices of unfoldings.

The second category above comprises those tensor ID/CUR factorizations in which the entries of the factor matrices are entries of the input tensor. The early algorithm of [77] to compute a rank-$(R_1, \ldots, R_d)$ "approximate tensor SVD" forms the factor matrices in a Tucker decomposition as randomly sampled columns of unfoldings; this factorization would later be termed a "higher-order ID" in [186], where the mode-j factor matrix in a Tucker decomposition is computed as column skeletons of the mode-j unfolding selected by greedy pivoting. We detail several recent algorithms that leverage randomization to compute tensor ID/CUR approximations efficiently in Sect. 6.

6 Randomized Low-Rank Tensor Decompositions

As in the matrix setting, we are interested in algorithms that compute low-rank tensor decompositions accurately and efficiently, with provable performance guarantees, and with demonstrated high practical speed. Because many tensor algorithms fundamentally rely on classical matrix algorithms, we can frequently fall back on the ideas introduced in Sect. 4 to leverage randomization for efficient tensor decompositions. However, given the rapidly increasing number of applications that involve large-scale tensorial data, it is often impractical or infeasible to rely on basic implementations of the randomized rangefinder, e.g., with Gaussian sketching.

In this section, we survey new developments in fast randomized algorithms for low-rank tensor decompositions in the CP and Tucker formats. Sect. 6.1 explores how CP-ALS can be accelerated with randomization, specifically through fast sampling methods that exploit the structure of the overdetermined least squares problems. Sect. 6.2 summarizes randomized methods to compute Tucker decompositions with orthogonal factors, which naturally extend the randomized rangefinder to the tensor setting. These algorithms are contrasted with those for tensor ID/CUR decompositions in Sect. 6.3, which includes both the CP and Tucker formats. Our goal is to broadly survey relevant literature and detail recent developments that may stimulate future work.

6.1 Randomized Algorithms for CP Decomposition

Over the last decade, many randomized algorithms have been proposed to compute CP decompositions more efficiently. We survey several notable algorithms, before summarizing the recent randomized CP-ALS procedures of [134] and [21], which have high-performance implementations in [20].

Many randomized algorithms for CP decompositions seek to exploit the structure of the Khatri–Rao product (KRP) in CP-ALS, cf. (41), for fast sketching and sampling. In [217], the input tensor is sketched without being explicitly formed, then the KRP is computed. The "randomized ALS" algorithm of [181] incorporates projections onto random tensors to improve the conditioning of the least squares problems in CP-ALS. A procedure to sample rows of the KRP near-optimally via fast leverage score approximations is presented in [50]. In [17], a Kronecker fast Johnson-Lindenstrauss transform (KFJLT, cf. [116]) is applied to the KRP to reduce the coherence for better sampling performance. The algorithm of [3] builds on [17] with options for adaptive sketch sizes and regularization terms. The CP-ARLS-LEV algorithm of [134] samples the KRP using an approximate leverage score distribution. In [144], the TNS-CP algorithm samples from the exact leverage score distribution and improves on the sampling complexity of CP-ARLS-LEV by avoiding exponential dependence on d through recursive sketching [4], though it is still more computationally expensive than the method of [21], known as STS-CP, which we detail in this section.

Before we discuss algorithmic details, we briefly summarize the problem. Recall from Sect. 5.3.1 that within each iteration of CP-ALS (Algorithm 5), we compute a rank-R CP decomposition that approximates our input tensor by solving d least squares problems in succession. More precisely, given input tensor $\mathcal{X} \in \mathbb{R}^{N_1 \times \cdots \times N_d}$, in each iteration of CP-ALS, we solve a least squares problem in each mode $j = 1, \ldots, d$, in which every factor matrix except $\mathbf{A}^{(j)}$ is fixed, and $\mathbf{A}^{(j)}$ is computed as the solution of

$$\min_{\widehat{\mathbf{A}} \in \mathbb{F}^{N_j \times R}} \left\| \widehat{\mathbf{A}} \mathbf{Z}^* - \mathbf{X}_{(j)} \right\|_{\mathrm{F}}^2 = \min_{\widehat{\mathbf{A}} \in \mathbb{F}^{N_j \times R}} \left\| \mathbf{Z} \widehat{\mathbf{A}}^* - \mathbf{X}_{(j)}^* \right\|_{\mathrm{F}}^2, \tag{45}$$

where $\mathbf{Z} = \mathbf{A}^{(d)} \odot \cdots \odot \mathbf{A}^{(j+1)} \odot \mathbf{A}^{(j-1)} \odot \cdots \odot \mathbf{A}^{(1)} \in \mathbb{R}^{N^{(-j)} \times R}$.

If we solve (45) without exploiting any inherent problem structure, the cost of computing the QR decomposition of $\mathbf{Z}$ is $O(R^2 N^{(-j)})$, and the cost of applying it N_j times brings the total cost to $O(R^2 N^{(-j)} + RN)$.

6.1.1 Randomized CP-ALS with Approximate Leverage Score Sampling

In [134], the CP-ARLS-LEV algorithm is developed to accelerate each least squares sub-problem (45). Namely, $S \ll N^{(-j)}$ rows of $\mathbf{Z}$ are sampled according to an approximate leverage score distribution. The corresponding S rows of $\mathbf{X}_{(j)}^*$ are then selected for a smaller "sketched" problem, of which the solution approximates the solution of (45). Sketching reduces the cost of solving the jth least squares problem to $O(R^2 S + RSN_j)$. However, forming $\mathbf{Z}$ or $\mathbf{X}$ and computing leverage scores would incur a cost of $O(R^2 N^{(-j)})$. To avoid this, in CP-ARLS-LEV, the leverage scores are estimated as the product of leverage scores of the respective factor matrices in the KRP, and only the S sampled rows of $\mathbf{Z}$ are formed explicitly.

To this end, let $j \in [d]$ be fixed. We first observe that there is a bijection mapping each row $i \in [N^{(-j)}]$ of $\mathbf{Z}$ to a $(d-1)$-tuple $(i_d, \ldots, i_{j+1}, i_{j-1}, \ldots, i_1)$, with $i_k \in [N_k]$, $k \neq j$, which indexes rows of the factor matrices in the KRP:

$$\mathbf{Z}(i,:) = \mathbf{A}^{(d)}(i_d,:) * \cdots * \mathbf{A}^{(j+1)}(i_{j+1},:) * \mathbf{A}^{(j-1)}(i_{j-1},:) * \cdots * \mathbf{A}^{(1)}(i_1,:). \quad (46)$$

Using [50, Theorem 3.3], the leverage scores of the rows of $\mathbf{Z}$ are bounded above by the product of the leverage scores of the rows of its constituent factor matrices:

$$\ell_i(\mathbf{Z}) \leq \hat{\ell}_i(\mathbf{Z}) := \prod_{k \neq j} \ell_{i_k}\left(\mathbf{A}^{(k)}\right). \quad (47)$$

From (47), we define a probability distribution on the rows of $\mathbf{Z}$:

$$p_i = \hat{\ell}_i(\mathbf{Z}) / R^{d-1}. \quad (48)$$

Computing the probabilities in (48) explicitly would cost $O(N^{(-j)})$ operations. However, [134, Lemma 9] establishes that each i_k in the $(d-1)$-tuple can be sampled independently, using the leverage scores of the rows of $\mathbf{A}^{(k)}$, $k \neq j$, which cost $O(N_k R^2)$ to compute. This gives a multi-index corresponding to the i-th row of $\mathbf{Z}$ without computing the Kronecker product of leverage scores, and $\mathbf{Z}(i,:)$ can be formed in $O(R(d-1))$ operations using (46).

There are additional modifications that can be made to CP-ARLS-LEV to further improve performance, cf. [134], and a distributed-memory implementation of CP-ARLS-LEV is developed in [20], along with a distributed-memory version of an algorithm that performs random walks on a binary tree for fast exact leverage score estimation instead, which we describe next.

6.1.2 Randomized CP-ALS with Fast Exact Leverage Score Sampling

The recent work of [21] provides a fast algorithm to exactly sample from the leverage score distribution of the KRP. This work builds upon the leverage score computations in [141, 185], as well as the approximate ridge leverage score sampling method of [218], which uses recursive sketching with CountSketch and Sparse Sign operators. Namely, the algorithm of [21], known as STS-CP, samples a row index from the KRP by sampling a row from each factor matrix in sequence and computing Hadamard products, with each subsequent draw conditioned on prior ones. Broadly speaking, the goal is to build an "easy-to-update" data structure in $O(\sum_{j=1}^{d-1} N_j R^2)$ time with $O(\sum_{j=1}^{d-1} N_j R)$ storage requirements that produces S samples from the KRP according to the exact leverage score distribution in $O(dR^3 + S\sum_{j=1}^{d-1} R^2 \log \max(N_j, R))$ time, with $O(R^3)$ intermediate storage requirements (cf. [21, Theorem 1.1]).

Recall that given a matrix $\mathbf{A} \in \mathbb{R}^{I\times R}$, we can compute the leverage score of row i via

$$\ell_i = \mathbf{A}(i,:)\left(\mathbf{A}^\top \mathbf{A}\right)^\dagger \mathbf{A}(i,:)^\top. \tag{49}$$

Without loss of generality, let $\mathbf{Z} = \mathbf{A}^{(1)} \circ \cdots \circ \mathbf{A}^{(d-1)}$ for notational convenience. We again index the rows of $\mathbf{Z}$ by the $(d-1)$-tuple $(i_1,\ldots,i_{d-1})$, with $i_k \in [N_k]$, $k \neq d$. Using (49), we compute the leverage scores of $\mathbf{Z}$ via

$$\ell_{i_1,\ldots,i_{d-1}} = \mathbf{Z}((i_1,\ldots,i_{d-1}),:)(\mathbf{Z}^\top\mathbf{Z})^\dagger\mathbf{Z}((i_1,\ldots,i_{d-1}),:)^\top.$$

Let

$$\mathbf{G}_j = (\mathbf{A}^{(j)})^\top\mathbf{A}^{(j)} \in \mathbb{R}^{R\times R}$$

for $j = 1,\ldots,d-1$, and define

$$\mathbf{G} = \mathbf{G}_1 * \cdots * \mathbf{G}_{d-1} \in \mathbb{R}^{R\times R},$$

so that $\mathbf{G} = \mathbf{Z}^\top\mathbf{Z}$. Note that $\mathbf{G}$ is positive semi-definite. To define probabilities, we use a generalized inner product given by $\langle \mathbf{A},\mathbf{B},\mathbf{C}\rangle = \sum_{i=1,j=1}^{m,n} \mathbf{A}(i,j)\mathbf{B}(i,j)\mathbf{C}(i,j)$ for $\mathbf{A},\mathbf{B},\mathbf{C} \in \mathbb{R}^{m\times n}$, or the sum of all entries in the Hadamard product.

Now, assume we have already sampled an index from each of $[N_1],\ldots,[N_{j-1}]$ for some $j = 1,\ldots,d-1$, denoted by $s_1,\ldots,s_{j-1}$. Let $\hat{s}_j$ be a random variable for a draw from $[N_j]$, with $\hat{s}_1 = s_1,\ldots,\hat{s}_{j-1} = s_{j-1}$; we abbreviate this condition by $\hat{s}_{<j} = s_{<j}$. Then to sample the next index s_j from $[N_j]$ conditioned on prior draws, we need the probability of $\hat{s}_j$ given $\hat{s}_{<j}$. By [21, Theorem 3.1], [141],

$$\begin{aligned} p(\hat{s}_j = s_j|\hat{s}_{<j} = s_{<j}) &= C^{-1}\langle h_{<j}h_{<j}^\top, \mathbf{A}^{(j)}(s_j,:)^\top\mathbf{A}^{(j)}(s_j,:), \mathbf{G}_{>j}\rangle \\ &=: q_{h_{<j},\mathbf{A}^{(j)},\mathbf{G}_{>j}}[s_j], \end{aligned}$$

where

$$\begin{aligned} C &:= \langle h_{<j}h_{<j}^\top, \mathbf{A}^{(j)\top}\mathbf{A}^{(j)}, \mathbf{G}_{>j}\rangle \text{ is nonzero,} \\ h_{<j} &:= \mathbf{A}^{(1)}(s_1,:)^\top * \ldots * \mathbf{A}^{(j-1)}(s_{j-1},:)^\top, \\ \mathbf{G}_{>j} &:= \mathbf{G}^\dagger * \mathbf{G}_{j+1} * \cdots * \mathbf{G}_d. \end{aligned}$$

Forming the data structure that samples efficiently from this distribution is reminiscent of the binary-tree inversion sampling scheme in [185], which we briefly describe.

For notational simplicity, let $q := q_{h_{<j},\mathbf{A}^{(j)},\mathbf{G}_{>j}}$ and $h := h_{<j}$. We partition the interval $[0,1]$ into N_j bins, where the i-th bin has width $q[i]$ for $i \in [N_j]$. We sample $\ell \sim \text{Uniform}[0,1]$ and return the index of the bin that contains it by performing a binary search. To control sampling complexity and space usage, the binary search terminates when a certain (prescribed) number of bins F remains in the search space. If the search reaches a node v of the tree with left child $L(v)$ containing a contiguous subset $I \subset [N_j]$, and the remaining search space is $[a,b] \subset [0,1]$, then the search branches left iff

$$\ell < a + \sum_{i \in I} q[i] = a + C^{-1}\langle hh^\top, \sum_{i \in I} \mathbf{A}^{(j)}(i,:)^\top \mathbf{A}^{(j)}(i,:), \mathbf{G}_{>j}\rangle. \tag{50}$$

Computing and caching $\sum_{i \in I} \mathbf{A}^{(j)}(i,:)^\top \mathbf{A}^{(j)}(i,:)$ for every node costs $O(IR^2)$ in time and $O(R^2\lceil I/F \rceil)$ in storage. Subsequent binary searches to evaluate (50) have runtime $O(R^2)$ for each of the $\log\lceil I/F \rceil$ internal tree nodes, and $O(FR^2)$ to evaluate q at the F indices in each leaf node. If, in addition, $\mathbf{G}_{>j}$ is rank-one, the cost further reduces to $O(FR)$; cf. [21, Appendix A.5]

Now, we note that in general, $\mathbf{G}_{>j}$ is not rank-one, but we can still enjoy sampling cost reduction by first computing a 1-dimensional subspace spanned by an eigenvector of $\mathbf{G}_{>j}$, then projecting relevant vectors onto it. Let $\mathbf{G}_{>j} = \mathbf{V}\mathbf{\Lambda}\mathbf{V}^\top$ be a symmetric eigenvalue decomposition, where columns of $\mathbf{V}$ are eigenvectors of $\mathbf{G}_{>j}$ and $\mathbf{\Lambda} = \mathrm{diag}((\lambda_\ell)_{\ell=1}^R)$ are eigenvalues. Then q can be written as

$$q[i] = C^{-1} \sum_{\ell=1}^{R} \lambda_\ell \langle hh^\top, \mathbf{A}^{(j)}(i,:)^\top \mathbf{A}^{(j)}(i,:), \mathbf{V}(:,\ell)\mathbf{V}(:,\ell)^\top\rangle. \tag{51}$$

Define $\mathbf{W} \in \mathbb{R}^{N_j \times R}$ by $\mathbf{W}(i,\ell) = \langle hh^\top, \mathbf{A}^{(j)}(i,:)^\top \mathbf{A}^{(j)}(i,:), \mathbf{V}(:,\ell)\mathbf{V}(:,\ell)^\top\rangle$, which has non-negative entries. Since $\lambda_\ell \geq 0$, we can express (51) in terms of probability mixtures defined by the columns of $\mathbf{W}$:

$$q = \sum_{\ell=1}^{R} w[\ell] \frac{\mathbf{W}(:,\ell)}{\|\mathbf{W}(:,\ell)\|_1}, \tag{52}$$

$$w[\ell] := C^{-1}\lambda_\ell \|\mathbf{W}(:,\ell)\|_1. \tag{53}$$

The expression in (52)-(53) reveals a sampling procedure that achieves the desired runtime and storage complexity using [21, Lemma 3.2]:

- Sample ℓ according to $p(\hat{\ell} = \ell) = C^{-1}\lambda_\ell \sum_{i=1}^{N_j} \mathbf{W}(i,\ell)$
- Sample s_j according to $p(\hat{s}_j = s_j | \hat{\ell} = \ell) := \mathbf{W}(s_j,\ell)/\|\mathbf{W}(:,\ell)\|_1$.

6.2 Randomized Algorithms for Tucker Decompositions

The Tucker decomposition provides a natural framework for data compression. As randomized algorithms, particularly the randomized SVD (Algorithm 3), have become mainstays for fast and reliable matrix compression, it is natural to consider how randomization can be leveraged for fast and reliable tensor compression.

Randomized algorithms for Tucker decompositions have been investigated in many problem settings, generally involving either randomized sampling or sketching. One of the earliest works on randomized HOSVD and HOOI algorithms is [205], called the MACH-HOSVD and MACH-HOOI. In MACH-HOSVD and MACH-HOOI, the input tensor is sparsified according to a coin flip for each nonzero entry; with probability p, a nonzero entry is kept and reweighted by $1/p$, and with probability $1-p$, the entry is set to 0. Leverage score sampling of matrix unfoldings to sparsify the tensor is explored in [175], and Tucker decompositions of sub-sampled tensors are also investigated in [162, 214]. A detailed survey of these methods can be found in [5].

As the randomized rangefinder gained traction in the numerical linear algebra community, many investigations began to explore randomized sketching for fast Tucker decompositions. In [223], the RandTucker algorithm is presented, which relies on the randomized rangefinder to compute an HOSVD from Gaussian sketches of mode unfoldings. Formal analysis of RandTucker is performed in [86], which incorporates power iteration into RandTucker for faster spectral decay; it is noted, though, that uniform sampling instead of Gaussian sketching performs similarly and may be preferable for large problems. Power iteration and randomized sketching are utilized in the related work of [41, 42] for randomized HOSVD and HOOI, which build on the rank-adaptive algorithm of [40] that computes a basis for the range of each mode unfolding with ε-accuracy.

In [193], a single-pass algorithm based on the HOSVD is developed that significantly improves on the cost of storing the random sketching matrices, via Khatri–Rao products of random dimension reduction maps (called tensor random projections); details can also be found in [204]. The Sketch-STHOSVD algorithm of [69] is inspired by [193] but incorporates power iteration, whereas the Tucker-TS and Tucker-TTMTS algorithms of [142] are single-pass variants of HOOI using TensorSketch [166]. Cost-effective sketching for Tucker decompositions is also investigated in [28], which replaces the randomized SVD step in computing approximate bases with generalized Nyström. In the recent work of [104], the Randomized Tucker with Single-Mode Sketching (RTSMS) algorithm is introduced as a rank-adaptive method that sequentially builds approximate bases of unfoldings by applying small random sketches one mode at a time, similar to the randomized ST-HOSVD of [153], which we describe next.

6.2.1 Randomized (ST-)HOSVD

The randomized HOSVD is a natural extension of the HOSVD, where we substitute the SVD of each unfolding with the randomized SVD (Algorithm 3). The R-HOSVD algorithm of [153] is precisely Algorithm 6, but with a sketch instead of the unfolding.

If $\widehat{\mathcal{X}}$ is the output of the randomized HOSVD, with Gaussian sketching of each mode-j unfolding, the expected error is within $\approx \sqrt{d(k+1)}$ of the optimal rank-$(k,\ldots,k)$ Tucker approximation error, cf. [153, Theorem 3.1]. Letting n be the uniform mode size again, the computational cost of R-STHOSVD is $O(\sum_{j=1}^{d} k^j n^{d-j+1} + k^j n^{d-j})$, where the first term is the cost of computing SVDs of mode unfoldings and the second term is the cost of forming the core tensor. By contrast, the R-HOSVD algorithm costs $O(dkn^d + \sum_{j=1}^{d} k^j n^{d-j+1})$ operations. In other words, the cost is reduced roughly by a factor of n with Gaussian sketching. While the resulting error does depend on processing order, the worst-case expected error does not, cf. [153, Sect. 3.3]. For more software implementation details, we refer to [213].

6.2.2 Randomized HOOI

Recall from Section 5.3.2 that HOOI computes Tucker decompositions iteratively by solving least-squares problems in each mode for the factor matrix, generally more accurate than the HOSVD but more expensive. Similar to our goal in CP-ALS, for a given input $\mathcal{X} \in \mathbb{F}^{N_1 \times \cdots \times N_d}$, we are seeking $\widehat{\mathcal{X}} = \mathcal{G} \times_1 \mathbf{A}^{(1)} \cdots \times_d \mathbf{A}^{(d)}$ that minimizes

$$\min_{\mathcal{G}, \mathbf{A}^{(1)}, \ldots, \mathbf{A}^{(d)}} \|\mathcal{X} - \mathcal{G} \times_1 \mathbf{A}^{(1)} \cdots \times_d \mathbf{A}^{(d)}\|,$$

where $\mathcal{G} \in \mathbb{F}^{R_1 \times \cdots \times R_d}$ and $\mathbf{A}^{(j)} \in \mathbb{F}^{N_j \times R_j}$. We can re-write the HOOI procedure in Algorithm 7 as two main steps that are repeated until convergence; cf. [142]:

1. For $j = 1, \ldots, d$, solve $\mathbf{A}^{(j)} = \underset{\mathbf{A} \in \mathbb{F}^{N_j \times R_j}}{\arg\min} \|\mathbf{Z}^{(j)} \mathbf{G}^*_{(j)} \mathbf{A}^* - \mathbf{X}^*_{(j)}\|^2$, where

$$\mathbf{Z}^{(j)} = \mathbf{A}^{(d)} \otimes \cdots \otimes \mathbf{A}^{(j+1)} \otimes \mathbf{A}^{(j-1)} \otimes \cdots \otimes \mathbf{A}^{(1)},$$

2. Update $\mathcal{G} = \underset{\mathcal{Y} \in \mathbb{F}^{R_1 \times R_d}}{\arg\min} \|\mathbf{Z}\mathbf{y} - \mathbf{x}\|^2$, where $\mathbf{Z} = \mathbf{A}^{(d)} \otimes \cdots \mathbf{A}^{(1)}$, and $\mathbf{y}, \mathbf{x}$ are (column) vectorizations of $\mathcal{Y}$, $\mathcal{X}$, respectively.

We observe that each of these two steps involves large overdetermined least squares problems, and similar to the strategy in CP-ARLS-LEV, we will exploit the matrix product structures to solve smaller sketched least squares problems efficiently, now with TensorSketch.

TensorSketch TensorSketch [166, 178, 11, 68] is a specialized version of CountSketch. A CountSketch operator can be defined as a linear map $\mathbf{S} : \mathbb{R}^I \to \mathbb{R}^J$ given by $\mathbf{S} = \mathbf{PD}$, where $\mathbf{P} \in \mathbb{R}^{J \times I}$ is a matrix with $\mathbf{P}(h(i), i) = 1$ and all other entries zero, for a random map $h : [I] \to [J]$ such that $\mathbb{P}[h(i) = j] = 1/J$ for all $i \in [I]$, $j \in [J]$. The matrix $\mathbf{D} \in \mathbb{R}^{I \times I}$ is a diagonal matrix of ± 1 with equal probability.

The CountSketch operator $\mathbf{S}$ can be rapidly applied to a matrix $\mathbf{A}$ in $O(\text{nnz}(\mathbf{A}))$ operations, without explicitly forming $\mathbf{S}$ or $\mathbf{A}$. If $\mathbf{S} \in \mathbb{R}^{J \times I}$ and $\mathbf{A} \in \mathbb{F}^{I \times L}$, typically $J \ll I$, we form $\mathbf{Y} = \mathbf{SA}$ by hashing each row i of $\mathbf{A}$ with an integer h sampled uniformly from $[J]$, assigning it a value s of ± 1 with equal probability, then summing the rows with the same hash value (i.e. adding $s\mathbf{A}_{i,:}$ to the hth row of $\mathbf{Y}$ if $\mathbf{A}_{i,:}$ has hash value h).

A TensorSketch operator is a specific type of CountSketch operator, which can be rapidly applied to Kronecker products, e.g., $\mathbf{Z} = \mathbf{A}^{(1)} \otimes \cdots \otimes \mathbf{A}^{(d)} \in \mathbb{F}^{n^d \times k^d}$. If $\mathbf{T} : \mathbb{R}^{n^d} \to \mathbb{R}^{\ell}$ is a TensorSketch operator, then the cost of sketching $\mathbf{TZ}$ is $O(\ell dnk + \ell k^d)$, suppressing log factors, instead of $O(\ell n^d k^d)$ for naive multiplication. This is accomplished by sketching each factor matrix with independent CountSketch operators, then convolving with Fast Fourier Transforms (FFTs).

Let $h_j : [n] \to [\ell]$ for $j = 1, \ldots, d$ be 3-wise independent hash functions, i.e. for any distinct $i_1, i_2, i_3 \in [n]$, the hash codes $h_j(i_1), h_j(i_2), h_j(i_3)$ are Uniform$\{[\ell]\}$. Let $s_j : [n] \to \{+1, -1\}$ be 4-wise independent sign functions (i.e. Uniform$\{\pm 1\}$). We form independent CountSketch matrices $\mathbf{S}^{(j)} = \mathbf{P}^{(j)} \mathbf{D}^{(j)}$ for $j = 1, \ldots, d$, using the hash functions h_j to form $\mathbf{P}^{(j)}$ and $s_j(i) = \mathbf{D}^{(j)}_{i,i}$ for $i \in [n]$. Applying $\mathbf{S}^{(j)}$ to each column $\mathbf{A}^{(j)}_{:,r_j}$, for $r_j = 1, \ldots, k$, can be represented by an $(\ell - 1)$-degree polynomial

$$\mathcal{P}^{(j)}_{r_j}(\omega) = \sum_{i=1}^{n} s_j(i)\mathbf{A}^{(j)}_{i,r_j}\omega^{h_j(i)-1} =: \sum_{l=1}^{\ell} c^{(j)}_{l,r_j}\omega^{j-1}, \tag{54}$$

where $\mathbf{c}^{(j)}_{r_j}$ are coefficients $(c^{(j)}_{1,r_j}, \ldots, c^{(j)}_{\ell,r_j})$, grouped in (54) by hash value.

The TensorSketch operator $\mathbf{T}$ is defined as the CountSketch operator comprised of hash function H and sign function S given by

$$H : [n]^d \to \ell, \;\; (i_1, \ldots, i_d) \mapsto \left(\sum_{j=1}^{d}(h_j(i_j) - 1) \mod \ell\right) + 1,$$

$$S : [n]^d \to \{+1, -1\}, \;\; (i_1, \ldots, i_d) \mapsto \prod_{j=1}^{d} s_j(i_j).$$

The r-th column of $\mathbf{T}(\mathbf{A}^{(1)} \otimes \cdots \otimes \mathbf{A}^{(d)}) = \mathbf{TZ}$, $r = 1, \ldots, k^d$, is the polynomial

$$\begin{aligned}
\mathcal{P}_r(\omega) &= \sum_{i=1}^{n^d} S(i_1, \ldots, i_d)\mathbf{Z}_{i,r}\omega^{H(i_1,\ldots,i_d)}, \\
&= \sum_{i=1}^{n^d} s_1(i_1)\cdots s_d(i_d)\mathbf{A}^{(1)}_{i_1,r_1}\cdots\mathbf{A}^{(d)}_{i_d,r_d}\omega^{(h_1(i_1)+\cdots+h_d(i_d)-d) \mod \ell} \\
&= \mathrm{FFT}^{-1}\left(\mathrm{FFT}(\mathbf{c}^{(1)}_{r_1}) * \cdots * \mathrm{FFT}(\mathbf{c}^{(d)}_{r_d})\right),
\end{aligned} \tag{55}$$

where $*$ denotes the Hadamard product, i in (55) corresponds to the tuple $(i_1, \ldots, i_d) \in [n]^d$, and r corresponds to the tuple $(r_1, \ldots, r_d) \in [k]^d$. In other words,

$$\mathbf{TZ} = \mathrm{FFT}^{-1}\left(\left(\bigodot_{j=1}^{d}\left(\mathrm{FFT}\left(\mathbf{S}^{(j)}\mathbf{A}^{(j)}\right)\right)^*\right)^*\right), \tag{56}$$

where $\odot$ is the KRP, and the FFT is applied column-wise. Instead of the usual $O(\ell n^d k^d)$ cost of matrix multiplication, the cost of $\mathbf{TZ}$ is $O(\ell dnk + \ell k^d \log(\ell))$ using (56).

Using TensorSketch in HOOI We now turn our attention back to the randomized HOOI algorithm of [142], called TUCKER-TS, which uses TensorSketch in each least squares subproblem in the two main steps above. In Step 1, the factor matrix sub-problems for modes $j = 1, \ldots, d$ are highly overdetermined, involving an $n^{d-1} \times k^{d-1}$ Kronecker product $\mathbf{Z}^{(j)}$. The core tensor sub-problem in Step 2 is also highly overdetermined, involving a $n^d \times k^d$ Kronecker product. For both steps, we form independent TensorSketches to solve smaller sketched least squares problems, as in Sect. 3.3. TUCKER-TS can also be made single-pass; cf. [142, Algorithm S3].

The dominant cost of randomized HOOI with TensorSketch (TUCKER-TS) is $O(dnk^2 + \ell dk^d + \ell_c k^d + k^{2d})$, or roughly $k^{O(d)}$, versus the dominant cost of $O(n^d)$ for classical HOOI (Algorithm 7), suppressing log factors and letting $\ell = \ell_1 = \cdots = \ell_d$. (Their numerical results suggest choosing $\ell = Ck^{d-1}$ and $\ell_c = Ck^d$ for constant $C > 4$, e.g., $C = 10$.) See [142, S3.2.6] for details on the complexity analysis. We note that the single-pass algorithm presented in [142] uses the same TensorSketch operators on the updated Kronecker products until some convergence criteria are met. For the theoretical guarantees of [68] to hold, it is technically necessary to form new independent TensorSketch operators in each iteration; however, in practice, it is noted in [142] that re-using TensorSketch operators is more efficient and works just as well.

6.2.3 Randomized Tucker with Single-Mode Sketching

We end this section by highlighting the recent work of [104] on Randomized Tucker decompositions with Single-Mode Sketching (RTSMS). The RTSMS algorithm sketches the input tensor one mode at a time, i.e. sketching on the left of the mode-j unfolding vs. sketching on the right, to avoid the computational bottleneck of multiplying along the larger dimension $N^{(-j)} \equiv n^{d-1}$. The algorithm is also single-pass and rank-adaptive.

To illustrate, suppose we want to compute a rank-(R_1, R_2, R_3) HOSVD approximating $\boldsymbol{\mathcal{X}} \in \mathbb{F}^{N_1 \times N_2 \times N_3}$. (We omit rank-adaptivity and iterative refinement; see [104] for practical implementation details.) We initialize tensors

$$\boldsymbol{\mathcal{G}}^{\text{old}} = \boldsymbol{\mathcal{X}} \text{ and } \boldsymbol{\mathcal{G}}^{\text{new}} = \boldsymbol{\mathcal{X}} \times_1 \mathbf{\Omega}^{(1)} \iff \mathbf{G}^{\text{new}}_{(1)} = \left(\mathbf{\Omega}^{(1)}\right)^* \mathbf{X}_{(1)} \in \mathbb{F}^{\tilde{R}_1 \times (N_2 N_3)}$$

for some sketching dimension $\tilde{R}_1 > R_1$. There exists a factor matrix $\mathbf{A}^{(1)} \in \mathbb{F}^{N_1 \times \tilde{R}_1}$ such that $\boldsymbol{\mathcal{G}}^{\text{new}} \approx \boldsymbol{\mathcal{G}}^{\text{old}} \times_1 \mathbf{A}^{(1)}$. In practice, this is computed as an approximate solution of the corresponding least squares problem, which is being sub-sampled from the right, cf. [104, Section 4.2]. We update $\boldsymbol{\mathcal{G}}^{\text{old}} = \boldsymbol{\mathcal{G}}^{\text{new}}$, and for the next mode, update $\boldsymbol{\mathcal{G}}^{\text{new}}$ via

$$\mathbf{G}^{\text{new}}_{(2)} = \left(\mathbf{\Omega}^{(2)}\right)^* \mathbf{G}^{\text{old}}_{(2)} \in \mathbb{F}^{\tilde{R}_2 \times (\tilde{R}_1 N_3)}$$

for some sketching dimension $\tilde{R}_2 > R_2$. To compute $\mathbf{A}^{(2)}$, we solve a smaller sketched least squares problem of size $\tilde{R}_2 \times (\tilde{R}_1 N_3)$, and repeat this for the final mode.

The computational complexity of RTSMS is dominated by the first TTM $\boldsymbol{\mathcal{X}} \times_1 \mathbf{\Omega}_1$. The dominant cost is $O(kn^d)$ to compute the first TTM, but the sketch size is only $k \times n$ vs. $k \times n^{d-1}$ for R-HOSVD. The least squares solves are also significantly reduced by sub-sampling and by computing a generalized Nyström decomposition vs. SVD. The full tensor does not need to be sketched again in later modes, and the core tensor is available immediately upon termination. Additional details can be found in [104].

6.3 Randomized Algorithms for Tensor ID/CUR Decompositions

As we observed in Sect. 5.3.4, the tensor ID/CUR decomposition has been formulated in many ways in the literature, with no one standard representation. However, there exists a natural categorization of tensor ID/CUR decompositions, based on whether the factor matrices or core tensors in the computed decompositions are comprised of entries of the input tensor. We discuss randomized algorithms to compute a Tucker ID/CUR decomposition from each category, as well as recent work on CP tensor ID/CUR.

6.3.1 Randomized Algorithms for Tucker Core ID/CUR

We first consider randomized algorithms for Tucker tensor ID/CUR factorizations of the first type, where the core tensor is a sub-tensor of the input tensor. In particular, we focus on the tensor ID/CUR decompositions that were recently characterized in [29], termed the "Chidori" and "Fiber" tensor CUR factorizations. This work was inspired by numerous investigations of tensor CUR decompositions, including [31, 77, 140].

In [29], given $\mathcal{X} \in \mathbb{F}^{N_1 \times \cdots \times N_d}$, the rank-$(R_1, \ldots, R_d)$ tensor CUR factorization is

$$\widehat{\mathcal{X}} = \mathcal{G} \times_1 (\mathbf{C}_1 \mathbf{U}_1^{\dagger}) \times_2 (\mathbf{C}_2 \mathbf{U}_2^{\dagger}) \times \cdots \times_d (\mathbf{C}_d \mathbf{U}_d^{\dagger}), \tag{57}$$

where the rank-$(R_1, \ldots, R_d)$ core tensor is given by

$$\mathcal{G} = \mathcal{X}(I_1, \ldots, I_d),$$

for index sets $I_j \subset [N_j]$, $j = 1, \ldots, d$. The mode-j factor matrix $\mathbf{C}_j \mathbf{U}_j^{\dagger}$ is defined by

$$\begin{aligned} \mathbf{C}_j &= \mathbf{X}_{(j)}(:, J_j), \quad \text{and} \\ \mathbf{U}_j &= \mathbf{C}_j(I_j, :), \end{aligned}$$

where $J_j \subset [N^{(-j)}]$. If indices J_j are computed independently from indices I_j, the decomposition in (57) is called a "Fiber" CUR decomposition in [29]. If $J_j = \otimes_{k \neq j} I_k$, then (57) is called a "Chidori" CUR decomposition.

By [29, Theorem 3.3], [101, Corollary 5.2], if the indices I_j and J_j are each sampled uniformly with $|I_j| = O(R_j \log N_j)$ and $|J_j| = O(R_j \log N^{(-j)})$, then the computed Tucker decomposition is a rank-$(R_1, \ldots, R_d)$ approximation with high probability, under certain incoherence assumptions on the input tensor; cf. [70]. With uniform sampling, the computational complexity of computing either CUR decomposition is dominated by the cost of $\mathbf{U}_j^{\dagger}$, which is $O(k^{d+1} \log^d(n))$ for Chidori and $O(k^2 \log^2(n))$ for Fiber, letting $n = N_1 = \cdots = N_d$ and $k = R_1 = \cdots R_d$ as before.

The recent work of [222] proposes a similar factorization to (57), called the "CoreID." The CoreID algorithm is reminiscent of the "structure-preserving" ST-HOSVD developed for sparse input tensors in [153]. Namely, in the CoreID, as each factor matrix is computed, the input tensor is compressed along that mode before proceeding to the next one, so that each subsequent unfolding is reduced in size. Through matricization, computing the CoreID is reduced to computing a randomized matrix row ID in each mode, cf. Algorithm 4. For $j = 1, \ldots, d$, the mode-j unfolding of the tensor is sketched by a random matrix $\mathbf{Y}_j = \mathbf{X}_{(j)}\mathbf{\Omega}_j$. Then a row ID of $\mathbf{Y}_j$ is computed by some skeleton selection method. The selected skeletons I_j form the core sub-tensor, and the mode-j factor matrix for the CoreID is defined as the associated interpolation matrix. The tensor is then compressed along mode j before proceeding to the next. Sketching is the dominant cost, so many of the numerical experiments in [222] consider structured input tensors for fast sketching, e.g., a sparse input tensor or a CP-factorized input tensor.

6.3.2 Randomized Algorithms for Tucker Factor ID/CUR

We now turn our attention to randomized algorithms for tensor ID/CUR factorizations in the Tucker format in which the factor matrices contain entries of the input tensor. We detail the "higher-order ID" of [186], contrasted with the "SatID" of [222].

For a given $\mathcal{X} \in \mathbb{F}^{N_1 \times \cdots \times N_d}$, the higher-order ID (HOID) of [186] is given by

$$\widehat{\mathcal{X}} = \mathcal{G} \times_1 \mathbf{C}_1 \times_2 \mathbf{C}_2 \times \cdots \times_d \mathbf{C}_d, \tag{58}$$

where the rank-$(R_1, \ldots, R_d)$ core tensor $\mathcal{G}$ is defined as

$$\mathcal{G} = \mathcal{X} \times_1 \mathbf{C}_1^\dagger \times_2 \mathbf{C}_2^\dagger \times \cdots \times_d \mathbf{C}_d^\dagger, \tag{59}$$

and each factor $\mathbf{C}_j \in \mathbb{F}^{N_j \times R_j}$ consists of column skeletons of the mode-j unfolding $\mathbf{X}_{(j)}$. The core tensor in (59) is optimal for the factorization in (58), cf. [58]. For the theoretical guarantees of [186], the strong rank-revealing QR algorithm of [97] is used for skeleton selection on the random sketch of $\mathbf{X}_{(j)}$. In practice, LUPP in the randomized matrix ID (i.e., Algorithm 4, with input $\mathbf{X}_{(j)}^*$) performs similarly and is less expensive.

For each mode j, the cost of sketching is $O(R_j N_j N^{(-j)}) \equiv O(kn^d)$. Skeleton selection as in [186] costs $O(R_j^2 N^{(-j)}) \equiv O(k^2 n^{d-1})$. The total cost is roughly $O(dkn^d + dk^{d-1}n^{d-1})$ after computing the core tensor. A sequentially-truncated version of the HOID (ST-HOID), analogous to the ST-HOSVD, is given in [186, Algorithm 4].

A very similar decomposition is developed in [222], termed the "SatID." For an n^d-sized tensor, a QR-based algorithm to compute skeletons for the mode-j factor matrix as in [186] requires n^{d-1} column norms, which is not optimal for structured tensors and can even have larger storage costs than the input. By contrast, the SatID algorithm is based on a randomized matrix ID with marginalized norm sampling, which can be viewed as an extension of the original tensor CUR of [77] that uses norm sampling.

We briefly summarize the marginalized norm sampling procedure. Recall that we can uniquely identify a column index $i \in [N^{(-j)}]$ of $\mathbf{X}_{(j)}$ with a $(d-1)$-tuple $(i_1, \ldots, i_{j-1}, i_{j+1}, \ldots, i_d)$. Similar to CP-ARLS-LEV in [134], the idea is to accelerate sampling of $i \in [N^{(-j)}]$, now based on column norm sampling of $\mathbf{X}_{(j)}$.

Suppose we have already selected some subset of column indices S of $\mathbf{X} = \mathbf{X}_{(j)}$ to determine the mode-j column skeletons comprising $\mathbf{C}_j$, and we want to select the next column. Norm sampling (cf. Sect. 3.2) requires the computation of column scores

$$d_i^{(S)} = \|\mathbf{X}_{:,i} - \mathbf{Q}_S\mathbf{Q}_S^*\mathbf{X}_{:,i}\|^2 = \min_{\mathbf{v}} \|\mathbf{X}_{:,S}\mathbf{v} - \mathbf{X}_{:,i}\|^2, \quad i \in [N^{(-j)}], \tag{60}$$

where $\mathbf{Q}_S$ is an orthonormal basis for the range of $\mathbf{X}_{:,S}$. If

$$X = (X_1, \ldots, X_{j-1}, X_{j+1}, \ldots, X_d)$$

is a random variable representing the next column index, from (60), $\mathbb{P}(X = i) \propto d_i^{(S)}$. The key observation is that

$$\mathbb{P}(X_1 = i_1) \propto \sum_{i_2,\ldots,i_d} d_{i_1,i_2,\ldots,i_d}^{(S)},$$

so that we can sample i_1 from its marginal distribution, then sample i_2 given i_1, and continue for all of $i = (i_1, i_2, \ldots, i_{j-1}, i_{j+1}, \ldots, i_d)$. This sum can be computed efficiently via the equation derived in [222, Eq. 18]: $\mathbb{P}(X_1 = i_1) \propto \|\widetilde{\mathbf{X}}_{:,i_1}\|^2$, where $\widetilde{\mathbf{X}} = \left(\mathcal{X} \times_j \mathbf{Q}_{S^c}\right)^*_{(1)}$ with $\mathbf{Q}_{S^c}$ an orthonormal basis for the range of $\mathbf{X}_{:,[N^{(-j)}]\setminus S}$. In other words, to sample i_1, we need only compute the N_1 column norms of $\widetilde{\mathbf{X}}$, which can be approximated efficiently using randomized sketching, cf. [222, Algorithm 6]. Once i_1 has been sampled, this process is repeated on the sliced tensor $\mathcal{X}(i_1, :, \ldots, :)$.

For structured inputs, this procedure can be modified for better performance. The total complexity (including randomized sketching) is $O(\mathrm{nnz}(\mathcal{X}) + \ell n^2 k^2)$ for a rank-$(k, \ldots, k)$ SatID of an n^d-sized tensor with sketch size ℓ; see [222, Algorithm 7].

6.3.3 Randomized Algorithms for CP Factor ID/CUR

We now consider tensor ID/CUR decompositions in the CP format, which necessarily fall into the category of decompositions in which the factor matrices contain entries of the input tensor. These decompositions can be interpreted as the result of pruning the input tensor, leaving only its most important rank-one components.

The seminal work of [22] introduces a method to obtain rank-k CP-tensor IDs from rank-R CP-factorized tensors, given by

$$\mathcal{X} = \sum_{r=1}^{R} \mathbf{a}_r^{(1)} \circ \cdots \circ \mathbf{a}_r^{(d)} \in \mathbb{F}^{N_1 \times \cdots \times N_d}. \tag{61}$$

To better illustrate the algorithm, we express (61) in a vectorized form, where $\mathbf{X}_r \in \mathbb{F}^{N \times 1}$ is a vectorization of the rank-one component $\mathbf{a}_r^{(1)} \circ \cdots \circ \mathbf{a}_r^{(d)}$. Then we can consider the matrix

$$\mathbf{X} = \left[\mathbf{X}_1 \middle| \cdots \middle| \mathbf{X}_R\right] \in \mathbb{F}^{N \times R}, \tag{62}$$

and within this framework, we are seeking a rank-k matrix column ID of $\mathbf{X}$, corresponding to a rank-k CP decomposition $\widehat{\mathcal{X}}$ denoted by

$$\widehat{\mathcal{X}} = \sum_{r=1}^{k} \mathbf{X}_{s_r}, \tag{63}$$

for k column skeletons of (62) indexed by $\{s_1, \ldots, s_k\} \subset [R]$. In particular, we want to select these k indices without explicitly forming the matrix $\mathbf{X}$.

To do this, the randomized CP compression algorithm of [22] uses rank-one random tensors $\mathbf{r}_r^{(1)} \circ \cdots \circ \mathbf{r}_r^{(d)}$, with vectorized form $\mathbf{R}_r$ for $r = 1, \ldots, \ell$ and $\ell = k + p$ for some small oversampling parameter p. An $\ell \times R$ random sample matrix is formed via

$$\mathbf{Y} = \begin{bmatrix} \langle \mathbf{R}_1, \mathbf{X}_1 \rangle & \langle \mathbf{R}_1, \mathbf{X}_2 \rangle & \cdots & \langle \mathbf{R}_1, \mathbf{X}_R \rangle \\ \vdots & \vdots & & \vdots \\ \langle \mathbf{R}_\ell, \mathbf{X}_1 \rangle & \langle \mathbf{R}_\ell, \mathbf{X}_2 \rangle & \cdots & \langle \mathbf{R}_\ell, \mathbf{X}_R \rangle \end{bmatrix} =: \mathbf{RX},$$

of which the rank-k column ID is then computed to obtain skeletons $\{s_1, \ldots, s_k\}$ as in (63). We note that the matrix $\mathbf{Y}$ can be computed efficiently without ever explicitly forming $\mathbf{R}$ or $\mathbf{X}$, since each $\mathbf{Y}(l, r) = \prod_{j=1}^{d} \langle \mathbf{r}_l^{(j)}, \mathbf{a}_r^{(j)} \rangle$. However, we observe that the elements of $\mathbf{R}$ are not independent, and so theoretical guarantees on the randomized matrix decomposition computed with $\mathbf{Y}$ are not given, though there is evidence that it performs well in practice [22].

To analyze the cost of this algorithm, let $n = N_1 = \cdots = N_d$. Forming the ℓ random tensors costs $O(t_R \ell dn)$, where t_R is the cost of generating a single random number. The cost of computing each entry of $\mathbf{Y}$ is $O(dn)$, thus the total cost to compute $\mathbf{Y}$ is $O(\ell R \cdot dn)$. The rank-k matrix ID of $\mathbf{Y}$ comes at a cost of $O(k\ell R)$, and forming the rank-k tensor ID from the rank-one terms of $\mathcal{X}$ requires $O(dkn)$ additional operations. In particular, the method of [22] is faster than a rank-k ALS procedure by a factor of $k \cdot n_{iter}$, where n_{iter} is the number of iterations of ALS.

The recent work of [143] constructs an identical rank-k CP approximation to a rank-R CP-factorized tensor $\mathcal{X}$, but uses the TensorSketch operator instead of rank-one random tensors. The complexity of the TensorSketch-based algorithm of [143] is dominated by the sketching step, which costs $O(d(nR + R\ell \log \ell))$; the ID of the $\ell \times R$ matrix $\mathbf{Y}$ costs the same as above. In addition to its theoretical performance guarantees, the algorithm's practical performance is demonstrated through numerical experiments, in which the TensorSketch-based method exhibits a runtime speed-up of roughly one order of magnitude over other randomized CP-tensor ID algorithms, including the random projection method of [22].

7 Concluding Remarks

The primary purpose of this survey was to review a set of recently developed randomized algorithms for processing large-scale tensor data. These new methods were in many cases inspired by analogous techniques for processing matrices that exploit randomized dimension reducing maps to accelerate computations. While such methods for the matrix case are reaching a certain state of maturity, the tensor environment is still evolving rapidly. In this survey, our focus was on techniques that are relatively straightforward generalizations of algorithms designed for matrices.

We did not attempt to describe the plethora of algorithms that are designed for more tensor-specific representations, such as, for instance, tensor trains. Nor did we cover techniques that are designed for tensors that are so large that they cannot be formed or stored explicitly.

We observe that much of the success of RNLA stems from the fact that randomized algorithms often involve less data movement, and therefore execute faster on modern hardware than many traditional algorithms that were designed to minimize the number of floating point operations. Such considerations have proven to be even more essential in the tensor case, given the often enormous size of the data sets under consideration, and the stronger tension between data being located close by in the tensor on the one hand, and being close in physical memory on the other.

Acknowledgments: The authors would like to express their gratitude to the editors and to the anonymous referees for providing remarkably detailed and helpful feedback on the manuscript. Additionally, we acknowledge financial support from the Office of Naval Research (N00014-18-1-2354), the National Science Foundation (DMS-2313434, DMS-2401889), and the Department of Energy ASCR (DE-SC0025312). PGM further expresses gratitude for an academic sabbatical in 2025/26 that was made possible through support provided by the Simons Foundation, and by the Swedish Royal Academy of Sciences through the Institute Mittag-Leffler.

References

1. E. Acar, D. M. Dunlavy, and T. G. Kolda. A scalable optimization approach for fitting canonical tensor decompositions. *Journal of Chemometrics*, 25(2):67–86, 2011.
2. D. Achlioptas and F. Mcsherry. Fast computation of low-rank matrix approximations. *Journal of the ACM*, 54(2):9, 2007.
3. K. S. Aggour, A. Gittens, and B. Yener. Adaptive sketching for fast and convergent Canonical Polyadic decomposition. In *Proceedings of the 37th International Conference on Machine Learning*, ICML'20. JMLR.org, 2020.
4. T. D. Ahle, M. Kapralov, J. B. Knudsen, R. Pagh, A. Velingker, D. P. Woodruff, and A. Zandieh. Oblivious sketching of high-degree polynomial kernels. In *Proceedings of the 14th Annual ACM-SIAM Symposium on Discrete Algorithms*, pages 141–160. SIAM, 2020.
5. S. Ahmadi-Asl, S. Abukhovich, M. G. Asante-Menash, A. Cichocki, A.-H. Phan, T. Tanaka, and I. Oseledets. Randomized algorithms for computation of Tucker decomposition and Higher Order SVD (HOSVD). *IEEE Access*, Volume 9:28684–28706, 2021.
6. A. Aidini, G. Tsagkatakis, N. D. Sidiropoulos, and P. Tsakalides. Few-shot classification using tensor completion. In *Proceedings of the 57th Annual Asilomar Conference on Signals, Systems, and Computers*, pages 1283–1287. IEEE, 2023.
7. A. E. Alaoui and M. W. Mahoney. Fast randomized kernel ridge regression with statistical guarantees. In *Advances in Neural Information Processing Systems 28*, NIPS 2015, pages 1–9, 2015.
8. N. Amsel, T. Chen, A. Greenbaum, C. Musco, and C. Musco. Nearly optimal approximation of matrix functions by the Lanczos method. In *Proceedings of the 38th International Conference on Neural Information Processing Systems*, NeurIPS 2024, pages 1–31, 2024.

9. A. Anandkumar, R. Ge, D. Hsu, S. M. Kakade, and M. Telgarsky. Tensor decompositions for learning latent variable models. *Journal of Machine Learning Research*, 15(1):2773–2832, 2014.
10. H. Avron, P. Maymounkov, and S. Toledo. Blendenpik: Supercharging LAPACK's least-squares solver. *SIAM Journal on Scientific Computing*, 32(3):1217–1236, 2010.
11. H. Avron, H. L. Nguyen, and D. P. Woodruff. Subspace embeddings for the polynomial kernel. In *Advances in Neural Information Processing Systems 27*. Curran Associates, Inc., 2014.
12. O. Balabanov and L. Grigori. Randomized Gram–Schmidt process with application to GMRES. *SIAM Journal on Scientific Computing*, 44(3):A1450–A1474, 2022.
13. O. Balabanov and L. Grigori. Randomized block Gram–Schmidt process for the solution of linear systems and eigenvalue problems. *SIAM Journal on Scientific Computing*, 47(1):A553–A585, 2025.
14. J. Ballani, L. Grasedyck, and M. Kluge. Black box approximation of tensors in hierarchical Tucker format. *Linear Algebra and Its Applications*, 438(2):639–657, 2013.
15. G. Ballard and T. G. Kolda. *Tensor Decompositions for Data Science*. Cambridge University Press, 2025.
16. K. Batselier, W. Yu, L. Daniel, and N. Wong. Computing low-rank approximations of large-scale matrices with the tensor network randomized SVD. *SIAM Journal on Matrix Analysis and Applications*, 39(3):1221–1244, 2018.
17. C. Battaglino, G. Ballard, and T. G. Kolda. A practical randomized CP tensor decomposition. *SIAM Journal on Matrix Analysis and Applications*, 39(2):876–901, 2018.
18. T. Bendory, Y. Khoo, J. Kileel, O. Mickelin, and A. Singer. Autocorrelation analysis for cryo-EM with sparsity constraints: Improved sample complexity and projection-based algorithms. *Proceedings of the National Academy of Sciences*, 120(18):e2216507120, 2023.
19. J. M. F. T. Berge and N. D. Sidiropoulos. On uniqueness in CANDECOMP/PARAFAC. *Psychometrika*, 67(3):399–409, 2002.
20. V. Bharadwaj, O. A. Malik, R. Murray, A. Buluç, and J. Demmel. Distributed-memory randomized algorithms for sparse tensor CP decomposition. In *Proceedings of the 36th ACM Symposium on Parallelism in Algorithms and Architectures*. ACM, 2024.
21. V. Bharadwaj, O. A. Malik, R. Murray, L. Grigori, A. Buluc, and J. Demmel. Fast exact leverage score sampling from Khatri-Rao products with applications to tensor decomposition. In *Thirty-Seventh Conference on Neural Information Processing Systems*, 2023.
22. D. J. Biagioni, D. Beylkin, and G. Beylkin. Randomized interpolative decomposition of separated representations. *The Journal of Computational Physics*, 281:116–134, 2015.
23. F. A. Bischoff, R. J. Harrison, and E. F. Valeev. Computing many-body wave functions with guaranteed precision: The first-order Møller-Plesset wave function for the ground state of helium atom. *The Journal of Chemical Physics*, 137(10):104103, 2012.
24. A. M. P. Boelens, D. Venturi, and D. M. Tartakovsky. Parallel tensor methods for high-dimensional linear PDEs. *The Journal of Computational Physics*, 375:519–539, 2018.
25. V. Braverman, A. Krishnan, and C. Musco. Sublinear time spectral density estimation. In *Proceedings of the 54th Annual ACM SIGACT Symposium on Theory of Computing*, STOC 2022, page 1144–1157, New York, NY, USA, 2022. Association for Computing Machinery.
26. S. Brin and L. Page. The anatomy of a large-scale hypertextual Web search engine. *Computer Networks and ISDN Systems*, 30(1–7):107–117, 1998.
27. R. Bro and C. A. Andersson. Improving the speed of multiway algorithms: Part ii: Compression. *Chemometrics and Intelligent Laboratory Systems*, 42(1-2):105–113, 1998.
28. A. Bucci and L. Robol. A multilinear Nyström algorithm for low-rank approximation of tensors in Tucker format. *SIAM Journal on Matrix Analysis and Applications*, 45(4):1929–1953, 2024.
29. H. Cai, K. Hamm, L. Huang, and D. Needell. Mode-wise tensor decompositions: Multi-dimensional generalizations of CUR decompositions. *Journal of Machine Learning Research*, 22(185):1–36, 2021.
30. C. F. Caiafa and A. Cichocki. Methods for factorization and approximation of tensors by partial fiber sampling. In *2009 3rd IEEE International Workshop on Computational Advances in Multi-Sensor Adaptive Processing*, pages 73–76, 2009.
31. C. F. Caiafa and A. Cichocki. Generalizing the column-row matrix decomposition to multi-way arrays. *Linear Algebra and Its Applications*, 433(3):557–573, 2010.

32. D. Calandriello, M. Dereziński, and M. Valko. Sampling from a k-DPP without looking at all items. In *Advances in Neural Information Processing Systems*, volume 33, pages 18688–18698, 2020.
33. C. Camaño, E. N. Epperly, R. Meyer, and J. A. Tropp. Faster linear algebra algorithms with structured random matrices, 08 2025. arXiv:2508.21189.
34. J. D. Carroll and J. J. Chang. Analysis of individual differences in multidimensional scaling via an N-way generalization of "Eckart-Young" decomposition. *Psychometrika*, 35(3):283–319, 1970.
35. J. D. Carroll, S. Pruzansky, and J. B. Kruskal. CANDELINC: A general approach to multidimensional analysis of many-way arrays with linear constraints on parameters. *Psychometrika*, 45(1):3–24, 1980.
36. A. Castillo, J. Haddock, I. Hartsock, P. Hoyos, L. Kassab, A. Kryshchenko, K. Larripa, D. Needell, S. Suryanarayanan, and K. Yacoubou-Djima. Randomized Kaczmarz methods for t-product tensor linear systems with factorized operators. *BIT Numerical Mathematics*, 65:38, 2025.
37. T. F. Chan. Rank-revealing QR factorizations. *Linear Algebra and Its Applications*, 88/89:67–82, 1987.
38. S. Chandrasekaran and I. C. F. Ipsen. On rank-revealing factorisations. *SIAM Journal on Matrix Analysis and Applications*, 15(2):592–622, 1994.
39. M. Charikar, K. Chen, and M. Farach-Colton. Finding frequent items in data streams. *Theoretical Computer Science*, 312(1):3–15, Jan. 2004.
40. M. Che and Y. Wei. Randomized algorithms for the approximations of Tucker and the tensor train decompositions. *Advances in Computational Mathematics*, 45:395–428, 2019.
41. M. Che, Y. Wei, and H. Yan. The computation of low multilinear rank approximations of tensors via power scheme and random projection. *SIAM Journal on Matrix Analysis and Applications*, 41(2):605–636, Jan. 2020.
42. M. Che, Y. Wei, and H. Yan. Randomized algorithms for the low multilinear rank approximations of tensors. *Journal of Computational and Applied Mathematics*, 390:113380, 2021.
43. M. Che, Y. Wei, and H. Yan. Randomized algorithms for computing the tensor train approximation and their applications, 2024. arXiv:2405.07147.
44. M. Che, Y. Wei, and H. Yan. Efficient algorithms for Tucker decomposition via approximate matrix multiplication. *Advances in Computational Mathematics*, 51(3), 2025.
45. H. Chen, F. Ahmad, S. Vorobyov, and F. Porikli. Tensor decompositions in wireless communications and MIMO radar. *IEEE Journal of Selected Topics in Signal Processing*, 15(3):438–453, 2021.
46. T. Chen, P. Niroula, A. Ray, P. Subrahmanya, M. Pistoia, and N. Kumar. GPU-parallelizable randomized sketch-and-precondition for linear regression using sparse sign sketches, 2025. arXiv:2506.03070.
47. Y. Chen, E. N. Epperly, J. A. Tropp, and R. J. Webber. Randomly pivoted Cholesky: Practical approximation of a kernel matrix with few entry evaluations. *Communications on Pure and Applied Mathematics*, 78(5):995–1041, 2025.
48. Y. Chen and Y. Yang. Fast statistical leverage score approximation in kernel ridge regression. In *Proceedings of the 24th International Conference on Artificial Intelligence and Statistics*, PMLR: Volume 130, 2021.
49. S. Chenakkod, M. Dereziński, X. Dong, and M. Rudelson. Optimal embedding dimension for sparse subspace embeddings. In *Proceedings of the 56th Annual ACM Symposium on Theory of Computing*, pages 1106–1117, New York, NY, USA, 2024. ACM.
50. D. Cheng, R. Peng, I. Perros, and Y. Liu. SPALS: Fast alternating least squares via implicit leverage scores sampling. In *Advances in Neural Information Processing Systems*, 2016.
51. A. Cichocki, D. Mandic, L. De Lathauwer, G. Zhou, Q. Zhao, C. Caiafa, and H. A. PHAN. Tensor decompositions for signal processing applications: From two-way to multiway component analysis. *IEEE Signal Processing Magazine*, 32(2):145–163, 2015.
52. M. B. Cohen. Nearly tight oblivious subspace embeddings by trace inequalities. In *Proceedings of the 27th Annual ACM-SIAM Symposium on Discrete Algorithms*, pages 278–287, Philadelphia, PA, USA, 2016. SIAM.

53. A. Cortinovis and D. Kressner. Low-rank approximation in the Frobenius norm by column and row subset selection. *SIAM Journal on Matrix Analysis and Applications*, 41(4):1651–1673, 2020.
54. A. Cortinovis and D. Kressner. Adaptive randomized pivoting for column subset selection, DEIM, and low-rank approximation. *SIAM Journal on Matrix Analysis and Applications*, 47(1):25–47, 2026.
55. A. Cortinovis, D. Kressner, and Y. Nakatsukasa. Speeding up Krylov subspace methods for computing $f(a)b$ via randomization. *SIAM Journal on Matrix Analysis and Applications*, 45(1):619–633, 2024.
56. W. F. de la Vega, M. Karpinski, R. Kannan, and S. Vempala. Tensor decomposition and approximation schemes for constraint satisfaction problems. In *Proceedings of the 37th Annual ACM Symposium on Theory of Computing*, page 747–754, New York, NY, USA, 2005. Association for Computing Machinery.
57. L. De Lathauwer, B. De Moor, and J. Vandewalle. A multilinear singular value decomposition. *SIAM Journal on Matrix Analysis and Applications*, 21(4):1253–1278, 2000.
58. L. De Lathauwer, B. D. Moor, and J. Vandewalle. On the best rank-1 and rank-$(r_1, r_2, \ldots, r_n)$ approximation of higher-order tensors. *SIAM Journal on Matrix Analysis and Applications*, 21(4):1324–1342, 2000.
59. A. Dektor, A. Rodgers, and D. Venturi. Rank-adaptive tensor methods for high-dimensional nonlinear PDEs. *Journal of Scientific Computing*, 88(36), 2021.
60. M. Dereziński, B. Bartan, M. Pilanci, and M. W. Mahoney. Debiasing distributed second order optimization with surrogate sketching and scaled regularization. *Advances in Neural Information Processing Systems*, 33:6684–6695, 2020.
61. M. Dereziński, D. Calandriello, and M. Valko. Exact sampling of determinantal point processes with sublinear time preprocessing. In *Advances in Neural Information Processing Systems*, volume 32, pages 11546–11558, 2019.
62. M. Dereziński, J. Lacotte, M. Pilanci, and M. W. Mahoney. Newton-LESS: Sparsification without trade-offs for the sketched Newton update. In *Advances in Neural Information Processing Systems*, volume 34, pages 2835–2847, 2021.
63. M. Dereziński and M. W. Mahoney. Recent and upcoming developments in randomized numerical linear algebra for machine learning. In *Proceedings of the 30th ACM SIGKDD Conference on Knowledge Discovery and Data Mining*, KDD '24, page 6470–6479, New York, NY, USA, 2024. Association for Computing Machinery.
64. M. Dereziński and M. W. Mahoney. Determinantal point processes in randomized numerical linear algebra. *Notices of the American Mathematical Society*, 68(1), Jan 2021.
65. A. Deshpande and L. Rademacher. Efficient volume sampling for row/column subset selection. In *Proceedings of the 42nd Annual ACM Symposium on Theory of Computing*. ACM, 2010.
66. A. Deshpande, L. Rademacher, S. S. Vempala, and G. Wang. Matrix approximation and projective clustering via volume sampling. In *Proceedings of the 17th Annual ACM–SIAM Symposium on Discrete Algorithms*, pages 1117–1126, 2006.
67. A. Deshpande and S. S. Vempala. Adaptive sampling and fast low-rank matrix approximation. *Electron. Colloquium Comput. Complex.*, TR06, 2006.
68. H. Diao, Z. Song, W. Sun, and D. P. Woodruff. Sketching for Kronecker product regression and P-splines. In A. Storkey and F. Perez-Cruz, editors, *Proceedings of the 21st International Conference on Artificial Intelligence and Statistics*, volume 84 of *Proceedings of Machine Learning Research*, pages 1299–1308. PMLR, April 2018.
69. W. Dong, G. Yu, L. Qi, and J. Cai. Practical sketching algorithms for low-rank Tucker approximation of large tensors. *Journal of Scientific Computing*, 95(52), 2023.
70. X. Dong, M. Yuan, C.-H. Zhang, and T. Zhang. Statistically optimal and computationally efficient low rank tensor completion from noisy entries. *The Annals of Statistics*, 49(1):76–99, 2021.
71. Y. Dong, C. Chen, P.-G. Martinsson, and K. Pearce. Robust blockwise random pivoting: Fast and accurate adaptive interpolative decomposition. *SIAM Journal on Matrix Analysis and Applications*, 46(3):1791–1815, 2025.
72. Y. Dong and P.-G. Martinsson. Simpler is better: A comparative study of randomized pivoting algorithms for CUR and interpolative decompositions. *Advances in Computational Mathematics*, 49(66), Aug 2023.

73. P. Drineas, R. Kannan, and M. W. Mahoney. Fast Monte Carlo algorithms for matrices I: Approximating matrix multiplication. *SIAM Journal on Computing*, 36(1):132–157, 2006.
74. P. Drineas, R. Kannan, and M. W. Mahoney. Fast Monte Carlo algorithms for matrices II: Computing a low-rank approximation to a matrix. *SIAM Journal on Computing*, 36(1):158–183, 2006.
75. P. Drineas, R. Kannan, and M. W. Mahoney. Fast Monte Carlo algorithms for matrices III: Computing a compressed approximate matrix decomposition. *SIAM Journal on Computing*, 36(1):184–206, 2006.
76. P. Drineas, M. Magdon-Ismail, M. W. Mahoney, and D. P. Woodruff. Fast approximation of matrix coherence and statistical leverage. *Journal of Machine Learning Research*, 13(1):3475–3506, 2012.
77. P. Drineas and M. W. Mahoney. A randomized algorithm for a tensor-based generalization of the singular value decomposition. *Linear Algebra and Its Applications*, 420(2-3):553–571, 2007.
78. P. Drineas, M. W. Mahoney, and S. Muthukrishnan. Sampling algorithms for ℓ_2 regression and applications. In *Proceedings of the 17th Annual ACM-SIAM Symposium on Discrete Algorithms*, pages 1127–1136, 2006.
79. C. Eckart and G. Young. The approximation of one matrix by another of lower rank. *Psychometrika*, 1(3):211–218, Sep 1936.
80. L. Eldén and B. Savas. A Newton–Grassmann method for computing the best multilinear rank-(r_1, r_2, r_3) approximation of a tensor. *SIAM Journal on Matrix Analysis and Applications*, 31(2):248–271, 2009.
81. E. N. Epperly. Fast and forward stable randomized algorithms for linear least-squares problems. *SIAM Journal on Matrix Analysis and Applications*, 45(4):1782–1804, 2024.
82. E. N. Epperly. *Make the Most of What You Have: Resource-Efficient Randomized Algorithms for Matrix Computations*. PhD thesis, California Institute of Technology, 2025.
83. E. N. Epperly, M. Meier, and Y. Nakatsukasa. Fast randomized least-squares solvers can be just as accurate and stable as classical direct solvers. *Communications on Pure and Applied Mathematics*, 79, 09 2025.
84. E. N. Epperly and J. A. Tropp. Efficient error and variance estimation for randomized matrix computations. *SIAM Journal on Scientific Computing*, 46(1):A508–A528, 2024.
85. E. N. Epperly, J. A. Tropp, and R. J. Webber. Embrace rejection: Kernel matrix approximation by accelerated randomly pivoted Cholesky. *SIAM Journal on Matrix Analysis and Applications*, pages 2527–2557, 2025.
86. N. B. Erichson, K. Manohar, S. L. Brunton, and J. N. Kutz. Randomized CP tensor decomposition. *Machine Learning: Science and Technology*, 1(2):025012, May 2020.
87. S. Friedland, A. Niknejad, M. Kaveh, and H. Zare. Fast Monte-Carlo low rank approximations for matrices. *2006 IEEE/SMC International Conference on System of Systems Engineering*, 2005.
88. S. Friedland and V. Tammali. Low-rank approximation of tensors. In P. Benner, M. Bollhöfer, D. Kressner, C. Mehl, and T. Stykel, editors, *Numerical Algebra, Matrix Theory, Differential-Algebraic Equations and Control Theory*, pages 377–411. Springer, Cham, 2015.
89. A. M. Frieze, R. Kannan, and S. Vempala. Fast Monte-Carlo algorithms for finding low-rank approximations. *Journal of the ACM*, 51:1025–1041, 2004.
90. T. Fukaya, R. Kannan, Y. Nakatsukasa, Y. Yamamoto, and Y. Yanagisawa. Shifted Cholesky QR for computing the QR factorization of ill-conditioned matrices. *SIAM Journal on Scientific Computing*, 42(1):A477–A503, 2020.
91. K. Gao, Z.-H. Huang, and Y. Xu. Tensor robust principal component analysis based on a two-layer Tucker rank minimization model. *SIAM Journal on Imaging Sciences*, 18(2):1522–1561, 2025.
92. S. Ghosh and P. Rigollet. Sparse multi-reference alignment: Phase retrieval, uniform uncertainty principles and the beltway problem. *Found. Comput. Math.*, 23(5):1851–1898, Aug. 2022.
93. L. Giraud, J. Langou, and M. Rozloznik. The loss of orthogonality in the Gram-Schmidt orthogonalization process. *Computers & Mathematics with Applications*, 50(7):1069–1075, 2005.
94. G. H. Golub and C. Reinsch. Singular value decomposition and least squares solutions. *Numerische Mathematik*, 14:403–420, 1970.
95. G. H. Golub and C. F. Van Loan. *Matrix Computations*. JHU Press, 2013.

96. L. Grasedyck, D. Kressner, and C. Tobler. A literature survey of low-rank tensor approximation techniques. *GAMM-Mitteilungen*, 36(1):53–78, 2013.
97. M. Gu and S. C. Eisenstat. Efficient algorithms for computing a strong rank-revealing QR factorization. *SIAM Journal on Scientific Computing*, 17(4):848–869, 1996.
98. N. Halko, P.-G. Martinsson, Y. Shkolnisky, and M. Tygert. An algorithm for the principal component analysis of large data sets. *SIAM Journal on Scientific Computing*, 33(5):2580–2594, 2011.
99. N. Halko, P.-G. Martinsson, and J. A. Tropp. Finding structure with randomness: Probabilistic algorithms for constructing approximate matrix decompositions. *SIAM Review*, 53(2):217–288, 2011.
100. E. Hallman, I. C. F. Ipsen, and A. K. Saibaba. Monte Carlo methods for estimating the diagonal of a real symmetric matrix. *SIAM Journal on Matrix Analysis and Applications*, 44(1):240–269, 2023.
101. K. Hamm and L. Huang. Stability of sampling for CUR decompositions. *Foundations of Data Science*, 2:83, 2020.
102. R. A. Harshman. Foundations of the PARAFAC procedure: Models and conditions for an "explanatory" multimodal factor analysis. *UCLA Working Papers in Phonetics*, 16:1–84, 1970.
103. R. A. Harshman and M. E. Lundy. Data preprocessing and the extended PARAFAC model. In H. G. Law, C. W. Snyder, J. Hattie, and R. K. McDonald, editors, *Research Methods for Multi-mode Data Analysis*, pages 216–284. Praeger Publishers, New York, 1984.
104. B. Hashemi and Y. Nakatsukasa. RTSMS: Randomized Tucker with single-mode sketching. *Electronic Transactions on Numerical Analysis*, 63:247–280, 2025.
105. H. He, D. Kressner, and B. Plestenjak. Randomized methods for computing joint eigenvalues, with applications to multiparameter eigenvalue problems and root finding. *Numerical Algorithms*, 100(3):861–892, 2025.
106. M. R. Henzinger, A. Heydon, M. Mitzenmacher, and M. Najork. Measuring index quality using random walks on the Web. *Computer Networks*, 31(11–16):1291–1303, 1999.
107. F. L. Hitchcock. The expression of a tensor or a polyadic as a sum of products. *Journal of Mathematics and Physics*, 6:164–189, 1927.
108. F. L. Hitchcock. Multiple invariants and generalized rank of a p-way matrix or tensor. *Journal of Mathematics and Physics*, 7:39–79, 1927.
109. J. Hough, M. Krishnapur, Y. Peres, and B. Virág. Determinantal processes and independence. *Probability Surveys*, 3:206–229, 04 2006.
110. J. Håstad. Tensor rank is NP-complete. *Journal of Algorithms*, 11(4):644–654, Dec. 1990.
111. Y. Hu. On the convergence of higher-order orthogonal iteration. *Linear and Multilinear Algebra*, 66(11):2247–2265, 2018.
112. B. Huber, R. Schneider, and S. Wolf. A randomized tensor train singular value decomposition. In *Compressed Sensing and Its Applications*, pages 261–290. Springer, Cham, 2017.
113. I. C. F. Ipsen and T. Wentworth. The effect of coherence on sampling from matrices with orthonormal columns, and preconditioned least squares problems. *SIAM Journal on Matrix Analysis and Applications*, 35(4):1490–1520, 2014.
114. Y. Jang and L. Grigori. Randomized orthogonalization process with reorthogonalization. *Numerical Linear Algebra with Applications*, 32(4):e70029, 2025.
115. Y. Ji, Q. Wang, X. Li, and J. Liu. A survey on tensor techniques and applications in machine learning. *IEEE Access*, 7:162950–162990, 2019.
116. R. Jin, T. G. Kolda, and R. Ward. Faster Johnson-Lindenstrauss transforms via Kronecker products. *Information and Inference: A Journal of the IMA*, 10(4):1533–1562, Dec. 2021.
117. W. Kahan. Numerical linear algebra. *Canadian Mathematical Bulletin*, 9(5):757–801, 1966.
118. D. M. Kane and J. Nelson. Sparser Johnson-Lindenstrauss transforms. *Journal of the ACM*, 61(1), Jan. 2014.
119. R. Kannan and S. Vempala. Randomized algorithms in numerical linear algebra. *Acta Numerica*, 26:95–135, 2017.
120. B. N. Khoromskij. Tensor numerical methods for multidimensional PDEs: Theoretical analysis and initial applications. In *CEMRACS 2013 – Modelling and Simulation of Complex Systems: Stochastic and Deterministic Approaches*, volume 48 of *ESAIM: Proceedings and Surveys*, pages 1–28. EDP Sciences, Les Ulis, France, 2015.

121. Y.-D. Kim and S. Choi. Nonnegative Tucker decomposition. In *Proceedings of the IEEE Conference on Computer Vision and Pattern Recognition*, pages 1–8. IEEE, 2007.
122. T. G. Kolda. Orthogonal tensor decompositions. *SIAM Journal on Matrix Analysis and Applications*, 23(1):243–255, 2001.
123. T. G. Kolda. A counterexample to the possibility of an extension of the Eckart-Young low-rank approximation theorem for the orthogonal rank tensor decomposition. *SIAM Journal on Matrix Analysis and Applications*, 24(3):762–767, 2003.
124. T. G. Kolda. Multilinear operators for higher-order decompositions. Technical Report SAND2006-2081, Sandia National Laboratories, Albuquerque, NM, Livermore, CA, 2006.
125. T. G. Kolda. Symmetric orthogonal tensor decomposition is trivial, 2015. arXiv:1503.01375.
126. T. G. Kolda and B. W. Bader. Tensor decompositions and applications. *SIAM Review*, 51(3):455–500, 2009.
127. D. Kressner and L. Periša. Recompression of Hadamard products of tensors in Tucker format. *SIAM Journal on Scientific Computing*, 39(5):A1879–A1902, 2017.
128. D. Kressner and B. Plestenjak. Analysis of eigenvalue condition numbers for a class of randomized numerical methods for singular matrix pencils. *BIT Numerical Mathematics*, 64(32), 2024.
129. P. M. Kroonenberg and J. De Leeuw. Principal component analysis of three-mode data by means of alternating least squares algorithms. *Psychometrika*, 45(1):69–97, 1980.
130. J. B. Kruskal. Three-way arrays: Rank and uniqueness of trilinear decompositions, with application to arithmetic complexity and statistics. *Linear Algebra and Its Applications*, 18:95–138, 1977.
131. J. B. Kruskal. Rank, decomposition, and uniqueness for 3-way and n-way arrays. In R. Coppi and S. Bolasco, editors, *Multiway Data Analysis*, pages 7–18. North-Holland, Amsterdam, 1989.
132. A. Kulesza and B. Taskar. Determinantal point processes for machine learning. *Foundations and Trends in Machine Learning*, 5(2–3):123–286, 2012.
133. J. Lacotte and M. Pilanci. Faster least squares optimization, 2021. arXiv:1911.02675.
134. B. W. Larsen and T. G. Kolda. Practical leverage-based sampling for low-rank tensor decomposition. *SIAM Journal on Matrix Analysis and Applications*, 43(3):1488–1517, 2022.
135. H. Li and S. Yin. Single-pass randomized algorithms for LU decomposition. *Linear Algebra and Its Applications*, 595:101–122, 2020.
136. L. Li, W. Yu, and K. Batselier. Faster tensor train decomposition for sparse data. *Journal of Computational and Applied Mathematics*, 405:113972, 2022.
137. E. Liberty, F. Woolfe, P.-G. Martinsson, V. Rokhlin, and M. Tygert. Randomized algorithms for the low-rank approximation of matrices. *Proceedings of the National Academy of Sciences*, 104(51):20167–20172, 2007.
138. L. Ma and E. Solomonik. Fast and accurate randomized algorithms for low-rank tensor decompositions. In *Advances in Neural Information Processing Systems 34*, pages 24299–24312, 2021.
139. M. W. Mahoney. Randomized algorithms for matrices and data. *Foundations and Trends in Machine Learning*, 3(2):123–224, 2011.
140. M. W. Mahoney, M. Maggioni, and P. Drineas. Tensor-CUR decompositions for tensor-based data. *SIAM Journal on Matrix Analysis and Applications*, 30(3):957–987, 2008.
141. O. A. Malik. More efficient sampling for tensor decomposition with worst-case guarantees. In *Proceedings of the 39th International Conference on Machine Learning, PMLR 162*, Baltimore, MD, 2022.
142. O. A. Malik and S. Becker. Low-rank Tucker decomposition of large tensors using Tensorsketch. In *Advances in Neural Information Processing Systems*, volume 31. Curran Associates, Inc., 2018.
143. O. A. Malik and S. Becker. Fast randomized matrix and tensor interpolative decomposition using CountSketch. *Advances in Computational Mathematics*, 46(76):1–28, 2020.
144. O. A. Malik, V. Bharadwaj, and R. Murray. Sampling-based decomposition algorithms for arbitrary tensor networks, October 2022. arXiv:2210.03828.
145. P.-G. Martinsson. Randomized methods for matrix computations. *The Mathematics of Data*, 25(4):187–231, 2019.

146. P.-G. Martinsson, V. Rokhlin, and M. Tygert. On interpolation and integration in finite-dimensional spaces of bounded functions. *Communications in Applied Mathematics and Computational Science*, pages 133–142, 2006.
147. P.-G. Martinsson, V. Rokhlin, and M. Tygert. A randomized algorithm for the approximation of matrices. Technical Report Yale CS research report YALEU/DCS/RR-1361, Yale University, Computer Science Department, 2006.
148. P.-G. Martinsson, V. Rokhlin, and M. Tygert. A randomized algorithm for the decomposition of matrices. *Applied and Computational Harmonic Analysis*, 30(1):47–68, 2011.
149. P.-G. Martinsson and J. A. Tropp. Randomized numerical linear algebra: Foundations and algorithms. *Acta Numerica*, 29:403–572, 2020.
150. M. Meier, Y. Nakatsukasa, A. Townsend, and M. Webb. Are sketch-and-precondition least squares solvers numerically stable? *SIAM Journal on Matrix Analysis and Applications*, 45(2):905–929, 2024.
151. X. Meng and M. W. Mahoney. Low-distortion subspace embeddings in input-sparsity time and applications to robust linear regression. In *Proceedings of the 45th Annual ACM Symposium on Theory of Computing*, pages 91–100. ACM, 2013.
152. X. Meng, M. A. Saunders, and M. W. Mahoney. LSRN: A parallel iterative solver for strongly over- or underdetermined systems. *SIAM Journal on Scientific Computing*, 36(2):C95–C118, 2014.
153. R. Minster, A. K. Saibaba, and M. E. Kilmer. Randomized algorithms for low-rank tensor decompositions in the Tucker format. *SIAM Journal on Mathematics of Data Science*, 2(1):189–215, 2020.
154. S. Miron, Y. Zniyed, R. Boyer, A. Lima Ferrer de Almeida, G. Favier, D. Brie, and P. Comon. Tensor methods for multisensor signal processing. *IET Signal Processing*, 14(10):693–709, 2020.
155. M. Mørup, L. K. Hansen, and S. M. Arnfred. Algorithms for sparse nonnegative Tucker decompositions. *Neural Computation*, 20(8):2112–2131, Aug. 2008.
156. R. Murray, J. Demmel, M. W. Mahoney, N. B. Erichson, M. Melnichenko, O. A. Malik, L. Grigori, P. Luszczek, M. Dereziński, M. E. Lopes, T. Liang, H. Luo, and J. Dongarra. Randomized numerical linear algebra: A perspective on the field with an eye to software. Technical Report UCB/EECS-2023-19, Electrical Engineering & Computer Sciences Department, University of California, Berkeley, February 2023.
157. C. Musco and C. Musco. Recursive sampling for the Nyström method. In *Proceedings of the 31st International Conference on Neural Information Processing Systems*, NIPS'17, page 3836–3848, Red Hook, NY, USA, 2017. Curran Associates Inc.
158. S. Na, M. Dereziński, and M. W. Mahoney. Hessian averaging in stochastic Newton methods achieves superlinear convergence. *Mathematical Programming*, 201(1):473–520, 2023.
159. Y. Nakatsukasa and J. A. Tropp. Fast and accurate randomized algorithms for linear systems and eigenvalue problems. *SIAM Journal on Matrix Analysis and Applications*, 45(2):1183–1214, 2024.
160. J. Nelson and H. L. Nguyên. OSNAP: Faster numerical linear algebra algorithms via sparser subspace embeddings. In *2013 IEEE 54th Annual Symposium on Foundations of Computer Science*, pages 117–126, 2013.
161. J. Oh, K. Shin, E. E. Papalexakis, C. Faloutsos, and H. Yu. S-HOT: Scalable high-order Tucker decomposition. In *Proceedings of the 10th ACM International Conference on Web Search and Data Mining*, WSDM '17, page 761–770, New York, NY, USA, 2017. Association for Computing Machinery.
162. I. V. Oseledets, D. V. Savostianov, and E. E. Tyrtyshnikov. Tucker dimensionality reduction of three-dimensional arrays in linear time. *SIAM Journal on Matrix Analysis and Applications*, 30(3):939–956, 2008.
163. A. I. Osinsky. Close to optimal column approximation using a single SVD. *Linear Algebra and Its Applications*, 725:359–377, 2025.
164. I. K. Ozaslan, M. Pilanci, and O. Arikan. Iterative Hessian sketch with momentum. In *Proceedings of the IEEE International Conference on Acoustics, Speech and Signal Processing*, pages 7470–7474, 2019.

165. L. Page, S. Brin, R. Motwani, and T. Winograd. The PageRank citation ranking: Bringing order to the Web. Technical Report 1999-66, Stanford University, Computer Science Department, 1998.
166. R. Pagh. Compressed matrix multiplication. *ACM Transactions on Computation Theory*, 5(3):9:1–9:17, Aug 2013.
167. C. Pan. On the existence and computation of rank-revealing LU factorizations. *Linear Algebra and Its Applications*, 316(1):199–222, 2000.
168. Y. Panagakis, J. Kossaifi, G. Chrysos, J. Oldfield, T. Patti, M. Nicolaou, A. Anandkumar, and S. Zafeiriou. Tensor methods in deep learning. In *Signal Processing and Machine Learning Theory*, pages 1009–1048. Academic Press, 01 2024.
169. C. H. Papadimitriou, P. Raghavan, H. Tamaki, and S. Vempala. Latent semantic indexing: A probabilistic analysis. *Journal of Computer and System Sciences*, 61(2):217–235, 2000.
170. E. E. Papalexakis, C. Faloutsos, and N. D. Sidiropoulos. Tensors for data mining and data fusion: Models, applications, and scalable algorithms. *ACM Transactions on Intelligent Systems and Technology*, 8(2), Oct. 2016.
171. T. Park and Y. Nakatsukasa. Accuracy and stability of CUR decompositions with oversampling. *SIAM Journal on Matrix Analysis and Applications*, 46(1):780–810, 2025.
172. D. Parker. Random butterfly transformations with applications in computational linear algebra. Technical Report CSD-950023, Computer Science Department, University of California, Los Angeles, Los Angeles, CA, 1995.
173. K. J. Pearce, C. Chen, Y. Dong, and P.-G. Martinsson. Adaptive parallelizable algorithms for interpolative decompositions via partially pivoted LU. *Numerical Linear Algebra with Applications*, 32(1):e70002, 2025.
174. C. Peng, C. Lewis, X. Wang, M. Clement, K. Pierce, V. Rishi, F. Pavošević, S. Slattery, J. Zhang, N. Teke, A. Kumar, C. Masteran, A. Asadchev, J. Calvin, and E. Valeev. Massively parallel quantum chemistry: A high-performance research platform for electronic structure. *The Journal of Chemical Physics*, 153:044120, 07 2020.
175. I. Perros, R. Chen, R. Vuduc, and J. Sun. Sparse hierarchical Tucker factorization and its application to healthcare. In *Proceedings of the 2015 IEEE International Conference on Data Mining*, ICDM '15, page 943–948, USA, 2015. IEEE Computer Society.
176. D. Persson, A. Cortinovis, and D. Kressner. Improved variants of the `Hutch++` algorithm for trace estimation. *SIAM Journal on Matrix Analysis and Applications*, 43(3):1162–1185, 2022.
177. D. Persson and D. Kressner. Randomized low-rank approximation of monotone matrix functions. *SIAM Journal on Matrix Analysis and Applications*, 44(2):894–918, 2023.
178. N. Pham and R. Pagh. Fast and scalable polynomial kernels via explicit feature maps. In *Proceedings of the 19th ACM SIGKDD International Conference on Knowledge Discovery and Data Mining*, KDD '13, page 239–247, New York, NY, USA, 2013. Association for Computing Machinery.
179. K. Pierce, V. Rishi, and E. Valeev. Robust approximation of tensor networks: Application to grid-free tensor factorization of the Coulomb interaction. *Journal of Chemical Theory and Computation*, 17, 03 2021.
180. N. Pritchard, T. Park, Y. Nakatsukasa, and P.-G. Martinsson. Fast rank adaptive CUR via a recycled small sketch, 2025. arXiv:2509.21963.
181. M. J. Reynolds, A. Doostan, and G. Beylkin. Randomized alternating least squares for canonical tensor decompositions: Application to a PDE with random data. *SIAM Journal on Scientific Computing*, 38(5):A2634–A2664, 2016.
182. E. Robeva. Orthogonal decomposition of symmetric tensors. *SIAM Journal on Matrix Analysis and Applications*, 37(1):86–102, 2016.
183. V. Rokhlin, A. Szlam, and M. Tygert. A randomized algorithm for principal component analysis. *SIAM Journal on Matrix Analysis and Applications*, 31(3):1100–1124, 2010.
184. V. Rokhlin and M. Tygert. A fast randomized algorithm for overdetermined linear least-squares regression. *Proceedings of the National Academy of Sciences*, 105(36):13212–13217, 2008.
185. F. A. Saad, C. E. Freer, M. C. Rinard, and V. K. Mansinghka. Optimal approximate sampling from discrete probability distributions. In *Proceedings of the ACM on Programming Languages*, volume 4, pages 1–31, 2020.

186. A. K. Saibaba. HOID: Higher order interpolatory decomposition for tensors based on Tucker representation. *SIAM Journal on Matrix Analysis and Applications*, 37(3):1223–1249, 2016.
187. T. Sarlós. Improved approximation algorithms for large matrices via random projections. In *The 47th Annual IEEE Symposium on Foundations of Computer Science*, pages 143–152. IEEE, 2006.
188. R. Schneider. *Pseudospectral Divide-and-Conquer for the Generalized Eigenvalue Problem.* PhD thesis, University of California San Diego, 2024.
189. N. D. Sidiropoulos and R. Bro. On the uniqueness of multilinear decomposition of n-way arrays. *Journal of Chemometrics*, 14(3):229–239, 2000.
190. N. D. Sidiropoulos, L. De Lathauwer, X. Fu, K. Huang, E. E. Papalexakis, and C. Faloutsos. Tensor decomposition for signal processing and machine learning. *IEEE Transactions on Signal Processing*, 65(13):3551–3582, 2017.
191. A. Sietsema, Z. Vural, J. Chapman, Y. Yaniv, and D. Needell. Stratified non-negative tensor factorization, 2024. arXiv:2411.18805.
192. L. Sorber, M. Van Barel, and L. De Lathauwer. Optimization-based algorithms for tensor decompositions: Canonical Polyadic decomposition, decomposition in rank-$(l_r, l_r, 1)$ terms, and a new generalization. *SIAM Journal on Optimization*, 23(2):695–720, 2013.
193. Y. Sun, Y. Guo, C. Luo, J. A. Tropp, and M. Udell. Low-rank Tucker approximation of a tensor from streaming data. *SIAM Journal on Mathematics of Data Science*, 2(4):1123–1150, 2020.
194. Y. Sun, Y. Guo, J. A. Tropp, and M. Udell. Tensor random projection for low memory dimension reduction, April 2021. arXiv:2105.00105.
195. D. A. Tarzanagh and G. Michailidis. Fast randomized algorithms for t-product based tensor operations and decompositions with applications to imaging data. *SIAM Journal on Imaging Sciences*, 11(4):2629–2664, 2018.
196. P. Tichavský and Z. Koldovský. Algorithms for nonorthogonal approximate joint block diagonalization. In *Proceedings of the 20th European Signal Processing Conference*, pages 2094–2098, Bucharest, Romania, 2012. IEEE.
197. P. Tichavský, A. H. Phan, and A. Cichocki. Non-orthogonal tensor diagonalization. *Signal Processing*, 138:313–320, 2017.
198. L. N. Trefethen and R. S. Schreiber. Average-case stability of Gaussian elimination. *SIAM Journal on Matrix Analysis and Applications*, 11(3):335–360, 1990.
199. J. A. Tropp. Improved analysis of the subsampled randomized Hadamard transform. *Advances in Adaptive Data Analysis*, 3(1-2):115–126, 2011.
200. J. A. Tropp. An introduction to matrix concentration inequalities. *Foundations and Trends in Machine Learning*, 8(1–2):1–230, 2015.
201. J. A. Tropp. Randomized block Krylov methods for approximating extreme eigenvalues. *Numerische Mathematik*, 150(1):217–255, 2022.
202. J. A. Tropp. Comparison theorems for the minimum eigenvalue of a random positive-semidefinite matrix. *Communications of the American Mathematical Society*, 2026. Accepted March 2026, awaiting publication.
203. J. A. Tropp, A. Yurtsever, M. Udell, and V. Cevher. More practical sketching algorithms for low-rank matrix approximation. Technical Report No. 2018-01, California Institute of Technology, October 2018.
204. J. A. Tropp, A. Yurtsever, M. Udell, and V. Cevher. Streaming low-rank matrix approximation with an application to scientific simulation. *SIAM Journal on Scientific Computing*, 41(4):A2430–A2463, 2019.
205. C. E. Tsourakakis. MACH: Fast randomized tensor decompositions. In *Proceedings of the 2010 SIAM International Conference on Data Mining*, pages 689–700. SIAM, 2010.
206. L. R. Tucker. Implications of factor analysis of three-way matrices for measurement of change. In C. W. Harris, editor, *Problems in Measuring Change*, pages 122–137. University of Wisconsin Press, Madison WI, 1963.
207. L. R. Tucker. The extension of factor analysis to three-dimensional matrices. In H. Gulliksen and N. Frederiksen, editors, *Contributions to Mathematical Psychology*, pages 110–127. Holt, Rinehart and Winston, New York, 1964.
208. L. R. Tucker. Some mathematical notes on three-mode factor analysis. *Psychometrika*, 31:279–311, 1966.

209. S. Ubaru and Y. Saad. Applications of trace estimation techniques. In T. Kozubek, M. Čermák, P. Tichý, R. Blaheta, J. Šístek, D. Lukáš, and J. Jaroš, editors, *High Performance Computing in Science and Engineering*, pages 19–33. Springer International Publishing, Cham, Switzerland, 2018.
210. N. Vannieuwenhoven, K. Meerbergen, and R. Vandebril. Computing the gradient in optimization algorithms for the CP decomposition in constant memory through tensor blocking. *SIAM Journal on Scientific Computing*, 37(3):C415–C438, 2015.
211. N. Vannieuwenhoven, R. Vandebril, and K. Meerbergen. A new truncation strategy for the higher-order singular value decomposition. *SIAM Journal on Scientific Computing*, 34(2):A1027–A1052, 2012.
212. N. Vervliet and L. De Lathauwer. A randomized block sampling approach to Canonical Polyadic decomposition of large-scale tensors. *IEEE Journal of Selected Topics in Signal Processing*, 10(2):284–295, 2016.
213. N. Vervliet, O. Debals, and L. De Lathauwer. Tensorlab 3.0 — Numerical optimization strategies for large-scale constrained and coupled matrix/tensor factorization. In *The 50th Asilomar Conference on Signals, Systems and Computers*, pages 1733–1738, 2016.
214. N. Vervliet, O. Debals, L. Sorber, and L. D. Lathauwer. Breaking the curse of dimensionality using decompositions of incomplete tensors: Tensor-based scientific computing in big data analysis. *IEEE Signal Processing Magazine*, 31:71–79, 2014.
215. M. Wang, H. Cui, and H. Li. SVD-based algorithms for tensor wheel decomposition. *Advances in Computational Mathematics*, 50:99, 2024.
216. Q. Wang, C. Cui, and D. Han. Accelerated doubly stochastic gradient descent for tensor CP decomposition. *Journal of Optimization Theory and Applications*, 197(2):665–704, 2023.
217. Y. Wang, H.-F. Tung, A. J. Smola, and A. Anandkumar. Fast and guaranteed tensor decomposition via sketching. In *Advances in Neural Information Processing Systems*, pages 991–999, 2015.
218. D. Woodruff and A. Zandieh. Leverage score sampling for tensor product matrices in input sparsity time. In *Proceedings of the 39th International Conference on Machine Learning, PMLR*, volume 162, pages 23933–23964, 2022.
219. D. P. Woodruff. Sketching as a tool for numerical linear algebra. *Foundational Trends in Theoretical Computer Science*, 10(1–2):1–157, Oct. 2014.
220. A. Zare, A. Ozdemir, M. A. Iwen, and S. Aviyente. Extension of PCA to higher order data structures: An introduction to tensors, tensor decompositions, and tensor PCA. *Proceedings of the IEEE*, 106(8):1341–1358, 2018.
221. A. Zhang, O. Mickelin, J. Kileel, E. J. Verbeke, N. F. Marshall, M. A. Gilles, and A. Singer. Moment-based metrics for molecules computable from cryogenic electron microscopy images. *Biological Imaging*, 4:e3, Feb 2024.
222. Y. Zhang, M. Fornace, and M. Lindsey. Fast and accurate interpolative decompositions for general, sparse, and structured tensors, 03 2025. arXiv:2503.18921.
223. G. Zhou, A. Cichocki, and S. Xie. Decomposition of big tensors with low multilinear rank, 12 2014. arXiv:1412.1885.
224. G. Zhou, A. Cichocki, Q. Zhao, and S. Xie. Nonnegative matrix and tensor factorizations: An algorithmic perspective. *IEEE Signal Processing Magazine*, 31(3):54–65, 2014.
225. G. Zhou, A. Cichocki, Q. Zhao, and S. Xie. Efficient nonnegative Tucker decompositions: Algorithms and uniqueness. *IEEE Transactions on Image Processing*, 24(12):4990–5003, Dec 2015.

Parallel Sparse and Data-Sparse Factorization-Based Linear Solvers

Xiaoye Sherry Li and Yang Liu

Abstract Efficient solutions of large-scale, ill-conditioned and indefinite algebraic equations are ubiquitously needed in numerous computational fields, including multiphysics simulations, machine learning, and data science. Because of their robustness and accuracy, direct solvers are crucial components in building a scalable solver toolchain. In this chapter, we will review recent advances of sparse direct solvers along two axes: 1) reducing communication and latency costs in both task- and data-parallel settings, and 2) reducing computational complexity via low-rank and other compression techniques such as hierarchical matrix algebra. In addition to algorithmic principles, we also illustrate the key parallelization challenges and best practices to deliver high speed and reliability on modern heterogeneous parallel machines.

1 Introduction

Efficient solutions of large-scale, ill-conditioned and indefinite algebraic equations are ubiquitously needed in numerous computational fields, including multiphysics modeling and simulations, machine learning, and data science. Because of their robustness and accuracy, direct solvers are crucial components in building a scalable solver toolchain. The field of sparse factorization-based solvers is vast and has been active since the 1970s. It evolved from classical Gaussian elimination to combinatorial techniques that control fill-in and memory use. Early ideas focused on ordering and factorization strategies to minimize fill (notably nested dissection and minimum-degree orderings) and on scalable factorization forms (LU, Cholesky, multifrontal and, later, supernodal approaches). These ideas were embodied in portable software packages, most notably pioneered in SPARSPACK [88] and Harwell Subroutine Libraries [206]. Since then, a vast amount of research has been conducted and many more portable and robust

Xiaoye Sherry Li
Lawrence Berkeley National Laboratory, xsli@lbl.gov
Yang Liu
Lawrence Berkeley National Laboratory, liuyangzhuan@lbl.gov

B. Koren and D. Roose (eds.), *Numerical Analysis and Scientific Computing: an Anthology*, Lecture Notes in Computational Science and Engineering 156,
https://doi.org/10.1007/978-3-032-33428-2_8

software packages were developed, see a partial list in section 6. For orientation, classic references include [87, 66, 71]. In this chapter, we will focus mostly on the developments in the distributed memory and GPU programming environments. We will review recent advances of sparse direct solvers along two axes: 1) reducing communication and latency costs in both task- and data-parallel settings, and 2) reducing computational complexity via low-rank and other compression techniques such as hierarchical matrix algebra. In addition to algorithmic principles, we also illustrate the key parallelization challenges and best practices to deliver high speed and reliability on modern heterogeneous parallel machines.

1.1 Problem Statement

A linear system is defined as $Ax = B$, where matrix A is of size $n \times n$ and the right-hand side B is of size $n \times n_{rhs}$. This chapter focuses on sparse linear systems in two scenarios. In the first case A is *structure-sparse* with many explicit zeros. This type of sparse systems typically arises from discretization of partial differential equations. In the second case A is dense but admits *data-sparse* representations, for example, it is numerically low-rank or has numerically low-rank blocks. This type of sparse systems often arises from integral equations, boundary value problems and many kernel matrices used in machine learning. Even though the solution techniques for the two types of sparse systems are distinct, we will illustrate that a hybrid method combining both types of sparsity-exploiting techniques is even more efficient than a traditional sparse direct solver. For the right-hand side, we focus on the most common case with dense B and $n_{rhs} = 1$.

In this chapter, by "direct solver" we mean more broadly the solvers or preconditioners that are based on certain matrix factorizations. That is, the core operations are factorizations and solutions using the factored matrices. In data-sparse cases, the factorizations are usually approximate, therefore they may not be used simply in direct solvers, instead, they can be used as effective preconditioners to accelerate convergence of a Krylov iterative solver. We note that the low-rank approximate factorization is very different from the conventional ILU-type approximate factorization, which will be clear in section 4.

We will focus on two main algorithmic advances in recent years: 1) asymptotic reduction of data movement on parallel machines, and 2) asymptotic reduction of arithmetic operations and memory footprint. Here, by asymptotic reduction we mean lowering the exponent in "big O" notation in analytical formulae. For example, in the first case, we will show orders of magnitude lower communication bandwidth and latency cost for some algorithms. In the second case, we will show almost linear complexity for arithmetic and memory for some algorithms. The complexity formulae will be summarized in tables 1 to 4. Guided by these algorithmic principles, a large number of implementation techniques were developed. We will focus on the distributed-memory algorithms for both CPUs and GPUs, targeting exascale.

For most practical problems, we first *transform* the original linear system into a more favorable form that can be solved more efficiently. A general description of the algorithm for direct solution of linear equations $AX = B$ is as follows:

1. Compute a *factorization*, for example, LU decomposition:

$$P_c P_r D_r A D_c P_c^T = LU. \qquad (1)$$

   Here D_r and D_c are diagonal matrices to equilibrate the system, which can often reduce the condition number. P_r and P_c are *permutation matrices* that reorder the rows and the columns of A. They serve the purpose of maintaining numerical stability and preserving sparsity in the sparse factored matrices.
2. Compute the *solution* of the transformed system: $(P_c P_r D_r A D_c P_c^T) P_c D_c^{-1} X = P_c P_r D_r B$. Thus, X can be evaluated as:

$$X = A^{-1} B = D_c P_c^T U^{-1} L^{-1} (P_c P_r D_r B). \qquad (2)$$

   Multiplying L^{-1} (or U^{-1}) means triangular solution in the case of LU decomposition. Or, it could mean some other solution methods for other types of factorizations, e.g., see section 4.

Remark. Computing D_r, D_c, P_r and P_c are preprocessing steps, which by themselves have large bodies of literature. Although they are not the focus of this chapter, they are critical components for a robust and efficient solver and we will give appropriate references throughout the chapter, see, for example, [136].

Most of the algorithms discussed in this chapter are based on Gaussian elimination (GE) and its sparse variants. In the GE algorithm the Schur complement updates can be performed in different orders, leading to different variants such as left-looking (fan-in) or right-looking (fan-out). These variants are mathematically equivalent under the assumption that the operations are associative, although this is not entirely true in floating point arithmetic. However, different variants have very different memory access and communication patterns. We mainly use the right-looking algorithm and its multifrontal variant to illustrate parallelism. Figure 1 gives a high level sketch of the two algorithms and compares them side-by-side. In the algorithms a node usually refers to a *supernode*. Loosely speaking, a supernode consists of several consecutive rows/columns having the same nonzero pattern. Supernodes help lift most of the submatrix operations to use Level-3 BLAS, leading to higher arithmetic intensity.

Remark. The multifrontal algorithm presented here works only for sparse matrices with *symmetric patterns*, hence, the computation can be organized in a tree structure, called assembly tree or elimination tree (etree) [152]. For nonsymmetric-pattern sparse matrices, the computation graph is a directed acyclic graph (DAG), the same as the right-looking or left-looking algorithm. See more explanation of a DAG in section 2.1. A good example of nonsymmetric-pattern multifrontal algorithm and code is UMFPACK [65].

In the algorithm description, we use the multifrontal term *extend-add*, with symbol $\updownarrow\!\!\leftrightarrow$, to denote the sparse operation of *scattering-and-adding* an element from the source array into an element of the destination array. The nonzero structure of the destination

Listing (1) right-looking LU

```
For each node K in DAG from sources to
      sink
  1. Factorize diagonal
       A(K,K) = L(K,K)U(K,K)
     (may involve pivoting)
  2. Compute off-diag panels L(R1,K),
       U(K,R1)
     (via triangular solves TRSM)
  3. In-place Schur complement update:
     A(R1,R1) ← A(R1,R1) +(scatter)
          (−L(R1,K) · U(K,R1))
     (via matrix multiplication GEMM)
Endfor
```

Listing (2) symmetric-pattern multifrontal LU

```
for each node K in etree from leaves
     to root
  0. Assemble children's Ts to frontal
        matrix:
     F ← [ A(K,K)   A(K,R1) ] +(scatter) T1 +(scatter)
         [ A(R1,K)  0       ]
        T2 +(scatter) ...
  1. Factorize diagonal
       F(K,K) = L(K,K)U(K,K)
     (may involve pivoting)
  2. Compute off-diag panels L(R1,K),
       U(K,R1)
     (via triangular solves TRSM)
  3. Compute and pass update T_K to
       parent:
     T_K = F(R1,R1) − L(R1,K) · U(K,R1)
     (via matrix multiplication GEMM)
Endfor
```

Figure 1 Sketch of the right-looking and multifrontal algorithms. We define $R \subseteq \{K{:}N\}$ consisting of indices corresponding to nonzeros of LU, and $R_1 = R \setminus \{K\}$. In Listing (2), T_i represents node i's update matrix corresponding to the partial Schur complement update. The two index sets R and R_1 change after each step of GE, reflecting the Schur complement sparsity after each elimination step.

is a superset of the nonzero structure of the source, hence the need for "scatter". Note that in the context of Schur complement update, the "add" operation should really be interpreted as "subtract". For notational convenience, we define two index sets $R \subseteq \{K{:}N\}$ referring to the remaining part of the matrix at the Kth elimination step, and $R_1 = R \setminus \{K\}$. These two index sets change after each step of GE, depending on sparsity. In the supernodal right-looking algorithm (Listing (1) in fig. 1), steps 1 and 2 involve dense matrix operations. Step 3 involves sparse scattering. In the multifrontal algorithm (Listing (2) in fig. 1), steps 1 and 2 are the same as the right-looking algorithm. The difference is in the treatment of Schur complement update. In right-looking, the Schur complement update is performed right away *in-place*, whereas in multifrontal, multiple partial Schur complement update matrices (denoted as T) are passed along the etree. The scatter operation occurs at step 0.

In [17], Amestoy et al. provided a comprehensive study and comparison of these two algorithms that were implemented in the distributed memory SuperLU_DIST [212] and MUMPS [18], respectively. Empirically, using a 512-processor Cray T3E and a set of matrices from real applications, the paper compared the performance characteristics of the two solvers with respect to uniprocessor speed, interprocessor communication, the memory requirement, and scalability.

The GE algorithms involving hierarchically low-rank matrices will be treated in section 4.

1.2 Challenges and Motivations

Parallel computing has become an increasingly indispensable tool in various computing disciplines, particularly in modeling physical phenomena in science and engineering simulations. Developing parallel algorithms and software has become a central task in the scientific community. Despite a long history of advances in parallelizing sparse linear algebra algorithms, there still exist many challenges, either architecture-related or application-related, which are summarized below.

1. Inter-process communication cost and load imbalance become an increasing scalability hindrance. This is due to the growing gap in hardware speed of data transfer and the speed of floating point operations. Therefore, it is essential to redesign algorithms to reduce communication and synchronization. We address this challenge in section 2.
2. Modern HPC architectures become increasingly heterogeneous. For the three exascale machines in the United States [215], more than 90% of the computing power comes from GPUs. Existing sparse direct solvers usually exploit only task parallelism, e.g., through OpenMP [187] to use on-node CPUs and MPI [179] to communicate across nodes. Novel methods are needed to make best use of massive data parallelism provided by GPUs. Worse yet, no common programming standard is developed for GPUs from different computer vendors, e.g., AMD, NVIDIA, Intel. We address this challenge in section 3.
3. Direct solvers based on classical LU decomposition and triangular solution are inherently not scalable from the viewpoint of algorithmic complexity, because the number of floating point operations and memory storage units grow super-linearly with respect to problem size. The growth rate is even higher for discretized PDEs in 3D geometry than in 2D. Hence, the number of computing nodes must increase more than the increase of problem size in order to be able to store the factored matrices in the aggregate memory of all nodes. This limits the size of the problems that are solvable given fixed hardware resources and time budget. A truly scalable algorithm should have a nearly linear time and memory complexity with respect to problem size. We address this algorithmic challenge in section 4.

1.3 Sources of Parallelism

Parallel architectures

Parallelism occurs at many levels of the hierarchy of modern processor architectures with varying granularity: bit level, instruction level, data, and task parallelism [110].

The vector machines with SIMD (Single Instruction, Multiple Data) instructions were the earliest parallel machines which are based on the pipeline principle. They provide high-level operations on arrays of elements. SIMD instructions perform exactly the same operations on multiple data objects, thus produce multiple results at the same time. For example, in the late 1990s, Intel introduced the SSE (Streaming SIMD Extensions) (and SSE2 extension for double precision) in the x86 instruction sets, which were later expanded to MMX, AVX, and AVX512. For the SIMD architectures, parallelization is usually achieved by a compiler which analyzes the program and generates the vector instructions. However, the SIMD instructions are not well suited for operations with irregular control structures.

The growth of multi-core CPUs with independent cores that can handle distinct threads of execution serves a much broader spectrum of applications that can take advantage of parallel computing. These so-called Multiple Instruction Multiple Data (MIMD) machines are suitable for general-purpose computing tasks, including servers, and high-performance computing (HPC) systems. In the MIMD architectures, shared memory machines are widely used now on desktops and laptops. Multiple processors access a common memory space and communicate through this shared memory. Programming such systems is made easier by using the standard OpenMP API that is portable to many platforms and programming languages. Distributed memory machines are needed to provide true scalability, where each processor has its own local memory. Processors communicate via an interconnect network. Programming such systems is significantly harder than a shared memory machine. The MPI standard brought a portable solution for message passing which quickly became the de facto standard. Nowadays, most HPC systems have a hybrid architecture which combines elements of both shared and distributed memory systems. Additionally, each node often has a heterogeneous organization with both multicore CPUs and GPU accelerators. It is insufficient to use MPI alone to program such systems. We often use a hybrid programming model such as MPI+X, where MPI is used to generate multiple processes across multiple nodes, and X is used within the address space of each MPI process. Here, X can be OpenMP and/or GPU-specific languages.

Parallel algorithms in matrix computation

Parallelization is to divide the whole computational task into multiple subtasks and map them to the underlying architecture. The art of designing an efficient parallel algorithm follows several key guiding principles: 1) dividing the subtasks evenly to maintain balanced workload across all the processes; 2) making subtasks as independent as possible to avoid information sharing and hence the waiting time; 3) increasing arithmetic intensity relative to memory access and interprocess communication; and 4) using high-level abstraction.

Matrix computation is one of the prime fields that could effectively utilize parallel computers at the dawn of parallel computing. Even though it is a mature field in parallel algorithms design, there are still ample opportunities for algorithm advances, which is mainly due to the frequent innovations in HPC architectures and novel applications that rely on numerical linear algebra. Over the years, the single most important abstraction is the Basic Linear Algebra Subprograms (BLAS) library [183] which contains low-level routines that perform basic vector and matrix operations, providing essential building blocks for linear algebra applications. Various optimized BLAS libraries are widely accessible, such as OpenBLAS, ATLAS and Intel MKL for CPUs, and NVIDIA cuBLAS and AMD rocBLAS for GPUs. The BLAS abstraction hides a lot of low level architectural optimization from the algorithm developers, such as vectorization or threads/warps on GPUs.

In the sparse matrix landscape, many algorithms are also restructured to express the operations in terms of BLAS. However, the dense blocks in sparse matrices are usually small, making an algorithm far from performing at the peak speed if it depends only on parallelizing BLAS. Therefore, in addition to data parallelism encompassed in BLAS, it is essential to exploit task parallelism as well. In sparse factorization-based solvers, two important data structures can greatly help design a good matrix-to-process mapping scheme and a parallel execution strategy. One data structure is a DAG as is used in the right-looking sparse LU. Another data structure is a tree as is used in the symmetric-pattern multifrontal LU and the hierarchically rank-structured algorithms in section 4. Both DAGs and trees encode the data and task dependencies among submatrix operations. There is a rich body of literature about both static and dynamic scheduling of a task graph described by a DAG or tree on a distributed memory machine to maintain load balance and to minimize communication.The parallel algorithms presented in this chapter are mostly based on static scheduling strategies. The challenge with dynamic scheduling is the overhead associated with the scheduler itself. It is cost effective only when the computational task has large granularity.

1.4 Analysis of Parallel Algorithms

We emphasize established performance bound models.

- *Amdahl's law* provides an upper bound on achievable speedup based on the fraction of sequential work, highlighting the diminishing returns of parallelization when sequential portions remain.
- The *roofline model* helps relate algorithmic arithmetic intensity to architectural bandwidth and peak performance, serving as a visual performance diagnostic.
- The *latency–bandwidth model* is discussed to quantify the per-message startup overheads and sustained transfer rates, guiding optimizations in communication patterns.

The rest of the chapter is organized as follows. Section 2 describes recent algorithms for mitigating communication cost for sparse LU factorization (SpLU) and sparse triangular solution (SpTRSV). This section also surveys analyses of the communication lower bounds for these operations. Section 3 describes various novel algorithms to exploit the massive fine-grain parallelism provided by the GPUs. Section 4 describes various data-sparse compression techniques to reduce both floating point and memory complexity. In addition to the parallel construction algorithms for compression, the section also illustrates how the GE algorithm works with the compressed formats and how they can be efficiently parallelized. Section 5 describes how the data-sparse formats can be embedded in the structure-sparse algorithm, leading to a nearly linear complexity fast solver. Section 6 gives a brief survey of existing parallel sparse solvers, using one or multiple algorithms discussed in this chapter. Finally, section 7 lists a number of open problems.

2 Mitigating Communication and Synchronization Costs

Communication cost occurs when transferring data between fast and slow memory or between two nodes across the network. This chapter focuses more on the latter. A commonly used cost model for communication performance is the α-β *model*, where α refers to the latency and β is the inverse of the bandwidth between two processes. The time to send a message of length n is roughly:

$$\text{Time} = \text{latency} + n/\text{bandwidth} = \alpha + n \times \beta \ .$$

This is a simplified and ideal model, without taking into account such practicalities as network congestion etc. For most parallel machines $\alpha \gg \beta \gg$ time_per_flop. For example, on the current Perlmutter node at NERSC [194], typical sparse matrix algorithms run at about 4% of the peak, i.e., the realistic time per flop is $10^{-5}\mu s$. According to the measurement by Lin [180] of typical message sizes, $\alpha = 1.7\mu s \approx 170,000$ flops, and $\beta = 4 \cdot 10^{-5}\mu s/\text{byte} \approx 30$ flops_per_double_word. Therefore, the fundamental principle is to organize the parallel algorithm so that it maintains high data locality and sends fewer long messages rather than sending many short messages. In the following subsections we will describe various algorithmic and implementation techniques to mitigate communication cost for SpLU and SpTRSV.

2.1 Synchronization-Avoiding via One-Sided Communication

Sparse computations in general have lower arithmetic intensity than the dense counterparts. The problem is exacerbated by a high degree of data and task dependency. For example, sparse triangular solution (SpTRSV) is notoriously difficult to parallelize – its arithmetic intensity is of $\mathcal{O}(1)$, similar to sparse matrix-vector multiplication (SpMV). Yet, it has a higher degree of dependency than SpMV, which imposes more sequentiality than SpMV.

The execution of factorization and triangular solution can be concisely prescribed by the DAG data structure, in which a vertex i represents a certain computation task T_i and a directed edge from vertex T_i to vertex T_j encodes the dependency of T_j on T_i, e.g., T_j needs some data produced by T_i. The vertex weight represents the amount of computation and the edge weight represents the amount of data to be transferred between two tasks. If an edge crosses the boundary between two MPI processes, inter-process communication needs to happen for data transfer.

The DAGs for sparse computations can be highly irregular in that both the vertex weights and the edge weights vary a lot. The communication links between pairs of processes also vary. It is particularly challenging to achieve good parallel efficiency for SpTRSV, which typically has limited task-level parallelism and relies on fine-grain synchronization to exploit this parallelism.

For SpTRSV on-node parallelism, extensive studies have led to efficient algorithms with satisfactory parallel efficiency. All these algorithms aim to schedule ready tasks as soon as possible, including level-set based methods [21, 153], level tile method [222], dynamic scheduling [46], and scheduling via the standard OpenMP-barrier type of loop [47]. Figure 2 shows an example of a sparse lower triangular matrix L and its associated task graph. Assuming a supernodal blocked algorithm is used, then each diagonal block L_{ii} defines a dense TRSV operation $x_i = L_{ii}^{-1}\bar{b}_i$ and each off-diagonal nonzero block L_{ij} defines a dense GEMV operation $L_{ij}x_j$. In a level-set algorithm, multiple processes perform parallel operations within a level, then synchronize globally before moving to the next level. Park et al. [191] developed a scheme to reduce the amount of synchronization. The algorithm uses local point-to-point synchronization for the two dependent tasks across the level sets if they are assigned to different processes. Dependent tasks assigned to the same process do not incur synchronization. They show that this reduced synchronization leads to a 1.6× speedup over the strictly level-based algorithm on a 12-core Intel Xeon processor.

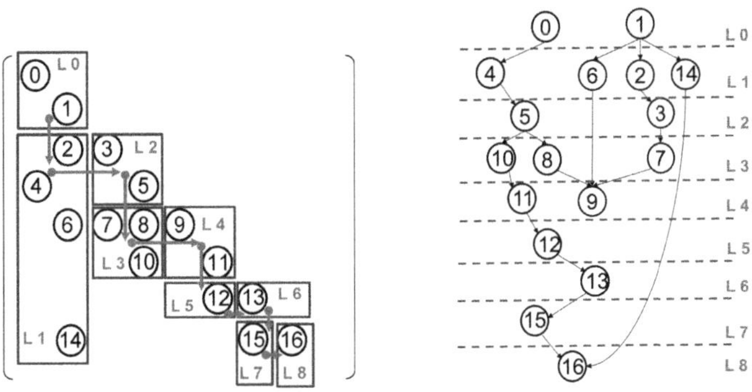

(a) nonzero pattern of L

(b) Nine level sets of task DAG

Figure 2 Illustration of level set in a lower triangular SpTRSV, $Lx = b$.

Implementing these techniques in a distributed manner is highly nontrivial, since the algorithms do not map well to MPI's two-sided handshaking paradigm, where both sender and receiver actively engage in the data exchange. In contrast, one-sided communication is an appealing alternative communication model, which was already established in the MPI standard since MPI-2 [99]. This model was inspired by the Bulk Synchronous Parallel (BSP) model, which separates communication from computation. The BSP model has been implemented in the BSPlib API [113]. In the one-sided communication paradigm communication between processes does not require explicit synchronization between the sender and receiver. Instead, one process can perform operations on the memory of another process without the receiving process's explicit participation. MPI-2 introduced the MPI_WIN and Remote Memory Access (RMA) operations to take advantage of the remote direct memory access (RDMA) features offered by modern interconnects. RDMA enables a process to directly access memory on remote processes without involvement of the activities at the remote side. The light-weight asynchronous one-sided communication primitives provide a new mechanism for efficient DAG execution at scale.

The first distributed-memory SpTRSV algorithm in SuperLU_DIST uses a *message-driven* approach [137]. Every process runs an asynchronous self-scheduling loop and triggers appropriate computations depending on one of the two types of messages received. This algorithm overcomes some synchronization overhead, but still uses MPI two-sided Isend/Recv communication protocol. This algorithm can hardly scale to 100 processes. The algorithm was improved by Liu et al. [158] by using asynchronous communication trees for collective communication among subsets of processes.

In [70], Ding et al. developed the first SpTRSV algorithm exploiting the one-sided communication paradigm. Firstly, they benchmarked a couple of one-sided MPI implementations on a Cray XC40 parallel machine with Intel Xeon E5-2698v3 processors, one is Cray MPI [120] and the other is foMPI [34]. The latter is a fast MPI-3.0 RDMA library on the Cray system. They found that on 16 nodes, the foMPI library achieved up to 3× more bandwidth than the Cray MPI two-side primitives for messages of size 64 to 256 bytes, which is the typical size in SpTRSV. Secondly, they profiled the earlier distributed SpTRSV code in SuperLU_DIST with communication tree optimization [158] and found that with 1000 MPI processes, as much as 90% of the parallel runtime is spent in communication.

Ding's algorithm uses a *passive target mode* in one-sided communication to implement two synchronization-free task queues to manage the messaging between producer-consumer pairs. One such task queue is dedicated to the solution components x_i after they are computed at the diagonal blocks and are remotely enqueued, via `MPI_Put`, to the column processes that need them for subsequent matvecs. Another task queue is dedicated to the partial sums of the matvecs at the diagonal blocks, via `MPI_Accumulate`. Ding et al. empirically showed that, using up to 4096 processes on the Cray XC40 system, their one-sided SpTRSV solver reduces communication time by up to 2.5× and improves runtime up to 2.4× compared to the SuperLU_DIST's two-sided MPI implementation.

Another contribution of Ding's work is the development of a critical path performance model for the DAG-based computational graph of SpTRSV. Ding showed that this performance model can accurately assess the observed performance relative to computation and communication speeds and provide advice on what shape of the process grid is optimal [70]. This model is generalizable to any parallel application with DAG as an underlying computation/communication graph.

Given the delicate task queue algorithm design and careful implementation, a natural question to ask is whether a generic DAG-based runtime system can achieve the same effect. In the last two decades, there has been tremendous progress in development and deployment of runtime frameworks for parallel algorithms portable to heterogeneous architectures. The two widely used systems in the linear algebra community are PaRSEC [41] and StarPU [26]. These two runtime systems were successfully used in parallel implementations of SpLU [133, 198]. However, it remains an open problem whether such a generic runtime scheduler is cost-effective for SpTRSV, because the task sizes in SpTRSV are much smaller than those in SpLU. The only work we are aware of in this vein is by Totoni et al. [216], where the message-passing parallel language and runtime system Charm++ [119] was used for load balancing and dynamic scheduling. They showed that the Charm++ assisted implementation can scale up to 512 cores of an IBM BlueGene/P machine.

To extend the one-sided communication paradigm to the GPU realm, Ding et al. leveraged the advantage of GPU-initiated data transfers of NVSHMEM to implement and evaluate a multi-GPU SpTRSV [69]. Ding's SpTRSV algorithm is based on a novel producer-consumer paradigm to manage the computation and communication using two separate CUDA streams. On two nodes of NVIDIA Tesla V100 GPUs, the algorithm achieved a 3.7× speedup when using two nodes of twelve GPUs relative to the baseline on a single GPU, and up to 6.1× compared to the cuSPARSE `csrsv2` solver over the range of one to eighteen GPUs.

2.2 Communication-Avoiding 3D Algorithm Framework

As the speed discrepancy between communication and floating point operation becomes increasingly large, communication time often dominates the parallel runtime of a parallel algorithm, especially for a sparse matrix algorithm. One primary innovation in parallel sparse direct solvers in the 1990s and 2000s was the 2D blocking algorithm for sparse LU decomposition and sparse triangular solution, such as those used in SuperLU_DIST [137, 138], WSMP [103]. This was motivated from the dense matrix algorithms in the ScaLAPACK library [38]. In these 2D algorithms, the processes are organized in a two-dimensional grid, with communication restricted within process rows or process columns. The sparse L and U factors are partitioned into 2D blocks, which generally are of non-uniform size and are sparsity-induced. These 2D blocks are cyclically mapped to the 2D process grid. Prior to these efforts, a more fine-grained 2D approach was developed by van der Stappen et al. for the sparse LU on a mesh of transputers [217], where they used the block size 1 by 1. This led to speedups of up to a factor 107 on 400 transputers. However, despite the improved scalability, when using thousands of processes, the 2D algorithm still spends more than 90% time in communication [204].

In last two decades, there is a growing body of research on novel algorithms that require asymptotically lower amount of communication than the earlier parallel algorithms (e.g., [27]). These algorithms are termed communication-avoiding algorithms, or CA algorithms for short. One notable class of CA algorithms trade off extra memory for reduced communication. The rationale behind this is based on the analysis that the communication lower bounds decrease as the size of local memory increases [27, Section 2]. Based on this principle, the 3D and 2.5D CA algorithms were developed for dense matrix-multiplication and LU decomposition [118, 209].[1]

Inspired by the success of 2.5D dense CA algorithms, Sao et al. developed the first 3D CA algorithm for SpLU [204] and SpTRSV [200]. The 3D CA algorithm framework uses a 3D logical process grid, instead of the 2D process grid that was the state-of-the-art as of 2018. The algorithm replicates some data to reduce both the number of messages and the communication volume. In addition, the 3D sparse algorithms can use the elimination tree to efficiently map the sparse matrix to the 3D process grid to further localize the communication among subsets of processes. On the other hand,

[1] The 3D process configuration is a special case of the 2.5D configuration.

the 2.5D dense LU algorithm cannot exploit this feature. As a result, the 3D sparse algorithm not only reduces communication but also shortens the critical path of SpLU. For matrices with planar graph structure (e.g., planar grids and meshes), the critical path length of Sao's 3D SpLU algorithm is $O(n / \log n)$ whereas the critical path of a typical 2D algorithm is $O(n)$.

Figure 3 illustrates the 3D processes grid, configured as $P = P_x \times P_y \times P_z$. It can be considered as P_z layers of 2D processes. The distribution of the sparse matrices is governed by the supernodal elimination tree: the standard etree is transformed into an etree which is binary at the top $\log_2(P_z)$ levels and has P_z subtrees at the bottom level. The transformation may make some subtrees disconnected, hence becoming forests. But for notational simplicity, we still use the term etree herein. The matrices A, L, and U corresponding to each subtree are assigned to one 2D process layer. The 2D layers are referred to as Grid-0, Grid-1, ..., for all P_z grids. Figure 4 shows the submatrix mapping to the four 2D process grids. The sparse LU on each 2D grid is performed using the same 2D algorithm locally within the 2D processes. Each 2D LU generates a Schur complement update to the common ancestor of the etree. The key feature of 3D LU is to replicate the Schur complement memory corresponding to the common ancestor tree among the P_z 2D process layers, so that each 2D grid can accumulate the Schur complement (partial sum) in its local memory, in parallel with the other 2D grids. Finally, the individual copies of partial sums are reduced (added) onto a single grid, where they are factorized in a 2D fashion. The communication reduction is mainly due to the independent 2D subgrids working on smaller subproblems. For example, in fig. 4a, let S_i^j denote the partial update of the Schur complement that is computed and residing in Grid-j but is destined to the etree node i. At the leaf level, Grid-0 and Grid-1 first factorize nodes 3 and 4 in parallel. Grid-0 produces the partial updates S_1^0 and S_0^0. Grid-1 produces the partial updates S_1^1 and S_0^1. Then the two grids reduce all the common ancestor nodes, namely nodes 1 and 0, to Grid-0. This amounts to subtracting $S_1^0 + S_1^1$ from A_1 and subtracting $S_0^0 + S_0^1$ from A_0, In parallel, similar operations occur for Grid-2 and Grid-3 along the etree nodes 5, 6, 2 and 0. At one level up, only Grid-0 and Grid-2 are active. They factorize nodes 1 and 2, producing S_0^0 and S_0^2, and reduce the updates on node 0 to Grid-0. At the root level, only Grid-0 factors node 0. The generalization of this process for any $P_z = 2^\ell$ is given as Algorithm 1 in [204]. In practice, the 3D SpLU algorithm improves strong scaling from 1000 cores to 32,000 cores, leading to speedups of up to 27× on planar and up to 3.3× on non-planar sparse matrices.

For a sparse matrix with arbitrary sparsity pattern, it is infeasible to quantify the communication reduction in "big O" sense. Fortunately, for a large class of discretized prototypical PDEs, there exist analytical closed forms of nonzero patterns in the LU factors. These include the finite difference discretization of PDEs defined on a 2D geometry giving rise to planar graphs, and PDEs defined on a 3D geometry giving rise to non-planar graphs. Sao et al. analyzed the memory-communication trade-off for these classes of problems. They assume that the matrices are first reordered by the nested dissection ordering [86] or recursive graph partitioning [150, 93], which is optimal for this class of problems [86, 150, 93]. With this ordering, the nonzero counts (and hence the memory) in the LU factors are $O(n \log n)$ for planar graphs and $n^{4/3}$ for non-planar

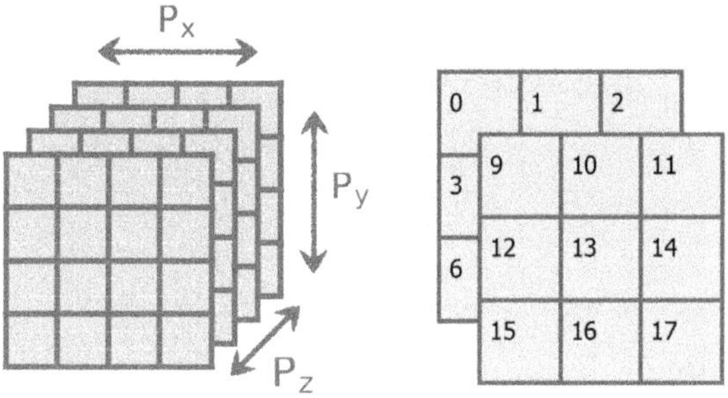

(a) (b)

Figure 3 The view of the logical 3D process grid and an example of 18 processes arranged as a $3 \times 3 \times 2$ process grid.

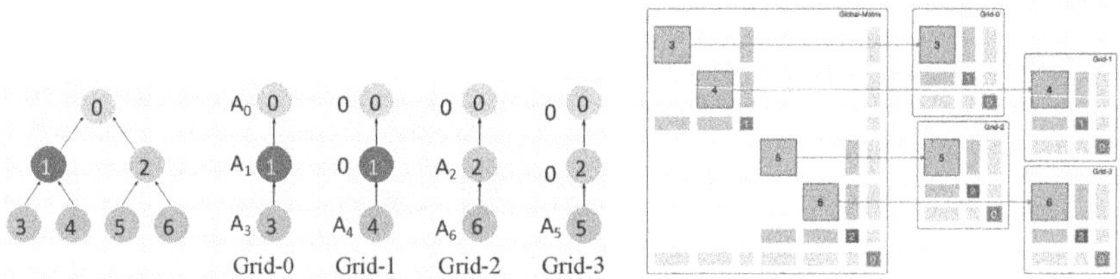

(a) (b)

Figure 4 Two-level etree partition and the matrix view of the submatrix mapping to four 2D process grids.

graphs. The flop counts are $n^{3/2}$ for planar graphs and n^2 for non-planar graphs. The etree corresponds to the *separator tree* where each internal node is a separator at a certain middle level of the dissection. The memory and communication analyses use the following two key facts:

- Each separator node in the etree corresponds to a dense submatrix;
- Communication time is determined by the amount of communication occurring on the *critical path* of the computational graph. Here, the critical path is a leaf-to-root path in the separator tree.

Table 1 summarizes the analysis results regarding the memory, bandwidth and latency costs, comparing the 2D and 3D sparse LU algorithms. The analysis calculates the memory and communication in two separate phases — first in the 2D algorithm for each subtree and second in the ancestor-reduction step. Thus, in the column "3D-SparseLU", two separate terms contribute to memory and communication. The calculation is only needed for Grid-0 as it is the only grid that participates in all levels and is on the critical path. The last column suggests that setting $P_z = O(\log n)$ would lead to the minimal memory and communication costs for the 3D algorithm, assuming $P \gg \log n$. Similarly, table 2 summarizes the memory, bandwidth and latency costs for the PDEs associated with the non-planar graph.

Parameter	2D PDE		
	2D-SparseLU	3D-SparseLU	3D-SparseLU (min) $P_z = O(\log n)$
FLOP count			
per process (F)	$\frac{O(n^{3/2})}{P}$	$\frac{O(n^{3/2})}{P}$	$\frac{O(n^{3/2})}{P}$
Memory			
per process (M)	$O\left(\frac{n}{P}\log n\right)$	$O\left(\frac{n}{P}(\log(\frac{n}{P_z}) + P_z)\right)$	$O\left(\frac{n}{P}\log n\right)$
Communication volume			
per process (W)	$O\left(\frac{n}{\sqrt{P}}\log n\right)$	$O\left(\frac{n}{\sqrt{P}}(2\sqrt{P_z} + \frac{\log n}{\sqrt{P_z}}) + \frac{2nP_z\log P_z}{P}\right)$	$O\left(\frac{n}{\sqrt{P}}\sqrt{\log n}\right)$
Latency	$O(n)$	$O\left(\frac{n}{P_z} + \sqrt{n}\right)$	$O\left(\frac{n}{\log n}\right)$

Table 1 [204, Table 2] Asymptotic memory, communication, and latency costs for 2D and 3D Sparse LU algorithm for 2D model PDEs.

Parameter	3D PDE	
	2D-SparseLU	3D-SparseLU
FLOP count		
per process (F)	$O\left(\frac{n^2}{P}\right)$	$O\left(\frac{n^2}{P}\right)$
Memory		
per process (M)	$O\left(\frac{n^{\frac{4}{3}}}{P}\right)$	$O\left(M_{2D}\left(P_z + \frac{1}{P_z^{4/3}}\right)\right)$
Communication volume		
per process (W)	$O\left(\frac{n^{4/3}}{\sqrt{P}}\right)$	$O\left(W_{2D}\left(\kappa_1\sqrt{P_z} + \frac{1-\kappa_1}{P_z{}^{5/6}}\right) + W_{2D}\left(\kappa_1\frac{P_z\log P_z}{\sqrt{P}}\right)\right)$ †
Latency	$O(n)$	$O\left(L_{2D}\left(\kappa_2 n^{-1/3} + \frac{1}{P_z}\right)\right)$ ‡

† The first and second term in the expression for W_{3D} represent W_{3D}^{xy} and W_{3D}^{z}, respectively. $\kappa_1 - \frac{2(\alpha-1)}{\alpha^4-1}$, whcrc $\alpha - 2^{1/3}$.
‡ $\kappa_2 = \frac{\alpha}{\alpha+1}$

Table 2 [204, Table 3] Asymptotic memory, communication, and latency costs for 2D and 3D Sparse LU algorithm for 3D model PDEs.

In both tables, the formulae for communication volumes in the 3D sparse LU are non-trivial functions depending on P_z. To provide visual perspective of the two functions, we plot these formulae with increasing P_z while fixing the matrix dimension n to be 1 million and total processes P to be 1024, see fig. 5. In both cases, there are some sweet spots for the P_z values that would minimize the communication volume. It is not always true that the larger P_z the less communication. Note that these formulae are just in "big O" sense and do not contain the constant prefactors in each case. Therefore, the relative communication volumes between the 3D and 2D SpLU are just notional magnitudes and should not be considered to be the actual differences on a real machine. In practice, Sao's paper showed that P_z ranging from 8 – 16 is usually good for both planar and non-planar problems.

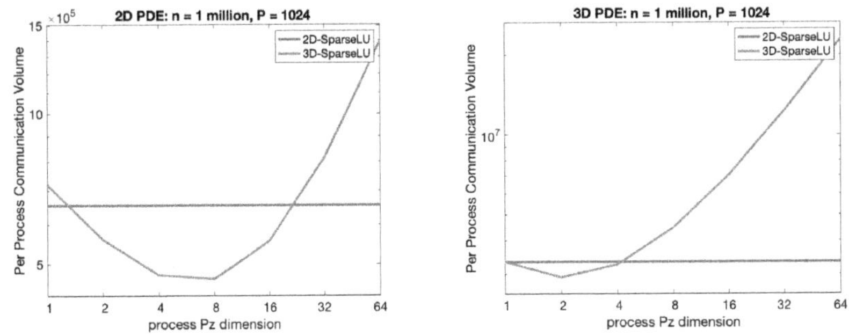

(a) (b)

Figure 5 Asymptotic per process communication volumes given in table 1 $\left(\frac{n}{\sqrt{P}}(2\sqrt{P_z} + \frac{\log n}{\sqrt{P_z}}) + \frac{2nP_z \log P_z}{P}\right)$ and table 2 $\left(W_{2D}\left(\kappa_1\sqrt{P_z} + \frac{1-\kappa_1}{{P_z}^{5/6}}\right) + W_{2D}\left(\kappa_1 \frac{P_z \log P_z}{\sqrt{P}}\right)\right)$, respectively, with different P_z settings.

Communication-avoiding 3D SpTRSV

Following the same 3D SpLU principle and the distribution of the LU factors, Sao et al. designed a new 3D SpTRSV algorithm [200]. Since the diagonal block of the L (or U) factor associated with a leaf subtree or an ancestor separator resides only on 2D process grids, the corresponding SpTRSV operations would still use the prior 2D algorithm. Similar to 3D LU, there is some replicated data on the common ancestor nodes at the higher levels of etree. Different from LU, where the replicated memory is for the partial Schur complement updates, here in SpTRSV, the replicated memory is for the partial updates to the right-hand side matrix B. Each 2D grid accumulates the local partial sum that is reduced to update a part of the right-hand side corresponding to the common ancestor nodes.

Consider solving the lower triangular system $LY = B$, using the same example as in fig. 4. The right-hand side B (and solution Y) is partitioned into 7 block rows, $B_0, B_1, \ldots, B_6$, corresponding to the 7 etree nodes. Let S_i^j denote the partial update that is computed and residing in Grid-j but is destined to the etree node i for block B_i. At the leaf level, Grid-0 and Grid-1 first solve nodes 3 and 4 in parallel, obtaining $Y_3 = L_3^{-1}B_3$ and $Y_4 = L_4^{-1}B_4$. Grid-0 produces the partial updates $S_1^0 = -L_{1,3} \cdot Y_3$ and $S_0^0 = -L_{0,3} \cdot Y_3$ destined to B_1 and B_0, respectively. Grid-1 produces the partial updates $S_1^1 = -L_{1,4} \cdot Y_4$ and $S_0^1 = -L_{0,4} \cdot Y_4$ destined to B_1 and B_0. The two grids reduce the updates at nodes 1 and 0, forming the updated $B_1 = B_1 + S_1^0 + S_1^1$ and the pending update $S_0^0 = S_0^0 + S_0^1$ on Grid-0. Grid-0 proceeds to solve $Y_1 = L_1^{-1} \cdot B_1$ using the 2D algorithm. In parallel, similar operations occur for Grid-2 and Grid-3 along the etree nodes 5, 6, 2 and 0. At one level up, only Grid-0 and Grid-2 are active. They solve for $Y_1 = L_1^{-1} \cdot B_1$ and $Y_2 = L_2^{-1} \cdot B_2$. They then reduce the merged updates $S_0^0 - L_{0,1} \cdot Y_1$ (Grid-0) and $S_0^2 - L_{0,2} \cdot Y_2$ (Grid-2) on node 0 to Grid-0. At the root of the etree, only Grid-0 solves for Y_0 using the 2D algorithm. The generalization of this process for any $P_z = 2^\ell$ is given in [200, Algorithm 2]. Table 3 summarizes the analysis results about the communication cost and per-process communication volume, comparing the 2D and 3D SpTRSV algorithms. The communication cost measures the number of words sent along the critical path of the computation. Liu et al. [155] further improved the 3D SpTRSV algorithm by allowing replicated computation on each grid by setting the corresponding RHS (Right-Hand Side) to 0. This greatly simplifies the 3D algorithm design, and removes the need for inter-grid communication and synchronization at each level, albeit with moderately increased inner-grid communication and computation. This approach achieved up to a 3.45× additional speedup compared with the above-mentioned algorithm [200] on up to 2048 CPU cores.

Problem type	Communication	2D SpTRSV	3D SpTRSV
Planar	Cost (W)	$O\left(\frac{n}{\sqrt{P}}+\sqrt{n}\right)$	$O\left(\frac{n}{\sqrt{P_z P}}+\sqrt{n}\right)$
(2D PDE)	Average Volume	$O\left(\frac{n}{\sqrt{P}}\right)$	$O\left(\frac{n}{\sqrt{P_z P}}\right)$
	Max Volume	$O\left(\frac{n}{\sqrt{P}}\right)$	$O\left(\frac{n}{\sqrt{P_z P}}+\frac{\sqrt{n P_z}}{\sqrt{P}}\right)$
Non-Planar	Cost (W)	$O\left(\frac{n}{\sqrt{P}}+n^{2/3}\right)$	$O\left(\frac{n}{\sqrt{P_z P}}+n^{2/3}\right)$
(3D PDE)	Average Volume	$O\left(\frac{n}{\sqrt{P}}\right)$	$O\left(\frac{n}{\sqrt{P_z P}}\right)$
	Max Volume	$O\left(\frac{n}{\sqrt{P}}\right)$	$O\left(\frac{n}{\sqrt{P_z P}}+n^{2/3}\frac{\sqrt{P_z}}{\sqrt{P}}\right)$

Table 3 [200, Table 2] Asymptotic communication cost and volume for 2D SpTRSV and 3D SpTRSV, on planar (2D PDE) and non-planar (3D-PDE) input problems

2.3 Communication Lower Bounds

Even though the 3D CA algorithm framework implemented in SuperLU_DIST has low communication complexity among the widely used parallel sparse direct solvers, a natural question arises whether this communication complexity reaches the lower bound, that is, whether this is a communication optimal algorithm and whether we can continue to seek better algorithms to close the gap from optimum. The answer to this question is not simple. Firstly, the communication cost consists of two parts: bandwidth (volume) and latency (number of messages). Some algorithms may minimize one cost but not both, or may be able to trade-off one for another. Secondly, some algorithms may be able to trade-off floating point operations and/or memory for lower communication cost. Thirdly, for sparse matrices, the optimality of an algorithm may depend on the sparsity pattern.

Let F denote the number of floating point operations per process, and let M denote the local memory size. For a broad class of matrix-multiplication-like linear algebra algorithms, dense or sparse, it was established that the lower bounds on communication bandwidths and communication latencies are:

$$W = \Omega\left(\frac{F}{\sqrt{M}}\right), \qquad S = \Omega\left(\frac{F}{M^{3/2}}\right), \tag{3}$$

see [27] and the historical notes therein. These formulae suggest that when increasing the local memory size, the lower bounds for both W and S can be decreasing. This intuition prompted the development of some CA algorithms that replicate some data for reduced communication. In [209], Solomonik and Demmel presented a 2.5D dense LU algorithm that arranges P processes as c layers of 2D process grids. The dense matrix is duplicated on each layer, thus, the algorithm uses $M \approx cn^2/P$ local memory. The algorithm reduces the bandwidth by a factor of $\sqrt{c}$ in comparison with standard 2D LU algorithms (when $c = 1$). However, it sends a factor of $\sqrt{c}$ more messages than the 2D algorithm.

For SpLU, the two available open-source implementations with low communication complexity are: the multifrontal approach with subtree-to-subcube mapping proposed by Gupta et al. [104] and the right-looking 3D mapping approach proposed by Sao et al. (see section 2.2 and [204]). Both of these approaches fully exploit the etree parallelism and do not suffer the latency cost increase as does the 2.5D dense LU. Gupta's algorithm partitions the 2D process grid into smaller 2D process grids that are assigned to the frontal matrices at different levels. Sao's 3D mapping approach partitions the 3D process grid into smaller 3D process grids. The essence of both approaches is to use multiple copies of the matrix to perform multiple Schur-complement updates in parallel. The low communication complexity of the multifrontal algorithm is also attributed to trading additional memory for reduced communication. This additional memory stores the multiple partial update matrices, i.e., the T matrices in fig. 1. For three-dimensional and higher-dimensional sparse matrices, e.g., those corresponding to PDEs on 2D and 3D spatial grids, the sparse LU operation is dominated by the dense LU at the top level separator, therefore, the behavior of both Gupta's and Sao's algorithms is optimal in terms of communication. But, for sparse matrices associated with planar graphs, both algorithms do not achieve the communication lower bound specified by eq. (3). Consider a model planar problem that comes from a finite difference discretization on a 2D grid of size $n = k \times k$, Gupta's algorithm has a gap of $\sqrt{\log n}$ from the bandwidth lower bound, and Sao's algorithm has a gap of $\log n$ (see last column of table 1, lower bound is $F/\sqrt{M} = O\left(n/(\sqrt{P}\sqrt{\log n})\right)$).

In [201], Sao et al. proposed a nearly-optimal algorithm in terms of communication, which combines the subtree-to-subcube mapping and the 3D mapping approach. Here, the lower level subtree of the etree uses subtree-to-subcube mapping and subtrees at higher levels use the 3D mapping algorithm. Thus, for higher level subtrees, data replication is used up to a factor of $\log n$, without increasing the asymptotic memory cost. This hybrid algorithm achieves the communication lower bound within a factor of $\log \log n$, a considerable improvement over the $\sqrt{\log n}$ and $\log n$ factors of existing approaches. However, it remains an open problem to implement this last communication-optimal algorithm.

The software details of the 3D sparse direct solver SuperLU_DIST are described in [139].

3 Uncovering Fine-Grain Parallelism for GPU Acceleration

In the exascale computing era, GPUs have become mainstream computing engines. In the November 2025 edition of the TOP500 supercomputer list, all four exascale machines are using various generations of GPUs from AMD, Intel and NVIDIA [215]. On those machines, more than 90% of the computing power comes from GPUs. Despite their increasing adoption in scientific computing, GPU architectures present unique challenges for sparse computations. Modern GPUs can have thousands of cores, enabling them to run a large number of threads concurrently. This high thread count allows for simultaneous execution of many operations, greatly increasing throughput. This type of SIMD data-parallel model is well-suited to large dense matrix operations. However, for sparse matrix operations, task parallelism is often more prevalent than data parallelism. Significant efforts have been made to expose fine-grained data parallelism.

Programming on GPUs often requires explicit management of threads and thread blocks. Developers need to understand how to organize computations into many small kernels and manage synchronization between threads, which can be non-intuitive. While GPUs can handle many threads in parallel, their architectures are not designed for handling complex control flows. Therefore, most GPU-capable sparse direct solvers still use CPUs to handle control flows and use GPUs only in the *offload mode*.

About fifteen years ago, several CPU-GPU SpLU algorithms were developed that mostly use GPU as a BLAS accelerator (i.e., dense LU, TRSM and GEMM in fig. 1) [164, 240, 89, 218]. All these earlier developments focused on single GPU or a few GPUs on a single node. Starting 2014, Sao et al. initiated the first line of work on large-scale distributed memory GPU accelerated sparse direct solvers with implementations in SuperLU_DIST. This required using the MPI+OpenMP+CUDA programming model with delicate coordinations of both inter-node communication, intra-node CPU-GPU communication and intra-node multithreading on each multicore node. In their first distributed memory GPU-enabled SpLU algorithm [203], the only part offloaded to GPUs is the GEMM computation that makes up the contribution to each Schur complement update. Sao et al. leveraged NVIDIA's cuBLAS library for the GEMM calls on the GPU side. To overcome the cost of data transfer between the CPU and GPU memories, they developed a sophisticated software pipelining for overlapping the GEMM operations on GPUs with the Scatter operations on CPU. This requires creating multiple CUDA streams to split a single large GEMM or Scatter operation into smaller pieces. This algorithm achieved up to 2× speedups on 8 nodes with 16 Tesla "Fermi" GPUs. However, for truly sparse matrices, the performance gain is marginal. Their second algorithm [202] also offloaded the Schur complement update to GPU, including both GEMM and Scatter operations. The algorithm is called HALO for Highly Asynchronous Lazy Offload, referring to combining asynchronous use of both CPU and GPU for each Schur complement update accounting for the GPU memory size and the Schur complement matrix size. The asynchronous and lazy updates serve to hide and reduce communication over the PCIe link between CPU and GPU and to overcome the small GPU memory capacity. This strategy delivered an additional 2.5× speedup over the GEMM-only-offload approach. In their subsequent effort [204], on top of the HALO strategy, Sao et al. added additional GPU offloading for the block diagonal dense

LU and the L and U panel factorizations, leveraging the cuBLAS TRSM routine and the cuSparse GETRF routine.

In [92], Ghysels and Synk developed the first distributed memory GPU accelerated symmetric-pattern multifrontal sparse direct solver with implementation in STRUMPACK. Their algorithm offloads both dense matrix operations and the sparse scatter–gather operations to GPUs. They leveraged various vendor libraries for the dense operations, such as NVIDIA's cuBLAS and cuSOLVER and AMD's rocBLAS and rocSOLVE. For the smaller frontal matrices they developed custom CUDA and HIP kernels to reduce kernel launch overhead. The multi-GPU setting uses SLATE [84] as a modern GPU-aware replacement for ScaLAPACK. On 4 nodes of the Summit machine at Oak Ridge National Laboratory, the code is $10\times$ faster when using all 24 V100 GPUs compared to when it only uses the 168 POWER9 CPU cores. On 8 Summit nodes, using 48 V100 GPUs, the sparse solver reaches over 50TFlop/s.

In a more recent effort in PanguLU [83], Fu et al. developed a different SpLU algorithm which does not use any dense supernode data structure; instead, they use a uniform 2D block partition of the sparse matrix, in which each block is generally sparse and represented in the standard Compressed Sparse Column (CSC) format. In order to accommodate a vast number of different sparsity patterns, they developed 17 different sparse kernels for block diagonal LU, panel lower and upper triangular solution and Schur complement update. Some of them are for CPU and some others for GPU. Based on the extensive microbenchmarks of various block patterns arising from different sparse matrices, they derived four decision trees to choose a specific kernel dynamically based on each block's sparsity pattern. Accompanied with an efficient asynchronous scheduling strategy for parallel block operations, PanguLU achieved strong scaling of $47\times$ and $74\times$ on 128 NVIDIA A100 GPUs and 128 AMD MPI50 GPUs, respectively, compared to single GPUs.

The new generation of GPUs has much larger memory sizes than earlier generations. For example, an NVIDIA GH200 (2024) Grace Hopper GPU has nearly 100 GB memory whereas an NVIDIA Tesla M2090 (2011) GPU had only 6 GB memory. As such, many applications were ported to run entirely on GPUs and there is a demand for the algebraic solvers to run entirely on the GPUs. The recent work of Caracal SpLU algorithm [198] epitomizes the "GPU-resident" paradigm. Two main techniques developed by Ren et al. greatly enhanced the performance of single GPU implementations: 1) a fine-grained static scheduling algorithm that utilizes multiple GPU streams to perform TRSM and GEMM on multiple sub-blocks simultaneously, and 2) a memory caching algorithm and clever use of the GPU memory pool to enhance data reuse and reduce the frequency of calling cudaMalloc/cudaFee that may cause GPU memory fragmentation. On a single NVIDIA A100 GPU, Caracal achieved up to 21% of the A100's theoretical peak performance, $7\times$ speedup over SuperLU_DIST, $94\times$ speedup over PanguLU, and $16\times$ speedup over PasTiX. Even though the Caracal approach eliminates the need for CPU-GPU data transfers during the LU factorization, the inter-GPU communication is still initiated on the CPU. That is, CPUs are not entirely out of the picture. From their limited scaling study using only 4 GPUs, their multi-GPU results are not superior to the other methods. Further improvement is needed for efficient use of large numbers of GPUs.

Batching

In the same vein as the GPU-resident solvers that intend to reduce the number of kernel launches, *batching* also emerged as an efficient algorithm paradigm for GPUs. When executing many small tasks, the overhead of launching kernels and managing resources can dominate the computation time. Batching reduces this overhead by executing a large number of operations within a single kernel call, increasing overall throughput. This leads to significant speedups compared to executing each task sequentially. The batched algorithm is particularly suitable for applications that involve a large number of independent and *small to medium sized* linear algebra problems. These problems arise from combustion modeling, bio-chemical modeling, fusion plasma modeling, and nuclear physics simulations, to name a few. The dimension of each matrix is of order hundreds to tens of thousands. Most of the GPU throughput would be wasted if we solve each of them individually. Well-known libraries supporting dense batched matrix operations include MAGMA [3, 1] and KBLAS [122]. The newer library Kokkos Kernels supports batched linear algebra operations for both dense and sparse matrices [127].

Batched dense operations are also critical components to enable the development of fast sparse direct solvers on GPUs. For example, in the GPU version of STRUMPACK, each level of the assembly tree involves multiple independent diagonal LU, off-diagonal TRSM and GEMM for the update matrices. These are all dense operations and can be organized as three batched routines across all the nodes on one level of the tree. However, the sub-matrices within a level are not of the same size. Abdelfattah et al. expanded the batched MAGMA library to handle irregular patterns of matrices in a batch and added new routines irrLU, irrTRSM and irrGEMM [2]. They showed that the new MAGMA enabled over 6× speedup for STRUMPACK running on a single AMD MI100 GPU.

The SuperLU team led the development of the batched sparse direct solver interface and the implementation in SuperLU_DIST [42]. A salient feature of the new interface is to allow a heterogeneous batch structure such that each linear system can have different size, sparsity pattern and numerical property. The interface allows each system in the batch to have different preprocessing: equilibration, and row and column reordering, depending on the numerical values and sparsity pattern of each individual system. The results with some combustion and fusion applications demonstrate that the batched sparse direct solver can be an order of magnitude faster than the batched banded solver or the batched dense solver in MAGMA.

There are several prominent efforts in developing batching sparse iterative solvers and preconditioners. The modern GPU-oriented Gingko library [23] supports batched iterative solvers for both symmetric and nonsymmetric sparse linear systems. The current version supports a fixed sparsity pattern across all the matrices in the batch. That is, their BatchCsr or BatchEll formats store only one copy of the meta-graph data structure, with different numerical values [121]. In addition, Ginkgo supports batched Block-Jacobi preconditioners for which the inversion of the diagonal block is performed via a batched variable size Gauss–Jordan elimination. The Kokkos Kernels library also supports a number of batched iterative solvers including CG and GMRES. Liegeois et al. described in detail how they optimized the batched sparse matrix vector product (SpMV), which is the performance-dominant operation in most sparse iterative solvers [146].

4 Acceleration via Data-Sparse Compression

Large-scale numerical linear algebra (NLA) algorithms, even possibly exploring exact structural sparsity and communication-avoiding techniques, may remain computationally expensive in terms of flop counts, storage units, communication and data movement. Moreover, modern HPC architectures are increasingly heterogeneous and compute-intensive, but the relatively smaller improvement on memory bandwidth, communication latency and energy efficiency is still limiting the size of problems that can be solved. Data-sparse compression, as a competitive complement, refers to compact representations where the effective amount of information is much smaller than the ambient size of a dense matrix/tensor/array. Data-sparse compression reduces the amount of data that must be stored, moved and processed while maintaining a controllable accuracy level, and has emerged as a fundamental enabler in NLA and scientific computing.

The data-sparse compression can be applied to the entire dense matrix of interest, its judiciously selected dense submatrices, or dense submatrices of a sparse matrix, leading to significant reduction in the asymptotic complexity and/or its large constant prefactors of the original exact NLA algorithm. Data-sparse compression can be categorized into two complementary families of techniques, i.e., low-rank compression and lossy floating-point compression, which are summarized as follows:

Low-rank compression exploits mathematical or algebraic structure in matrices to approximate them efficiently and accurately. Many matrices arising from PDEs [74, 54, 76, 78], integral equations [30, 107], kernel methods in machine learning [197, 56, 59, 58, 11, 151, 177, 9] and optimization [81, 6, 8] are not arbitrary dense matrices: they exhibit latent separability, long-range separability, smoothness, or decay, which implies that submatrices of the matrix have rapidly decaying singular values and can be approximated by products of small matrices of rank far less than the dimension of the matrix. Low-rank compression becomes essential in direct solvers and factorization-based preconditioners (compressing dense frontal matrices and Schur complements), iterative solvers (compressing operator evaluations and Krylov bases), fast transforms, and inverse problems (accelerating Jacobian/Hessian actions and PDE-constrained optimization). In the context of direct solution of large-scale sparse and dense linear systems, low-rank compression includes single low-rank compression [227, 117], rank-structured matrix algorithms leveraging single-level low-rank compression [207, 5] or multilevel compression such as hierarchical matrices [106, 55, 132, 64, 178, 9], and their higher-order tensor extensions [160, 140]. These methods can effectively reduce the $O(n^3)$ computational complexity for factorizing a dense $n \times n$ matrix (or lower but still superlinear complexities for sparse matrices) to quasi-linear, linear or even sublinear complexities. Moreover, many modern low-rank compression methods can reduce the constant pre-factors, fully utilize parallel computers, and maintain low asymptotic complexity, making them capable of solving record-breaking large-scale dense and sparse linear systems.

Lossy floating-point compression exploits bit-level redundancy in numerical arrays. Rather than exploiting algebraic structure by low-rank compression, it reduces cost by quantizing or transforming data to a compressed representation with controlled errors. Matrices and multi-dimensional arrays, arising from PDE solutions and operators, machine learning features, fast transforms (e.g, FFT), and numerical algorithms (e.g., AMG) can exhibit local smoothness/correlation, and small dynamic range or low entropy, whose bit-level representation can be efficiently transformed or quantized to exploit the redundancy. Existing lossy compressors such as ZFP [148], SZ [67], and FPZIP [149] can significantly reduce the storage cost of large scientific data and leverage shared-memory and GPU parallelism. To the best of our knowledge, lossy floating-point compression has been mostly used in reducing the memory requirement or communication volume, albeit incurring manageable computational overheads during compressing and decompressing the data.

It is worth emphasizing that low-rank compression and lossy floating-point compression are two complimentary families of data-sparse compression algorithms: the former exploits mathematical structure (long-range smoothness and spectral decay) and the latter exploits bit-level redundancy (local smoothness and correlation). In practice, one can often attain optimal compression by hybridizing the two algorithms such as in [130, 61, 186, 126] and we will see a few such examples in section 5.

In the rest of this section, we will focus on the description of the fundamentals of rank-structured matrices for the direct solution of dense matrices.

4.1 Overview of Rank-Structured Matrix Algorithms

Large-scale dense linear systems $AX = B$ arise from many computational science and engineering, machine learning and statistics problems, including matrices in boundary element methods and volume integral equations, Fourier integral operators, fractional Laplacian operators, kernel matrices and covariance matrices in machine learning, Hessian matrices in PDE-constrained optimization, and effectively dense matrices due to fill-in of a sparse direct factorization for PDEs. Exact arithmetics-based solvers for these large linear systems can be computationally very demanding even leveraging modern parallel HPC architectures. This is particularly the case when the system becomes ill-conditioned and/or involves many right-hand side vectors B, as in multi-physics, multi-scale simulations and machine learning pipelines. That said, these matrices often exhibit rich mathematical and physical structures that can be exploited to significantly reduce the solver complexities. We will see a few such examples in section 4.2.

Rank-structured matrix algorithms [172] aim at solving these large $n \times n$ linear systems with less than the conventional methods' $O(n^2)$ memory and $O(n^3)$ computational requirements. Taking hierarchical matrices as an example, they leverage the ideas of multilevel matrix partitioning and low-rank compression. They can be viewed as the algebraic version of the fast multipole methods with additional support for matrix inversion. After a proper *preprocessing* phase that reorders the matrix rows/columns and hierarchically clusters them into groups, certain off-diagonal blocks at each level ℓ of the reordered matrix reveal geometric structure or smooth interactions, leading to rapidly decaying singular values. This means these blocks can be approximated, up to a user-specified accuracy tolerance ϵ, as low-rank matrices:

$$A_{st}^{\ell} \approx U_{st} V_{st}^{T}. \tag{4}$$

Here A_{st}^{ℓ} is the $n_s \times n_t$ block corresponding to row group s and column group t, U_{st} and V_{st}^{T} are the low-rank factors, and the rank $r \ll \min(n_s, n_t)$ depends on the specified tolerance ϵ and the nature of the underlying kernel. As a result, one needs far less than $O(n_s n_t)$ flop operations and storage units for this block. After compressing all off-diagonal blocks at each level of the hierarchical clustering (which is referred to as the *construction* phase), one ends up with a compressed representation $H \approx A$ with almost $O(nr)$ memory requirement. Here we assume all blocks have the same numerical rank r for simplicity of explanation.

Once the construction of H is finished, many basic operations on A can be approximated and accelerated using H instead of A. These include *factorization* of H into H^{-1} [9, 97] or its equivalents (e.g., ULV [55, 53], LU [97, 207, 213], QR [128], Cholesky [228, 97] and telescoping factorization [174, 132]), *application* of H or H^{-1} to given vectors (also called the *solve* phase), recompression of the composition of H and other operators into a new rank-structured matrix, etc.

There are a wide variety of rank-structured matrices developed in applied mathematics, computer science and engineering (e.g., acoustics, seismics, electromagnetics and fluid dynamics) societies. Multilevel partitioning-based algorithms, i.e., hierarchical matrices, can be roughly classified into four categories depending on two criteria, i.e., the admissibility condition and the shared basis property. The admissibility condition refers to whether the low-rank compression of eq. (4) is applied to all off-diagonal blocks (dubbed weak admissibility) or only to those that are "well-separated" from each other using certain distance metrics (dubbed strong admissibility); the shared basis property refers to whether the U_{st} (and V_{st}) factors for two row (and column) groups in eq. (4) share common column (and row) bases with other groups. Weak-admissible and independent-basis algorithms typically refer to hierarchically off-diagonal low-rank (HODLR) matrices [9], and strong-admissible and independent-basis algorithms include $\mathcal{H}$ matrices [106] and their variants [125]. Weak-admissible and shared-basis algorithms include hierarchically semi-separable (HSS) matrices [55, 228, 226, 95], hierarchical interpolative factorization (HIF) [132], and strong-admissible and shared-basis algorithms include $\mathcal{H}^2$ matrices [106, 165], inverse fast multipole methods (IFMM) [64], and strong recursive skeletonization factorization (RS-S) [178], etc. In general, strong admissibility, as opposed to weak admissibility, leads to smaller ranks for high-dimensional problems (e.g., solving PDEs on 3D manifolds or in 3D spaces, or performing kernel regression in high-dimensional feature spaces), but results in higher computational costs for low-dimensional problems; shared-basis based algorithms are more sophisticated but computationally less expensive than independent-basis based algorithms. A summary of the four major classes of hierarchical matrix algorithms is provided in table 4.

When dealing with matrices arising from oscillatory problems such as high-frequency wave equations and oscillatory integral transforms, the rank r in eq. (4) can still remain high even for strong-admissibility-based hierarchical matrix algorithms. In this regime, one can replace the low-rank compression A^l_{st} in eq. (4) with multilevel compression algorithms such as the directional fast multipole method [73] or butterfly compression [176, 163], which leads to efficient extensions of hierarchical matrices capable of solving oscillatory problems with quasi-linear computational complexity. These algorithms include strong-admissible formats e.g., directional $\mathcal{H}^2$ matrices [48] and hierarchical block LU-based solvers [102, 109], as well as weak-admissible formats e.g., hierarchical off-diagonal butterfly matrices (HODBF) [157, 205] and directional preconditioners [239].

As opposed to the hierarchical matrix algorithms, single-level partitioning based algorithms, named block low-rank (BLR) matrices or tile low-rank matrices [207, 5], only compress off-diagonal blocks of A^l_{st} in eq. (4) at a fixed level ℓ. These algorithms are simpler than hierarchical matrix algorithms but can also provide significant computational and memory savings. Just like the hierarchical matrix algorithms, BLR algorithms also have variants that use shared bases or different admissibility conditions [193, 24, 16].

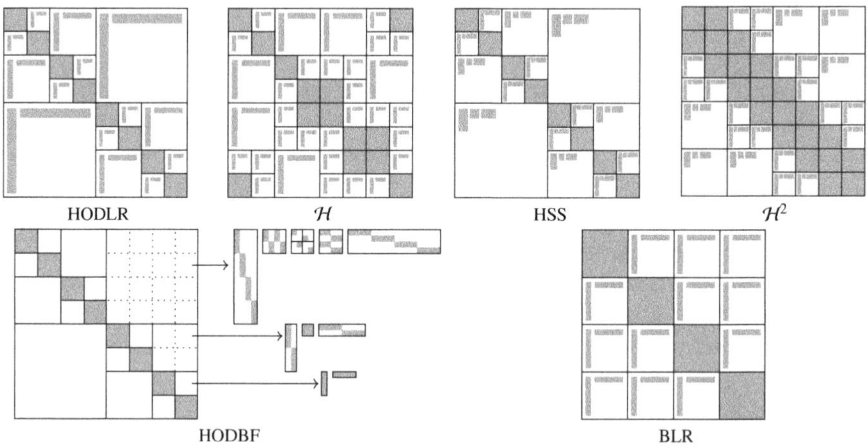

Figure 6 Illustration of several types of hierarchical matrices: HODLR, $\mathcal{H}$, HSS and $\mathcal{H}^2$, HODBF (the butterfly extension of HODLR), and BLR (single-level partitioning-based). In HODLR, $\mathcal{H}$ and BLR, each admissible block A^ℓ_{st} is represented as $A^\ell_{st} \approx U_{st} V^T_{st}$ where U_{st} and V^T_{st} are plotted. In HSS and $\mathcal{H}^2$, each admissible block A^ℓ_{st} is represented as $A^\ell_{st} \approx U_s B_{st} V^T_t$ where U_s and V^T_t are not explicitly formed and can be constructed from the transfer matrices via eq. (12). At non-leaf levels, the transfer matrices E_s, E^T_t and the coupling matrices B_{st} are plotted.

Format	Storage	Construction	Factorization	Apply	Adm	Shared Basis	Dim
HODLR	$O(nr\log n)$	$O\left(nr^{\alpha}\log^{\beta} n\right)$	$O\left(nr^2\log^2 n\right)$	$O(nr\log n)$	Weak	No	$d \le 2$
$\mathcal{H}$	$O(nr\log n)$	$O\left(nr^{\alpha}\log^{\beta} n\right)$	$O\left(nr^2\log^2 n\right)$	$O(nr\log n)$	Strong	No	$d \ge 3$
HSS	$O(nr)$	$O(nr^{\alpha})$	$O(nr^2)$	$O(nr)$	Weak	Yes	$d \le 2$
$\mathcal{H}^2$	$O(nr)$	$O(nr^{\alpha})$	$O(nr^2)$	$O(nr)$	Strong	Yes	$d \ge 3$

Table 4 Summary of four major classes of hierarchical matrix representation of an $n \times n$ matrix. r denotes the maximum numerical rank in the off-diagonal blocks, d denotes the dimensionality of the underlying scientific or machine learning kernels. $\alpha = 1, 2$, $\beta = 1, 2$. "Adm" means "admissibility", "Dim" means "dimensionality", and "Apply" represents application of the matrix or its inverse to a single vector. Note that $\mathcal{H}$ and $\mathcal{H}^2$ matrices also support weak admissibility, but their major advantage over other hierarchical matrix algorithms is attributed to their support for strong admissibility for high-dimensional problems.

4.2 Existence of Hierarchical Matrix Structure in Practical Applications

The efficiency of the rank-structured matrix algorithms heavily relies on, after certain preprocessing, the existence of the low-rank structure for the matrix operators appearing in many scientific computing, physics, engineering, and machine learning applications. Here we list a few application examples where rank-structured matrix algorithms are known to be effective.

4.2.1 Integral Equations

For the simulation of wave phenomena (e.g., in acoustics, elastics, electromagnetics and optics [169, 213, 207, 205, 174, 62, 107, 100, 159]), fluid flow (e.g. Stokes flow [29]) and plasma dynamics (e.g., MHD [124], Poisson solvers [37]), one oftentimes solves a linear system with matrix elements A_{ij} in the form of

$$A_{ij} = \int_{D_i}\int_{D_j} \psi_i(y)\, G(x, y)\, \phi_j(y)\, dx\, dy. \tag{5}$$

Here ψ_i and ϕ_j are local basis functions (often $\psi_i = \phi_i$), D_i and D_j are the domain of integration, and G is the Green's function for Helmholtz, Laplace, Yukawa or Stokes kernels, etc. For example, the Green's function of the integral equation (IE) for the 3D Laplace kernel is

$$G(x, y) = \frac{1}{4\pi|x - y|}. \tag{6}$$

For off-diagonal blocks representing Green's function interaction between well-separated spatial groups, one can prove the algebraic compressibility for analytical [238, 74] and semi-analytical [162] Green's function for both smooth and oscillatory kernels. On the other hand, the compressibility of off-diagonal blocks of A^{-1}, essential for preserving the computational advantage of fast direct solvers, remains an active research area [22, 77, 75].

4.2.2 Kernel Matrices

Many machine learning and statistics algorithms, e.g., kernel ridge regression and Gaussian processes, require the inversion of kernel matrices in the form of a Gaussian kernel with elements

$$A_{ij} = \exp - \frac{\|x_i - x_j\|^2}{2\sigma^2} \tag{7}$$

with hyperparameter σ, or many other kernels. Due to smoothness or rapid decay (with respect to $\|x_i - x_j\|$) of the kernels, the kernel matrix A exhibits low numerical ranks in the off-diagonal blocks and rank-structured matrix algorithms have been widely used [197, 56, 59, 58, 11, 151, 177, 9]. Moreover, when optimizing hyperparameters in Gaussian processes with gradient-based methods, one often constructs a hierarchical matrix representation for $\partial_\sigma A$ as well [85].

4.2.3 PDEs and Schur Complements

Discretization of many elliptic PDEs (e.g., Poisson, Helmholtz, Yukawa, convection-diffusion, convection-reaction, and incompressible Stokes, etc.) leads to a sparse linear system with matrix A. In supernodal or multifrontal sparse direct solvers, block elimination of the properly permuted system leads to Schur complements in the form of

$$S = A_{ss} - A_{sr} A_{rr}^{-1} A_{rs}, \quad A = \begin{bmatrix} A_{rr} & A_{rs} \\ A_{sr} & A_{ss} \end{bmatrix}. \tag{8}$$

The Schur complement can be viewed as the numerical Green's function for the underlying PDE [74, 54, 76, 78] and hence exhibits low-rank structures. Here we assume A is 2×2 block partitioned for simplicity. Note that in reality, the block elimination process often leads to Schur complements in the form of

$$S = F_{ss} - F_{sr} F_{rr}^{-1} F_{rs}. \tag{9}$$

Here F are the dense frontal matrices where F_{rr} and F_{ss} are represented as hierarchical matrices, and F_{rs} and F_{sr} are presented as either hierarchical matrices or single low-rank compression [61], depending on the admissibility condition.

4.2.4 PDE-Constrained Inverse Problem and Optimization

In PDE-constrained optimization and inverse problems [81, 6, 8], one optimizes some misfit term $f(u(m))$ with respect to model parameter m subject to a PDE constraint $A(m)u - b = 0$, which leads to a Hessian matrix in the form of

$$G = A^{-T}\partial^2_{mm} f A^{-1} + S. \tag{10}$$

Here A is the discretized PDE operator, A^{-1} can be computed via hierarchical matrix algorithms, $\partial^2_{mm} f$ and S are local operators. One can show that G also inherits the hierarchical matrix structure of A^{-1} [97, 8, 32] and can be efficiently inverted by hierarchical matrix algorithms.

In addition to the above-mentioned classes of matrices, many other special matrices such as Toeplitz matrix, Hankel matrix, Fourier integral operator, and spherical harmonic transform also exhibit hierarchical rank structures.

4.3 Algorithm Description with an $\mathcal{H}^2$ Matrix Example

4.3.1 Preprocessing

The preprocessing phase is typically required for all existing rank-structured matrix algorithms to reveal the proper hierarchical structure to achieve optimal compressibility and computational efficiency. The preprocessing phase determines (a) clustering: permutation information that block partitions the matrix into, typically, multiple levels and (b) admissibility assignment: which blocks at each level of the permuted matrix can be algebraically compressed. When geometrical information is available for each row/column, e.g., Cartesian coordinates x_i in eq. (7), or any geometrical point in the support of local basis functions ψ_i, ϕ_j in eq. (5), clustering algorithms such as KD tree, cobblestone-like [207] or random projection tree [56] are commonly used. For dense frontal matrices arising from sparse multifrontal solvers with no geometrical information, the frontal matrices are clustered using graph partitioning algorithms such as recursive bisection [156, 91].

Similarly, admissibility assignment needs to estimate how far away a pair of block rows and block columns are from each other. When geometrical information is available, the distance and diameters of two geometrical groups can be easily computed via geometrical distance (Note that even frontal matrices for PDEs can inherit the geometrical information [245] when the PDE is discretized on e.g., a 3D Cartesian grid). However, for (SPD) kernel matrices algebraically constructed without geometrical information, kernel distances have been exploited to determine the admissibility [245]. Similarly, for frontal matrices without geometrical information, one can rely on graph distance [156], or adaptive algorithms (i.e., attempting compressing each block but further partitioning it if the rank is too high) [60]. In addition to admissibility, sometimes it suffices to compute a set of nearest neighbors for each row/column. This can be computed with k-nearest neighbor search [36], approximate nearest neighbor search [56] or graph-based algorithms [156].

As an example, consider an $\mathcal{H}^2$ representation of the kernel matrix in eq. (7) with geometrical information x_i. A KD tree-based clustering will generate a cluster tree I of $L = O(\log n)$ levels (see fig. 7(b)). Each pair of clusters (I_s, I_t) (or simply (s,t)) at each level ℓ are deemed admissible;

$$\texttt{adm}(s,t) = 1,\ \texttt{if}\ \frac{\texttt{Diam}(s) + \texttt{Diam}(t)}{2} \leq \eta\, \texttt{Dist}(s,t), \tag{11}$$

where Diam and Dist are the diameters and distance of the two clusters. The block A_{st} for the admissible pair, i.e., admissible block, is algebraically compressed and the inadmissible block is kept as dense. Figure 7(a) shows the matrix structure with a set of $n = 2^{15}$ 3D geometry points using the admissibility parameter $\eta = 0.7$ (green and red indicate admissible and inadmissible blocks, respectively). The set of clusters that form inadmissible pairs with a cluster t are denoted as $\mathcal{N}_t$. The set of clusters (a) that form admissible pairs with cluster t and (b) whose parents form inadmissible pairs with the parent of t, are denoted as $\mathcal{F}_t$. In $\mathcal{H}^2$ matrix format, each admissible block is compressed as $A_{st} = U_s B_{st} V_t^T$ where B_{st} is a $r \times r$ coupling matrix, and U_s and V_t^T are column and row bases. The basis for leaf nodes in the cluster tree are stored explicitly, and the basis for an inner node t of the cluster tree is defined in terms of the basis of its children t_1 and t_2 using transfer matrices E_{t_1} and E_{t_2}, resulting in a nested and shared basis:

$$U_t = \begin{bmatrix} U_{t_1} & \\ & U_{t_2} \end{bmatrix} \begin{bmatrix} E_{t_1} \\ E_{t_2} \end{bmatrix}. \tag{12}$$

See the $\mathcal{H}^2$ illustration in fig. 6 where the leaf-level inadmissible blocks, coupling matrices, leaf-level basis matrices, and higher-level transfer matrices are shown.

In addition to $\mathcal{H}^2$ matrices, different choices of admissibility parameter ($\eta \geq 1$ (weak admissibility) or $\eta \leq 0.5$ (strong admissibility)) and shared basis property lead to all the other hierarchical matrix formats in table 4 and fig. 6. It is worth noting that in independent basis-based formats such as $\mathcal{H}$ or HODLR matrices, the low-rank compression of each admissible block becomes $A_{st} \approx U_{st} V_{st}^T$ of eq. (4) instead of $A_{st} \approx U_s B_{st} V_t^T$ (again, see fig. 6).

4.3.2 Construction

The construction phase essentially computes the low-rank compression $A_{st} \approx U_{st} V_{st}^T$ of eq. (4) for all admissible blocks to form a computationally efficient approximation $H \approx A$. In order to achieve lower than $O(n^2)$ computational complexity for large matrices, one needs to use alternative algorithms to truncated SVD or rank-revealing QR that forms a full $n \times n$ admissible block. Instead, optimal-complexity algorithms rely on the assumption of either (a) any entry of the block A_{st} can be efficiently computed on the fly, or (b) the block and its transpose can be rapidly applied to any random vector Ω as $A_{st}\Omega$ and $A_{st}^T\Omega$. The former leads to purely algebraic algorithms such as adaptive cross approximation (ACA) or Nyström method [30, 31, 242, 161, 227], and proxy point or Chebyshev interpolation-based methods [234, 232] requiring geometrical information. These algorithms have been extended for butterfly compression as well [50, 142, 188]. On the other hand, the latter leads to random sketching-type algorithms such as randomized SVD [145], QR [231, 82], UTV [173] and butterfly [163, 142] algorithms. These faster algorithms can achieve up to an optimal $O(nr)$ or $O(nr^2)$ computational complexity.

When it comes to efficient construction of rank structured matrix representations, the above assumptions become (a) entry-evaluation: any entry of A can be efficiently computed, (b) fully sketching: $A\Omega$ and $A^T\Omega$ can be rapidly computed, or (c) partially sketching: a hybrid of (a) and (b). In practical applications, entry-evaluation applies to integral equations eq. (5) and kernel matrices eq. (7), etc.; randomly sketching applies to frontal matrices eq. (9) and Hessian matrices eq. (10), etc; partially sketching is often applied to frontal matrices (when the update term $F_{sr}F_{rr}^{-1}F_{rs}$ in eq. (9) is formed as a low-rank compression first).

In the context of $\mathcal{H}^2$ matrices, the adaptation of fast low-rank compression algorithms becomes more sophisticated due to the shared-basis property. In other words, one needs to compute a single low-rank basis U_s of A_{st}, $t \notin \mathcal{N}_s$ for each cluster s and a single low-rank basis V_t of A_{st}, $s \notin \mathcal{N}_t$ for each cluster t via entry evaluation or sketching. Fast entry evaluation-based $\mathcal{H}^2$ algorithms include proxy surface-based [116, 177, 174, 232], polynomial interpolation-based [107, 35, 241], and ACA-type [33, 243] algorithms. On the other hand, fast sketching-based $\mathcal{H}^2$ algorithms include fully sketching-based top-down algorithms [147, 135, 241], fully sketching-based bottom-up algorithm [237], and partially sketching-based bottom-up algorithm [44].

Here we give a brief description of the partially sketching-based bottom-up algorithm [44] using the example of fig. 7(b). Note that admissible blocks only exist at the lower two levels. Assuming A is real and symmetric for simplicity of explanation, the algorithm uses interpolative decomposition (ID) for admissible blocks such that $A_{st} \approx U_s B_{st} U_t^T$, and $B_{st} = A(\widetilde{I}_s, \widetilde{I}_t)$ for some skeleton sets $\widetilde{I}_s \subseteq I_s$ and $\widetilde{I}_t \subseteq I_t$. From the sketching data $Y = A\Omega$ with some random matrix $\Omega \in \mathbb{R}^{n\times d}$, $d > r$, the algorithm first computes $Y_s = Y(I_s, :) - \sum_{b\in \mathcal{N}_s} D_{sb}\Omega_b$ with $\Omega_b = \Omega(I_b, :)$ for each leaf-level cluster s. Note that D_{sb} denotes leaf-level inadmissible blocks and is explicitly computed via entry evaluation. Y_s effectively sketches the column space of all admissible blocks at block row s (see fig. 7(c)), and U_s and $\widetilde{I}_s$ (also denoted as I_{sS}) can be computed from Y_s via ID. Note that A in fig. 7(c) has been reordered such that the redundant indices $I_{sR} = I_s \setminus I_{sS}$ appear before skeleton indices I_{sS} for each cluster s. After applying a transformation associated with U_s, only the skeletonized rows $A(\widetilde{I}_s, :)$ (i.e., rows of A corresponding to skeleton sets $\widetilde{I}_s$) and columns $A(:, \widetilde{I}_s)$ remain (see fig. 7(d)(e)). At the next level, one computes $Y_s = \begin{bmatrix} U_{s_1} Y_{s_1} \\ U_{s_2} Y_{s_2} \end{bmatrix} - \begin{bmatrix} \sum_{b\in\mathcal{F}_{s_1}} B_{s_1 b} U_b^T \Omega_b \\ \sum_{b\in\mathcal{F}_{s_2}} B_{s_2 b} U_b^T \Omega_b \end{bmatrix}$ for each level-2 cluster s with s_1, s_2 being its children. Note that the coupling matrices $B_{s_1 b}$ and $B_{s_2 b}$ are explicitly computed via entry-evaluation (see the blocks in red in fig. 7(f)) and Ω_b and T_b for a level-1 cluster b have been scaled by U_b. Y_s effectively sketches the column space of all admissible blocks shown in green in fig. 7(h) from which transfer matrices E_{s_1} and E_{s_2} can be computed via ID. Finally, the coupling matrices B_{st} at level 2 are explicitly formed via entry evaluation. As can be seen, this algorithm leverages both sketching and entry evaluation, and can achieve the optimal $O(nr^2)$ computational complexity. Numerical results in [44] report up to a 1000× speedup over an existing GPU implementation of the top-down sketching-based algorithm [241] and a 660× speedup over a CPU implementation of the algorithm [135] (see fig. 8).

In addition to $\mathcal{H}^2$ matrices, fast construction algorithms have also been applied in many other rank-structured matrix algorithms, including entry-evaluation-based algorithms for $\mathcal{H}$ [97], HODLR [11], HODBF [205], HSS [56] and BLR [207], as well as sketching-based algorithms for $\mathcal{H}$, HODLR [147, 135], HSS [96, 170, 134] and uniform BLR [193], etc.

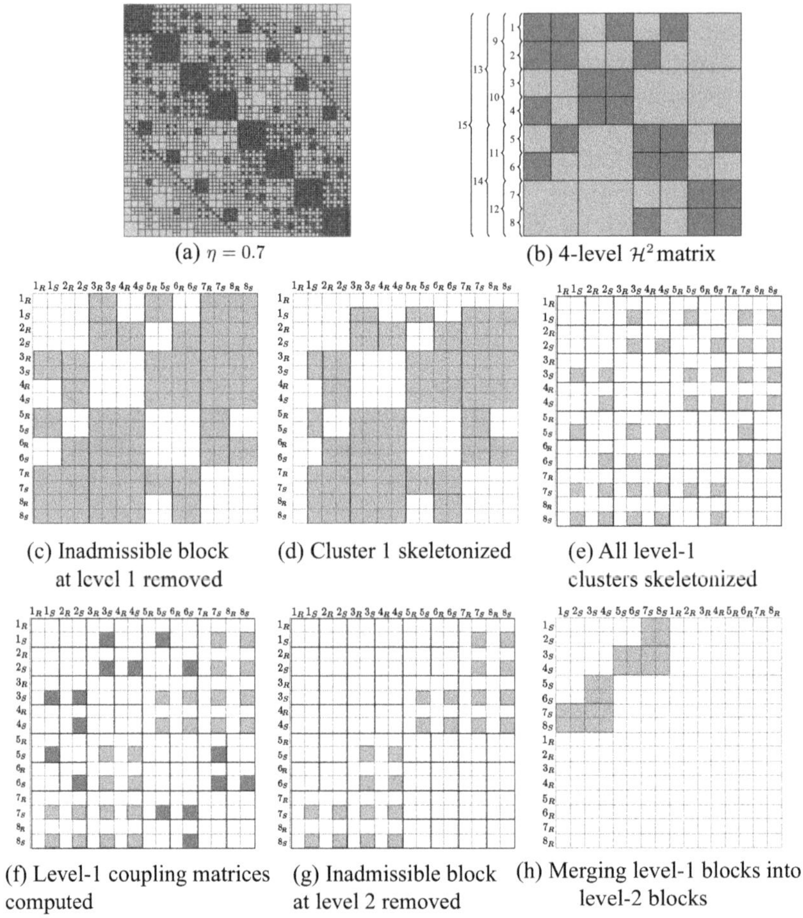

(a) $\eta = 0.7$ (b) 4-level $\mathcal{H}^2$ matrix

(c) Inadmissible block at level 1 removed (d) Cluster 1 skeletonized (e) All level-1 clusters skeletonized

(f) Level-1 coupling matrices computed (g) Inadmissible block at level 2 removed (h) Merging level-1 blocks into level-2 blocks

Figure 7 [44, figs. 1 and 4] (a) Block partitioning of a hierarchical matrix for a 3D problem of size $N = 2^{15}$ with different η. (b) Illustration of a 4-level $\mathcal{H}^2$ matrix. (c)-(h) First two levels of the partially sketching-based algorithm in [44].

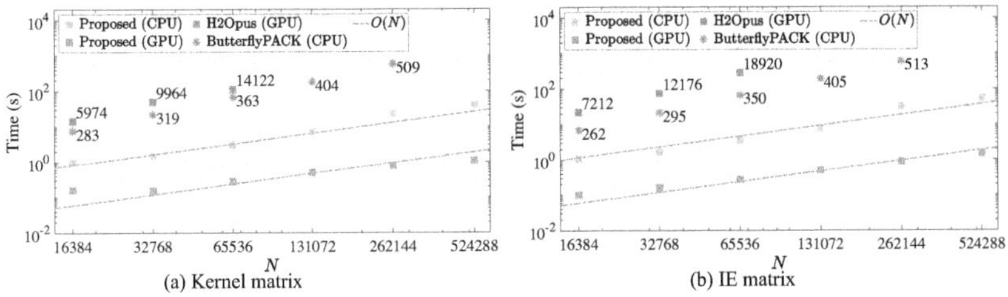

Figure 8 [44, fig. 5] The execution time for the CPU and GPU implementations of the construction algorithm [44] for the kernel and IE matrices. Also shown is the top-down construction algorithm implemented in ButterflyPACK [135] and H2Opus [241], with its data points labeled with the total number of random samples d. In comparison, algorithm [44] requires only $d = 256$.

4.3.3 Factorization and Solve

The factorization phase efficiently computes the inverse of the rank-structured matrix H^{-1} or its equivalent, by leveraging arithmetic operations in the compressed formats such as block addition and multiplication [97], Sherman–Morrison–Woodbury formula [9], and skeletonization [174]. Once H^{-1} has been computed, it can be used in the solve phase to compute $X = H^{-1}B$ for any RHS B leveraging fast arithmetic operations as well. It's worth noting that the factorization (and solve) algorithms can significantly differ from each other based on different rank-structured matrix formats. These include recursive block LU [39, 97, 102, 109] for $\mathcal{H}$ or $\mathcal{H}^2$ matrices, single-level block LU for BLR matrices [207], block ULV or Cholesky for HSS [55, 53, 228, 95], multiplicative inversion for HODLR [12, 11, 9] and HODBF [157, 205], recursive skeletonization for HSS and $\mathcal{H}^2$ [174, 178]. These algorithms can achieve an up to $O\left(nr^2\right)$ optimal computational complexity as shown in table 4. While many factorization algorithms for weak-admissible formats such as HODLR [9] or HSS [174] can guarantee a constant growth factor in r during factorization, these formats are less preferable for high-dimensional problems as r does not stay independent of n due to weak admissibility. Although there exist specialized algorithms [62, 132] for PDEs on structured grids, strong-admissible formats such as $\mathcal{H}$ or $\mathcal{H}^2$ matrices exhibit much smaller ranks.

As efficient and parallel $\mathcal{H}^2$ factorization algorithms represent one state-of-the-art rank-structured matrix algorithm research area, we summarize existing $\mathcal{H}^2$ algorithms in the following. The algorithms of Börm et al. [40] and Ma et al. [165] represent the early $\mathcal{H}^2$ factorization algorithms based on recursive block addition/multiplication and local update, which can attain $O\left(nr^2\right)$ computational complexity, albeit with high constant prefactors. Later, inverse fast multipole (IFMM) methods [64, 10, 196, 53] were developed which rewrite the dense system as an extended sparse system using FMM bases and perform block LU elimination. More recently, the recursive skeletonization with strong admissibility (RS-S) [178] was proposed leveraging both interpolative decomposition and block LU elimination. RS-S exhibits much smaller prefactors and is more amenable to efficient parallelization. A sequence of follow-up works exists [144, 167, 166, 237, 213, 43].

Here we give a high-level overview of the $\mathcal{H}^2$ factorization algorithm in [43] using the 4-level $\mathcal{H}^2$ matrix of fig. 7(b). For simplicity of explanation, we write $H = M+D+F$ where M and D contain respectively all admissible and inadmissible blocks at level 1 and F contains the fill-ins to be generated (see fig. 9(a)). For cluster 1, let $Q_1 = \mathtt{diag}(\widetilde{Q}_1, I, \cdots, I)$ where $\widetilde{Q}_1 = [U_1^{\perp}, U_1]$, $\bar{H} = Q_1^T H Q_1$ skeletonizes cluster 1. Next, LU elimination of $\bar{H}_{1_R 1_R}$ by L_1 and U_1 modifies blocks in D (shown in blue in fig. 9(b)) and introduces fill-ins in F (shown in black in fig. 9(b)). Let A_s^{ℓ} denote the matrix after LU elimination of s_R at level ℓ by L_s and U_s, we have $A_1^1 = L_1 Q_1^T H Q_1 U_1$. With a slight abuse of notation, we still split A_1^1 (and the results of all subsequent A_s^{ℓ}) into $A_1^1 = M + D + F$. We repeat these operations for cluster 2 with $A_2^1 = L_2 Q_2^T A_1^1 Q_2 U_2$ (see fig. 9(c)) and for all the remaining level-1 clusters until $A_8^1 = L_8 Q_8^T A_7^1 Q_8 U_8$ (see fig. 9(d)). Note that the fill-in blocks (in black) appearing at the row block of cluster s are used to augment the original U_s. The blocks in M and F corresponding to inadmissible blocks at level 2, marked in red in fig. 9(e), are then moved into D. Finally, a permutation P^1 is applied to permute all the diagonal blocks of redundant indices to the lower right corner (see fig. 9(f)), followed by an LU factorization $L_r^1 U_r^1$ that eliminates these redundant blocks. The above procedure is repeated for level $2, 3, \ldots, L$, until at the highest level L there are no more admissible blocks left and a final LU factorization $L_L U_L$ is performed. This LU-like factorization can be summarized as

$$H = \left(\prod_{\ell=1}^{L}\left(\prod_{s\,\mathrm{at}\,\ell} Q_s L_s^{-1}\right) P^{\ell^T} L_r^{\ell}\right) L_L U_L \left(\prod_{\ell=L}^{1}\left(U_r^{\ell} P^{\ell} \prod_{s} U_s^{-1} Q_s^T\right)\right). \tag{13}$$

It is worth noting that unlike weak-admissible formats such as HODLR and HSS, the rank growth of H^{-1} for $\mathcal{H}$ or $\mathcal{H}^2$ matrices is due to the fill-in blocks (in black in fig. 9), which hasn't been fully understood to date [22, 77, 75].

Once the factorization in eq. (13) has been computed, H^{-1} can be applied to any RHS vectors B by applying each factor in eq. (13) one by one. Note that the inverses of Q_s, P^{ℓ}, L_r^{ℓ} and U_r^{ℓ} are readily available now. As a few examples, fig. 10 demonstrates the factorization time, factorization memory and solve time when applying the algorithm of [43] to 2D and 3D, IE and kernel matrices of varying sizes on a shared-memory CPU architecture. As can be clearly seen, the $O(n)$ complexities have been validated.

Needless to say, depending on the accuracy of U_s and condition number of A, the factorization in eq. (13) can be either used as a direct solution operator or efficient preconditioner.

4.4 Parallelization of Rank-Structured Matrix Algorithm

Beyond the superior asymptotic complexities of rank-structured matrix algorithms shown in table 4, their efficient implementation on modern parallel computers represents a core enabler for real-world large-scale multi-physics, multi-scale, and high-fidelity simulations. That said, these algorithms can face significant parallel scalability challenges as (a) different parallel data layouts can introduce vastly different communication

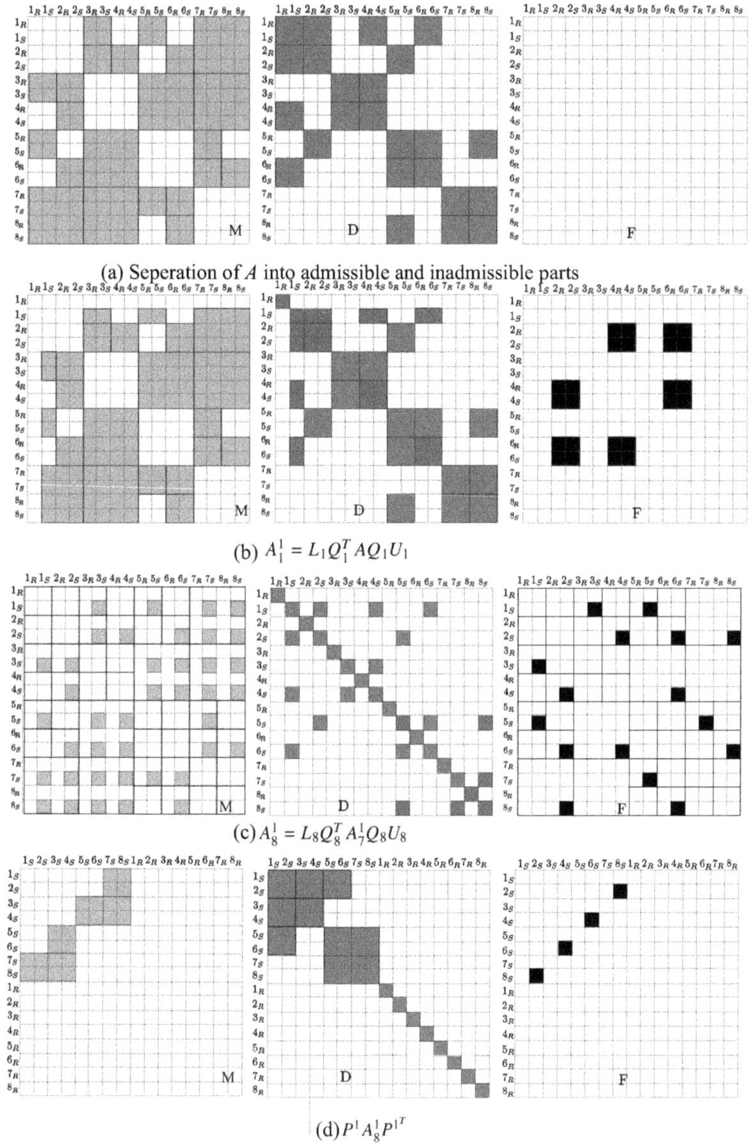

(a) Seperation of A into admissible and inadmissible parts

(b) $A_1^1 = L_1 Q_1^T A Q_1 U_1$

(c) $A_8^1 = L_8 Q_8^T A_7^1 Q_8 U_8$

(d) $P^1 A_8^1 P^{1^T}$

Figure 9 [43, figs. 4-12] Illustration of the factorization algorithm of [43] on the $\mathcal{H}^2$ example of fig. 7(b). (a) The splitting of $A = M + D + F$ where M and D contain respectively all admissible and inadmissible blocks at level 1 and F contains the fill-ins to be generated. Note that A has been reordered such that for each cluster t, its index set $I_t = [t_R, t_S]$ consists of the redundant indices t_R and skeletonization indices t_S, respectively. (b) Skeletonization using Q_1 and elimination using L_1, U_1 for cluster 1. After the skeletonization, the elimination of block row/column 1_R introduces modifications to existing inadmissible blocks (shown in blue) and new fill-ins which are admissible (shown in black). (c) Skeletonization using Q_8 and elimination using L_8, U_8 for cluster 8. (d) Permutation P^1 is applied to permute the diagonal blocks of redundant indices to the lower right corner, which is then LU factorized as $L_r^1 U_r^1$. Note that certain level-1 admissible blocks (green and black) now become level-2 inadmissible blocks.

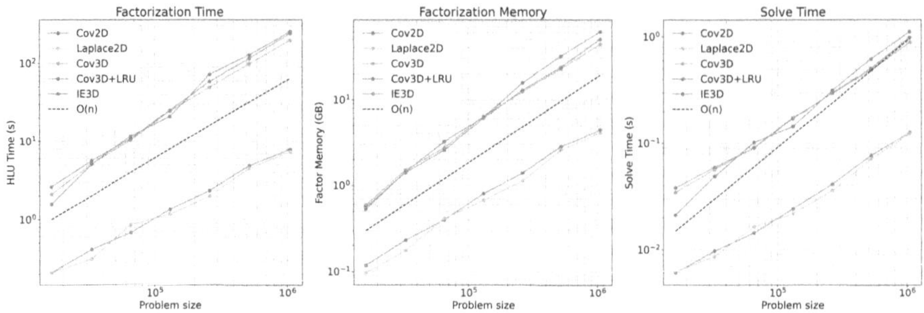

Figure 10 [43, figs. 13 and 16] The factorization time, memory requirement and solve time of the shared-memory parallel $\mathcal{H}^2$ algorithm of [43].

patterns, (b) many algorithms feature sequences of small and irregular-sized numerical linear algebra operations making them hard to achieve peak flop performance of modern hardware, and (c) many factorization algorithms involve complex data dependencies and recursive computations that are sequential in nature. In what follows, we review existing parallelization strategies for rank-structured matrices to tackle these challenges.

4.4.1 Distributed-Memory Process Layout

When the matrix size n is large enough such that the entire rank-structured matrix cannot fit onto a single shared-memory machine, distributed-memory parallelism becomes essential to fully unlock the potential of these algorithms. The optimal parallel data layout should be able to simultaneously achieve (a) good load balance for matrices with general sizes and non-uniform ranks on a non-power-of-two number of processes, and (b) minimal communication during construction, factorization and solve. Here we summarize a few existing parallel data layouts using examples of fig. 11.

For a hierarchical matrix, the most common parallel layout is perhaps the 1D block layout of fig. 11(a), which has been used for HODLR and HODBF matrices [59, 205], $\mathcal{H}$ matrices [141, 101], and $\mathcal{H}^2$ matrices [35, 241] for the construction phase and matrix-vector multiplication. This layout exhibits good load balance, but may require extra data redistribution in matrix-vector multiplication when the input vectors also follow the 1D block layout. Moreover, although this layout leads to manageable communication in the factorization phase for weak-admissible formats such as HODLR and HODBF matrices [205] (see fig. 12 for the strong scaling performance of HODLR, HODBF and HSS-BF algorithms in the ButterflyPACK package [154]), they post significant challenges for recursive computation-based $\mathcal{H}$ and $\mathcal{H}^2$ factorization algorithms. In [233], a so-called lattice-$\mathcal{H}$ matrix algorithm representing a hybrid algorithm between BLR and $\mathcal{H}$ uses a 2D block-cyclic layout as shown in fig. 11(b), which improves parallel efficiency of the factorization phase at the cost of introducing more flop operations. See fig. 12 for the performance difference between 1D block and 2D block-cyclic layouts for the lattice-$\mathcal{H}$ algorithm in ButterflyPACK [154]. In addition, the HSS algorithm of [199] uses a tree-based parallel layout that simplifies the communication pattern during the HSS-ULV factorization and some strong scaling results are shown in fig. 12 for HSS construction. Other parallel layouts have been used in parallel HIF [143], $\mathcal{H}^2$ [144] and BLR [61] factorization algorithms, etc.

When it comes down to the parallel layout of each individual off-diagonal block (e.g., the top-level off-diagonal block of fig. 11(a)), one can use the 2D block-cyclic [199] or 1D block [141] layout for each low-rank factor, and a parallel FFT-like layout for each butterfly representation [195, 163]. See fig. 11(d)-(e) for an illustration.

4.4.2 Fine-Grained Parallelism on CPU and GPU

Efficient parallel implementation of an algorithm (or its local workload of each MPI process) on shared-memory CPU and GPU architectures must take into consideration the irregular data structures in each phase of a rank-structured matrix algorithm. For example, the matrix-vector multiplication of an $\mathcal{H}$ matrix involves an $O\left(n/2^{\ell}\right)$ number of fine-grained GEMV operations of non-uniform sizes at each tree level ℓ. As another example, the factorization of a HODLR matrix involves an $O\left(n/2^{\ell}\right)$ number of fine-grained LU decompositions of variable-size diagonal blocks at each level ℓ. Common implementation strategies are to flatten and marshal the tree metadata so that pertinent operations can be batched on GPUs or CPUs, and the CPU-GPU data transfer can be minimized. This is particularly true for shared-basis algorithms such as HSS and $\mathcal{H}^2$ where even at higher levels of the tree, operations only involve small-sized blocks. These strategies have been leveraged in GPU implementation of BLR factorization [60, 123], HODLR factorization [57], $\mathcal{H}^2$ multiplication [45, 123, 241, 235], sketching-based $\mathcal{H}^2$ construction [44], and shared-memory CPU-based $\mathcal{H}^2$ factorization [43]. Needless to say, the same strategies implemented on GPUs can behave much better than those implemented on CPUs, see the example of $\mathcal{H}^2$ construction [44] in fig. 8.

4.4.3 Strategies to Tackle Data Dependency

The other parallelization challenge is the data dependency, particularly in the factorization phase, of many rank-structured matrix algorithms, e.g., the recursive computation-based $\mathcal{H}$ and $\mathcal{H}^2$ factorization algorithms [39, 97], the skeletonization-based $\mathcal{H}^2$ factorization algorithms [178], and the sketching-based $\mathcal{H}$ matrix construction algorithms [147]. Common practices to exploit parallelism within data dependency are threefold: (a) One can leverage graph-based algorithms such as graph coloring to explore operations that can be simultaneously executed. This has been demonstrated for $\mathcal{H}^2$ factorization [144] and construction [135]. (b) Runtime scheduling systems such as StarPU [26], PARSEC [51] and OmpSs can be leveraged based on the DAG information of e.g., BLR [5, 51] and $\mathcal{H}$ matrix [233, 129, 7] algorithms. (c) One can potentially modify the original factorization algorithm to remove, to some extent, the data dependency. Examples include the lattice-$\mathcal{H}$ [233] algorithm that behaves as BLR (which is easier to parallelize) at the higher tree levels, and the $\mathcal{H}^2$/BLR/HSS algorithms of [167, 210, 168] which precompute the augmented basis matrices due to fill-ins associated with inadmissible blocks.

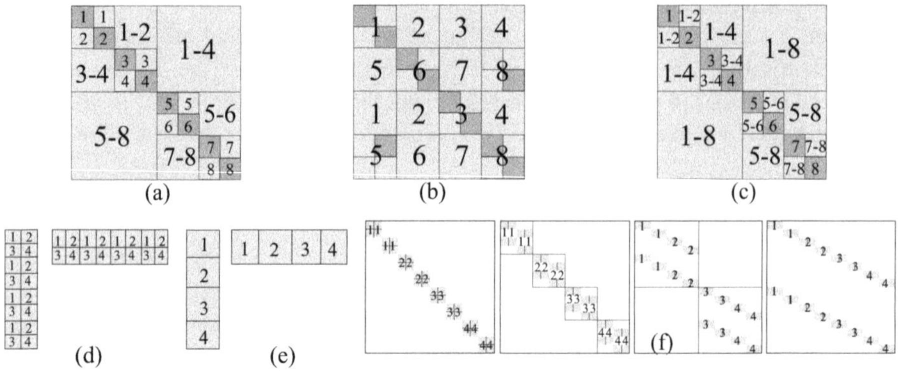

Figure 11 (Top) Parallel process layouts for distributing a hierarchical matrix on 8 MPI processes: (a) 1D block layout for HODLR, HODBF matrices [59, 205], $\mathcal{H}$ matrices [141], and $\mathcal{H}^2$ matrices [35, 241], (b) 2D block-cyclic layout for lattice-$\mathcal{H}$ matrix algorithm [233] and BLR algorithm of [61], and (c) tree-based layout for HSS [199]. (Bottom) Parallel process layouts for distributing a single low-rank or butterfly compression on 4 MPI processes: (d) 2D block-cyclic layout for a low-rank representation used in [199], (e) 1D block layout for a low-rank representation used in [141], and (f) A parallel FFT-like layout for a butterfly representation used in [163, 205, 156].

5 Combining Structure-Sparse and Data-Sparse Formats

Many sparse matrices arising from PDE discretization on structured and unstructured space and time grids (see a few examples in section 4.2.3) exhibit rank structures in their inverse or equivalent factorization [32, 54, 78, 76, 74], hence the rank-structured matrix algorithms introduced in section 4 can be combined with sparse supernodal or multifrontal solvers to construct fast direct solvers or efficient preconditioners. Moreover, other data-sparse compression algorithms such as lossy floating-point compression can also be combined with rank-structured matrix algorithms and sparse direct solvers. In what follows, we focus on reviewing multifrontal sparse solvers with data-sparse compression. That said, there are also attempts to augment supernodal sparse solvers with algebraic compression, such as the integration of HODLR [79] and BLR [185] in the PaStiX package [111], and the use of other low-rank compression techniques in a supernodal-based Cholesky solver [52].

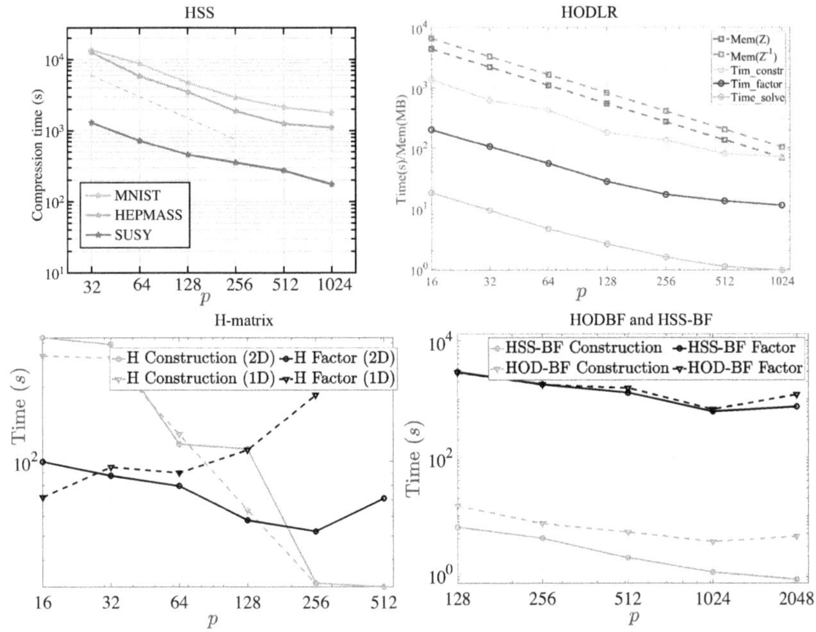

Figure 12 Strong scaling (on distributed-memory systems) of various phases of several hierarchical matrix algorithms: HSS ([56, fig. 7]), HODLR, $\mathcal{H}$ matrix, HODBF and HSSBF. In general, weak-admissible algorithms can attain much better parallel efficiencies than strong-admissible algorithms (e.g., $\mathcal{H}$ and $\mathcal{H}^2$). For $\mathcal{H}$ matrix algorithms, the 2D block-cyclic process layout can achieve a higher parallel efficiency than the 1D block process layout.

5.1 Overview of Data-Sparse Multifrontal Sparse Solvers and Preconditioners

Data-sparse multifrontal solvers have a similar algorithmic workflow to exact multifrontal solvers, except that the dense frontal matrices and their arithmetic are compressed using various compression techniques depending on their sizes and underlying PDE characteristics.

The overall workflow is described in Algorithm 1. Just like exact multifrontal and supernodal solvers, the matrix A is first preprocessed as eq. (1) with diagonal scaling matrices D_r and D_c, and permutation matrices P_r and P_c. For example, a typical choice for the fill-in reduction permutation P_c is nested dissection, from which the frontal matrices can be viewed as graphs between the separators consisting of subsets of vertices of the original graph A. As explained in section 4.3.1, the separators need to be further clustered to reveal rank structures inside each frontal matrix, via graph partitioning algorithms such as recursive bisection [156, 91] or geometrical information if available [245]. This permutation is denoted $\widehat{P}$ in Algorithm 1. Based on P_c, a structure called the assembly tree is constructed consisting of dense frontal matrices as the nodes (see fig. 14 for an assembly tree example). For each non-root node, the frontal matrix has

Algorithm 1 Sparse rank-structured multifrontal factorization using ZFP, BLR, HSS and HODBF matrix compression, followed by a GMRES iterative solve using the multifrontal factorization as an efficient preconditioner.

Input: $A \in \mathbb{R}^{n\times n}$, $b \in \mathbb{R}^n$
Output: $x \approx A^{-1}b$

$A \leftarrow P(D_r A D_c Q_c)P^\top$ ▹ Scaling, and permutation for stability and fill reduction
$A \leftarrow \widehat{P} A \widehat{P}^\top$ ▹ Rank-reducing separator reordering
Build assembly tree: define I_τ^{s} and I_τ^{u} for every frontal matrix F_τ
for nodes τ in assembly tree in topological order **do**
  if no compression **then**
    construct F_τ and perform arithmetic as a dense matrix
  else if ZFP compression **then**
    construct F_τ and perform arithmetic as a dense matrix
    compress as zfp matrix
  else if BLR compression **then**
    $F_\tau \leftarrow \begin{bmatrix} A(I_\tau^{\mathrm{s}}, I_\tau^{\mathrm{s}}) & A(I_\tau^{\mathrm{s}}, I_\tau^{\mathrm{u}}) \\ A(I_\tau^{\mathrm{u}}, I_\tau^{\mathrm{s}}) & 0 \end{bmatrix} \updownarrow\!\!\!\!\leftrightarrow T_{\tau_1} \updownarrow\!\!\!\!\leftrightarrow T_{\tau_2}$
    $P_\tau L_\tau U_\tau \leftarrow F_{11}$ ▹ LU with partial pivoting
    $F_{11}, F_{12}, F_{21} \leftarrow$ BLR compress(F_{11}, F_{12}, F_{21})
    $F_{12} \leftarrow L_\tau^{-1} P_\tau^\top F_{12}$
    $F_{21} \leftarrow F_{21} U_\tau^{-1}$
    $T_\tau \leftarrow F_{22} - F_{21} F_{12}$ ▹ Schur update
  else if HODBF compression **then**
    $F_{11} \leftarrow$ HODBF compress $\left(A(I_\tau^{\mathrm{s}}, I_\tau^{\mathrm{s}}) \updownarrow\!\!\!\!\leftrightarrow T_{\tau_1} \updownarrow\!\!\!\!\leftrightarrow T_{\tau_2}\right)$
    $F_{11}^{-1} \leftarrow$ HODBF invert (F_{11})
    $F_{12} \leftarrow$ butterfly compress $\left(A(I_\tau^{\mathrm{s}}, I_\tau^{\mathrm{u}}) \updownarrow\!\!\!\!\leftrightarrow T_{\tau_1} \updownarrow\!\!\!\!\leftrightarrow T_{\tau_2}\right)$
    $F_{21} \leftarrow$ butterfly compress $\left(A(I_\tau^{\mathrm{u}}, I_\tau^{\mathrm{s}}) \updownarrow\!\!\!\!\leftrightarrow T_{\tau_1} \updownarrow\!\!\!\!\leftrightarrow T_{\tau_2}\right)$
    $S \leftarrow$ butterfly compress $\left(F_{21} F_{11}^{-1} F_{12}\right)$ ▹ Sketching-based construction
    $T_\tau \leftarrow$ HODBF compress $\left(T_{\tau_1} \updownarrow\!\!\!\!\leftrightarrow T_{\tau_2} - S\right)$
  else if HSS compression **then**
    $F_\tau \leftarrow$ HSS compress $\left(\begin{bmatrix} A(I_\tau^{\mathrm{s}}, I_\tau^{\mathrm{s}}) & A(I_\tau^{\mathrm{s}}, I_\tau^{\mathrm{u}}) \\ A(I_\tau^{\mathrm{u}}, I_\tau^{\mathrm{s}}) & 0 \end{bmatrix} \updownarrow\!\!\!\!\leftrightarrow T_{\tau_1} \updownarrow\!\!\!\!\leftrightarrow T_{\tau_2}\right)$ ▹ Sketching-based construction
    $U_\tau L_\tau V_\tau \leftarrow F_{11}$ ▹ HSS ULV
    $F_{12} \leftarrow L_\tau^{-1} U_\tau^T F_{12}$
    $F_{21} \leftarrow F_{21} V_\tau^T$
    $T_\tau \leftarrow F_{22} - F_{21} F_{12}$ ▹ Low-rank update
  end if
end for
$x \leftarrow \mathrm{GMRES}(A, b, M : u \leftarrow D_c Q_c P^\top \widehat{P}^\top$ bwd-solve(fwd-solve$(\widehat{P} P D_r v)))$

a 2×2 block structure $F_\tau = \begin{bmatrix} F_{11} & F_{12} \\ F_{21} & F_{22} \end{bmatrix}$, where F_{11} corresponds to fully summed variables I_τ^s, and F_{22} with indices I_τ^u corresponds the trailing submatrix to be Schur complement updated by F_{11} and will contribute to the parent frontal matrix via extend-add. The data-sparse multifrontal solvers will then traverse the assembly tree in the topological order, and for each (non-leaf) frontal matrix perform factorization of F_{11}, update panels F_{12}/F_{21}, Schur complement update F_{22} as T, and propagate T into its parent using extend-add. For the root frontal matrix, only $F_{11} = F_\tau$ is factorized. Each frontal matrix can be compressed using any existing data-sparse compression algorithm such as lossy floating-point compression including ZFP [148] and SZ [67], or rank-structured matrix algorithms including $\mathcal{H}$, $\mathcal{H}^2$, HSS, BLR, HODLR, HODBF and their other variants. Most existing data-sparse multifrontal solvers use a single compression algorithm, but there also exist solvers that leverage more than one compression algorithm (i.e., hybrid or composite compression algorithms). Since lossy floating-point compression only works well for very small frontal matrices or the inadmissible blocks inside each rank-structured matrix algorithm, we refer to data-sparse multifrontal sparse solvers simply as rank-structured multifrontal sparse solvers in the rest of this section. Note that as the frontal matrix arithmetic is approximated by the compression methods using a prescribed tolerance, the multifrontal solver only computes $M \approx A^{-1}$, which can be used either as a direct solver or a preconditioner (in e.g., GMRES) depending on the tolerance and application needs. In what follows, we review existing rank-structured multifrontal sparse solvers in literature.

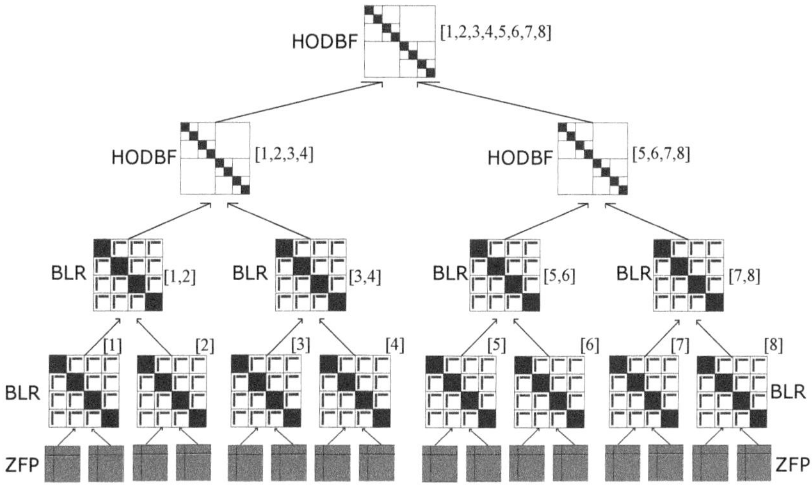

Figure 13 [61, fig. 5] The assembly tree of rank-structured multifrontal solvers using different compression algorithms. The solver is parallelized using 8 MPI processes.

$\mathcal{H}$ and $\mathcal{H}^2$ represent the first hierarchical matrix algorithms applied to frontal matrices. Earlier works consider recursive computation-based factorization algorithms [98, 244, 30] which exhibit large computational overheads and difficulty in efficient parallelization [98], albeit with promising complexity claims. Modern $\mathcal{H}^2$ algorithms and variants such as IFMM and skeletonization-based algorithms have been recently introduced to multifrontal solvers [236, 237, 196, 245], representing an active research area. In addition to these strong-admissible algorithms, weak-admissible algorithms such as HSS [90, 91, 229, 225, 221, 219, 230] represent one of the foundational algorithms in rank-structured multifrontal solver package STRUMPACK [211]. As shown in Algorithm 1, HSS is used to compress F_τ, typically using sketching-based construction algorithms to ensure compression accuracy in fully algebraic settings. Next, F_{11} (which is also a HSS matrix) is factorized using HSS-ULV, and F_{22} can be low-rank

problem	dim	rank $r(m)$ BF	HSS	BLR	$\mathcal{H}^2$	factor flops BF	HSS	BLR	$\mathcal{H}^2$	solve flops/factor memory BF	HSS	BLR	$\mathcal{H}^2$
Helmholtz	2	$\log m$	m	m	m	n	$n^{3/2}$	$n^{3/2}$	$n^{3/2}$	n	$n\log n$	$n\log n$	$n\log n$
	3	$m^{1/4}$	m	m	m	$n\log^2 n$	n^2	n^2	n^2	n	$n^{4/3}$	$n^{4/3}$	$n^{4/3}$
Poisson	2	$\log m$	$\log m$	1	1	n	n	$n\log n$	n	n	n	n	n
	3	$m^{1/4}$	$m^{1/2}$	1	1	$n\log^2 n$	$n^{4/3}$	$n^{4/3}$	n	n	n	$n\log n$	n

Table 5 Asymptotic complexity of rank-structured matrix (HODBF (denoted as "BF" in the table), HSS, BLR and $\mathcal{H}^2$) enhanced multifrontal solvers for 2D and 3D, Helmholtz and Poisson equations. Other variants like HIF [115], HSS2D [226], and multilevel BLR [16] are not shown here. The "big O" notation $O(\cdot)$ has been dropped. Here m denotes the size of a front and n is the size of the global sparse system.

updated into T_τ in an efficient manner. As HSS turns out to be less efficient for high-dimensional problems (e.g. 3D PDEs), its high-dimensional variants such as HSS2D [226] and HIF [115, 80, 49] have been integrated into multifrontal solvers as well. Other weak-admissible formats such as HODLR and HODBF have also been integrated into multifrontal solvers [156] where the HODBF algorithm is well-suited to compress large frontal matrices arising from high-frequency wave equations. As shown in Algorithm 1, HODBF is used to compress F_{11}, F_{12}, F_{21} and T_τ separately, and the Schur complement update S is compressed using sketching-based algorithms [163]. Finally, BLR has also been a popular method for rank-structured multifrontal solvers [175, 13, 20, 185, 15], which is available in both MUMPS [18] and STRUMPACK [211]. BLR has several variants [14, 61] each with different characteristics in flop performance, memory usage and communication complexity. We show the right-looking (RL)-BLR in Algorithm 1, which requires forming the frontal matrix in dense format and performs the arithmetic in compressed representations. Other more memory efficient BLR variants such as hybrid BLR [61] have also been integrated into STRUMPACK. It is worth mentioning that in addition to standard finite difference or finite element discretizations, rank-structured multifrontal solvers have also been developed for higher order discretizations such as the hierarchical merging of Poincaré–Steklov (HPS) solvers [171, 108, 94] and spectral methods [208].

Remark on numerical stability. When one frontal matrix is dense without low-rank compression, F_{11} can be factorized as LU with partial pivoting P_τ which is numerically stable for ill-conditioned frontal matrices. However, rank-structured matrix algorithms can only allow partial pivoting on each leaf-level diagonal block, yielding these algorithms more prone to numerical instability. One exception is the BLR algorithm [175], which allows permutation of the blocks during factorization. In practice, a frontal matrix may need to be scaled or preconditioned before performing the hierarchical matrix arithmetic as demonstrated in [230, 80, 49].

Remark on computational complexity. The asymptotic complexity of rank-structured multifrontal solvers is closely related to the numerical rank of each frontal matrix compressed by a specific rank-structured matrix algorithm. As the frontal matrix represents numerical Green's functions of the governing PDEs, acting on separators that are orthogonal or parallel to each other, the rank behavior can be estimated to derive the overall sparse solver complexity. In table 5 we summarize the asymptotic complexities for the factor flops, solve flops and factor memory of different rank-structured multifrontal solvers using two classes of 2D and 3D PDEs, namely Poisson and high-frequency Helmholtz. We list the rank of each frontal matrix of size $m \times m$, $r(m)$, for HODBF, HSS, BLR and $\mathcal{H}^2$ algorithms as well as the overall solver complexity in terms of the sparse matrix size n. It is worth noting that for 3D high-frequency Helmholtz, the theoretical rank estimate is $r(m) = O(m)$ for HSS, BLR, and $\mathcal{H}/\mathcal{H}^2$, etc. [156, 74], which leads to no theoretical asymptotic complexity reduction compared to $O(n^2)$. However, one often empirically observes $r(m) = O(m^{0.5})$ when $n \leq 400^3$ [219, 221, 220], leading to an empirically reduced solver complexity. While the optimal asymptotic complexity depends on the PDE type, dimensionality, problem size and compression algorithm, we remark that in practice there does not necessarily exist a single compression algorithm that yields the best compression performance for all frontal matrices in a multifrontal solver. As such, hybrid or composite compression-based [61] multifrontal solvers may become necessary. As an example, we illustrate a composite multifrontal solver [61] in fig. 13 where small-sized frontal matrices are compressed using ZFP, small-to-medium-sized ones are compressed using BLR (hybrid BLR or RL BLR), and large ones are compressed using HODBF. Figure 14 demonstrates that the fully composite solver (labeled as "HODBF_BLR(Hybrid)_ZFP") is capable of handling $n = 420^3 = 74,088,000$ for 3D high-frequency wave equations and $n = 512^3 = 134,217,728$ for 3D singularly perturbed reaction-diffusion equations, using 2048 NERSC Cori CPU cores.

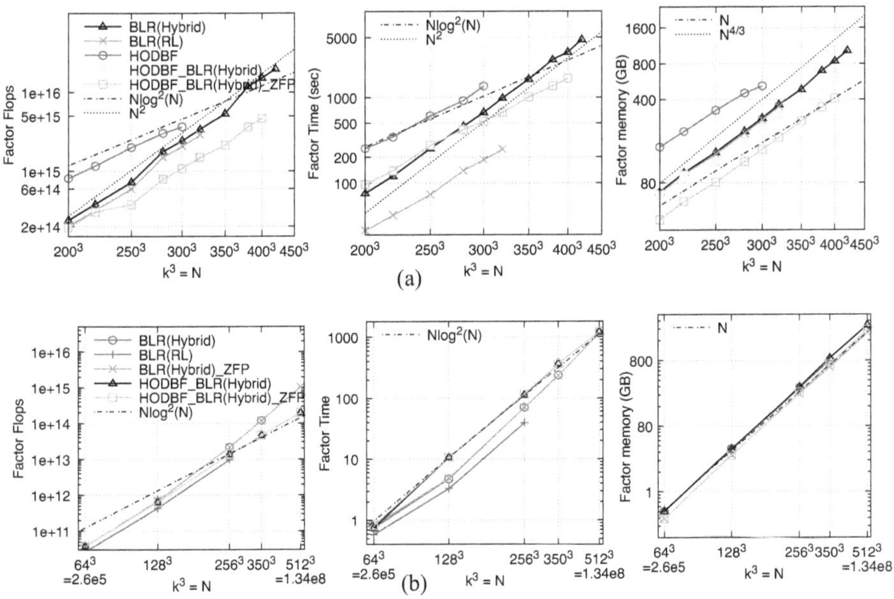

Figure 14 [61, figs. 8 and 9] Characteristics of the factorization phase including flops, CPU time and memory for a multifrontal sparse solver with composite compression algorithms using 64 NERSC Cori CPU nodes.

Remark on distributed-memory and fine-grained parallelization. Distributed-memory parallelization of rank-structured multifrontal solvers represents a core technique to fully unlock the potential of these solvers for large-scale PDE simulations and machine learning computation. Most distributed-memory parallel multifrontal solvers follow a nested parallelization design pattern, which is briefly explained as follows: At the outer layer, the assembly tree is distributed using a tree parallelism approach [19]. The root node frontal matrix is computed on all processes in the root MPI communicator; the root MPI communicator is split into two smaller ones, each handling one of the children frontal matrices at the next level; this procedure continues until a frontal matrix (and the corresponding subtree) is fully occupied by one MPI process. Figure 13 shows the distribution of the assembly tree with 8 MPI processes. Note that each parent MPI communicator does not need to be split into two approximately equal-sized communicators. One can estimate the workload of its child (and the associated subtree) and use the so-called proportional mapping to assign communicator sizes proportional to the estimated workloads [19, 221]. At the inner layer of the nested parallelization, each rank-structured frontal matrix occupied by more than one MPI process can be parallelized using the data layouts explained in section 4.4.1, and each dense frontal matrix can be parallelized using e.g., a 2D block-cyclic layout. The above-described nested parallelization strategy has been utilized in several rank-structured multifrontal solvers, including BLR-based [60, 175], HSS-based [90, 221, 219], HODBF-based [156] and composite compression-based [61] solvers. Figure 15(a) shows the strong scaling performance of different phases of a HSS-based multifrontal solver in STRUMPACK [90] for a sparse matrix tdr455k from accelerator simulations using up to 1152 CPU cores. Fine-grained parallelism inside each frontal matrix can naturally leverage the techniques summarized in section 4.4.2. Beyond that, this kind of parallelism can also be exploited across dense [92] or rank-structured frontal matrices, particularly at lower levels of the assembly tree consisting of large numbers of small-sized frontal matrices. Examples include shared-memory HSS-multifrontal solvers [91] and GPU BLR-multifrontal solvers [60]. In addition, GPUs have been exploited for several two-level multifrontal solvers such as [236, 131]. For large-sized frontal matrices, it is worth mentioning that distributed-memory multi-GPU parallelization of a single frontal matrix has also been recently developed leveraging batched vendor GPU BLAS libraries and GPU-aware MPI libraries [60]. Figure 15(b) demonstrates the performance of the BLR-multifrontal solver [60] compared with exact supernodal and multifrontal solvers on 2048 CPU cores or 128 GPUs on the NERSC Perlmutter.

6 Available Software for Distributed Memory Parallel Computers

In this section, we list in table 6 several direct solver packages (open-source for most of them) leveraging structure-sparse and/or data-sparse algorithms. Here we only mention a few representative packages designed with distributed-memory parallelism, and implemented with HPC programming languages such as C, C++, Fortran or Julia. Note that several of them also provide support for Python or Matlab interfaces. A more complete list can be found at `https://portal.nersc.gov/project/sparse/superlu/`

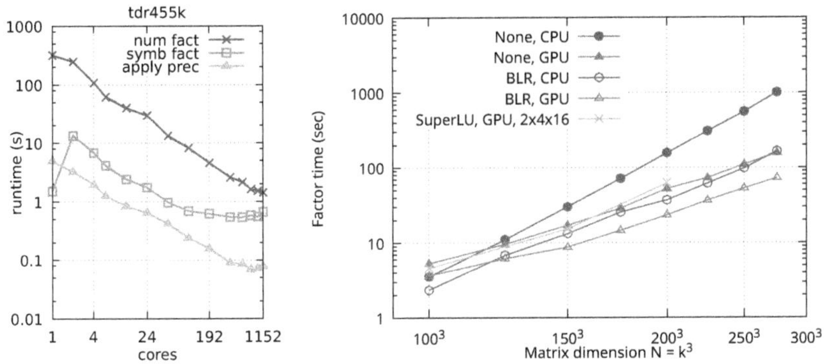

Figure 15 (a) [90, fig. 6] Strong scaling of the parallel HSS-multifrontal solver of STRUMPACK. (b) [60, fig. 1] Performance comparison of the exact multifrontal solver and BLR-multifrontal solver in STRUMPACK and the supernodal solver SuperLU_DIST on multiple CPU and GPU nodes.

`SparseDirectSurvey.pdf` for structure-sparse solvers and `https://github.com/gchavez2/awesome_hierarchical_matrices` for data-sparse solvers.

Software	GPU	Matrix Type	Data-sparse	Programming Language	Typical Application
ASKIT [25, 59]	No	Dense	Yes	C++	Kernel
AHMED [4]	No	Dense/Sparse	Yes	C++	IE/FEM
ButterflyPACK [154]	No	Dense	Yes	F/C++/PY	IE/Kernel/Optimization
DINFMM [68, 144]	No	Dense	Yes	Julia	IE
h2opus [105, 241]	Yes	Dense	Yes	C++	IE/Kernel/Optimization
HiCMA [112]	Yes	Dense	Yes	C/C++	IE/Kernel
HLibPro [114]	No	Dense/Sparse	Yes	C++	IE/Kernel/PDE
MUMPS [181, 18]	Yes	Sparse	Yes	C/F	Any sparse
mxc [182, 168]	Yes	Dense	Yes	C++	IE/Kernel
PaStiX [192, 111]	Yes	Sparse	Yes	C/F/PY/Julia	Any sparse
PARDISO [190]	No	Sparse	No	C/C++/F/PY/Matlab	Any sparse
STRUMPACK [211]	Yes	Dense/Sparse	Yes	C++/F/PY	IE/Kernel/Optimization/Any sparse
SuperLU_DIST [212, 139]	Yes	Sparse	No	C/C++/F/PY/Julia/Matlab	Any sparse

Table 6 List (in alphabetical order) of some distributed-memory direct solver packages leveraging structure sparsity and/or data sparsity."PY" stands for "Python" and "F" stands for "Fortran".

Among the packages in table 6, SuperLU_DIST and PARDISO are exact direct solvers that leverage only data-sparse algorithms and can be applied to any sparse matrix. Other such packages not listed here include PanguLU [189], symPACK [214] and WSMP [223], etc. On the other hand, the other packages in table 6 leveraging data-sparse algorithms for dense matrices implement a variety of rank-structured matrix algorithms described in section 4.1 and usually work as efficient preconditioners. Some of these packages are designed for a particular application such as kernel methods in machine learning and statistics, integral equation (IE) or finite-element method (FEM) matrices, and thus provide additional support at the application layer (e.g., PDE discretization, quadrature, hyperparameter optimization, etc.). Note that there are several dense direct solver packages that only support construction, rather than factorization, of a rank-structured matrix, which we don't list here. Finally, MUMPS,

PaStiX, STRUMPACK, AHMED and HLibPro in the table combine structure-sparse and data-sparse algorithms for sparse matrices exhibiting low-rank structures, often arising from PDE or optimization problems. Note that MUMPS, PaStiX and STRUMPACK can behave as exact direct solvers when their data-sparse algorithms are turned off.

7 Open Problems

Despite tremendous progress in parallel algorithms for both structure-sparse and data-sparse direct solvers, there remain many open problems, both theoretical and practical. We list some of them in the following.

- *GPU-resident solvers.* GPU memory capacity today is orders of magnitude larger than a decade ago. As such, many application codes are moved entirely to GPUs, avoiding the data transfer cost between CPU and GPU. This requires the linear solvers to be running on GPUs as well. The open problems are mostly in the preprocessing steps that are largely on CPUs. These are all combinatorial algorithms and much less progress is achieved on GPUs.
- *Distributed memory symmetric indefinite sparse direct solvers.* This capability is sorely needed by the optimization community. The symmetric LDL^T factorization is highly desirable in numerical optimization methods that use second-order information, as it exposes matrix inertia needed to assess local convexity and KKT optimality. As of now, a robust and widely used algorithm is MA57 [72], but there is no reliable distributed memory code.
- *Communication analysis for data-sparse parallel algorithms.* While communication analysis is well-established for exact dense and structure-sparse direct solvers, there is a dearth of research concerning data-sparse direct solvers. In fact, there exists only a limited body of latency–bandwidth model-based analyses for the construction phase [141, 163], with none available for the factorization phase. Accordingly, communication avoiding algorithms remain unexplored for rank-structured matrix algorithms. Moreover, distributed-memory parallel algorithms and dynamical runtime scheduling techniques have not been reported for sketching-based construction and factorization algorithms for strong-admissible rank-structured matrix algorithms.
- *Data-sparse in tensor computations.* The efficiency of rank-structured matrix algorithms for high-dimensional applications—such as 3D integral equations, kernel matrices, and quantum chemistry—remains a significant challenge with substantial potential for optimization. Tensor-based extensions of these algorithms represent a promising frontier for reducing both the constant prefactors and the asymptotic complexities of existing fast direct solvers. Recent efforts include the development of tensor-structured $\mathcal{H}$-matrices [140] and butterfly algorithms [160, 126], tensor-train-based solvers for integral equations [63, 184], as well as tensor-based approaches for wave equations with varying frequencies or coefficients. However, these methods primarily target structured data, such as PDEs discretized on regular grids; their applicability to unstructured data remains an open problem. Furthermore, while several

parallel algorithms for individual tensor compressions exist [28], fully parallelized tensor-based direct solvers have yet to be developed.

- *Data-sparse rank analysis for matrix inverse.* While the rank behavior of frontal matrices in sparse solvers for PDEs is relatively well understood, the rank behavior of inverses for integral equation matrices, particularly using strong-admissible-based rank structured algorithms, has been addressed primarily through empirical studies. Theoretical results have only been established for Laplacian kernels with quasi-uniform discretization [77, 75]. Comparing with the rank structure of the integral equation matrices, the ranks in the inverse often exhibit mild growth. This phenomenon is attributed to both the accumulation of numerical errors and the increasingly complex physics (e.g., scattering matrices in wave problems) encoded within the Schur complements. Consequently, rigorous analytical tools for characterizing rank growth in the inverses of integral equation matrices remain largely unavailable.
- *Error and spectral analysis for data-sparse matrices.* Error analysis for rank-structured matrices represents an indispensable tool for understanding the reliability of these algorithms and providing practical guidance on algorithm parameter choices. Backward stability analysis exists for certain weak-admissible formats such as HSS [170, 224], but remains largely unavailable for most rank-structured matrix algorithms. Moreover, the effectiveness of rank-structured preconditioners for ill-conditioned systems, such as recent $\mathcal{H}^2$ preconditioners [168, 210], requires rigorous spectral analysis tools.

Acknowledgements This work was supported by the U.S. Department of Energy, Office of Science, Office of Advanced Scientific Computing Research's Applied Mathematics Competitive Portfolios program and Scientific Discovery through Advanced Computing (SciDAC) program through the FASTMath Institute, under contract number DE-AC02-05CH11231.

References

1. A. Abdelfattah, N. Beams, R. Carson, P. Ghysels, T. Kolev, T. Stitt, A. Vargas, S. Tomov, and J. Dongarra. MAGMA: Enabling exascale performance with accelerated BLAS and LAPACK for diverse GPU architectures. *Int. J. High Perform. Comput.*, 38(5):468–490, 2024.
2. A. Abdelfattah, P. Ghysels, W. Boukaram, S. Tomov, X. S. Li, and J. Dongarra. Addressing irregular patterns of matrix computations on GPUs and their impact on applications powered by sparse direct solvers. In *SC22: International Conference for High Performance Computing, Networking, Storage and Analysis*, pages 1–14. IEEE, 2022.
3. E. Agullo, J. Demmel, J. Dongarra, B. Hadri, J. Kurzak, J. Langou, H. Ltaief, P. Luszczek, and S. Tomov. Numerical linear algebra on emerging architectures: The PLASMA and MAGMA projects. *J. Phys.: Conf. Ser.*, 180(1), 2009.
4. AHMED. `https://www.wr.uni-bayreuth.de/en/software/ahmed/index.html`.
5. N. Al-Harthi, R. Alomairy, K. Akbudak, R. Chen, H. Ltaief, H. Bagci, and D. Keyes. Solving acoustic boundary integral equations using high performance tile low-rank LU factorization. In *International Conference on High Performance Computing*, pages 209–229. Springer, 2020.
6. N. V. Alger. *Data-scalable Hessian preconditioning for distributed parameter PDE-constrained inverse problems.* PhD thesis, The University of Texas at Austin, 2019.

7. J. I. Aliaga, R. Carratalá-Sáez, R. Kriemann, and E. S. Quintana-Ortí. Task-parallel LU factorization of hierarchical matrices using OmpSs. In *2017 IEEE International Parallel and Distributed Processing Symposium Workshops (IPDPSW)*, pages 1148–1157. IEEE, 2017.
8. I. Ambartsumyan, W. Boukaram, T. Bui-Thanh, O. Ghattas, D. Keyes, G. Stadler, G. Turkiyyah, and S. Zampini. Hierarchical matrix approximations of Hessians arising in inverse problems governed by PDEs. *SIAM J. Sci. Comput.*, 42(5):A3397–A3426, 2020.
9. S. Ambikasaran and E. Darve. An $O(N \log N)$ fast direct solver for partial hierarchically semi-separable matrices: with application to radial basis function interpolation. *J. Sci. Comput.*, 57(3):477–501, 2013.
10. S. Ambikasaran and E. Darve. The inverse fast multipole method. *arXiv preprint arXiv:1407.1572*, 2014.
11. S. Ambikasaran, D. Foreman-Mackey, L. Greengard, D. W. Hogg, and M. O'Neil. Fast direct methods for Gaussian processes. *IEEE Trans. Pattern Anal. Mach. Intell.*, 38(2):252–265, 2015.
12. S. Ambikasaran, M. O'Neil, and K. R. Singh. Fast symmetric factorization of hierarchical matrices with applications. *arXiv preprint arXiv:1405.0223*, 2014.
13. P. Amestoy, C. Ashcraft, O. Boiteau, A. Buttari, J.-Y. L'Excellent, and C. Weisbecker. Improving multifrontal methods by means of block low-rank representations. *SIAM J. Sci. Comput.*, 37(3):A1451–A1474, 2015.
14. P. Amestoy, A. Buttari, J.-Y. L'Excellent, and T. Mary. On the complexity of the block low-rank multifrontal factorization. *SIAM J. Sci. Comput.*, 39(4):A1710–A1740, 2017.
15. P. R. Amestoy, A. Buttari, J.-Y. L'Excellent, and T. Mary. Performance and scalability of the block low-rank multifrontal factorization on multicore architectures. *ACM Trans. Math. Softw.*, 45(1), Feb. 2019.
16. P. R. Amestoy, A. Buttari, J.-Y. L'Excellent, and T. A. Mary. Bridging the gap between flat and hierarchical low-rank matrix formats: the multilevel block low-rank format. *SIAM J. Sci. Comput.*, 41(3):A1414–A1442, 2019.
17. Amestoy, P. R. and Duff, I. S. and L'Excellent, J.-Y. and Li, X. S. Analysis and comparison of two general sparse solvers for distributed memory computers. *ACM Trans. Math. Softw.*, 27(4):388–421, Dec. 2001.
18. Amestoy, P. R. and Duff, I. S. and L'Excellent, J-Y and Koster, J. MUMPS: a general purpose distributed memory sparse solver. In *International Workshop on Applied Parallel Computing*, pages 121–130. Springer, 2000.
19. Amestoy, P. R. and Duff, I. S. and L'Excellent, J.-Y. and Koster, J. A fully asynchronous multifrontal solver using distributed dynamic scheduling. *SIAM Journal on Matrix Anal. Appl.*, 23:15–41, 2001.
20. A. Aminfar, S. Ambikasaran, and E. Darve. A fast block low-rank dense solver with applications to finite-element matrices. *J. Comput. Phys.*, 304:170–188, 2016.
21. E. Anderson and Y. Saad. Solving sparse triangular linear systems on parallel computers. *Int. J. High Speed Comput.*, 1(1):73–95, Apr. 1989.
22. N. Angleitner, M. Faustmann, and J. M. Melenk. $\mathcal{H}$-inverses for RBF interpolation. *Adv. Comput. Math.*, 49:1–46, 2021.
23. H. Anzt, T. Cojean, G. Flegar, F. Göbel, T. Grützmacher, P. Nayak, T. Ribizel, Y. M. Tsai, and E. S. Quintana-Ortí. Ginkgo: a modern linear operator algebra framework for high performance computing. *ACM Trans. Math. Softw.*, 48(1):2:1–2:33, Feb. 2022.
24. C. Ashcraft, A. Buttari, and T. Mary. Block low-rank matrices with shared bases: potential and limitations of the BLR^2 format. *SIAM J. Matrix Anal. Appl.*, 42(2):990–1010, 2021.
25. ASKIT. `https://padas.oden.utexas.edu/libaskit`.
26. C. Augonnet, S. Thibault, R. Namyst, and P.-A. Wacrenier. StarPU: a unified platform for task scheduling on heterogeneous multicore architectures. *Concurrency Computat.: Pract. Exper.*, 23(2):187–198, 2011.
27. G. Ballard, E. Carson, J. Demmel, M. Hoemmen, N. Knight, and O. Schwartz. Communication lower bounds and optimal algorithms for numerical linear algebra. *Acta Numer.*, 23:1–155, 2014.
28. G. Ballard and T. G. Kolda. Tensor decompositions for data science. *Cambridge University Press*, 2025.

29. A. Barnett, B. Wu, and S. Veerapaneni. Spectrally accurate quadratures for evaluation of layer potentials close to the boundary for the 2D Stokes and Laplace equations. *SIAM J. Sci. Comput.*, 37(4):B519–B542, 2015.
30. M. Bebendorf. Approximation of boundary element matrices. *Numer. Math.*, 86(4):565–589, Oct 2000.
31. M. Bebendorf and R. Grzhibovskis. Accelerating Galerkin BEM for linear elasticity using adaptive cross approximation. *Math. Methods Appl. Sci.*, 29(14):1721–1747, 2006.
32. M. Bebendorf and W. Hackbusch. Existence of $\mathcal{H}$-matrix approximants to the inverse FE-matrix of elliptic operators with $L\infty$-coefficients. *Numer. Math.*, 95(1):1–28, 2003.
33. M. Bebendorf and R. Venn. Constructing nested bases approximations from the entries of non-local operators. *Numer. Math.*, 121(4):609–635, 2012.
34. R. Belli and T. Hoefler. Notified access: Extending remote memory access programming models for producer-consumer synchronization. In *IEEE International Parallel and Distributed Processing Symposium*, pages 871–881. IEEE, 2015.
35. J. Bendoraityte and S. Börm. Distributed $\mathcal{H}^2$-matrices for non-local operators. *Comput. Vis. Sci*, 11:237–249, 2008.
36. N. Bhatia and Vandana. Survey of nearest neighbor techniques. *arXiv preprint arXiv:1007.0085*, 2010.
37. C. K. Birdsall and A. B. Langdon. *Plasma Physics via Computer Simulation*. CRC Press, 2018.
38. L. S. Blackford, J. Choi, E. D'Azevedo, J. Demmel, I. Dhillon, J. Dongarra, S. Hammarling, G. Henry, A. Petitet, K. Stanley, D. Walker, and R. C. Whaley. *ScaLAPACK users' guide*. SIAM, Philadelphia, 1997. `http://www.netlib.org/scalapack`.
39. S. Börm, L. Grasedyck, and W. Hackbusch. Introduction to hierarchical matrices with application. *Eng. Anal. Bound. Elem.*, 27:405–422, 2003.
40. S. Börm and K. Reimer. Efficient arithmetic operations for rank-structured matrices based on hierarchical low-rank updates. *Comput. Vis. Sci.*, 16:247–258, 2013.
41. G. Bosilca, A. Bouteiller, A. Danalis, T. Herault, P. Lemarinier, and J. Dongarra. DAGuE: A generic distributed DAG engine for High Performance Computing. *Parallel Comput.*, 38(1–2):37–51, Jan. 2012.
42. W. Boukaram, Y. Hong, Y. Liu, T. Shi, and X. S. Li. Batched sparse direct solver design and evaluation in SuperLU_DIST. *Int. J. High Perform. Comput.*, 38(6):585–598, 2024.
43. W. Boukaram, D. Keyes, S. Li, Y. Liu, and G. Turkiyyah. Linear complexity $\mathcal{H}^2$ direct solver for fine-grained parallel architectures. *arXiv preprint arXiv:2509.11152*, 2025.
44. W. Boukaram, Y. Liu, P. Ghysels, and X. S. Li. Adaptive sketching based construction of H2 matrices on GPUs. In *The 26th IEEE International Workshop on Parallel and Distributed Scientific and Engineering Computing (PDSEC 2025)*, pages 433–442. IEEE, 06 2025.
45. W. H. Boukaram, G. Turkiyyah, and D. Keyes. Hierarchical matrix operations on GPUs: Matrix-vector multiplication and compression. *ACM Trans. Math. Softw.*, 45(1):1–28, Feb. 2019.
46. A. M. Bradley. A hybrid multithreaded direct sparse triangular solver. In *Proceedings of the seventh SIAM Workshop on Combinatorial Scientific Computing (CSC)*, pages 13–22, 2016.
47. T. Böhnlein, P. A. Papp, R. S. Steiner, C. K. Matzoros, and A. N. Yzelman. Efficient parallel scheduling for sparse triangular solvers. *arXiv preprint arXiv:2503.05408*, 2025.
48. S. Börm. Directional $\mathcal{H}^2$-matrix compression for high-frequency problems. *Numer. Linear Algebra Appl.*, 24(6):e2112, 2017.
49. L. Cambier, C. Chen, E. G. Boman, S. Rajamanickam, R. S. Tuminaro, and E. Darve. An algebraic sparsified nested dissection algorithm using low-rank approximations. *SIAM J. Matrix Anal. Appl.*, 41(2):715–746, 2020.
50. E. Candès, L. Demanet, and L. Ying. A fast butterfly algorithm for the computation of Fourier integral operators. *SIAM Multiscale Model. Simul.*, 7(4):1727–1750, 2009.
51. Q. Cao, Y. Pei, T. Herault, K. Akbudak, A. Mikhalev, G. Bosilca, H. Ltaief, D. Keyes, and J. Dongarra. Performance analysis of tile low-rank Cholesky factorization using PARSEC instrumentation tools. In *2019 IEEE/ACM International Workshop on Programming and Performance Visualization Tools (ProTools)*, pages 25–32. IEEE, 2019.
52. J. N. Chadwick and D. S. Bindel. An efficient solver for sparse linear systems based on rank-structured Cholesky factorization. *arXiv preprint arXiv:1507.05593*, 2015.

53. S. Chandrasekaran, P. Dewilde, M. Gu, W. Lyons, and T. Pals. A fast solver for HSS representations via sparse matrices. *SIAM J. Matrix Anal. Appl.*, 29(1):67–81, 2007.
54. S. Chandrasekaran, P. Dewilde, M. Gu, and N. Somasunderam. On the numerical rank of the off-diagonal blocks of Schur complements of discretized elliptic PDEs. *SIAM Journal on Matrix Anal. Appl.*, 31:2261–2290, 2010.
55. S. Chandrasekaran, M. Gu, and T. Pals. A fast ULV decomposition solver for hierarchically semiseparable representations. *SIAM J. Matrix Anal. Appl.*, 28(3):603–622, 2006.
56. G. Chávez, Y. Liu, P. Ghysels, X. S. Li, and E. Rebrova. Scalable and memory-efficient kernel ridge regression. In *2020 IEEE International Parallel and Distributed Processing Symposium (IPDPS)*, pages 956–965. IEEE, 2020.
57. C. Chen and P.-G. Martinsson. Solving linear systems on a GPU with hierarchically off-diagonal low-rank approximations. In *SC22: International Conference for High Performance Computing, Networking, Storage and Analysis*, pages 1–15. IEEE, 2022.
58. D. Y. Chenhan, W. B. March, and G. Biros. An $n \log n$ parallel fast direct solver for kernel matrices. In *2017 IEEE International Parallel and Distributed Processing Symposium (IPDPS)*, pages 886–896. IEEE, 2017.
59. D. Y. Chenhan, W. B. March, B. Xiao, and G. Biros. INV-ASKIT: a parallel fast direct solver for kernel matrices. In *2016 IEEE International Parallel and Distributed Processing Symposium (IPDPS)*, pages 161–171. IEEE, 2016.
60. L. Claus, P. Ghysels, W. H. Boukaram, and X. S. Li. A graphics processing unit accelerated sparse direct solver and preconditioner with block low rank compression. *Int. J. High Perform. Comput.*, 39(1):18–31, 2025.
61. L. Claus, P. Ghysels, Y. Liu, T. A. Nhan, R. Thirumalaisamy, A. P. S. Bhalla, and S. Li. Sparse approximate multifrontal factorization with composite compression methods. *ACM Trans. Math. Softw.*, 49(3):1–28, 2023.
62. E. Corona, P.-G. Martinsson, and D. Zorin. An $O(N)$ direct solver for integral equations on the plane. *Appl. Comput. Harmon. Anal.*, 38(2):284–317, 2015.
63. E. Corona, A. Rahimian, and D. Zorin. A tensor-train accelerated solver for integral equations in complex geometries. *J. Comput. Phys.*, 334:145–169, 2017.
64. P. Coulier, H. Pouransari, and E. Darve. The inverse fast multipole method: using a fast approximate direct solver as a preconditioner for dense linear systems. *SIAM J. Sci. Comput.*, 39(3):A761–A796, 2017.
65. T. A. Davis. Algorithm 832: UMFPACK V4. 3—an unsymmetric-pattern multifrontal method. *ACM Trans. Math. Softw.*, 30(2):196–199, 2004.
66. T. A. Davis. *Direct Methods for Sparse Linear Systems*. Society for Industrial and Applied Mathematics, Philadelphia, 2006.
67. S. Di and F. Cappello. Fast error-bounded lossy HPC data compression with SZ. In *IEEE International Parallel and Distributed Processing Symposium (IPDPS)*, pages 730–739. IEEE, 2016.
68. DINFMM. https://github.com/Tianyu-Liang/DINFMM/tree/main.
69. N. Ding, Y. Liu, S. Williams, and X. S. Li. A message-driven, multi-GPU parallel sparse triangular solver. In *Proceedings of the SIAM Conference on Applied and Computational Discrete Algorithms (ACDA21)*, pages 147–159, 2021.
70. N. Ding, S. Williams, Y. Liu, and X. S. Li. Leveraging one-sided communication for sparse triangular solvers. In *Proceedings of the SIAM Conference on Parallel Processing for Scientific Computing*, pages 93–105, 2020.
71. I. Duff, A. Erisman, and J. Reid. *Direct Methods for Sparse Matrices, Second Edition*. Oxford University Press, London, 2017.
72. I. S. Duff. Ma57—a code for the solution of sparse symmetric definite and indefinite systems. *ACM Trans. Math. Softw.*, 30(2):118–144, jun 2004.
73. B. Engquist and L. Ying. Fast directional multilevel algorithms for oscillatory kernels. *SIAM J. Sci. Comput.*, 29(4):1710–1737, 2007.
74. B. Engquist and H. Zhao. Approximate separability of the Green's function of the Helmholtz equation in the high frequency limit. *Commun. Pure Appl. Math.*, 71(11):2220–2274, 2018.
75. M. Faustmann, J. Melenk, and D. Praetorius. Existence of $\mathcal{H}$-matrix approximants to the inverses of BEM matrices: The simple-layer operator. *Math. Comp.*, 85(297):119–152, 2016.

76. M. Faustmann, J. M. Melenk, and D. Praetorius. A new proof for existence of H-matrix approximants to the inverse of FEM matrices: the Dirichlet problem for the Laplacian. In *Spectral and High Order Methods for Partial Differential Equations-ICOSAHOM 2012: Selected papers from the ICOSAHOM conference, June 25-29, 2012, Gammarth, Tunisia*, pages 249–259. Springer, 2013.
77. M. Faustmann, J. M. Melenk, and D. Praetorius. Existence of $\mathcal{H}$-matrix approximants to the inverse of BEM matrices: the hyper-singular integral operator. *IMA J. Numer. Anal.*, 37:1211–1244, 2015.
78. M. Faustmann, J. M. Melenk, and D. Praetorius. $\mathcal{H}$-matrix approximability of the inverses of FEM matrices. *Numer. Math.*, 131(4):615–642, Dec. 2015.
79. M. Faverge, G. Pichon, P. Ramet, and J. Roman. On the use of H-matrix arithmetic in PaStiX: a preliminary study. In *Workshop on Fast Direct Solvers*, Toulouse, France, June 2015.
80. J. Feliu-Fabà, K. L. Ho, and L. Ying. Recursively preconditioned hierarchical interpolative factorization for elliptic partial differential equations. *Commun. Math. Sci.*, 18(1):91–108, 2020.
81. J. Feliu-Fabà and L. Ying. Approximate inversion of discrete Fourier integral operators. *J. Comput. Phys.*, 446:110654, 2021.
82. Y. Feng, J. Xiao, and M. Gu. Flip-flop spectrum-revealing QR factorization and its applications to singular value decomposition. *Electron. Trans. Numer. Anal.*, 51:469–494, 2019.
83. X. Fu, B. Zhang, T. Wang, W. Li, Y. Lu, E. Yi, J. Zhao, X. Geng, F. Li, J. Zhang, Z. Jin, and W. Liu. PanguLU: a scalable regular two-dimensional block-cyclic sparse direct solver on distributed heterogeneous systems. In *Proceedings of the International Conference for High Performance Computing, Networking, Storage and Analysis*, SC '23, New York, NY, USA, 2023. Association for Computing Machinery.
84. M. Gates, J. Kurzak, A. Charara, A. YarKhan, and J. Dongarra. SLATE: design of a modern distributed and accelerated linear algebra library. In *Proceedings of the International Conference for High Performance Computing, Networking, Storage and Analysis*, SC '19, New York, NY, USA, 2019. Association for Computing Machinery.
85. C. J. Geoga, M. Anitescu, and M. L. Stein. Scalable Gaussian process computations using hierarchical matrices. *J. Comput. Graph. Stat.*, 29(2):227–237, 2020.
86. A. George. Nested dissection of a regular finite element mesh. *SIAM J. Numer. Anal.*, 10:345–363, 1973.
87. A. George and J. W. H. Liu. *Computer Solution of Large Sparse Positive Definite Systems.* Prentice Hall, Englewood Cliffs, NJ, 1981.
88. A. George and E. Ng. A brief description of SPARSPAK Waterloo sparse linear equations package. *SIGNUM Newsl.*, 16(2):17–20, 1981.
89. T. George, V. Saxena, A. Gupta, A. Singh, and A. Choudjury. Multifrontal factorization of sparse SPD matrices on GPUs. In *Proc. of IEEE International Parallel and Distributed Processing Symposium (IPDPS 2011)*, Anchorage, Alaska, May 16-20 2011.
90. P. Ghysels, X. S. Li, C. Gorman, and F.-H. Rouet. A robust parallel preconditioner for indefinite systems using hierarchical matrices and randomized sampling. In *2017 IEEE International Parallel and Distributed Processing Symposium (IPDPS)*, pages 897–906. IEEE, 2017.
91. P. Ghysels, X. S. Li, F.-H. Rouet, S. Williams, and A. Napov. An efficient multicore implementation of a novel HSS-structured multifrontal solver using randomized sampling. *SIAM J. Sci. Comput.*, 38(5):S358–S384, 2016.
92. P. Ghysels and R. Synk. High performance sparse multifrontal solvers on modern GPUs. *Parallel Computing*, 110:102897, 2022.
93. J. R. Gilbert and R. E. Tarjan. The analysis of a nested dissection algorithm. *Numer. Math.*, pages 377–404, 1987.
94. A. Gillman and P.-G. Martinsson. An $O(N)$ algorithm for constructing the solution operator to 2D elliptic boundary value problems in the absence of body loads. *Adv. Comput. Math.*, 40(4):773–796, 2014.
95. A. Gillman, P. M. Young, and P.-G. Martinsson. A direct solver with $O(N)$ complexity for integral equations on one-dimensional domains. *Front. Math. China*, 7(2):217–247, 2012.
96. C. Gorman, G. Chávez, P. Ghysels, T. Mary, F.-H. Rouet, and X. S. Li. Robust and accurate stopping criteria for adaptive randomized sampling in matrix-free hierarchically semiseparable construction. *SIAM J. Sci. Comput.*, 41(5):S61–S85, 2019.

97. L. Grasedyck and W. Hackbusch. Construction and arithmetics of H-matrices. *Computing*, 70(4):295–334, 2003.
98. L. Grasedyck, R. Kriemann, and S. Le Borne. Parallel black box $\mathcal{H}$-LU preconditioning for elliptic boundary value problems. *Comput. Vis. Sci.*, 11(4):273–291, 2008.
99. W. Gropp, E. Lusk, and R. Thakur. *Using MPI-2: Advanced Features of the Message-Passing Interface*. MIT Press, Cambridge, MA, USA, 1999.
100. H. Guo, J. Hu, and E. Michielssen. On MLMDA/butterfly compressibility of inverse integral operators. *IEEE Antennas Wirel. Propag. Lett.*, 12:31–34, 2013.
101. H. Guo, J. Hu, and Z. Nie. An MPI-OpenMP hybrid parallel $\mathcal{H}$-LU direct solver for electromagnetic integral equations. *Int. J. Antennas Propag.*, 2015(1):615743, 2015.
102. H. Guo, Y. Liu, J. Hu, and E. Michielssen. A butterfly-based direct integral-equation solver using hierarchical LU factorization for analyzing scattering from electrically large conducting objects. *IEEE Trans. Antennas Propag.*, 65(9):4742–4750, 2017.
103. A. Gupta. Recent advances in direct methods for solving unsymmetric sparse systems of linear equations. *ACM Trans. Math. Softw.*, 28(3):301–324, Sept. 2002.
104. A. Gupta, G. Karypis, and V. Kumar. Highly scalable parallel algorithms for sparse matrix factorization. *IEEE Trans. Parallel Distrib. Syst.*, 8(5):502–520, 1997.
105. H2Opus. `https://github.com/ecrc/h2opus`.
106. W. Hackbusch. *Hierarchical Matrices: Algorithms and Analysis*. Springer, 2015.
107. W. Hackbusch and S. Börm. H2-matrix approximation of integral operators by interpolation. *Appl. Numer. Math.*, 43(1-2):129–143, 2002.
108. S. Hao and P.-G. Martinsson. A direct solver for elliptic PDEs in three dimensions based on hierarchical merging of Poincaré–Steklov operators. *J. Comput. Appl. Math.*, 308:419–434, 2016.
109. A. Heldring, E. Ubeda, and J. Rius. Fast direct high frequency MoM solver using Butterfly algebra. *IEEE Trans. Antennas Propag.*, 47(1):703–712, 2026.
110. J. Hennessy, D. Patterson, and C. Kozyrakis. *Computer Architecture: A Quantitative Approach*. The Morgan Kaufmann Series in Computer Architecture and Design. Elsevier Science, 2025.
111. P. Hénon, P. Ramet, and J. Roman. PaStiX: a high-performance parallel direct solver for sparse symmetric positive definite systems. *Parallel Computing*, 28(2):301–321, 2002.
112. HiCMA. `https://github.com/ecrc/hicma`.
113. J. M. Hill, B. McColl, D. C. Stefanescu, M. W. Goudreau, K. Lang, S. B. Rao, T. Suel, T. Tsantilas, and R. H. Bisseling. BSPlib: The BSP programming library. *Parallel Computing*, 24(14):1947–1980, 1998.
114. HLibPro. `https://www.hlibpro.com`.
115. K. L. Ho and L. Ying. Hierarchical interpolative factorization for elliptic operators: differential equations. *Commun. Pure Appl. Math.*, 69(8):1415–1451, 2016.
116. H. Huang, X. Xing, and E. Chow. H2Pack: High-performance H2 matrix package for kernel matrices using the proxy point method. *ACM Trans. Math. Softw. (TOMS)*, 47(1):1–29, 2020.
117. H. Huang, T. Xu, Y. Xi, and E. Chow. HiGP: A high-performance Python package for Gaussian process. *arXiv preprint arXiv:2503.02259*, 2025.
118. D. Irony and S. Toledo. Trading replication for communication in parallel distributed-memory dense solvers. *Parallel Process. Lett.*, 12(01):79–94, 2002.
119. L. V. Kale and G. Zheng. Charm++ and AMPI: adaptive runtime strategies via migratable objects. In *Advanced Computational Infrastructures for Parallel and Distributed Adaptive Applications*, chapter 13, pages 265–282. John Wiley & Sons, Ltd, 2009.
120. K. Kandalla, D. Knaak, K. McMahon, N. Radcliffe, and M. Pagel. Optimizing Cray MPI and Cray SHMEM for current and next generation Cray-XC supercomputers. *Cray User Group (CUG)*, 2015, 2015.
121. A. Kashi, P. Nayak, D. Kulkarni, A. Scheinberg, P. Lin, and H. Anzt. Batched sparse iterative solvers on GPU for the collision operator for fusion plasma simulations. In *2022 IEEE International Parallel and Distributed Processing Symposium (IPDPS)*, pages 157–167, 2022.
122. KBLAS-GPU. `https://github.com/ecrc/kblas-gpu`.
123. D. Keyes, H. Ltaief, and G. Turkiyyah. Hierarchical algorithms on hierarchical architectures. *Philos. Trans. R. Soc. A*, 378(2166):20190055, Jan. 2020.

124. B. Khalichi, A. Cerfon, D. Malhotra, and M. O'Neil. Taylor relaxations with stepped-pressure profile: integral equation solver (BIEST) for computation of multi-region relaxed MHD equilibria. In *2025 International Applied Computational Electromagnetics Society Symposium (ACES)*, pages 1–2. IEEE, 2025.
125. R. Khan, V. Kandappan, and S. Ambikasaran. HODLRdD: A new black-box fast algorithm for N-body problems in d-dimensions with guaranteed error bounds: Applications to integral equations and support vector machines. *J. Comput. Phys.*, 501:112786, 2024.
126. P. M. Kielstra, T. Shi, H. Luo, J. Qian, and Y. Liu. A linear-complexity tensor butterfly algorithm for compressing high-dimensional oscillatory integral operators. *SIAM Multiscale Model. Simul.*, 23(2):864–893, 2025.
127. Kokkos Kernels. `https://github.com/kokkos/kokkos-kernels`.
128. D. Kressner and A. Susnjara. Fast QR decomposition of HODLR matrices. *arXiv preprint arXiv:1809.10585*, 2018.
129. R. Kriemann. H-LU factorization on many-core systems. *Comput. Vis. Sci.*, 16(3):105–117, 2013.
130. R. Kriemann. Hierarchical low-rank arithmetic with floating point compression. *SIAM J. Sci. Comput.*, 47(4):B763–B784, 2025.
131. J. Kump, A. Yesypenko, and P.-G. Martinsson. A two-level direct solver for the hierarchical Poincaré-Steklov method. *arXiv preprint arXiv:2503.04033*, 2025.
132. K. L. Ho and L. Ying. Hierarchical interpolative factorization for elliptic operators: integral equations. *Commun. Pure Appl. Math.*, 69(7):1314–1353, 2016.
133. X. Lacoste, M. Faverge, G. Bosilca, P. Ramet, and S. Thibault. Taking advantage of hybrid systems for sparse direct solvers via task-based runtimes. In *2014 IEEE International Parallel & Distributed Processing Symposium Workshops*, pages 29–38, 2014.
134. J. Levitt and P.-G. Martinsson. Linear-complexity black-box randomized compression of rank-structured matrices. *SIAM J. Sci. Comput.*, 46(3):A1747–A1763, 2024.
135. J. Levitt and P.-G. Martinsson. Randomized compression of rank-structured matrices accelerated with graph coloring. *J. Comput. Appl. Math.*, 451:116044, 2024.
136. X. Li. Factorization-based sparse solvers and preconditioners. In Z. Bai, W. Gao, and Y. Su, editors, *Series in Contemporary Applied Mathematics: Vol 19, Matrix Functions and Matrix Equations*, pages 109–137. World Scientific, 2015. Extended Summary for 4th Gene Golub SIAM Summer School, July 22-Aug. 9, 2013, Shanghai.
137. X. S. Li and J. Demmel. Making sparse Gaussian elimination scalable by static pivoting. In *SC '98: Proceedings of the 1998 ACM/IEEE Conference on Supercomputing*, pages 34–34, 1998.
138. X. S. Li and J. W. Demmel. SuperLU_DIST: A scalable distributed-memory sparse direct solver for unsymmetric linear systems. *ACM Trans. Math. Softw.*, 29(2):110–140, June 2003. `https://portal.nersc.gov/project/sparse/superlu/`.
139. X. S. Li, P. Lin, Y. Liu, and P. Sao. Newly released capabilities in the distributed-memory SuperLU sparse direct solver. *ACM Trans. Math. Softw.*, 49(1), mar 2023.
140. Y. Li and J. Liu. Hierarchical Tucker low-rank matrices: construction and matrix-vector multiplication. *arXiv preprint arXiv:2508.05958*, 2025.
141. Y. Li, J. Poulson, and L. Ying. Distributed-memory $\mathcal{H}$-matrix algebra I: data distribution and matrix-vector multiplication. *arXiv preprint arXiv:2008.12441*, 2020.
142. Y. Li, H. Yang, E. R. Martin, K. L. Ho, and L. Ying. Butterfly factorization. *SIAM Multiscale Model. Simul.*, 13(2):714–732, 2015.
143. Y. Li and L. Ying. Distributed-memory hierarchical interpolative factorization. *Research in the Mathematical Sciences*, 4(1):12, 2017.
144. T. Liang, C. Chen, P.-G. Martinsson, and G. Biros. An $O(N)$ distributed-memory parallel direct solver for planar integral equations. In *2024 IEEE International Parallel and Distributed Processing Symposium (IPDPS)*, pages 440–452. IEEE, 2024.
145. E. Liberty, F. Woolfe, P.-G. Martinsson, V. Rokhlin, and M. Tygert. Randomized algorithms for the low-rank approximation of matrices. *Proceedings of the National Academy of Sciences*, 104(51):20167–20172, 2007.
146. K. Liegeois, S. Rajamanickam, and L. Berger-Vergiat. Performance portable batched sparse linear solvers. *IEEE Trans. Parallel Distrib. Syst.*, 34(5):1524–1535, 2023.

147. L. Lin, J. Lu, and L. Ying. Fast construction of hierarchical matrix representation from matrix–vector multiplication. *J. Comput. Phys.*, 230(10):4071–4087, 2011.
148. P. Lindstrom, J. Hittinger, J. Diffenderfer, A. Fox, D. Osei-Kuffuor, and J. Banks. ZFP: A compressed array representation for numerical computations. *Int. J. High Perform. Comput.*, 39(1):104–122, 2025.
149. P. Lindstrom and M. Isenburg. Fast and efficient compression of floating-point data. *IEEE Trans. Vis. Comput. Graph.*, 12(5):1245–1250, 2006.
150. R. J. Lipton, D. J. Rose, and R. E. Tarjan. Generalized nested dissection. *SIAM J. Numer. Anal.*, 16:346–358, 1979.
151. A. Litvinenko, R. Kriemann, M. G. Genton, Y. Sun, and D. E. Keyes. HLIBCov: Parallel hierarchical matrix approximation of large covariance matrices and likelihoods with applications in parameter identification. *MethodsX*, 7:100600, 2020.
152. J. W. H. Liu. The role of elimination trees in sparse factorization. *SIAM J. Matrix Anal. Appl.*, 11(1):134–172, January 1990.
153. W. Liu, A. Li, J. Hogg, I. S. Duff, and B. Vinter. A synchronization-free algorithm for parallel sparse triangular solves. In *Proceedings of the 22nd International Conference on Euro-Par 2016: Parallel Processing - Volume 9833*, page 617–630, Berlin, Heidelberg, 2016. Springer-Verlag.
154. Y. Liu. ButterflyPACK, 2018. `https://portal.nersc.gov/project/sparse/butterflypack/`.
155. Y. Liu, N. Ding, P. Sao, S. Williams, and X. S. Li. Unified communication optimization strategies for sparse triangular solver on CPU and GPU clusters. In *Proceedings of the International Conference for High Performance Computing, Networking, Storage and Analysis*, SC '23, New York, NY, USA, 2023. Association for Computing Machinery.
156. Y. Liu, P. Ghysels, L. Claus, and X. S. Li. Sparse approximate multifrontal factorization with butterfly compression for high-frequency wave equations. *SIAM J. Sci. Comput.*, 43(5):S367–S391, 2021.
157. Y. Liu, H. Guo, and E. Michielssen. An HSS matrix-inspired butterfly-based direct solver for analyzing scattering from two-dimensional objects. *IEEE Antennas Wirel. Propag. Lett.*, 16:1179–1183, 2017.
158. Y. Liu, M. Jacquelin, P. Ghysels, and X. Li. Highly scalable distributed-memory sparse triangular solution algorithms. In *Proceedings of the SIAM Workshop on Combinatorial Scientific Computing*, Bergen, Norway, June 6-8, 2018 2018.
159. Y. Liu, T. Luo, A. Rani, H. Luo, and X. S. Li. Detecting resonance of radio-frequency cavities using fast direct integral equation solvers and augmented Bayesian optimization. *IEEE J. Multiscale Multiphysics Comput. Tech.*, 2023.
160. Y. Liu, T. Shi, P. M. Kielstra, H. Luo, and J. Qian. A block diagonal preconditioned and tensor butterfly accelerated volume integral equation solver for 3D high-frequency scattering analysis. In *2025 International Applied Computational Electromagnetics Society Symposium (ACES)*, pages 1–1. IEEE, 2025.
161. Y. Liu, W. Sid-Lakhdar, E. Rebrova, P. Ghysels, and X. S. Li. A parallel hierarchical blocked adaptive cross approximation algorithm. *Int. J. High Perform. Comput.*, 34(4):394–408, 2020.
162. Y. Liu, J. Song, R. Burridge, and J. Qian. A fast butterfly-compressed Hadamard-Babich integrator for high-frequency Helmholtz equations in inhomogeneous media with arbitrary sources. *SIAM Multiscale Model. Simul.*, 21(1):269–308, 2023.
163. Y. Liu, X. Xing, H. Guo, E. Michielssen, P. Ghysels, and X. Li. Butterfly factorization via randomized matrix-vector multiplications. *SIAM. J. Sci. Comput.*, 43(2):A883–A907, 2021.
164. R. Lucas, G. Wagenbreth, D. Davis, and R. Grimes. Multifrontal computations on GPUs and their multi-core hosts. In *VECPAR'10: Proc. 9th Intl. Meeting for High Performance Computing for Computational Science*, Berkeley, CA, 2010. `http://vecpar.fe.up.pt/2010/papers/5.php`.
165. M. Ma and D. Jiao. Accuracy directly controlled fast direct solution of general $\mathcal{H}^2$-matrices and its application to solving electrodynamic volume integral equations. *IEEE Trans. Microw. Theory Techn.*, 66(1):35–48, 2018.
166. M. Ma and D. Jiao. Direct solution of general $\mathcal{H}^2$ -matrices with controlled accuracy and concurrent change of cluster bases for electromagnetic analysis. *IEEE Trans. Microw. Theory Techn.*, 67(6):2114–2127, 2019.

167. Q. Ma, S. Deshmukh, and R. Yokota. Scalable linear time dense direct solver for 3-d problems without trailing sub-matrix dependencies. In *SC22: International Conference for High Performance Computing, Networking, Storage and Analysis*, pages 1–12. IEEE, 2022.
168. Q. Ma and R. Yokota. An inherently parallel $\mathcal{H}^2$-ULV factorization for solving dense linear systems on GPUs. *Int. J. High Perform. Comput.*, 38(4):314–336, Apr. 2024.
169. A. B. Manić, A. P. Smull, F.-H. Rouet, X. S. Li, and B. M. Notaroš. Efficient scalable parallel higher order direct MoM-SIE method with hierarchically semiseparable structures for 3-D scattering. *IEEE Trans. Antennas Propag.*, 65(5):2467–2478, 2017.
170. P.-G. Martinsson. A fast randomized algorithm for computing a hierarchically semiseparable representation of a matrix. *SIAM J. Matrix Anal. Appl.*, 32(4):1251–1274, 2011.
171. P.-G. Martinsson. A direct solver for variable coefficient elliptic PDEs discretized via a composite spectral collocation method. *J. Comput. Phys.*, 242:460–479, 2013.
172. P.-G. Martinsson and M. O'Neil. Fast direct solvers. *arXiv preprint arXiv:2511.07773*, 2025.
173. P. G. Martinsson, G. Quintana-Ortí, and N. Heavner. RandUTV: a blocked randomized algorithm for computing a rank-revealing UTV factorization. *ACM Trans. Math. Softw.*, 45(1), Mar. 2019.
174. P.-G. Martinsson and V. Rokhlin. A fast direct solver for boundary integral equations in two dimensions. *J. Comput. Phys.*, 205(1):1–23, 2005.
175. T. Mary. *Block low-rank multifrontal solvers: complexity, performance, and scalability*. PhD thesis, Université Paul Sabatier-Toulouse III, 2017.
176. E. Michielssen and A. Boag. A multilevel matrix decomposition algorithm for analyzing scattering from large structures. *IEEE Trans. Antennas Propag.*, 44(8):1086–1093, 1996.
177. V. Minden, A. Damle, K. L. Ho, and L. Ying. Fast spatial Gaussian process maximum likelihood estimation via skeletonization factorizations. *SIAM Multiscale Model. Simul.*, 15(4):1584–1611, 2017.
178. V. Minden, K. L. Ho, A. Damle, and L. Ying. A recursive skeletonization factorization based on strong admissibility. *SIAM Multiscale Model. Simul.*, 15(2):768–796, 2017.
179. MPI Forum. `https://www.mpi-forum.org/`.
180. Private Communicatioon, 2026.
181. MUMPS. `http://mumps.enseeiht.fr/`.
182. mxc. `https://github.com/QxGbM/mxc`.
183. BLAS. `https://netlib.org/blas/`.
184. C. Nguyen, A. Boyko, and V. Okhmatovski. Tensor train accelerated method of moment solution of volume integral equations for arbitrary objects with logarithmic complexity. *IEEE Trans. Microw. Theory Tech.*, 73(12), 2025.
185. R. Nies and M. Hoelzl. Testing performance with and without block low rank compression in MUMPS and the new PaStiX 6.0 for JOREK nonlinear MHD simulations. *arXiv preprint arXiv:1907.13442*, 2019.
186. R. Nuca, M. Parsani, and G. Turkiyyah. An adaptive two-stage algorithm for error-bounded scientific data compression. In *2025 IEEE International Parallel and Distributed Processing Symposium (IPDPS)*, pages 987–997. IEEE, 2025.
187. OpenMP. `https://www.openmp.org/`.
188. Q. Pang, K. L. Ho, and H. Yang. Interpolative decomposition butterfly factorization. *SIAM J. Sci. Comput.*, 42(2):A1097–A1115, 2020.
189. PanguLU. `https://github.com/SuperScientificSoftwareLaboratory/PanguLU`.
190. PARDISO. `https://www.intel.com/content/www/us/en/docs/onemkl/developer-reference-c/2023-0/pardiso.html`.
191. J. Park, M. Smelyanskiy, N. Sundaram, and P. Dubey. Sparsifying synchronization for high-performance shared-memory sparse triangular solver. In J. M. Kunkel, T. Ludwig, and H. W. Meuer, editors, *Supercomputing*, pages 124–140, Cham, 2014. Springer International Publishing.
192. PaStiX. `https://gitlab.inria.fr/solverstack/pastix`.
193. K. J. Pearce, A. Yesypenko, J. Levitt, and P.-G. Martinsson. Randomized rank-structured matrix compression by tagging. *arXiv preprint arXiv:2501.05528*, 2025.
194. Perlmutter System. `https://www.nersc.gov/what-we-do/computing-for-science/perlmutter`.
195. J. Poulson, L. Demanet, N. Maxwell, and L. Ying. A parallel butterfly algorithm. *SIAM J. Sci. Comput.*, 36(1):C49–C65, 2014.

196. H. Pouransari, P. Coulier, and E. Darve. Fast hierarchical solvers for sparse matrices using extended sparsification and low-rank approximation. *SIAM J. Sci. Comput.*, 39(3):A797–A830, 2017.
197. E. Rebrova, G. Chávez, Y. Liu, P. Ghysels, and X. S. Li. A study of clustering techniques and hierarchical matrix formats for kernel ridge regression. In *IEEE International Parallel and Distributed Processing Symposium Workshops (IPDPSW)*, pages 883–892. IEEE, 2018.
198. J. Ren, T. Zhong, Y. Hong, G. Feng, X. Wang, W. Jia, H. Ltaief, and D. E. Keyes. Caracal: A GPU-resident sparse LU solver with lightweight fine-grained scheduling. In *Proceedings of the International Conference for High Performance Computing, Networking, Storage and Analysis*, SC '25, page 1477–1494, New York, NY, USA, 2025. Association for Computing Machinery.
199. F.-H. Rouet, X. S. Li, P. Ghysels, and A. Napov. A distributed-memory package for dense hierarchically semi-separable matrix computations using randomization. *ACM Trans. Math. Softw. (TOMS)*, 42(4):1–35, 2016.
200. P. Sao, R. Kannan, X. S. Li, and R. Vuduc. A communication-avoiding 3D sparse triangular solver. In *Proceedings of the ACM International Conference on Supercomputing*, ICS '19, page 127–137, New York, NY, USA, 2019. Association for Computing Machinery.
201. P. Sao and X. S. Li. Brief announcement: communication optimal sparse LU factorization for planar matrices. SPAA '23, page 427–430, New York, NY, USA, 2023. Association for Computing Machinery.
202. P. Sao, X. Liu, R. Vuduc, and X. Li. A sparse direct solver for distributed memory Xeon Phi-accelerated systems. In *29th IEEE International Parallel & Distributed Processing Symposium (IPDPS)*, Hyderabad, India, May 25-29 2015.
203. P. Sao, R. Vuduc, and X. Li. A distributed CPU-GPU sparse direct solver. In *Proc. of Euro-Par 2014, LNCS Vol. 8632, pp. 487-498*, Porto, Portugal, August 25-29 2014.
204. P. Sao, R. Vuduc, and X. Li. A communication-avoiding 3D algorithm for sparse LU factorization on heterogeneous systems. *J. Parallel and Distributed Computing*, September 2019.
205. S. B. Sayed, Y. Liu, L. J. Gomez, and A. C. Yucel. A butterfly-accelerated volume integral equation solver for broad permittivity and large-scale electromagnetic analysis. *IEEE Trans. Antennas Propag.*, 70(5):3549–3559, 2022.
206. J. A. Scott. HSL60: a brief history of the HSL mathematical software library. Technical Report STFC-TR-2023-002, STFC Rutherford Appleton Laboratory, May 2023.
207. J. Shaeffer. Direct solve of electrically large integral equations for problem sizes to 1 M unknowns. *IEEE Trans. Antennas Propag.*, 56(8):2306–2313, 2008.
208. J. Shen, Y. Wang, and J. Xia. Fast structured direct spectral methods for differential equations with variable coefficients, I. the one-dimensional case. *SIAM J. Sci. Comput.*, 38(1):A28–A54, 2016.
209. E. Solomonik and J. Demmel. Communication-optimal parallel 2.5D matrix multiplication and LU factorization algorithms. In *Euro-Par 2011 Parallel Processing*, pages 90–109. Springer, 2011.
210. T. Spendlhofer, Q. Ma, Y. Matsumoto, and R. Yokota. On the interplay between precision, rank and admissibility for iterative hierarchical low-rank matrix solvers. In *ISC High Performance 2025 Research Paper Proceedings (40th International Conference)*, pages 1–12. Prometeus GmbH, 2025.
211. STRUMPACK: STRUctured Matrices PACKages, last accessed 2025. `http://portal.nersc.gov/project/sparse/strumpack/`.
212. SuperLU: sparse direct solver, Last accessed: 2025. `http://portal.nersc.gov/project/sparse/superlu/`.
213. D. Sushnikova, L. Greengard, M. O'Neil, and M. Rachh. FMM-LU: a fast direct solver for multiscale boundary integral equations in three dimensions. *SIAM Multiscale Model. Simul.*, 21(4):1570–1601, 2023.
214. symPACK. `https://github.com/symPACK/symPACK`.
215. TOP 500, 2025. `https://www.top500.org/lists/top500/2025/11/`.
216. E. Totoni, M. T. Heath, and L. V. Kale. Structure-adaptive parallel solution of sparse triangular linear systems. *Parallel Comput.*, 40(9):454–470, Oct. 2014.
217. A. F. van der Stappen, R. H. Bisseling, and J. G. van de Vorst. Parallel Sparse LU Decomposition on a Mesh Network of Transputers. *SIAM J. Matrix Anal. Appl.*, 14(3):853–879, 1993.

218. R. Vuduc, A. Chandramowlishwaran, J. Choi, M. Guney, and A. Shringarpure. On the limits of GPU acceleration. In *Proc. of the 2nd USENIX Conference on Hot Topics in Parallelism, HotPar'10*, Berkeley, CA, 2010.
219. S. Wang, M. V. de Hoop, and J. Xia. On 3D modeling of seismic wave propagation via a structured parallel multifrontal direct Helmholtz solver. *Geophysical Prospecting*, 59:857–873, 2011.
220. S. Wang, M. V. de Hoop, J. Xia, and X. S. Li. Massively parallel structured multifrontal solver for time-harmonic elastic waves in 3-D anisotropic media. *Geophys. J. Int.*, 191(1):346–366, 2012.
221. S. Wang, X. S. Li, F.-H. Rouet, J. Xia, and M. V. De Hoop. A parallel geometric multifrontal solver using hierarchically semiseparable structure. *ACM Trans. Math. Softw.*, 42(3), May 2016.
222. X. Wang, W. Liu, W. Xue, and L. Wu. swSpTRSV: a fast sparse triangular solve with sparse level tile layout on sunway architectures. *SIGPLAN Not.*, 53(1):338–353, Feb. 2018.
223. WSMP. `https://dominoweb.draco.res.ibm.com/95ce158b896a010d85256a57006aadaf.html`.
224. Y. Xi and J. Xia. On the stability of some hierarchical rank structured matrix algorithms. *SIAM J. Matrix Anal. Appl.*, 37(3):1279–1303, 2016.
225. J. Xia. Randomized sparse direct solvers. *SIAM J. Matrix Anal. Appl.*, 34(1):197–227, 2013.
226. J. Xia. $O(N)$ complexity randomized 3D direct solver with HSS2D structure. *Proceedings of the Project Review, Geo-Mathematical Imaging Group*, pages 317–325, 2014.
227. J. Xia. Making the Nyström method highly accurate for low-rank approximations. *SIAM J. Sci. Comput.*, 46(2):A1076–A1101, 2024.
228. J. Xia, S. Chandrasekaran, M. Gu, and X. S. Li. Fast algorithms for hierarchically semiseparable matrices. *Numer. Linear Algebra Appl.*, 17(6):953–976, 2010.
229. J. Xia, S. Chandrasekaran, M. Gu, and X. S. Li. Superfast multifrontal method for large structured linear systems of equations. *SIAM J. Matrix Anal. Appl.*, 31(3):1382–1411, 2010.
230. J. Z. Xia. Effective and robust preconditioning of general SPD matrices via structured incomplete factorization. *SIAM J. Matrix Anal. Appl.*, 38(4):1298–1322, 2017.
231. J. Xiao, M. Gu, and J. Langou. Fast parallel randomized QR with column pivoting algorithms for reliable low-rank matrix approximations. In *2017 IEEE 24th International Conference on High Performance Computing (HiPC)*, pages 233–242, Dec 2017.
232. X. Xing and E. Chow. Interpolative decomposition via proxy points for kernel matrices. *SIAM J. Matrix Anal. Appl.*, 41(1):221–243, 2020.
233. I. Yamazaki, A. Ida, R. Yokota, and J. Dongarra. Distributed-memory lattice H-matrix factorization. *Int. J. High Perform. Comput.*, 33(5):1046–1063, 2019.
234. X. Ye, J. Xia, and L. Ying. Analytical low-rank compression via proxy point selection. *SIAM J. Matrix Anal. Appl.*, 41(3):1059–1085, 2020.
235. A. Yesypenko, C. Chen, and P.-G. Martinsson. A simplified fast multipole method based on strong recursive skeletonization. *J. Comput. Phys.*, 524:113707, 2025.
236. A. Yesypenko and P.-G. Martinsson. SlabLU: a two-level sparse direct solver for elliptic PDEs. *Adv. Comput. Math.*, 50(4):90, 2024.
237. A. Yesypenko and P.-G. Martinsson. Randomized strong recursive skeletonization: simultaneous compression and LU factorization of hierarchical matrices using matrix-vector products. *arXiv preprint arXiv:2311.01451*, 2025.
238. L. Ying. A pedestrian introduction to fast multipole methods. *Science China Mathematics*, 55(5):1043–1051, 2012.
239. L. Ying. Directional preconditioner for 2D high frequency obstacle scattering. *SIAM Multiscale Model. Simul.*, 13(3):829–846, 2015.
240. C. Yu, W. Wang, and D. Pierce. A CPU-GPU hybrid approach for the unsymmetric multifrontal method. *Parallel Computing*, 37:759–770, 2011.
241. S. Zampini, W. Boukaram, G. Turkiyyah, O. Knio, and D. Keyes. H2Opus: a distributed-memory multi-GPU software package for non-local operators. *Adv. Comput. Math.*, 48(3):31, 2022.
242. K. Zhao, M. N. Vouvakis, and J.-F. Lee. The adaptive cross approximation algorithm for accelerated method of moments computations of EMC problems. *IEEE Trans. Electromagn. Compat.*, 47(4):763–773, Nov 2005.
243. Y. Zhao, D. Jiao, and J. Mao. Fast nested cross approximation algorithm for solving large-scale electromagnetic problems. *IEEE Trans. Microw. Theory Techn.*, 67(8):3271–3283, 2019.

244. B. Zhou and D. Jiao. Direct finite-element solver of linear complexity for large-scale 3-D electromagnetic analysis and circuit extraction. *IEEE Trans. Microw. Theory Techn.*, 63(10):3066–3080, 2015.
245. X. Zhu and J. Lai. Recursive sparse LU decomposition based on nested dissection and low rank approximations. *J. Comput. Phys.*, 539:114231, 2025.

Ensemble Kalman–Bucy Filtering for Nonlinear Model Predictive Control

Sebastian Reich

Abstract We consider the problem of optimal control for partially observed dynamical systems. Despite its prevalence in practical applications, there are still very few algorithms available that take uncertainties in the current state estimates and future observations into account. In other words, most current approaches separate state estimation from the optimal control or planning problem. In this paper, we extend the popular ensemble Kalman filter to receding horizon optimal control problems in the spirit of nonlinear model predictive control. We provide an interacting particle approximation to the forward-backward stochastic differential equations arising from Pontryagin's maximum principle with the forward stochastic differential equation provided by the time-continuous ensemble Kalman–Bucy filter equations. The receding horizon control laws are approximated as linear and are continuously updated as in nonlinear model predictive control. We illustrate the performance of the proposed methodology for an inverted pendulum example.

1 Introduction

Optimal control for partially observed dynamical systems was first considered in [18] and theoretical foundations have been established, for example, in [3] and [16]. It is well-known [10, 14] that, for linear systems, the involved optimal control and state estimation problems can be solved separately. However, such a separation or uncertainty equivalence principle no longer applies to nonlinear dynamical systems. Still, such a principle is widely applied in practice due to the computational burden of combined estimation/data assimilation and optimal control/planning problems.

Sebastian Reich
Institut für Mathematik, Universität Potsdam, Karl-Liebknecht-Str. 24/25, 14476 Potsdam,
sebastian.reich@uni-potsdam.de

B. Koren and D. Roose (eds.), *Numerical Analysis and Scientific Computing: an Anthology*, Lecture Notes in Computational Science and Engineering 156,
https://doi.org/10.1007/978-3-032-33428-2_9

At the same time, advances in uncertainty quantification, reduced order modeling, data assimilation, and optimal control and the increasing importance of digital twins in engineering, personalized medicine, and weather forecasting have opened the door to new approaches and new challenges [15, 21].

In this paper, we revisit the problem of optimal control for partially observed diffusion processes by combining nonlinear model predictive control (NMPC) [19] with ensemble Kalman filtering (EnKF) [8, 5] for data assimilation and state estimation; thus providing an alternative approach to the currently available output feedback NMPC techniques [9, 19]. Our approach is related to the covariance steering approach to belief space planning [23]. However, in contrast to splitting the problem into a nonlinear nominal problem and subsequently solving a linearized optimal control and estimation problem as used in [23], we solve an ensemble of nonlinear data assimilation problems and subsequently reduce the arising stochastic optimal control problem in the resulting belief states to linear feedback control laws; as also at the core of the EnKF for data assimilation [8, 5].

To simplify the exposition, we assume a data assimilation setting, where partial and noisy observations are taken continuously in time [12, 20] and processed by the ensemble Kalman–Bucy filter (EnKBF) [4, 5]. The novelty of our approach arises from the fact that we use the EnKBF for (i) estimating model states from actual observations and (ii) predicting uncertainties in the optimal control problem arising from the statistics of future observations and uncertainties in the estimated model states. For the latter task, we formulate an appropriate set of forward-backward stochastic differential equations (FBSDEs) based on Pontryagin's maximum principle [6]. In the spirit of the EnKBF, we also utilize a linear approximation *ansatz* when solving the associated backward stochastic differential equation (BSDE), giving rise to linear mean-field control laws in the estimated model states. While linear control laws become insufficient for highly nonlinear infinite horizon optimal control problems, the receding horizon approach of NMPC allows for a continuous adaptation of linear approximations, very much in line with modern data assimilation techniques such as the EnKF. As for NMPC, the success of the proposed methodology depends on suitable choices of the optimal control cost function as well as sufficiently informative observations to apply EnKF-based data assimilation.

In this paper, we assume a deterministic model dynamics. Model errors can be taken into account through stochastic perturbations in the EnKBF or robust NPMC schemes [9, 19].

The remainder of the paper is structured as follows. A mathematical problem formulation is provided in Section 2. We first develop our novel methodology in Section 3 in the context of partially observed linear control problems. In particular, we derive an interacting particle formulation in terms of FBSDEs which exactly reproduces the known state estimation and the optimal control solution [10, 14]. The proposed methodology is then extended to nonlinear problems in the spirit of EnKF-based methodologies [5] in Section 4 with the addition that we also need to approximate solutions to a BSDE. We introduce the acronym EnKBF-NMPC for the newly proposed method. Section 5 provides a numerical illustration in the context of an inverted pendulum. The paper closes with some final remarks in Section 6.

2 Mathematical Problem Formulation

We consider controlled ordinary differential equations (ODEs) of the form

$$\dot{X}_t = f(X_t) + G(X_t)U_t \tag{1}$$

subject to uncertain initial conditions $X_0 \in \mathbb{R}^{d_x}$, which are distributed according to a known probability measure $\nu_0(\mathrm{d}x)$. The control applied at time $t \geq 0$ is denoted by $U_t \in \mathbb{R}^{d_u}$. The drift function $f : \mathbb{R}^{d_x} \to \mathbb{R}^{d_x}$ and the matrix-valued function $G : \mathbb{R}^{d_x} \to \mathbb{R}^{d_x \times d_u}$ are assumed to be given.

We also assume that the process is partially observed continuously in time with the forward model

$$\mathrm{d}\mathcal{Y}_t = h(X_t)\mathrm{d}t + R^{1/2}\mathrm{d}W_t, \tag{2}$$

where $h : \mathbb{R}^{d_x} \to \mathbb{R}^{d_y}$ denotes the forward model, $R \in \mathbb{R}^{d_y \times d_y}$ is the measurement error covariance matrix, and W_t is standard Brownian motion (BM) of dimension d_y.

Actual observations are denoted by $\mathcal{Y}_t^\dagger$, $t \geq 0$, and its associated filtering distribution at time $\tau > 0$ by

$$\nu_\tau(\mathrm{d}x) := \nu_\tau(\mathrm{d}x|\mathcal{Y}_{0:\tau}^\dagger). \tag{3}$$

Here we have introduced the notation $\mathcal{Y}_{0:\tau}^\dagger$ to indicate the dependence on observations $\mathcal{Y}_t^\dagger$ from the time interval $t \in (0, \tau]$.

At each time instant $\tau_n = n\Delta\tau$, $n \geq 0$ and for available observations $\mathcal{Y}_{0:\tau_n}^\dagger$, we solve a finite horizon stochastic optimal control problem with cost function

$$\mathcal{L}(U_{\tau_n:\tau_n+T}) = \mathbb{E}\left[\int_{\tau_n}^{\tau_n+T} \left(\frac{1}{2}\|U_t\|^2 + c(X_t)\right)\mathrm{d}t + \psi(X_{\tau_n+T})\right], \tag{4}$$

$\tau_n + T \geq \tau_{n+1}$, with expectations taken over initial conditions

$$X_{\tau_n} \sim \nu_{\tau_n} \tag{5}$$

and all possible future observation paths $\mathcal{Y}_{\tau_n:\tau_n+T}$ that satisfy the model (2). Here, $c(x)$ and $\psi(x)$ are given cost functions.

Since the initial conditions are uncertain and obey (5), the optimal feedback control law becomes a function of the distributions $\nu(\mathrm{d}x)$, which we denote by $u_t^*(\nu)$. Then $U_t^* = u_t^*(\nu_t)$ is applied over the time interval $t \in (\tau_n, \tau_{n+1}]$ along the filtering distribution ν_t for given data $\mathcal{Y}_{0:t}^\dagger$. Hence, we obtain the controlled ODE

$$\dot{X}_t = f(X_t) + G(X_t)u_t^*(\nu_t) \tag{6}$$

for $t \in (\tau_n, \tau_{n+1}]$ and initial $X_{\tau_n} \sim \nu_{\tau_n}$.

Receding horizon optimal control is at the heart of NMPC [19]. NMPC typically replaces the uncertain X_{τ_n} by its expectation value under the filtering distribution ν_{τ_n}; i.e. by

$$\bar{X}_{\tau_n} := \mathbb{E}[X_{\tau_n}]. \tag{7}$$

This approximation reduces the stochastic optimal control problem to a standard deterministic optimal control problem with cost

$$\mathcal{L}(u_{\tau_n:\tau_n+T}) = \int_{\tau_n}^{\tau_n+T} \left(\frac{1}{2}\|U_t\|^2 + c(\bar{x}_t)\right) \mathrm{d}t + \psi(\bar{x}_{\tau_n+T}) \tag{8}$$

subject to the ODE

$$\dot{\bar{x}}_t = f(\bar{x}_t) + G(\bar{x}_t)U_t \tag{9}$$

with initial condition $\bar{x}_{\tau_n} = \bar{X}_{\tau_n}$. We note that the optimal control U_t^* becomes open-loop.

In this paper, we develop an alternative approach that takes the uncertainties in the initial conditions and future observations into account and results in feedback control laws of the form

$$u_t^*(\nu_t) = -G(\bar{X}_t)^{\mathrm{T}} \left(\Lambda_t^* \bar{X}_t + \lambda_t^*\right) \tag{10}$$

for suitable matrices $\Lambda_t^* \in \mathbb{R}^{d_x \times d_x}$ and vectors $\lambda_t^* \in \mathbb{R}^{d_x}$, $t \in (\tau_{t_n}, \tau_{t_n} + T]$. The feedback control law (10) is linear for constant G.

Let us embed the discussion so far into a digital twin setting [11]. The optimal control $U_t^\dagger = u_t^*(\nu_t)$; i.e. (10), would be applied to the physical twin, which could be of the form

$$\mathrm{d}X_t^\dagger = f(X_t^\dagger)\mathrm{d}t + G(X_t^\dagger)U_t^\dagger \mathrm{d}t + \sigma \mathrm{d}B_t^\dagger. \tag{11}$$

Here $B_t^\dagger$ stands for random perturbations, which mimic the difference between the physical and digital twin and which could take the form of standard BM for simplicity of exposition, and $\sigma > 0$. The physical twin (11) produces a reference trajectory, which we denote by $X_t^\dagger$ and which is partially observed according to the observation model (2); i.e., the actual observations $\mathcal{Y}_{0:t}^\dagger$ satisfy

$$\mathrm{d}\mathcal{Y}_t^\dagger = h(X_t^\dagger)\mathrm{d}t + R^{1/2}\mathrm{d}W_t^\dagger \tag{12}$$

with $W_t^\dagger$ a particular realization of BM. Given the data, the digital twin evolves according to the EnKBF mean-field equation

$$\begin{aligned}\mathrm{d}X_t = &\left(f(X_t) - G(X_t)G(\bar{X}_t)^{\mathrm{T}} \left(\Lambda_t^* \bar{X}_t + \lambda_t^*\right)\right) \mathrm{d}t \\ &- \frac{1}{2} C_t^{xh} R^{-1}(h(X_t) + \bar{h}_t)\mathrm{d}t + C_t^{xh} R^{-1} \mathrm{d}\mathcal{Y}_t^\dagger,\end{aligned} \tag{13}$$

which depends on the law ν_t of X_t via its mean $\bar{X}_t$, the expectation value $\bar{h}_t$ of $h(X_t)$, and the covariance matrix C_t^{xh} between X_t and $h(X_t)$.

Since our approach is based on the continuous-time EnKBF approximation [5] to filtering distributions (3), we abbreviate our approach EnKBF-NMPC. In order to implement EnKBF-NMPC, it remains to formulate an algorithm for computing the required Λ_t^* and λ_t^*. In the following section, our approach is motivated by first considering linear control problems.

3 Optimal Control for Partially Observed Linear ODEs

We start from the linear control problem

$$\dot{X}_t = AX_t + b + GU_t \tag{14}$$

in state space $\mathbb{R}^{d_x}$ subject to continuous-time observations

$$\mathrm{d}\mathcal{Y}_t = HX_t \mathrm{d}t + R^{1/2}\mathrm{d}W_t. \tag{15}$$

An actual observation is denoted by $\mathcal{Y}_t^\dagger \in \mathbb{R}^{d_y}$ and the initial X_0 is Gaussian distributed with mean m_0 and covariance C_0. The ensemble-based EnKBF evolution equations are given by

$$\mathrm{d}X_t^{(i)} = \left(AX_t^{(i)} + b + GU_t\right)\mathrm{d}t - \frac{1}{2}C_t H^{\mathrm{T}} R^{-1} H(X_t^{(i)} + \bar{X}_t)\mathrm{d}t + C_t H^{\mathrm{T}} R^{-1/2} \mathrm{d}\mathcal{Y}_t^\dagger, \tag{16}$$

$i = 1, \ldots, M$ [5]. Here $M > d_x$ is the size of the ensemble,

$$\bar{X}_t = \frac{1}{M}\sum_{j=1}^{M} X_t^{(j)} \tag{17}$$

denotes the empirical mean of the ensemble $\{X_t^{(j)}\}$ and

$$C_t = \frac{1}{M}\sum_{j=1}^{M}(X_t^{(j)} - \bar{X}_t)(X_t^{(j)} - \bar{X}_t)^{\mathrm{T}} \tag{18}$$

its empirical covariance matrix.

The initial ensemble is chosen such that $\bar{X}_0 = m_0$ and $C_0 = C_0$ for given mean m_0 and covariance matrix C_0. Given these initial conditions, the EnKBF formulation (16) is equivalent to the standard Kalman–Bucy filter equations in terms of the time evolution of their mean and covariance matrix; i.e., $\bar{X}_t = m_t$ and $C_t = C_t$ for all $t \geq 0$ regardless of the ensemble size $M > d_x$. However, note that the EnKBF is often used with $M < d_x$ in combination with localization and that the EnKBF easily generalizes to non-Gaussian and nonlinear evolution settings.

We set $\tau_n = 0$ in (4) for simplicity of exposition and consider the time window $[0, T]$ with the optimal control U_t^* now chosen to minimize the cost functional

$$\mathcal{V}(U_{0:T}) = \mathbb{E}\left[\int_0^T \left(\frac{1}{2}\|U_t\|^2 + \frac{1}{M}\sum_{j=1}^{M} c(X_t^{(j)})\right)\mathrm{d}t + \frac{1}{2M}\sum_{j=1}^{M}\psi(X_T^{(j)})\right], \tag{19}$$

where

$$c(x) = \frac{1}{2}(x - c)^{\mathrm{T}} V (x - c) \tag{20}$$

is the running cost and

$$\psi(x) = \frac{1}{2}(x - c_T)^{\mathrm{T}} V_T (x - c_T) \tag{21}$$

the terminal cost, respectively. The symmetric positive-definite matrices $V \in \mathbb{R}^{d_x \times d_x}$ and $V_T \in \mathbb{R}^{d_x \times d_x}$ as well as the vectors $c \in \mathbb{R}^{d_x}$ and $c_T \in \mathbb{R}^{d_x}$ are assumed to be given. The expectation in (19) is taken over all possible future observations $\mathcal{Y}_t$, $t > 0$, and initial conditions $X_0 \sim \mathrm{N}(m_0, C_0)$.

In order to apply the stochastic Pontryagin maximum principle [6], we first rewrite the EnKBF (16) in the form

$$\mathrm{d}X_t^{(i)} = \left(AX_t^{(i)} + b + GU_t\right)\mathrm{d}t - \frac{1}{2}C_t H^{\mathrm{T}} R^{-1} H(X_t^{(i)} - \bar{X}_t)\mathrm{d}t + C_t H^{\mathrm{T}} R^{-1/2}\mathrm{d}\mathcal{W}_t \tag{22}$$

$t > 0$, where $\mathcal{W}_t$ is BM of dimension d_y. This reformulation is justified because the innovation process $\mathcal{I}_t$ defined by $\mathrm{d}\mathcal{I}_t = \mathrm{d}\mathcal{Y}_t - H\bar{X}_t \mathrm{d}t$ behaves like BM [1]. It should be noted that noise $\mathcal{W}_t$ only affects the time evolution of the mean $\bar{X}_t$.

For simplicity, let us assume that observations $\mathcal{Y}_t$ in (15) are scalar-valued, i.e., $d_y = 1$. Following [6], the Hamiltonian appearing in Pontryagin's maximum principle is given by

$$\begin{aligned}\mathcal{H}(\{x_j\}, \{p_j\}, \{z_j\}, u) &= \frac{M}{2}\|u\|^2 + \sum_{j=1}^{M} c(x_j) + \\ &\sum_{j=1}^{M} \left\{ \left(Ax_j + b + Gu - \frac{1}{2}CH^{\mathrm{T}} R^{-1} H(x_j - \bar{x}) \right)^{\mathrm{T}} p_j + R^{-1/2} HCz_j \right\}\end{aligned} \tag{23}$$

with $x_i \in \mathbb{R}^{d_x}$, $p_i \in \mathbb{R}^{d_x}$, and $z_i \in \mathbb{R}^{d_x}$. The FBSDEs in the associated time-dependent random variables $X_t^{(i)} \in \mathbb{R}^{d_x}$, $P_t^{(i)} \in \mathbb{R}^{d_x}$, and $Z_t^{(i)} \in \mathbb{R}^{d_x}$ become

$$\mathrm{d}X_t^{(i)} = \nabla_{p_i} H(\{X_t^{(j)}\}, \{P_t^{(j)}\}, \{Z_t^{(j)}\}, U_t^*)\mathrm{d}t + C_t H^{\mathrm{T}} R^{-1/2} \mathrm{d}\mathcal{W}_t, \tag{24a}$$

$$\mathrm{d}P_t^{(i)} = -\nabla_{x_i} H(\{X_t^{(j)}\}, \{P_t^{(j)}\}, \{Z_t^{(j)}\}, U_t^*)\mathrm{d}t + Z_t^{(i)} \mathrm{d}\mathcal{W}_t \tag{24b}$$

for $i = 1, \ldots, M$, subject to the terminal condition

$$P_T^{(i)} = \nabla_x \psi(X_T^{(i)}) \tag{25}$$

and the initial ensemble $\{X_0^{(j)}\}$. The desired optimal control U_t^* is provided by

$$U_t^* = \arg\min_u H(\{X_t^{(j)}\}, \{P_t^{(j)}\}, \{Z_t^{(j)}\}, u) = -\frac{1}{M}\sum_{j=1}^{M} G^{\mathrm{T}} P_t^{(j)}. \tag{26}$$

Upon substitution into (23), we obtain the reduced Hamiltonian

$$\mathcal{H}_{\text{red}}(\{x_j\},\{p_j\},\{z_j\}) = -\frac{M}{2}\|G^{\mathrm{T}}\bar{p}\|^2 + \sum_{j=1}^{M} c(x_j) + \sum_{j=1}^{M}\left\{\left(Ax_j + b - \frac{1}{2}CH^{\mathrm{T}}R^{-1}H(x_j-\bar{x})\right)^{\mathrm{T}} p_j + R^{-1/2}HC\bar{z}\right\} \tag{27}$$

with empirical means

$$\bar{p} = \frac{1}{M}\sum_{j=1}^{M} p_j, \qquad \bar{z} = \frac{1}{M}\sum_{j=1}^{M} z_j. \tag{28}$$

The computation of $\nabla_{x_i}\mathcal{H}_{\text{red}}$ is rather complex due to the appearance of the empirical covariance matrix C and the empirical mean

$$\bar{x} = \frac{1}{M}\sum_{j=1}^{M} x_j \tag{29}$$

in the Hamiltonian $\mathcal{H}_{\text{red}}$:

$$\begin{aligned}\nabla_{x_i}\mathcal{H}_{\text{red}} &= A^{\mathrm{T}}p_i - \frac{1}{2}H^{\mathrm{T}}R^{-1}HC(p_i-\bar{p}) + V(x_i - c) + \\ &- \frac{1}{2}H^{\mathrm{T}}R^{-1}HC^{xp}(x_i-\bar{x}) - \frac{1}{2}C^{px}H^{\mathrm{T}}R^{-1}H(x_i-\bar{x}) + \\ &+ H^{\mathrm{T}}R^{-1/2}\bar{z}^{\mathrm{T}}(x_i-\bar{x}) + \bar{z}R^{-1/2}H(x_i-\bar{x}).\end{aligned} \tag{30}$$

Here we have introduced the empirical covariance matrices

$$C^{px} = \frac{1}{M}\sum_{j=1}^{M}(p_j-\bar{p})(x_j-\bar{x})^{\mathrm{T}} = (C^{xp})^{\mathrm{T}}. \tag{31}$$

We next introduce the ensemble deviations

$$\delta X_t^{(i)} := X_t^{(i)} - \bar{X}_t \tag{32}$$

for $i = 1,\ldots,M$ and make the *ansatz*

$$P_t^{(i)} = \Lambda_t\bar{X}_t + \Omega_t\delta X_t^{(i)} + \lambda_t \tag{33}$$

for suitable matrices $\Lambda_t \in \mathbb{R}^{d_x\times d_x}$, $\Omega_t \in \mathbb{R}^{d_x\times d_x}$ and vectors $\lambda_t \in \mathbb{R}^{d_x}$. It then follows from [6] eq. (4.25) that

$$\bar{Z}_t = Z_t^{(j)} = \Lambda_t C_t H^{\mathrm{T}} R^{-1/2} \tag{34}$$

for all $j = 1, \dots, M$. Also note that

$$C_t^{px} = \Omega_t C_t. \tag{35}$$

The optimal control is defined in terms of the ensemble mean

$$\bar{P}_t = \frac{1}{M}\sum_{j=1}^{M} P_t^{(j)} = \Lambda_t \bar{X}_t + \lambda_t \tag{36}$$

and, therefore, we only consider the evolution equation in $\bar{P}_t$ from now on, which is given by the BSDE

$$-\mathrm{d}\bar{P}_t = A^{\mathrm{T}}\bar{P}_t \mathrm{d}t + V(\bar{X}_t - c)\mathrm{d}t - \bar{Z}_t \mathrm{d}\mathcal{W}_t. \tag{37}$$

Furthermore, the forward stochastic differential equation (FSDE) (24a) becomes

$$\mathrm{d}X_t^{(i)} = \left(AX_t^{(i)} + b - GG^{\mathrm{T}}\bar{P}_t\right)\mathrm{d}t - \frac{1}{2}C_t H^{\mathrm{T}} R^{-1} H(X_t^{(i)} - \bar{X}_t)\mathrm{d}t + C_t H^{\mathrm{T}} R^{-1/2}\mathrm{d}\mathcal{W}_t \tag{38}$$

and the desired optimal control law is provided by

$$U_t^* = u_t^*(\nu_t) := -G^{\mathrm{T}}\bar{P}_t = -G^{\mathrm{T}}\left(\Lambda_t \bar{X}_t + \lambda_t\right). \tag{39}$$

Taking the ensemble mean in (24) yields the following FSDE in $\bar{X}_t$:

$$\mathrm{d}\bar{X}_t = \left(A\bar{X}_t + b - GG^{\mathrm{T}}\bar{P}_t\right)\mathrm{d}t + C_t H^{\mathrm{T}} R^{-1/2}\mathrm{d}\mathcal{W}_t. \tag{40}$$

Hence, (40) and (37) together with (33) lead to the following evolution equations for Λ_t and λ_t:

$$-\dot{\Lambda}_t = \Lambda_t A + A^{\mathrm{T}}\Lambda_t - \Lambda_t GG^{\mathrm{T}}\Lambda_t + V, \tag{41a}$$

$$-\dot{\lambda}_t = A^{\mathrm{T}}\lambda_t - Vc + \Lambda_t(b - GG^{\mathrm{T}}\lambda_t) \tag{41b}$$

with terminal conditions $\Lambda_T = V_T$ and $\lambda_T = -V_T c_T$, respectively.

The formulations (41a) and (41b)coincide with the optimal control equations for linear systems [10, 14]. We have derived them here from a perspective based on McKean (interacting particle) reformulation (16) of the standard Kalman–Bucy filter equations [5] and an application of the stochastic Pontryagin maximum principle [6].

4 The EnKBF-NMPC Algorithm

The attractiveness of formulations (22) and (37) stems from the fact that they generalize beyond the linear Gaussian setting in the spirit of EnKF [5]. We now also return to vector-valued observations; i.e. $d_y \geq 1$. For example, a generalized formulation of FBSDE is provided by

$$\mathrm{d}X_t^{(i)} = \left(f(X_t^{(i)}) + G(X_t^{(i)})U_t^* - \frac{1}{2}C_t^{xh}R^{-1}(h(X_t^{(i)}) - \bar{h}_t)\right)\mathrm{d}t + C_t^{xh}R^{-1/2}\mathrm{d}\mathcal{W}_t, \tag{42a}$$

$$-\mathrm{d}\bar{P}_t = \left(Df(\bar{X}_t) + DG(\bar{X}_t)U_t^*\right)^{\mathrm{T}}\bar{P}_t\mathrm{d}t + \nabla_x c(\bar{X}_t)\mathrm{d}t + \bar{Z}_t\mathrm{d}\mathcal{W}_t, \tag{42b}$$

where

$$\bar{h}_t = \frac{1}{M}\sum_{j=1}^{M} h(X_t^{(j)}) \tag{43}$$

and C_t^{xh} denotes the empirical covariance matrix between $X_t \in \mathbb{R}^{d_x}$ and $h(X_t) \in \mathbb{R}^{d_y}$, respectively. The desired optimal control is approximated by

$$U_t^* = u_t^*(\nu_t) := -G(\bar{X}_t)^{\mathrm{T}}\bar{P}_t. \tag{44}$$

Here, $Df(x) \in \mathbb{R}^{d_x \times d_x}$ denotes the Jacobian of $f(x)$ and $DG(x)U_t^* \in \mathbb{R}^{d_x \times d_x}$ the Jacobian of $G(x)U_t^*$.

The formulation (42) is now used as a basis to extend NMPC [19] to partially observed diffusion processes and EnKBF-based data assimilation. In particular, we have to simulate (42) over multiple realizations of the noise $\mathcal{W}_t$, which we denote by $\mathcal{W}_{t,k}$, $k = 1, \ldots, K$. We also apply statistical linearization [5] in order to avoid computing the Jacobian $Df(x)$ and $DG(x)U_{t,k}^*$. These approximations yield the FBSDEs

$$\mathrm{d}X_{t,k}^{(i)} = \left(f(X_{t,k}^{(i)}) + G(X_{t,k}^{(i)})U_{t,k}^* - \frac{1}{2}C_{t,k}^{xh}R^{-1}(h(X_{t,k}^{(i)}) - \bar{h}_{t,k})\right)\mathrm{d}t + C_{t,k}^{xh}R^{-1/2}\mathrm{d}\mathcal{W}_{t,k}, \tag{45a}$$

$$-\mathrm{d}\bar{P}_{t,k} = C_{t,k}^{-1}\left(C_{t,k}^{xf} + C_{t,k}^{xG}U_{t,k}^*\right)\bar{P}_{t,k}\mathrm{d}t + \nabla_x c(\bar{X}_{t,k})\mathrm{d}t - \bar{Z}_{t,k}\mathrm{d}\mathcal{W}_{t,k}, \tag{45b}$$

for $i = 1, \ldots, M$ and $k = 1, \ldots, K$, together with optimal control $U_{t,k}^* = -G(\bar{X}_{t,k})^{\mathrm{T}}\bar{P}_{t,k}$. Empirical covariance matrices are defined by

$$C_{t,k}^{xh} = \frac{1}{M}\sum_{j=1}^{M}(X_{t,k}^{(j)} - \bar{X}_{t,k})(h(X_{t,k}^{(j)}) - \bar{h}_{t,k})^{\mathrm{T}} \tag{46}$$

with related expressions for $C_{t,k} = C_{t,k}^{xx}$, $C_{t,k}^{xf}$, and $C_{t,k}^{xG}$. Furthermore,

$$\bar{X}_{t,k} = \frac{1}{M}\sum_{j=1}^{M} X_{t,k}^{(j)}, \qquad \bar{h}_{t,k} = \frac{1}{M}\sum_{j=1}^{M} h(X_{t,k}^{(j)}). \tag{47}$$

BSDE (45b) is now approximated using the linear regression *ansatz*

$$\bar{P}_{t,k} \approx \Lambda_t(\bar{X}_{t,k} - \bar{m}_t) + \mu_t, \tag{48}$$

where $\bar{\Lambda}_t$, μ_t, and $\bar{Z}_{t,k}$ are chosen such that the resulting approximation error is minimized [6] and

$$\bar{m}_t = \frac{1}{K}\sum_{k=1}^{K} \bar{X}_{t,k}. \tag{49}$$

We note that the choice

$$U^*_{t,k} = -G(\bar{m}_t)^{\mathrm{T}}\mu_t \tag{50}$$

corresponds to an open-loop control in NMPC.

More precisely, given a time-discretization at time levels $t_{n+1} = t_n + \Delta t$, $n = 0,\ldots,N-1$ for given step-sizes $\Delta t = T/N > 0$ and $t_0 = 0$, we define the least squares cost function

$$\mathcal{L}_n(\Lambda_{t_n},\mu_{t_n}) = \frac{1}{K}\sum_{k=1}^{K} \left\|\tilde{P}_{t_n,k} - \gamma_{t_n,k}\right\|^2, \tag{51}$$

with

$$\gamma_{t_n,k} = \left(I + \Delta t C^{-1}_{t_{n+1},k}\left(C^{xf}_{t_{n+1},k} + C^{xGU}_{t_{n+1},k}\right)\right)\bar{P}_{t_{n+1},k} + \nabla_x c(\bar{X}_{t_{n+1},k})\Delta t \tag{52}$$

subject to

$$\tilde{P}_{t_n,k} = \Lambda_{t_n}(\bar{X}_{t_n,k} - \bar{m}_{t_n}) + \mu_{t_n}. \tag{53}$$

We further simplify this formulation by explicitly removing the noise in $\bar{X}_{t_n,k}$ using the backward integration step

$$\tilde{X}^{(i)}_{t_n,k} = X^{(i)}_{t_{n+1},k} - \Delta t\left(f(X^{(i)}_{t_{n+1},k}) - G(X^{(i)}_{t_{n+1},k})G(\bar{X}_{t_{n+1},k})^{\mathrm{T}}\bar{P}_{t_{n+1},k}\right). \tag{54}$$

Then replace (53) with

$$\tilde{P}_{t_n,k} = \Lambda_{t_n}(\tilde{X}_{t_n,k} - \tilde{m}_{t_n}) + \mu_{t_n}, \qquad \tilde{X}_{t_n,k} = \frac{1}{M}\sum_{j=1}^{M}\tilde{X}^{(i)}_{t_n,k}, \quad \tilde{m}_{t_n} = \frac{1}{K}\sum_{k=1}^{K}\tilde{X}_{t_n,k}, \tag{55}$$

and (51) becomes

$$\mathcal{L}_n(\Lambda_{t_n},\mu_{t_n}) = \frac{1}{K}\sum_{k=1}^{K}\left\|\Lambda_{t_n}(\tilde{X}_{t_n,k} - \tilde{m}_{t_n}) + \mu_{t_n} - \gamma_{t_n,k}\right\|^2. \tag{56}$$

Denote the minimizer of (56) by $(\Lambda^*_{t_n},\mu^*_{t_n})$. See the Appendix for explicit solution formulas. Then we set

$$\bar{P}_{t_n,k} := \Lambda^*_{t_n}(\bar{X}_{t_n,k} - \tilde{m}_{t_n}) + \mu^*_{t_n} = \Lambda^*_{t_n}\bar{X}_{t_n,k} + \lambda^*_{t_n} \tag{57}$$

with

$$\lambda^*_{t_n} = \mu^*_{t_n} - \Lambda^*_{t_n}\tilde{m}_{t_n}. \tag{58}$$

The terminal condition is provided by

$$\bar{P}_{T,k} = \nabla_x \psi(\bar{X}_{T,k}). \tag{59}$$

The forward and backward evolution equations are iterated until convergence for fixed realizations of the noise $\mathcal{W}_{t,k}$.

Formulation (54) is justified in our setting, since we assume linear McKean control laws and hence second-order terms, which would arise from the stochastic contributions, do not appear in the control. We also note that Λ_t^* is a symmetric matrix in the linear setting, which is no longer the case when used as an approximation in the nonlinear setting. A symmetric approximation could still be obtained using $(\Lambda_{t_n}^* + (\Lambda_{t_n}^*)^{\mathrm{T}})/2$ instead of $\Lambda_{t_n}^*$ in (60) below.

The sequence $\{\Lambda_{t_n}^*, \lambda_{t_n}^*\}_{n=0}^N$ is stored and linearly interpolated to provide linear control

$$U_t^* = u_t^*(\nu_t) := -G(\bar{X}_t)^{\mathrm{T}} \left(\Lambda_t^* \bar{X}_t + \lambda_t^*\right) \tag{60}$$

for the physical twin (11) and for the forward prediction step of its digital twin

$$\mathrm{d}X_t^{(i)} = \left(f(X_t^{(i)}) + G(X_t^{(i)})U_t^* - \frac{1}{2}C_t^{xh}R^{-1}(h(X_t^{(i)}) + \bar{h}_t)\right)\mathrm{d}t + C_t^{xh}R^{-1/2}\mathrm{d}\mathcal{Y}_t^{\dagger}. \tag{61}$$

Let us briefly comment on the relation of the proposed EnKBF-NMPC scheme to available NMPC for partially observed processes. NMPC for partially observed processes is called the output NMPC in Section 4 of [19] and the output feedback NMPC in [9]. Most existing approaches rely on what is called the nominal solution in [19]; which in our context amounts to setting $\mathcal{W}_{t,k} \equiv 0$ and $X_{0,k}^{(i)} = \bar{X}_0$ and thus reduces the associated optimal control problem to a deterministic finite-horizon control problem or, equivalently, to an open-loop control of the form (50). Instead, EnKBF-NMPC takes into account uncertainties, which are induced by uncertain initial conditions as well as uncertain future observations. The related approach of [23, 24] also builds upon a nominal solution, as in NMPC, and computes Gaussian approximations of the uncertainties using control theory for partially observed linear processes. Our approach, in contrast, avoids explicit linearization about a nominal solution and represents uncertainties by evolving particles.

5 Numerical Example

A simple nonlinear example is provided by the controlled inverted pendulum

$$\dot{v}_t = \sin\phi_t - \gamma v_t + \cos(\phi_t)u_t, \qquad \dot{\phi}_t = v, \tag{62}$$

with $\gamma = 5$ [14]. The state variable is $x = (\phi, v)^{\mathrm{T}} \in \mathbb{R}^2$. We note that the equilibrium $(0, 0)$ is unstable, while $(\pi, 0)$ is stable. The control u_t should stabilize the unstable equilibrium given uncertainty in the initial conditions and noisy observations of the

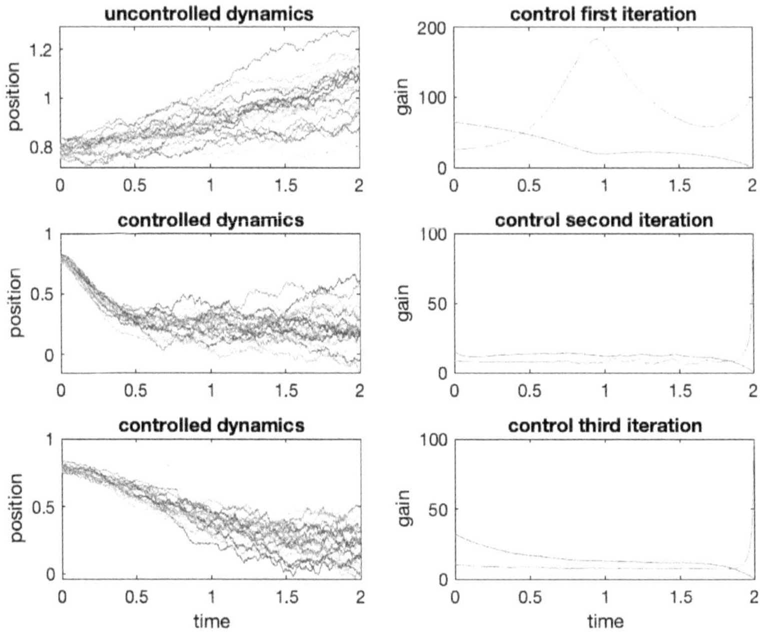

Figure 1 Left panels: We display the frequentist forecast uncertainties obtained from simulating (45a) using the optimal control obtained from the previous iteration of (45) with the initial control being $u \equiv 0$. We perform three iterations with results displayed sequentially from top to bottom. Right panels: We display the two entries from the second column of $\Lambda_t^* \in \mathbb{R}^{2\times 2}$, which have been obtained by minimizing (56). The control has essentially converged after three iteration steps.

angle ϕ_t; i.e.,

$$\mathrm{d}\mathcal{Y}_t = \phi_t \mathrm{d}t + R^{1/2}\mathrm{d}W_t. \tag{63}$$

The initial positions ϕ_0 are Gaussian distributed with mean $\pi/4$ and the initial velocities $\dot{\phi}_0$ are Gaussian distributed with mean zero. The variance is $\sigma_0 = 0.05$ for both variables. We set $R = 1$ unless otherwise stated. Observations $\mathcal{Y}_t^\dagger$, $t \geq 0$, are generated in an ideal twin experimental setting; i.e., we set $\sigma = 0$ in the corresponding physical twin (11).

We first implement a fixed-horizon version of the EnKBF-NMPC subject to final cost

$$\psi(\phi, v) = \frac{\lambda_T}{2}(\phi^2 + v^2) \tag{64}$$

and running cost

$$c(\phi, v) = \frac{\lambda}{2}(\phi^2 + v^2) \tag{65}$$

with $\lambda_T = \lambda = 100$. The EnKBF-NMPC scheme implemented starts from $u_t \equiv 0$ in the FSDE and then iterates forward and backward three times over the fixed time interval $[0, 2]$; i.e. $T = 2$. The results can be found in Figure 1. The ensemble sizes are

$M = K = 20$ and the step-size is set to $\Delta t = 10^{-3}$. The effect of the control can be clearly seen, and also the (frequentist) uncertainty due to possible future observations.

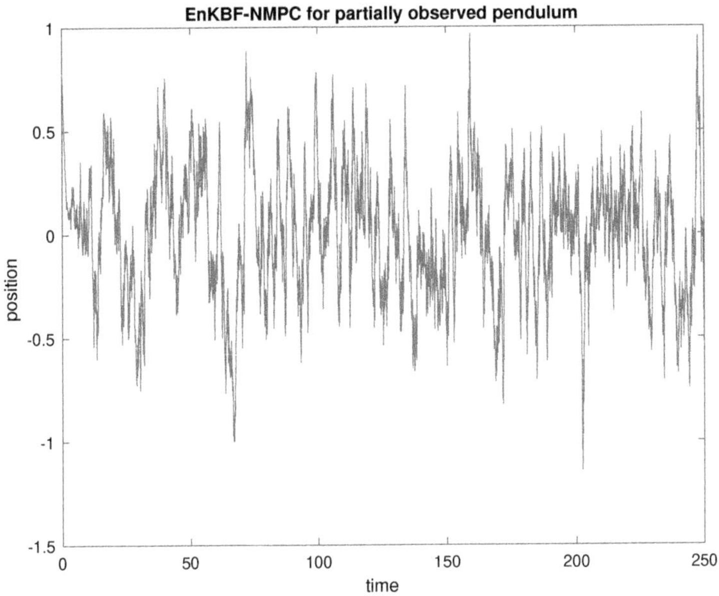

Figure 2 Ensemble mean from an EnKBF filter implementation with controls computed using the proposed EnKBF-NMPC methodology. The control laws were computed over time intervals of length $T = 0.5$ with the computed control then applied over intervals of length $\Delta\tau = 0.05$. The process was repeated 5000 times. One can clearly identify the stabilizing effect of the control.

In a second experiment, we repeatedly solved the receding NMPC problem over time intervals of length $T = 0.5$ with final cost $\psi(x)$ and running cost $c(x)$ as stated before. The computed control is then applied over time intervals $\Delta\tau = 0.05$ subject to a noisy observation of the angle of its physical twin. The resulting time evolution of the mean of the ensemble from the associated controlled EnKBF (61) with the entire process repeated 5000 times can be found for $R = 1$ in Figure 2 and for $R = 0.1$ in Figure 3.

It should be noted that the stated formulation does not allow one to move the pendulum from stable to unstable equilibrium positions. This is likely due to the choice of $\psi(x)$ in (64).

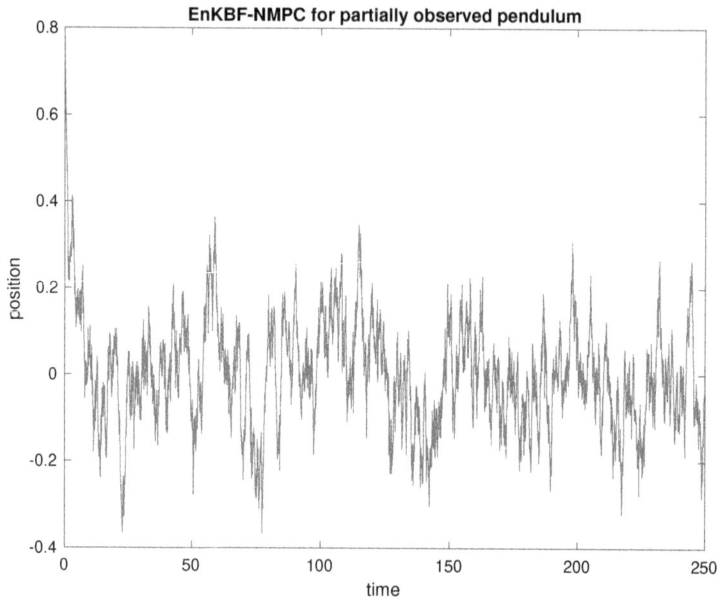

Figure 3 Same setting as used for Figure 2 except for a reduced measurement error with $R = 0.1$. The reduction in variance of the computed solution relative to the unstable equilibrium $\phi = 0$ is clearly visible.

6 Conclusions

A combination of NMPC and the EnKBF has been proposed in order to represent the effect of future observations and uncertainties in current state estimates on the control of partially observed dynamical systems. Related computational problems arise within the digital twin paradigm with partial and noisy measurements taken from its physical twin and digital twin based control laws in turn being applied to the physical twin.

Although this paper has followed a purely algorithmic approach, it remains to investigate the stability of the EnKBF-NMPC method under idealized digital and physical twin scenarios and its dependence on the chosen terminal cost function ψ. See, for example, [22] for a related accuracy and stability study for the EnKBF alone.

A possible extension is to replace the EnKBF by the feedback particle filter [13]. Note that [13] relies on the value function and its associated Hamilton–Jacobi–Bellman equation approach, while this paper follows Pontryagin's maximum principle [6]. A mean-field extension of Pontryagin's maximum principle, which can be applied to both the EnKBF and the feedback particle filter, has been proposed in [17]. Such an extension would allow one to consider global and nonlinear feedback control laws.

It would be of interest to also investigate Conditional Value at Risk (CVaR) measures and their sample approximations as a cost function; i.e., replace $\mu_T[\psi]$ by

$$\tilde{\psi}_\beta(\mu_T) := \arg\min_s \{s + \frac{1}{1-\beta}\mu_T\left[(\psi - s)_+\right]\} \tag{66}$$

for given $\beta \in (0, 1)$ [7]. Alternatively, one could consider a terminal cost

$$\tilde{\psi}_\beta(\mu_T) = \frac{1}{\beta}\log\left(\mu_T\left[\exp(\beta\psi)\right]\right), \tag{67}$$

for $\beta \in \mathbb{R}$ [6].

We finally mention that the proposed approach can be extended to discrete-time observations along the lines of [2]; again replacing the Hamilton–Jacobi–Bellman equation by Pontryagin's maximum principle.

In closing, let me mention that I gave an overview of the then emerging field of ensemble data assimilation and the EnKF at the Woudschoten Conference in 2009. In 2023, I returned to the Woudschoten conference, highlighting advances at the interface of interacting particle systems, data assimilation and control. A second set of lectures, presented by Jean Francois Chassagneux, covered numerical solutions of stochastic optimal control problems and forward-backward SDE in particular. The material presented in this paper has been inspired by those advances in combination with the topics presented in my 2023 Woudschoten lectures. I wish to thank and congratulate the Woudschoten numerical analysis community for having run such a wonderful series of annual meetings.

Competing Interests This work has been partially funded by Deutsche Forschungsgemeinschaft (DFG) - Project-ID 318763901 - SFB1294.

Appendix

The minimizer of a functional of the form

$$\mathcal{L}(\Lambda, \mu) = \sum_{k=1}^{K} \|\Lambda(\bar{X}_k - \bar{m}) + \mu - \gamma_k\|^2, \qquad \bar{m} = \frac{1}{K}\sum_{k=1}^{M} \bar{X}_k, \tag{68}$$

is provided by

$$\mu^* = \frac{1}{K}\sum_{k=1}^{K} \gamma_k \tag{69}$$

and

$$\Lambda^* = \left(\frac{1}{K}\sum_{k=1}^{K}(\gamma_k - \mu^*)(\bar{X}_k - \bar{m})^{\mathrm{T}}\right)\left(\frac{1}{K}\sum_{k=1}^{K}(\bar{X}_k - \bar{m})(\bar{X}_k - \bar{m})^{\mathrm{T}}\right)^{-1} \tag{70a}$$

$$= C^{\gamma x}(C^{xx})^{-1}, \tag{70b}$$

which implies a form of statistical linearization as also commonly used in ensemble Kalman filtering [5].

References

1. Bain, A., and Crisan, D. *Fundamentals of Stochastic Filtering*, vol. 60 of *Stoch. Model. Appl. Probab.* Springer, New York, 2009.
2. Bayer, C., Djehiche, B., Rezvanova, E., and Tempone, R. F. Continuous time stochastic optimal control under discrete time partial observations. Tech. rep., arXiv:2407.18018, 2024.
3. Bensoussan, A. *Stochastic Control of Partially Observable Systems.* Cambridge University Press, Cambridge, 1992.
4. Bergemann, K., and Reich, S. An ensemble Kalman-Bucy filter for continuous data assimilation. *Meteorologische Zeitschrift 21*, 3 (2012), 213.
5. Calvello, E., Reich, S., and Stuart, A. M. Ensemble Kalman methods: A mean field perspective. *Acta Numerica 34* (2025), 123–291.
6. Carmona, R. *Lectures on BSDEs, Stochastic Control, and Stochastic Differential Games with Financial Applications.* SIAM, Philadelphia, 2016.
7. Chaudhuri, A., Kramer, B., Norton, M., Royset, J. O., and Willcox, K. Certifiable risk-based engineering design optimization. *AIAA Journal 60*, 2 (2022), 551–565.
8. Evensen, G., Vossepoel, F., and van Leeuwen, P. *Data Assimilation Fundamentals: A Unified Formulation of the State and Parameter Estimation Problem.* Springer Nature Switzerland AG, Cham, Switzerland, 2022.
9. Findeisen, R., and Allgöwer, F. An introduction to nonlinear model predictive control. In *Systems and Control : 21th Benelux Meeting 2002* (2002), A. de Jager and H. Zwart, Eds., Technische Universiteit Eindhoven, pp. 119–141.
10. van Handel, R. Stochastic calculus, filtering, and stochastic control. *Lecture Notes, Caltech* (2007).
11. Kapteyn, M. G., Pretorius, J. V., and Willcox, K. E. A probabilistic graphical model foundation for enabling predictive digital twins at scale. *Nature Computational Science 1* (2021), 337–347.
12. Law, K., Stuart, A., and Zygalakis, K. *Data Assimilation.* Springer, Cham, Switzerland, 2015.
13. Mehta, P., and Meyn, S. A feedback particle filter-based approach to optimal control with partial observations. In *52nd IEEE Conference on Decision and Control* (2013), IEEE, pp. 3121–3127.
14. Meyn, S. *Control Systems & Reinforcement Learning.* Cambridge University Press, Cambridge, 2022.
15. National Academies of Sciences, Engineering, and Medicine. Foundational research gaps and future directions for digital twins. Tech. rep., Washington, DC: The National Academies Press, 2024.
16. Nisio, M. *Stochastic Control Theory*, 2nd ed. Springer, Tokyo, 2015.
17. Opper, M., and Reich, S. On a mean-field Pontryagin minimum principle for stochastic optimal control. Tech. rep., arXiv:2506.10506, 2025.
18. Åström, K. J. Optimal control of Markov processes with incomplete state information. *Journal of Mathematical Analysis and Applications 10* (1965), 174–205.
19. Rawlings, J. B., Mayne, D. Q., and Diehl, M. M. *Model Predictive Control: Theory, Computation, and Design*, 2nd ed. Nob Hill Publishing, Madison, 2018.
20. Reich, S., and Cotter, C. *Probabilistic Forecasting and Bayesian Data Assimilation.* Cambridge University Press, Cambridge, 2015.

21. Society for Industrial and Applied Mathematics. Report on the future of computational science. Tech. rep., SIAM Publishing, Philadelphia, 2024.
22. de Wiljes, J., Reich, S., and Stannat, W. Long-time stability and accuracy of the ensemble Kalman–Bucy filter for fully observed processes and small measurement noise. *SIAM Journal on Applied Dynamical Systems 17*, 2 (2018), 1152–1181.
23. Zheng, D., Ridderhof, J., Tsiotras, P., and Agha-Mohammadi, A.-A. Belief space planning: A covariance steering approach. In *2022 International Conference on Robotics and Automation (ICRA)* (2022), pp. 11051–11057.
24. Zheng, D., Ridderhof, J., Zhang, Z., Tsiotras, P., and Agha-Mohammadi, A.-A. CS-BRM: A probabilistic roadmap for consistent belief space planning with reachability guarantees. *IEEE Transactions on Robotics* (2024), 1630–1649.

Computing Stationary Distribution via Dirichlet-Energy Minimization by Coordinate Descent

Konstantin Avrachenkov, Lorenzo Gregoris and Nelly Litvak

Abstract We present an optimization-based interpretation of the Red Light Green Light (RLGL) algorithm for computing stationary distributions of large Markov chains. For reversible chains, RLGL is shown to be a block coordinate descent method for minimizing the Dirichlet energy associated with the symmetrized Laplacian. This connection yields exponential convergence guarantees under mild assumptions on the update schedule. The variational viewpoint also motivates new Gauss–Southwell–Dirichlet (GSD) coordinate selection rules, which provably maximize the instantaneous energy decrease. We further extend the analysis to nearly reversible chains, treated as perturbations of the reversible case. By controlling the antisymmetric part of the transition matrix through explicit irreversibility parameters, we obtain sufficient conditions for exponential convergence to the stationary distribution. The new GSD coordinate selection rules perform well in numerical experiments, even in the general irreversible case.

1 Introduction

Computing the stationary distribution of a Markov chain is a fundamental task with applications ranging from queueing systems, performance evaluation, and chemical networks, to large-scale problems such as PageRank [7], semi-supervised learning [3], and graph neural networks [11, 5].

Konstantin Avrachenkov
Centre Inria d'Université Côte d'Azur, NEO team, Sophia Antipolis, France, k.avrachenkov@inria.fr

Lorenzo Gregoris
Eindhoven University of Technology, Eindhoven, The Netherlands, l.gregoris@tue.nl

Nelly Litvak
Eindhoven University of Technology, Eindhoven, The Netherlands, n.v.litvak@tue.nl

B. Koren and D. Roose (eds.), *Numerical Analysis and Scientific Computing: an Anthology*, Lecture Notes in Computational Science and Engineering 156,
https://doi.org/10.1007/978-3-032-33428-2_10

Formally, the task is to solve the eigenvector problem $\pi P = \pi$ for a transition probability matrix P. In many real-world settings, the state space may be extremely large, making direct numerical methods infeasible and leaving iterative algorithms as the only practical option.

Iterative methods start from an initial approximation $\widehat{\pi}_0$ and generate a sequence $\{\widehat{\pi}_t\}_{t\geq 0}$ by progressively refining the estimate. In large-scale problems, it is often essential that each iteration is inexpensive, which motivates algorithms that update only a subset of coordinates at each step.

A large class of iterative methods for stationary distribution computation is based on the *residual*

$$r_t = \widehat{\pi}_t(P - I), \tag{1}$$

and aim to drive $\|r_t\|$ to zero. Examples include OPIC [1], Gauss–Southwell-type PageRank algorithms [18], and more generally residual-elimination methods [14].

The RLGL ("Red Light Green Light") algorithm [2] is a recent unifying framework for many iterative approaches. In RLGL, only a subset of coordinates is updated at each iteration according to a chosen schedule (following the terminology in [2], the updated coordinates receive a "green light", thus the name of the algorithm). Different schedules and initializations correspond to many of the classical algorithms cited above. RLGL is remarkably simple and performs extremely well in practice, but proving convergence guarantees for the best performing schedules has been challenging. Nonetheless, in the experiments, RLGL consistently outperforms many known algorithms, including state-of-the-art Krylov subspace methods such as GMRES [24].

For general linear systems, including overdetermined ones, a classical example of the residual minimization approach is the Kaczmarz method [15], which updates the solution by projecting onto one equation at a time. While Kaczmarz-type methods are very general, they do not exploit the specific structure of the problem at hand. In fact, we applied Algorithm 1 in [27] for computing the stationary distribution, and it converged significantly slower than simple power iterations. Indeed, the convergence rate of classical Kaczmarz methods depends on the square of the condition number of the system matrix [27, 4], whereas the first-order methods we consider scale with the spectral gap.

Another classical approach for stationary distribution computation is reformulating the eigenvector problem as a minimization problem of the least-squares objective

$$f(x) = \tfrac{1}{2}\|x(P - I)\|^2, \tag{2}$$

for some norm $\|\cdot\|$. Then f is a convex function minimized at the solutions of the linear system, and we can apply the well-established theory for this class of optimization problems. The simplest iterative optimization algorithms are first-order methods [28], which use gradient information to update the current estimate.

There are drawbacks when applying first-order methods to minimize the least-squares objective function (2). For instance, the gradient reads

$$\nabla f(x) = x(P - I)(P - I)^\top, \tag{3}$$

which depends not only on P, but also depends on $PP^\top$ and $P^\top$. This could be a problem because even if P is sparse, $PP^\top$ may not be (an effect known as *fill-in* [23, Section 3.3.2]). A way to address this is to avoid forming $PP^\top$ explicitly and instead compute $(xP)P^\top$ through two sparse matrix–vector multiplications. While this only doubles the cost per iteration, the condition number gets squared. This is relevant because the convergence rate of gradient-based methods depends on the condition number of the objective, so this squaring can significantly slow down convergence (see, e.g., [20, 6]).

Finally, in applications like PageRank where accessing out-neighbors is cheaper than in-neighbors, approaches whose gradients (or updates) depend only on P and not on $P^\top$, are operationally attractive. Indeed, web crawlers discover the graph by following outgoing links, so the outlinks of a page are immediately available, whereas identifying all pages linking to a given page requires (near-)global knowledge of the crawl [9].

Another formulation as an optimization problem applies to the special case when the matrix A of a linear system $Ax = b$ is symmetric positive semi-definite. In this case, one can construct a convex objective function, of which the gradient depends linearly on A:

$$\Phi(x) = \tfrac{1}{2}x^\top Ax - b^\top x. \tag{4}$$

This is usually referred to as the *energy function* of the linear system. Minimizing Φ is equivalent to minimizing the error in the A-norm, also called the energy norm. This formulation is the starting point of the conjugate-gradient method and other techniques for solving positive semi-definite linear systems [12]. The power of this approach is in the fact that the gradient of the energy function is

$$\nabla\Phi(x_t) = Ax_t - b = -r_t. \tag{5}$$

Relation (5) bridges the optimization algorithms based on the gradient, and the iterative algorithms based on the residual.

The limitation of this approach for stationary distribution computations is that, in general, transition matrices are not symmetric (in this trivial case the stationary distribution is uniform). Therefore, the conjugate-gradient framework is not directly applicable. The RLGL algorithm however, uses linear updates in the transition matrix P, suggesting there may exist an objective function of the form (4), for which a coordinate descent method produces similar updates and converges exponentially fast.

Contributions

In this chapter, we provide a partial theoretical explanation for the strong empirical performance of RLGL. Our main results are:

1. **Variational formulation.** We show that when P is similar to a symmetric matrix—for example, for reversible Markov chains—the stationary distribution problem admits an energy formulation based on the *Dirichlet energy*. In this setting, RLGL corresponds to a block coordinate descent method that minimizes this energy.

2. **Exponential convergence for nearly reversible chains.** Using this energy viewpoint, we prove that RLGL converges *exponentially fast* for a class of chains we call *nearly reversible*, with minimal assumption on the update schedule. This extends the previously known special cases in [2].
3. **Energy inspired heuristics.** The variational formulation yields new coordinate selection rules that provably maximize the decrease in energy at each step, which we call *Gauss-Southwell-Dirichlet* heuristics. We show that these rules outperform the previously known heuristics.

Outline of the chapter

In Section 2 we start by introducing the notation, and give brief preliminaries on Markov chains, the RLGL algorithm, the Dirichlet energy of a Markov chain, and coordinate descent methods to minimize convex functions. In Section 3 we prove the equivalence of RLGL to coordinate descent in the case when the transition matrix is similar to a symmetric matrix. In Section 4 we extend these ideas by viewing a general irreversible Markov chain as a perturbed reversible chain. By doing so, we obtain spectral conditions that, when satisfied, guarantee exponential convergence. Finally, in Section 5, we propose new energy-inspired heuristics, and we compare them with baseline heuristics for computing the stationary distribution and PageRank on both real and synthetic networks.

2 Preliminaries

This section recollects the main tools used in the chapter, linking the probabilistic and optimization viewpoints. We first fix notation and recall basic facts on Markov chains, stationary distribution and reversibility. We then introduce the RLGL algorithm, a block-coordinate analogue of power iteration. Finally, we discuss the graph Laplacian and the Dirichlet energy, which provide the variational framework to interpret RLGL as a coordinate descent method.

2.1 Notation and Conventions

We adopt the row vector convention for Markov chains: probability distributions are represented as row vectors $\pi \in \mathbb{R}^n$ satisfying $\pi \mathbf{1}^\top = 1$, where $\mathbf{1} \in \mathbb{R}^n$ denotes the all-ones row vector. A matrix $P \in \mathbb{R}^{n \times n}$ is called *row stochastic* if $P\mathbf{1}^\top = \mathbf{1}^\top$ and $P_{ij} \geq 0$ for all i, j.

For a positive integer n, we write $[n] := \{1, 2, \ldots, n\}$. We denote by $\{e_i\}_{i=1}^n$ the standard basis of $\mathbb{R}^n$, where e_i is the row vector with a 1 in position i and 0 elsewhere. Given a subset $B \subseteq [n]$, we let $I_B \in \mathbb{R}^{n \times n}$ be the diagonal *projector* onto the coordinates in B, defined by

$$(I_B)_{ij} = \delta_{ij}\,\mathbb{I}(i \in B),$$

where δ_{ij} is the Kronecker delta and $\mathbb{I}(\cdot)$ denotes the indicator function. For a matrix $M \in \mathbb{R}^{n\times n}$ and a subset $B \subseteq [n]$, we denote by M_{BB} the *principal submatrix* of M with rows and columns indexed by B. Equivalently, $M_{BB} = (M_{ij})_{i,j\in B} \in \mathbb{R}^{|B|\times|B|}$.

Time or iteration indices are denoted by $t \in \mathbb{N} = \{0, 1, 2, \dots\}$, starting at $t = 0$. For a differentiable function $f : \mathbb{R}^n \to \mathbb{R}$, we denote by $\nabla f \in \mathbb{R}^n$ its gradient, and by $\partial_i f := (\nabla f)_i$.

2.2 Markov Chains and Stationary Distribution

A discrete-time Markov chain on the state space $V = [n]$ is specified by an $n \times n$ row-stochastic matrix P, where P_{ij} denotes the probability of transitioning from state i to state j. We always assume that the chain is irreducible, so that by the Perron–Frobenius theorem there exists a unique stationary distribution vector π, which is the left eigenvector of P corresponding to eigenvalue 1, i.e.,

$$\pi = \pi P, \qquad \sum_{i=1}^{n} \pi_i = 1, \qquad \pi_i > 0 \text{ for all } i \in V.$$

A Markov chain is called *reversible* with respect to π if it satisfies the *detailed balance equations*

$$\pi_i P_{ij} = \pi_j P_{ji} \qquad \text{for all } i, j \in V. \tag{6}$$

Given a Markov chain with stationary distribution π, we define its *time-reversal* as the chain with transition matrix

$$P^* = \Pi^{-1} P^\top \Pi,$$

where $\Pi = \operatorname{diag}(\pi)$. Both P and P^* share the same spectrum and stationary distribution, and the chain is reversible if and only if $P = P^*$. Equivalently, P is reversible if and only if it is self-adjoint in the Hilbert space $L^2(\pi)$, equipped with the rescaled inner product $\langle\cdot,\cdot\rangle_\pi$, defined for $f, g : [n] \to \mathbb{R}$ as

$$\langle f, g\rangle_\pi := \sum_{x=1}^{n} \pi_x f(x) g(x).$$

For our purposes, it is more convenient to work with the standard Euclidean inner product. This can be achieved by applying the isometry $x \mapsto x\Pi^{-1/2}$. In these new coordinates, which we denote by y, a reversible transition matrix P with stationary distribution π becomes symmetric.

It is often convenient to separate a transition matrix into its symmetric and antisymmetric components with respect to π. Indeed, any irreducible P admits the decomposition

$$P = \tfrac{1}{2}(P + P^*) \;+\; \tfrac{1}{2}(P - P^*), \tag{7}$$

where the first term is the *reversible part* of P, itself a stochastic matrix, and the second term is a matrix with zero row sums that captures the deviation from reversibility.

2.3 RLGL Algorithm

The Red-Light-Green-Light (RLGL) algorithm [2] works as follows. Given an iterate x_t and residual vector $r_t = x_t(P - I)$, the update is

$$x_{t+1} = x_t - x_t(I - P)I_{B_t} = x_t + r_t I_{B_t}. \tag{8}$$

Instead of recomputing the residual at every iteration, one can update it recursively as

$$\begin{aligned} r_{t+1} &= x_{t+1}(P - I) \\ &= (x_t - x_t(I - P)I_{B_t})(P - I) \\ &= r_t + r_t I_{B_t}(P - I) = r_t P(B_t), \end{aligned}$$

where $P(B_t) = I + I_{B_t}(P - I)$ is the matrix such that its rows $i \in B_t$ are the same as in P and the rows $i \notin B_t$ are as in the identity matrix. Thus, RLGL can be viewed as a block-coordinate variant of the power iterations. In other words, instead of updating all coordinates simultaneously as in power iteration, RLGL in (8) updates only a subset $B_t \subseteq [n]$; when $B_t = [n]$ for all t, the method reduces to the classical power iteration.

The total residual is conserved and equals zero:

$$r_t \mathbf{1}^\top = x_t(P - I)\mathbf{1}^\top = x_t(P\mathbf{1}^\top - \mathbf{1}^\top) = 0.$$

The terminology in [1, 2] emphasizes this by alluding to the residual as *cash*, and interpreting the updates as cash movements or transactions.

At each step, the residual may decrease due to cancellations, when cash of opposite sign meets at a node. Consequently, the ℓ_1 norm of the residual is non-increasing:

$$\|r_{t+1}\|_1 \leq \|r_t\|_1.$$

The approximation of the stationary distribution is obtained by normalizing vector x_t, which in turn is expressed as the sum of updates at each step, as seen from (8):

$$x_t = x_0 + \sum_{k=1}^{t-1} r_k I_{B_k}, \quad \widehat{\pi}_t = \frac{1}{\|x_t\|_1} x_t.$$

In other words, x_t records the cumulative *transaction history*, that is, the total amount of cash that each node has distributed up to time t. When the residual r_t goes to zero, $\widehat{\pi}_t$ converges to the stationary distribution.

Thus, the crux of RLGL updates is to drain the cash using an as small as possible number of transactions. However, identifying an optimal sequence of transactions is a difficult optimization problem, intractable even for relatively small graphs. Consequently, practical implementations rely on heuristics, both randomized and deterministic. Several of these heuristics are summarized in Table 2.

Heuristic	Selection Rule
RR	$B_t = \{i \mid i \equiv t \pmod n\}$
Rand	$B_t = \{i\}$ with probability $\frac{1}{n}$
Greedy	$B_t = \arg\min_i \|r_t + r_t I_{\{i\}}(P - I)\|_1$
MaxC	$B_t = \{i \mid i = \operatorname{argmax}_j \lvert r_{t,j}\rvert\}$
PC	$B_t = \{i\}$ with probability proportional to $\lvert r_{t,i}\rvert$
Theta	$B_t = \{i \mid i \equiv t \pmod n \text{ and } \lvert r_{t,i}\rvert \geq \theta_t(r)\}, \quad \theta_t(r) = \left(\frac{1}{n}\sum_j \lvert r_{t',j}\rvert^r\right)^{1/r}$

Table 1 Summary of RLGL heuristics considered in [2].

Numerical evidence suggests that residual-dependent heuristics proposed in [2], such as Theta, achieve state-of-the-art performance, often outperforming classical methods in practice.

2.4 Dirichlet Energy and the Graph Laplacian

The normalized Laplacian [8] associated with an irreducible transition matrix P is defined as

$$\mathcal{L}_{\text{norm}} = I - \Pi^{1/2} P \Pi^{-1/2}, \qquad \Pi = \operatorname{diag}(\pi),$$

where π is the stationary distribution of P. If $\lambda_1 = 1, \lambda_2, \ldots, \lambda_n$ are the eigenvalues of P then $1 - \lambda_i$ are the eigenvalues of $\mathcal{L}_{\text{norm}}$. In particular, the kernel of $\mathcal{L}_{\text{norm}}$ is one-dimensional, spanned by $\sqrt{\pi}$. Since $\mathcal{L}_{\text{norm}}$ is in general non-symmetric, its eigenvalues may be complex numbers. A standard way to handle this is to consider the *additive symmetrization*, or *symmetrized Laplacian*:

$$\mathcal{L}_{\text{sym}} = \frac{1}{2}(\mathcal{L}_{\text{norm}} + \mathcal{L}_{\text{norm}}{}^{\top}) = I - \Pi^{1/2}\Big(\frac{P + P^*}{2}\Big)\Pi^{-1/2}.$$

The symmetrized Laplacian is symmetric and positive semi-definite, so its eigenvalues, denoted by $\mu_1, \ldots, \mu_n$, are real and contained in $[0, 2]$. Its kernel coincides with that of $\mathcal{L}_{\text{norm}}$, spanned by $\sqrt{\pi}$.

The *Dirichlet energy* [17, Sec. 13.2] associated with P is the quadratic form induced by the symmetrized Laplacian:

$$\mathcal{E}(y) = \tfrac{1}{2} y \mathcal{L}_{\text{sym}} y^{\top}. \tag{9}$$

This function is convex, with the only flat direction being the one spanned by the kernel of the Laplacian. More precisely, restricted to the orthogonal complement of $\sqrt{\pi}$, the

Dirichlet energy is μ_2-strongly convex, where μ_2 is the smallest non-zero eigenvalue of $\mathcal{L}_{\text{sym}}$. The constant μ_2 is often called the *Poincaré constant* of P [25]. In the reversible and lazy case, one has $\mu_2 = 1 - \lambda_2(P)$, so the Poincaré constant coincides with the usual spectral gap.

For brevity, we will henceforth write $\mathcal{L}$ for the symmetrized Laplacian and denote by μ its smallest nonzero eigenvalue.

The Dirichlet energy plays a central role in the analysis of mixing times: it controls the decay of variance and underlies functional inequalities (such as Poincaré and log-Sobolev inequalities) that quantify the speed of convergence to equilibrium [17]. The definition of the Dirichlet energy in this chapter differs slightly from the classical one, as we work in Euclidean coordinates rather than in the weighted space $\ell^2(\pi)$, and use the transpose of the symmetrized Laplacian; however, the essential properties remain unchanged.

The next proposition establishes an important relation between the squared ℓ_2-norm of the residual and the Dirichlet energy.

Proposition 1. *There exist constants* $0 < c_1 \leq c_2 < +\infty$ *such that*

$$c_1 \mathcal{E}(x) \leq \|x\Pi^{1/2}(P - I)\|_2^2 \leq c_2 \mathcal{E}(x).$$

Proof. Apply Lemma 9 in the Appendix to the Dirichlet energy and $x\Pi^{1/2}(P - I)(P - I)^\top \Pi^{1/2} x^\top$. □

In particular, Proposition 1 implies that if either the energy or the residual goes to zero, then both these quantities go to zero together with the same rate.

2.5 Optimization with Coordinate Descent

We begin by recalling standard results on coordinate descent methods for minimizing a differentiable function $f : \mathbb{R}^n \to \mathbb{R}$. A coordinate descent method generates a sequence of iterates via updates of the form:

$$x_{t+1} = x_t - \alpha_t \partial_{i_t} f(x_t) e_{i_t}, \tag{10}$$

where i_t is the chosen coordinate at step t and $\alpha_t \in \mathbb{R}$ is the step-size.

It is known that coordinate descent methods converge exponentially if f satisfies the μ-Polyak–Lojasiewicz (PL) condition (see Definition 1) and has a Lipschitz continuous gradient.

Definition 1 (μ-PL function, [28])**.** *Let* f *be a differentiable function, and let* $f^* = \inf_{x \in \mathbb{R}^n} f(x)$. *We say* f *is* μ*-PL for some* $\mu > 0$ *if it satisfies*

$$f(x) - f^* \leq \frac{1}{2\mu} \|\nabla f(x)\|_2^2, \qquad \forall x \in \mathbb{R}^n.$$

The PL condition is strictly weaker than strong convexity: it does not require positive curvature in every direction and therefore allows for flat directions. In particular, any μ-strongly convex function (32) is μ-PL (see Lemma 6 in the Appendix). This is, for instance, the case for quadratic objectives whose Hessian is only positive semidefinite. This makes the PL condition a natural assumption in our setting: the Dirichlet energy (9) is not strongly convex due to the flat direction corresponding to the stationary distribution, but it still satisfies the PL inequality, with μ equal to the Poincaré constant; see Lemma 7 in the Appendix.

Let $L_{max} := \max_i L_i$, where L_i's are the Lipschitz constants:

$$|\partial_i f(x + he_i) - \partial_i f(x)| \leq L_i |h|, \qquad i = 1, \ldots, n. \tag{11}$$

Functions that satisfy (11) have quadratically bounded growth as stated in Lemma 1 below.

Lemma 1 *Let $f : \mathbb{R}^n \to \mathbb{R}$ be differentiable with coordinate-wise Lipschitz constants $L_1, \ldots, L_n$. Then,*

$$f(x + he_i) \leq f(x) + h\partial_i f(x) + \frac{L_i}{2} h^2, \qquad \forall h \in \mathbb{R}.$$

Proof. See Appendix 6. □

Using Lemma 1 with $h = -\alpha_t \partial_{i_t} f(x_t)$, one can show that the step size that maximizes the guaranteed one-step progress, i.e., the decrease of the objective function in one step, is $\alpha_t = 1/L_{i_t}$. This step size yields a guaranteed decrease of:

$$f(x_t) - f(x_{t+1}) \geq \frac{1}{2L_{i_t}} |\partial_{i_t} f(x_t)|^2. \tag{12}$$

From Eq. (12), we see that updating a coordinate with a large gradient component seems like a good idea. If this component has a big enough fraction of the full gradient norm, then we can use PL-inequality (Definition 1) to relate the one-step progress to the current sub-optimality $f(x_t) - f^*$, and obtain exponential convergence[1].

Let us assume that our selection rule selects coordinates i_t with at least a fraction $\beta \in (0, 1]$ of the full gradient ℓ_2-norm squared:

$$|\partial_{i_t} f(x_t)|^2 \geq \beta \, \|\nabla f(x_t)\|_2^2. \tag{13}$$

For random update rules, this should hold in expectation. If i_t satisfies this condition, we call it a *good* update. To guarantee exponential convergence, we need the number of good updates to grow linearly with t, i.e., the density of good updates is non-vanishing. Let $\mathcal{G}(t)$ be the number of good updates up to iteration t, then the long-run density of

[1] In the numerical analysis literature this is called *linear convergence*, since the decrease of the objective function at a given step is linear in the current sub-optimality.

the good updates is

$$\liminf_{t\to\infty} \frac{\mathcal{G}(t)}{t} = \rho > 0. \tag{14}$$

With the Gauss–Southwell rule, i.e., updating the coordinate with greatest absolute value gradient, we can guarantee that $\beta \geq 1/n$ at least for every update. Another common rule is updating a uniformly random coordinate, which achieves this in expectation $\mathbb{E}[\beta] = 1/n$.

The following theorem and its block version are standard results in optimization theory.

Theorem 1. *[28, Thm. 1] Let f be μ-PL and admit coordinate-wise Lipschitz constants L_i. Consider the coordinate descent method* (10) *with $\alpha_t = 1/L_{i_t}$ and using a selection rule that satisfies* (13). *Then*

$$f(x_{t+1}) - f^* \leq \left(1 - \frac{\beta\mu}{L_{i_t}}\right)(f(x_t) - f^*).$$

With assumption (14) we obtain exponential convergence by iterating Theorem 1. We know that by time t we have at least ρt updates for which the theorem applies, so we obtain at least a rate

$$-\rho \log\left(1 - \frac{\beta\mu}{L_{\max}}\right).$$

We can extend Theorem 1 to *block descent*, in which we update a subset of coordinates $B_t \subseteq [n]$ simultaneously at each iteration:

$$x_{t+1} = x_t - \alpha_t \odot \nabla f(x_t) I_{B_t}, \tag{15}$$

where $\odot$ denotes the Hadamard product and α_t is a row vector, allowing for non-homogeneous step sizes.

For each block $B \subseteq [n]$, we define the *block Lipschitz constant* L_B as

$$\|\nabla_B f(x + hI_B) - \nabla_B f(x)\|_2 \leq L_B \|hI_B\|_2 \qquad \forall x, h \in \mathbb{R}^n.$$

In the case of $f(x) = xAx^\top$ with A positive semi-definite, the block Lipschitz coefficients are simply

$$L_B = \lambda_{max}(A_{BB}).$$

Theorem 2. *[21, Sec. 3.5] Suppose f satisfies the PL-inequality with parameter μ and admits block Lipschitz constants $\{L_B\}$. Consider a block descent method with $\alpha_t = 1/L_{B_t}$ and where the selected block B_t satisfies the following block version of* (13)*: $\beta \leq \frac{\|\nabla f(x) I_{B_t}\|_2^2}{\|\nabla f(x)\|_2^2}$. Then*

$$f(x_{t+1}) - f^* \leq \left(1 - \frac{\beta\mu}{L_{B_t}}\right)(f(x_t) - f^*).$$

3 Coordinate Descent for Reversible Chains

3.1 Finding an Energy Function for RLGL

We would like to interpret the RLGL update (8) as a gradient descent step for some energy function. This means that the residual vector $r(x) = x(I-P)$ should be a gradient field. It is clear that this is not possible in general, since the Jacobian of a gradient field needs to be symmetric, and this is true if and only if P itself is symmetric.

To recover a gradient interpretation, we look for an appropriate coordinate transformation. Since the RLGL update is linear, the transformation must also be linear. Hence we seek an invertible matrix $M \in \mathbb{R}^{n\times n}$ such that, in the new coordinates $y = xM$, the update becomes the gradient of a quadratic energy.

Lemma 2 *The RLGL dynamics can be represented as gradient descent on a quadratic energy if and only if P is similar to a symmetric matrix.*

Proof. Suppose $y = xM$ with M invertible. Then in the y–coordinates the residual becomes

$$\widetilde{r}(y) = r(yM^{-1})M = y(I - M^{-1}PM).$$

If $M^{-1}PM$ is symmetric, then $\widetilde{r}(y)$ is a gradient field with potential

$$\Phi(y) = \tfrac{1}{2}y(I - M^{-1}PM)y^{\top}.$$

Thus RLGL admits a gradient interpretation when P is similar to a symmetric matrix. □

In particular, if P is reversible with stationary distribution π, then the linear transformation

$$M = \Pi^{-1/2}, \qquad \Pi = \mathrm{diag}(\pi),$$

yields $\Pi^{1/2}P\Pi^{-1/2}$, which is symmetric by detailed balance (6). In this case, the quadratic energy is the Dirichlet energy (9).

3.2 RLGL as Block Descent

As established above, for reversible chains, the coordinate transformation $y = x\Pi^{-1/2}$ allows us to rigorously interpret the RLGL dynamics as a gradient descent process on the Dirichlet energy. We can now use this optimization framework to formally connect RLGL to block coordinate descent and analyze its convergence.

As in the single coordinate update, where the absence of self-loops ensures that RLGL has the optimal step size, in the case of block updates this happens when the updated block is an *independent set*. Here, by independent set we mean that no two vertices in the block are connected by an edge in the transition graph, that is, for all $x, y \in B_t$ we have $P_{xy} = 0$. This ensures that, after the update, the cash in the block is zero.

Theorem 3. *Let P be an irreducible, reversible transition matrix. If the updated block B_t is an independent set, the RLGL update matches a block descent update with an optimal step size in the symmetrized coordinates.*

Proof. Let $y = x\Pi^{-1/2}$. In these coordinates, the associated energy is the Dirichlet energy $\mathcal{E}(y) = \frac{1}{2} y\mathcal{L}y^\top$, whose gradient is $\nabla_y\mathcal{E}(y) = y\mathcal{L}$. Furthermore, the rescaled residual is given by $\widetilde{r} = -y\mathcal{L}$, so that $\widetilde{r} = -\nabla_y\mathcal{E}(y)$.

Consider any update u supported on B_t (so $u_i = 0$ for $i \notin B_t$). The new iterate is $y_{t+1} = y_t + u$, and the energy after the update reads

$$\begin{aligned}\mathcal{E}(y_{t+1}) &= \tfrac{1}{2}(y_t + u)\mathcal{L}(y_t + u)^\top \\ &= \mathcal{E}(y_t) - \widetilde{r}_{B_t}u_{B_t}^\top + \tfrac{1}{2}u_{B_t}\mathcal{L}_{B_tB_t}u_{B_t}^\top,\end{aligned}$$

where $\widetilde{r}_{B_t}$ is the restriction of $\widetilde{r}$ to B_t and $\mathcal{L}_{B_tB_t}$ is the principal submatrix of $\mathcal{L}$ on B_t. This expression is a strictly convex quadratic in u_{B_t}, whose unique minimizer gives the optimal block update. The first-order condition is

$$-\widetilde{r}_{B_t} + u_{B_t}\mathcal{L}_{B_tB_t} = 0,$$

so the optimal block update is

$$u^*_{B_t} = \widetilde{r}_{B_t}\mathcal{L}^{-1}_{B_tB_t}.$$

Since B_t is an independent set and the chain is irreducible, $\mathcal{L}_{B_tB_t} = I$, so the optimal update simplifies to $u^*_{B_t} = \widetilde{r}_{B_t}$, meaning the new iterate is obtained simply by adding the residual on the updated coordinates:

$$y_{t+1} = y_t + \widetilde{r}_t I_{B_t}.$$

Comparing this to (15), we see that the per-coordinate step size is exactly 1.

Transforming back into x coordinates, we obtain

$$\begin{aligned}x_{t+1}\Pi^{-1/2} &= x_t\Pi^{-1/2} - (x_t\Pi^{-1/2})\mathcal{L}I_{B_t} \\ &= x_t\Pi^{-1/2} - x_t(I - P)\Pi^{-1/2}I_{B_t}.\end{aligned}$$

Multiplying on the right by $\Pi^{1/2}$ yields

$$x_{t+1} = x_t - x_t(I - P)I_{B_t},$$

which is exactly the RLGL update (8). □

Since with these assumptions RLGL is a block descent method and the Dirichlet energy (9) satisfies the hypothesis of Theorem 1, we obtain exponential convergence.

Theorem 4. *If P is reversible with Poincaré constant μ, then one β-good update yields*

$$\mathcal{E}(y_{t+1}) \leq \left(1 - \frac{\beta\mu}{L_{B_t}}\right)\mathcal{E}(y_t). \tag{16}$$

In particular, if the update sequence uses independent sets and has a fraction ρ of β-good updates, then RLGL converges exponentially, with a rate of at least

$$-\rho \log(1 - \beta\mu).$$

Remark 1 If the updated block B_t is not an independent set, then $\mathcal{L}_{B_t B_t}$ is not the identity. In this case, RLGL still decreases the energy, but the step size is generally suboptimal. For example, if $B_t = \{i\}$ with $P_{ii} > 0$, i.e., i has a self-loop, then the optimal step size on coordinate i would be $(1 - P_{ii})^{-1}$ rather than 1. Thus exponential convergence is preserved, but the guaranteed per-step decrease is smaller.

From Theorem 3 we have a new interpretation of RLGL: one update minimizes the Dirichlet energy in the subspace of the updated coordinates when the updated block is an independent set. This makes it look like an Orthogonal-subspace method [23].

When the updated block B_t is an independent set, the optimal block update yields an exact energy decrease of

$$\mathcal{E}(y_t) - \mathcal{E}(y_{t+1}) = \tfrac{1}{2}\|\widetilde{r}_{B_t}\|_2^2. \tag{17}$$

So the Dirichlet energy strictly decreases whenever the block residual is nonzero. Crucially, this statement holds irrespective of whether there is any local residual cancellation of the original "raw" residual vector in the x-coordinates.

3.3 When Coordinate Descent Outperforms Power Iteration

We now compare the exponential convergence rates of coordinate descent (CD) from Theorem 4 and power iteration (PI). Since our focus is on computational efficiency, we measure progress per unit of arithmetic cost. Updating a single coordinate of the residual in CD requires work proportional to the out-degree of the chosen node, so that n coordinate updates correspond roughly to one matrix–vector multiplication in PI. This normalization is exact on average when the coordinate is selected uniformly at random.

Recall that PI converges at a rate governed by the second eigenvalue of P: the ℓ_2–error decays as λ_2^t, or equivalently at an exponential rate of $-\log\lambda_2$. For CD, Theorem 1 guarantees that after n coordinate updates the Dirichlet energy decreases at a rate of at least

$$-n\rho \log\bigl(1 - \beta(1 - \lambda_2)\bigr).$$

Under uniform random selection, we have $\rho = 1$ and $\mathbb{E}[\beta] = 1/n$. We use Proposition 1 to compare the two.

In the asymptotic regime $n \gg 1$, CD with the Gauss–Southwell or random selection is always slower than PI, since

$$-\log \lambda_2 \geq \frac{1-\lambda_2}{2} \qquad \text{for all } \lambda_2 \in [0,1].$$

However, CD can outperform PI if the coordinate rule achieves a larger β. Specifically, for $\rho = 1$, CD beats PI whenever

$$\beta \geq \frac{1-\lambda_2^{2/n}}{1-\lambda_2} = \frac{2}{n} + \frac{(n-2)(1-\lambda_2)}{n^2} + o\big((1-\lambda_2)^2\big).$$

Thus, surpassing PI requires that the chosen coordinate captures more than the average $2/n$ share of the total residual.

The Gauss–Southwell rule (after the first iteration) achieves a β that is strictly greater than $1/n$. Empirically, we consistently observe that the selected coordinate accounts for more than $2/n$ of the total residual. The same empirical observation was noted in [1], where the residual vector is non-negative.

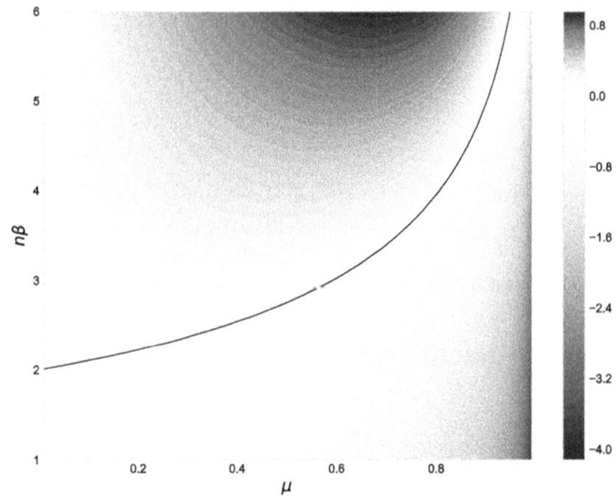

Figure 1 Comparison of convergence rates of coordinate descent and power iteration, for $n = 100$. Plotted difference $r_{CD} - r_{PI}$. The blue region is where n iterations of CD beat one PI, in red the converse. The black line shows when $r_{CD} = r_{PI}$.

These results suggest a natural interpretation: to surpass PI, the algorithm must avoid diffusing the residual evenly across all coordinates. Instead, it is advantageous to let the residual accumulate so that a few coordinates carry a significant portion of the error, allowing each update to make more substantial progress. In sparse networks this concentration occurs naturally, which explains why CD can outperform PI in practice even when theoretical upper bounds based on the spectral gap μ are pessimistic.

4 Extension to Nearly Reversible Chains

In this section we extend Theorem 1 beyond the reversible setting. Our goal is to identify conditions, under which exponential convergence can still be guaranteed for general, possibly non-reversible, Markov chains. To this end, we decompose an arbitrary transition matrix into a reversible component and a perturbation term.

We then analyze the behavior of coordinate-descent dynamics in the presence of this linear perturbation, and derive conditions, under which exponential convergence is preserved. When specialized to our setting, these conditions involve the row norm of the irreversible part of the transition matrix and the Poincaré constant of the chain. As we will show, for single-coordinate update rules—such as the Greedy Gauss–Southwell and the Random rule—to guarantee a one-step decrease in the energy, the irreversibility of the chain must be vanishingly small, in a precise sense described below.

4.1 Irreversible Chains as Perturbed Coordinate Descent

We now relax the reversibility assumption. In this setting, convergence of the RLGL updates is no longer guaranteed in general. For example, on the directed cycle one can keep pushing the residual along the cycle indefinitely without ever reducing its norm, and even some aperiodic chains exhibit this phenomenon [2]. Thus, in the irreversible setting we must impose additional structural assumptions on the chain, the update sequence, and their interaction.

Our approach is to interpret RLGL as coordinate descent with a perturbed gradient. The perturbation arises from the antisymmetric part of the dynamics, and we will derive spectral conditions, under which it remains small enough to preserve convergence.

Let P be an arbitrary (not necessarily reversible) Markov chain with stationary distribution π. Define the symmetric and antisymmetric components of P in the π-weighted coordinates by

$$S = \tfrac{1}{2}\Pi^{1/2}(P + P^*)\Pi^{-1/2}, \qquad A = \tfrac{1}{2}\Pi^{1/2}(P - P^*)\Pi^{-1/2},$$

so that $S^\top = S$ and $A^\top = -A$. With this notation, the normalized Laplacian decomposes as

$$\mathcal{L}_{\text{norm}} = \Pi^{1/2}(I - P)\Pi^{-1/2} = (I - S) - A = \mathcal{L} - A.$$

Writing the updates in normalized coordinates $y_t = x_t\Pi^{-1/2}$, the RLGL rule (8) becomes

$$y_{t+1} = y_t - y_t\big[(I - S) - A\big]I_{B_t} = \underbrace{y_t - y_t\mathcal{L}I_{B_t}}_{\text{descent step}} + \underbrace{y_tAI_{B_t}}_{\text{perturbation}}. \tag{18}$$

The first term is a coordinate descent step for the Dirichlet energy, while the second term is the perturbation induced by the irreversible dynamics. If this perturbation remains sufficiently small, in a sense made precise below, RLGL still exhibits exponential convergence.

To quantify how much each coordinate deviates from reversibility, we introduce the following *local irreversibility* coefficient for each state:

$$\kappa_i := \|A_{i,:}\|_2 = \frac{1}{2\sqrt{\pi_i}} \left(\sum_{j=1}^{n} \frac{1}{\pi_j} (\pi_j p_{ji} - \pi_i p_{ij})^2 \right)^{1/2} . \tag{19}$$

The quantity κ_i captures the local irreversibility at state i. It measures how strongly the chain violates detailed balance at i, by weight averaging the squared net probability currents with respect to the stationary distribution. When $\kappa_i = 0$, state i satisfies detailed balance with all other states. Thus, κ_i quantifies the extent to which the dynamics around i fail to be reversible.

We now introduce two global irreversibility coefficients: one based on the ℓ_∞-norm and the other based on the ℓ_2-norm.

$$\eta_\infty := \max_{i \in [n]} \frac{\kappa_i}{\mu}, \qquad \eta_2 := \frac{1}{\mu} \left(\sum_{i=1}^{n} \kappa_i^2 \right)^{1/2}, \tag{20}$$

where μ is the Poincaré constant of P. Note that we have the bound $\eta_2 \leq \sqrt{n}\, \eta_\infty$.

Definition 2 (Nearly reversible chains). *We say that a Markov chain with transition matrix P is* nearly reversible *if*

$$\eta_\infty < \frac{1}{2n + \sqrt{n}}.$$

This threshold is chosen because it is precisely the regime in which single–coordinate update rules remain effective: as we will see in Corollary 1, such a bound on η_∞ guarantees that both the Greedy Gauss–Southwell rule and the Random rule produce a one-step decrease in energy and therefore achieve exponential convergence.

4.2 Coordinate Descent with Linear Perturbation

As observed in (18), an RLGL update can be decomposed into a descent step plus an additional perturbation term. This perturbation is linear in the current iterate and arises from the irreversible component of the transition matrix (7). If this perturbation is sufficiently small, one may still expect convergence.

To make this precise, we analyze a generic coordinate descent update in the presence of a linear perturbation and derive explicit conditions under which convergence is preserved. Additive perturbations for full gradient descent have been studied in prior work (see, e.g., [10]); here we adapt the core argument to the coordinate-wise setting.

Motivated by the updates (18), we will study the case when, for each coordinate i, the perturbation is a linear function of the current iterate:

$$\xi_i(x_t) = \langle x_t, v^{(i)} \rangle,$$

for some fixed $v^{(i)} \in \mathbb{R}^n$. Set $\kappa_i := \|v^{(i)}\|_2$ to be the ℓ_2-norm of these vectors, and let $\kappa := \max_i \kappa_i$. (We will show in Section 4.3 that these are the same κ's as defined in (19).)

At iteration t we pick coordinate i_t and perform the update

$$x_{t+1} = x_t - \alpha_t \, \partial_{i_t} f(x_t) \, e_{i_t} - \xi_{i_t}(x_t) \, e_{i_t}. \tag{21}$$

We impose a natural consistency condition that the perturbation must vanish at the minimizer:

$$\langle v^{(i)}, x^* \rangle = 0, \qquad \forall i \in [n], \tag{22}$$

which ensures that x^* remains a fixed point of the dynamics (21). The next theorem gives a sufficient condition for exponential convergence of perturbed coordinate descent.

Theorem 5. *Let $f : \mathbb{R}^n \to \mathbb{R}$ be differentiable and μ-strongly convex, and assume each partial derivative $\partial_i f$ is L_i-Lipschitz. Let $L_{\max} := \max_i L_i$. Suppose the selection rule guarantees updates that are* good *with parameter $\beta \in (0, 1]$ (as above). Let the perturbations satisfy the above Assumption 22.*

If

$$\kappa < \frac{\mu\sqrt{\beta}}{L_{\max}}, \tag{23}$$

then for a fixed step size $\alpha_t = 1/L_{i_t}$, the iterates converge exponentially to x^*. Specifically, for any good update i_t, the per-step decrease is at least:

$$f(x_{t+1}) - f^* \le \left(1 - \frac{\beta\mu}{L_{i_t}} + \frac{L_{i_t}\kappa_{i_t}^2}{\mu}\right)(f(x_t) - f^*),$$

leading to an overall rate of at least

$$-\log\left(1 - \frac{\beta\mu}{L_{max}} + \frac{\kappa^2}{\mu}\right).$$

Proof. The update along coordinate i_t is $\Delta_t = -\alpha_t \partial_{i_t} f(x_t) + \xi_{i_t}(x_t)$. Lemma 1 gives

$$f(x_{t+1}) \le f(x_t) + \partial_{i_t} f(x_t)\Delta_t + \frac{L_{i_t}}{2}\Delta_t^2.$$

We choose the standard fixed step size $\alpha_t = 1/L_{i_t}$. Substituting this into the update Δ_t and then into the inequality above yields:

$$
\begin{aligned}
f(x_{t+1}) &\leq f(x_t) - \partial_{i_t} f(x_t)\left(\frac{1}{L_{i_t}}\partial_{i_t} f(x_t) - \xi_{i_t}(x_t)\right) + \frac{L_{i_t}}{2}\left(\frac{1}{L_{i_t}}\partial_{i_t} f(x_t) - \xi_{i_t}(x_t)\right)^2 \\
&= f(x_t) - \frac{1}{L_{i_t}}|\partial_{i_t} f(x_t)|^2 - \partial_{i_t} f(x_t)\xi_{i_t}(x_t) + \\
&\qquad \frac{1}{2L_{i_t}}|\partial_{i_t} f(x_t)|^2 + \partial_{i_t} f(x_t)\xi_{i_t}(x_t) + \frac{L_{i_t}}{2}|\xi_{i_t}(x_t)|^2 \\
&= f(x_t) - \frac{1}{2L_{i_t}}|\partial_{i_t} f(x_t)|^2 + \frac{L_{i_t}}{2}|\xi_{i_t}(x_t)|^2.
\end{aligned}
$$

Now, we bound the last two terms. Since f is μ-strong convex, by Lemma 6, f is also μ-PL. Thus, for a β-good update (13), we obtain

$$
|\partial_{i_t} f(x_t)|^2 \geq 2\beta\mu(f(x_t) - f(x^*)).
$$

For the perturbation, we use the μ-strong convexity of f and the consistency condition (22):

$$
|\xi_{i_t}(x_t)|^2 = |\langle x_t, v^{(i_t)}\rangle|^2 = |\langle x_t - x^*, v^{(i_t)}\rangle|^2 \leq \kappa_{i_t}^2 \|x_t - x^*\|_2^2 \leq \frac{2\kappa_{i_t}^2}{\mu}(f(x_t) - f(x^*)).
$$

Substituting these bounds into our descent inequality, we get:

$$
f(x_{t+1}) \leq f(x_t) - \frac{1}{2L_{i_t}}(2\beta\mu(f(x_t) - f^*)) + \frac{L_{i_t}}{2}\left(\frac{2\kappa_{i_t}^2}{\mu}(f(x_t) - f^*)\right).
$$

Subtracting the optimal value f^* from both sides and rearranging gives the one-step contraction:

$$
f(x_{t+1}) - f^* < \left(1 - \frac{\beta\mu}{L_{i_t}} + \frac{L_{i_t}\kappa_{i_t}^2}{\mu}\right)(f(x_t) - f^*).
$$

Condition (23) ensures that this is a contraction. By considering the worst-case values $L_{\max}$ and κ, we obtain the desired exponential convergence rate. □

Computing the optimal step size α_t^* for the perturbed update yields

$$
\alpha_t^* = \frac{1}{L_{i_t}} + \frac{\xi_{i_t}(x_t)}{\partial_{i_t} f(x_t)}, \tag{24}
$$

which is the same as the unperturbed case plus the ratio of the perturbation and the partial derivative with respect to the updated coordinate. Thus if the latter is small, the unperturbed step size is close to the optimal one.

4.3 RLGL on Nearly Reversible Chains

Our goal is to show that if the Markov chain is nearly reversible, and the update sequence satisfies a condition of the form (13), then the RLGL algorithm converges exponentially. We achieve this by applying Theorem 5, since in the irreversible case we can view the RLGL update as perturbed coordinate descent (18).

The perturbation term in (18) takes the form

$$\xi_{i_t}(y_t) = (y_t A)e_{i_t} = \langle y_t, A_{:,i_t} \rangle,$$

i.e., the perturbation vector is the i_t-th column of the anti-symmetric matrix A, so $v^{(i)} = A_{:,i}$, meaning $\kappa_i = \|A_{:,i}\|_2$, which is precisely our measure of local irreversibility (19).

We define two residuals:

$$s_t = y_t \mathcal{L}_{\text{sym}}, \qquad r_t = y_t \mathcal{L}_{\text{norm}}.$$

The residual s_t corresponds to the gradient of the Dirichlet energy, and is the quantity that appears in the convergence theorems for coordinate descent. However, the residual actually observed in RLGL during execution is r_t. In the reversible case the two coincide, but in the irreversible setting this distinction becomes essential: the update sequence must be β–good with respect to s_t, while selection happens using r_r. We therefore define

$$|s_{t,i}|^2 \geq \beta_{\text{sym}} \|s_t\|_2^2, \qquad |r_{t,i}|^2 \geq \beta_{\text{norm}} \|r_t\|_2^2.$$

Later, in Proposition 2, we will quantify how far the two residuals can differ under the nearly-reversible assumption and show how we can relate β_{sym} and β_{norm}.

We now prove a bound for the perturbation term, which we will need to apply Theorem 5 to the updates based on r_t. Recall that the Dirichlet energy (9) is not strongly convex, but it is μ-PL.

Lemma 3 *For each coordinate i,*

$$|(yA)_i|^2 \leq \frac{2\kappa_i^2}{\mu} \mathcal{E}(y), \qquad \forall y \in \mathbb{R}^n.$$

Proof. Let $y^\perp$ be the projection of y in $Ker(\mathcal{L})^\perp$. By the consistency condition (22) and Cauchy–Schwarz,

$$|\xi_i(y)| = |\xi_i(y^\perp)| \leq \kappa_i \|y^\perp\|_2.$$

In $Ker(\mathcal{L})^\perp$ the Dirichlet energy is μ-strongly convex, and the unique minimizer is 0, hence

$$\mathcal{E}(y^\perp) \geq \frac{\mu}{2} \|y^\perp\|_2^2.$$

Noting that $\mathcal{E}(y) = \mathcal{E}(y^{\perp})$, we obtain

$$|\xi_i(y)|^2 \leq \frac{2\kappa_i^2}{\mu}\mathcal{E}(y). \qquad \square$$

Theorem 6. *Let P be a Markov chain with Poincaré constant μ. Then one β_{sym}-good update yields:*

$$\mathcal{E}(y_{t+1}) \leq \left(1 - \frac{\beta_{sym}\mu}{L_{i_t}} + \frac{L_{i_t}\kappa_{i_t}^2}{\mu}\right)\mathcal{E}(y_t).$$

If P is such that

$$\eta_\infty < \frac{\sqrt{\beta_{sym}}}{L_{max}}, \tag{25}$$

and P has no self-loops ($p_{ii} = 0$ for all i), and the update sequence $(i_t)_{t\geq 0}$ has only β_{sym}-good updates, then the RLGL algorithm converges exponentially with a rate of at least

$$-\log\left(1 - \beta_{sym}\,\mu + \frac{\kappa^2}{\mu}\right).$$

Proof. The only step in the proof of Theorem 5 that required μ-strong convexity is the bound on the perturbation term. In our case, we can use Lemma 3.

The condition for convergence from Theorem 5 is $\kappa < \mu\sqrt{\beta_{\text{sym}}}/L_{\max}$. This condition is ensured by the definition of $\eta_\infty = \kappa/\mu$ in (20) and (25). Applying Theorem 5, for a good update we have a one-step contraction of at least:

$$\mathcal{E}(y_{t+1}) \leq \left(1 - \frac{\beta_{\text{sym}}\,\mu}{L_{i_t}} + \frac{L_{i_t}\kappa_{i_t}^2}{\mu}\right)\mathcal{E}(y_t).$$

The Lipschitz constants are $L_i = 1 - p_{ii}$. For a chain with no self-loops, $p_{ii} = 0$, so $L_i = 1$ for all i, and $L_{\max} = 1$. Substituting $L_{i_t} = 1$ and taking the worst-case bounds over all coordinates we obtain the stated convergence rate. $\square$

Computing the energy decrease per-step in the case of no self-loops, we find:

$$\mathcal{E}(y_t) - \mathcal{E}(y_{t+1}) = \frac{1}{2}|s_t|^2 - \frac{1}{2}|\xi_{i_t}(x_t)|^2,$$

which shows that if the perturbation is smaller than the residual, the energy decreases.

Our objective is now to guarantee that a β_{norm}-good coordinate remains β_{sym}-good for some positive, possibly smaller, β_{sym}. To achieve this, we need to compare the two residuals.

Lemma 4 *For every coordinate i,*

$$\frac{|r_{t,i} - s_{t,i}|}{\|s_t\|_2} \leq \eta_\infty, \qquad \frac{\|r_t - s_t\|_2}{\|s_t\|_2} \leq \eta_2.$$

Proof. From Lemma 3 we have

$$|r_{t,i} - s_{t,i}|^2 = |(y_t A)_i|^2 \le \frac{2\kappa_i^2}{\mu} \mathcal{E}(y_t).$$

Using the PL inequality (1),

$$\mathcal{E}(y_t) \le \frac{1}{2\mu} \|s_t\|_2^2,$$

so we obtain

$$\frac{|r_{t,i} - s_{t,i}|^2}{\|s_t\|_2^2} \le \frac{\kappa_i^2}{\mu^2}.$$

Using the definition of η_∞ and taking square roots yields the first bound. Summing over the coordinates yields the second. □

Proposition 2. *Let P be nearly reversible. If a coordinate i is β_{norm}-good, such that*

$$\gamma_\infty^\star := \frac{\sqrt{n}\,\eta_\infty}{1 - \sqrt{n}\,\eta_\infty} \le \sqrt{\beta_{\text{norm}}}, \tag{26}$$

then the same coordinate is β_{sym}-good with

$$\beta_{\text{sym}} \ge \left(\frac{\sqrt{\beta_{\text{norm}}} - \gamma_\infty^\star}{1 + \gamma_\infty^\star} \right)^2.$$

Proof. Define $\delta_t = r_t - s_t = y_t A$. From the coordinate-wise bound of Lemma 3 and the definition $\kappa := \max_i \kappa_i$ we get

$$\|\delta_t\|_2 \le \frac{\sqrt{n}\,\kappa}{\mu} \|s_t\|_2 \le \sqrt{n}\,\eta_\infty \|s_t\|_2.$$

Set $\gamma_t := \|\delta_t\|_2 / \|r_t\|_2$. We have

$$\|r_t\|_2 = \|s_t + \delta\|_2 \ge \|s_t\|_2 - \|\delta_t\|_2 \ge (1 - \sqrt{n}\,\eta_\infty)\|s_t\|_2,$$

and hence

$$\gamma_t = \frac{\|\delta_t\|_2}{\|r_t\|_2} \le \frac{\sqrt{n}\,\eta_\infty}{1 - \sqrt{n}\,\eta_\infty} = \gamma_\infty^\star.$$

Furthermore, by Definition 2, it holds that $0 < \gamma_t \le 1$.

Since by assumption the update is β_{norm}-good, we have $|r_{t,i}| \ge \sqrt{\beta_{\text{norm}}}\,\|r_t\|_2$, hence

$$|s_{t,i}| \ge |r_{t,i}| - |\delta_i| \ge (\sqrt{\beta_{\text{norm}}} - \gamma_t)\|r_t\|_2.$$

Also $\|s_t\|_2 \le \|r_t\|_2 + \|\delta\|_2 = (1 + \gamma_t)\|r_t\|_2$. Therefore

$$\frac{|s_{t,i}|^2}{\|s_t\|_2^2} \ge \frac{(\sqrt{\beta_{\text{norm}}} - \gamma_t)^2}{(1 + \gamma_t)^2}.$$

The right-hand side is decreasing in γ_t for $0 \le \gamma_t \le \sqrt{\beta_{\text{norm}}}$, so substituting $\gamma_t = \gamma_\infty^\star$ yields the stated lower bound for β_{sym}. □

We can obtain an analogous bound using η_2.

Proposition 3. *Let P be nearly reversible. If a coordinate i is β_{norm}-good, such that*

$$\gamma_2^\star := \frac{\eta_2}{1-\eta_2} \le \sqrt{\beta_{\text{norm}}}, \tag{27}$$

then the same coordinate is β_{sym}-good with

$$\beta_{\text{sym}} = \left(\frac{\sqrt{\beta_{\text{norm}}} - \gamma_2^\star}{1+\gamma_2^\star}\right)^2.$$

Proof. We start by summing up the inequalities of Lemma 3. The rest of the proof follows. □

Theorem 7. *Let P be nearly reversible. If the update schedule satisfies*

$$\eta_\infty^2 < \left(\frac{\sqrt{\beta_{\text{norm}}} - \gamma^\star}{1+\gamma^\star}\right)^2 =: \widetilde{\beta}, \tag{28}$$

where $\gamma^\star$ is defined as in (26) or (27). Then the RLGL algorithm converges exponentially, with a rate of at least

$$-\log\left(1 - \widetilde{\beta}\mu + \frac{\kappa^2}{\mu}\right).$$

Proof. By Proposition 2 or Proposition 3, $\beta_{\text{sym}} \ge \widetilde{\beta}$. We then apply Theorem 6. □

Corollary 1 *Let P be nearly reversible. Then the RLGL algorithm converges exponentially for the greedy Gauss-Southwell and random rules.*

Proof. See Appendix 6 □

We want to highlight that near-reversibiltiy is a strong condition because we require the irreversibility term to scale as $O(1/n)$. However, under this condition, we also obtain the strong result that the fraction by which the energy is reduced, is $O(1/n)$ as well, while updating only one coordinate. Intuitively, we see the perturbation as noise, and we require the noise to be smaller than the update.

When the chain is not nearly reversible, updating a single coordinate is insufficient: the method cannot guarantee a descent step. Instead, one must update a block of coordinates so that the update rule has a sufficiently large β to tolerate the higher irreversibility. This might explain the empirical success of RLGL on irreversible chains: although a single-coordinate update may fail to decrease the energy, multiple or block updates can do so. A purely one-step analysis, which is what we have carried out, is insufficient to capture this phenomenon.

4.4 Nearly Reversible Convex Combinations

We analyze the nearly reversible property (Definition 2) for convex combination of reversible and irreversible chains. Consider

$$P_\varepsilon = (1-\varepsilon)R + \varepsilon Q, \qquad \varepsilon \in [0,1], \tag{29}$$

where R is irreducible and reversible with respect to π, and Q is irreversible. The local irreversibility constants κ_i, the Poincaré constant μ, and the corresponding η_∞ will be functions of ε.

Consider first the simple case where Q has the same stationary distribution π. Since $A(R) = 0$, the anti-symmetric part satisfies $A(P_\varepsilon) = \varepsilon A(Q)$, so that

$$\kappa(P_\varepsilon) = \varepsilon\, \kappa(Q).$$

By Weyl's inequality (see, e.g., [13, Thm. 4.3.1]), it follows that

$$\mu(P_\varepsilon) \geq (1-\varepsilon)\mu(R).$$

Hence

$$\eta_\infty(P_\varepsilon) = \frac{\kappa(P_\varepsilon)}{\mu(P_\varepsilon)} \leq \frac{\varepsilon}{1-\varepsilon}\frac{\kappa(Q)}{\mu(R)}.$$

Thus, P_ε is nearly reversible for ε small enough.

The restriction that R and Q share the same stationary distribution can be relaxed.

Proposition 4. *Let P_ε be of the form* (29). *Let $\pi(\varepsilon)$ be the stationary distribution of P_ε. Then there are factors*

$$c_1(\varepsilon) := \frac{(1-\varepsilon)\|R\|_2}{\pi_{\min}(\varepsilon)}, \qquad c_2(\varepsilon) := \frac{\|Q\|_2\, \pi_{\max}(\varepsilon)}{\pi_{\min}(\varepsilon)},$$

such that

$$\kappa(\varepsilon) \leq c_1(\varepsilon)\, \|\pi(\varepsilon) - \pi\|_\infty + \varepsilon\, c_2(\varepsilon). \tag{30}$$

In particular $\kappa(\varepsilon) = O(\varepsilon)$ as $\varepsilon \downarrow 0$.

Proof. Set

$$M(\varepsilon) := \Pi(\varepsilon)P_\varepsilon - P_\varepsilon^\top \Pi(\varepsilon), \qquad A(\varepsilon) = \tfrac{1}{2}\,\Pi(\varepsilon)^{-1/2} M(\varepsilon)\Pi(\varepsilon)^{-1/2}.$$

Since $\|\Pi(\varepsilon)^{-1/2}\|_2 = 1/\sqrt{\pi_{\min}(\varepsilon)}$ we have

$$\kappa(\varepsilon) \leq \|A(\varepsilon)\|_2 \leq \frac{\|M(\varepsilon)\|_2}{2\pi_{\min}(\varepsilon)}.$$

Using (29) and the reversibility relation $\Pi R - R^\top \Pi = 0$, we expand

$$\begin{aligned} M(\varepsilon) &= (1-\varepsilon)\big(\Pi(\varepsilon)R - R^\top \Pi(\varepsilon)\big) + \varepsilon\big(\Pi(\varepsilon)Q - Q^\top \Pi(\varepsilon)\big) \\ &= (1-\varepsilon)\big(\Delta(\varepsilon)R - R^\top \Delta(\varepsilon)\big) + \varepsilon\big(\Pi(\varepsilon)Q - Q^\top \Pi(\varepsilon)\big), \end{aligned}$$

where $\Delta(\varepsilon) = \Pi(\varepsilon) - \Pi$ is diagonal. Taking operator norms and using $\|\Delta(\varepsilon)\|_2 = \|\pi(\varepsilon) - \pi\|_\infty$ and $\|\Pi(\varepsilon)\|_2 = \pi_{\max}(\varepsilon)$ gives

$$\|M(\varepsilon)\|_2 \le 2(1-\varepsilon)\|R\|_2\, \|\pi(\varepsilon) - \pi\|_\infty + 2\varepsilon\, \|Q\|_2\, \pi_{\max}(\varepsilon).$$

Combining with the previous inequality yields

$$\kappa(\varepsilon) \le \frac{(1-\varepsilon)\|R\|_2}{\pi_{\min}(\varepsilon)} \|\pi(\varepsilon) - \pi\|_\infty + \frac{\varepsilon\|Q\|_2\, \pi_{\max}(\varepsilon)}{\pi_{\min}(\varepsilon)},$$

which is exactly (30) with the factors $c_1(\varepsilon), c_2(\varepsilon)$ defined above.

To obtain the $O(\varepsilon)$ claim we use the Schweitzer updating formula [26]

$$\|\pi(\varepsilon) - \pi\|_\infty \le \frac{\varepsilon\, \|(Q-R)D\|_\infty}{1 - \varepsilon\, \|(Q-R)D\|_\infty} \quad \text{whenever} \quad \varepsilon\|(Q-R)D\|_\infty < 1,$$

where D is the deviation matrix, given by

$$D = (I - R + \mathbf{1}^\top \pi)^{-1} - \mathbf{1}^\top \pi.$$

Substituting this into (30) shows both terms on the right-hand side are $O(\varepsilon)$ as $\varepsilon \downarrow 0$, hence $\kappa(\varepsilon) = O(\varepsilon)$. □

As an example, let Q be the transition matrix of the directed cycle on n vertices, and let $R = \frac{1}{n}\mathbf{1}^\top\mathbf{1}$. Both Q and R have the uniform stationary distribution $\pi = \frac{1}{n}\mathbf{1}$. Consider P_ε as in (29). The antisymmetric part is

$$A(P_\varepsilon) = \tfrac{1}{2}(P_\varepsilon - P_\varepsilon^\top) = \tfrac{\varepsilon}{2}\,(Q - Q^\top),$$

therefore the local irreversibility coefficients are

$$\kappa(P_\varepsilon) = \max_i \|A_{i,:}\|_2 = \frac{\varepsilon}{\sqrt{2}}.$$

The symmetric part is

$$S(P_\varepsilon) = \tfrac{1}{2}(P_\varepsilon + P_\varepsilon^\top) = (1-\varepsilon)R + \varepsilon\, \tfrac{Q+Q^\top}{2}.$$

Its eigenvalues are

$$\lambda_0(S) = 1, \qquad \lambda_k(S) = \varepsilon \cos\frac{2\pi k}{n}, \quad k = 1, \ldots, n-1,$$

so the Poincaré constant of S is

$$\mu = 1 - \varepsilon \cos\frac{2\pi}{n}.$$

Thus the irreversibility ratio is

$$\eta_\infty = \frac{\varepsilon/\sqrt{2}}{1 - \varepsilon\cos(2\pi/n)}.$$

The threshold $\varepsilon^*(n)$ at which the chain is nearly reversible, i.e., $\eta_\infty \le 1/(2n + \sqrt{n})$, is

$$\varepsilon^*(n) = \frac{1}{(\sqrt{2}n + \sqrt{n/2})(1 + \cos(2\pi/n))}.$$

For $n \ge 4$, $\varepsilon^*(n) \le 1/(\sqrt{2}n)$, in particular $\varepsilon^*(n) = O(1/n)$.

This example has direct implications for PageRank: the PageRank transition matrix is precisely of the form (29), where the reversible component is the teleportation distribution. Proposition 4 shows that by choosing a sufficiently large teleportation factor, one can always ensure that the resulting PageRank chain becomes nearly reversible. The directed-cycle example illustrates a worst-case scenario: in practice, we expect that for many real-world networks the amount of teleportation required to achieve near-reversibility is substantially smaller than what is needed for directed cycles of the same size.

5 Heuristics and Numerical Experiments

Inspired by the equivalence of RLGL to coordinate descent in the reversible case, we come up with new heuristics. We show that they perform well even in the irreversible case, when we employ them to compute the stationary distribution and PageRank of real world and synthetic networks.

5.1 Gauss–Southwell–Dirichlet (GSD) Heuristics

The insights of Theorem 3 suggest the following new class of heuristics. Indeed, intuitively, one should prioritize coordinate updates that yield the largest one-step decrease of the Dirichlet energy in (17). This suggests heuristics that choose the updated coordinates based on the residual rescaled by $\sqrt{\pi_i}$ rather than the actual residual as was done in [2], or rescaled by π_i as proposed in [18].

Specifically, assume that P has no self-loops, so that any singleton $\{i\}$ is an independent set. By Theorem 3, the one-step decrease in the symmetrized coordinates from updating coordinate i with the optimal step is

$$\mathcal{E}(y_t) - \mathcal{E}(y_{t+1}) = \tfrac{1}{2}\left|(y_t\mathcal{L})_i\right|^2. \tag{31}$$

Hence the greedy (Gauss–Southwell) rule in the y–coordinates is

$$i_t = \mathrm{argmax}_i \ |(y_t \mathcal{L})_i|.$$

Translating back to the original coordinates x, this becomes

$$i_t = \mathrm{argmax}_i \ \left|\left(x_t(I-P)\Pi^{-1/2}\right)_i\right|,$$

i.e., one should apply Gauss–Southwell to the residuals *rescaled* by $\sqrt{\pi_i}$. We call this the *Gauss–Southwell–Dirichlet* (GSD) rule.

The stationary distribution π is not known a priori, so in practice one must use its proxies $\widehat{\pi}$. A natural choice is to use the current iterate x_t. Note that the proxy does not need to be normalized.

To exploit parallel hardware, one wants to update many coordinates at each iteration. Theorem 3 suggests updating an *independent set* that carries large total rescaled residual. A simple local scheme that produces such a set is the following:

1. each node computes its rescaled residual.
2. node i is *updated* if it holds the largest rescaled residual in its neighborhood.

Due to the locality, this heuristic is distributed, and yields an independent set B_t that contains the coordinate with maximum rescaled residuals; hence, it preserves the descent property from Theorem 3.

The heuristics above are robust to multiplicative approximation errors in the rescaling factor. The following lemma quantifies this statement.

Lemma 5 *Let π be the true stationary distribution and let $\widehat{\pi}$ be a proxy satisfying*

$$c^{-1}\pi_i \le \widehat{\pi}_i \le c\,\pi_i \qquad \text{for all } i \in [n],$$

for some $c \ge 1$. Denote by $\widetilde{r}_i = |(x_t(I-P)\Pi^{-1/2})_i|$ the true rescaled residuals and by $\widehat{r}_i = |(x_t(I-P)\widehat{\Pi}^{-1/2})_i|$ the residuals rescaled using the proxy. Then

$$c^{-1/2}\widetilde{r}_i \le \widehat{r}_i \le c^{1/2}\widetilde{r}_i \qquad \text{for all } i \in [n].$$

In particular, the coordinate selected by the proxy rule satisfies

$$\max_i \widehat{r}_i \ge c^{-1/2} \max_i \widetilde{r}_i,$$

and the corresponding one-step energy decrease is at least c^{-1} times the optimal single-coordinate decrease.

Proof. Since $\widehat{\Pi}^{-1/2}\Pi^{1/2}$ is diagonal with entries in $[c^{-1/2}, c^{1/2}]$, the claimed multiplicative bounds on $\widehat{r}_i$ follow immediately. Squaring the scores yields the c^{-1} factor for the energy decrease. □

Thus a fixed multiplicative proxy degrades the greedy decrease by at most a factor c^{-1} that depends only on the quality of the proxy.

Finally, we quantify the costs as the cumulative number of edges used, divided by the number of edges in the graph: one full power iteration has a normalized cost of 1, while an iteration where a single node i is updated has a normalized cost of $d_i^+/|E|$. Therefore, it is natural to include some weight that takes into account the cost of updating a node, which is its out-degree d_i^+, and choose the node i that maximizes the decrease in energy (31) per cost. We call this heuristic `GSD-deg` and its local version `LocalGSD-deg`.

Table 2 summarizes all the new heuristics described above.

Heuristic	**Selection Rule**
`GSD`	$B_t = \{ i \mid i = \text{argmax}_j \; \lvert r_{t,j} \rvert / \sqrt{x_{t,j}} \}$
`GSD-deg`	$B_t = \{ i \mid i = \text{argmax}_j \; \lvert r_{t,j} \rvert / \sqrt{d_j^+ x_{t,j}} \}$
`LocalGSD`	$B_t = \{ i \mid i = \text{argmax}_{j \in N(i)} \; \lvert r_{t,j} \rvert / \sqrt{x_{t,j}} \}$
`LocalGSD-deg`	$B_t = \{ i \mid i = \text{argmax}_{j \in N(i)} \; \lvert r_{t,j} \rvert / \sqrt{d_j^+ x_{t,j}} \}$

Table 2 Here $r_t = x_t(I - P)$, x_t is the current estimate of the stationary distribution, $N(i)$ denotes the neighborhood of node i, including itself, d_i^+ its the out-degree.

Selecting the maximum-score coordinate at each iteration can be costly. Moreover, in sparse settings it is reasonable to expect that a single update affects only a small number of scores. This motivates a batched strategy in which scores are recomputed only after several updates: the scores are evaluated, ranked, and then the next batch of coordinates is updated sequentially or in parallel.

We compare our new heuristics to the baseline heuristics summarized in Table 3. The `Theta` heuristic is the best performing heuristic in [2].

Baseline heuristic	**Selection Rule**
`Gauss-Southwell`	$B_t = \{ i \mid i = \arg\max_j \lvert r_{t,j} \rvert \}$
`PI`	$B_t = [n]$
`PCash`	$B_t = \{ i \}$ with probability $\propto \lvert r_{t,i} \rvert$
`Theta`	$B_t = \{ i \mid i \equiv t \pmod{n}$ and $\lvert r_{t,i} \rvert \geq \theta_t(r) \}$, $\theta_t(r) = \left(\frac{\sum_j \lvert r_{t_0,j} \rvert^r}{n} \right)^{1/r}, \quad r \geq 1, \quad t_0 = \lfloor \frac{t}{n} \rfloor$

Table 3 Baseline heuristics.

5.2 Numerical Results

In this section, we compare the existing and the newly proposed RLGL heuristics on web graphs, and two synthetic graphs: a stochastic block model and a scale-free preferential attachment model. More details on the graphs used can be found in Table 4.

For each graph, we compute both the stationary distribution on their largest strongly connected component (LSCC) and the PageRank [7], which is the stationary distribution of P_ε in (29), where Q is a simple random walk on a (directed) graph, $R = \frac{1}{n}\mathbf{1}^\top\mathbf{1}$ and the damping factor is $\varepsilon = 0.85$. We report the ℓ_1-norm of the residual as a function of the *normalized cost*.

We start with a comparison of the Gauss–Southwell rule with different rescalings of the residuals, on the LSCC of the Harvard500 graph. As we observe in Figure 2, rescaling by $\sqrt{x_{t,i}}$, which is our proposed GSD heuristic, results in the fastest convergence.

Across all the graphs we tested, our new heuristics, particularly GSD-deg, outperform the previously best-known methods, including the Theta heuristic of [2]. It is noteworthy that the local heuristics also perform consistently well, almost always surpassing the prior state of the art, despite using only local information.

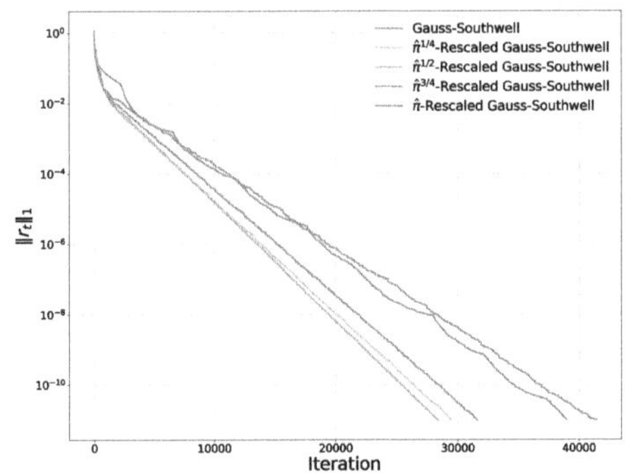

Figure 2 Stationary distribution computation of the Harvard500 LSCC. Comparison of different residual rescalings for the Gauss-Southwell heuristic. On the y-axis the ℓ_1 norm of the residual, on the x-axis the iteration number. Rescaling by $\sqrt{\hat{\pi}}$ results in faster convergence.

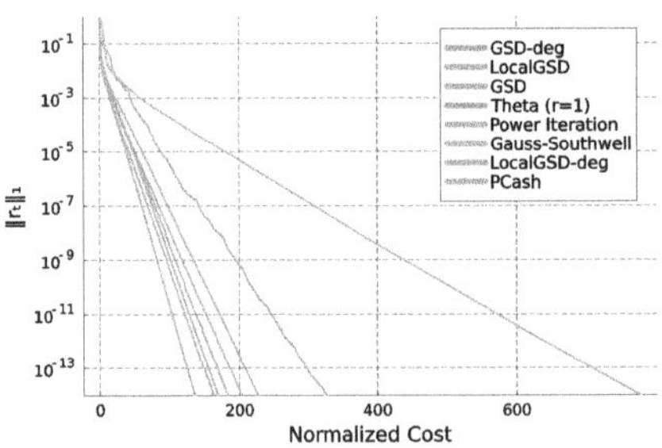

Figure 3 Stationary distribution on Harvard500 LSCC.

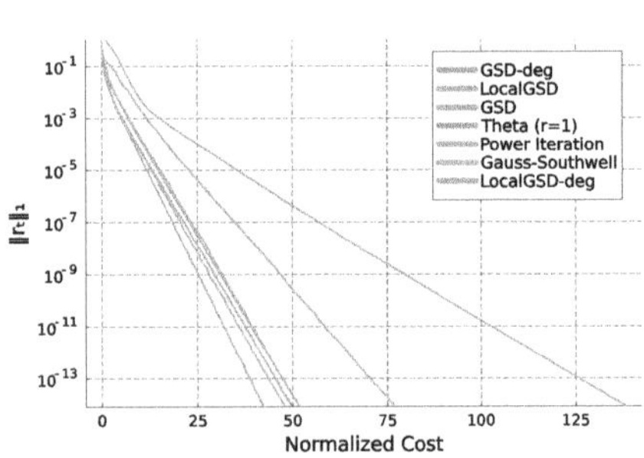

Figure 4 PageRank on Harvard500 LSCC.

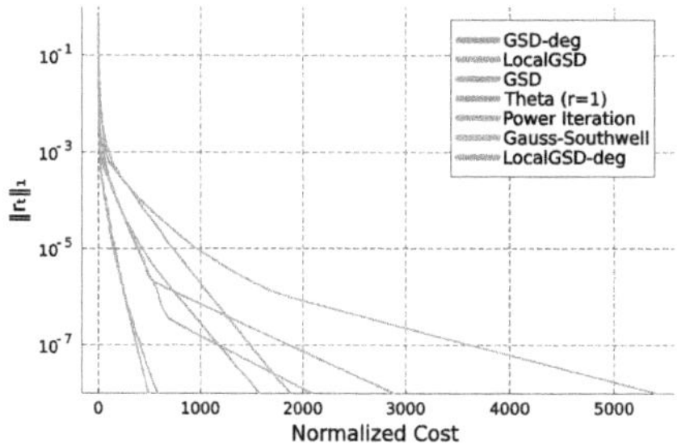

Figure 5 Stationary distribution computation on web-edu.

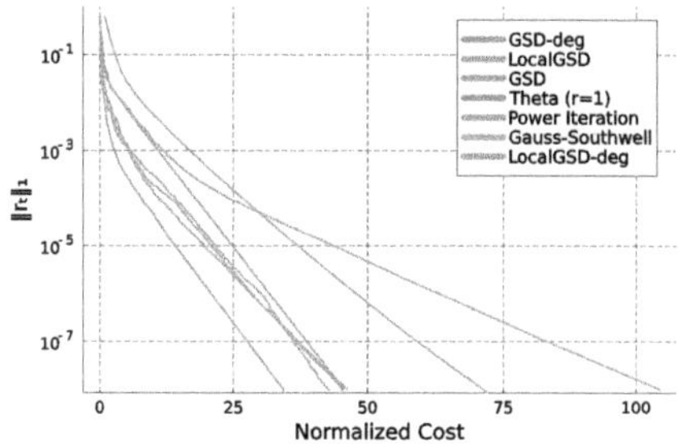

Figure 6 PageRank computation on web-edu.

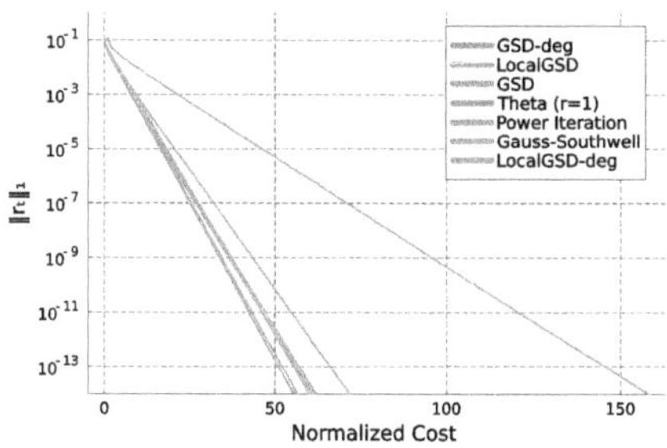

Figure 7 Stationary distribution computation SBM.

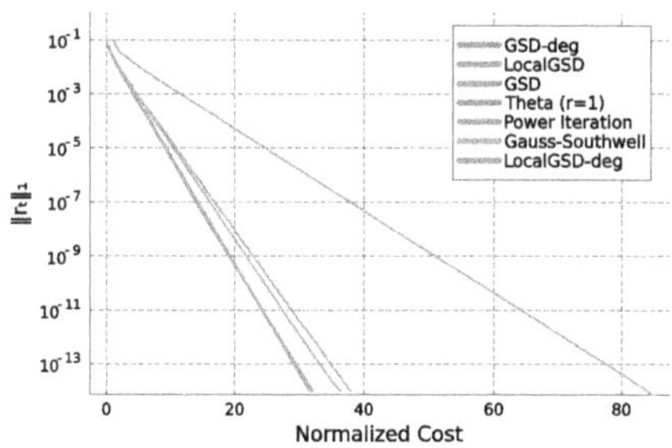

Figure 8 PageRank computation on SBM.

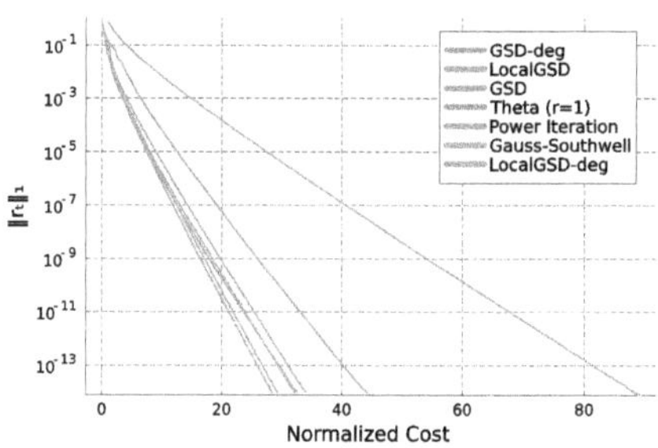

Figure 9 Stationary distribution computation on scale-free.

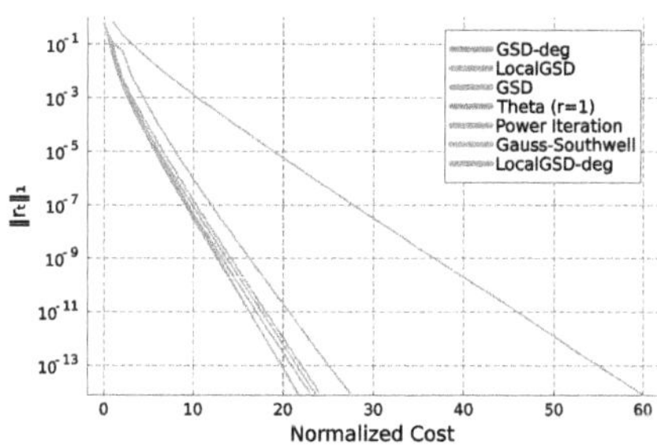

Figure 10 PageRank on computation scale-free.

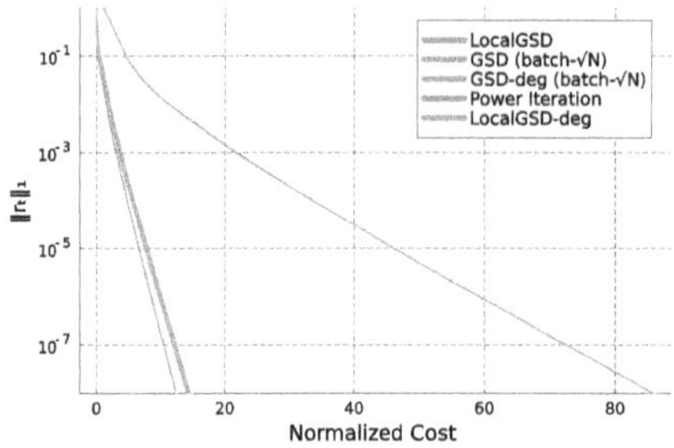

Figure 11 PageRank computation on web-stanford LSCC.

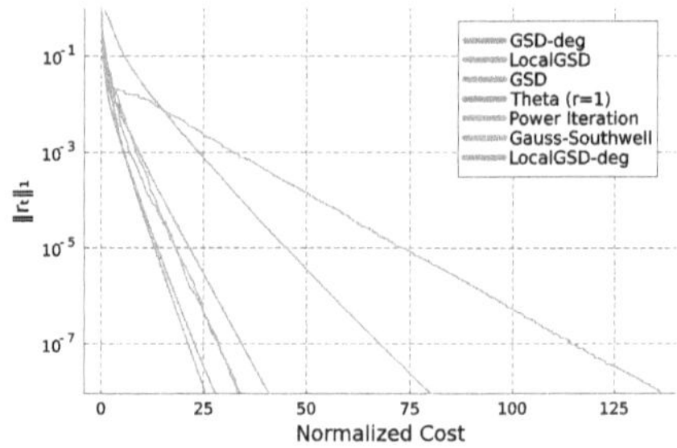

Figure 12 PageRank computation on wb-cs-stanford LSCC.

Graph	Nodes	Edges	Description
Harvard500 LSCC†	335	1963	web graph
SBM ‡	800	27852	synthetic block model
scale-free ‡	1000	2991	synthetic scale-free network
web-cs-stanford LSCC*	2759	13895	web graph
web-edu†	3031	6474	web graph
web-stanford LSCC*	150532	1576314	web graph

Table 4 Graphs used in the numerical experiments. †: dataset from the Network Repository [22]; *: dataset from SNAP [16]; ‡: synthetic graphs generated by the authors.

6 Conclusion and Further Research

We have found the connection between the RLGL algorithm and coordinate descent method. Specifically, our Theorem 1 says that when P is similar to a symmetric matrix, then the RLGL update is a gradient update when minimizing a specific energy function. When P is reversible, we have shown that this energy function is the Dirichlet energy of the Markov chain. Moreover, if updated coordinates form an independent set with respect to P, then the RLGL update matches the optimal step size of the coordinate descent.

In light of computing stationary distributions of large Markov chains, which is the original goal of the RLGL algorithm, reversibility is a restrictive assumption because the stationary distribution of a reversible chain can be easily computed by other methods. To address this, in Theorem 5, we have extended our results to a class of so-called 'nearly reversible' Markov chains, using a perturbation approach.

Even though the nearly reversibility condition is still rather stringent, in practice, the heuristics motivated by the Dirichlet energy minimization, improve the performance of the RLGL method. In our numerical experiments on both synthetic and real-world networks, these new heuristics consistently outperform the baselines. In particular,

`GSD-deg` always surpasses the previous state of the art. Moreover, the local variant, `LocalGSD-deg`, almost always outperforms the prior SOTA, despite relying only on local information.

A natural next step is to look for conditions on P that are weaker than near reversibility, and yet allow for an energy–minimization interpretation of the RLGL updates, or at least guarantee some form of descent. It would also be of interest to better understand the role of block updates in the irreversible setting, where single-coordinate updates may fail, and to characterize structural properties of directed graphs that enable efficient coordinate-based algorithms. Finally, in our experiments, the rescaling of coordinates derived for reversible Markov chains, speeds up the convergence for general Markov chains as well. It will be very interesting to find a mathematical justification for this phenomenon beyond the nearly-reversible case.

Acknowledgments

The work of NL was supported in part by the Dutch Research Council (NWO) through Gravitation-grant NETWORKS-024.002.003. The work of LG is supported by the EAISI institute at Eindhoven University of Technology under the project "Distributed solutions for large Markov chains". KA has been supported by the French government, through the France 2030 investment plan managed by the Agence Nationale de la Recherche, as part of the Université Côte d'Azur's Initiative of Excellence, reference ANR-15-IDEX-0001 and of the "UCA DS4H" project, reference ANR-17-EURE-0004.

Appendix

Definition 3. *A differentiable function* $f : \mathbb{R} \to \mathbb{R}$ *is said to be* μ-strongly convex *if there exists a* $\mu > 0$*, such that*

$$f(y) \geq f(x) + \nabla f(y)^\top (y-x) + \frac{\mu}{2}\|y-x\|_2^2, \qquad \forall x, y \in \mathbb{R}. \tag{32}$$

Lemma 6 *[19, Sec. 2.1.3] Let* f *be a differentiable* μ*-strongly convex function, and let* $f^* = \inf_{x\in\mathbb{R}^n}$*. Then,*

$$\frac{1}{2\mu}\|\nabla f(x)\|_2^2 \geq f(x) - f^*.$$

Proof. Minimize both sides of the inequality (32). □

The next lemma follows immediately from the spectral theorem for symmetric matrices; see [13].

Lemma 7 *Let* $Q \in \mathbb{R}^{n\times n}$ *be symmetric and positive semi-definite. Then the quadratic form*

$$f(x) = \tfrac{1}{2} x Q x^\top$$

satisfies the PL inequality with $\mu = \min\{\lambda > 0 \mid \lambda \text{ eigenvalue of } Q\}$.

Proof. Since the gradient is $\nabla f(x) = xQ$, we see that

$$\tfrac{1}{2}\|\nabla f(x)\|^2 = \tfrac{1}{2} x Q^2 x^\top,$$

thus the PL-inequality is equivalent to

$$xQ^2x^\top \geq \mu\, xQx^\top.$$

We can show this diagonalizing $Q = U^\top \Lambda U$ with $\Lambda = \operatorname{diag}(\lambda_1, \ldots, \lambda_n)$, and let μ be the smallest non-zero eigenvalue. Let $y = xU^\top$. Then

$$xQ^2x^\top = \sum_{i=1}^{n} \lambda_i^2 y_i^2 \geq \mu \sum_{i=1}^{n} \lambda_i y_i^2 = \mu x Q x^\top.$$ □

We define the Lipschitz constants for each coordinate

$$|\partial_i f(x + \lambda e_i) - \partial_i f(x)| \leq L_i |\lambda|, \qquad i = 1, \ldots, n,$$

and define $L_{max} := \max_i L_i$.

Lipschitzianity of the gradient allows us to bound the growth of a function, the next result is a component version of the *descent lemma.*

Lemma 8 *[19, Lemma 1.2.3] Let* $f : \mathbb{R}^n \to \mathbb{R}$ *be differentiable and with gradient* $L_1, L_2, \ldots, L_n$ *Lipschitz constant. Then*

$$f(x + \lambda e_i) \leq f(x) + \lambda \partial_i f(x) + \frac{L_i}{2}\lambda^2.$$

Proof. Define the function $g : \mathbb{R} \to \mathbb{R}$, $g(t) = \partial_i f(x + te_i)$. Then,

$$\begin{aligned} g(\lambda) - g(0) &= \int_0^\lambda g'(t)dt \\ &\leq \int_0^\lambda |g'(t) - g'(0)|dt + \int_0^\lambda g'(0)dt \\ &\leq L_i \int_0^\lambda |t|dt + \lambda g'(0) \\ &\leq \frac{L_i}{2}\lambda^2 + \lambda g'(0). \end{aligned}$$

□

The next lemma follows from basic properties of symmetric positive semidefinite matrices and the generalized Rayleigh quotient; see [13].

Lemma 9 *Let $A, B \in \mathbb{R}^{n\times n}$ be symmetric positive semidefinite matrices such that* $\ker(A) = \ker(B)$. *Then there exist constants $0 < c_1 \leq c_2 < \infty$ such that for all $x \in \mathbb{R}^n$,*

$$c_1 xBx^\top \;\leq\; xAx^\top \;\leq\; c_2 xBx^\top.$$

Proof. Let $K := \ker(A) = \ker(B)$ and $V = K^\perp$. For $x \in \mathbb{R}^n$ write $x = u + k$ with $u \in V$, $k \in K$. Since A and B vanish on K, we have

$$x^\top Ax = u^\top Au, \qquad x^\top Bx = u^\top Bu.$$

On V, both A and B are positive definite, so the generalized Rayleigh quotient

$$\phi(u) := \frac{u^\top Au}{u^\top Bu}, \quad u \in V \setminus \{0\},$$

is strictly positive. Restricting ϕ to the compact unit sphere in V, $S := \{u \in V : \|u\| = 1\}$, the continuous function ϕ attains a minimum $c_1 > 0$ and a maximum $c_2 < \infty$. Hence,

$$c_1 u^\top Bu \leq u^\top Au \leq c_2 u^\top Bu, \quad \forall u \in V,$$

and therefore the same inequality holds for all $x \in \mathbb{R}^n$. □

Proof of Corollary 1. We seek conditions under which Theorem 6 guarantees convergence. This requires two things:

1. We must effectively transfer the goodness from the observed residual to the symmetric gradient (Proposition 2).
2. The perturbation must be small enough to satisfy the convergence condition (Theorem 5).

Theorem 6 guarantees convergence if:

$$\kappa < \mu\sqrt{\beta_{\text{sym}}}.$$

Using the relation $\kappa \leq \mu\sqrt{n}\eta_\infty$, this condition becomes:

$$\sqrt{n}\,\eta_\infty < \sqrt{\beta_{\text{sym}}}. \tag{33}$$

We use Proposition 2 to substitute $\sqrt{\beta_{\text{sym}}}$. Let $B = \sqrt{\beta_{\text{norm}}} = 1/\sqrt{n}$. The proposition states:

$$\sqrt{\beta_{\text{sym}}} = \frac{B - \gamma_\infty^\star}{1 + \gamma_\infty^\star}, \quad \text{where} \quad \gamma_\infty^\star = \frac{\sqrt{n}\,\eta_\infty}{1 - \sqrt{n}\,\eta_\infty}.$$

To simplify notation, let $\omega = \sqrt{n}\,\eta_\infty$. Then $\gamma_\infty^\star = \frac{\omega}{1-\omega}$. Substituting this into the expression for $\sqrt{\beta_{\text{sym}}}$:

$$\sqrt{\beta_{\text{sym}}} = B(1 - \omega) - \omega.$$

We now substitute this simplified form back into our convergence condition (33):

$$2\omega < B(1 - \omega).$$

Thus, we require:

$$\omega < \frac{B}{2 + B}.$$

Substituting back gives the desired bound. □

References

1. S. Abiteboul, M. Preda, and G. Cobena. Adaptive on-line page importance computation. In *Proceedings of the 12th International Conference on World Wide Web*, pages 280–290, 2003.
2. K. Avrachenkov, P. Brown, and N. Litvak. Red Light Green Light method for solving large Markov chains. *Journal on Scientific Computing*, 93(1):18.
3. K. Avrachenkov, A. Mishenin, P. Gonçalves, and M. Sokol. Generalized optimization frame work for graph-based semi-supervised learning. In *Proceedings of the 2012 SIAM International Conference on Data Mining*, pages 966–974, 2012.
4. Z.-Z. Bai and W.-T. Wu. On greedy randomized Kaczmarz method for solving large sparse linear systems. *SIAM Journal on Scientific Computing*, 40(1):A592–A606, 2018.
5. A. Bojchevski, J. Gasteiger, B. Perozzi, A. Kapoor, M. Blais, B. Rózemberczki, M. Lukasik, and S. Günnemann. Scaling graph neural networks with approximate PageRank. In *Proceedings of the 26th ACM SIGKDD International Conference on Knowledge Discovery & Data Mining*, pages 2464–2473, 2020.
6. S. Boyd and L. Vandenberghe. *Convex Optimization*. Cambridge University Press, 2004.
7. S. Brin and L. Page. The anatomy of a large-scale hypertextual web search engine. *Computer Networks and ISDN Systems*, 30(1-7):107–117, 1998.
8. F. R. Chung. *Spectral Graph Theory*, volume 92. American Mathematical Society, 1997.
9. T. K. N. Dang, D. Bucur, B. Atil, G. Pitel, F. Ruis, H. Kadkhodaei, and N. Litvak. Look back, look around: A systematic analysis of effective predictors for new outlinks in focused web crawling. *Knowledge-Based Systems*, 260:110126, 2023.
10. M. P. Friedlander and M. Schmidt. Hybrid deterministic-stochastic methods for data fitting. *SIAM Journal on Scientific Computing*, 34(3):A1380–A1405, 2012.
11. J. Gasteiger, A. Bojchevski, and S. Günnemann. Predict then propagate: Graph neural networks meet personalized PageRank. *arXiv preprint arXiv:1810.05997*, 2018.
12. M. R. Hestenes and E. Stiefel. The conjugate gradient method for solving linear systems. *Journal of Research of the National Bureau of Standards*, 49:409–436, 1952.

13. R. A. Horn and C. R. Johnson. *Matrix Analysis*. Cambridge University Press, 2 edition, 2013.
14. G. Jeh and J. Widom. Scaling personalized web search. In *Proceedings of the 12th International Conference on World Wide Web*, pages 271–279, 2003.
15. S. Kaczmarz. Angenäherte Auflösung von Systemen Linearer Gleichungen (English translation by Jason Stockmann): Bulletin International de l'Académie Polonaise des Sciences et des Lettres. 1937.
16. J. Leskovec and A. Krevl. SNAP Datasets: Stanford large network dataset collection. `http://snap.stanford.edu/data`, June 2014.
17. D. A. Levin and Y. Peres. *Markov Chains and Mixing Times*. American Mathematical Society, 2017.
18. F. McSherry. A uniform approach to accelerated PageRank computation. In *Proceedings of the 14th International Conference on World Wide Web - WWW'05*, page 575. ACM Press.
19. Y. Nesterov. *Introductory Lectures on Convex Optimization: A Basic Course*, volume 87. Springer Science & Business Media, 2013.
20. Y. Nesterov. *Lectures on Convex Optimization*. Springer, 2018.
21. J. Nutini, I. Laradji, and M. Schmidt. Let's make block coordinate descent converge faster: Faster greedy rules, message-passing, active-set complexity, and superlinear convergence. *arXiv preprint arXiv:1712.08859*, 2017.
22. R. A. Rossi and N. K. Ahmed. The network data repository with interactive graph analytics and visualization. In *Proceedings of the Twenty-Ninth AAAI Conference on Artificial Intelligence*, 2015.
23. Y. Saad. *Iterative Methods for Sparse Linear Systems*. SIAM, 2003.
24. Y. Saad and M. H. Schultz. GMRES: A generalized minimal residual algorithm for solving nonsymmetric linear systems. *SIAM Journal on Scientific and Statistical Computing*, 7(3):856–869, 1986.
25. J. Salez. Modern aspects of Markov chains: Entropy, curvature and the cutoff phenomenon. *arXiv preprint arXiv:2508.21055*, 2025.
26. P. J. Schweitzer. Perturbation theory and finite Markov chains. *Journal of Applied Probability*, 5(2):401–413, 1968.
27. T. Strohmer and R. Vershynin. A randomized Kaczmarz algorithm with exponential convergence. *Journal of Fourier Analysis and Applications*, 15(2):262–278, 2009.
28. S. J. Wright. Coordinate descent algorithms. *arXiv preprint arXiv:1502.04759*, 2015.

Bayesian Perspectives in Scientific Machine Learning: Methods, Challenges, and Opportunities

Dario Coscia and Gianluigi Rozza

Abstract Scientific Machine Learning (SciML) is an interdisciplinary field combining data-driven machine learning with physics-based modeling, for accurate predictions across scientific and engineering applications. Despite the great successes of SciML, the widespread deployment in real-world applications remains challenged by model uncertainty, limited data, and potential model misspecification, which can undermine reliability and trustworthiness. Bayesian learning offers a principled solution by providing probabilistic predictions, quantifying uncertainty, and supporting continual adaptation as new data becomes available. However, the integration of Bayesian methods into SciML, particularly Bayesian Deep Learning, remains relatively underexplored. In this review, we provide a comprehensive overview of Bayesian perspectives in SciML, focusing on methodological advances, current challenges, and emerging opportunities. We highlight how Bayesian approaches can enhance uncertainty quantification, continual learning, interpretability, and robustness in scientific applications, bridging the gap between statistical rigor and the practical demands of large-scale applications and high-dimensional models. By systematically examining these contributions, we outline promising research directions and opportunities at the intersection of Bayesian methods and Scientific Machine Learning.

1 Introduction

The foundation of scientific discovery lies in the collection, transformation, and interpretation of data—an ongoing process shaped by advances in methodology, technology, and human social contexts. In recent decades, the rise of Artificial Intelligence (AI) has

Dario Coscia
SISSA Mathematics Area, mathLab, via Bonomea, 265, Trieste 34136, Italy, dcoscia@sissa.it

Gianluigi Rozza
SISSA Mathematics Area, mathLab, via Bonomea, 265, Trieste 34136, Italy, grozza@sissa.it

B. Koren and D. Roose (eds.), *Numerical Analysis and Scientific Computing: an Anthology*, Lecture Notes in Computational Science and Engineering 156,
https://doi.org/10.1007/978-3-032-33428-2_11

introduced a paradigm shift, with AI increasingly serving as an integrated and, in some cases, indispensable tool [166].

Machine Learning (ML) is widely used in scientific discovery to extract patterns from large datasets, guide experimentation, validate existing theories, explore new data-compatible hypotheses, and automate many aspects of the discovery process. However, purely data-driven ML models often struggle to respect underlying physical laws, while traditional physics-based models can be computationally intensive or limited in scope. The interdisciplinary field that connects ML with physics-based computational models is Scientific Machine Learning [130] (SciML). By combining data-driven methods with first-principles physics, physical constraints, correlations, and structural information directly into ML models, SciML enables more accurate, efficient, and physically consistent modeling across a wide range of scientific applications.

Notable successes of SciML span a wide range of scientific domains. In climate modeling [18,128], hybrid ML-physics models enhance prediction accuracy and enable the early forecasting of extreme weather events. In protein folding [1,79], ML models have proved capable of predicting the joint structures of protein complexes, nucleic acids, and small molecules, accelerating our understanding of biomolecular interactions. In materials science [119,158,181], ML has been instrumental in predicting material properties with high precision and speeding up molecular dynamics simulations, thereby accelerating the discovery of novel materials. These examples illustrate how SciML not only complements traditional modeling approaches but also opens new avenues for exploration that were previously computationally or experimentally intractable.

However, while SciML has demonstrated remarkable successes across many domains, its widespread adoption in real-world scientific applications requires careful consideration of model reliability [166]. In practice, even the most accurate ML models can be affected by uncertainty, limited data, or model misspecification, which may lead to unreliable predictions if not properly accounted for [129]. Bayesian learning provides a principled framework to address these challenges by representing model parameters and predictions probabilistically, allowing for explicit quantification of uncertainty and more. This probabilistic perspective enables scientists to assess the reliability of model outputs, identify potential errors arising from model misspecification, and make informed decisions before applying models to real-world scenarios. Moreover, Bayesian methods naturally adapt to new data while maintaining calibrated uncertainty, making them particularly suitable for the dynamic and iterative nature of scientific discovery.

Despite these advantages, Bayesian methods remain underutilized in SciML, leaving significant potential to enhance model reliability, interpretability, and adaptability. Systematically incorporating Bayesian principles could empower SciML models to rigorously quantify uncertainty, address model misspecification, and continuously update as new data emerge, ultimately enabling more robust and trustworthy scientific discovery. However, a key reason for their limited adoption is computational cost: exact Bayesian inference requires integrating over high-dimensional parameter spaces, which is analytically intractable and prohibitively expensive for modern deep neural networks, necessitating approximate methods that trade off accuracy for scalability.

1.1 Scope and Objectives of this Review

The goal of this review is to provide a comprehensive overview of Bayesian perspectives in Scientific Machine Learning, with a particular emphasis on Bayesian Deep Learning, a paradigm that remains relatively underexplored in the SciML community. We focus on methodological developments, current challenges, and emerging opportunities, aiming to bridge the gap between the statistical rigor of Bayesian inference and the practical demands of large-scale, high-dimensional Deep Learning models in scientific applications. In particular, we examine how Bayesian Deep Learning can contribute to three key aspects of scientific modeling:

- **Uncertainty Quantification:** Bayesian models provide principled estimates of confidence in predictions, which is crucial for guiding experiments, simulations, or data collection in scientific workflows.
- **Continual Learning:** By updating knowledge as new data become available, Bayesian approaches enable models to adapt to evolving scientific problems while retaining insights from prior tasks.
- **Interpretability and Robustness:** Bayesian methods help identify relevant model components and reduce sensitivity to errors, enhancing the reliability and transparency of learned models in complex scientific applications.

Throughout this review, we highlight state-of-the-art methods that balance computational efficiency and scalability with the rigor of Bayesian inference, discuss practical challenges in applying these methods to SciML problems, and outline emerging opportunities that promise to advance the integration of Bayesian techniques into scientific modeling and simulation.

2 The Many Faces of Learning: Deep, Bayesian, and Scientific

To navigate the rich landscape of modern machine learning, it helps to have a clear map of the main approaches shaping research and applications. In this work, we focus on three complementary perspectives: Deep Learning, Bayesian Learning, and Scientific Machine Learning. It is important to stress that these are not competing approaches; rather, they represent different modalities that can collaborate, fuse, and interact to address diverse problems.

Deep learning has transformed how we discover patterns in large, complex datasets, driving advances from computer vision and speech recognition to natural language processing and, more recently, the natural sciences. Yet finding patterns alone is often insufficient—real-world problems also require reasoning under uncertainty. Bayesian learning provides a principled framework for probabilistic reasoning, allowing models to quantify uncertainty and more. Finally, Scientific Machine Learning fuses data-driven methods with scientific principles. By embedding physical laws, symmetries, correlations, and structural insights, SciML produces models that are both accurate and faithful to the phenomena they represent. In the subsection below we review the fundamentals for each learning modality.

2.1 Deep Learning

Over the past two decades, the emergence of Deep Learning has driven significant progress in the field of Machine Learning. It has revolutionised numerous domains, including computer vision [43, 62, 91, 134], speech recognition [4, 66], and natural language processing [11, 22, 131, 156, 161]. Today, fields like robotics, wireless communication, and the natural sciences are also undergoing substantial transformation, thanks also to the great improvement in artificial neural network architectures [9, 57, 76, 88, 143, 148, 161].

Most of the breakthroughs in Deep Learning have been driven by *supervised learning*. In this paradigm [13], a model is trained to capture the *conditional probability distribution* $p(\boldsymbol{y} \mid \boldsymbol{x})$, which represents the likelihood of producing an output $\boldsymbol{y} \in \mathbb{Y}$ given an input $\boldsymbol{x} \in \mathbb{X}$. The "learning process" starts with a labeled dataset $\mathcal{D} = \{(\boldsymbol{x}_i, \boldsymbol{y}_i)\}_{i=1}^N$, and a parametrized model $f_{\boldsymbol{\theta}} : \mathbb{X} \to \mathbb{Y}$ (e.g. a neural network). The model is trained by optimizing the parameters $\boldsymbol{\theta} \in \Theta$ by minimizing a loss function $\mathcal{L} : \Theta \to \mathbb{R}$ over the observed input-output pairs:

$$\mathcal{L}(\boldsymbol{\theta}) = \mathbb{E}_{(\boldsymbol{x}_i, \boldsymbol{y}_i) \sim \mathcal{D}}[l(f_{\boldsymbol{\theta}}(\boldsymbol{x}_i), \boldsymbol{y}_i)], \tag{1}$$

where $l : \mathbb{Y} \times \mathbb{Y} \to \mathbb{R}$ is a quality metric, measuring the mismatch between the predicted output by the model $f_{\boldsymbol{\theta}}(\boldsymbol{x}_i)$, and the true one $\boldsymbol{y}_i$.

Application: Regression

A classic example of supervised learning is the regression problem, where the goal is to predict a continuous target value $\boldsymbol{y}$ from an input $\boldsymbol{x}$. In this setting:

- $f_{\boldsymbol{\theta}}(\boldsymbol{x}_i)$ is the model's prediction for input $\boldsymbol{x}_i$, e.g., the output of a neural network given the input features.
- $l(f_{\boldsymbol{\theta}}(\boldsymbol{x}_i), \boldsymbol{y}_i)$ is a pointwise loss function that measures the error between the predicted value and the true target y_i. For regression, a common choice is the *squared error*:
$$l(f_{\boldsymbol{\theta}}(\boldsymbol{x}_i), \boldsymbol{y}_i) = \|f_{\boldsymbol{\theta}}(\boldsymbol{x}_i) - \boldsymbol{y}_i\|_2^2.$$
- $\mathcal{L}(\boldsymbol{\theta})$ is the overall loss (or objective) function that aggregates the errors over the dataset, typically as an expectation or empirical average:
$$\mathcal{L}(\boldsymbol{\theta}) = \frac{1}{N} \sum_{i=1}^{N} l(f_{\boldsymbol{\theta}}(\boldsymbol{x}_i), \boldsymbol{y}_i),$$
which is then minimized with respect to the parameters $\boldsymbol{\theta}$ to train the model.

Through this process, the model learns to approximate the mapping from inputs $\boldsymbol{x}$ to outputs $\boldsymbol{y}$, effectively capturing the conditional relationship $p(\boldsymbol{y} \mid \boldsymbol{x})$ in the supervised learning sense.

Although supervised learning has proven highly effective, acquiring labelled data $(\boldsymbol{x}, \boldsymbol{y})$ can be costly, time-consuming, or even impractical in some domains. Consequently, the field has increasingly recognized the importance of unsupervised or self-supervised methods as a pathway to scale learning [58, 87, 105, 138, 149, 150]. In unsupervised or self-supervised methods, the final goal is to learn a probability distribution $p_{\boldsymbol{\theta}}(z)$ given unlabeled data $\mathcal{D} = \{z_i\}_{i=1}^{N}$. In this setting, the model seeks to capture the underlying structure, regularities, or generative mechanisms of the data without relying on explicit target labels. Depending on the approach, this may involve maximising the likelihood of the observed samples [39, 40], learning a latent representation that explains the data [58, 67], or specifying a generative process capable of producing realistic samples from the learned distribution [87, 149]. Through these techniques, unsupervised and self-supervised learning enable the exploitation of vast amounts of unlabelled data, ultimately serving as a foundation for more scalable and generalizable learning systems [131, 156].

2.2 Bayesian Learning

Bayesian models have a long-standing role in Deep Learning [16, 53, 61, 68, 176], providing a principled framework for modelling uncertainty in neural networks through probabilistic inference. The foundation of Bayesian learning is Bayes' theorem [8], which provides a systematic way to update beliefs about unknown quantities based on observed data. In this framework, the model parameters are treated as random variables,

$$\omega \sim p(\omega), \tag{2}$$

where $p(\omega)$ is the prior distribution encoding our initial beliefs about the parameters. Note that we use ω (rather than $\boldsymbol{\theta}$ as in Section 2.1) to emphasize that in the Bayesian setting, parameters are treated as random variables rather than fixed point estimates. After observing a dataset $\mathcal{D}$ we can update this prior to obtain the *posterior distribution* over parameters, $p(\omega \mid \mathcal{D})$, using Bayes' theorem:

$$p(\omega \mid \mathcal{D}) = \frac{p(\mathcal{D} \mid \omega)\, p(\omega)}{\int p(\mathcal{D} \mid \omega)\, p(\omega)\, d\omega}, \tag{3}$$

where $p(\mathcal{D} \mid \omega)$ is the likelihood of the data given the parameters. Once the posterior distribution over weights, $p(\omega \mid \mathcal{D})$, is obtained, *Bayesian Model Averaging* (BMA) can be used to make predictions by averaging over all plausible models:

$$f^*(\star) = \mathbb{E}_{p(\omega \mid \mathcal{D})}\left[f_\omega(\star))\right] = \int f_\omega(\star)\, p(\omega \mid \mathcal{D})\, d\omega. \tag{4}$$

In the context of Deep Learning, Bayesian learning has been applied by placing probability distributions over neural network weights rather than relying on deterministic point (maximum likelihood) estimates.

Application: Bayesian Regression

A Bayesian extension of the regression problem is done by treating the model parameters ω as random variables and placing a prior distribution over them. In this setting:

- $p(\omega)$ is the prior distribution, representing our beliefs about the parameters before seeing any data.
- $p(\boldsymbol{y}_i \mid \boldsymbol{x}_i, \omega)$ is the likelihood, which models the probability of the observed target $\boldsymbol{y}_i$ given the input $\boldsymbol{x}_i$ and parameters ω. For example, for Gaussian noise in regression:

$$p(\boldsymbol{y}_i \mid \boldsymbol{x}_i, \omega) = \mathcal{N}(\boldsymbol{y}_i \mid f_\omega(\boldsymbol{x}_i), \sigma^2).$$

- $p(\omega \mid \mathcal{D})$ is the posterior distribution over the model parameters given the dataset $\mathcal{D} = \{(\boldsymbol{x}_i, \boldsymbol{y}_i)\}_{i=1}^N$, obtained via Bayes' theorem:

$$p(\omega \mid \mathcal{D}) = \frac{p(\mathcal{D} \mid \omega)\, p(\omega)}{p(\mathcal{D})},$$

where $p(\mathcal{D} \mid \omega) = \prod_{i=1}^N p(\boldsymbol{y}_i \mid \boldsymbol{x}_i, \omega)$ is the likelihood of the full dataset.
- Predictions for a new input $\boldsymbol{x}_*$ are made by marginalizing over the posterior:

$$p(\boldsymbol{y}_* \mid \boldsymbol{x}_*, \mathcal{D}) = \int p(\boldsymbol{y}_* \mid \boldsymbol{x}_*, \omega)\, p(\omega \mid \mathcal{D})\, d\omega.$$

If the prior $p(\omega)$ is uniform (non-informative), the posterior mean coincides with the Maximum Likelihood Estimate (MLE):

$$\hat{\omega}_{\text{MLE}} = \arg\max_{\omega} p(\mathcal{D} \mid \omega).$$

This shows that deterministic regression (Section 2.1) is a special case of Bayesian regression: when the prior is non-informative, minimizing the loss in Equation (1) is equivalent to maximizing the likelihood.

At first glance, Bayesian inference may appear conceptually straightforward: one specifies a prior distribution, observes data, and updates beliefs via Bayes' theorem. In practice, however, executing this update in the context of modern neural networks is profoundly challenging. A primary difficulty concerns the specification of an appropriate prior [49, 163, 180]. The prior plays a decisive role in determining the posterior distribution and, consequently, the model's inductive behaviour. Its selection is far from arbitrary and raises important questions regarding expressiveness, regularisation, and robustness, issues highlighted in classical consistency results [42]. An even more fundamental obstacle arises from the computation of the posterior itself. The Bayesian update involves evaluating a normalising constant obtained by integrating the product

$p(\mathcal{D} \mid \omega)p(\omega)$ over the full parameter space. For contemporary deep architectures, this space is extremely high-dimensional and exhibits complex, multimodal structure, rendering the integral analytically intractable and computationally prohibitive. These challenges render exact Bayesian inference in Deep Learning settings practically infeasible, necessitating approximate methods.

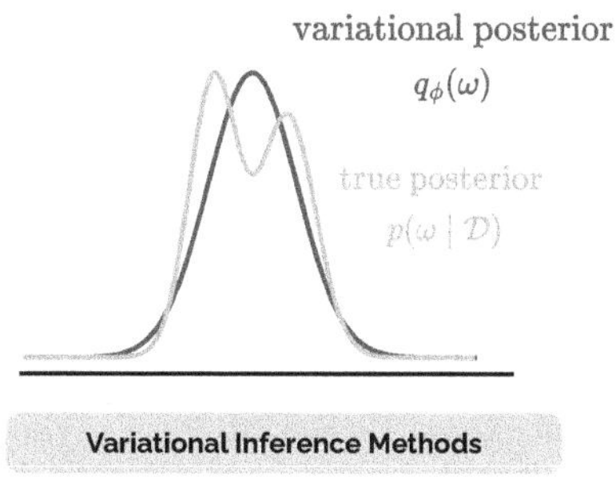

Figure 1 Variational Inference reformulates posterior estimation as an optimization problem: it approximates the true posterior $p(\omega \mid \mathcal{D})$ with a tractable distribution $q_\phi(\omega)$ by maximizing the ELBO, equivalent to minimizing the KL divergence between $q_\phi(\omega)$ and $p(\omega \mid \mathcal{D})$.

Variational inference [61, 78, 165] has emerged as a popular approach by reframing the inference problem into an optimisation task (Figure 1). The final goal of variational inference is to approximate the true posterior $p(\omega \mid \mathcal{D})$ with a tractable parametric distribution $q_\phi(\omega)$, typically a Gaussian or a more expressive distribution [109, 110, 141], by minimizing their Kullback–Leibler (KL) divergence:

$$\mathrm{KL}[q_\phi(\omega) \mid p(\omega \mid \mathcal{D})] = \mathbb{E}_{q_\phi(\omega)}[\log q_\phi(\omega) - \log p(\omega \mid \mathcal{D})]. \tag{5}$$

Since the posterior distribution is not accessible, it can be shown that the KL minimisation is mathematically equivalent to maximising the Evidence Lower BOund (ELBO):

$$\mathrm{ELBO}[q_\phi(\omega)] = \mathbb{E}_{q_\phi(\omega)}[\log p(\mathcal{D} \mid \omega)] - \mathrm{KL}[q_\phi(\omega) \mid p(\omega)]. \tag{6}$$

Despite its efficiency and scalability, variational inference often suffers from bias introduced by the choice of the approximating family [15], potentially leading the inferred posterior to concentrate on overly simple or unimodal regions of the parameter space. Recent advances, such as stochastic variational inference [69] and amortized inference [87, 137], attempt to mitigate these limitations by making the approximation more flexible and computationally efficient.

Markov Chain Monte Carlo (MCMC) methods [21] provide asymptotically exact posterior inference by generating dependent samples that, in the limit, recover the true distribution. In Figure 2, we illustrate the MCMC sampling pipeline. Blue boxes denote the prior parameter space; red squares represent proposed samples rejected by the Metropolis–Hastings acceptance criterion; and orange squares indicate accepted samples comprising the Markov chain trajectory converging toward the true posterior distribution. Classical MCMC algorithms such as Metropolis–Hastings or Hamiltonian Monte Carlo (HMC) are theoretically appealing but often impractical for deep networks due to the high dimensionality of parameter spaces, slow mixing, and the need

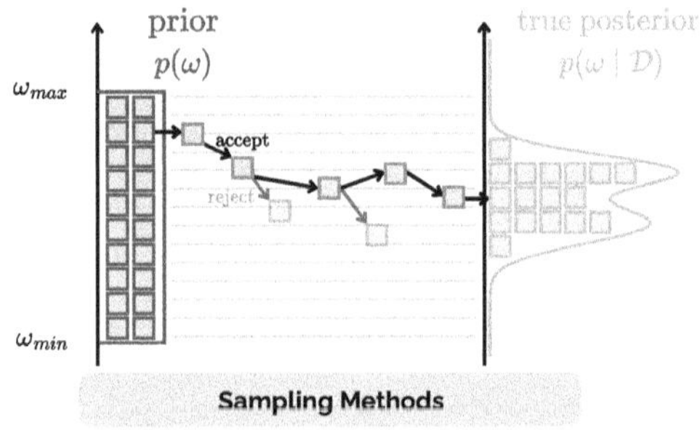

Figure 2 Markov Chain Monte Carlo (MCMC) methods estimate the posterior $p(\omega \mid \mathcal{D})$ by generating dependent samples that asymptotically recover the true distribution. Unlike variational inference, MCMC treats inference as a sampling problem rather than an optimisation problem, allowing exploration of complex, multimodal posteriors.

for careful tuning of step sizes. Gradient-based variants, such as Langevin dynamics or stochastic gradient Langevin dynamics (SGLD) [176], or Stochastic Gradient Markov Chain Monte Carlo (SGMCMC) [27, 118], improve scalability by using gradient information to guide exploration while injecting noise to preserve the correct stationary distribution. These methods can efficiently navigate complex, multimodal landscapes and are compatible with mini-batch training, making them more suitable for Deep Learning. However, they still require careful hyperparameter tuning (e.g., step size and noise scale), and convergence guarantees are typically asymptotic, which may limit their practical fidelity to the true posterior in very high-dimensional or highly non-convex settings.

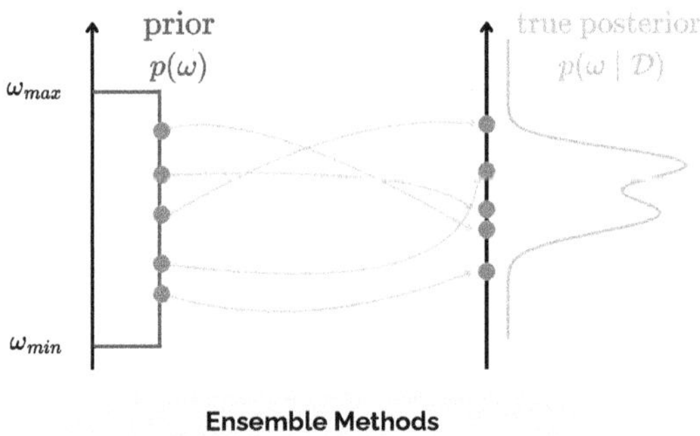

Figure 3 Ensemble methods approximate posterior uncertainty by training multiple independent models and aggregating their predictions, capturing variability in model outputs to represent epistemic uncertainty and improve generalization.

Ensemble methods [73, 96, 112] provide a simple approach to representing epistemic uncertainty by capturing the variability in model predictions (Figure 3). By training multiple independent models and aggregating their predictions, ensembles can empirically approximate posterior uncertainty [75, 180] and improve generalisation performance. Unlike MCMC methods—which generate correlated samples that asymptotically converge to the true posterior via a Markov chain—ensemble methods rely on the diversity introduced by different random initializations and the loss landscape, without a principled sampling scheme or posterior convergence guarantees. As a result, ensembles only implicitly and approximately capture posterior uncertainty. While straightforward to implement and often highly effective in practice, ensembles come at the cost of substantial computational and memory overhead, especially when the number of models grows or when each model is itself large.

2.3 Scientific Machine Learning

> "*Scientific Machine Learning is an interdisciplinary field empowered by the synergy of physics-based computational models with machine-learning algorithms for scientific and engineering applications*" [130].

Traditional scientific computing relies on first principles, such as partial differential equations or dynamical systems, to simulate complex phenomena [45,97]. While being highly interpretable and grounded in theory, first-principle calculations often come with high computational costs, limited scalability, or difficulties in handling incomplete or noisy data. In contrast, purely data-driven ML methods excel at capturing patterns from large datasets but may ignore the underlying physical laws, leading to models that are untrustworthy or physically inconsistent. Scientific Machine Learning (SciML) [5, 130] is an interdisciplinary field bridging the physics-based modelling and data-driven learning approaches into a unified framework.

In Scientific Machine Learning, a defining feature is the incorporation of scientific domain knowledge directly into machine learning models. Such knowledge can appear in several forms (Figure 4), including ab-initio principles, physical constraints (e.g., equivariance, conservation laws), correlations (e.g., space–time or parameter correlations), and structural information (e.g., discrete or graphical structure).

Ab-initio principles are often used to embed first-principles physics directly into learning algorithms. For example, Hermann et al. [64] develop highly flexible neural-network wavefunctions that surpass state-of-the-art variational quantum chemistry methods. Similar approaches have been applied in quantum Density Functional Theory (DFT) [102, 153] and classical DFT [37, 133], in gene prediction [71], and in biomolecular dynamics simulations [170].

Physical constraints are another common source of domain knowledge, especially in learning or solving Partial Differential Equations (PDEs). Physics-Informed Neural Networks (PINNs) [32,41,82,132] incorporate the governing equations through automatic differentiation, enforcing them as soft constraints during training. Equivariant Neural Networks [20,29,34,173,174] encode symmetry properties—such as rotational or translational invariance—directly into the architecture, ensuring that model predictions respect the underlying physical symmetries.

Correlations leverage known relationships in the system, such as space–time patterns or parameter dependencies. Reduced-order modeling (ROM) [65,98,111,140] is a common approach that captures the dominant modes of a complex system, reducing computational cost while retaining essential dynamics. Deep learning methods can enhance the above methodologies by learning non-linear correlations from simulation or

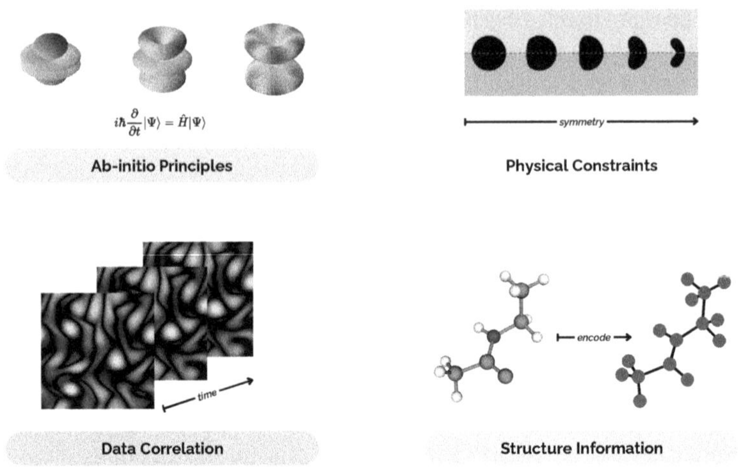

Figure 4 Taxonomy of scientific domain knowledge in machine learning. Scientific domain knowledge can be incorporated into ML models in four complementary — and sometimes overlapping — ways. *Ab-initio principles* build the model architecture around a first-principles formulation, so that the physics *defines* the model (e.g., neural-network wavefunctions [64], deep DFT [102]). *Physical constraints* regularise or restrict an otherwise general architecture so that predictions respect governing equations or symmetries (e.g., PINNs [132], equivariant networks [29, 173]). *Correlations* exploit known space–time or parameter relationships to build compact surrogate models, often via reduced-order modelling [65, 140] or nonlinear latent representations [51]. *Structural information* leverages the discrete or topological anatomy of scientific data — sequences, meshes, or graphs — to preserve connectivity and topology in the learned model [7, 142].

experimental data [51,99,127,162], obtaining a richer low-dimensional representation of the data.

Finally, structural information exploits the inherent anatomy of scientific data. Autoregressive models [10, 159] capture sequential dependencies, for example, in time-dependent PDE modelling [3, 19, 142], or molecular physical simulations [12, 147]. Another prominent example of scientific data structures is graphs. In Machine Learning Interatomic Potentials (MLIPs) [7,52,119,139,181], for instance, graphs represent the discrete arrangement of atoms and their interactions, enabling fast and accurate predictions of energies and forces while preserving the underlying chemical structure. Graphs are also widely used in learning over meshes, where they capture the connectivity and topology of spatial discretisations in simulations, with a wide spectrum of applications, ranging from object deformation [142] to biomedical modeling [44], to cite a few.

3 Why Bayesian in Scientific Machine Learning

Having introduced Deep Learning, Bayesian methods, and the principles of Scientific Machine Learning, it becomes natural to ask: why integrate Bayesian approaches specifically into SciML? The answer lies in the unique demands of scientific applications. Unlike conventional settings, scientific modeling often involves scarce or costly data, high-dimensional parameter spaces, and complex physical constraints. In these contexts, understanding the confidence and reliability of predictions is as important as the predictions themselves. However, most models still rely on deterministic Deep Learning, and Bayesian learning seems less common in practice.

In this section, we then examine the potential of Bayesian approaches to enhance Scientific Machine Learning models. First, Bayesian methods provide principled uncertainty quantification, allowing models to indicate how confident they are in their predictions and to identify regions where additional data or careful analysis may be needed. Second, they naturally support continual learning, enabling models to incorporate new information as it becomes available while retaining knowledge from previous tasks—a key advantage when data arrive sequentially, or experiments are costly. Third, Bayesian techniques improve interpretability and robustness: methods like model averaging combine predictions from multiple models to reduce the impact of misspecification, while Bayesian pruning identifies non-essential parameters to simplify models without compromising performance.

Across fields such as materials science, chemistry, fluid dynamics, and weather forecasting, these capabilities make Bayesian approaches a powerful tool for building SciML models that are not only accurate but also reliable, adaptable, and transparent. By integrating predictive power with principled uncertainty estimates and interpretable structures, Bayesian methods are increasingly central to advancing Scientific Machine Learning.

3.1 Uncertainty Quantification

In Scientific Machine Learning, understanding and managing uncertainty is just as important as achieving high predictive accuracy. Uncertainty can originate from many sources—such as noisy data, overparameterized models, or physical-model misspecification [129]—and failing to account for it can undermine the reliability of scientific predictions. Conventional Deep Learning methods generally return only point estimates, which may provide a single plausible prediction but offer no insight into the confidence or robustness of that prediction. In contrast, Bayesian Deep Learning produces full probability distributions over model parameters (see Figure 5), allowing the model to express its confidence—or lack thereof—when making predictions [125].

In scientific domains—such as chemistry, materials science, fluid dynamics, weather forecasting, chaotic systems (e.g., turbulent flows), and multi-scale models (e.g., molecular-to-continuum coupling)—where experimental data collection is costly or constrained, predictions are sensitive to small perturbations, or models span vastly

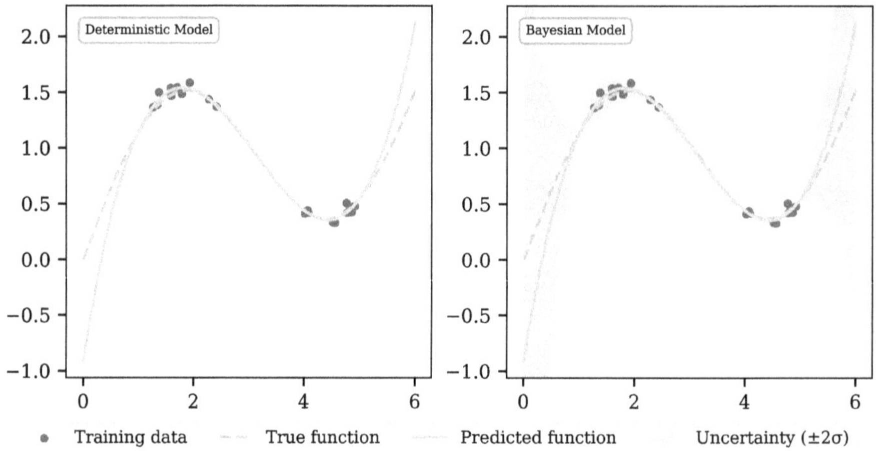

Figure 5 Illustrative comparison of deterministic and Bayesian approaches in Scientific Machine Learning. While *Deterministic Models* provide single-point predictions, *Bayesian Models* capture uncertainty by producing predictive distributions, allowing one to know when model predictions are less confident. This is particularly important in Scientific Machine Learning, where understanding uncertainty is as critical as achieving high predictive accuracy [125, 129].

different scales, Bayesian Deep Learning excels at quantifying uncertainty [125]. For instance, in chemistry and materials science, Bayesian uncertainty quantification is widely used to assess model reliability when computing physical quantities such as energies, forces, and stresses [30, 152, 154]; guiding active learning to improve models while reducing the cost of DFT labeling [77,93]; and steering molecular dynamics simulations toward regions of high uncertainty to accelerate scientific discovery [92]. In fluid dynamics, Bayesian models have been widely employed to quantify PDE surrogates' errors compared to true solvers in both forward and inverse problems, particularly within physics-informed neural networks [106, 129, 185] and Neural Operators [31, 113, 114]. More recently, large-scale foundation models have also recognized the value of Bayesian learning—both in weather forecasting [2], where it helps capture the many sources of uncertainty inherent in atmospheric prediction, and in PDE solvers [190], where it provides a principled framework for modelling uncertainty in complex physical systems.

Finally, it is worth mentioning that in addition to Bayesian Deep Learning, Gaussian Processes (GPs) [179] have long served as a foundational Bayesian approach for uncertainty quantification in scientific applications [36, 122]. GPs provide closed-form predictive distributions and well-calibrated uncertainty estimates, making them particularly attractive in low-data regimes. However, their computational complexity and limited scalability to high-dimensional inputs have motivated the growing adoption of Bayesian Deep Learning methods, which retain probabilistic reasoning while scaling to large datasets and complex systems.

3.2 Continual Learning

A key property of the Bayesian theorem in Equation (3) is that the posterior can dynamically update beliefs as new data become available. This continual refinement enables the model to retain and reuse relevant information from previously learned tasks while still adapting to new evidence. Consequently, Bayesian updating offers a principled mechanism for knowledge transfer, helping prevent forgetting and improving performance when learning sequentially.

In Deep Learning, updating a model across shifting data distributions is studied in the field of continual learning [169]. A major challenge in this setting is that deterministic neural networks suffer from catastrophic forgetting, meaning they tend to overwrite or discard previously learned knowledge when trained on new tasks. To address this, [101] proposes a Bayesian continual learning approach that adapts to new tasks while preserving information from earlier ones. Similarly, [86] propagates the posterior distribution as the prior for subsequent tasks by approximating the posterior using a density estimator fitted to Hamiltonian Monte Carlo (HMC) samples. Finally, [100] introduces a meta-continual-learning framework in which neural networks are meta-learned offline and then kept fixed during continual deployment; only the statistical (Bayesian) components are updated with new data, demonstrating strong resistance to forgetting.

In classical Bayesian inference, continual learning is most commonly used in the field of *data assimilation* [47, 135], where the goal is to sequentially update the state estimate of a dynamical system as new observational data arrive. Classical methods such as the Ensemble Kalman Filter [46] and variational approaches (3D-Var, 4D-Var) [107] are fundamentally Bayesian: they maintain a prior over the system state and update it into a posterior each time new measurements become available. Recent work has shown that neural network surrogates can be embedded within these assimilation cycles, combining the expressive power of deep learning with the principled sequential updating of Bayesian inference [17, 25], making data assimilation a natural bridge between classical Bayesian inference and modern SciML continual learning methods

In the SciML context, the continual learning approach is particularly relevant when physical surrogate models encounter streaming or evolving data distributions [187]. For example, in fluid dynamics [25], continual learning enables neural PDE solvers to update their solutions sequentially. This can be particularly advantageous when the full discretized grid of the domain is too large to store in memory, or when data arrive from sensors in a streaming fashion and are not available all at once. Another promising direction involves finetuning large pretrained models on vast amounts of scientific data for specialized downstream tasks. In [30], for instance, the authors finetune a universal machine-learning interatomic potential on an out-of-distribution, real-world silica glass dataset, demonstrating that Bayesian finetuning not only outperforms the pretrained model by preventing forgetting, but also surpasses standard finetuning using the deterministic network counterpart.

3.3 Interpretability and Avoiding Misspecification

One of the enduring criticisms of deep neural networks — including those used in Scientific Machine Learning — is their "black-box" nature: although they often deliver accurate predictions, it can be challenging to interpret how these predictions are made or to extract scientific insight from the learned representations. A practical approach to improve interpretability is Bayesian pruning, which identifies and removes redundant parameters based on posterior uncertainty [108]. By zeroing out non-essential neurons, this method not only reduces the number of parameters but also demonstrates that overparameterized models are not necessary for effective inference. Such pruning is particularly useful for developing lightweight SciML models that can run efficiently on limited hardware [193]. More precisely, Bayesian pruning can be understood as a form of model selection. By placing sparsity-promoting priors over network weights, the posterior naturally concentrates on simpler, more parsimonious model structures. This is directly analogous to Bayesian model selection via the marginal likelihood, where more complex models are penalized unless the data explicitly support them. In this view, pruning does not merely reduce parameter count but also identifies the minimal model structure sufficient to explain the data, which is a principled and interpretable outcome in itself.

Another key advantage of Bayesian inference in SciML is its ability to average predictions across multiple models. Bayesian model averaging (BMA) mitigates the effects of model misspecification, providing a robust framework that accounts for uncertainty in both parameters and model structures, ultimately leading to more reliable and interpretable predictions [129]. For instance, in fluid dynamics, [74] applied a weighted BMA approach to an ensemble of Reduced Order Models (ROMs), where the weights were optimized to reflect each model's relative importance for specific spatial variables. More broadly, the systematic gap between a computational model and reality is formalized in the statistical literature under the terms *model discrepancy* or *model error* [83, 84], providing a principled framework for accounting for structural inadequacies beyond parameter uncertainty alone.

4 Challenges of Bayesian Scientific Machine Learning

Having introduced why Bayesian methods are needed for Scientific Machine Learning, it is natural to consider the challenges that arise when combining these approaches as well. While Bayesian Deep Learning offers principled uncertainty quantification, enhanced interpretability, and robustness, it also introduces substantial computational and methodological complexities. Modern neural networks are high-dimensional, highly non-linear, and often multimodal, making exact Bayesian inference intractable. Furthermore, the practical implementation of Bayesian methods, such as model averaging, can be computationally expensive, particularly in SciML applications, where each model evaluation may need to be performed multiple times, such as in molecular dynamics simulations for materials science.

In this section, we examine these challenges in detail. We discuss the scalability and efficiency limitations of standard Bayesian approaches, including Monte Carlo-based predictive inference and ensemble methods. We then explore issues related to inference, highlighting the computational overhead of obtaining predictive moments. Finally, we address prior selection, emphasizing the difficulties of specifying meaningful priors in high-dimensional parameter spaces and reviewing approaches that leverage functional priors or data-informed weight priors.

4.1 Scalability and Efficiency

One of the main challenges of Bayesian Deep Learning is the high computational cost [75]. We refer the reader to Section 2.2 for a description of the main Bayesian inference methods; here we focus on the scalability challenges that arise when applying them to large SciML models. More specifically, inference in modern neural networks poses a substantial challenge because, in principle, it requires integrating over an enormously high-dimensional posterior distribution to compute BMA —one that spans millions of parameters, contains numerous modes, and exhibits complex geometric structures such as mode connectivity [55]. These properties make the posterior exceptionally difficult to explore or approximate with precision. Hamiltonian Monte Carlo [75, 129, 191] has long been considered the gold standard for sampling from the posterior, but its applicability to large-scale models is limited, as it requires orders of magnitude more computation than standard deep neural network training. As a result, much contemporary work focuses on developing approximations, heuristics, and alternative strategies for tractable inference.

Deep Ensembles [96] and more computationally efficient variants such as Snapshot Ensembles [73] or SWAG [112] offer strong generalization performance. However, their predictive distributions still differ substantially from the HMC posterior, even though they are significantly more accurate than variational inference methods [75]. This has led practitioners to rely on deep ensembles for uncertainty quantification. However, as noted in [125], when working with large and computationally expensive Deep Learning models—such as Large Language Models [22, 156]—or when repeatedly invoking models in scientific workflows (e.g., running molecular dynamics simulations), the use of deep ensembles can become impractical due to their substantial computational demands.

Recently, several approaches have been proposed to address computational efficiency while remaining suitable for large SciML models. In [113, 114], the Laplace approximation is used to construct Bayesian Neural Operators. Although applying the Laplace approximation over the full parameter space can be challenging, the authors demonstrate that even a last-layer Laplace approximation can yield strong uncertainty estimates for PDE solvers. A major advantage of this methodology is that it can be applied to any pretrained model. However, the last-layer approximation may underperform compared to deep ensembles, especially for out-of-distribution data, which is a quite common scenario in SciML. Another line of research is the Adaptive Variational Dropout approach introduced in [31] in the context of autoregressive models for PDE solvers and molecular generation, and later extended in [30] to graph-based modeling, particularly for Machine-Learned Interatomic Potentials (MLIPs). The major advantage of this method is that it provides a fully Bayesian treatment of all network weights, resulting in calibrated uncertainties even for out-of-distribution data, while remaining scalable by choosing an efficient, structured variational distribution that avoids the prohibitive costs typically associated with full Bayesian inference. The authors in [30] further show that the method can surpass ensembles while using a single model with only about twice the FLOPs of standard training, and it also extends effectively to fine-tuning.

4.2 Inference

We have already stated that one key advantage of Bayesian Deep Learning is its ability to marginalize over model uncertainty [180]. However, computing the predictive distribution via model averaging—necessary for uncertainty quantification and downstream tasks—typically relies on Monte Carlo estimation for both ensemble and variational inference methods, introducing significant computational overhead [125]. Indeed, deriving statistical moments (e.g., mean and variance) from the predictive distribution requires K repeated model evaluations. For a model with $O(C)$ FLOPs per forward pass, the total computational cost scales as $O(K \cdot C)$, which can become prohibitive in practice, particularly for large models or high values of K, even when GPU vectorization is possible. This is especially problematic in Scientific Machine Learning tasks, where each model evaluation may involve expensive simulations. It is important to clarify that the computational concern here is not the cost of a single model evaluation — which SciML models reduce dramatically compared to first-principles calculations, but to a *simulation pipeline*. A prototypical example is that of Machine-Learning Interatomic Potentials (MLIPs) [119, 181]: a single Density Functional Theory (DFT) calculation, which may take hours, is replaced by a MLIP query that takes milliseconds. However, when such surrogate models are deployed within large-scale scientific pipelines — such as molecular dynamics simulations, nudged elastic band calculations, or free energy perturbation methods — the model may be queried millions of times over the course of a single simulation. In this setting, the $O(K \cdot C)$ cost of Monte Carlo uncertainty estimation, where K forward passes are required per query, becomes non-negligible even for individually cheap surrogate models, and can render Bayesian inference computationally prohibitive in practice. This highlights that the challenge of scalable Bayesian inference in SciML is not simply about the cost of individual evaluations, but about the cumulative cost incurred when models are embedded in iterative scientific workflows [30].

An interesting alternative, recently proposed in [103], is to perform local linearization of activation functions and local Gaussian approximations at linear layers. This approach allows approximate predictive moments to be computed analytically, significantly reducing the need for repeated Monte Carlo forward passes and the associated computational cost.

4.3 Prior Selection

In Bayesian Deep Learning, the prior defined over network parameters induces a corresponding prior over functions, and [180] argued that this function prior influences generalization. Neural network architectures naturally encode useful structure into this functional prior—for instance, convolutional networks inherently enforce translation equivariance. However, specifying meaningful priors directly in the high-dimensional parameter space is difficult due to its complexity and lack of interpretability [125]. Therefore, an important objective is to design priors on neural network weights that are both computationally tractable and encourage solutions with desirable model behaviors [50, 117, 157].

In the context of Scientific Machine Learning for PDE solvers, functional priors [157] have recently been used to provide more informative priors in Bayesian neural networks. Unlike traditional priors over network weights, functional priors are defined directly in function space. For example, [115, 129] employ generative modeling and neural operator techniques to learn functional priors from historical data and underlying physical principles. Another promising strategy, focusing on priors over weights, is proposed in [31] in the context of autoregressive models. Their approach encodes historical data into a prior that is a time-dependent, aggregated mixture of posteriors. The authors show that for temporal data, this prior allows the network to learn more effectively and generalize better than standard time-independent priors, such as Gaussian or log-uniform distributions.

5 Proposed Future Directions and Opportunities

Building on the challenges discussed in previous sections, this section highlights promising research directions and opportunities at the intersection of Bayesian methods and Scientific Machine Learning. These approaches provide principled ways to use uncertainty, incorporate prior knowledge, and optimize computational resources, enabling more efficient and reliable scientific models.

5.1 Active Learning

Active learning [136, 155] is a framework that leverages model uncertainty to reduce the amount of labeled data required for training. Let $\mathcal{D}_L = \{(\boldsymbol{x}_i, \boldsymbol{y}_i)\}_{i=1}^{N_L}$ denote the labeled dataset and $\mathcal{D}_U = \{\boldsymbol{x}_j\}_{j=1}^{N_U}$ the unlabeled dataset. Instead of labeling all examples in $\mathcal{D}_U$ randomly, the model identifies the most informative samples $\boldsymbol{x}^* \in \mathcal{D}_U$ to label next. This selection is typically performed using an acquisition function $\mathcal{A}(\boldsymbol{x})$ [172], which assigns a score to each unlabeled example based on criteria such as predictive uncertainty, diversity, or expected improvement. Formally, the next sample is chosen as

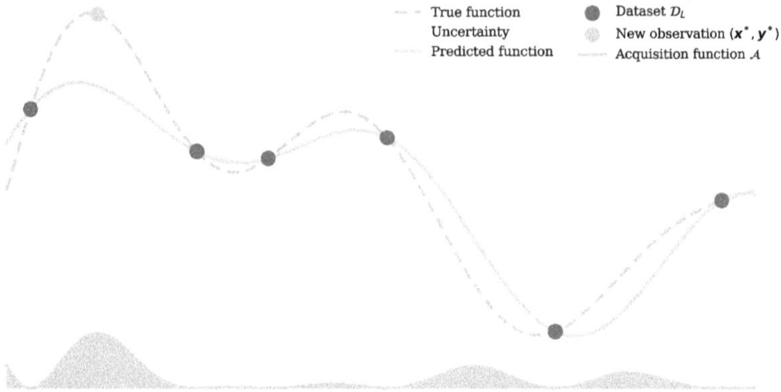

Figure 6 Schematic of the active learning workflow. A model is initially trained on a small labeled dataset $\mathcal{D}_L$. A sampler proposes new unlabeled samples, and an acquisition function $\mathcal{A}$ scores these samples. The sample with a higher acquisition function value $\boldsymbol{x}^*$ is selected and its label $\boldsymbol{y}^*$ is obtained. The pair $(\boldsymbol{x}^*, \boldsymbol{y}^*)$ is added to the labeled set $\mathcal{D}_L$, and the process is iterated, allowing the model to improve efficiently while minimizing labeling costs.

$$\boldsymbol{x}^* = \arg\max_{x \in \mathcal{D}_U} \mathcal{A}(\boldsymbol{x}), \tag{7}$$

and its label $\boldsymbol{y}^*$ is obtained and added to the labeled set, $\mathcal{D}_L \leftarrow \mathcal{D}_L \cup \{(\boldsymbol{x}^*, \boldsymbol{y}^*)\}$. By iteratively repeating this process, active learning allows the model to achieve higher performance with fewer labeled samples (see Figure 6).

In Scientific Machine Learning, obtaining labeled data is often expensive or time-consuming [166]. This makes active learning particularly valuable, as the uncertainty-driven selection of data points—introduced above—can directly reduce the number of expensive simulations or experiments needed. Bayesian inference further enhances this process [54], providing principled uncertainty estimates that guide acquisition functions to select the most informative samples, ensuring that each labeled point contributes maximally to improving the model.

A prominent application of active learning arises in the generation of training data for machine-learned interatomic potentials (MLIPs) [92]. The process begins with a small bootstrap dataset used to train an initial (Bayesian) MLIP. After training, a sampler generates new configurations, and the model's predictive uncertainty is evaluated for each one. The acquisition function—driven primarily by this uncertainty—identifies configurations for which the model is least confident. These high-uncertainty samples are then recalculated with a QM calculator to obtain accurate labels (energy and forces), and the resulting data is added to the training set to improve the model's reliability [77, 189]. Interestingly, models' uncertainties can also be used directly during sampling—for example, to steer molecular dynamics simulations toward regions where the model is highly uncertain [160].

The active learning strategy extends far beyond materials science. In PDE learning, for example, active learning can be applied when solutions of a parametrized PDE are available at different parameter values [116] or fidelity levels [35]. By targeting parameter regions where the model is less confident, one can selectively compute new high-fidelity PDE solutions, thereby improving the surrogate while keeping the overall computational cost manageable.

5.2 Transfer Learning

In the era of big data, training a model from scratch typically demands substantial computational resources, which are out of reach for most practitioners. Transfer learning [124, 175] helps address this challenge by allowing a model built for one task to be repurposed as the starting point for a different but related task.

Common approaches for transfer learning include feature extraction [188], where a pretrained model is used as a fixed representation and only a lightweight task-specific head is trained; full fine-tuning in which all parameters of the pretrained model are updated for the downstream task; and partial fine-tuning [89], where only a subset of layers (often the upper layers) are adapted to balance computational cost and performance. Recent work has introduced a family of parameter-efficient fine-tuning (PEFT) methods [38] that significantly reduce memory and compute requirements by updating only a small fraction of the model's parameters. A particularly effective PEFT strategy is Low-Rank Adaptation [72], which injects trainable low-rank matrices into existing weight matrices while freezing the original parameters, offering strong performance with minimal parameter overhead.

In the context of SciML, many works have focused on transfer learning for PDE modelling, for ROMs [80], PINNs [24, 104], and Neural Operators [59, 192]. Another application is the use of transfer learning techniques to generalize Neural PDE solutions to different geometries [59, 167, 182]. Very recent works [151, 184] analyzed how neural operator models scale and transfer across multiple PDE systems, considering model size, dataset size, task physics, and pre-training adaptation, highlighting that developing general SciML Foundation models [28] with shared weights is effective, even under small to moderate distribution shifts in downstream tasks.

In the context of transfer learning, Bayesian approaches in SciML offer complementary advantages. By explicitly modeling uncertainty in the parameters, Bayesian methods can mitigate overfitting, which is particularly important when downstream datasets are small or noisy [30, 60]. Additionally, continual posterior learning provides a natural framework for transfer learning: as new tasks or data are introduced, the posterior can be updated, enabling principled adaptation while preserving knowledge from previous tasks. Although scaling Bayesian Deep Learning to large models remains computationally challenging [6, 146], recent advances in parameter-efficient fine-tuning open promising avenues for integrating Bayesian methods with modern SciML models [121, 171, 183].

5.3 Reinforcement Learning

Reinforcement learning (RL) and control share strong connections with Bayesian learning [56, 164]. Combining these approaches provides a principled framework for action selection, allowing agents to balance exploration and exploitation based on uncertainty, while also offering a systematic way to incorporate prior knowledge into the learning and control process.

In Scientific Machine Learning, RL and control techniques are increasingly employed, and their link to Bayesian inference opens up promising research directions. For instance, in transition path sampling, [70] uses path-integral cross entropy [81] to identify optimal control functions by maximizing a regularized expected reward. This method can be interpreted from a Bayesian perspective, where the prior corresponds to the uncontrolled dynamics and the likelihood is represented by a Boltzmann distribution over the negative expected reward.

Similarly, RL from human feedback—commonly used to fine-tune pretrained models, including large language models, for task-specific behaviors [123]—can also be understood within this Bayesian framework [90]. In molecular design, RL has been applied to generate molecules with desired properties [14, 120]. Beyond molecular applications, RL has been used to model turbulence via data-driven closure models [94,95], although Bayesian approaches remain largely underexplored in this context.

5.4 Model compression

The online–offline paradigm [140] amortizes the high training cost of machine learning models over many low-cost inference queries. However, for large-scale models or computationally intensive SciML applications, inference can still be expensive: the large number of parameters directly increases the number of FLOPs required per evaluation. To mitigate this cost, Bayesian approaches offer a principled framework for model compression, enabling reductions in model size while preserving predictive performance. A principled strategy is to employ sparsity-promoting priors [108], which enable the pruning of substantial portions of a neural network's weights. Alternatively, the prior can be used as an entropy model [186], achieving similar compression effects. At test time, these strategies enable models to operate at varying levels of precision, allowing practitioners to balance predictive accuracy against computational cost.

5.5 Probabilistic Numerics

Probabilistic numerics [63] refers to a class of numerical algorithms in linear algebra, integration, optimization, and the solution of differential equations, where Bayesian methods are used to quantify uncertainty in their results. Within the context of SciML, probabilistic numerics has been applied to solving ordinary differential equations [85, 144], partial differential equations [168], and optimization problems [126, 177, 178].

5.6 Differentiable Physics Models

Differentiable physics models are a central component of modern scientific machine learning, where physical systems are constructed to be end-to-end differentiable, often via automatic differentiation [33, 145]. This includes, for example, neural differential equations [26] and machine-learned interatomic potentials, which provide direct access to gradients of the underlying physical quantities. Recently, molecular dynamics applications have leveraged differentiable models to compute forces as gradients of learned energy surfaces, as in conventional MD, while augmenting the potential with Bayesian uncertainty estimates to bias sampling toward high-uncertainty regions, thereby improving exploration of complex energy landscapes [189]. While these connections are increasingly explored in the SciML literature, they remain a promising direction for further integration of differentiable modeling and Bayesian methods.

6 Conclusion

Scientific Machine Learning has demonstrated transformative potential across diverse domains, from climate modeling to protein folding and materials science, by integrating data-driven machine learning with physics-based modeling. Despite these advances, challenges related to uncertainty, model misspecification, and continual adaptation remain critical barriers to reliable and trustworthy deployment in real-world scientific applications. Bayesian learning, particularly Bayesian Deep Learning, provides a principled framework to address these challenges, enhancing uncertainty quantification, continual learning, interpretability, robustness, and more. Looking forward, further integration of Bayesian techniques into SciML workflows, combined with advances in scalable inference, will be essential to deliver reliable and adaptable AI tools that support confident decision-making in complex scientific problems.

Yet a balanced perspective requires acknowledging that Bayesian methods are not without limitations, and that alternative approaches exist which may be preferable in certain settings. From a practical standpoint, Bayesian inference — even in its approximate forms — introduces additional computational overhead compared to standard deterministic training. In resource-constrained settings, or when uncertainty estimates are not critical to the downstream task, this overhead may not be justified. Moreover, the quality of Bayesian uncertainty estimates depends heavily on the choice of prior and the quality of the posterior approximation: poorly specified priors can bias results significantly, and approximate posteriors obtained via variational inference or the Laplace approximation may be overconfident or fail to capture the true multimodal structure of the posterior.

It is also worth noting that principled alternatives to Bayesian uncertainty quantification exist. Conformal prediction [48], for instance, provides distribution-free, finite-sample coverage guarantees without requiring a probabilistic model, and has recently been extended to regression and classification settings relevant to SciML [23]. Cross-validation-based uncertainty estimates offer practical robustness with fewer modeling assumptions. These methods may be preferable when computational budgets are tight or when the Bayesian modeling assumptions are difficult to justify [13].

Bayesian methods are most valuable when the uncertainty structure of the problem is scientifically meaningful — for example, when uncertainty estimates directly inform experimental design, guide active learning, or quantify the reliability of safety-critical predictions. In these settings, the principled probabilistic framework that Bayesian learning provides remains unmatched. As Scientific Machine Learning continues to mature, the integration of Bayesian principles will be essential to bridging the gap between predictive power and the rigorous uncertainty awareness that responsible scientific discovery demands.

7 Ethical Statement

This paper presents work aimed at advancing machine learning and computational sciences, with the potential to accelerate scientific discovery in the natural sciences. As these methods continue to evolve, they carry numerous potential societal implications. Although no specific issues are highlighted here, we stress the importance of ongoing discussions on ethical considerations and best practices to ensure the responsible development and application of these technologies for the benefit of society.

References

1. J. Abramson, J. Adler, J. Dunger, R. Evans, T. Green, A. Pritzel, O. Ronneberger, L. Willmore, A. J. Ballard, J. Bambrick, et al. Accurate structure prediction of biomolecular interactions with AlphaFold 3. *Nature*, 630(8016):493–500, 2024.
2. F. Alet, I. Price, A. El-Kadi, D. Masters, S. Markou, T. R. Andersson, J. Stott, R. Lam, M. Willson, A. Sanchez-Gonzalez, et al. Skillful joint probabilistic weather forecasting from marginals. *arXiv preprint arXiv:2506.10772*, 2025.
3. B. Alkin, A. Fürst, S. Schmid, L. Gruber, M. Holzleitner, and J. Brandstetter. Universal physics transformers: A framework for efficiently scaling neural operators. *Advances in Neural Information Processing Systems*, 37:25152–25194, 2024.
4. A. Baevski, Y. Zhou, A. Mohamed, and M. Auli. wav2vec 2.0: A framework for self-supervised learning of speech representations. *Advances in Neural Information Processing Systems*, 33:12449–12460, 2020.
5. N. Baker, F. Alexander, T. Bremer, A. Hagberg, Y. Kevrekidis, H. Najm, M. Parashar, A. Patra, J. Sethian, S. Wild, et al. Basic Research Needs Workshop for Scientific Machine Learning: Core Technologies for Artificial Intelligence. *Document prepared for Department of Energy Advanced Scientific Computing Research, USA*, 10, 2019.

6. N. Band, T. G. Rudner, Q. Feng, A. Filos, Z. Nado, M. W. Dusenberry, G. Jerfel, D. Tran, and Y. Gal. Benchmarking Bayesian deep learning on diabetic retinopathy detection tasks. *arXiv preprint arXiv:2211.12717*, 2022.
7. I. Batatia, D. P. Kovacs, G. Simm, C. Ortner, and G. Csányi. MACE: Higher Order Equivariant Message Passing Neural Networks for Fast and Accurate Force Fields. *Advances in Neural Information Processing Systems*, 35:11423–11436, 2022.
8. T. Bayes. An Essay towards Solving a Problem in the Doctrine of Chances. By the late Rev. Mr. Bayes, FRS communicated by Mr. Price, in a letter to John Canton, A.M.F.R.S. *Philosophical Transactions of the Royal Society of London*, 1763.
9. M. Beck, K. Pöppel, M. Spanring, A. Auer, O. Prudnikova, M. Kopp, G. Klambauer, J. Brandstetter, and S. Hochreiter. xLSTM: Extended long short-term memory. *Advances in Neural Information Processing Systems*, 37:107547–107603, 2024.
10. Y. Bengio, R. Ducharme, and P. Vincent. A Neural Probabilistic Language Model. *Advances in Neural Information Processing Systems*, 2000.
11. Y. Bengio, R. Ducharme, P. Vincent, and C. Jauvin. A neural probabilistic language model. *Journal of Machine Learning Research*, 3(Feb):1137–1155, 2003.
12. F. Bigi, S. Chong, A. Kristiadi, and M. Ceriotti. FlashMD: long-stride, universal prediction of molecular dynamics. *arXiv preprint arXiv:2505.19350*, 2025.
13. C. M. Bishop and N. M. Nasrabadi. *Pattern Recognition and Machine Learning*. Springer, 2006.
14. T. Blaschke, J. Arús-Pous, H. Chen, C. Margreitter, C. Tyrchan, O. Engkvist, K. Papadopoulos, and A. Patronov. REINVENT 2.0: an AI tool for de novo drug design. *Journal of Chemical Information and Modeling*, 60(12):5918–5922, 2020.
15. D. M. Blei, A. Kucukelbir, and J. D. McAuliffe. Variational inference: A review for statisticians. *Journal of the American Statistical Association*, 112(518):859–877, 2017.
16. C. Blundell, J. Cornebise, K. Kavukcuoglu, and D. Wierstra. Weight Uncertainty in Neural Networks. In *International Conference on Machine Learning*, pages 1613–1622. PMLR, 2015.
17. M. Bocquet, J. Brajard, A. Carrassi, and L. Bertino. Data assimilation as a learning tool to infer ordinary differential equation representations of dynamical models. *Nonlinear Processes in Geophysics*, 26(3):143–162, 2019.
18. C. Bodnar, W. P. Bruinsma, A. Lucic, M. Stanley, A. Allen, J. Brandstetter, P. Garvan, M. Riechert, J. A. Weyn, H. Dong, et al. A foundation model for the Earth system. *Nature*, pages 1–8, 2025.
19. J. Brandstetter, D. E. Worrall, and M. Welling. Message Passing Neural PDE Solvers. In *International Conference on Learning Representations*, 2022.
20. M. M. Bronstein, J. Bruna, Y. LeCun, A. Szlam, and P. Vandergheynst. Geometric deep learning: going beyond Euclidean data. *IEEE Signal Processing Magazine*, 34(4):18–42, 2017.
21. S. Brooks, A. Gelman, G. Jones, and X.-L. Meng. *Handbook of Markov Chain Monte Carlo*. CRC press, 2011.
22. T. Brown, B. Mann, N. Ryder, M. Subbiah, J. D. Kaplan, P. Dhariwal, A. Neelakantan, P. Shyam, G. Sastry, A. Askell, et al. Language models are few-shot learners. *Advances in Neural Information Processing Systems*, 33:1877–1901, 2020.
23. F. Cairoli, N. Paoletti, and L. Bortolussi. Conformal quantitative predictive monitoring of STL requirements for stochastic processes. In *Proceedings of the 26th ACM International Conference on Hybrid Systems: Computation and Control*, pages 1–11, 2023.
24. S. Chakraborty. Transfer learning based multi-fidelity physics informed deep neural network. *Journal of Computational Physics*, 426:109942, 2021.
25. P. Chen, T. Meng, Z. Zou, J. Darbon, and G. E. Karniadakis. Leveraging Hamilton-Jacobi PDEs with time-dependent Hamiltonians for Continual Scientific Machine Learning. In *6th Annual Learning for Dynamics & Control Conference*, pages 1–12. PMLR, 2024.
26. R. T. Chen, Y. Rubanova, J. Bettencourt, and D. K. Duvenaud. Neural ordinary differential equations. *Advances in Neural Information Processing Systems*, 31, 2018.
27. T. Chen, E. Fox, and C. Guestrin. Stochastic Gradient Hamiltonian Monte Carlo. In *International Conference on Machine Learning*, pages 1683–1691. PMLR, 2014.
28. Y. Choi, S. W. Cheung, Y. Kim, P.-H. Tsai, A. N. Diaz, I. Zanardi, S. W. Chung, D. M. Copeland, C. Kendrick, W. Anderson, et al. Defining Foundation Models for Computational Science: A Call for Clarity and Rigor. *arXiv preprint arXiv:2505.22904*, 2025.

29. T. Cohen and M. Welling. Group equivariant convolutional networks. In *International Conference on Machine Learning*, pages 2990–2999. PMLR, 2016.
30. D. Coscia, P. de Haan, and M. Welling. BLIPs: Bayesian Learned Interatomic Potentials. *arXiv preprint arXiv:2508.14022*, 2025.
31. D. Coscia, M. Welling, N. Demo, and G. Rozza. BARNN: A Bayesian Autoregressive and Recurrent Neural Network. In *International Conference on Machine Learning*, 2025.
32. S. Cuomo, V. S. Di Cola, F. Giampaolo, G. Rozza, M. Raissi, and F. Piccialli. Scientific machine learning through physics–informed neural networks: Where we are and what's next. *Journal of Scientific Computing*, 92(3):88, 2022.
33. F. de Avila Belbute-Peres, K. Smith, K. Allen, J. Tenenbaum, and J. Z. Kolter. End-to-end differentiable physics for learning and control. *Advances in Neural Information Processing Systems*, 31, 2018.
34. P. De Haan, M. Weiler, T. Cohen, and M. Welling. Gauge Equivariant Mesh CNNs: Anisotropic Convolutions on Geometric Graphs. *arXiv preprint arXiv:2003.05425*, 2020.
35. N. Demo, M. Tezzele, and G. Rozza. A DeepONet multi-fidelity approach for residual learning in reduced order modeling. *Advanced Modeling and Simulation in Engineering Sciences*, 10(1):12, 2023.
36. V. L. Deringer, A. P. Bartók, N. Bernstein, D. M. Wilkins, M. Ceriotti, and G. Csányi. Gaussian process regression for materials and molecules. *Chemical Reviews*, 121(16):10073–10141, 2021.
37. J. Dijkman, M. Dijkstra, R. Van Roij, M. Welling, J.-W. van de Meent, and B. Ensing. Learning neural free-energy functionals with pair-correlation matching. *Physical Review Letters*, 134(5):056103, 2025.
38. N. Ding, Y. Qin, G. Yang, F. Wei, Z. Yang, Y. Su, S. Hu, Y. Chen, C.-M. Chan, W. Chen, et al. Parameter-efficient fine-tuning of large-scale pre-trained language models. *Nature machine intelligence*, 5(3):220–235, 2023.
39. L. Dinh, D. Krueger, and Y. Bengio. Nice: Non-linear independent components estimation. *arXiv preprint arXiv:1410.8516*, 2014.
40. L. Dinh, J. Sohl-Dickstein, and S. Bengio. Density Estimation using Real NVP. *arXiv preprint arXiv:1605.08803*, 2016.
41. M. G. Dissanayake and N. Phan-Thien. Neural-network-based approximations for solving partial differential equations. *Communications in Numerical Methods in Engineering*, 10(3):195–201, 1994.
42. J. L. Doob. Application of the theory of Martingales. *Le Calcul des Probabilités et ses Applications*, pages 23–27, 1949.
43. A. Dosovitskiy, L. Beyer, A. Kolesnikov, D. Weissenborn, X. Zhai, T. Unterthiner, M. Dehghani, M. Minderer, G. Heigold, S. Gelly, et al. An image is worth 16×16 words: Transformers for Image Recognition at Scale. *arXiv preprint arXiv:2010.11929*, 2020.
44. G. A. D'Inverno, S. Moradizadeh, S. Salavatidezfouli, P. C. Africa, and G. Rozza. Mesh-informed Reduced Order Models for aneurysm rupture risk prediction. *Journal of Computational and Applied Mathematics*, page 116727, 2025.
45. L. C. Evans. *Partial Differential Equations*, volume 19. American Mathematical Society, 2022.
46. G. Evensen. Sequential data assimilation with a nonlinear quasi-geostrophic model using Monte Carlo methods to forecast error statistics. *Journal of Geophysical Research: Oceans*, 99(C5):10143–10162, 1994.
47. G. Evensen. *Data Assimilation: the Ensemble Kalman Filter*. Springer, 2009.
48. M. Fontana, G. Zeni, and S. Vantini. Conformal prediction: a unified review of theory and new challenges. *Bernoulli*, 29(1):1–23, 2023.
49. V. Fortuin. Priors in Bayesian Deep Learning: A Review. *International Statistical Review*, 90(3):563–591, 2022.
50. V. Fortuin, A. Garriga-Alonso, S. W. Ober, F. Wenzel, G. Rätsch, R. E. Turner, M. van der Wilk, and L. Aitchison. Bayesian neural network priors revisited. *arXiv preprint arXiv:2102.06571*, 2021.
51. S. Fresca, L. Dede', and A. Manzoni. A comprehensive deep learning-based approach to reduced order modeling of nonlinear time-dependent parametrized PDEs. *Journal of Scientific Computing*, 87(2):61, 2021.

52. X. Fu, B. M. Wood, L. Barroso-Luque, D. S. Levine, M. Gao, M. Dzamba, and C. L. Zitnick. Learning smooth and expressive interatomic potentials for physical property prediction. *arXiv preprint arXiv:2502.12147*, 2025.
53. Y. Gal and Z. Ghahramani. Dropout as a Bayesian Approximation: Representing Model Uncertainty in Deep Learning. In *International Conference on Machine Learning*, pages 1050–1059. PMLR, 2016.
54. Y. Gal, R. Islam, and Z. Ghahramani. Deep Bayesian Active Learning with image data. In *International Conference on Machine Learning*, pages 1183–1192. PMLR, 2017.
55. T. Garipov, P. Izmailov, D. Podoprikhin, D. P. Vetrov, and A. G. Wilson. Loss Surfaces, Mode Connectivity, and Fast Ensembling of DNNs. *Advances in Neural Information Processing Systems*, 31, 2018.
56. M. Ghavamzadeh, S. Mannor, J. Pineau, A. Tamar, et al. Bayesian Reinforcement Learning: A Survey. *Foundations and Trends® in Machine Learning*, 8(5-6):359–483, 2015.
57. J. Gilmer, S. S. Schoenholz, P. F. Riley, O. Vinyals, and G. E. Dahl. Neural Message Passing for Quantum Chemistry. In *International Conference on Machine Learning*, pages 1263–1272. PMLR, 2017.
58. I. J. Goodfellow, J. Pouget-Abadie, M. Mirza, B. Xu, D. Warde-Farley, S. Ozair, A. Courville, and Y. Bengio. Generative Adversarial Nets. *Advances in Neural Information Processing Systems*, 27, 2014.
59. S. Goswami, K. Kontolati, M. D. Shields, and G. E. Karniadakis. Deep transfer operator learning for partial differential equations under conditional shift. *Nature Machine Intelligence*, 4(12):1155–1164, 2022.
60. F. Grasselli, S. Chong, V. Kapil, S. Bonfanti, and K. Rossi. Uncertainty in the Era of Machine Learning for Atomistic Modeling. *Digital Discovery*, 4(10):2654–2675, 2025.
61. A. Graves. Practical Variational Inference for Neural Networks. In *Advances in Neural Information Processing Systems*, volume 24, 2011.
62. K. He, X. Zhang, S. Ren, and J. Sun. Deep residual learning for image recognition. In *Proceedings of the IEEE Conference on Computer Vision and Pattern Recognition*, pages 770–778, 2016.
63. P. Hennig, M. A. Osborne, and M. Girolami. Probabilistic numerics and uncertainty in computations. *Proceedings of the Royal Society A: Mathematical, Physical and Engineering Sciences*, 471(2179):20150142, 2015.
64. J. Hermann, Z. Schätzle, and F. Noé. Deep-neural-network solution of the electronic Schrödinger equation. *Nature Chemistry*, 12(10):891–897, 2020.
65. J. S. Hesthaven, G. Rozza, B. Stamm, et al. *Certified Reduced Basis Methods for Parametrized Partial Differential Equations*, volume 590. Springer, 2016.
66. G. Hinton, L. Deng, D. Yu, G. E. Dahl, A.-r. Mohamed, N. Jaitly, A. Senior, V. Vanhoucke, P. Nguyen, T. N. Sainath, et al. Deep neural networks for acoustic modeling in speech recognition: The shared views of four research groups. *IEEE Signal Processing Magazine*, 29(6):82–97, 2012.
67. G. E. Hinton and R. R. Salakhutdinov. Reducing the dimensionality of data with Neural Networks. *Science*, 313(5786):504–507, 2006.
68. G. E. Hinton and D. Van Camp. Keeping the Neural Networks Simple by Minimizing the Description Length of the Weights. In *Conference on Computational Learning Theory*, 1993.
69. M. D. Hoffman, D. M. Blei, C. Wang, and J. Paisley. Stochastic variational Inference. *Journal of Machine Learning Research*, 14(1):1303–1347, 2013.
70. L. Holdijk, Y. Du, F. Hooft, P. Jaini, B. Ensing, and M. Welling. Stochastic optimal control for collective variable free sampling of molecular transition paths. *Advances in Neural Information Processing Systems*, 36:79540–79556, 2023.
71. F. Holst, A. M. Bolger, F. Kindel, C. Günther, J. Maß, S. Triesch, N. Kiel, N. Saadat, O. Ebenhöh, B. Usadel, et al. Helixer: ab initio prediction of primary eukaryotic gene models combining deep learning and a hidden Markov model. *Nature Methods*, pages 1–8, 2025.
72. E. J. Hu, Y. Shen, P. Wallis, Z. Allen-Zhu, Y. Li, S. Wang, L. Wang, W. Chen, et al. Lora: Low-Rank Adaptation of Large Language Models. *ICLR*, 1(2):3, 2022.
73. G. Huang, Y. Li, G. Pleiss, Z. Liu, J. E. Hopcroft, and K. Q. Weinberger. Snapshot ensembles: Train 1, get M for free. *arXiv preprint arXiv:1704.00109*, 2017.
74. A. Ivagnes, N. Tonicello, P. Cinnella, and G. Rozza. Enhancing non-intrusive reduced-order models with space-dependent aggregation methods. *Acta Mechanica*, pages 1–30, 2024.

75. P. Izmailov, S. Vikram, M. D. Hoffman, and A. G. G. Wilson. What are Bayesian neural network posteriors really like? In *International Conference on Machine Learning*, pages 4629–4640. PMLR, 2021.
76. A. Jaegle, S. Borgeaud, J.-B. Alayrac, C. Doersch, C. Ionescu, D. Ding, S. Koppula, D. Zoran, A. Brock, E. Shelhamer, et al. Perceiver IO: A general architecture for structured inputs & outputs. *arXiv preprint arXiv:2107.14795*, 2021.
77. R. Jinnouchi, K. Miwa, F. Karsai, G. Kresse, and R. Asahi. On-the-fly active learning of interatomic potentials for large-scale atomistic simulations. *Journal of Physical Chemistry Letters*, 11(17):6946–6955, 2020.
78. M. I. Jordan, Z. Ghahramani, T. S. Jaakkola, and L. K. Saul. An introduction to variational methods for graphical models. *Machine Learning*, 37(2):183–233, 1999.
79. J. Jumper, R. Evans, A. Pritzel, T. Green, M. Figurnov, O. Ronneberger, K. Tunyasuvunakool, R. Bates, A. Žídek, A. Potapenko, et al. Highly accurate protein structure prediction with AlphaFold. *Nature*, 596(7873):583–589, 2021.
80. T. Kadeethum, D. O'Malley, Y. Choi, H. S. Viswanathan, and H. Yoon. Progressive transfer learning for advancing machine learning-based reduced-order modeling. *Scientific Reports*, 14(1):15731, 2024.
81. H. J. Kappen and H. C. Ruiz. Adaptive importance sampling for control and inference. *Journal of Statistical Physics*, 162(5):1244–1266, 2016.
82. G. E. Karniadakis, I. G. Kevrekidis, L. Lu, P. Perdikaris, S. Wang, and L. Yang. Physics-informed machine learning. *Nature Reviews Physics*, 3(6):422–440, 2021.
83. M. C. Kennedy and A. O'Hagan. Predicting the output from a complex computer code when fast approximations are available. *Biometrika*, 87(1):1–13, 2000.
84. M. C. Kennedy and A. O'Hagan. Bayesian calibration of computer models. *Journal of the Royal Statistical Society: Series B (Statistical Methodology)*, 63(3):425–464, 2001.
85. H. Kersting and P. Hennig. Active Uncertainty Calibration in Bayesian ODE Solvers. In *32nd Conference on Uncertainty in Artificial Intelligence (UAI 2016)*, pages 309–318. Curran Associates, Inc., 2016.
86. S. Kessler, A. Cobb, T. G. Rudner, S. Zohren, and S. J. Roberts. On sequential Bayesian inference for continual learning. *Entropy*, 25(6):884, 2023.
87. D. P. Kingma and M. Welling. Auto-encoding Variational Bayes. *arXiv preprint arXiv:1312.6114*, 2013.
88. T. Kipf. Semi-supervised classification with graph convolutional networks. *arXiv preprint arXiv:1609.02907*, 2016.
89. P. Kirichenko, P. Izmailov, and A. G. Wilson. Last Layer Re-Training is Sufficient for Robustness to Spurious Correlations. In *The Eleventh International Conference on Learning Representations*, 2023.
90. T. Korbak, E. Perez, and C. Buckley. RL with KL penalties is better viewed as Bayesian inference. In *Findings of the Association for Computational Linguistics: EMNLP 2022*, pages 1083–1091, 2022.
91. A. Krizhevsky, I. Sutskever, and G. E. Hinton. Imagenet classification with deep convolutional neural networks. *Advances in Neural Information Processing Systems*, 25, 2012.
92. M. Kulichenko, K. Barros, N. Lubbers, Y. W. Li, R. Messerly, S. Tretiak, J. S. Smith, and B. Nebgen. Uncertainty-driven dynamics for active learning of interatomic potentials. *Nature Computational Science*, 3(3):230–239, 2023.
93. M. Kulichenko, B. Nebgen, N. Lubbers, J. S. Smith, K. Barros, A. E. Allen, A. Habib, E. Shinkle, N. Fedik, Y. W. Li, et al. Data generation for machine learning interatomic potentials and beyond. *Chemical Reviews*, 124(24):13681–13714, 2024.
94. M. Kurz, A. Beck, and B. Sanderse. Harnessing equivariance: Modeling turbulence with graph neural networks. *arXiv preprint arXiv:2504.07741*, 2025.
95. M. Kurz, P. Offenhäuser, and A. Beck. Deep Reinforcement Learning for Turbulence Modeling in Large Eddy Simulations. *International Journal of Heat and Fluid Flow*, 99:109094, 2023.
96. B. Lakshminarayanan, A. Pritzel, and C. Blundell. Simple and Scalable Predictive Uncertainty Estimation using Deep Ensembles. *Advances in Neural Information Processing Systems*, 30, 2017.

97. L. D. Landau and E. M. Lifshitz. *Course of Theoretical Physics*. Elsevier, 2013.
98. T. Lassila, A. Manzoni, A. Quarteroni, and G. Rozza. Model order reduction in fluid dynamics: challenges and perspectives. *Reduced Order Methods for Modeling and Computational Reduction*, pages 235–273, 2014.
99. K. Lee and K. T. Carlberg. Model reduction of dynamical systems on nonlinear manifolds using deep convolutional autoencoders. *Journal of Computational Physics*, 404:108973, 2020.
100. S. Lee, H. Jeon, J. Son, and G. Kim. Learning to continually learn with the Bayesian principle. *arXiv preprint arXiv:2405.18758*, 2024.
101. H. Li, P. Barnaghi, S. Enshaeifar, and F. Ganz. Continual learning using Bayesian neural networks. *IEEE Transactions on Neural Networks and Learning Systems*, 32(9):4243–4252, 2020.
102. H. Li, Z. Wang, N. Zou, M. Ye, R. Xu, X. Gong, W. Duan, and Y. Xu. Deep-learning density functional theory Hamiltonian for efficient ab initio electronic-structure calculation. *Nature Computational Science*, 2(6):367–377, 2022.
103. R. Li, M. Klasson, A. Solin, and M. Trapp. Streamlining prediction in Bayesian deep learning. In *International Conference on Learning Representations*, 2024.
104. Z. Li, H. Zheng, N. Kovachki, D. Jin, H. Chen, B. Liu, K. Azizzadenesheli, and A. Anandkumar. Physics-informed neural operator for learning partial differential equations. *ACM/IMS Journal of Data Science*, 1(3):1–27, 2024.
105. Y. Lipman, R. T. Chen, H. Ben-Hamu, M. Nickel, and M. Le. Flow Matching for Generative Modeling. In *International Conference on Learning Representations*, 2022.
106. X. Liu, W. Yao, W. Peng, and W. Zhou. Bayesian physics-informed extreme learning machine for forward and inverse PDE problems with noisy data. *Neurocomputing*, 549:126425, 2023.
107. A. C. Lorenc. Analysis methods for numerical weather prediction. *Quarterly Journal of the Royal Meteorological Society*, 112(474):1177–1194, 1986.
108. C. Louizos, K. Ullrich, and M. Welling. Bayesian compression for deep learning. *Advances in Neural Information Processing Systems*, 30, 2017.
109. C. Louizos and M. Welling. Structured and efficient variational deep learning with matrix Gaussian posteriors. In *International Conference on Machine Learning*, pages 1708–1716. PMLR, 2016.
110. C. Louizos and M. Welling. Multiplicative normalizing flows for variational bayesian neural networks. In *International Conference on Machine Learning*, pages 2218–2227. PMLR, 2017.
111. D. J. Lucia, P. S. Beran, and W. A. Silva. Reduced-order modeling: new approaches for computational physics. *Progress in Aerospace Sciences*, 40(1-2):51–117, 2004.
112. W. J. Maddox, P. Izmailov, T. Garipov, D. P. Vetrov, and A. G. Wilson. A simple baseline for Bayesian uncertainty in deep learning. *Advances in Neural Information Processing Systems*, 32, 2019.
113. E. Magnani, N. Krämer, R. Eschenhagen, L. Rosasco, and P. Hennig. Approximate Bayesian neural operators: Uncertainty quantification for parametric PDEs. *arXiv preprint arXiv:2208.01565*, 2022.
114. E. Magnani, M. Pförtner, T. Weber, and P. Hennig. Linearization turns neural operators into function-valued Gaussian processes. *arXiv preprint arXiv:2406.05072*, 2024.
115. X. Meng, L. Yang, Z. Mao, J. del Águila Ferrandis, and G. E. Karniadakis. Learning functional priors and posteriors from data and physics. *Journal of Computational Physics*, 457:111073, 2022.
116. D. Musekamp, M. Kalimuthu, D. Holzmüller, M. Takamoto, and M. Niepert. Active Learning for Neural PDE Solvers. In *International Conference on Learning Representations*, 2025.
117. E. T. Nalisnick. *On priors for Bayesian neural networks*. PhD thesis, University of California, Irvine, 2018.
118. C. Nemeth and P. Fearnhead. Stochastic Gradient Markov Chain Monte Carlo. *Journal of the American Statistical Association*, 116(533):433–450, 2021.
119. M. Neumann, J. Gin, B. Rhodes, S. Bennett, Z. Li, H. Choubisa, A. Hussey, and J. Godwin. ORB: A fast, scalable neural network potential. *arXiv preprint arXiv:2410.22570*, 2024.
120. M. Olivecrona, T. Blaschke, O. Engkvist, and H. Chen. Molecular De-Novo Design through Deep Reinforcement Learning. *Journal of Cheminformatics*, 9(1):48, 2017.

121. E. Onal, K. Flöge, E. Caldwell, A. Sheverdin, and V. Fortuin. Gaussian stochastic weight averaging for Bayesian low-rank adaptation of large language models. *arXiv preprint arXiv:2405.03425*, 2024.
122. G. Ortali, N. Demo, and G. Rozza. A Gaussian process regression approach within a data-driven POD framework for engineering problems in fluid dynamics. *Mathematics in Engineering*, 4(3):1, 2022.
123. L. Ouyang, J. Wu, X. Jiang, D. Almeida, C. Wainwright, P. Mishkin, C. Zhang, S. Agarwal, K. Slama, A. Ray, et al. Training language models to follow instructions with human feedback. *Advances in Neural Information Processing Systems*, 35:27730–27744, 2022.
124. S. J. Pan and Q. Yang. A survey on Transfer Learning. *IEEE Transactions on Knowledge and Data Engineering*, 22(10):1345–1359, 2009.
125. T. Papamarkou, M. Skoularidou, K. Palla, L. Aitchison, J. Arbel, D. Dunson, M. Filippone, V. Fortuin, P. Hennig, J. M. Hernández-Lobato, et al. Position: Bayesian Deep Learning is Needed in the Age of Large-Scale AI. In *International Conference on Machine Learning*, pages 39556–39586. PMLR, 2024.
126. M. Pförtner, I. Steinwart, P. Hennig, and J. Wenger. Physics-informed Gaussian process regression generalizes linear PDE solvers. *arXiv preprint arXiv:2212.12474*, 2022.
127. F. Pichi, B. Moya, and J. S. Hesthaven. A graph convolutional autoencoder approach to model order reduction for parametrized PDEs. *Journal of Computational Physics*, 501:112762, 2024.
128. I. Price, A. Sanchez-Gonzalez, F. Alet, T. R. Andersson, A. El-Kadi, D. Masters, T. Ewalds, J. Stott, S. Mohamed, P. Battaglia, et al. Probabilistic weather forecasting with machine learning. *Nature*, 637(8044):84–90, 2025.
129. A. F. Psaros, X. Meng, Z. Zou, L. Guo, and G. E. Karniadakis. Uncertainty quantification in scientific machine learning: Methods, metrics, and comparisons. *Journal of Computational Physics*, 477:111902, 2023.
130. A. Quarteroni, P. Gervasio, and F. Regazzoni. Combining physics-based and data-driven models: advancing the frontiers of research with scientific machine learning. *arXiv preprint arXiv:2501.18708*, 2025.
131. A. Radford, K. Narasimhan, T. Salimans, and I. Sutskever. Improving Language Understanding by Generative Pre-Training. *OpenAI Blog*, 2018.
132. M. Raissi, P. Perdikaris, and G. E. Karniadakis. Physics-informed neural networks: A deep learning framework for solving forward and inverse problems involving nonlinear partial differential equations. *Journal of Computational physics*, 378:686–707, 2019.
133. K. Ram, J. Dijkman, R. Van Roij, J.-W. van de Meent, B. Ensing, M. Welling, and D. Cremers. Learned free-energy functionals from pair-correlation matching for dynamical density functional theory. *Physical Review Letters*, 112(4):045314, 2025.
134. J. Redmon, S. Divvala, R. Girshick, and A. Farhadi. You only look once: Unified, real-time object detection. In *Proceedings of the IEEE Conference on Computer Vision and Pattern Recognition*, pages 779–788, 2016.
135. S. Reich and C. Cotter. *Probabilistic forecasting and Bayesian data assimilation*. Cambridge University Press, 2015.
136. P. Ren, Y. Xiao, X. Chang, P.-Y. Huang, Z. Li, B. B. Gupta, X. Chen, and X. Wang. A survey of deep active learning. *ACM Romputing Surveys (CSUR)*, 54(9):1–40, 2021.
137. D. Rezende and S. Mohamed. Variational inference with normalizing flows. In *International Conference on Machine Learning*, pages 1530–1538. PMLR, 2015.
138. D. J. Rezende, S. Mohamed, and D. Wierstra. Stochastic backpropagation and approximate inference in deep generative models. In *International Conference on Machine Learning*, pages 1278–1286. PMLR, 2014.
139. B. Rhodes, S. Vandenhaute, V. Šimkus, J. Gin, J. Godwin, T. Duignan, and M. Neumann. ORB-v3: atomistic simulation at scale. *arXiv preprint arXiv:2504.06231*, 2025.
140. G. Rozza, G. Stabile, and F. Ballarin. *Advanced Reduced Order Methods and Applications in Computational Fluid Dynamics*. SIAM, 2022.
141. T. G. Rudner, Z. Chen, Y. W. Teh, and Y. Gal. Tractable function-space variational inference in Bayesian neural networks. *Advances in Neural Information Processing Systems*, 35:22686–22698, 2022.

142. A. Sanchez-Gonzalez, J. Godwin, T. Pfaff, R. Ying, J. Leskovec, and P. Battaglia. Learning to Simulate Complex Physics with Graph Networks. In *International Conference on Machine Learning*, 2020.
143. F. Scarselli, M. Gori, A. C. Tsoi, M. Hagenbuchner, and G. Monfardini. The graph neural network model. *IEEE Transactions on Neural Networks*, 20(1):61–80, 2008.
144. M. Schober, S. Särkkä, and P. Hennig. A probabilistic model for the numerical solution of initial value problems. *Statistics and Computing*, 29(1):99–122, 2019.
145. S. Schoenholz and E. D. Cubuk. JAX MD: a framework for differentiable physics. *Advances in Neural Information Processing Systems*, 33:11428–11441, 2020.
146. F. Seligmann, P. Becker, M. Volpp, and G. Neumann. Beyond deep ensembles: A large-scale evaluation of Bayesian deep learning under distribution shift. *Advances in Neural Information Processing Systems*, 36:29372–29405, 2023.
147. F. Sestak, A. Toshev, A. Fürst, G. Klambauer, A. Mayr, and J. Brandstetter. LaM-SLidE: Latent Space Modeling of Spatial Dynamical Systems via Linked Entities. *arXiv preprint arXiv:2502.12128*, 2025.
148. K. Simonyan and A. Zisserman. Very deep convolutional networks for large-scale image recognition. *arXiv preprint arXiv:1409.1556*, 2014.
149. J. Sohl-Dickstein, E. Weiss, N. Maheswaranathan, and S. Ganguli. Deep unsupervised learning using nonequilibrium thermodynamics. In *International Conference on Machine Learning*, pages 2256–2265. pmlr, 2015.
150. Y. Song, J. Sohl-Dickstein, D. P. Kingma, A. Kumar, S. Ermon, and B. Poole. Score-Based Generative Modeling through Stochastic Differential Equations. In *International Conference on Learning Representations*, 2020.
151. S. Subramanian, P. Harrington, K. Keutzer, W. Bhimji, D. Morozov, M. W. Mahoney, and A. Gholami. Towards foundation models for scientific machine learning: Characterizing scaling and transfer behavior. *Advances in Neural Information Processing Systems*, 36:71242–71262, 2023.
152. A. R. Tan, S. Urata, S. Goldman, J. C. Dietschreit, and R. Gómez-Bombarelli. Single-model uncertainty quantification in neural network potentials does not consistently outperform model ensembles. *npj Computational Materials*, 9(1):225, 2023.
153. Z. Tang, H. Li, P. Lin, X. Gong, G. Jin, L. He, H. Jiang, X. Ren, W. Duan, and Y. Xu. A deep equivariant neural network approach for efficient hybrid density functional calculations. *Nature Communications*, 15(1):8815, 2024.
154. S. Thaler, G. Doehner, and J. Zavadlav. Scalable Bayesian uncertainty quantification for neural network potentials: promise and pitfalls. *Journal of Chemical Theory and Computation*, 19(14):4520–4532, 2023.
155. A. Tharwat and W. Schenck. A survey on active learning: State-of-the-art, practical challenges and research directions. *Mathematics*, 11(4):820, 2023.
156. H. Touvron, T. Lavril, G. Izacard, X. Martinet, M.-A. Lachaux, T. Lacroix, B. Rozière, N. Goyal, E. Hambro, F. Azhar, et al. Llama: Open and efficient foundation language models. *arXiv preprint arXiv:2302.13971*, 2023.
157. B.-H. Tran, S. Rossi, D. Milios, and M. Filippone. All you need is a good functional prior for Bayesian deep learning. *Journal of Machine Learning Research*, 23(74):1–56, 2022.
158. O. T. Unke, S. Chmiela, H. E. Sauceda, M. Gastegger, I. Poltavsky, K. T. Schutt, A. Tkatchenko, and K.-R. Muller. Machine Learning Force Fields. *Chemical Reviews*, 121(16):10142–10186, 2021.
159. B. Uria, M.-A. Côté, K. Gregor, I. Murray, and H. Larochelle. Neural Autoregressive Distribution Estimation. *Journal of Machine Learning Research*, 2016.
160. C. van der Oord, M. Sachs, D. P. Kovács, C. Ortner, and G. Csányi. Hyperactive learning for data-driven interatomic potentials. *npj Computational Materials*, 9(1):168, 2023.
161. A. Vaswani, N. Shazeer, N. Parmar, J. Uszkoreit, L. Jones, A. N. Gomez, L. Kaiser, and I. Polosukhin. Attention is all you need. *Advances in Neural Information Processing Systems*, 30, 2017.
162. R. Vinuesa and S. L. Brunton. Enhancing computational fluid dynamics with machine learning. *Nature Computational Science*, 2(6):358–366, 2022.

163. M. Vladimirova, J. Verbeek, P. Mesejo, and J. Arbel. Understanding priors in Bayesian neural networks at the unit level. In *International Conference on Machine Learning*, pages 6458–6467. PMLR, 2019.
164. N. Vlassis, M. Ghavamzadeh, S. Mannor, and P. Poupart. Bayesian Reinforcement Learning. *Reinforcement Learning: State-of-the-Art*, pages 359–386, 2012.
165. M. J. Wainwright, M. I. Jordan, et al. Graphical models, exponential families, and variational inference. *Foundations and Trends® in Machine Learning*, 1(1–2):1–305, 2008.
166. H. Wang, T. Fu, Y. Du, W. Gao, K. Huang, Z. Liu, P. Chandak, S. Liu, P. Van Katwyk, A. Deac, et al. Scientific discovery in the age of artificial intelligence. *Nature*, 620(7972):47–60, 2023.
167. H. Wang, R. Planas, A. Chandramowlishwaran, and R. Bostanabad. Mosaic flows: A transferable deep learning framework for solving PDEs on unseen domains. *Computer Methods in Applied Mechanics and Engineering*, 389:114424, 2022.
168. J. Wang, J. Cockayne, O. Chkrebtii, T. J. Sullivan, and C. J. Oates. Bayesian numerical methods for nonlinear partial differential equations. *Statistics and Computing*, 31(5):55, 2021.
169. L. Wang, X. Zhang, H. Su, and J. Zhu. A Comprehensive Survey of Continual Learning: Theory, Method and Application. *IEEE Transactions on Pattern Analysis and Machine Intelligence*, 46(8):5362–5383, 2024.
170. T. Wang, X. He, M. Li, Y. Li, R. Bi, Y. Wang, C. Cheng, X. Shen, J. Meng, H. Zhang, et al. Ab initio characterization of protein molecular dynamics with AI2BMD. *Nature*, 635(8040):1019–1027, 2024.
171. X. Wang, L. Aitchison, and M. Rudolph. LoRA Ensembles for Large Language Model Fine-tuning. *arXiv preprint arXiv:2310.00035*, 2023.
172. X. Wang, Y. Jin, S. Schmitt, and M. Olhofer. Recent advances in Bayesian optimization. *ACM Computing Surveys*, 55(13s):1–36, 2023.
173. M. Weiler, P. Forré, E. Verlinde, and M. Welling. Coordinate Independent Convolutional Networks–Isometry and Gauge Equivariant Convolutions on Riemannian Manifolds. *arXiv preprint arXiv:2106.06020*, 2021.
174. M. Weiler, P. Forré, E. Verlinde, and M. Welling. Equivariant and coordinate independent convolutional networks. *A Gauge Field Theory of Neural Networks*, page 110, 2023.
175. K. Weiss, T. M. Khoshgoftaar, and D. Wang. A survey of transfer learning. *Journal of Big Data*, 3(1):9, 2016.
176. M. Welling and Y. W. Teh. Bayesian Learning via Stochastic Gradient Langevin Dynamics. In *International Conference on Machine Learning (ICML-11)*, pages 681–688, 2011.
177. J. Wenger and P. Hennig. Probabilistic linear solvers for machine learning. *Advances in Neural Information Processing Systems*, 33:6731–6742, 2020.
178. J. Wenger, G. Pleiss, M. Pförtner, P. Hennig, and J. P. Cunningham. Posterior and computational uncertainty in Gaussian processes. *Advances in Neural Information Processing Systems*, 35:10876–10890, 2022.
179. C. Williams and C. Rasmussen. Gaussian processes for regression. *Advances in neural information processing systems*, 8, 1995.
180. A. G. Wilson and P. Izmailov. Bayesian deep learning and a probabilistic perspective of generalization. *Advances in Neural Information Processing Systems*, 33:4697–4708, 2020.
181. B. M. Wood, M. Dzamba, X. Fu, M. Gao, M. Shuaibi, L. Barroso-Luque, K. Abdelmaqsoud, V. Gharakhanyan, J. R. Kitchin, D. S. Levine, et al. UMA: A Family of Universal Models for Atoms. *arXiv preprint arXiv:2506.23971*, 2025.
182. C. Xu, B. T. Cao, Y. Yuan, and G. Meschke. Transfer learning based physics-informed neural networks for solving inverse problems in engineering structures under different loading scenarios. *Computer Methods in Applied Mechanics and Engineering*, 405:115852, 2023.
183. A. X. Yang, M. Robeyns, X. Wang, and L. Aitchison. Bayesian Low-Rank Adaptation for Large Language Models. *arXiv preprint arXiv:2308.13111*, 2023.
184. L. Yang, S. Liu, T. Meng, and S. J. Osher. In-context operator learning with data prompts for differential equation problems. *Proceedings of the National Academy of Sciences*, 120(39):e2310142120, 2023.
185. L. Yang, X. Meng, and G. E. Karniadakis. B-PINNs: Bayesian physics-informed neural networks for forward and inverse PDE problems with noisy data. *Journal of Computational Physics*, 425:109913, 2021.

186. Y. Yang, S. Mandt, L. Theis, et al. An introduction to neural data compression. *Foundations and Trends® in Computer Graphics and Vision*, 15(2):113–200, 2023.
187. D. Ye, W. Yan, C. Brune, and M. Guo. PDE-constrained Gaussian process surrogate modeling with uncertain data locations. *Advanced Modeling and Simulation in Engineering Sciences*, 12(1):33, 2025.
188. J. Yosinski, J. Clune, Y. Bengio, and H. Lipson. How transferable are features in deep neural networks? *Advances in Neural Information Processing Systems*, 27, 2014.
189. V. Zaverkin, D. Holzmüller, H. Christiansen, F. Errica, F. Alesiani, M. Takamoto, M. Niepert, and J. Kästner. Uncertainty-biased molecular dynamics for learning uniformly accurate interatomic potentials. *npj Computational Materials*, 10(1):83, 2024.
190. B. J. Zhang, S. Liu, S. J. Osher, and M. A. Katsoulakis. Probabilistic operator learning: generative modeling and uncertainty quantification for foundation models of differential equations. *arXiv preprint arXiv:2509.05186*, 2025.
191. R. Zhang, C. Li, J. Zhang, C. Chen, and A. G. Wilson. Cyclical stochastic gradient MCMC for Bayesian deep learning. *arXiv preprint arXiv:1902.03932*, 2019.
192. M. Zhu, H. Zhang, A. Jiao, G. E. Karniadakis, and L. Lu. Reliable extrapolation of deep neural operators informed by physics or sparse observations. *Computer Methods in Applied Mechanics and Engineering*, 412:116064, 2023.
193. Y. Zhu and N. Zabaras. Bayesian deep convolutional encoder–decoder networks for surrogate modeling and uncertainty quantification. *Journal of Computational Physics*, 366:415–447, 2018.

Numerical Methods for Optimal Decumulation of a Defined Contribution Pension Plan

Peter A. Forsyth and George Labahn

Abstract The decumulation of a defined contribution (DC) pension plan is well known to be one of the hardest problems in finance. We model this decumulation challenge as an optimal stochastic control problem. The control problem is solved, at each rebalancing date, by alternatively solving a linear partial-integro differential equation (PIDE) followed by an optimization step. We solve the PIDE by using a δ-monotone Fourier method, which ensures that monotonicity holds to $O(\delta)$. We allow for the use of leverage (i.e., borrowing to invest in stocks), as well as minimum constraints on bond holdings. We pay particular attention to minimizing wrap-around error, an issue which is endemic for Fourier methods and central to the effective use of these methods for optimal control problems. Rather unexpectedly, we find that restricting the portfolio equity fraction to a maximum of 50% does not reduce portfolio efficiency noticeably. This may be a useful strategy for risk-averse retirees.

1 Introduction

In developed countries, there is a strong and growing shift from Defined Benefit (DB) pension plans to Defined Contribution (DC) plans. The private sector, and increasingly government institutions, are unwilling to take on the balance sheet liabilities of a DB plan (Thinking Ahead Institute, 2024).

Under a DC plan, the employer and employee contribute to a (usually) tax advantaged account, which is typically invested in a mix of stocks and bonds. Upon retirement, the employee is now responsible for both choosing an investment strategy and deciding on a

Peter A. Forsyth
David R. Cheriton School of Computer Science, University of Waterloo, Waterloo ON, Canada N2L 3G1, paforsyt@uwaterloo.ca

George Labahn
David R. Cheriton School of Computer Science, University of Waterloo, Waterloo ON, Canada N2L 3G1, glabahn@uwaterloo.ca

B. Koren and D. Roose (eds.), *Numerical Analysis and Scientific Computing: an Anthology*, Lecture Notes in Computational Science and Engineering 156,
https://doi.org/10.1007/978-3-032-33428-2_12

withdrawal schedule. Although it is often advocated that retirees should buy annuities, these are unpopular with DC plan holders (Peijnenburg et al., 2016) for numerous valid reasons (MacDonald et al., 2013). In particular, annuities are generally overpriced, lack true inflation protection and retirees have no access to capital in the event of emergencies.

Since it is usually challenging to fund a reasonable lifestyle investing solely in riskless assets, the retiree is now exposed to both investment risk and longevity risk. This problem of funding retirement from accumulated DC capital, also known as the decumulation problem, has been termed *"the nastiest, hardest problem in finance,"* by Nobel Laureate William Sharpe (quoted in (Ritholz, 2017)).

There have been several suggestions for DC plan *spending rules* (Anarkulova et al., 2025). Probably the most ubiquitous rule is the *four per cent rule* (Bengen, 1994). This rule suggests that retirees can withdraw four per cent of their initial capital (at age 65) annually, adjusted for inflation. Based on historical US data, an investor who invested in a portfolio of 50% bonds and 50% stocks (rebalanced annually), and followed the four per cent rule, would have never run out of savings over any thirty year historical period.

However, the four per cent rule has many critics. For example, rolling thirty year periods have high correlations. In addition, a fixed withdrawal schedule, as well as a constant weight stock allocation is somewhat simplistic. In order to better stress test spending rules, Anarkulova et al. (2025) use block bootstrap resampling of historical data, to generate a distribution of outcomes. We will also use block bootstrap resampling to test our results in this chapter.

Decumulation strategies are naturally posed as a problem in optimal stochastic control. The controls in this case are the stock/bond split on each rebalancing date and the (real) annual withdrawal amounts. In our case we impose maximum and minimum constraints on the withdrawal amounts and also impose constraints on the maximum stock allocation.

1.1 Our Contributions

In this chapter, we consider the same scenario posed in Bengen (1994). Our objective function is to maximize total withdrawals over a thirty year period (with minimum/maximum annual constraints) and minimize the risk of running out of savings before year thirty.

An optimal stochastic control problem used for decumulation is solved at each rebalancing date, by alternatively solving a linear partial-integro differential equation (PIDE) followed by an optimization step. The PIDE is solved by using a δ-monotone Fourier method, which ensures that monotonicity holds to $O(\delta)$, an important property for control problems. The $\delta-$monotone technique generalizes the method described in Forsyth and Labahn (2019). Our basic dynamic programming technique is similar to the high-level description in Forsyth (2022), but here we delve much deeper in describing the numerical methods. In particular, we focus on such issues as wrap-around errors in Fourier methods for Partial-Integro Differential Equations (PIDEs), and the 2-d δ-monotone Green's function approximation.

Previous studies for decumulation have imposed no-shorting and no-leverage constraints on stock investments. In this chapter, we relax the no-leverage constraint on the stock investment and also investigate more restrictive investments in stocks. For example, a risk averse retiree may wish to restrict the maximum fraction invested in equities to be significantly less than one.

Rather surprisingly, we find that restricting the maximum fraction in stocks to be 50% does not lower the efficient frontier by a large amount. The most significant effect of the 50% restriction, compared to allowing a maximum stock fraction of up to 130% (leverage), is that the time for the median withdrawals to reach the maximum is delayed by one year. This may be an acceptable trade-off for risk averse investors.

From the numerical point of view, when solving optimal control problems using Fourier methods, wrap-around error can be significant, due to the possible instantaneous movement to domain boundaries at reallocation times (Lippa, 2013; Ignatieva et al., 2018; Alonso-Garcia et al., 2018). We show that our technique reduces wrap-around error to almost machine level precision.

1.2 Organization of this Chapter

The rest of this chapter is organized as follows. In the next section we state our problem setting followed by a mathematical formulation of our model in Section 3. The latter section includes information on the stochastic processes followed by bonds and stocks along with a description of our control set. Section 4 describes our conflicting risk and reward measures, Expected Shortfall and Expected Withdrawals, while Section 5 looks at how one maximizes these conflicting measures, something which is initially set up as a pre-commitment problem which is not implementable. Section 6 then discusses how one can formulate our problem into an (implementable) dynamic program, one which is solved by a PIDE between withdrawal time intervals. Section 7 then shows how one can solve these PIDEs using Green's functions and Fourier analysis. The wrap-around problem that comes with using Fourier analysis and periodicity is addressed in the next section. That section also includes the resulting numerical algorithm for solving our optimal control decumulation problem. Section 9 then presents an example along with its numerical solution, followed in the next section by tests of robustness of our numerical example. The paper ends with a conclusion along with an Appendix providing extra details regarding the wrap-around error discussed earlier.

2 Problem Setting

As noted in Anarkulova et al. (2025), retirees have a revealed preference for spending rules, such as that advocated by Bengen (1994). Spending rules (such as the four per cent rule) are popular with retirees as they are simple to implement, have an intuitive interpretation, and the rolling 30 year backtests in Bengen (1994) are easy to understand.

In this paper we restrict attention to the scenario outlined in Bengen (1994), that is, we consider a 65-year old retiree who desires fixed minimum annual (real) cash flows over a 30 year time horizon. We also impose an upper limit (a cap) on maximum withdrawals in any year. From the CPM2014 table from the Canadian Institute of Actuaries[1], the probability that a 65-year old Canadian male attains the age of 95 is about 0.13. In spite of this relatively low probability of achieving the age of 95, assuming a 30 year time horizon is considered a prudent approach for minimizing the risk of exhausting savings. In addition, observe that we will not mortality weight future cash flows, as is done when averaging over a population for pricing annuities. Mortality weighting may not be meaningful for an individual investor, since it reflects population averages, rather than the circumstances of an individual retiree.

Since we allow investing in risky assets, with a minimum cash withdrawal each year, it is possible to exhaust savings.[2] In this case, we continue to withdraw funds from the portfolio, which is equivalent to borrowing cash. In such circumstances, we assume that withdrawals continue, something which can be viewed as borrowing against other resources. This debt accumulates at the borrowing rate. Essentially, we are assuming that the investor has other assets, for example real estate, which can be used as a hedge of last resort. In this case, this debt could then be funded using a reverse mortgage, with real estate as collateral (Pfeiffer et al., 2013).

It is important to note that an implicit assumption here is that real estate is not fungible with other financial assets, except as a last resort. This mental separation of assets is a common aspect of behavioral finance (Shefrin and Thaler, 1988). Real estate in this case has a dual role: if market returns are exceptionally strong, or the retiree passes away earlier than expected, this property can be a bequest. If market returns are poor, or in the case of extreme longevity, then real estate can be used to fund expenses if the retirement account is exhausted.

3 Mathematical Formulation

We assume that the investor has access to two funds: a broad market stock index fund and a constant maturity bond index fund. Let $S(t)$ and $B(t)$ denote the real (inflation adjusted) *amounts* invested in the stock index and the bond index, respectively, at time t with T being the investment horizon. These amounts will evolve over time, depending on the investor's asset allocation, and changes in the real unit prices of the assets. In the absence of an investor determined control such as cash withdrawals or rebalancing, any changes in $S(t)$ and $B(t)$ result from asset price evolution. We model the stock index as following a jump diffusion, which permits modelling both continuous stochastic processes and market jumps.

[1] `www.cia-ica.ca/docs/default-source/2014/214013e.pdf`.

[2] Ignoring the trivial and unlikely case where investing in riskless assets can fund 30 year minimum withdrawals.

We model the real returns of a constant maturity bond index as a stochastic process (see for example, Lin et al., 2015; MacMinn et al., 2014). This avoids the need to model bond prices and inflation separately. As done in MacMinn et al. (2014), we assume that the constant maturity bond index follows a jump diffusion process, with justification being found in Forsyth et al. (2022, Appendix A).

3.1 The Jump Diffusion Processes

For any time t let

$$S(t^-) \equiv \lim_{\epsilon \to 0^+} S(t-\epsilon) \quad \text{and} \quad S(t^+) \equiv \lim_{\epsilon \to 0^+} S(t+\epsilon)$$

denote the values of S the instant before and the instant after time t. Such notation will also be used for any other time-dependent function. We let ξ^s be a random number representing a jump multiplier, that is, when a jump occurs, $S(t) = \xi^s S(t^-)$. The use of a jump process accounts for non-normal asset returns. As in (Kou, 2002; Kou and Wang, 2004), we assume that $\log(\xi^s)$ follows a double exponential distribution with u^s the probability of an upward jump and $1 - u^s$ the probability of a downward jump. The density function for $y = \log \xi^s$ is

$$f^s(y) = u^s \eta_1^s e^{-\eta_1^s y} \mathbf{1}_{y \geq 0} + (1-u^s)\eta_2^s e^{\eta_2^s y} \mathbf{1}_{y<0} \quad ; \quad \mathbf{1}_C = \begin{cases} 1, & C \text{ is true.} \\ 0, & \text{otherwise.} \end{cases} \tag{1}$$

We also define

$$\gamma_\xi^s = E[\xi^s - 1] = \frac{u^s \eta_1^s}{\eta_1^s - 1} + \frac{(1-u^s)\eta_2^s}{\eta_2^s + 1} - 1\,, \tag{2}$$

where $E[\cdot]$ is the expectation. In the absence of control, $S(t)$ evolves according to

$$\frac{dS(t)}{S(t^-)} = \left(\mu^s - \lambda_\xi^s \gamma_\xi^s\right) dt + \sigma^s \, dZ^s + d\left(\sum_{i=1}^{\pi_t^s} (\xi_i^s - 1)\right), \tag{3}$$

where μ^s is the (uncompensated) drift rate, σ^s is the volatility, dZ^s is the increment of a Wiener process, π_t^s is a Poisson process with positive intensity parameter λ_ξ^s, and the ξ_i^s are i.i.d. positive random variables having distribution (1). Moreover, ξ_i^s, π_t^s, and Z^s are assumed to all be mutually independent.

Similarly, let the amount in the bond index the instant before t be $B(t^-)$. In the absence of investor intervention, $B(t)$ evolves as

$$\frac{dB(t)}{B(t^-)} = \left(\mu^b - \lambda_\xi^b \gamma_\xi^b + \mu_c^b \mathbf{1}_{\{B(t^-)<0\}}\right) dt + \sigma^b \, dZ^b + d\left(\sum_{i=1}^{\pi_t^b} (\xi_i^b - 1)\right), \tag{4}$$

where the terms in equation (4) are defined analogously to equation (3). In particular, π_t^b is a Poisson process with positive intensity parameter λ_ξ^b, and the density function for $y = \log \xi^b$ is

$$f^b(y) = u^b \eta_1^b e^{-\eta_1^b y} \mathbf{1}_{y \geq 0} + (1 - u^b) \eta_2^b e^{\eta_2^b y} \mathbf{1}_{y<0} \,,$$

and $\gamma_\xi^b = E[\xi^b - 1]$. Again ξ_i^b, π_t^b, and Z^b are assumed to all be mutually independent. The term $\mu_c^b \mathbf{1}_{\{B(t^-)<0\}}$ in equation (4) represents the extra cost of borrowing, that is, the spread.

The diffusion processes are correlated, that is, $dZ^s \cdot dZ^b = \rho_{sb} \, dt$. However, contrary to common belief, an analysis of historical data suggests that the stock and bond jump processes are essentially uncorrelated (see Forsyth (2020b) for empirical justification). We make this assumption in this work.

We define the investor's total wealth at time t as

$$\text{Total wealth} \equiv W(t) = S(t) + B(t).$$

In case of insolvency, the portfolio is liquidated, trading stops and any outstanding debt accrues interest at the borrowing rate.

3.2 The Set of Controls

Consider a set of discrete times

$$\mathcal{T} = \{t_0 = 0 < t_1 < t_2 < \ldots < t_M = T\},$$

where we assume that $t_i - t_{i-1} = \Delta t = T/M$ is constant. Here $t_0 = 0$ is the inception time of the investment and $\mathcal{T}$ is the set of withdrawal/rebalancing times, as defined in equation (3.2). At each rebalancing time t_i, $i = 0, 1, \ldots, M-1$, the investor first withdraws an amount of cash q_i from the portfolio, and then afterwards rebalances the portfolio. At $t_M = T$ the portfolio is liquidated and no cash flow occurs. This is enforced by specifying $q_M = 0$.

Let

$$W(t_i^-) = S(t_i^-) + B(t_i^-)$$

denote the total portfolio wealth the instant before withdrawals and rebalancing at t_i, we then have that $W(t_i^+)$ is given by

$$W(t_i^+) = W(t_i^-) - q_i \ ; \ i \in \mathcal{T} \ ,$$

with $W(t_M^+) = W(t_M^-)$ since $q_M \equiv 0$.

Typically, DC plan savings are held in a tax-advantaged account, with no taxes triggered by rebalancing. With infrequent (e.g., yearly) rebalancing, we also expect other transaction costs to be small, and hence they can be ignored. It is possible to include transaction costs, but at the expense of increased computational cost (Van Staden et al., 2018).

Let $X(t) = (S(t), B(t))$, $t \in [0,T]$ denote the multi-dimensional controlled underlying process and $x = (s, b)$ the realized state of the system. The rebalancing control $p_i(\cdot)$ is the fraction invested in the stock index at the rebalancing date t_i, that is,

$$p_i\left(X(t_i^-)\right) = p\left(X(t_i^-), t_i\right) = \frac{S(t_i^+)}{S(t_i^+) + B(t_i^+)} \ .$$

Let $q_i(\cdot)$ be the amount withdrawn at time t_i, that is, $q_i\left(X(t_i^-)\right) = q\left(X(t_i^-), t_i\right)$. Formally, the controls depend on the state of the investment portfolio, before the rebalancing occurs, that is, $p_i(\cdot) = p\left(X(t_i^-), t_i\right)$ and $q_i(\cdot) = q\left(X(t_i^-), t_i\right)$, $t_i \in \mathcal{T}$, where $\mathcal{T}$ is the set of rebalancing times. However, it will be convenient to note that in our case, we find the optimal control $p_i(\cdot)$ amongst all strategies with constant wealth (after withdrawal of cash). Hence, with some abuse of notation, we will now consider $p_i(\cdot)$ to be a function of wealth after withdrawal of cash, where we use the shorthand notation W_i^- and W_i^+ for the variables representing wealth the instant before and instant after t_i. Using a similar shorthand notation for S_i and B_i we then have

$$\begin{aligned} p_i(\cdot) &= p(W(t_i^+), t_i) = p_i(W_i^+) \ , \\ S_i^+ &= p_i(W_i^+) \ W_i^+ \ , \\ B_i^+ &= (1 - p_i(W_i^+)) \ W_i^+ \ . \end{aligned}$$

Note that the control for $p_i(\cdot)$ depends only on W_i^+. Since $p_i(\cdot) = p_i(W_i^- - q_i)$, it follows that

$$q_i(\cdot) = q_i(W_i^-) \ ,$$

which we discuss further in Section 6.

A control at time t_i is then given by the pair $(q_i(\cdot), p_i(\cdot))$ where the notation $(\cdot)$ denotes that the control is a function of the state. Let $\mathcal{Z}$ represent the set of admissible values of the controls $(q_i(\cdot), p_i(\cdot))$. We impose bounded leverage constraints (assuming solvency) along with maximum and minimum values for the withdrawals. In addition we apply the constraint that in the event of insolvency due to withdrawals ($W(t_i^+) < 0$), or in the case of leverage, trading ceases and debt (negative wealth) accumulates at the appropriate borrowing rate of return (that is, a spread over the bond rate). We also specify that the stock assets are liquidated at $t = t_M$.

We can then define our controls by

$$\mathcal{Z}_q(W_i^-, t_i) = \begin{cases} [q_{\min}, q_{\max}] & t_i \in \mathcal{T} \; ; \; t_i \neq t_M \; ; \; W_i^- \geq q_{\max} \; , \\ [q_{\min}, \max(q_{\min}, W_i^-)] & t_i \in \mathcal{T} \; ; \; t \neq t_M \; ; \; W_i^- < q_{\max} \; , \\ \{0\} & t_i = t_M \; , \end{cases} \tag{5}$$

$$\mathcal{Z}_p(W_i^+, t_i) = \begin{cases} [0, p_{\max}] & t_i \in \mathcal{T} \; ; \; t_i \neq t_M \; ; \; W_i^+ > 0 \; . \\ \{0\} & t_i \in \mathcal{T} \; ; \; t_i \neq t_M \; ; \; W_i^+ \leq 0 \; . \\ \{0\} & t_i = t_M \; . \end{cases} \tag{6}$$

The rather complicated expression in equation (5) imposes the assumption that as wealth becomes small, the retiree first tries to avoid insolvency as much as possible and in the event of insolvency, withdraws only $q_{\min}$.

The set of admissible values for $(q_i, p_i), t_i \in \mathcal{T}$, can then be written as

$$(q_i, p_i) \in \mathcal{Z}(W_i^-, W_i^+, t_i) = \mathcal{Z}_q(W_i^-, t_i) \times \mathcal{Z}_p(W_i^+, t_i) \; . \tag{7}$$

For implementation purposes, we have written equation (7) in terms of the wealth after withdrawal of cash. However, we remind the reader that since $W_i^+ = W_i^- - q_i$, the controls are formally a function of the state $X(t_i^-)$ before the control is applied.

The admissible control set $\mathcal{A}$ can then be written as

$$\mathcal{A} = \left\{ (q_i, p_i)_{0 \leq i \leq M} : (p_i, q_i) \in \mathcal{Z}(W_i^-, W_i^+, t_i) \right\} ,$$

with an admissible control $\mathcal{P} \in \mathcal{A}$ written as

$$\mathcal{P} = \{ (q_i(\cdot), p_i(\cdot)) \; : \; i = 0, \ldots, M \} \; .$$

We also define $\mathcal{P}_n \equiv \mathcal{P}_{t_n} \subset \mathcal{P}$ as the tail of the set of controls in $[t_n, t_{n+1}, \ldots, t_M]$, that is,

$$\mathcal{P}_n = \{ (q_n(\cdot), p_n(\cdot)), \ldots, (q_M(\cdot), p_M(\cdot)) \}$$

and $\mathcal{A}_n$, the tail of the admissible control set, as

$$\mathcal{A}_n = \left\{ (q_i, p_i)_{n \leq i \leq M} : (q_i, p_i) \in \mathcal{Z}(W_i^-, W_i^+, t_i) \right\} ,$$

so that $\mathcal{P}_n \in \mathcal{A}_n$.

4 Risk and Reward

In this section we consider our measures of risk and reward for our retiree. In this case the retiree is primarily concerned with the risk of depleting savings while at the same time hoping to maximize the cash withdrawals from his plan.

4.1 Risk: Expected Shortfall (ES)

Two typical measures of financial tail risk of an investment portfolio are Value at Risk (VAR) and Conditional Value of risk (CVAR), with the latter also known as Expected Shortfall. Expected Shortfall is an alternative to value at risk that is more sensitive to the shape of the tail of the loss distribution.

Suppose

$$\int_{-\infty}^{W_\alpha^*} g(W_T)\, dW_T = \alpha,$$

where $g(W_T)$ is the probability density function of wealth W_T at $t = T$. Then W_α^* can be viewed as the Value at Risk (VAR) at level α since the probability of wealth being more than W_α^* is $1 - \alpha$. For example, if $\alpha = .01$, then 99% of the outcomes have $W_T > W_\alpha^*$. If W_α^* is sufficiently large, this suggests very low risk of running out of savings. The Expected Shortfall (ES) at level α is then

$$\mathrm{ES}_\alpha = \frac{1}{\alpha} \int_{-\infty}^{W_\alpha^*} W_T\, g(W_T)\, dW_T\ , \tag{8}$$

which is the mean of the worst α fraction of outcomes. Typically $\alpha \in \{.01, .05\}$. The definition of ES in equation (8) uses the probability density of the final wealth distribution, not the density of *loss*. Hence in our case, a larger value of ES, that is, a larger value of average worst case terminal wealth, is desired. [3]

Define $X_0^- = X(t_0^-)$. Given an expectation under control $\mathcal{P}$, $E_{\mathcal{P}}[\cdot]$, then as noted by Rockafellar and Uryasev (2000), the expected shortfall at level α has the alternate formulation as

$$\mathrm{ES}_\alpha(X_0^-, t_0^-) = \sup_{W^*} E_{\mathcal{P}_0}^{X_0^-, t_0^-}\left[W^* + \frac{1}{\alpha} \min(W_T - W^*, 0) \right] . \tag{9}$$

The admissible set for W^* in equation (9) is over the set of possible values for W_T.

The notation $\mathrm{ES}_\alpha(X_0^-, t_0^-)$ emphasizes that ES_α is as seen at (X_0^-, t_0^-), that is, the pre-commitment ES_α. A strategy based purely on optimizing the pre-commitment value of ES_α at time zero is *time-inconsistent*. As such it has been termed by many as *non-implementable*, since the investor has an incentive to deviate from the time zero pre-commitment strategy at $t > 0$. However, we consider the pre-commitment strategy merely as a device to determine an appropriate level of W^* in equation (9). If we fix W^* for all positive t, then this strategy is the induced time-consistent strategy (Strub et al., 2019; Forsyth, 2020a; Cui et al., 2022) and hence is implementable. For further discussion of the relationship between time consistent and pre-commitment strategies, see (Vigna, 2014; Menoncin and Vigna, 2017; Vigna, 2022; Strub et al., 2019; Forsyth, 2020a; Bjork et al., 2021; Cui et al., 2022). In particular, see Forsyth (2020a) for discussion of the induced time consistent policy resulting from the use of ES risk.

[3] In practice, the negative of W_α^* is often the reported VAR. and the negative of ES is commonly referred to as Conditional Value at Risk (CVAR).

An alternative measure of risk could be based on variability of withdrawals (Forsyth et al., 2020). However, we note that we have constraints on the minimum and maximum withdrawals, so that variability is mitigated. We also assume that given these constraints, the retiree is primarily concerned with the risk of depleting savings, something which is well measured by ES.

4.2 Reward: Expected Total Withdrawals (EW)

We will use expected total withdrawals as a measure of reward in the following. More precisely, we define EW (expected withdrawals) as

$$\text{EW}(X_0^-, t_0^-) = E_{\mathcal{P}_0}^{X_0^+, t_0^+}\left[\sum_{i=0}^{M} q_i\right], \tag{10}$$

where we assume that the investor survives for the entire decumulation period. This is consistent with the scenario in Bengen (1994) .

We remark that there is no discounting term in equation (10) as all quantities are real, that is, inflation-adjusted. It is straightforward to introduce discounting, but we view setting the real discount rate to zero to be a reasonable and conservative choice. See Forsyth (2022) for further comments.

5 Maximizing Conflicting Measures: Problem EW-ES

Expected withdrawals (EW) and expected shortfall (ES) are two conflicting measures. We handle this by using a scalarization technique to find the Pareto points for this bi-objective optimization problem, that is, the set of points where one objective cannot be improved without worsening the other. For a given scalarization parameter $\kappa > 0$, the goal is then to find the control $\mathcal{P}_0$ that maximizes

$$\text{EW}(X_0^-, t_0^-) + \kappa\, \text{ES}_\alpha(X_0^-, t_0^-) .$$

More precisely, we define the pre-commitment EW-ES problem $(PCES_{t_0}(\kappa))$ in terms of the value function $J(s, b, t_0^-)$ derived from (9) and (10):

$$J\left(s,b,t_0^-\right) = \sup_{\mathcal{P}_0\in\mathcal{A}} \sup_{W^*} \left\{ E_{\mathcal{P}_0}^{X_0^-,t_0^-}\left[\sum_{i=0}^{M} q_i + \kappa\left(W^* + \frac{1}{\alpha}\min(W_T - W^*,0)\right) + \epsilon W_T \middle| X(t_0^-) = (s,b) \right]\right\}, \tag{11}$$

$$\text{subject to } \begin{cases} (S(t),B(t)) \text{ follows processes (3) and (4); } t \notin \mathcal{T} . \\ W_\ell^+ = W_\ell^- - q_\ell \,;\; X_\ell^+ = (S_\ell^+, B_\ell^+) . \\ W_\ell^- = \left(S_\ell^- + B_\ell^-\right) . \\ S_\ell^+ = p_\ell(\cdot)W_\ell^+ \,;\; B_\ell^+ = (1-p_\ell(\cdot))W_\ell^+ . \\ (q_\ell(\cdot),p_\ell(\cdot)) \in \mathcal{Z}(W_\ell^-, W_\ell^+, t_\ell) . \\ \ell = 0,\ldots,M \,;\; t_\ell \in \mathcal{T} . \end{cases} \tag{12}$$

Note that we have added an extra term $E_{\mathcal{P}_0}^{X_0^-,t_0^-}[\epsilon W_T]$ to equation (11). If we have a maximum withdrawal constraint, and if $W_t \gg W^*$ as $t \to T$, then the controls become ill-posed. In this fortunate state for the investor, we can break investment policy ties either by setting $\epsilon < 0$, which will force investments in bonds, or by setting $\epsilon > 0$, which will force investments into stocks. Choosing $|\epsilon| \ll 1$ ensures that this term only has an effect if $W_t \gg W^*$ and $t \to T$. See Forsyth (2022) for more discussion of this.

We can interchange the $\sup\sup(\cdot)$ [4] in equation (11) to represent the value function as

$$J\left(s,b,t_0^-\right) = \sup_{W^*} \sup_{\mathcal{P}_0\in\mathcal{A}} \left\{ E_{\mathcal{P}_0}^{X_0^+,t_0^+}\left[\sum_{i=0}^{M} \mathfrak{q}_i + \kappa\left(W^* + \frac{1}{\alpha}\min(W_T - W^*,0)\right) + \epsilon W_T \middle| X(t_0^-) = (s,b) \right]\right\}. \tag{13}$$

Since the inner supremum in equation (13) is a continuous function of W^* and the optimal value of W^* in equation (13) is bounded, [5] we can define

$$\mathcal{W}^*(s,b) = \arg\max_{W^*} \left\{ \sup_{\mathcal{P}_0\in\mathcal{A}} \left\{ E_{\mathcal{P}_0}^{X_0^+,t_0^+}\left[\sum_{i=0}^{M} \mathfrak{q}_i + \kappa\left(W^* + \frac{1}{\alpha}\min(W_T - W^*,0)\right) + \epsilon W_T \middle| X(t_0^-) = (s,b) \right]\right\}\right\}. \tag{14}$$

[4] Let $F = \sup_{(a,b)\in A\times B} f(a,b)$, then $\forall \epsilon > 0$, $\exists (a^*,b^*) \in A \times B$, s.t. $f(a^*,b^*) > F - \epsilon$. Then $F \ge \sup_{a\in A}\sup_{b\in B} f(a,b) \ge \sup_{b\in b} f(a^*,b) \ge f(a^*,b^*) > F - \epsilon$. Hence, $\epsilon \to 0$ implies $\sup_{a\in A}\sup_{b\in B} f(a,b) = F$. Similarly $\sup_{b\in B}\sup_{a\in A} f(a,b) = F$.

[5] This is the same as noting that a finite value at risk exists.

Given $\mathcal{W}^*(s,b)$ from equation (14), then the optimal control $\mathcal{P}^*$ which solves Problem (13) is the same control which solves

$$\hat{J}\left(s,b,t_0^-\right) = \sup_{\mathcal{P}_0 \in \mathcal{A}} \left\{ E_{\mathcal{P}_0}^{X_0^+,t_0^+} \left[\sum_{i=0}^{M} \mathfrak{q}_i + \frac{\kappa}{\alpha} \min(W_T - \mathcal{W}^*(s,b),0) + \epsilon W_T \middle| X(t_0^-) = (s,b) \right] \right\}. \tag{15}$$

Hence Problem (15) is the induced time consistent policy for Problem (13). We refer the reader to Forsyth (2020a) for an extensive discussion concerning pre-commitment and time consistent expected shortfall strategies.

6 Formulation as a Dynamic Program

We can use the method in Forsyth (2020a) to determine our value function. Write (13) as

$$J(s,b,t_0^-) = \sup_{W^*} \tilde{v}(s,b,W^*,0^-) \ ,$$

where $\tilde{v}(s,b,W^*,t)$ is defined as

$$\tilde{v}(s,b,W^*,t_n^-) = \sup_{\mathcal{P}_n \in \mathcal{A}_n} \left\{ E_{\mathcal{P}_n}^{\hat{X}_n^+,t_n^+} \left[\sum_{i=n}^{M} q_i + \kappa\left(W^* + \frac{1}{\alpha} \min((W_T - W^*),0) \right) + \epsilon W_T \;\middle|\; \hat{X}(t_n^-) = (s,b,W^*) \right] \right\}, \tag{16}$$

$$\text{subject to} \begin{cases} (S(t),B(t)) \text{ follows processes (3) and (4)}; \;\; t \notin \mathcal{T} \, . \\ W_\ell^+ = W_\ell^- - q_\ell \, ; \; X_\ell^+ = (S_\ell^+, B_\ell^+, W^*) \, . \\ W_\ell^- = \left(S_\ell^- + B_\ell^- \right) . \\ S_\ell^+ = p_\ell(\cdot) W_\ell^+ \, ; \; B_\ell^+ = (1 - p_\ell(\cdot)) W_\ell^+ \, . \\ (q_\ell(\cdot), p_\ell(\cdot)) \in \mathcal{Z}(W_\ell^-, W_\ell^+, t_\ell) \, . \\ \ell = n, \ldots, M \, ; \; t_\ell \in \mathcal{T} \, . \end{cases} \tag{17}$$

The original problem (13) has therefore been decomposed into two steps:

1. For given initial cash W_0, and a fixed value of W^*, solve problem (16) using dynamic programming in order to determine $\tilde{v}(0,W_0,W^*,0^-)$.
2. Solve problem (13) by maximizing over W^*

$$J(0,W_0,0^-) = \sup_{W^*} v(0,W_0,W^*,0^-) \, . \tag{18}$$

6.1 Dynamic Programming Solution of Problem (16)

We give a brief overview of the method described in detail in Forsyth (2022). We apply the dynamic programming principle to $t_n \in \mathcal{T}$

$$\tilde{v}(s,b,W^*,t_n^-) = \sup_{q \in \mathcal{Z}_q(w^-,t_n)} \left\{ \sup_{p \in \mathcal{Z}_p(w^+,t_n)} \left[q + \tilde{v}(w^+ p, w^+(1-p), W^*, t_n^+) \right] \right\},$$
$$\text{where } w^- = (s+b)\ ;\ w^+ = w^- - q\ . \tag{19}$$

If we set

$$h(w,t_n,W^*) = \left[\sup_{p \in \mathcal{Z}_p(w,t_n)} \tilde{v}(wp, w(1-p), W^*, t_n^+) \right], \tag{20}$$

then equation (19) becomes

$$\tilde{v}(s,b,W^*,t_n^-) = \sup_{q \in \mathcal{Z}_q(w^-,t_n)} \left\{ q + \left[h((w^- - q), W^*, t_n^+) \right] \right\},$$
$$\text{with } w^- = (s+b). \tag{21}$$

This approach effectively replaces a two-dimensional optimization for (q_n, p_n), by two sequential one-dimensional optimizations. From equations (20) and (21), the optimal pair (q_n, p_n) satisfies

$$q_n = q_n(w^-, W^*)\ , \qquad \text{where} \quad w^- = (s+b)\ ,$$
$$p_n = p_n(w^+, W^*)\ , \qquad \text{where} \quad w^+ = w^- - q_n\ .$$

In other words, the optimal withdrawal control q_n is only a function of total wealth before withdrawals while the optimal control p_n is only a function of total wealth after withdrawals. If a withdrawal results in $W^+ > 0$, then the optimal control for p is determined along lines of constant wealth in the (s, b) plane while if a withdrawal results in $W^+ < 0$, then stocks are liquidated ($p = 0$). This is illustrated in Figure 1.

At $t = T$, we have

$$\tilde{v}(s,b,W^*,T^+) = \kappa\left(W^* + \frac{\min((s+b-W^*),0)}{\alpha} \right) + \epsilon(s+b)\ ,$$

while at points in between rebalancing times, that is when $t \notin \mathcal{T}$, standard arguments from SDEs (3-4), and Forsyth (2022) give

$$\tilde{v}_t + \frac{(\sigma^s)^2 s^2}{2}\tilde{v}_{ss} + (\mu^s - \lambda_\xi^s \gamma_\xi^s) s\tilde{v}_s + \lambda_\xi^s \int_{-\infty}^{+\infty} \tilde{v}(e^y s, b, t) f^s(y)\, dy + \frac{(\sigma^b)^2 b^2}{2}\tilde{v}_{bb}$$
$$+ (\mu^b + \mu_c^b \mathbf{1}_{\{b<0\}} - \lambda_\xi^b \gamma_\xi^b) b\tilde{v}_b + \lambda_\xi^b \int_{-\infty}^{+\infty} \tilde{v}(s, e^y b, t) f^b(y)\, dy - (\lambda_\xi^s + \lambda_\xi^b)\tilde{v}$$
$$+ \rho_{sb}\sigma^s\sigma^b sb\tilde{v}_{sb} = 0\ . \tag{22}$$

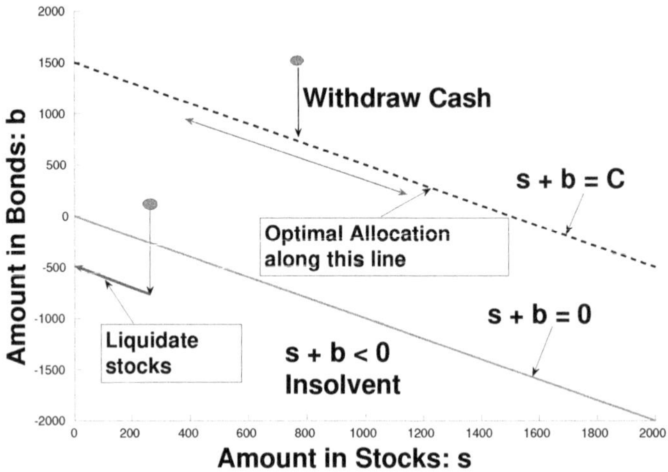

Figure 1 Schematic of withdrawal controls.

It is convenient to consider the two cases $b \geq 0$ and $b \leq 0$ separately. For $b \geq 0$, we solve

$$\tilde{v}_t + \frac{(\sigma^s)^2 s^2}{2}\tilde{v}_{ss} + (\mu^s - \lambda^s_\xi \gamma^s_\xi) s\tilde{v}_s + \lambda^s_\xi \int_{-\infty}^{+\infty} \tilde{v}(e^y s, b, t) f^s(y)\, dy + \frac{(\sigma^b)^2 b^2}{2}\tilde{v}_{bb}$$
$$+ (\mu^b - \lambda^b_\xi \gamma^b_\xi) b\tilde{v}_b + \lambda^b_\xi \int_{-\infty}^{+\infty} \tilde{v}(s, e^y b, t) f^b(y)\, dy - (\lambda^s_\xi + \lambda^b_\xi)\tilde{v}$$
$$+ \rho_{sb}\sigma^s\sigma^b sb\tilde{v}_{sb} = 0\,,$$
$$b \geq 0, s \geq 0\,. \tag{23}$$

When $b < 0$, it is convenient to rewrite equation (22) in terms of debt $\hat{b} = -b$. Letting $\hat{v}(s, \hat{b}, t) = \tilde{v}(s, b, t), b < 0, \hat{b} = -b$ in equation (22) we obtain

$$\hat{v}_t + \frac{(\sigma^s)^2 s^2}{2}\hat{v}_{ss} + (\mu^s - \lambda^s_\xi \gamma^s_\xi) s\hat{v}_s + \lambda^s_\xi \int_{-\infty}^{+\infty} \hat{v}(e^y s, \hat{b}, t) f^s(y)\, dy + \frac{(\sigma^b)^2 \hat{b}^2}{2}\hat{v}_{\hat{b}\hat{b}}$$
$$+ (\mu^b + \mu^b_c - \lambda^b_\xi \gamma^b_\xi) \hat{b}\hat{v}_{\hat{b}} + \lambda^b_\xi \int_{-\infty}^{+\infty} \hat{v}(s, e^y \hat{b}, t) f^b(y)\, dy - (\lambda^s_\xi + \lambda^b_\xi)\hat{v}$$
$$+ \rho_{sb}\sigma^s\sigma^b s\hat{b}\hat{v}_{s\hat{b}} = 0\,,$$
$$\hat{b} > 0, s \geq 0\,. \tag{24}$$

Note that equation (24) is now amenable to a transformation of the form $\hat{x} = \log \hat{b}$ since $\hat{b} > 0$, something required when using a Fourier method (Forsyth and Labahn, 2019; Forsyth, 2022) to solve equation (24).

After rebalancing, if $b \geq 0$, then b cannot become negative, since $b = 0$ is a barrier in equation (22). However, b can become negative after withdrawals, in which case b remains in the state $b < 0$. Then equation (24) applies, unless there is an injection of cash to move to a state with $b > 0$. The terminal condition for equation (24) is

$$\hat{v}(s, \hat{b}, W^*, T^+) = \kappa\left(W^* + \frac{\min((s - \hat{b} - W^*), 0)}{\alpha}\right) + \epsilon(s - \hat{b}) \; ; \; \hat{b} > 0 \, .$$

7 The δ-Monotone Fourier Method

The δ-monotone Fourier method originates from the work of (Forsyth and Labahn, 2019), where it was applied to one-dimensional problems in stochastic control. In the following, we give a sketch showing how to extend those methods to work for two-dimensional problems.

We illustrate this method by focusing on equation (23). Since $b > 0, s > 0$ we can let $x_1 = \log s, x_2 = \log b, \tau = T - t$ with $v(x_1, x_2, \tau) = \tilde{v}(e^{x_1}, e^{x_2}, T - \tau)$. Then (23) converts to

$$v_\tau = \frac{(\sigma^s)^2}{2} v_{x_1 x_1} + (\mu^s - \lambda_\xi^s \gamma_\xi^s) v_{x_1} + \lambda_\xi^s \int_{-\infty}^{\infty} v(x_1 + y, x_2, \tau) f^s(y)\, dy + \frac{(\sigma^b)^2}{2} v_{x_2 x_2}$$
$$+ (\mu^b - \lambda_\xi^b \gamma_\xi^b) v_{x_2} + \lambda_\xi^b \int_{-\infty}^{\infty} v(x_1, x_2 + y, t) f^b(y)\, dy - (\lambda_\xi^s + \lambda_\xi^b) v + \rho_{sb} \sigma^s \sigma^b v_{x_1 x_2} . \tag{25}$$

The exact solution of equation (25) can be written as

$$v(x_1, x_2, \tau + \Delta\tau) = \int_{-\infty}^{\infty} \int_{-\infty}^{\infty} g(x_1 - z_1, x_2 - z_2, \Delta\tau)\, v(z_1, z_2, \tau)\, dz_1\, dz_2 \, , \tag{26}$$

where $g(x_1, x_2, \Delta\tau)$ is the Green's function of equation (23) (Garroni and Menaldi, 1992).

Remark 1 (Form of Green's function) Note that the Green's function for equation (23) is of the form $g(x_1, x_2, z_1, z_2) = g(x_1 - z_1, x_2 - z_2)$, that is a function of the differences. Intuitively, we can view the Green's function as a scaled probability density f, and this means that $f(x_1, x_2)|(z_1, z_2)$ depends only on the difference, $f(x_1, x_2)|(z_1, z_2) = f(x_1 - z_1, x_2 - z_2)$. See Forsyth and Labahn (2019) for more discussion of this observation.

The Fourier transform pair for the Green's function is given as

$$G(\omega_1, \omega_2, \Delta\tau) = \int_{-\infty}^{\infty} \int_{-\infty}^{\infty} g(x_1, x_2, \Delta\tau) e^{-2\pi i \omega_1 x_1} e^{-2\pi i \omega_2 x_2}\, dx_1\, dx_2 \, ,$$
$$g(x_1, x_2, \Delta\tau) = \int_{-\infty}^{\infty} \int_{-\infty}^{\infty} G(\omega_1, \omega_2, \Delta\tau) e^{2\pi i \omega_1 x_1} e^{2\pi i \omega_2 x_2}\, d\omega_1\, d\omega_2 \, . \tag{27}$$

where $i = \sqrt{-1}$. Standard techniques then give the Fourier Transform of the Green's function $G(\omega_1, \omega_2, \Delta\tau)$ for equation (25) as

$$G(\omega_1, \omega_2, \Delta\tau) = e^{\Psi(\omega_1, \omega_2)\Delta\tau} , \tag{28}$$

where

$$\begin{aligned}\Psi(\omega_1, \omega_2) = & -\frac{(\sigma^s)^2}{2}(2\pi\omega_1)^2 + \left(\mu^s - \lambda_\xi^s \gamma_\xi^s - \frac{(\sigma^s)^2}{2}\right)(2\pi i\omega_1) + \lambda_\xi^s \overline{F^s}(\omega_1) \\ & -\frac{(\sigma^b)^2}{2}(2\pi\omega_2)^2 + \left(\mu^b - \lambda_\xi^b \gamma_\xi^b - \frac{(\sigma^b)^2}{2}\right)(2\pi i\omega_2) + \lambda_\xi^b \overline{F^b}(\omega_1) \\ & -(\lambda_\xi^s + \lambda_\xi^b) - \rho_{sb}\sigma^s\sigma^b(2\pi\omega_1)(2\pi\omega_2). \end{aligned} \tag{29}$$

Here $\overline{F^b}(\omega_1), \overline{F^s}(\omega_1)$ are the complex conjugates of the Fourier transforms of the density functions f^s, f^b (see e.g., equation (1)):

$$\begin{aligned}\overline{F^s}(\omega_1) &= \frac{u^s}{1 - \frac{2\pi i\omega_1}{\eta_1^s}} + \frac{1 - u^s}{1 + \frac{2\pi i\omega_1}{\eta_2^s}} , \\ \overline{F^b}(\omega_2) &= \frac{u^b}{1 - \frac{2\pi i\omega_2}{\eta_1^b}} + \frac{1 - u^b}{1 + \frac{2\pi i\omega_2}{\eta_2^b}} .\end{aligned}$$

7.1 Localization

Define

$$\Omega = [(x_1)_{\min}, (x_1)_{\max}] \times [(x_2)_{\min}, (x_2)_{\max}] .$$

As is typically the case, we assume that the Green's function $g(x_1, x_2, \Delta\tau)$ decays to zero as $|x_1|, |x_2| \to \infty$. More precisely, we assume that

Assumption 1 (Decay of Green's Function). *For some $A > 0$, $g(x_1, x_2, \Delta\tau)$ is negligible if* $\min(|x_1|, |x_2|) > A$.

We choose our region Ω so that we can assume that the Green's function is negligible for $(x_1,x_2) \notin \Omega$. As such we replace the Fourier transform pair (27) by their Fourier series equivalent

$$\begin{aligned} G(\omega_1^k, \omega_2^j, \Delta\tau) &\simeq \int_{(x_1)_{\min}}^{(x_1)_{\max}} \int_{(x_2)_{\min}}^{(x_2)_{\max}} g(x_1, x_2, \Delta\tau)\, e^{-2\pi i\omega_1^k x_1}\, e^{-2\pi i\omega_2^j x_2}\, dx_1\, dx_2 , \\ g(x_1, x_2, \Delta\tau) &\simeq \frac{1}{P_1 P_2} \sum_{k=-\infty}^{\infty} \sum_{j=-\infty}^{\infty} G(\omega_1^k, \omega_2^j, \Delta\tau)\, e^{2\pi i\omega_1^k x_1}\, e^{2\pi i\omega_2^j x_2} , \end{aligned} \tag{30}$$

with

$$P_1 = (x_1)_{\max} - (x_1)_{\min} \ ; \ P_2 = (x_2)_{\max} - (x_2)_{\min}; \quad \omega_1^k = \frac{k}{P_1} \ ; \ \omega_2^j = \frac{j}{P_2} \ ,$$

along with a localized version of equation (26)

$$v(x_1, x_2, \tau + \Delta\tau) \simeq \int_{(x_1)_{\min}}^{(x_1)_{\max}} \int_{(x_2)_{\min}}^{(x_2)_{\max}} g(x_1 - z_1, x_2 - z_2, \Delta\tau) \, v(z_1, z_2, \tau) \, dz_1 \, dz_2 \ . \tag{31}$$

Scaling factors in equation (30) are selected to correspond to a finite domain form of equation (27).

7.2 Periodic Extension

We have informally derived the expression for the Green's function by localizing the infinite domain Green's function. Localizing the problem on Ω, and using a Fourier series representation of the Green's function makes the implicit assumption of periodic extension.

Assumption 2. *Periodicity*

1. *The control problem is defined on the finite domain* Ω*.*
2. *The solution* $v(x_1, x_2, \tau)$ *is extended periodically*

$$v(x_1 \pm P_1, x_2, \tau) = v(x_1, x_2 \pm P_2, \tau) = v(x_1, x_2, \tau) \ . \tag{32}$$

3. *The jump size density functions* $f^s(y_1), f^b(y_2)$ *are defined on* $y_1 \in [(x_1)_{\min}, (x_1)_{\max}]$ *and* $y_2 \in [(x_2)_{\min}, (x_2)_{\max}]$*, with periodic extension*

$$f^s(y_1 \pm P_1) = f^s(y_1) \quad \textit{and} \quad f^b(y_2 \pm P_2) = f^b(y_2) \ . \tag{33}$$

4. *Periodic boundary conditions (32)-33) are specified for the PIDE problem (25).*
5. *From equation (30) we can also see that*

$$g(x_1 \pm P_1, x_2, \tau) = g(x_1, x_2 \pm P_2, \tau) = g(x_1, x_2, \tau).$$

It is more rigorous to make Assumption 2, and then derive the Green's function. It is straightforward to verify that this yields the Fourier series representation (30), with $G(\omega_1^k, \omega_2^j)$ given by equations (28)-(29). In particular, we have that (Garroni and Menaldi, 1992)

$$g(x_1, x_2, \Delta\tau) \geq 0 \ ,$$
$$\int_\Omega g(x_1, x_2, \Delta\tau) dx_1 \, dx_2 = 1 \ .$$

7.3 Discretization

We discretize equation (31) on a grid $\{(x_1, x_2)_{j,k}\}$, $\{(z_1, z_2)_{j,k}\}$, by setting

$$x_1^j = \hat{x}_1^0 + j\Delta x_1 \; ; \; x_2^k = \hat{x}_2^0 + k\Delta x_2 \; ; \; j = -\frac{N_1}{2}, \ldots \frac{N_1}{2} - 1 \; ; \; k = -\frac{N_2}{2}, \ldots \frac{N_2}{2} - 1 \,,$$
$$\Delta x_1 = \frac{P_1}{N_1} \; ; \; \Delta x_2 = \frac{P_2}{N_2} \,,$$
$$P_1 = (x_1)_{\max} - (x_1)_{\min} \; ; \; P_2 = (x_2)_{\max} - (x_2)_{\min} \,,$$
$$v_{j,k}(\tau) \equiv v(x_1^j, x_2^k, \tau).$$

Define linear basis functions as

$$\phi_1^j(x_1) = \begin{cases} \frac{x_1 - x_1^{j-1}}{\Delta x_1} & x_1^{j-1} \le x_1 \le x_1^j \,. \\ \frac{x_1^{j+1} - x_1}{\Delta x_1} & x_1^j \le x_1 \le x_1^{j+1} \,. \\ 0 & \text{otherwise} \,. \end{cases}$$

with a similar definition for $\phi_2^k(x_2)$. We can then represent the solution $v(x_1, x_2, \tau)$ as a linear combination of the basis functions

$$v(x_1, x_2, \tau) \simeq \sum_{j=-N_1/2}^{N_1/2-1} \sum_{k=-N_2/2}^{N_2/2-1} \phi_1^j(x_1)\phi_2^k(x_2) v_{j,k}(\tau) \,. \tag{34}$$

Substituting equation (34) into equation (31) then gives

$$\begin{aligned} &v_{\ell,m}(\tau + \Delta\tau) \\ &= \sum_{j=-N_1/2}^{N_1/2-1} \sum_{k=-N_2/2}^{N_2/2-1} v_{j,k}(\tau) \int_{(x_1)_{\min}}^{(x_1)_{\max}} \int_{(x_2)_{\min}}^{(x_2)_{\max}} \phi_1^j(x_1)\, \phi_2^k(x_2)\, g(x_1^\ell - x_1, x_2^m - x_2, \Delta\tau)\, dx_1\, dx_2 \,. \end{aligned} \tag{35}$$

For a given pair (j, k) the double integrals then simplify to

$$\begin{aligned} &\int_{(x_1)_{\min}}^{(x_1)_{\max}} \int_{(x_2)_{\min}}^{(x_2)_{\max}} \phi_1^j(x_1)\, \phi_2^k(x_2)\, g(x_1^\ell - x_1, x_2^m - x_2, \Delta\tau)\, dx_1\, dx_2 \\ &\quad = \int_{x_1^j - \Delta x_1}^{x_1^j + \Delta x_1} \int_{x_2^k - \Delta x_2}^{x_2^k + \Delta x_2} \phi_1^j(x_1)\, \phi_2^k(x_2)\, g(x_1^\ell - x_1, x_2^m - x_2, \Delta\tau)\, dx_1\, dx_2 \\ &\quad = \int_{x_1^\ell - x_1^j - \Delta x_1}^{x_1^\ell - x_1^j + \Delta x_1} \int_{x_2^m - x_2^k - \Delta x_2}^{x_2^m - x_2^k + \Delta x_2} \phi_1^j(x_1^\ell - x_1)\, \phi_2^k(x_2^m - x_2)\, g(x_1, x_2, \Delta\tau)\, dx_1\, dx_2 \\ &\quad = \int_{x_1^\ell - x_1^j - \Delta x_1}^{x_1^\ell - x_1^j + \Delta x_1} \int_{x_2^m - x_2^k - \Delta x_2}^{x_2^m - x_2^k + \Delta x_2} \phi_1^{\ell-j}(x_1) \phi_2^{m-k}(x_2)\, g(x_1, x_2, \Delta\tau)\, dx_1\, dx_2 \,. \end{aligned}$$

Here the last line follows from the property of linear basis functions,

$$\phi_1^j(x_1^\ell - x_1) = \phi_1^{\ell-j}(x_1), \quad \phi_2^k(x_2^k - x_2) = \phi_2^{m-k}(x_2).$$

Defining

$$\tilde{g}_{\ell-j,m-k}(\Delta\tau) \equiv \frac{1}{\Delta x_1 \Delta x_2} \int_{x_1^\ell - x_1^j - \Delta x_1}^{x_1^\ell - x_1^j + \Delta x_1} \int_{x_2^m - x_2^k - \Delta x_2}^{x_2^m - x_2^k + \Delta x_2} \phi_1^{\ell-j}(x_1)\phi_2^{m-k}(x_2)\, g(x_1,x_2,\Delta\tau)\, dx_1\, dx_2,$$

then implies that for each $\ell = -N_1/2, \ldots, N_1/2 - 1$, $m = -N_2/2, \ldots, N_2/2 - 1$, the convolution (35) can be written as

$$v_{\ell,m}(\tau + \Delta\tau) = \sum_{j=-N_1/2}^{N_1/2-1} \sum_{k=-N_2/2}^{N_2/2-1} v_{j,k}(\tau)\, \tilde{g}_{\ell-j,m-k}(\tau)\, \Delta x_1\, \Delta x_2. \tag{36}$$

Furthermore using the Fourier series form for $g(\cdot)$ from equation (30) we have

$$\begin{aligned}
&\tilde{g}_{\ell-j,m-k}(\Delta\tau) \\
&= \frac{1}{\Delta x_1\, \Delta x_2} \int_{x_1^\ell - x_1^j - \Delta x_1}^{x_1^\ell - x_1^j + \Delta x_1} \int_{x_2^m - x_2^k - \Delta x_2}^{x_2^m - x_2^k + \Delta x_2} \phi_1^{\ell-j}(x_1)\, \phi_2^{m-k}(x_2)\, g(x_1, x_2, \tau)\, dx_1\, dx_2 \\
&= \frac{1}{P_1 P_2} \sum_{u=-\infty}^{\infty} \sum_{p=-\infty}^{\infty} \frac{\sin^2 \pi\omega_1^u \Delta x_1}{(\pi\omega_1^u \Delta x_1)^2} \frac{\sin^2 \pi\omega_2^p \Delta x_2}{(\pi\omega_2^p \Delta x_2)^2} G(\omega_1^u, \omega_2^p, \Delta\tau)\, e^{2\pi i \omega_1^u (x_1^\ell - x_1^j)}\, e^{2\pi i \omega_2^p (x_2^m - x_2^k)} \\
&= \frac{1}{P_1 P_2} \sum_{u=-\infty}^{\infty} \sum_{p=-\infty}^{\infty} \frac{\sin^2 \pi\omega_1^u \Delta x_1}{(\pi\omega_1^u \Delta x_1)^2} \frac{\sin^2 \pi\omega_2^p \Delta x_2}{(\pi\omega_2^p \Delta x_2)^2} G(\omega_1^u, \omega_2^p, \Delta\tau)\, e^{2\pi i u(\ell-j)/N_1}\, e^{2\pi i p(m-k)/N_2}.
\end{aligned} \tag{37}$$

Here for the inside integrals in (37) we have used the following formula for linear basis functions

$$\frac{1}{\Delta x_1} \int_{x_1^j - \Delta x_1}^{x_1^j + \Delta x_1} e^{2\pi i \omega x_1} \phi_1^j(x) dx_1 = e^{2\pi i \omega x_1^j} \frac{\sin^2 \pi\omega\Delta x_1}{(\pi\omega\Delta x_1)^2},$$

with a similar result for the integral involving ϕ_2. We note that on a uniform grid $x_1^j = \hat{x}_1^0 + j\Delta x_1$ and $x_2^k = \hat{x}_2^0 + k\Delta x_2$, with the result that $x_1^{\ell-j} = x_1^\ell - x_1^j = (\ell - j)\Delta x_1$, and $x_2^{m-k} = x_2^m - x_2^k = (m - k)\Delta x_2$. Consequently, equation (37) is independent of $\hat{x}_1^0, \hat{x}_2^0$.[6]

For future reference, we note the DFT pair for any grid function[7] $h(x_1^j, x_2^k) = h_{jk}$ is given by

[6] For the case of $u = 0$, $p = 0$ in equation (37) we take the limiting value as $\omega_1^0 \to 0$, $\omega_2^0 \to 0$.

[7] Note that the DFT here is independent of any physical coordinates.

$$H(\omega_1^\ell, \omega_2^m, \Delta\tau) = \frac{P_1}{N_1}\frac{P_2}{N_2}\sum_{j,k} e^{-2\pi i \ell j/N_1} e^{-2\pi i m k/N_2} h_{jk} \,, \tag{38}$$

$$h_{pn}(\Delta\tau) = \frac{1}{P_1 P_2}\sum_{\ell,m} e^{2\pi i \ell p/N_1} e^{2\pi i m n/N_2} H(\omega_1^\ell, \omega_2^m, \Delta\tau) \,, \tag{39}$$

which can be verified by substituting equation (38) into equation (39)

$$h_{pn}(\Delta\tau) = \frac{1}{N_1 N_2}\sum_{j,k} h_{jk} \sum_{\ell,m} e^{2\pi i \ell (p-j)/N_1} e^{2\pi i m (n-k)/N_2} = h_{pn} \,,$$

since

$$\sum_{\ell,m} e^{2\pi i \ell (p-j)/N_1} e^{2\pi i m (n-k)/N_2} = \begin{cases} N_1 N_2 & p = j \text{ and } n = k \,. \\ 0 & \text{otherwise} \,. \end{cases} \tag{40}$$

We denote this pair by $H = \mathrm{DFT}(h), h = \mathrm{IDFT}(H)$.

7.3.1 Discrete Index Domain

Although we have defined the solution on the physical domain Ω, note that from equation (37) $\tilde{g}_{\ell-j,m-n}$ is independent of $(\hat{x}_1^0, \hat{x}_2^0)$. In addition, the DFT pair is defined independently of the original grid. Consequently, in order to avoid confusion concerning the various physical and grid domains, we will refer only to discrete index domains. Define

$$\mathcal{D} = \left\{ (\ell, j) \Big| -\frac{N_1}{2} \le \ell \le \frac{N_1}{2} - 1 \,, \ -\frac{N_2}{2} \le j \le \frac{N_2}{2} - 1 \right\}.$$

Then $\tilde{g}_{\ell,j}, v_{\ell,j}$ are both in the index domain $\mathcal{D}$.

7.4 Efficient Computation of the Discrete Convolution

The discrete convolution (36) is performed at each rebalancing date, hence an efficient evaluation is desirable. The convolution can be written as

$$\begin{aligned} v_{\ell,m}(\tau + \Delta\tau) &= \sum_{j=-N_1/2}^{N_1/2-1} \ \sum_{k=-N_2/2}^{N_2/2-1} v_{j,k}(\tau) \tilde{g}_{\ell-j,m-k}(\Delta\tau) \, \Delta x_1 \, \Delta x_2 \\ &= \frac{P_1}{N_1}\frac{P_2}{N_2}\sum_{j,k} \tilde{g}_{\ell-j,m-k}(\Delta\tau) v_{j,k}(\tau) \ . \end{aligned} \tag{41}$$

Using equations (38) and (39) to write $\tilde{g}(\Delta\tau) = \text{IDFT}(\tilde{G}(\Delta\tau)), v = \text{IDFT}(V)$ in equation (41) gives

$$
\begin{aligned}
& v_{\ell,m}(\tau + \Delta\tau) \\
&= \frac{1}{N_1 P_1}\frac{1}{N_2 P_2}\sum_{j,k}\sum_{p,u}\tilde{G}_{pu}(\Delta\tau)e^{2\pi i p(\ell-j)/N_1}e^{2\pi i u(m-k)/N_2}\sum_{q,r}V_{q,r}e^{2\pi i q j/N_1}e^{2\pi i r k/N_2} \\
&= \frac{1}{N_1 P_1}\frac{1}{N_2 P_2}\sum_{p,u}\sum_{q,r}\tilde{G}_{pu}(\Delta\tau)V_{qr}e^{2\pi i p\ell/N_1}e^{2\pi i u m/N_2}\sum_{j,k}e^{2\pi i j(q-p)/N_1}e^{2\pi i k(r-u)/N_2} \\
&= \frac{1}{P_1}\frac{1}{P_2}\sum_{p,u}\tilde{G}_{pu}(\Delta\tau)V_{pu}e^{2\pi i p\ell/N_1}e^{2\pi i u m/N_2} ,
\end{aligned}
$$

where we have used equation (40) in the last step. More compactly we can write

$$
\tilde{v}(\tau + \Delta\tau) = \text{IDFT}\big(\, \tilde{G}(\Delta\tau) \circ \text{DFT}\,(v(\tau))\,\big) , \tag{42}
$$

where $y \circ z$ is the Hadamard product of vectors y, z.

7.5 δ-Monotonicity

Using the definition for $\tilde{g}$ from equation (37), then this represents an exact integration over linear basis functions of the exact Green's function (with periodic boundary conditions) on Ω. Hence

$$
\tilde{g}_{j,k} \geq 0 \; ; \; \forall (j,k) \in \mathcal{D} . \tag{43}
$$

This gives rise to an important *monotonicity* property (Forsyth and Labahn, 2019). Suppose we have two discrete solutions $v_{\ell,m}(\tau), u_{\ell,m}(\tau)$ such that

$$
v_{\ell,m}(\tau) > u_{\ell,m}(\tau) \; ; \; \forall (\ell,m) \in \mathcal{D} . \tag{44}
$$

Then, using the time advance algorithm (36), it follows that the *discrete comparison principle* holds:

$$
v_{\ell,m}(\tau + \Delta\tau) > u_{\ell,m}(\tau + \Delta\tau) \; ; \; \forall (\ell,m) \in \mathcal{D}.
$$

The positivity condition (43) ensures that the order property (44) is preserved by the discretized algorithm. This is, of course, a property of the exact solution. This property is important since we compare the values of the discrete solution in order to determine the optimal control. See Forsyth and Labahn (2019) for more discussion of this.

However, note that the dimension of $\tilde{g}$ is (N_1, N_2), but we need to sum an infinite series to determine the discrete values of $\tilde{g}_{m,u}$ from equation (36). In practice, we truncate the series in equation (37), that is,

$$\tilde{g}_{\ell-j,m-k}(\Delta\tau)$$
$$\simeq \frac{1}{P_1 P_2} \sum_{u=-N_1^g/2}^{N_1^g/2-1} \sum_{p=-N_2^g/2}^{N_2^g/2-1} \left(\frac{\sin^2 \pi\omega_1^u \Delta x_1}{(\pi\omega_1^u \Delta x_1)^2} \right) \left(\frac{\sin^2 \pi\omega_2^p \Delta x_2}{(\pi\omega_2^p \Delta x_2)^2} \right) G(\omega_1^u, \omega_2^p, \Delta\tau)\, e^{2\pi i \frac{u(\ell-j)}{N_1}}\, e^{2\pi i \frac{p(m-k)}{N_2}} , \tag{45}$$

where the grid sizes (N_1^g, N_2^g) can be chosen independently from (N_1, N_2). However, truncating the Fourier series means that the positivity condition (43) may not hold any longer. In Forsyth and Labahn (2019), it is suggested that (N_1^g, N_2^g) be selected so that

$$\sum_{m=-N_1^g/2}^{N_1^g/2+1} \sum_{n=-N_2^g/2}^{N_2^g/2-1} \Delta x_1 \Delta x_2 \, |\min(\tilde{g}_{m,n}, 0)| < \delta \frac{\Delta\tau}{T} . \tag{46}$$

This ensures that the cumulative effect (after $(T/(\Delta\tau)$ steps) of non-positivity means that the discrete comparison principle holds to $O(\delta)$. Note that if the timestep $(\Delta\tau)$ is constant (which is usually the case), then equation (45) needs to be computed only once. Hence it is inexpensive to select sizes (N_1^g, N_2^g) which satisfy condition (46). Equation (45) can also be computed efficiently using FFTs, assuming $N_1^g/N_1, N_2^g/N_2$ are powers of two (see Forsyth and Labahn (2019).

8 Periodicity and the Problem of Wrap-Around Error

Localization of the Green's function effectively implies a periodic extension of the Green's function and the solution (see Assumption 2). In option pricing applications, this *wrap-around* error does not cause difficulty. However, in control applications, impulse controls (such as rebalancing) may require extensive use of solution values near the edges of the grid. For example, derisking by investing in an all bond portfolio results in an impulse control which drives $s \to 0$ instantaneously. This can cause significant wrap-around pollution (see (Lippa, 2013; Ruijter et al., 2013; Ignatieva et al., 2018)).

Given the localized problem defined on $\Omega = [(x_1)_{\min}, (x_1)_{\max}] \times [(x_2)_{\min}, (x_2)_{\max}]$, with widths

$$P_1 = (x_1)_{\max} - (x_1)_{\min} \quad \text{and} \quad P_2 = (x_2)_{\max} - (x_2)_{\min} ,$$

we construct an auxiliary grid with $N_1^\dagger = 2N_1$ nodes in the x_1 direction, and $N_2^\dagger = 2N_2$ nodes in the x_2 direction, on the domain $\Omega^\dagger = [(x_1)^\dagger_{\min}, (x_1)^\dagger_{\max}] \times [(x_2)^\dagger_{\min}, (x_2)^\dagger_{\max}]$, with

$$(x_1)^\dagger_{\min} = (x_1)_{\min} - \frac{P_1}{2} \; ; \quad (x_1)^\dagger_{\max} = (x_1)_{\max} + \frac{P_1}{2} \, ,$$
$$(x_2)^\dagger_{\min} = (x_2)_{\min} - \frac{P_2}{2} \; ; \quad (x_2)^\dagger_{\max} = (x_2)_{\max} + \frac{P_2}{2} \, ,$$

$$P_1^\dagger = (x_1)^\dagger_{\max} - (x_1)^\dagger_{\min} = 2P_1 \; ; \quad P_2^\dagger = (x_2)^\dagger_{\max} - (x_2)^\dagger_{\min} = 2P_2 \, .$$

Although we have increased the size of the physical domain Ω, we remind the reader of the discussion in Section 7.3.1. It will be more convenient in the following to refer to the extended index domain

$$\mathcal{D}^\dagger = \left\{(j,k) \mid \; -N_1^\dagger/2 \le j \le N_1^\dagger/2 - 1 \, , \; -N_2^\dagger/2 \le k \le N_2^\dagger/2 - 1\right\} \, .$$

Denote the projected Green's function for grid indices in $\mathcal{D}^\dagger$ by $\tilde{g}^\dagger$. Note from equation (37) that $\tilde{g}^\dagger$ does not depend on the actual physical domain, but only on $(\Delta x_1, \Delta x_2)$. We construct and store the DFT of the projection of the Green's function $\tilde{G}^\dagger(\Delta\tau) = \text{DFT}(\tilde{g}^\dagger(\Delta\tau))$ on this auxiliary index grid $\mathcal{D}^\dagger$.

Before applying the time advance algorithm (42) we form the padded array $v^\dagger(\tau)$:

$$v^\dagger(x_1^j, x_2^k, \tau) = \begin{cases} v(x_1^j, x_2^k, \tau) \, , & (j,k) \in \mathcal{D} \, , \\ v(x_1^{-N_1/2}, x_2^k, \tau) \, , & j \in [-N_1^\dagger/2, -N_1/2 - 1], \;\; k \in [-N_2/2, N_2/2 - 1] \, , \\ v(x_1^{-N_1/2}, x_2^{-N_2/2}, \tau) \, , & j \in [-N_1^\dagger/2, -N_1/2 - 1], \;\; k \in [-N_2^\dagger/2, -N_2/2 - 1] \, , \\ v(x_1^j, x_2^{-N_2/2}, \tau) \, , & j \in [-N_1/2, N_1/2 - 1], \;\; k \in [-N_2^\dagger/2, -N_2/2 - 1] \, , \\ A(x_1^j, x_2^k, \tau) \, , & j > N_1/2 - 1 \;\text{ or }\; k > N_2/2 - 1 \, , \end{cases} \tag{47}$$

where $A(x_1^j, x_2^k, \tau)$ is an asymptotic form of the solution which we assume to be available from financial reasoning. On the auxiliary grid, the points where $x_1 < (x_1)_{min}$ or $x_2 < (x_2)_{\min}$ correspond to the points where $s \to 0$ or $b \to 0$, with very small grid spacing. Hence extending the solution by constant values to the left and bottom is expected to generate a very small error.

We then modify the time advance algorithm (42) as shown in Algorithm 1. Note that the step (3) discards the computed solution in the padded areas $\{v^\dagger(\tau + \Delta\tau)_{j,k} | (j,k) \in \mathcal{D}^\dagger - \mathcal{D}\}$, since these points may be contaminated by wrap-around errors.

Require: $v(\tau)$; $\widetilde{G}^\dagger$
1: Form $v^\dagger(\tau)$ using (47)
2: $v^\dagger(\tau + \Delta\tau) = \text{IDFT}\big(\, \tilde{G}^\dagger \circ \text{DFT}\left(v^\dagger(\tau)\right)\big)$ {IDFT (Hadamard product)}
3: $v_{j,k}(\tau + \Delta\tau) = v^\dagger_{j,k}(\tau + \Delta\tau) \, , \, (j,k) \in \mathcal{D}$ {Discard values in $\mathcal{D}^\dagger - \mathcal{D}$.}

Algorithm 1 Advance time $v(\tau) \to v(\tau + \Delta\tau)$.

8.1 Error Due to Wrap-Around

Let $\tau^n = n\Delta\tau$ and $\mathcal{N}\Delta\tau = T$, with $\Delta\tau$ fixed. We can then write the discrete convolution (41) in the physical domain, compactly as

$$v_{\ell,m}^{n+1} = \Delta x_1 \Delta x_2 \sum_{(j,k)\in\mathcal{D}^\dagger} \tilde{g}^\dagger_{\ell-j,m-k} v^\dagger_{j,k}(\tau^n) \; , \; (\ell,m) \in \mathcal{D} \; . \tag{48}$$

Note that the righthand side of equation (48) uses the padded values for $v^\dagger_{j,k}$, but generates the unpadded $v^{n+1}_{\ell,m}$ on the left-hand side. In addition, $\tilde{g}^\dagger$ is periodic with period $P_1^\dagger$ in the x_1 direction, and period $P_2^\dagger$ in the x_2 direction.

The use of Algorithm 1 means that the periodic extension of $\tilde{g}$ is used in equation (48) for any terms of the sum such that

$$(\ell - j, m - k) \notin \mathcal{D}^\dagger \; .$$

This periodic extension *wraps around* the solution domain. Points with indices $\ell - j < -N_1^\dagger/2$ map to points with index $\ell - j + N_1^\dagger$. These mapped points are at the opposite side of the grid, and are a source of wrap-around error. See also Appendix for further discussion of this.

Definition 1 (Wraparound error.) Assume $\tilde{g}^\dagger_{\ell,m}$ for $(\ell, m) \in \mathcal{D}^\dagger$ is periodic

$$\tilde{g}^\dagger(\ell \pm N_1^\dagger, m) = \tilde{g}^\dagger(\ell, m \pm N_2^\dagger) = \tilde{g}^\dagger(\ell, m) \; . \tag{49}$$

Suppose $v^\dagger_{j,k}(\tau^n)$ is determined by boundary data for $(x_1^j, x_2^m) \in (\Omega^\dagger - \Omega)$, which we assume to be exact,[8] with $(N_1^\dagger, N_2^\dagger) = (2N_1, 2N_2)$. Then, the wrap-around error for convolution (48) evaluated using DFTs, at timestep $n\Delta\tau$, denoted by ϵ^n_{wrap} is

$$\varepsilon^n_{\text{wrap}} = \Delta x_1 \Delta x_2 \max_{(\ell,m)\in\mathcal{D}} \sum_{(j,k)\in\mathcal{D}^\dagger} \left| \tilde{g}^\dagger_{\ell-j,m-k} v^\dagger_{j,k}(\tau^n) \right| \mathbf{1}_{(\ell-j,m-k)\notin\mathcal{D}^\dagger} \; . \tag{50}$$

We now state a theorem on the effectiveness of our padding method. See Appendix for a proof.

Theorem 1 (Wraparound error) *Suppose $\tilde{g}^\dagger_{\ell,m}$ for $(\ell, m) \in \mathcal{D}^\dagger$ is periodic as defined in equation (49), and that $v^\dagger_{\ell,m}(\tau^n)$ be determined by boundary data for $(\ell, m) \in (\mathcal{D}^\dagger - \mathcal{D})$, with $(N_1^\dagger, N_2^\dagger) = 2(N_1, N_2)$. Assume further that there exists a constant C such that*

$$\left| v^\dagger_{\ell,m}(\tau^n) \right| \le C \; , \quad (\ell, m) \in \mathcal{D}^\dagger \; , \quad \forall\, n \; \textit{such that} \; (n\Delta\tau) \le T \; ,$$

[8] In practice, we use the method in equation (47), which we expect will generate a small error.

and that we choose (N_1, N_2) *sufficiently large so that*

$$\Delta x_1 \Delta x_2 \sum_{(j,k)\in\mathcal{D}^\dagger - \mathcal{D}} |\tilde{g}^\dagger_{j,k}| \leq \varepsilon_b \, \Delta\tau \,. \tag{51}$$

Then

$$\varepsilon^n_{wrap} \leq C\varepsilon_b \, \Delta\tau \,, \tag{52}$$

and the wrap-around error after N *steps is bounded by*

$$\varepsilon^N \leq T\epsilon_b C \quad where\; T = N\Delta\tau \,.$$

Remark 2 (Asymptotic Form of Green's Function) For a pure diffusion in one dimension, the Green's function for the Black–Scholes equation behaves like

$$g(x - x', \Delta\tau) = O\left(e^{-(x-x')^2/(\sigma^2\Delta\tau)}\right) \text{ as } |x - x'| \to \infty \,.$$

Similar asymptotic exponential decay as $\max(|x_1|,|x_2|) \to \infty$ is also seen for the full 2-d Green's function (Garroni and Menaldi, 1992). Hence ensuring that condition (51) holds is not onerous.

8.2 Numerical Method for Optimal Control

As shown in equation (16), we need the optimal pair (q, p) at each rebalancing date. The optimal controls are obtained by first discretizing the controls and thereafter determining the controls by exhaustive search. Recall that the PIDE grid spacing is $\Delta x_1, \Delta x_2$. Let N_q and N_p denote the number of points in the discretizations of q and p, respectively. We will also need to discretize the possible values of total wealth $w = s + b$, with N_w denoting the number of discrete wealth values. This approach (discretization and search), is then guaranteed to converge to the viscosity solution of the optimal control problem in the limit as

$$\Delta x_1, \Delta x_2 \to 0 \,;\, N_q, N_p, N_w \to \infty \,;\, \delta \to 0 \,, \tag{53}$$

which is discussed in Section 8.3.

Given an array of discrete values, that is, $v_{j,k}(\tau_n^-)$, we denote the linear interpolant of the grid values at an arbitrary point (z_1, z_2) by

$$v(\mathcal{I}(z_1, z_2), \tau_n^-) \,.$$

The numerical analogue of equation (20) requires a temporary array, $h(w_k), k = 1, \ldots, N_w$, where $w_k = s + b$, the total wealth. This array is determined by

$$\underbrace{p^*(w_k)}_{k=1,\ldots,N_w} = \underset{\substack{j=1,\ldots,N_p \\ p_j \in \mathcal{Z}_p(w_k,\tau_n^-)}}{\arg\max} \; v\Big(\mathcal{I}\left(\log(w_k p_j) \,,\, \log(w_k(1-p_j)) \right), \tau_n^- \Big) \,,$$

$$\underbrace{h(w_k)}_{k=1,\ldots,N_w} = v\Big(\mathcal{I}\Big(\log(\, w_k p^*(w_k)\,) \,,\, \log(\, w_k(1-p^*(w_k))\,) \Big), \tau_n^- \Big) \,. \tag{54}$$

In order to compute (21), we break this down into two steps, first using

$$\underbrace{q^*(w_k)}_{k=1,\ldots,N_w} = \underset{\substack{j=1,\ldots,N_q \\ q_j \in \mathcal{Z}_q(w_k,\tau_n)}}{\arg\max} \left\{ q_j + h\left(\mathcal{I}(w_k - q_j) \right) \right\} , \tag{55}$$

where $h(\mathcal{I}(\cdot))$ refers to one-dimensional interpolation for the one-dimensional array $h(w_k)$. Then

$$\underbrace{v(x_1^j, x_2^k, \tau_n^+)}_{\substack{j=-N_1/2,\ldots,N_1/2-1 \\ k=-N_2/2,\ldots,N_2/2-1 \\ w_{j,k}=e^{x_1^j}+e^{x_2^k}}} = q^*\left(\mathcal{I}(w_{j,k}) \right) + h\Big(\mathcal{I}\left(w_{j,k} - q^*\left(\mathcal{I}(w_{j,k}) \right) \right) \Big) \,. \tag{56}$$

8.3 Convergence to the Viscosity Solution

Convergence of a numerical method for the HJB equations is ensured provided that the complete numerical scheme is $\delta-$monotone, consistent and ℓ_∞ stable, (see Barles and Souganidis (1991); Forsyth and Labahn (2019)). We should clarify that each of these properties must be verified for the complete numerical scheme, including the time advance and the impulse control update.

Stability is easily established using the techniques in Forsyth and Labahn (2019). The time advance step is δ-monotone by construction. The optimization and projection steps in equations (54-56) are monotone, since linear interpolation is used at all off-grid points. Note that extrapolation is never used (which ensures stability).

Proving consistency is a bit more involved. The key is to note that consistency, in the viscosity sense, is defined in terms of smooth, infinitely differentiable test functions. This means that proof of consistency involves replacing all the functions in equations (54-56) by smooth test functions. Discretizing the control and using linear interpolation means that the maximum of the smooth test functions in equations (54-56) will converge to the global maximum in the limit.

While the discretization of the control and use of exhaustive search may appear to be crude, this technique (coupled with linear interpolation) is the only method which is guaranteed to converge to the global maximum of a (possibly) non-convex smooth test function. This is not merely a point of academic interest. Initial attempts to use standard optimization methods sometimes resulted in convergence to a different solution (compared to the solution obtained by exhaustive search). Since the viscosity solution is unique, we can deduce that the solution obtained without using exhaustive search was spurious.

However, numerical experiments show that the discretization of the control can be quite coarse, and this does not appear to impede convergence. A non-rigorous explanation of this effect is that a smooth function has a vanishing first derivative at extreme points.

8.4 Summary: Algorithm for Problem (16).

Algorithm 2 summarizes the final algorithm. The initial set-up cost is computing the Fourier weights on the extended domain $\widetilde{G}^{\dagger}$. Each timestep consists of first determining the optimal control, then followed by advancing to the next rebalancing time using the precomputed weights $\widetilde{G}^{\dagger}$.

Require: Weights $\widetilde{G}^{\dagger}$ in padded Fourier domain satisfying tolerance δ (see equations (45-46))

Input: number of timesteps $\mathcal{N}$, initial solution $(v^0)^-$
for $n = 1, \ldots, \mathcal{N}$ **do** {Timestep loop}
  Optimal control $v(\tau_{n-1}^-) \to v(\tau_{n-1}^+)$ equations (54-56)
  Advance time $v(\tau_{n-1}^+) \to v(\tau_n^-)$ using Algorithm 1
end for{End timestep loop}

Algorithm 2 Monotone Fourier method.

Remark 3 (Use of linear interpolation) Equations (54-56) make heavy use of linear interpolation. As discussed in Forsyth and Labahn (2019)) this is the only interpolation method (in general) which will preserve δ-monotonicity. This means that the convergence rate cannot exceed second order in the grid spacing. This is in contrast to the methods in (Fang and Oosterlee, 2008; 2009; Ruijter et al., 2013), where, in some circumstances, high-order rates of convergence can be achieved, but at the cost of possible non-monotonicity.

Remark 4 (Insolvency Between Rebalancing Times) For $b > 0$, insolvency cannot occur between rebalancing dates. However, for $b < 0$, it is possible that a large drop in the value of stocks could result in $s + b < 0$ between rebalancing times. Numerical tests using subtimesteps between rebalancing times (and checking for insolvency) did not show any significant changes to the final values at $\tau = T$. Hence, in all our reported numerical tests, we only check for insolvency at rebalancing times.

8.5 Final Algorithm: Problem EW-ES (18).

The final step for the complete solution of the EW-ES problem is the optimization step (18). We solve Problem EW-ES on a sequence of grids, with increasing number of grid nodes. On the coarsest grid, we discretize W^* and find the maximum of equation (18) by exhaustive search. Note that each evaluation of the objective function in (18) requires a solution of Problem 16. Using the coarse grid value of W^* as a starting value, we use a one-dimensional optimization algorithm to maximize the objective function (18) on each finer grid.

9 Numerical Example and Results

We consider a 65 year old retiree who has \$1,000K (one million) in pension savings. The retiree needs to withdraw a minimum of \$30K per year, but has no use for more than \$60K per year. All amounts are inflation adjusted. We assume that the retiree has other sources of income (work pension, government benefits) which, when added to the \$30K per year of withdrawals, accumulate to a satisfactory income level.

We consider a 30 year time horizon, since this is consistent with the Bengen (1994) scenario. Recall that the probability of a 65 year old Canadian male attaining the age of 95 is about 0.13, so this is a fairly conservative assumption. Similar probabilities are also true for other Western countries.

The pension savings are assumed to be held in a tax advantaged account, so that there are no tax consequences for rebalancing. We consider the withdrawals to be the desired amount before any income taxes.

Note that we also consider that the retiree has mortgage free real estate worth \$400K. We can regard this as a hedge of last resort.

Remark 5 (Reverse Mortgage Hedge) If we assume that a reverse mortgage can be used to obtain a non-recourse loan of half the value of the real estate (\$200K), then from a risk management point of view, any strategy which results in an expected shortfall $> -\$200$K is probably acceptable. All quantities are assumed inflation adjusted.

Other scenario details are listed in Table 1. As a point of comparison, the Bengen (1994) strategy would withdraw a fixed amount of \$40K (real) per year (4% of the initial \$1,000K), and rebalance to a constant weight of 50% in stocks at each rebalancing date.

Investment horizon T (years)	30.0
Equity market index	CRSP Cap-weighted index (real)
Bond index	30-day T-bill (US) (real)
Initial portfolio value W_0	1000
Mortgage free real estate	400
Cash withdrawal/rebalancing times	$t = 0, 1.0, 2.0, \ldots, 29.0$
Maximum withdrawal (per year)	$q_{\max} = 60$
Minimum withdrawal (per year)	$q_{\min} = 30$
Equity fraction range	$[0, p_{\max}]$
	$p_{\max} = 0.5, 0.8, 1.0, 1.3$
Borrowing spread μ_c^b	0.03
Rebalancing interval (years)	1.0
α (Expected shortfall parameter)	.05
Stabilization ϵ (see equation (11))	-10^{-4}
Market parameters	See Table 2

Table 1 Input data for examples. Monetary units: thousands of dollars. All amounts inflation adjusted.

For the computational study in this paper, we use data from the Center for Research in Security Prices (CRSP) on a monthly basis from 1926:1 to 2024:12.[9] The specific indices used are the CRSP 30 day U.S. T-bill index for the bond asset, and the CRSP cap-weighted total return index for the stock asset[10]. Retirees are, naturally, concerned with preserving real (not nominal) spending power. Hence, we use the US CPI index (from CSRP) to adjust these indices for inflation. We use the above market data in two different ways in subsequent investigations.

We use the threshold technique (Mancini, 2009; Cont and Mancini, 2011; Dang and Forsyth, 2016) to estimate the parameters for the parametric stochastic process models. Since the index data is in real terms, all parameters reflect real returns. Table 2 shows the results of calibrating the models to the historical data. The correlation ρ_{sb} is computed by removing any returns which occur at times corresponding to jumps in either series, and then using the sample covariance. Further discussion of the validity of assuming that the stock and bond jumps are independent is given in Forsyth (2020b).

We first compute and store the optimal controls in the *synthetic market*, that is, the stock market processes follow the parametric models (3) and (4). We then use Monte Carlo simulation, coupled with the stored optimal controls, to analyze the properties of the optimal allocation/withdrawal.

---

[9] More specifically, results presented here were calculated based on data from Historical Indexes, ©2024 Center for Research in Security Prices (CRSP), The University of Chicago Booth School of Business. Wharton Research Data Services was used in preparing this article. This service and the data available thereon constitute valuable intellectual property and trade secrets of WRDS and/or its third-party suppliers.

[10] The stock index includes all distributions for all domestic stocks trading on major U.S. exchanges.

CRSP	μ^s	σ^s	λ^s	u^s	η_1^s	η_2^s	ρ_{sb}
	0.088241	0.147361	0.31313	0.22581	4.3608	5.5309	0.096279
30-day T-bill	μ^b	σ^b	λ^b	u^b	η_1^b	η_2^b	ρ_{sb}
	0.0034	0.0139	0.3838	0.3947	61.510	53.356	0.096279

Table 2 Parameters for parametric market models (3) and (4), fit to CRSP data (inflation adjusted) for 1926:1 to 2024:12.

Finally, as a check on robustness, we test the controls (computed in the synthetic market), in the *historical market*, using block bootstrap resampled historical data.

9.1 Results: Convergence

We first check on the convergence of our method. The numerical parameters are given in Table 3.

Grid sizes ($b > 0$)	512×512
	1024×1024
	2048×2048
Extended grid $(N_1^\dagger, N_2^\dagger)$	$(2N_1, 2N_2)$
$\hat{x}_1^0, \hat{x}_2^0$	$\log(100)$
$(x_1)_{\min}, (x_2)_{\min}$	$-7.5 + \hat{x}_1^0$
$(x_1)_{\max}, (x_2)_{\max}$	$+10 + \hat{x}_2^0$
Monotonicity condition δ (equation(46))	10^{-6}
Number of points in w grid N_w (equation (54))	$4N_1$
Number of points in p grid N_p (equation (54))	$N_1/10$
Number of points in q grid N_q (equation (55))	$N_1/10$

Table 3 Numerical parameters

Remark 6 (Choice of $(x)_{\min}, (x)_{\max}$*)* Consider the grid with $b > 0$. In this case, we choose $x_{\min}$ (for both x_1, x_2) so that

$$e^{x_{\min}} W_0 < 10^{-4} W_0 \tag{57}$$

which is essentially zero allocation to bonds or stocks for practical purposes (compared to the original investment). For example, if the original investment was \$1000, then we assume that an allocation of less than \$0.10 is negligible.

In the case of $x_{\max}$ (for both x_1, x_2) we assume that the largest possible wealth would occur with an all-stock portfolio having zero withdrawals. A rough estimate of $x_{\max}$ is then given by assuming geometric Brownian motion with drift μ and volatility σ. We choose $x_{\max}$ such that

$$e^{x_{\max}} W_0 > e^{\mu T} e^{4\sigma\sqrt{T}} W_0 \,. \tag{58}$$

Equation (58) postulates a four sigma event, which is extremely unlikely. Note that we use $(x_1)_{\max} = (x_2)_{\max} = x_{\max}$, since the impulse control can (in theory) move the entire stock allocation into bonds instantaneously.

We have tested our results by increasing the ranges of $x_{\min}, x_{\max}$ and verified that the numerical results were unaffected to six digits.

Figure 2 shows the EW-ES efficient frontiers, computed using various grid sizes. The grid here refers to the grid for $b > 0$. There is an additional grid (of the same size) for $b < 0$. The optimal control is computed and stored, using the δ-monotone PIDE method. Statistics are then generated using Monte Carlo (MC) simulations, using the stored optimal controls, with 2.56×10^6 MC simulations being used. More precisely, Monte Carlo methods are used only as a forward policy evaluation tool, with a known policy. The optimal policies are obtained as part of the backward dynamic programming algorithm.

The curves for all grid sizes essentially overlap, indicating convergence for practical purposes. All subsequent results use the 2048×2048 grid. For comparison, we also show the results for the Bengen (1994) policy, which is clearly much less efficient than the optimal policy.

Table 4 shows detailed convergence results for a single point on the EW-ES frontier ($p_{\max} = 1.0, \kappa = 0.866$). The optimal controls computed using Algorithm 2 are stored, and then used in Monte Carlo (MC) simulations. The value function appears to converge smoothly (for the PIDE method).

	Algorithm in 2			Monte Carlo	
Grid	ES	$E[\sum_i q_i]/M$	Value Function (13)	ES	$E[\sum_i q_i]/M$
512×512	-14.6176	51.1162	1520.941	-10.302	51.1507
1024×1024	-9.16685	51.0706	1524.251	-7.5173	51.0781
2048×2048	-4.84764	50.9780	1525.179	-3.8866	50.9762

Table 4 EW-ES convergence test. Real stock index: deflated real capitalization weighted CRSP, real bond index: deflated 30 day T-bills. Scenario in Table 1. Parameters in Table 2. Maximum fraction in stocks $p_{\max} = 1.0$. $\kappa = 0.8660$. Other information is given in the caption of Figure 2.

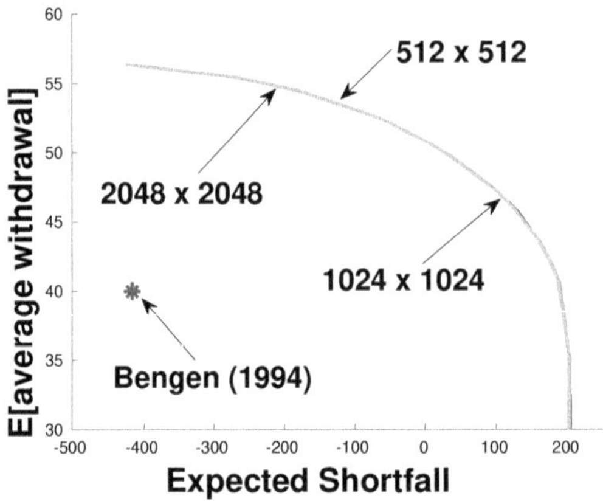

Figure 2 EW-ES convergence test. Real stock index: deflated real capitalization weighted CRSP, real bond index: deflated 30 day T-bills. Scenario in Table 1. Parameters in Table 2. The optimal control is determined by using the method in Algorithm 2. Grid refers to the grid used in the PIDE solve, $n_s \times n_b$, where n_s is the number of nodes in the $\log s$ direction, and n_b is the number of nodes in the $\log b$ direction. Units: thousands of dollars (real). The controls are stored, and then the final results are obtained using a Monte Carlo method, with 2.56×10^6 simulations. Maximum fraction in stocks $p_{\max} = 1.0$. Bengen (1994) refers to a constant weight in stocks $p = 0.5$, rebalanced annually, and constant yearly withdrawals of \$40K per year. All amounts are inflation adjusted.

It is also interesting to note that, in all our tests using the parameters in Table 3, we find that the wraparound error condition (51) is bounded by

$$\Delta x_1 \Delta x_2 \sum_{(j,k)\in \mathcal{D}^\dagger - \mathcal{D}} |\tilde{g}^\dagger_{j,k}| < 10^{-14} . \tag{59}$$

Note that

$$\Delta x_1 \Delta x_2 \sum_{(j,k)\in \mathcal{D}^\dagger} |\tilde{g}^\dagger_{j,k}| \leq 1 + \delta \simeq 1 ,$$

with δ given in Table 3. Consequently, the numerical result (59) indicates that the wrap-around error is only slightly larger than double precision machine epsilon.

9.2 Efficient Frontiers: Synthetic Market

Figure 3 shows the efficient frontiers computed using various values of $p_{\max}$ (labelled *pmax* on the Figure), the maximum values of the fraction in stocks. As must be true mathematically, the curves with higher values of $p_{\max}$ are above curves with smaller $p_{\max}$. However, a striking result is that the curves are all fairly tightly clustered, in comparison to the Bengen (1994) strategy. In particular, it seems reasonable to target an $ES \simeq 0$. The expected annual withdrawal for all values of $p_{\max}$ are very close for $ES = 0$.[11] However, this is not quite a free lunch, as we will see when we examine the results more closely.

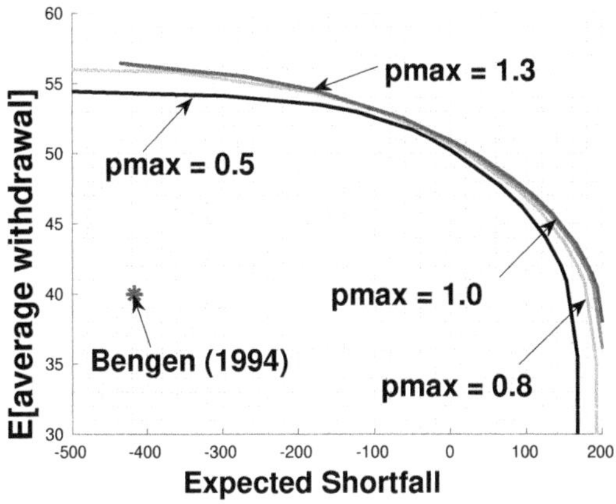

Figure 3 Comparison of maximum leverage constraint $p_{\max}$, synthetic market. Real stock index: deflated real capitalization weighted CRSP, real bond index: deflated 30 day T-bills. Scenario in Table 1. Parameters in Table 2. The optimal control is determined using Algorithm 2. The controls are stored (using 2048×2048 grid), and then the final results are obtained using a Monte Carlo method, with 2.56×10^6 simulations. Bengen (1994) refers to a constant weight in stocks $p = 0.5$, rebalanced annually, and constant yearly withdrawals of \$40K per year. All amounts are inflation adjusted.

9.3 Heat Maps: Synthetic Market

In order to gain some insight into the optimal controls, we plot the heat maps of the optimal asset allocation and the optimal withdrawals, for the cases $p_{\max} = 1.3, 1.0, 0.5$ in Figures 4, 5, 6. For each case, we choose the point on the efficient frontier so that $ES \simeq 0$, since this is an *interesting* point on the efficient frontier.

First, note that Figures 4(b), 5(b) and 6(b) show that the optimal withdrawal strategies for a wide range of $p_{\max}$ are very similar. In all cases, the withdrawal control is approximately *bang-bang*, that is, the optimal policy is only to withdraw the maximum or minimum amounts. In the continuous withdrawal limit, the withdrawal control can be proven to be bang-bang, Forsyth (2022). We have been unable to prove that this property holds for discrete withdrawals. However, numerical studies appear to show that the transition zone between $q_{\min}$ and $q_{\max}$ becomes smaller as the grid is refined. We conjecture that the true converged withdrawal control is bang-bang even for discrete rebalancing/withdrawals. The lefthand plots of Figures 4, 5, 6 show the

[11] Although from Remark 5, even $ES \simeq -200K$ is also acceptable. Hence a target of $ES = 0$ is very conservative.

optimal allocation strategies. We use the same colour scale for all plots, with the maximum allocation to stocks being 130%. It is interesting to see that the allocation strategies are quite similar as long as we restrict attention to the areas in the heat maps above the 5th wealth percentile. Large differences in the allocation appear below the 5th wealth percentile. This, of course, is due to the differing values of $p_{\max}$.[12]

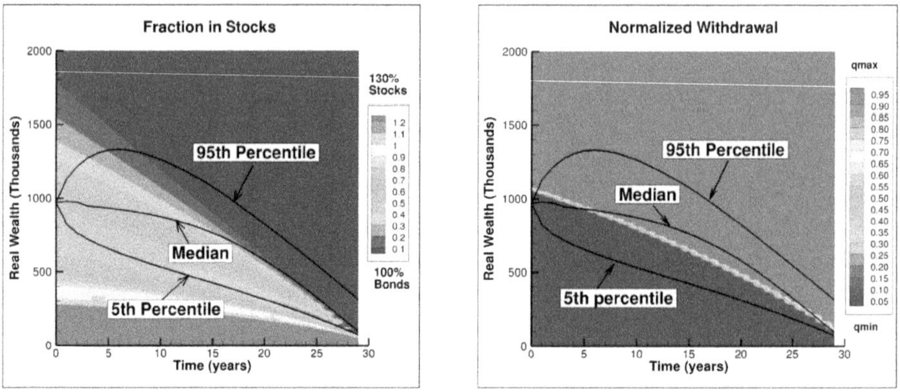

((a)) Fraction in stocks ((b)) Withdrawals

Figure 4 $p_{\max} = 1.3$. Heat map of the controls: fraction in stocks and withdrawals, computed using Algorithm 2. Real capitalization weighted CRSP index, and real 30-day T-bills. Scenario given in Table 1. Control computed and stored from the Algorithm 2 in the synthetic market. $q_{min} = 30$, $q_{\max} = 60$ (per year). $\kappa = 0.8583$, $EW \simeq 50.9$, $ES \simeq 0.96$. Percentiles from bootstrapped historical market. Normalized withdrawal $(q - q_{\min})/(q_{\max} - q_{\min})$. Units: thousands of dollars.

9.4 Percentiles Versus Time: Synthetic Market

Figures 7, 8 and 9 show the percentiles of fraction in stocks, wealth, and withdrawals versus time. We select the point on the efficient frontier for each case so that $ES \simeq 0$.

Rather surprisingly, the percentiles of fraction in stocks and percentiles of wealth are very similar, for all values of $p_{\max}$. However, we do see some differences in the withdrawal percentiles (the rightmost panel in each plot). The median withdrawal is slower to increase to the maximum for $p_{\max} = 0.5$ compared to the case with $p_{\max} = 1.0$.

In addition, we can see that even for $p_{\max} = 1.3$, the 95th percentile for stock allocation is less than 0.6 (Figure 7(a)), suggesting that with this level of ES risk, the optimal policy rarely allocates a large fraction to stocks.

[12] In ice hockey, if your team is down by one goal with one minute left to play, a common strategy is to *pull the goalie* (Asness and Brown, 2018). This gives you an extra attacker, but with an empty net. This is a risky strategy, but increases your chances of at least a tie. Below the 5th percentile of wealth, you have to invest the maximum amount in stocks in order to have a chance of satisfying the objective function.

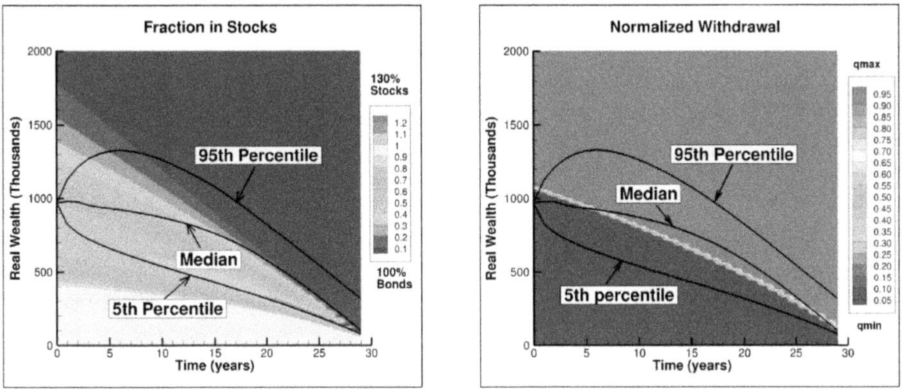

((a)) Fraction in stocks ((b)) Withdrawals

Figure 5 $p_{\max} = 1.0$. Heat map of the controls. $\kappa = 0.8860$, $EW = 50.7$, $ES = 4.6$. Other information as in caption for Figure 4.

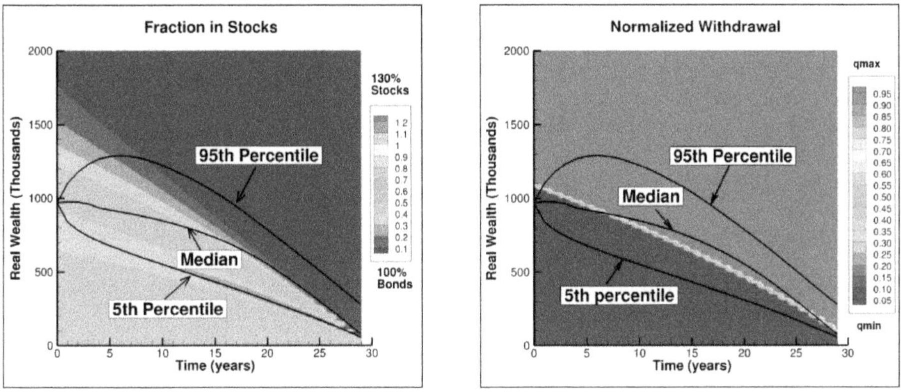

((a)) Fraction in stocks ((b)) Withdrawals

Figure 6 $p_{\max} = 0.5$. Heat map of the controls. $\kappa = 1.0$, $EW \simeq 50.2$, $ES = 1.25$. Other information as in caption for Figure 4.

10 Bootstrap Results

10.1 Historical Market

In order to check on the robustness of the above results, we proceed as follows. We compute and store the optimal controls based on the parametric model (3-4) as for the synthetic market case. However, we compute statistical quantities using the stored controls, but using bootstrapped historical return data directly. In this case, we make no assumptions concerning the stochastic processes followed by the stock and bond indexes. We remind the reader that all returns are inflation-adjusted. We use the stationary block

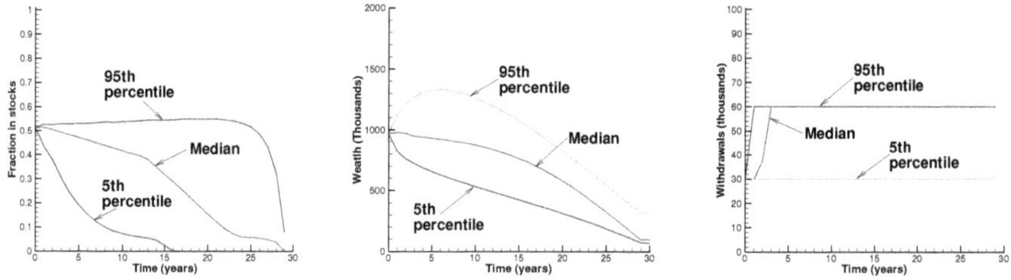

((a)) Percentiles fraction in stocks ((b)) Percentiles wealth ((c)) Percentiles withdrawals

Figure 7 $p_{\max} = 1.3$. Percentiles of the controls. Scenario in Table 1. Parameters based on the real CRSP index, and real 30-day T-bills (see Table 2). Control computed and stored from the Algorithm 2 in the synthetic market. $q_{min} = 30$, $q_{\max} = 60$ (per year), $EW \simeq 50.9$, $ES = 0.96$ ($\kappa = 0.8583$). Units: thousands of dollars.

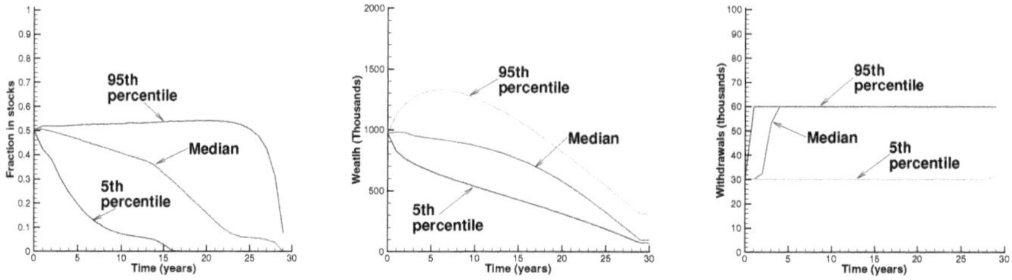

((a)) Percentiles fraction in stocks ((b)) Percentiles wealth ((c)) Percentiles withdrawals

Figure 8 $p_{\max} = 1.0$. Percentiles of the controls. $EW \simeq 50.7$, $ES = 4.6$ ($\kappa = 0.8860$). Other information as in caption for Figure 7.

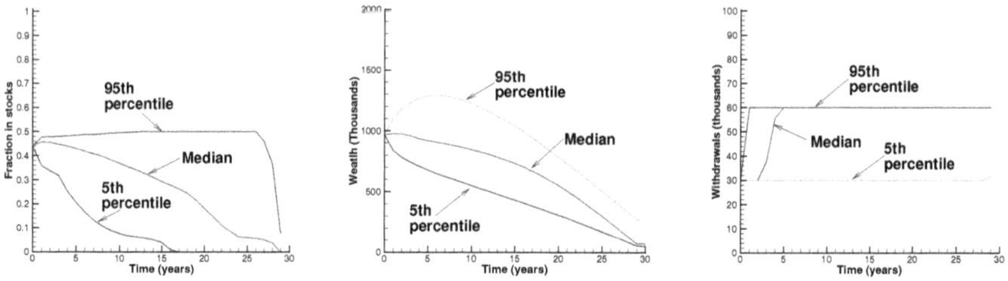

((a)) Percentiles fraction in stocks ((b)) Percentiles wealth ((c)) Percentiles withdrawals

Figure 9 $p_{\max} = 0.5$. Percentiles of the controls. $EW = 50.2$, $ES = 1.25$ ($\kappa = 1.0$). Other information as in caption of Figure 7.

bootstrap method (Politis and Romano, 1994; Politis and White, 2004; Patton et al., 2009; Cogneau and Zakalmouline, 2013; Dichtl et al., 2016; Cavaglia et al., 2022; Simonian and Martirosyan, 2022; Anarkulova et al., 2022).

A key parameter is the expected blocksize. Sampling the data in blocks accounts for serial correlation in the data series. We use the algorithm in Patton et al. (2009) to determine the optimal blocksize for the bond and stock returns separately, see Table 5. However, in our simulations, we use a paired sampling approach to simultaneously draw returns from both time series. In this case, a reasonable estimate for the blocksize for the paired resampling algorithm would be about 2.0 years. We will give results for a range of blocksizes as a check on the robustness of the bootstrap results. Detailed pseudo-code for block bootstrap resampling is given in Forsyth and Vetzal (2019).

Data	Optimal expected block size $\hat{b}$ (months)
30-day T-bill	51
CRSP cap weighted index	4.0

Table 5 Optimal expected blocksize $\hat{b} = 1/v$, from Patton et al. (2009). Range of historical data is between 1926:1 and 2024:12. The blocksize is a draw from a geometric distribution with $Pr(b = k) = (1 - v)^{k-1}v$. Expected blocksize estimate algorithm from Politis and White (2004); Patton et al. (2009) using market data from CRSP.

Figure 10 plots (i) the synthetic efficient frontier and (ii) bootstrap efficient frontier, computed using the synthetic market controls, tested in the historical market, for various blocksizes. Figure 10(a) shows the results for $p_{\max} = 1.3$. In this case, all the frontiers are quite close, especially compared to the Bengen (1994) strategy, tested in the historical market. The effect of using different expected blocksizes in the block bootstrap algorithm is quite small. Overall, this plot suggests that the optimal controls in this case are robust to model parameter misspecification.

The historical test for $p_{\max} = 0.5$ is shown in Figure 10(b). Again, the historical frontiers using different blocksizes are quite close. However, the historical efficient frontiers do deviate somewhat from the synthetic market frontier, for values of $ES > 100$. This indicates that there is some effect of model parameter misspecification, for large values of ES in the case of $p_{\max} = 0.5$. However, for values of $ES < 100$, the synthetic and historical market frontiers are very close. So, unless the retiree is extremely risk averse, this may not be a problem of practical concern.

Figures 11 and 12 show the percentiles of fraction in stocks, wealth, and optimal withdrawals, tested in the historical market. The two extreme cases: $p_{max} = 1.3$ and $p_{\max} = 0.5$ are shown.

The percentiles in stocks and wealth are fairly close for both cases. However, a comparison of Figures 11(c) and Figure 12(c) indicates that the median withdrawal increases from the minimum to the maximum over about three years for the case with $p_{max} = 1.3$. The comparable time frame for $p_{\max} = 0.5$ is about four years.

Although most retirees would prefer larger withdrawals during the early years of retirement, slightly lower initial withdrawals in exchange for never having more than 50% stock allocation might be seen as a reasonable tradeoff.

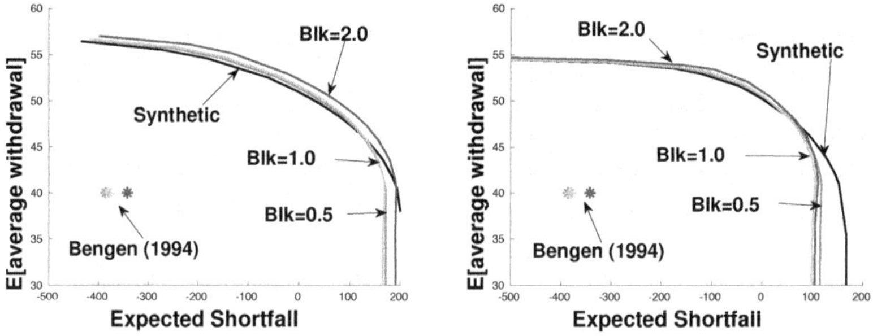

((a)) $p_{\max} = 1.3$ ((b)) $p_{\max} = 0.5$

Figure 10 Optimal controls were computed using the synthetic market model. These controls tested using bootstrapped historical data. Expected blocksizes (years) shown. 10^6 bootstrap resamples. Real stock index: deflated real capitalization weighted CRSP, real bond index: deflated 30 day T-bills. Scenario in Table 1. Parameters in Table 2. The Bengen control withdraws 40 per year (units: thousands of dollars), and rebalances annually to 50% bonds and 50% stocks. The Bengen results are also shown for expected blocksizes of 0.5, 1.0, 2.0 years.

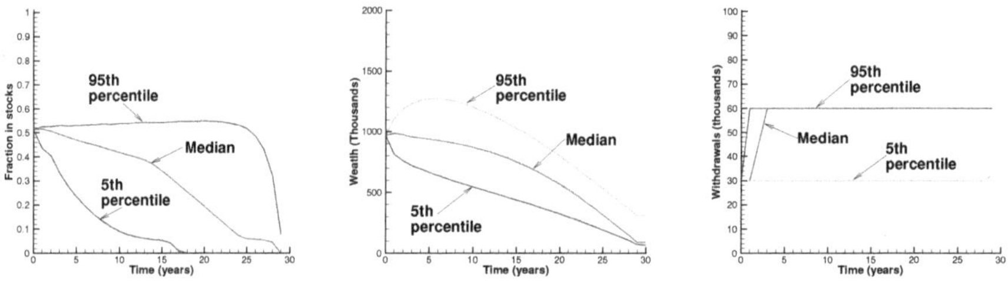

((a)) Percentiles fraction in stocks ((b)) Percentiles wealth ((c)) Percentiles withdrawals

Figure 11 $p_{\max} = 1.3$. Percentiles of the controls. $EW = 51.7$, $ES = 19$ ($\kappa = 0.8583$). Median withdrawal at $q_{\max}$ at year 3.0. Bootstrap results. Scenario in Table 1 Strategy computed in synthetic market. Tested in the bootstrapped historical market, expected blocksize 1.0 years. 10^6 bootstrap simulations. Bootstrap date based on the real CRSP index, and real 30-day T-bills 1926:1-2024:12. Control computed and stored using Algorithm 2 in the synthetic market. $q_{min} = 30$, $q_{\max} = 60$ (per year). Units: thousands of dollars.

11 Limitations of the Methods Used

We should note the following limitations of the PIDE-Fourier methods used in this Chapter.

- PIDE methods (in general) are limited to problems with no more than three stochastic factors. This excludes a number of problems of interest, for example portfolios having more than three risky assets, and adding stochastic mortality.
- Fourier-based approaches require stochastic processes which have known, closed-form expressions for the Fourier transform of the Green's function (equivalently the characteristic function).

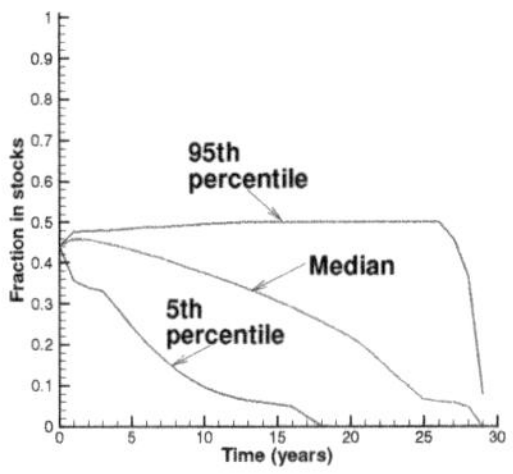

((a)) Percentiles fraction in stocks

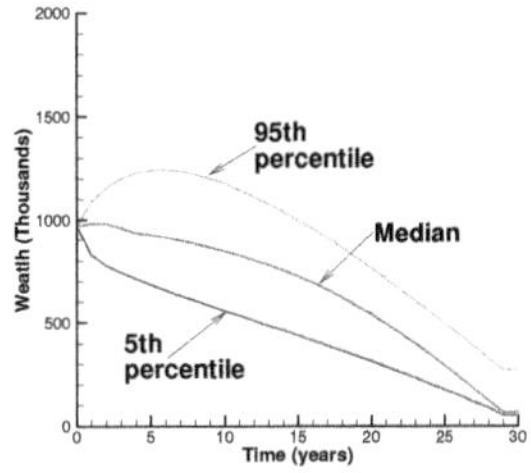

((b)) Percentiles wealth

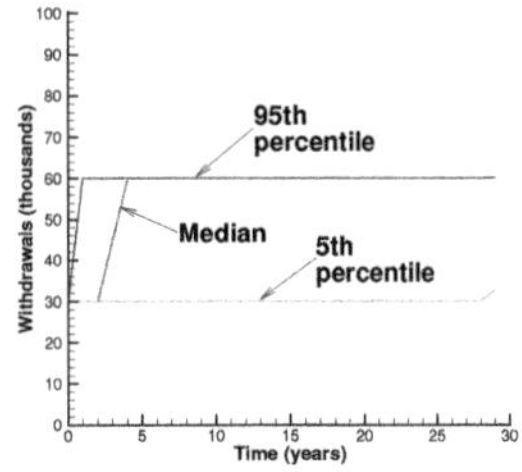

((c)) Percentiles withdrawals

Figure 12 $p_{\max} = 0.5$. Percentiles of the controls. Bootstrap results. $EW = 50.3$, $ES = 6.7$ ($\kappa = 1.0$). Median withdrawal at $q_{\max}$ at year 4.0. Other information as in caption to Figure 11.

- In order to ensure δ-monotonicity, piecewise linear basis functions are used. This caps the rate of convergence (for smooth functions) at second order in the grid spacing. This contrasts with the Chebyshev (Gaßet al., 2018; Glau et al., 2019) or Fourier-based COS method (Fang and Oosterlee, 2008; 2009; Ruijter et al., 2013) which, assuming smoothness, can achieve exponential convergence in some cases. But these techniques are not guaranteed to satisfy the δ-monotonicity property needed to prove convergence. We also note that it is non-trivial to obtain high-order convergence even for American options (Ruijter et al., 2013), due to the jump in the second derivative across the exercise boundary. Careful tracking of the exercise boundary is necessary. These methods would be difficult to implement for general optimal control problems.

At the same time these limitations are counterbalanced by the ability to obtain provably convergent solutions to complex control problems, with realistic constraints.

Finally we should mention a recent approach for high-dimensional stochastic control which is based on approximating the control by a neural network. This machine learning technique (Li and Forsyth, 2019; Van Staden et al., 2023; 2024; Chen et al., 2025) is purely data-driven, and makes no assumptions about the underlying stochastic processes. However, this also comes with the *caveat* that convergence proofs require determination of the global minimum of a non-convex function, which is problematic in practice.

12 Conclusions

In this paper, we have developed a technique for solving the optimal control problem associated with decumulation of a defined benefit (DC) pension plan. The basic control algorithm at each rebalancing time consists of (i) solution of an optimization problem and (ii) advancing the solution to the next rebalancing time (going backwards). We use a δ-monotone scheme, based on Fourier methods, for step (ii). This method preserves order relations to $O(\delta)$, which is an important property for numerical solution of optimal control. We pay particular attention to ensuring that the Fourier wrap-around error can be made arbitrarily small.

From a practical point of view, we have verified that the controls are robust by testing the controls using block bootstrap resampling of historical data, which makes no assumptions about stochastic processes for the underlying stock and bond indexes.

Perhaps the most interesting result, in terms of investment strategies for retirees, is the following. If the maximum fraction in equities is restricted to be less than 50%, this strategy is only slightly less efficient than allowing a maximum fraction of 130% (that is, using leverage). This suggests that for most retirees, there is no need to undertake risky strategies during the decumulation of a DC account.

These optimal control strategies are far more efficient than the typical four percent rule (Bengen, 1994). The optimal policy has an expected average withdrawal (real) of more than 5% of initial capital over thirty years, with approximately zero expected shortfall at age 95 (based on historical data). In contrast, the strategy suggested in (Bengen, 1994), withdraws 4% of initial capital annually, but has an expected shortfall of more than 35% of the initial capital at age 95.

Appendix

Wrap Around Error: Details

To gain some insight into the wrap-around problem, we consider a highly simplified, one-dimensional problem. To avoid subscript clutter, in this section, we use the notation

$$\tilde{g}(m-\ell) \equiv \tilde{g}_{m-\ell} \; ; \; u^n(m) \equiv u^n_m .$$

Using the above notation, consider the discrete time advance convolution

$$u^n(m) = \Delta x \sum_{\ell=-N^\dagger/2}^{N^\dagger/2-1} \tilde{g}(m-\ell)\, u^{n-1}(\ell) \,, \qquad m = -N/2, \ldots, N/2-1 \,, N \le N^\dagger \,, \tag{60}$$

where we assume periodic extension of $\tilde{g}$

$$\tilde{g}(\ell \pm N^\dagger) = \tilde{g}(\ell) \,.$$

In the above, if $N^\dagger > N$, nodes corresponding to $m < -N/2$ and $m > N/2 - 1$ are the padded nodes. In this case, we assume that the padded nodal values $u^n(m)$, $m < -N/2$ and $m > N/2 - 1$, are determined by boundary data.

As an example of wrap-around error, we examine a worst case term in equation (60). Consider the term in (60) corresponding to $m = -N/2$, and $l = N^\dagger/2 - 1$, namely

$$\Delta x \, \tilde{g}(-N/2 - N^\dagger/2 + 1)\, u^{n-1}(N^\dagger/2 - 1). \tag{61}$$

By periodic extension, we shift the argument of $\tilde{g}(\cdot)$ by $N^\dagger$, resulting in

$$\tilde{g}(-N/2 - N^\dagger/2 + 1) = \tilde{g}(-N/2 - N^\dagger/2 + 1 + N^\dagger) = \tilde{g}(-N/2 + N^\dagger/2 + 1), \quad (62)$$

and hence, the term (61) becomes

$$\Delta x\, \tilde{g}(-N/2 + N^\dagger/2 + 1)\, u^{n-1}(N^\dagger/2 - 1). \quad (63)$$

Hence, in this extreme case, equation (60) becomes

$$u^n(-N/2) = \Delta x\, \tilde{g}(-N/2 + N^\dagger/2 + 1)\, u^{n-1}(N^\dagger/2 - 1) + \sum_{l=-N^\dagger/2}^{N^\dagger/2-2} (\text{ remaining terms }). \quad (64)$$

Example 1 (No padding: $N^\dagger = N$) Suppose we do not use any padding, so that $N^\dagger = N$. In this case, equation (64) becomes

$$u^n(-N/2) = \Delta x\, \tilde{g}(1)\, u^{n-1}(N/2 - 1) + \sum_{l=-N/2}^{N/2-2} (\text{ remaining terms }).$$

Since, in general, $\tilde{g}(1)$ is not small, we can see that the term $u^{n-1}(N/2 - 1)$ has a considerable effect on $u^n(-N/2)$, which should not be the case. We can see here that the periodic extension of $\tilde{g}$ causes a *wrap-around* effect.

Example 2 (Padding: $N^\dagger = 2N$) If $N^\dagger = 2N$, then equation (64) becomes

$$u^n(-N/2) = \Delta x\, \tilde{g}(N/2 + 1)\, u^{n-1}(N^\dagger/2 - 1) + \sum_{l=-N^\dagger/2}^{N^\dagger/2-2} (\text{ other terms }). \quad (65)$$

If we select N sufficiently large so that $\tilde{g}(\pm N/2) \simeq 0$, then the leading term in equation (65) is small, and hence, the wrap-around error is reduced.

Wrap-Around: A Formal Result

From our previous examples, we can see that wrap-around may occur in equation (60) if

$$(m - \ell) < -N^\dagger/2 \text{ or } (m - \ell) > N^\dagger/2 - 1.$$

This leads us to the following formal definition of wrap-around error. We show that, with $N^\dagger = 2N$, the wrap-around error is sufficiently reduced.

Definition 2 (Wrap-around error) Assume that $\{\tilde{g}(\cdot)\}$ is periodic with period $N^\dagger$ and that the $u^n(m)$, for $m < -N/2$ or $m > N/2 - 1$, are determined by boundary data with $N^\dagger = 2N$. Then the wrap-around error for equation (60), at timestep n, denoted by e^n_{wrap}, is

$$e^n_{\text{wrap}} = \max_m \left\{ \Delta x \sum_{\ell \in \mathbb{N}^\dagger} \left| \tilde{g}(m-\ell)\, u^{n-1}(l) \right| \left(\mathbf{1}_{\{(m-\ell) < -N^\dagger/2\}} + \mathbf{1}_{\{(m-\ell) > N^\dagger/2-1\}} \right) \right\} . \tag{66}$$

We can now state the following result.

Theorem 2 *Let $\tilde{g}(\cdot)$ be periodic with period $N^\dagger$ and $u^n(m)$, for $m < -N/2$ or $m > N/2 - 1$, be determined by boundary data with $N^\dagger = 2N$. Assume further that $u^n(m)$ is bounded, so for $0 \le n \le M$, there exists a positive constant C such that*[13]

$$|u^n(m)| \le C, \quad m \in \mathbb{N}^\dagger, \forall n . \tag{67}$$

If N is selected sufficiently large so that

$$\Delta x \sum_{\ell=-N^\dagger/2}^{-N/2-1} |\tilde{g}(\ell)| + \Delta x \sum_{\ell=N/2}^{N^\dagger/2-1} |\tilde{g}(\ell)| \le \epsilon_e \Delta\tau , \tag{68}$$

then the wrap-around error after N steps is bounded by $TC\epsilon_e$.

Proof. Applying property (67) to equation (66) gives

$$e^n_{\text{wrap}} \le C \max_m \left\{ \Delta x \sum_{\ell=-N^\dagger/2}^{N^\dagger/2-1} |\tilde{g}(m-\ell)| \left(\mathbf{1}_{\{(m-\ell) < -N^\dagger/2\}} + \mathbf{1}_{\{(m-\ell) > N^\dagger/2-1\}} \right) \right\} . \tag{69}$$

Recall that $m \in \{-N/2, \ldots, N/2-1\}$, hence the worst case values of m on the righthand side of equation (69) are $m = -N/2$ and $m = N/2 - 1$. Thus equation (69) gives

$$\begin{aligned} e^n_{\text{wrap}} \le{}& C\Delta x \sum_{\ell=-N^\dagger/2}^{N^\dagger/2-1} |\tilde{g}(N/2-1-\ell)| \mathbf{1}_{\{(N/2-1-\ell) > N^\dagger/2-1\}} \\ &+ C\Delta x \sum_{\ell=-N^\dagger/2}^{N^\dagger/2-1} |\tilde{g}(-N/2-\ell)|\, \mathbf{1}_{\{(-N/2-\ell) < -N^\dagger/2\}} . \end{aligned} \tag{70}$$

[13] This is essentially a stability condition. See Forsyth and Labahn (2019) for a proof of stability for the δ-monotone method.

Also, since $N = N^\dagger/2$ equation (70) becomes

$$e^n_{\text{wrap}} \le C\Delta x \sum_{\ell=-N^\dagger/2}^{N^\dagger/2-1} |\tilde{g}(N^\dagger/4 - 1 - \ell)| \, \mathbf{1}_{\{(N^\dagger/4-1-\ell)>N^\dagger/2-1\}}$$
$$+ C\Delta x \sum_{\ell=-N^\dagger/2}^{N^\dagger/2-1} |\tilde{g}(-N^\dagger/4 - l)| \, \mathbf{1}_{\{(-N^\dagger/4-\ell)<-N^\dagger/2\}}, \tag{71}$$

and eliminating the indicator functions gives

$$e^n_{\text{wrap}} \le C\Delta x \sum_{\ell=-N^\dagger/2}^{-N^\dagger/4-1} |\tilde{g}(N^\dagger/4 - 1 - \ell)| + C\Delta x \sum_{\ell=N^\dagger/4+1}^{N^\dagger/2-1} |\tilde{g}(-N^\dagger/4 - \ell)|.$$

Shifting $\tilde{g}(\cdot)$ by $\pm N^\dagger$ so that the argument of $\tilde{g}(\cdot)$ is in the range $[-N^\dagger/2, N^\dagger/2 - 1]$, implies

$$e^n_{\text{wrap}} \le C\Delta x \sum_{\ell=-N^\dagger/2}^{-N^\dagger/4-1} |\tilde{g}(N^\dagger/4 - 1 - \ell - N^\dagger)| + C\Delta x \sum_{\ell=N^\dagger/4+1}^{N^\dagger/2-1} |\tilde{g}(-N^\dagger/4 - \ell + N^\dagger)|$$
$$= C\Delta x \sum_{\ell=-N^\dagger/2}^{-N^\dagger/4-1} |\tilde{g}(-3N^\dagger/4 - 1 - \ell)| + C\Delta x \sum_{\ell=N^\dagger/4+1}^{N^\dagger/2-1} |\tilde{g}(3N^\dagger/4 - \ell)|. \tag{72}$$

Rearranging the indices, gives

$$e^n_{\text{wrap}} \le C\Delta x \sum_{\ell=-N^\dagger/2}^{-N^\dagger/4-1} |\tilde{g}(\ell)| + C\Delta x \sum_{\ell=N^\dagger/4+1}^{N^\dagger/2-1} |\tilde{g}(\ell)|, \tag{73}$$

which, since $N = N^\dagger/2$, implies that equation (73) satisfies

$$e^n_{\text{wrap}} \le C\Delta x \sum_{\ell=-N^\dagger/2}^{-N/2-1} |\tilde{g}(\ell)| + C\Delta x \sum_{\ell=N/2+1}^{N^\dagger/2-1} |\tilde{g}(\ell)|. \tag{74}$$

Since

$$\Delta x \sum_{\ell=N/2+1}^{N^\dagger/2-1} |\tilde{g}(\ell)| \le \Delta x \sum_{\ell=N/2}^{N^\dagger/2-1} |\tilde{g}(\ell)| \, , \tag{75}$$

then

$$e^n_{\text{wrap}} \le C\Delta x \sum_{\ell=-N^\dagger/2}^{-N/2-1} |\tilde{g}(\ell)| + C\Delta x \sum_{\ell=N/2}^{N^\dagger/2-1} |\tilde{g}(\ell)| \,, \tag{76}$$

$$= C\epsilon_e \Delta\tau \tag{77}$$

where the last step follows from condition (68). Applying equation (77) recursively gives the bound $TC\epsilon_e$. □

It might at first sight seem odd to weaken the error bound using equation (75). However, this makes the final result easy to interpret. Let

$$\mathcal{S}^\dagger = \{-N^\dagger/2, \ldots, +N^\dagger/2 - 1\} \; ; \; \mathcal{S} = \{-N/2, \ldots, +N/2 - 1\} \,, \tag{78}$$

so that $\mathcal{S}^\dagger$ is the set of node indices for the padded domain, and $\mathcal{S}$ is the set of node indices for the original domain. Consequently, condition (76) can be written compactly as

$$e^n_{\text{wrap}} \le C\Delta x \sum_{\ell \in (\mathcal{S}^\dagger - \mathcal{S})} |\tilde{g}(\ell)| \;\le\; C\epsilon_e \Delta\tau \,. \tag{79}$$

Proving Theorem 1 basically follows the previous proof of the simpler case. Divide the region $\mathcal{D}' - \mathcal{D}$ outside $\mathcal{D}$ into right and left A_r, A_ℓ and upper and lower A_u, A_d regions as follows.

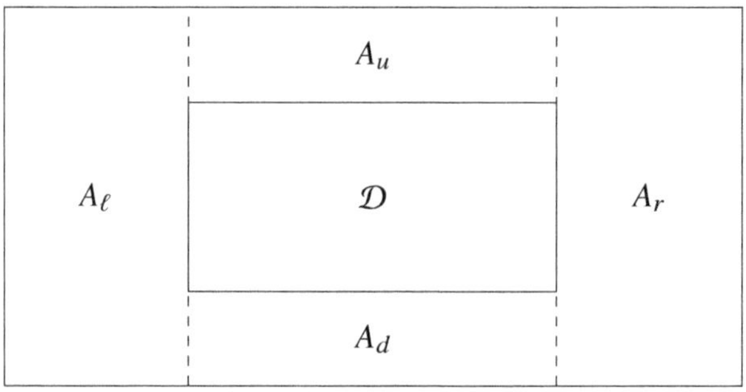

Our goal is to give an upper bound to

$$\varepsilon^n_{\text{wrap}} = \Delta x_1 \Delta x_2 \max_{(\ell,m)\in\mathcal{D}} \sum_{(j,k)\in\mathcal{D}^\dagger} \left|\tilde{g}^\dagger_{\ell-j,m-k} v^\dagger_{j,k}(\tau^n)\right| \mathbf{1}_{(\ell-j,m-k)\notin\mathcal{D}^\dagger} \,,$$

which by the boundedness assumptions of Theorem 1 is the same as bounding

$$\Delta x_1 \Delta x_2 \max_{(\ell,m)\in\mathcal{D}} \sum_{(j,k)\in\mathcal{D}^\dagger} \left|\tilde{g}^\dagger_{\ell-j,m-k}\right| \mathbf{1}_{(\ell-j,m-k)\notin\mathcal{D}^\dagger} \,.$$

As in the one-dimensional wraparound error, our worst-case values occur at borders, in this case the four corners of $\mathcal{D}$. In the horizontal direction when the first components are $-N_1/2$ and $N_1/2 - 1$ we can use the same manipulations as done in the one-dimensional case to obtain the wraparound error contribution from A_ℓ and A_r and use an identical argument when the second argument is $-N_2/2$ and $N_2/2 - 1$ to get the wraparound error contribution from A_u and A_d. These manipulations reduce to

$$e^n_{\text{wrap}} \leq C \cdot \Delta x_1 \Delta x_2 \sum_{(\ell,m)\in(\mathcal{D}^\dagger - \mathcal{D})} |\tilde{g}(\ell,m)| \;\leq\; C\epsilon_e \Delta\tau \,,$$

for each time step and hence give the error bound $e^{\mathcal{N}}_{\text{wrap}} \leq T\epsilon_e C$ after $\mathcal{N}$ steps.

References

Alonso-Garcia, J., O. Wood, and J. Ziveyi (2018). Pricing and hedging guaranteed minimum withdrawal benefits under a general Lévy framework using the COS method. *Quantitative Finance 18*, 1049–1075.

Anarkulova, A., S. Cederburg, and M. S. O'Doherty (2022). Stocks for the long run? Evidence from a broad sample of developed markets. *Journal of Financial Economics 143:1*, 409–433.

Anarkulova, A., S. Cederburg, M. S. O'Doherty, and R. W. Sias (2025). The safe withdrawal rate: Evidence from a broad sample of developed markets. *Journal of Pension Economics and Finance 24:3*, 464–500.

Asness, C. and A. Brown (2018). Pulling the goalie: Hockey and investment implications. `https://content.lesaffaires.com/LAF/lacom/goallie.pdf`.

Barles, G. and P. Souganidis (1991). Convergence of approximation schemes for fully nonlinear equations. *Asymptotic Analysis 4*, 271–283.

Bengen, W. (1994). Determining withdrawal rates using historical data. *Journal of Financial Planning 7*, 171–180.

Bjork, T., M. Khapko, and A. Murgoci (2021). *Time inconsistent control theory with finance applications*. New York: Springer Finance.

Cavaglia, S., L. Scott, K. Blay, and S. Hixon (2022). Multi-asset class factor premia: A strategic asset allocation perspective. *The Journal of Portfolio Management 48:9*, 14–32.

Chen, M., M. Shirazi, P. A. Forsyth, and Y. Li (2025). Machine learning and Hamilton-Jacobi-Bellman equation for optimal decumulation: a comparison study. *Journal of Computational Finance 29:1*, 77–118.

Cogneau, P. and V. Zakalmouline (2013). Block bootstrap methods and the choice of stocks for the long run. *Quantitative Finance 13*, 1443–1457.

Cont, R. and C. Mancini (2011). Nonparametric tests for pathwise properties of semimartingales. *Bernoulli 17*, 781–813.

Cui, X., D. Li, X. Qiao, and M. Strub (2022). Risk and potential: An asset allocation framework with applications to robo-advising. *Journal of the Operations Research Society of China 10*, 529–558.

Dang, D.-M. and P. A. Forsyth (2016). Better than pre-commitment mean-variance portfolio allocation strategies: a semi-self-financing Hamilton-Jacobi-Bellman equation approach. *European Journal of Operational Research 250*, 827–841.

Dichtl, H., W. Drobetz, and M. Wambach (2016). Testing rebalancing strategies for stock-bond portfolios across different asset allocations. *Applied Economics 48*, 772–788.

Fang, F. and C. Oosterlee (2008). A novel pricing method for European options based on Fourier cosine series expansions. *SIAM Journal on Scientific Computing 31*, 826–848.

Fang, F. and C. Oosterlee (2009). Pricing early-exercise and discrete barrier options by Fourier cosine series expansions. *Numerische Mathematik 114*, 27–62.

Forsyth, P. and G. Labahn (2019). ϵ-Monotone Fourier methods for optimal stochastic control in finance. *Journal of Computational Finance 22:4*, 25–71.

Forsyth, P. A. (2020a). Multi-period mean CVAR asset allocation: Is it advantageous to be time consistent? *SIAM Journal on Financial Mathematics 11:2*, 358–384.

Forsyth, P. A. (2020b). Optimal dynamic asset allocation for DC plan accumulation/decumulation: Ambition-CVAR. *Insurance: Mathematics and Economics 93*, 230–245.

Forsyth, P. A. (2022). A stochastic control approach to defined contribution plan decumulation: "The Nastiest, Hardest Problem in Finance". *North American Actuarial Journal 26:2*, 227–252.

Forsyth, P. A. and K. R. Vetzal (2019). Optimal asset allocation for retirement savings: deterministic vs. time consistent adaptive strategies. *Applied Mathematical Finance 26:1*, 1–37.

Forsyth, P. A., K. R. Vetzal, and G. Westmacott (2020). Optimal asset allocation for a DC pension decumulation with a variable spending rule. *ASTIN Bulletin 50:2*, 419–447.

Forsyth, P. A., K. R. Vetzal, and G. Westmacott (2022). Optimal control of the decumulation of a retirement portfolio with variable spending and dynamic asset allocation. *ASTIN Bulletin 51:3*, 905–938.

Garroni, M. G. and J. L. Menaldi (1992). *Green functions for second order parabolic integro-differential problems*. New York: Longman Scientific.

Gaß, M., K. Grau, M. Mahlstedt, and M. Mair (2018). Chebyshev interpolation for parametric option pricing. *Finance and Stochastics 22*, 701–731.

Glau, K., M. Mahlstedt, and C. Potz (2019). A new approach for American option pricing: the dynamic Chebyshev method. *SIAM Journal on Scientific Computing 41*, B153–B180.

Ignatieva, K., A. Song, and J. Ziveyi (2018). Fourier space timestepping algorithm for valuing guaranteed minimum withdrawal benefits in variable annuities under regime switching and stochastic mortality. *ASTIN Bulletin 48*, 139–169.

Kou, S. G. (2002). A jump-diffusion model for option pricing. *Management Science 48*, 1086–1101.

Kou, S. G. and H. Wang (2004). Option pricing under a double exponential jump diffusion model. *Management Science 50*, 1178–1192.

Li, Y. and P. Forsyth (2019). A data-driven neural network approach to optimal asset allocation for target based defined contribution pension plans. *Insurance: Mathematics and Economics 86*, 189–204.

Lin, Y., R. MacMinn, and R. Tian (2015). De-risking defined benefit plans. *Insurance: Mathematics and Economics 63*, 52–65.

Lippa, J. (2013). A Fourier space time-stepping approach to problems in finance. MMath thesis, University of Waterloo.

MacDonald, B.-J., B. Jones, R. J. Morrison, R. L. Brown, and M. Hardy (2013). Research and reality: A literature review on drawing down retirement financial savings. *North American Actuarial Journal 17*, 181–215.

MacMinn, R., P. Brockett, J. Wang, Y. Lin, and R. Tian (2014). The securitization of longevity risk and its implications for retirement security. In O. S. Mitchell, R. Maurer, and P. B. Hammond (Eds.), *Recreating Sustainable Retirement*, pp. 134–160. Oxford: Oxford University Press.

Mancini, C. (2009). Non-parametric threshold estimation models with stochastic diffusion coefficient and jumps. *Scandinavian Journal of Statistics 36*, 270–296.

Menoncin, F. and E. Vigna (2017). Mean-variance target based optimisation for defined contribution pension schemes in a stochastic framework. *Insurance: Mathematics and Economics 76*, 172–184.

Patton, A., D. Politis, and H. White (2009). Correction to: automatic block-length selection for the dependent bootstrap. *Econometric Reviews 28*, 372–375.

Peijnenburg, K., T. Nijman, and B. J. Werker (2016). The annuity puzzle remains a puzzle. *Journal of Economic Dynamics and Control 70*, 18–35.

Pfeiffer, S., J. R. Salter, and H. E. Evensky (2013). Increasing the sustainable withdrawal rate using the standby reverse mortgage. *Journal of Financial Planning 26:12*, 55–62.

Politis, D. and J. Romano (1994). The stationary bootstrap. *Journal of the American Statistical Association 89*, 1303–1313.

Politis, D. and H. White (2004). Automatic block-length selection for the dependent bootstrap. *Econometric Reviews 23*, 53–70.

Ritholz, B. (2017). Tackling the 'nastiest, hardest problem in finance'. `www.bloomberg.com/view/articles/2017-06-05/tackling-the-nastiest-hardest-problem-in-finance`.

Rockafellar, R. T. and S. Uryasev (2000). Optimization of conditional value-at-risk. *Journal of Risk 2*, 21–42.

Ruijter, M. J., C. W. Oosterlee, and R. Albers (2013). On the Fourier Cosine series expansion method for stochastic control problems. *Numerical Linear Algebra and Applications 20*, 598–625.

Shefrin, H. M. and R. H. Thaler (1988). The behavioral life-cycle hypothesis. *Economic Inquiry 26*, 609–643.

Simonian, J. and A. Martirosyan (2022). Sharpe parity redux. *The Journal of Portfolio Management 48:9*, 183–193.

Strub, M., D. Li, X. Cui, and J. Gao (2019). Discrete-time mean-CVaR portfolio selection and time-consistency induced term structure of the CVaR. *Journal of Economic Dynamics and Control 108*. Article 103751 (electronic).

Thinking Ahead Institute (2024). Global pension assets study 2024. `https://www.thinkingaheadinstitute.org/content/uploads/2024/02/GPAS-2024.pdf`.

Van Staden, P. M., D.-M. Dang, and P. Forsyth (2018). Time-consistent mean-variance portfolio optimization: a numerical impulse control approach. *Insurance: Mathematics and Economics 83*, 9–28.

Van Staden, P. M., P. A. Forsyth, and Y. Li (2023). Beating a benchmark: dynamic programming may not be the right numerical approach. *SIAM Journal on Financial Mathematics 14*(2), 407–451.

Van Staden, P. M., P. A. Forsyth, and Y. Li (2024). A global-in-time neural network approach to dynamic portfolio optimization. *Applied Mathematical Finance 31:3*, 131–163.

Vigna, E. (2014). On efficiency of mean-variance based portfolio selection in defined contribution pension schemes. *Quantitative Finance 14*, 237–258.

Vigna, E. (2022). Tail optimality and preferences consistency for intertemporal optimization problems. *SIAM Journal on Financial Mathematics 13*(1), 295–320.

Computational Optics

Martijn Anthonissen, Pieter Braam, Lisa Kusch, Koondanibha Mitra,
Jan ten Thije Boonkkamp and Wilbert IJzerman

Abstract The key problem in computational optics is to find the shapes of optical surfaces that transform a given light source into a desired target distribution. We present a mathematical model based on the principles of geometrical optics and formulated in terms of the optical map. This map connects the source and target domains and satisfies energy conservation, leading to an optimal transport problem expressed as a nonlinear partial differential equation of Monge–Ampère type. We use a least-squares method to compute the optical map and to construct the shapes of the optical surfaces. The computation is an iterative process with fast convergence. The method can handle complicated source and target distributions.

1 Introduction

We present mathematical models for optical systems with a parallel or point source and a parallel, point, near-field or far-field target, see Fig. 1. If we use reflectors only and take the minimum number of freeform surfaces required, this gives eight systems. Similarly, we get eight basic lens systems if we only use freeform lens surfaces. The surfaces are referred to as freeform since they have no imposed symmetries. Our goal is to find the shapes and locations of the freeform surfaces that convert a given source distribution into a desired target distribution. Our models are based on Hamilton's characteristic functions and energy conservation.

The mathematical models are formulated as the standard Monge–Ampère equation (SMA), the generalized Monge–Ampère equation (GMA) and a so-called generated Jacobian equation (GJE). An overview can be found in [2, 53]. The number of circles

Martijn Anthonissen, Pieter Braam, Lisa Kusch, Koondanibha Mitra, Jan ten Thije Boonkkamp
Eindhoven University of Technology, PO box 513, 5600 MB Eindhoven, The Netherlands,
m.j.h.anthonissen@tue.nl

Wilbert IJzerman
Signify Research, High Tech Campus 7, 5656 AE Eindhoven, The Netherlands

B. Koren and D. Roose (eds.), *Numerical Analysis and Scientific Computing: an Anthology*, Lecture Notes in Computational Science and Engineering 156,
https://doi.org/10.1007/978-3-032-33428-2_13

that surround a system in Fig. 1 indicates the model required: one circle indicates standard Monge–Ampère, two circles generalized Monge–Ampère and three circles generated Jacobian equation.

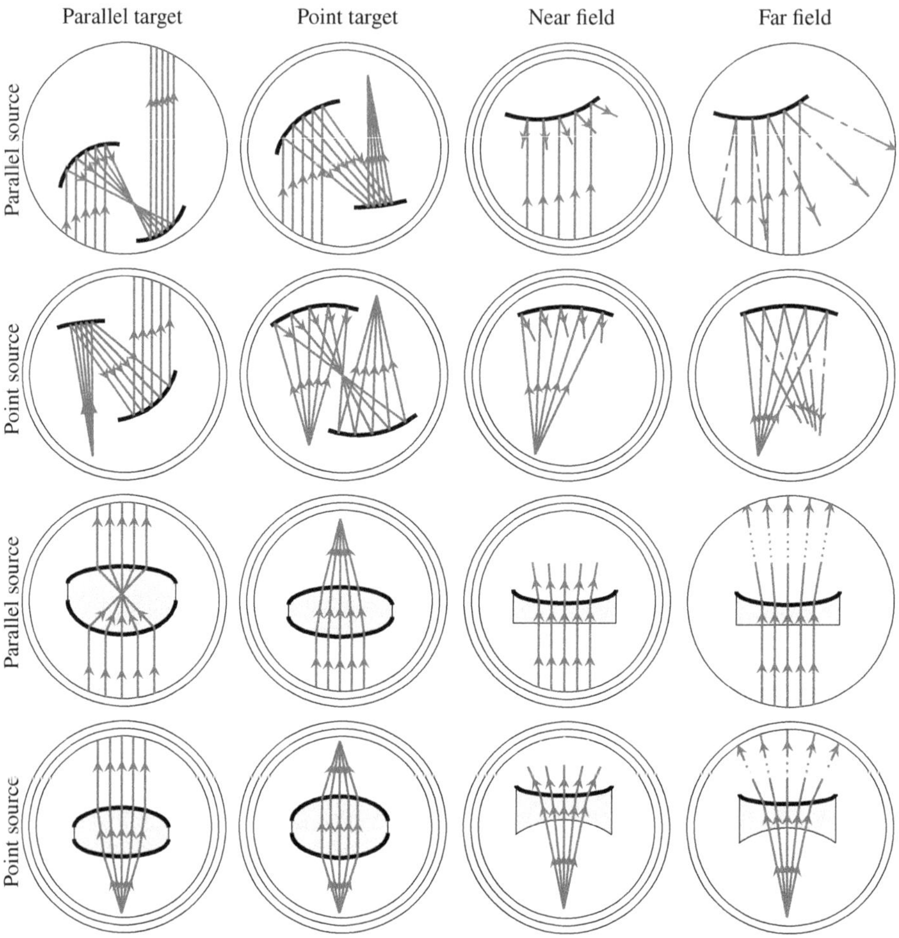

Figure 1 Schematic representations of the basic optical systems. The top eight are reflector systems and the bottom eight are lens systems. Freeform optical surfaces are plotted as thick black lines.

Generally, papers in freeform illumination optics cover only one particular optical system. Here we derive models for all sixteen using one framework. We give a short overview of literature—a more detailed survey can be found in [53, Ch. 5]. We outline three approaches: (1) methods that directly solve the partial differential equation (PDE) for the optical surface using finite difference or finite element discretizations, (2) optimization strategies for the Monge–Kantorovich optimal transport problem and (3) methods based on ray mapping. The first approach is followed by Benamou, Froese and Oberman [5, 6, 28–31, 42, 43], Brix et al. [13, 14], Caboussat et al. [16, 17], Dean and Glowinski [20–22], Prins et al. [49–51], Romijn et al. [53–57], Wu et al. [18,71–74,79,80], and Yadav et al. [76–78]. Lakkis et al. [35,39,40] introduce a Galerkin-type finite element method that can solve PDEs of Monge-Ampère type.

The second approach is to use optimization techniques for the Monge–Kantorovich transport problem that corresponds to the optical design problem. Examples are Angenent et al. [1], Benamou et al. [4], and Haber et al. [33]. Doskolovich et al. reduce the transport problem to a linear assignment problem [15, 23, 24]. Oliker et al. have introduced the methods of supporting paraboloids and ellipsoids, now known as the supporting quadric method (SQM) [32, 36–38, 44–48]. SQM combines geometrical methods with calculus of variations. Instead of dealing directly with the complicated nonlinear PDEs, it produces weak solutions that are the envelope of a set of, e.g., paraboloids of revolution. The surfaces found are not necessarily differentiable. The method provides both proofs of existence of solutions as well as a way to compute them numerically.

Third are ray-mapping techniques. In these methods a ray mapping is computed, which is subsequently used to compute the surface using the law of reflection or Snell's law. This method is used by Bösel and Gross [9, 10] and Feng et al. [25–27].

We have organized this contribution as follows. In Section 2, we derive models for three representative examples that cover the SMA, GMA and GJE models; the examples are a parallel-to-parallel reflector system, a parallel-to-point reflector system and a parallel-to-near-field lens system. All sixteen basic systems can be modeled similarly to these three systems. We will also go beyond the basic systems by outlining the model for a parallel-to-two-targets reflector system. We present our least-squares algorithm to numerically solve the mathematical models in Section 3 and present a challenging numerical example in Section 4. We draw conclusions in Section 5.

2 Mathematical Models for Optical Systems

We formulate mathematical models for a parallel-to-parallel reflector system, a parallel-to-point reflector system and a parallel-to-near-field lens system. We first outline the models in 2D for ease of presentation and then sketch the generalization to 3D. Our numerical method solves the full 3D problems.

2.1 Model 1: Standard Monge–Ampère Equation

In this section we present the mathematical model for the parallel-to-parallel reflector system, the top left basic system in Fig. 1. We will find that this system can be modeled by the standard Monge–Ampère equation.

We first consider the two-dimensional problem as sketched in Fig. 2. Given are a source domain S located at $z = 0$ and a target domain T at $z = L$. Both are assumed to be compact sets. We denote the spatial coordinates at the source by x and at the target by y.

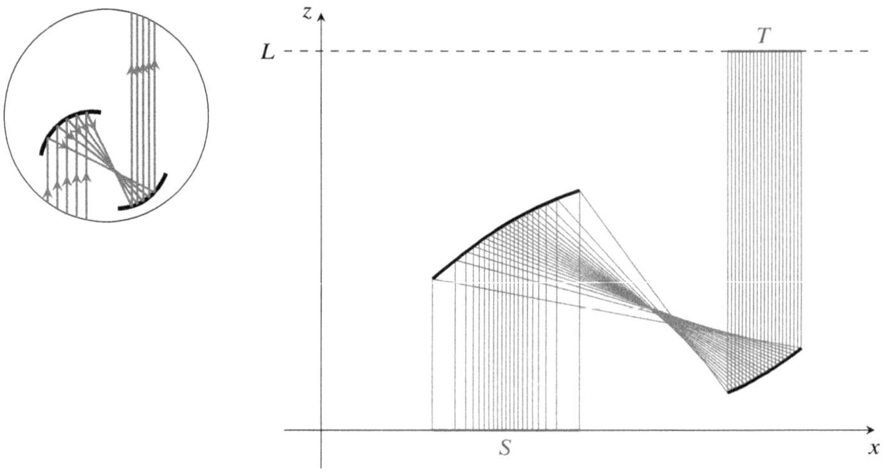

Figure 2 Two-dimensional parallel-to-parallel reflector system.

The source emits a parallel beam with light distribution (emittance) $f(x) > 0, x \in S$. The emittance is the luminous flux emitted by the source per unit length (2D) or area (3D). At the target we want to have a parallel beam with the desired light distribution (illuminance) $g(y) > 0, y \in T$, which is also the flux per unit length/area, see Fig. 3. The goal is to compute two freeform reflector surfaces that manage to do this conversion. We assume all light emitted from S reaches T, so that the areas of the green regions in Fig. 3 (total flux emitted by S and total flux reaching T) are the same.

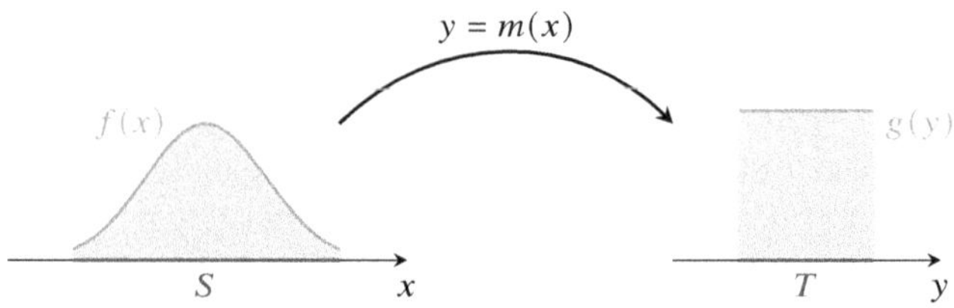

Figure 3 Emittance $f(x)$ of the source S and illuminance $g(y)$ at the target T. The optical map m gives the landing position $y \in T$ of a light ray starting at $x \in S$.

2.1.1 Cost-Function Relation

We introduce some notation in the sketch of the reflector system in Fig. 4. The surface of the first reflector is parameterized as $z = u(x)$ with $x \in S$; for the second reflector we set $L - z = w(y)$ with $y \in T$. Our goal is to find the functions u and w.

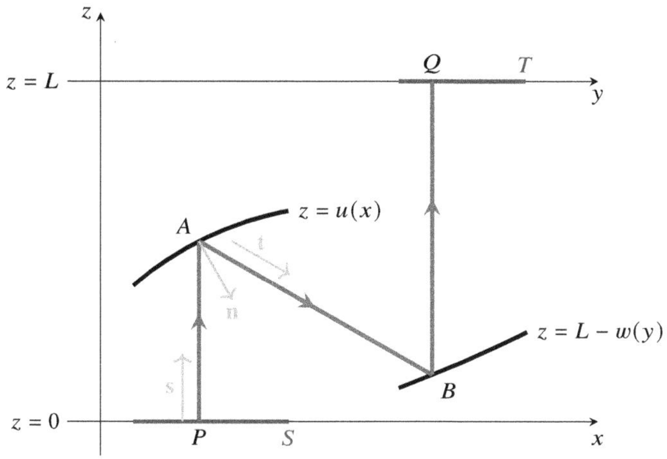

Figure 4 Sketch of the parallel-to-parallel reflector system.

Consider a light ray starting at the point $P = (x, 0)$ with $x \in S$ with direction $\mathbf{s} = (0, 1)^T$, that hits the first reflector at $A = (x, u(x))$, see Fig. 4. The law of reflection states that the ray is reflected in the direction $\mathbf{t} = \mathbf{s} - 2(\mathbf{s} \cdot \mathbf{n})\mathbf{n}$ with $\mathbf{n}$ the unit normal of the surface at A. We choose the normal in the direction towards the source, so

$$\mathbf{n} = \frac{1}{\sqrt{u'(x)^2 + 1}} \begin{pmatrix} u'(x) \\ -1 \end{pmatrix}, \tag{1}$$

which leads to

$$\mathbf{t} = \begin{pmatrix} t_1 \\ t_2 \end{pmatrix} = \frac{1}{u'(x)^2 + 1} \begin{pmatrix} 2u'(x) \\ u'(x)^2 - 1 \end{pmatrix}.$$

If we denote the distance between the points A, where the ray hits the first reflector, and B, where it hits the second reflector, by d, then we find that B has coordinates $(x + dt_1, u(x) + dt_2) = (y, L - w(y))$ where $y = m(x) = x + dt_1$. Finally, the ray hits the target at $Q = (y, L)$. The mapping $y = m(x)$, relating a position in the source to a position in the target, is called the *optical mapping*. Any light ray traveling from source to target has an optical path length [34, Sec. 4.5] $V = u(x) + d + w(y)$. This length is the same for all rays (this is known as the principle of Malus and Dupin in geometrical optics [8, Sec. 3.3.3]). Together with $u(x) + dt_2 = L - w(y)$, we find $d = (V - L)/(1 - t_2)$. Substituting this expression into the equation for the optical mapping and using that $t_1/(1 - t_2) = u'(x)$, we find that the optical mapping is

$$m(x) = x + (V - L)u'(x) = \varphi'(x), \tag{2}$$

with the auxiliary function $\varphi(x) = \frac{1}{2}x^2 + (V - L)u(x)$. Using the coordinates of A and B, we also have $(y - x)^2 + (u(x) + w(y) - L)^2 = d^2$. Replacing d by $V - u(x) - w(y)$, we can rewrite this as

$$u(x) + w(y) = c(x, y), \tag{3a}$$

with

$$c(x, y) = \tfrac{1}{2}(V + L) - \frac{(y - x)^2}{2(V - L)}. \tag{3b}$$

Eq. (3a) is known as an *optimal transport formulation* and the function $c(x, y)$ is called a *cost function*. For the optical system considered here, $c(x, y)$ is a quadratic function.

2.1.2 Optimal Transport

The cost-function relation Eq. (3a) is well known in optimal transport. It was originally studied by Monge in the context of moving soil from one location to another to construct walls and ditches for military fortification [41]. In our optical problem, we aim to design a reflector system that transports all light emitted by the source to the target. The shapes of the reflectors are such that $f(x) > 0$, $x \in S$, is redistributed into $g(y) > 0$, $y \in T$. As such, the problem of inverse optical design corresponds to that of building a castle from a pile of sand on the beach, see Fig. 5. Here, $u(x)$ and $w(y)$ model the shapes of the pile and of the castle, whereas $c(x, y)$ is the cost of moving sand from one place to another.

Figure 5 Sending light from a source to a target distribution corresponds to building a sand castle. Original watercolor by Lara Diamand. Used with permission.

Eq. (3a) has many solution pairs $(u(x), w(y))$. A common choice in optimal transport theory is given below. It is motivated by Villani [70, Ch. 5] by considering an application in economics, where the unknown functions $u(x)$ and $w(y)$ are prices that need to be maximized for competitiveness. The special choice

$$u(x) = \max_{y \in T} \left(c(x,y) - w(y) \right), \quad x \in S, \tag{4a}$$

$$w(y) = \max_{x \in S} \left(c(x,y) - u(x) \right), \quad y \in T, \tag{4b}$$

gives a c-convex pair. It leads to the necessary condition

$$\frac{\partial c}{\partial x}(x,y) - u'(x) = 0. \tag{5}$$

Substituting Eq. (3b) into Eq. (5) and rewriting gives $y = m(x) = x + (V - L)u'(x)$, which recovers Eq. (2).

2.1.3 Energy Conservation

To complete the mathematical model of this reflector system, we consider energy conservation. Let $S = [x_{\min}, x_{\max}]$ and $T = [y_{\min}, y_{\max}]$. Conservation of energy implies that for any $x \in S$, we have

$$\int_{x_{\min}}^{x} f(\xi)\, d\xi = \pm \int_{m(x_{\min})}^{m(x)} g(y)\, dy = \pm \int_{x_{\min}}^{x} g(m(\xi)) m'(\xi)\, d\xi, \tag{6}$$

where we need the plus sign on the right-hand side if $m(x) \geq m(x_{\min})$ and the minus sign otherwise. Eq. (6) states that all energy emitted from $(x_{\min}, x), x \in S$, at the source is received at the interval $\mathrm{int}(m(x_{\min}), m(x))$ at the target. Here, $\mathrm{int}(a,b)$ is the interval (a,b) if $a \leq b$ and (b,a) otherwise. Differentiating Eq. (6), we find

$$m'(x) = \pm \frac{f(x)}{g(m(x))}. \tag{7}$$

Substituting $m(x) = \varphi'(x)$, we find the second-order differential equation

$$\varphi''(x) = \pm \frac{f(x)}{g(\varphi'(x))}. \tag{8}$$

We impose the so-called *transport boundary condition* $m(\partial S) = \partial T$. It states that all light from the boundary of the source is mapped to the boundary of the target. The condition follows from the implicit condition $m(S) = T$, stating that all light from the source S must be transferred to the target T [52]. This leads to $m(x_{\min}) = \varphi'(x_{\min}) = y_{\min}$ and $\varphi'(x_{\max}) = y_{\max}$ if we take the plus sign in Eq. (6), and to $\varphi'(x_{\min}) = y_{\max}$ and $\varphi'(x_{\max}) = y_{\min}$ if we take the minus sign. Note that φ is determined up to a constant. The value of this constant varies the distance from the source to the first reflector; it does not change the shape of the reflector. We can fix a solution by prescribing the distance from the source to the reflector along a specific ray or by prescribing the average height of the reflector. Once we have found $\varphi(x)$ by solving the differential equation (8), we know the first reflector's surface, since $u(x) = (\varphi(x) - \frac{1}{2}x^2)/(V - L)$. The second reflector surface $w(y)$ follows from Eq. (3a).

We show an example in Fig. 6, where S is chosen as $-1 \leq x \leq 1, z = 0$ and T as $3 \leq y \leq 4, z = 5$. Both optical systems convert a Gaussian source distribution to a uniform target distribution. Here we take

$$f(x) = \exp\left(-\frac{x^2}{2\sigma^2}\right),$$

with $\sigma = \sqrt{0.2}$. Furthermore, $g(y) = C$ with the constant C chosen such that global energy conservation holds, i.e., Eq. (6) holds with $x = x_{\max}$. The optical path length is chosen as $V = 11$. The height of the first reflector is fixed by imposing $u(-1) = 2$. For the left system in Fig. 6, we have chosen the plus sign in Eq. (7). The right system corresponds to the minus sign.

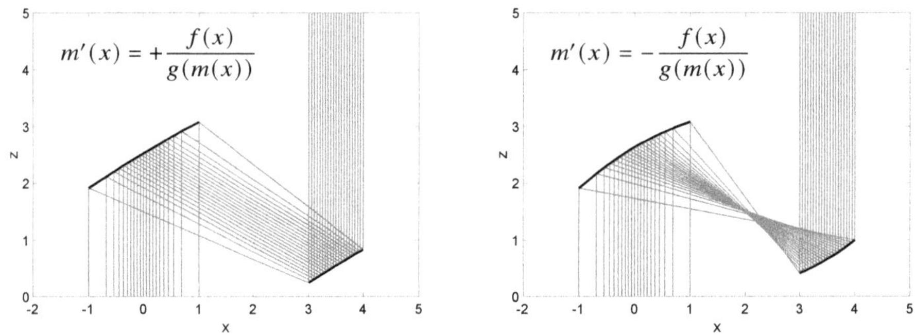

Figure 6 Two-dimensional reflector systems that convert a parallel light beam with a Gaussian distribution into a parallel beam with a uniform distribution. The light source is located at $z = 0$ and the target at $z = 5$.

2.1.4 Generalization to Three-Dimensional Optics

Equations (2)–(8) can be generalized to three-dimensional reflector systems (see [77] for details). A similar approach as above gives

$$\mathbf{m}(\mathbf{x}) = \mathbf{x} + (V - L)\nabla u(\mathbf{x}) = \nabla\left(\tfrac{1}{2}|\mathbf{x}|^2 + (V - L)u(\mathbf{x})\right) = \nabla\varphi(\mathbf{x}),$$

where the source coordinates $\mathbf{x} = (x_1, x_2)^T$ and the optical mapping $\mathbf{m}(\mathbf{x}) = (m_1(\mathbf{x}), m_2(\mathbf{x}))^T$ are vectors now. The optimal transport formulation becomes:

$$u(\mathbf{x}) + w(\mathbf{y}) = c(\mathbf{x}, \mathbf{y}), \tag{9a}$$

with target coordinates $\mathbf{y} = (y_1, y_2)^T$ and cost function

$$c(\mathbf{x}, \mathbf{y}) = \tfrac{1}{2}(V + L) - \frac{\|\mathbf{y} - \mathbf{x}\|^2}{2(V - L)}. \tag{9b}$$

As in Section 2.1.2, we choose the c-convex solution pair, cf. Eq. (4),

$$u(\mathbf{x}) = \max_{\mathbf{y}\in T} \left(c(\mathbf{x}, \mathbf{y}) - w(\mathbf{y})\right), \quad \mathbf{x} \in S, \tag{10a}$$

$$w(\mathbf{y}) = \max_{\mathbf{x}\in S} \left(c(\mathbf{x}, \mathbf{y}) - u(\mathbf{x})\right), \quad \mathbf{y} \in T. \tag{10b}$$

This leads to the necessary condition

$$\nabla_{\mathbf{x}} \left(c(\mathbf{x}, \mathbf{y}) - u(\mathbf{x})\right) = \mathbf{0}. \tag{11}$$

Substituting the optical mapping $\mathbf{y} = \mathbf{m}(\mathbf{x})$ into Eq. (11) and subsequently differentiating to $\mathbf{x}$, we find

$$\mathrm{D}_{\mathbf{xx}}c(\mathbf{x}, \mathbf{m}(\mathbf{x})) + \mathrm{D}_{\mathbf{xy}}c(\mathbf{x}, \mathbf{m}(\mathbf{x}))\,(\mathrm{D}\mathbf{m})(\mathbf{x}) = \mathrm{D}^2 u(\mathbf{x}). \tag{12}$$

Here, $\mathrm{D}_{\mathbf{xx}}c$ and $\mathrm{D}^2 u$ are Hessian matrices, $\mathrm{D}_{\mathbf{xy}}c$ is the matrix with mixed second-order derivatives and $\mathrm{D}\mathbf{m}(\mathbf{x})$ is the Jacobian matrix of $\mathbf{m}$. Eq. (12) can be simplified using $\mathrm{D}_{\mathbf{xx}}c = -1/(V - L) \cdot \mathbf{I}$, $\mathrm{D}_{\mathbf{xy}}c = 1/(V - L) \cdot \mathbf{I}$ and $\mathrm{D}^2\varphi = \mathbf{I} + (V - L)\mathrm{D}^2 u$ as

$$(\mathrm{D}\mathbf{m})(\mathbf{x}) = \mathrm{D}^2\varphi(\mathbf{x}). \tag{13}$$

Energy conservation becomes

$$\iint_A f(\mathbf{x})\, dA(\mathbf{x}) = \iint_{\mathbf{m}(A)} g(\mathbf{y})\, dA(\mathbf{y}) = \iint_A g(\mathbf{m}(\mathbf{x}))\, |\det(\mathrm{D}\mathbf{m}(\mathbf{x}))|\, dA(\mathbf{x}),$$

with A any subset of the source domain S. Because this equation needs to hold for any A, we find

$$\det(\mathrm{D}\mathbf{m}(\mathbf{x})) = \pm \frac{f(\mathbf{x})}{g(\mathbf{m}(\mathbf{x}))}. \tag{14}$$

Substituting Eq. (13) into Eq. (14), we find the nonlinear second-order partial differential equation

$$\frac{\partial^2\varphi(\mathbf{x})}{\partial x_1^2}\frac{\partial^2\varphi(\mathbf{x})}{\partial x_2^2} - \left(\frac{\partial^2\varphi(\mathbf{x})}{\partial x_1\,\partial x_2}\right)^2 = \pm \frac{f(\mathbf{x})}{g(\mathbf{m}(\mathbf{x}))}. \tag{15}$$

Eq. (15) is called the standard Monge–Ampère equation. As in Section 2.1.3, we impose the transport boundary condition

$$\mathbf{m}(\partial S) = \partial T, \tag{16}$$

stating that the boundary of the source domain S is mapped to the boundary of the target domain T. In 2D, there are just two options to satisfy Eq. (16), but in 3D there are many more. Bertens et al. [7, Sec. 2.3] discuss three methods to impose the transport boundary condition. In the first method, the images of points on ∂S are projected on ∂T during the iterative solution algorithm. The second method cuts both ∂S and ∂T in segments and applies the projection method to the individual segments. Finally, the third method also uses segmentations of the boundaries, but it generates parameterizations of boundary segments and uses these to map segments of ∂S to ∂T.

Table 1 lists the equations for the two- and three-dimensional problems. It is clear that these are very similar, but in 2D the resulting ordinary differential equations can be solved using standard numerical methods. In 3D, we need to solve a nonlinear partial differential equation. We present an iterative least-squares method to do so in Section 3.

Table 1 Equations for the parallel-to-parallel reflector, 2D vs 3D.

	2D	3D
Optical mapping	$y = m(x)$	$\mathbf{y} = \mathbf{m}(\mathbf{x})$
	$= x + (V - L)u'(x) = \phi'(x)$	$= \mathbf{x} + (V - L)\nabla u(\mathbf{x}) = \nabla \phi(\mathbf{x})$
Optimal transport	$u(x) + w(y) = c(x, y)$	$u(\mathbf{x}) + w(\mathbf{y}) = c(\mathbf{x}, \mathbf{y})$
	$c(x, y) = \frac{1}{2}(V + L) - \frac{(y-x)^2}{2(V-L)}$	$c(\mathbf{x}, \mathbf{y}) = \frac{1}{2}(V + L) - \frac{\lVert\mathbf{y}-\mathbf{x}\rVert^2}{2(V-L)}$
Energy conservation	$m'(x) = \phi''(x)$	$\det(\mathrm{D}\mathbf{m}(\mathbf{x})) = \frac{\partial^2 \phi}{\partial x_1^2}\frac{\partial^2 \phi}{\partial x_2^2} - \left(\frac{\partial^2 \phi}{\partial x_1 \partial x_2}\right)^2$
	$= \pm \frac{f(x)}{g(m(x))}$	$= \pm \frac{f(\mathbf{x})}{g(\mathbf{m}(\mathbf{x}))}$

2.2 Model 2: Generalized Monge–Ampère Equation

Next we model the parallel-to-point reflector system, the second basic system in Fig. 1. This system can be modeled by the generalized Monge–Ampère equation.

As before, we first consider the two-dimensional problem, sketched in Fig. 7. Given are a parallel source S located at $z = -L$ and a point target $T = (0, 0)$. We denote the spatial coordinates at the source by x.

The source emits a parallel beam with light distribution $f(x) > 0$, $x \in S$. At the target we specify the desired light distribution in terms of the directions of the light rays arriving at T, so it is the luminous flux received by the target per radian (2D) or steradian (3D). We denote the unit direction vector of an arriving ray by $\mathbf{t} = (t_1, t_2)^T$, cf. Fig. 9. It turns out to be convenient to use the stereographic projection from the south pole $P_{\text{south}} = (0, -1)$ as a parameterization for $\mathbf{t}$. This projection, which we denote by y, is defined as the intersection of the horizontal axis with the line segment that connects P_{south} with the point (t_1, t_2) on the unit circle, viz.

$$y = \frac{t_1}{t_2 + 1}, \tag{17a}$$

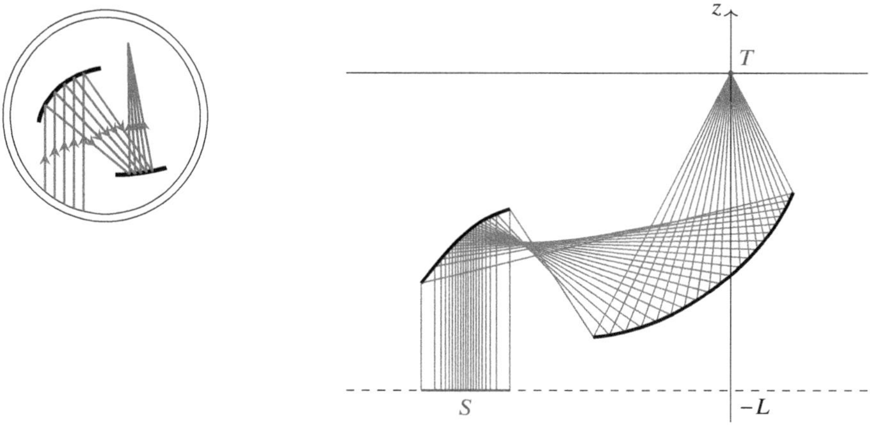

Figure 7 Two-dimensional parallel-to-point reflector system.

see Fig. 8. For the inverse stereographic projection we have

$$\mathbf{t} = \begin{pmatrix} t_1 \\ t_2 \end{pmatrix} = \frac{1}{y^2+1} \begin{pmatrix} 2y \\ 1-y^2 \end{pmatrix}. \tag{17b}$$

As in the first model, we write $g(y)$ for the desired target light distribution.

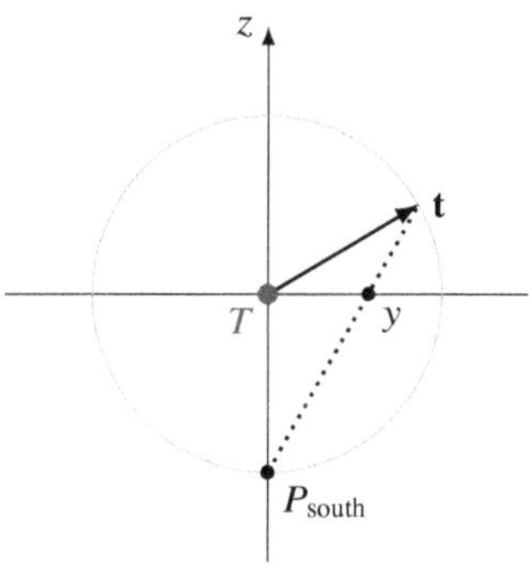

Figure 8 Stereographic projection from the south pole.

2.2.1 Cost-Function Relation

The notation for the current system is shown in Fig. 9. The surface of the first reflector is parameterized as $z = -L + u(x)$ with $x \in S$. The second reflector is parameterized as $\mathbf{r} = -w(y)\mathbf{t}$ with $y \in Y$. Here, Y is the stereographic target domain.

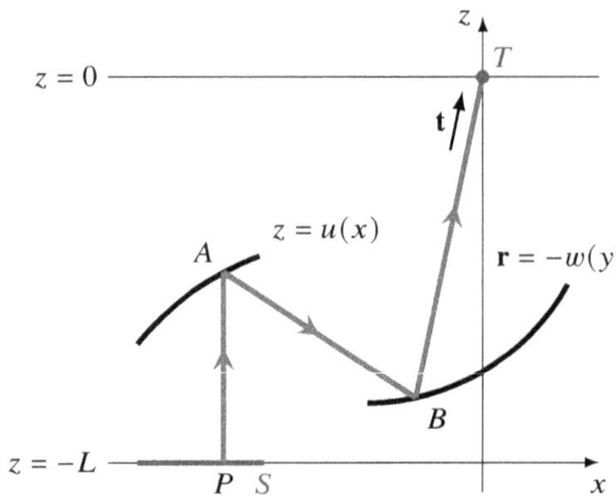

Figure 9 Sketch of the parallel-to-point reflector system.

Consider a light ray starting at the point $P = (x, -L)$ with $x \in S$ with direction $\mathbf{s} = (0, 1)^T$, that hits the first reflector at $A = (x, -L + u(x))$, see Fig. 9. The ray is reflected and hits the second reflector at $B = (-w(y)t_1, -w(y)t_2)$. After reflection at B, the ray is sent in the direction $\mathbf{t}$ and will arrive at the target point T.

Any light ray traveling from source to target has an optical path length $V = u(x) + d + w(y)$, where d is the distance between A and B. V is the same for all rays (principle of Malus and Dupin, cf. Section 2.1.1). Combining $d = V - u - w$ with

$$d^2 = (x + w(y)t_1)^2 + (-L + u(x) + w(y)t_2)^2,$$

we find, using $t_1^2 + t_2^2 = 1$,

$$V^2 - 2Vu - 2Vw + 2uw = x^2 + 2xwt_1 + L^2 - 2Lu - 2Lwt_2 + 2uwt_2. \tag{18}$$

Introducing the reduced optical path length $\beta = V - L$, moving all terms in Eq. (18) to the left-hand side and dividing by $2\beta w$, we get

$$\frac{V^2 - L^2}{2\beta w} - \frac{u}{w} - 1 - \frac{L}{\beta}(1 - t_2) + \frac{u}{\beta}(1 - t_2) - \frac{x^2}{2\beta w} - \frac{xt_1}{\beta} = 0. \tag{19}$$

We can express t_1 and t_2 in the stereographic coordinate y using Eq. (17b) and factorize Eq. (19) to obtain the geometric relation

$$\left(-\frac{u}{\beta} - \frac{x^2}{2\beta^2} + \frac{V + L}{2\beta}\right) \cdot \left(\frac{\beta}{w}(1 + y^2) - 2y^2\right) = \left(1 + \frac{xy}{\beta}\right)^2. \tag{20}$$

The factors in the left-hand side of Eq. (20) are positive [69, Sec. 3.3], so we can take the logarithm and rewrite it as

$$u_1(x) + u_2(y) = c(x, y), \tag{21a}$$

with

$$u_1(x) = \log\left(-\frac{u}{\beta} - \frac{x^2}{2\beta^2} + \frac{V+L}{2\beta}\right), \tag{21b}$$

$$u_2(y) = \log\left(\frac{\beta}{w}(1+y^2) - 2y^2\right), \tag{21c}$$

$$c(x,y) = \log\left(\left(1 + \frac{xy}{\beta}\right)^2\right). \tag{21d}$$

We conclude that we have again found an optimal transport formulation. The cost function is not quadratic however. We will discuss the consequences for the model and solution method in Sections 2.4 and 3.

2.2.2 Energy Conservation

Conservation of energy now reads, using Eq. (17b),

$$\begin{aligned}\int_{x_{\min}}^{x} f(\xi)\,d\xi &= \pm\int_{m(x_{\min})}^{m(x)} g(y)\,\|\mathbf{t}'(y)\|\,dy = \pm\int_{m(x_{\min})}^{m(x)} g(y)\,\frac{2}{1+y^2}\,dy \\ &= \pm\int_{x_{\min}}^{x} g(m(\xi))\frac{2}{1+m(\xi)^2}m'(\xi)\,d\xi. \end{aligned} \tag{22}$$

Differentiating Eq. (22), we find

$$m'(x) = \pm\frac{f(x)}{g(m(x))}\frac{1+m(x)^2}{2}. \tag{23}$$

Imposing the transport boundary condition gives $m(x_{\min}) = y_{\min}$ if we take the plus sign in Eq. (23), and $m(x_{\min}) = y_{\max}$ if we take the minus sign. Once we have found $m(x)$ by solving the differential equation, we use the necessary condition

$$\frac{\partial c}{\partial x}(x,y) - u_1'(x) = 0, \tag{24}$$

cf. Eq (5), substitute $y = m(x)$ and solve the resulting ODE for $u_1(x)$. This gives the shape of the first reflector. Finally, Eq. (21a) gives $u_2(m(x)) = c(x, m(x)) - u_1(x)$, from which the second reflector surface $w(y)$ follows, cf. Eq. (21c).

We show an example in Fig. 10, where S is chosen as $-7 \le x \le -5, z = -7$ and $T = (0,0)$. The stereographic target domain is $Y = (-0.25, 0.25)$. The optical path length is $V = 14$. The height of the first reflector is fixed by imposing $u(-5) = -3$. Both optical systems convert a Gaussian source distribution to a uniform (as a function of direction) target distribution. We take

$$f(x) = \exp\left(-\frac{(x+6)^2}{2\sigma^2}\right),$$

with $\sigma = \sqrt{0.2}$ and $g(y) = C$ with the constant C chosen such that global energy conservation holds, i.e., Eq. (22) holds with $x = x_{\max}$. For the left system in Fig. 10, we have chosen the plus sign in Eq. (23). The right system corresponds to the minus sign.

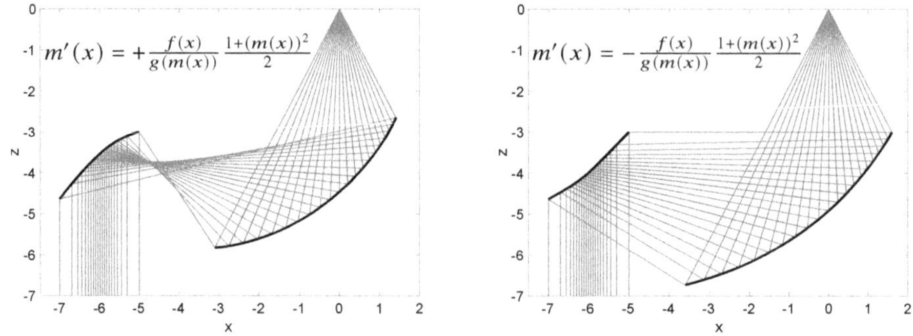

Figure 10 Two-dimensional reflector systems that send a parallel light beam with a Gaussian distribution to a point target with a uniform distribution. The light source is located at $z = -7$ and the target at $(0, 0)$.

2.2.3 Generalization to Three-Dimensional Optics

Equations (21a)–(21d) and (23) can be generalized to three-dimensional reflector systems, see [69, Ch. 7] or, alternatively, [66] for details on the equivalent point-to-parallel reflector system. One can show that the following equations hold

$$u_1(\mathbf{x}) + u_2(\mathbf{y}) = c(\mathbf{x}, \mathbf{y}), \tag{25a}$$

$$u_1(\mathbf{x}) = \log\left(-\frac{u}{\beta} - \frac{\|\mathbf{x}\|^2}{2\beta^2} + \frac{V+L}{2\beta}\right), \tag{25b}$$

$$u_2(\mathbf{y}) = \log\left(\frac{\beta}{w}(1 + \|\mathbf{y}\|^2) - 2\|\mathbf{y}\|^2\right), \tag{25c}$$

$$c(\mathbf{x}, \mathbf{y}) = \log\left(1 + \frac{2\mathbf{x}\cdot\mathbf{y}}{\beta} + \frac{\|\mathbf{x}\|^2\|\mathbf{y}\|^2}{\beta^2}\right), \tag{25d}$$

with source coordinates $\mathbf{x} = (x_1, x_2)^T$ and target stereographic coordinates $\mathbf{y} = (y_1, y_2)^T$, defined similarly as in 2D, cf. Eq. (17). As before, we take the c-convex solution pair, cf. Eq. (10), and find the zero-gradient condition, cf. Eq. (11),

$$\nabla_{\mathbf{x}}\left(c(\mathbf{x}, \mathbf{y}) - u_1(\mathbf{x})\right) = \mathbf{0}. \tag{26}$$

Substituting the optical mapping $\mathbf{y} = \mathbf{m}(\mathbf{x}) = (m_1(\mathbf{x}), m_2(\mathbf{x}))^T$ into Eq. (26) and subsequently differentiating to $\mathbf{x}$, we find

$$\mathrm{D}_{\mathbf{xx}}c(\mathbf{x}, \mathbf{m}(\mathbf{x})) + \mathrm{D}_{\mathbf{xy}}c(\mathbf{x}, \mathbf{m}(\mathbf{x}))\,(\mathrm{D}\mathbf{m})(\mathbf{x}) = \mathrm{D}^2 u_1(\mathbf{x}). \tag{27}$$

Introducing $\mathbf{C} = \mathrm{D}_{\mathbf{xy}}c$ and $\mathbf{P} = \mathrm{D}^2 u_1 - \mathrm{D}_{\mathbf{xx}}c$, we can write Eq. (27) as $\mathbf{C}\,\mathrm{D}\mathbf{m} = \mathbf{P}$. Energy conservation reads

$$\det(\mathrm{D}\mathbf{m}(\mathbf{x})) = \pm\frac{f(\mathbf{x})}{g(\mathbf{m}(\mathbf{x}))}\frac{\left(1 + \|\mathbf{m}(\mathbf{x})\|^2\right)^2}{4}. \tag{28}$$

We can combine Eqs. (27) and (28) to find

$$\det(\mathrm{D}^2 u_1(\mathbf{x}) - \mathrm{D}_{\mathbf{xx}}c(\mathbf{x})) = \pm\det(\mathbf{C})\frac{f(\mathbf{x})}{g(\mathbf{m}(\mathbf{x}))}\frac{\left(1 + \|\mathbf{m}(\mathbf{x})\|^2\right)^2}{4}. \tag{29}$$

Eq. (29) is called the generalized Monge–Ampère equation (GMA). We again impose the transport boundary condition (16).

Table 2 lists the equations for the two- and three-dimensional problems. As for the first model, we find ODEs in 2D, whereas we need to solve a nonlinear PDE in 3D.

Table 2 Equations for the parallel-to-point reflector, 2D vs 3D.

	2D	3D
Optimal transport	$u_1(x) + u_2(y) = c(x, y)$	$u_1(\mathbf{x}) + u_2(\mathbf{y}) = c(\mathbf{x}, \mathbf{y})$
	$c(x, y) = \log\left(\left(1 + \frac{xy}{\beta}\right)^2\right)$	$c(\mathbf{x}, \mathbf{y}) = \log\left(1 + \frac{2\mathbf{x}\cdot\mathbf{y}}{\beta} + \frac{\|\mathbf{x}\|^2\|\mathbf{y}\|^2}{\beta^2}\right)$
Energy conservation	$m'(x) = \pm\frac{f(x)}{g(m(x))}\frac{1+(m(x))^2}{2}$	$\det(\mathrm{D}\mathbf{m}(\mathbf{x})) = \pm\frac{f(\mathbf{x})}{g(\mathbf{m}(\mathbf{x}))}\frac{\left(1+\|\mathbf{m}(\mathbf{x})\|^2\right)^2}{4}$

2.3 Model 3: Generated Jacobian Equation

Next we model the parallel-to-near-field lens system, the eleventh basic system in Fig. 1. This system can be modeled by a generated Jacobian equation.

We first consider the two-dimensional problem, sketched in Fig. 11. Given are a parallel source S located at $z = 0$ and a near-field target T located at $z = L$. We denote the spatial coordinates at the source by x and at the target by y.

The source emits a parallel beam with emittance $f(x) > 0$, $x \in S$. At the target we have a desired illuminance $g(y) > 0$, $y \in T$. Note that we require only one freeform surface. For this reason we choose the first lens surface to be flat and it does not change the directions of incoming rays.

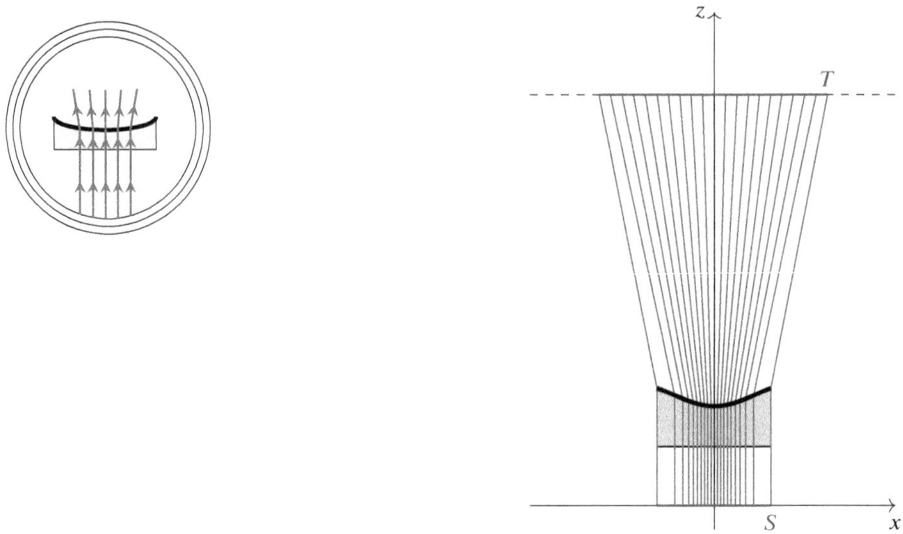

Figure 11 Two-dimensional parallel-to-near-field lens system.

2.3.1 Geometric Relation

The notation for the current system is shown in Fig. 12. Because the first lens surface is flat, we extend the lens to the source in our model. The second lens surface is parameterized as $z = u(x)$ with $x \in S$.

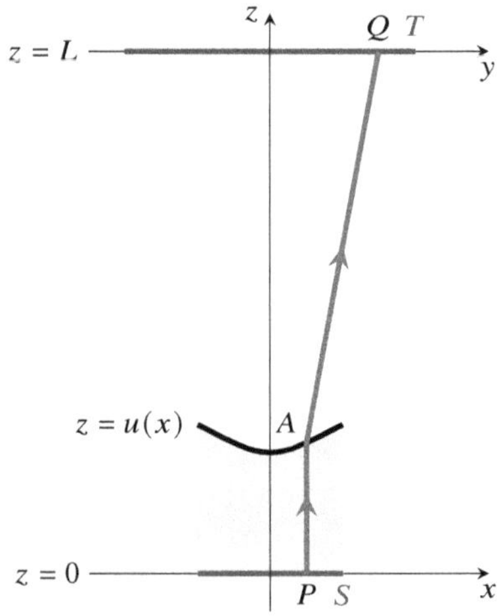

Figure 12 Sketch of the parallel-to-near-field lens system.

Consider a light ray starting at the point $P = (x, 0)$ with $x \in S$ with direction $\mathbf{s} = (0, 1)^T$, that hits the freeform lens surface at $A = (x, u(x))$, see Fig. 12. The ray is refracted and hits the target at the point $Q = (y, L)$. The optical path length equals $V(y) = n \cdot u(x) + d$ where n is the (constant) refractive index of the lens and d is the distance between A and Q. Note that $V(y)$ is no longer a constant. Together with

$$d^2 = (y - x)^2 + (L - u(x))^2,$$

we find the geometric relation

$$V(y) = n \cdot u(x) + \sqrt{(y - x)^2 + (L - u(x))^2}. \tag{30}$$

Eq. (30) differs from the geometric relations we found in Models 1 and 2. The unknown functions $V(y)$ and $u(x)$ cannot be separated and Eq. (30) cannot be written in the form of a cost-function relation as before.

2.3.2 Energy Conservation

As in Model 1, we get

$$m'(x) = \pm \frac{f(x)}{g(m(x))}. \tag{31}$$

Imposing the transport boundary condition gives $m(x_{\min}) = y_{\min}$ if we take the plus sign in Eq. (31), and $m(x_{\min}) = y_{\max}$ if we take the minus sign.

Once we have found $m(x)$ by solving the differential equation, we differentiate Eq. (30) with respect to x and substitute $y = m(x)$. This results in the following ODE for $u(x)$:

$$u'(x) = \frac{m(x) - x}{n\sqrt{(m(x) - x)^2 + (L - u(x))^2} + u(x) - L}. \tag{32}$$

Solving this ODE gives the shape of the freeform lens surface.

We show an example in Fig. 13, where S is chosen as $-1 \leq x \leq 1, z = 0$ and T as $-2 \leq y \leq 2, z = 7$. The height of the freeform lens surface is fixed by $u(-1) = 2$. The refractive index of the lens equals $n = 1.5$, which is the refractive index of glass. Both optical systems convert a Gaussian source distribution to a uniform target distribution; here we take

$$f(x) = \exp\left(-\frac{x^2}{2\sigma^2}\right),$$

with $\sigma = \sqrt{0.2}$ and $g(y)$ is constant. For the left system in Fig. 13, we have chosen the plus sign in Eq. (31). The right system corresponds to the minus sign.

2.3.3 Generalization to Three-Dimensional Optics

Equations (30) and (31) can be generalized to three-dimensional reflector systems. One can show that the following equations hold

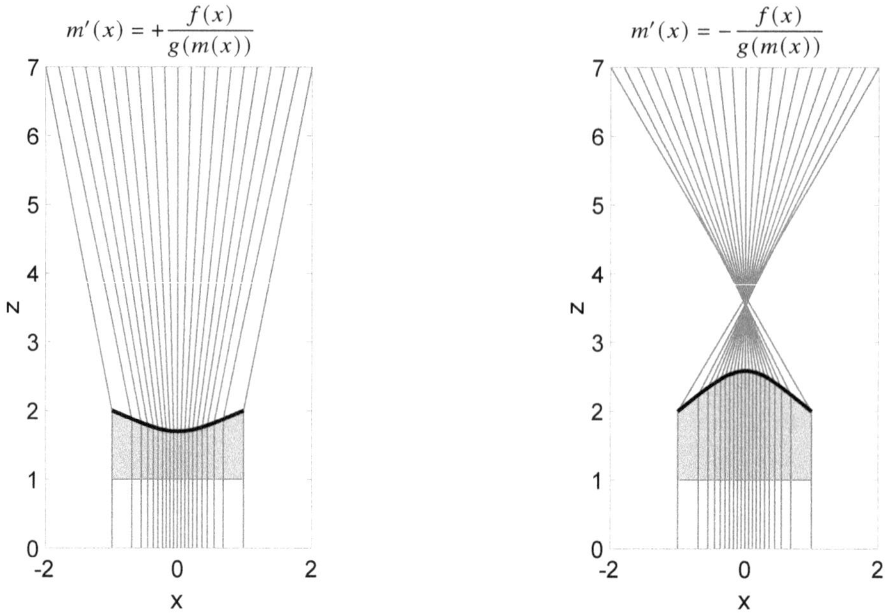

Figure 13 Two-dimensional lens systems that send a parallel light beam with a Gaussian distribution to a near-field target with a uniform illuminance. The light source is located at $z = 0$ and the target at $z = 7$.

$$V(\mathbf{y}) = n \cdot u(\mathbf{x}) + \sqrt{\|\mathbf{y}-\mathbf{x}\|^2 + (L - u(\mathbf{x}))^2}, \tag{33}$$

$$\det(\mathrm{D}\mathbf{m}(\mathbf{x})) = \pm\frac{f(\mathbf{x})}{g(\mathbf{m}(\mathbf{x}))}, \tag{34}$$

with source coordinates $\mathbf{x} = (x_1, x_2)^T$, target coordinates $\mathbf{y} = (y_1, y_2)^T$ and optical mapping $\mathbf{m}(\mathbf{x}) = (m_1(\mathbf{x}), m_2(\mathbf{x}))^T$. We define

$$H(\mathbf{x}, \mathbf{y}, v) = n \cdot v + \sqrt{\|\mathbf{y}-\mathbf{x}\|^2 + (L - v)^2}, \tag{35}$$

so that Eq. (33) becomes $V(\mathbf{y}) = H(\mathbf{x}, \mathbf{y}, u(\mathbf{x}))$. Because $H_v(\mathbf{x}, \mathbf{y}, \cdot) > 0$ for all $\mathbf{x} \in S$ and $\mathbf{y} \in T$, the function has an inverse $G(\mathbf{x}, \mathbf{y}, \cdot) = H^{-1}(\mathbf{x}, \mathbf{y}, \cdot)$. In a similar way as in the other models, we choose the solution pair, cf. Eq. (10),

$$u(\mathbf{x}) = \max_{\mathbf{y}\in T} G(\mathbf{x}, \mathbf{y}, V(\mathbf{y})), \quad \mathbf{x} \in S, \tag{36a}$$

$$V(\mathbf{y}) = \min_{\mathbf{x}\in S} H(\mathbf{x}, \mathbf{y}, u(\mathbf{x})), \quad \mathbf{y} \in T. \tag{36b}$$

We introduce $\widetilde{H}(\mathbf{x}, \mathbf{y}) = H(\mathbf{x}, \mathbf{y}, u(\mathbf{x}))$ and find, using Eq. (36b), the zero-gradient condition

$$\nabla_{\mathbf{x}}\widetilde{H}(\mathbf{x}, \mathbf{y}) = \nabla_{\mathbf{x}} H(\mathbf{x}, \mathbf{y}, u(\mathbf{x})) + H_v(\mathbf{x}, \mathbf{y}, u(\mathbf{x}))\nabla u(\mathbf{x}) = \mathbf{0}. \tag{37}$$

Substituting $\mathbf{y} = \mathbf{m}(\mathbf{x})$ into Eq. (37) and subsequently differentiating to $\mathbf{x}$, we find

$$\mathrm{D}_{\mathbf{xx}}\widetilde{H}(\mathbf{x}, \mathbf{m}(\mathbf{x})) + \mathrm{D}_{\mathbf{xy}}\widetilde{H}(\mathbf{x}, \mathbf{m}(\mathbf{x}))\,(\mathrm{D}\mathbf{m})(\mathbf{x}) = \mathbf{O}, \tag{38}$$

with $\mathbf{O}$ the zero matrix. Combining Eqs. (34) and (38) leads to

$$\det\big(\mathrm{D}_{\mathbf{xx}}\widetilde{H}(\mathbf{x}, \mathbf{m}(\mathbf{x}))\big) = \pm\det(\mathbf{C})\frac{f(\mathbf{x})}{g(\mathbf{m}(\mathbf{x}))}, \tag{39}$$

where we have introduced $\mathbf{C} = \mathrm{D}_{\mathbf{xy}}\widetilde{H}(\mathbf{x}, \mathbf{m}(\mathbf{x}))$. Eq. (39) is called a generated Jacobian equation (GJE) [65]. We again impose the transport boundary condition (16).

Table 3 lists the equations for the two- and three-dimensional problems. We again find ODEs in 2D, whereas we need to solve a nonlinear PDE in 3D.

Table 3 Equations for the parallel-to-near-field lens, 2D vs 3D.

	2D	3D
Geometric relation	$V(y) = n \cdot u(x) + d$	$V(\mathbf{y}) = n \cdot u(\mathbf{x}) + d$
	$d = \sqrt{(y-x)^2 + (L-u(x))^2}$	$d = \sqrt{\lVert\mathbf{y}-\mathbf{x}\rVert^2 + (L-u(\mathbf{x}))^2}$
Energy conservation	$m'(x) = \pm\frac{f(x)}{g(m(x))}$	$\det(\mathrm{D}\mathbf{m}(\mathbf{x})) = \pm\frac{f(\mathbf{x})}{g(\mathbf{m}(\mathbf{x}))}$

2.4 Unified Mathematical Framework

For all three models, SMA, GMA and GJE, energy conservation can be written in the form (cf. Eqs. (14), (28) and (34))

$$\det(\mathrm{D}\mathbf{m}) = F(\mathbf{x}, \mathbf{m}(\mathbf{x})), \tag{40}$$

with known right-hand side function $F(\mathbf{x}, \mathbf{m}(\mathbf{x}))$. We impose the transport boundary condition (16).

We have the following matrix equation for the optical map (cf. Eqs. (13), (27) and (38))

$$\mathbf{C}\,\mathrm{D}\mathbf{m} = \mathbf{P}, \tag{41}$$

where the matrices $\mathbf{C}$ and $\mathbf{P}$ are defined as follows:

- Model 1: **Standard Monge–Ampère** (SMA)—Optimal transport formulation, $c(\mathbf{x}, \mathbf{y})$ quadratic

$$\mathbf{C} = \mathbf{I}, \qquad \mathbf{P} = \mathrm{D}^2\varphi, \tag{42}$$

- Model 2: **Generalized Monge–Ampère** (GMA)—Optimal transport formulation, $c(\mathbf{x}, \mathbf{y})$ non-quadratic

$$\mathbf{C} = \mathrm{D}_{\mathbf{xy}}c, \qquad \mathbf{P} = \mathrm{D}^2u_1 - \mathrm{D}_{\mathbf{xx}}c, \tag{43}$$

- Model 3: **Generated Jacobian equation** (GJE)—No optimal transport formulation

$$\mathbf{C} = \mathrm{D}_{\mathbf{xy}}\widetilde{H}, \qquad \mathbf{P} = -\mathrm{D}_{\mathbf{xx}}\widetilde{H}. \tag{44}$$

In Models 1 and 2, we first compute the optical map from Eqs. (40) and (41) together with the transport boundary condition (16). Next we compute the location of the first optical surface from the zero-gradient condition Eq. (11) (for Model 1) or Eq. (26) (for Model 2). In Model 3, we will compute the mapping and the first surface simultaneously. We will use Eq. (37) for computing the surface. The numerical scheme is presented in Section 3.

2.5 Beyond the Basic Systems

In this section we outline an optical system that does not belong to the sixteen basic systems. We consider a system with a light source that emits a parallel beam with given emittance $f(\mathbf{x})$, where $\mathbf{x} \in S$ are spatial coordinates in the source domain S. Also given are *two* target illuminances $g(\mathbf{y})$, $\mathbf{y} \in T_1$, and $h(\mathbf{z})$, $\mathbf{z} \in T_2$, where $\mathbf{y}$ and $\mathbf{z}$ are the coordinates in the target domains T_1 and T_2. See Fig. 14 for sketches of the optical system. Our goal is to find the shapes of the two reflectors that transform light distribution f at the source into g at the first target and h at the second.

We consider the propagation of a light ray from source to first target, described by the optical mapping $\mathbf{m}_1 : S \to T_1$ that gives the coordinates $\mathbf{y} \in T_1$ as a function of the source coordinates $\mathbf{x} \in S$, so $\mathbf{y} = \mathbf{m}_1(\mathbf{x})$. In a similar way, we use a second optical mapping $\mathbf{m}_2 : T_1 \to T_2$ for propagation from the first to the second target, so $\mathbf{z} = \mathbf{m}_2(\mathbf{y})$.

Since there are no optical surfaces between the two targets, we can immediately find the optical path length V_2 for any light ray traveling from T_1 to T_2. Combining this with conservation of luminous flux and transport boundary conditions, we can compute the optical mapping $\mathbf{m}_2$. At this stage, we have found the direction vector of the ray at T_1

$$\mathbf{t} = \begin{pmatrix} \mathbf{p}_t \\ t_3 \end{pmatrix} = \frac{1}{\sqrt{|\mathbf{m}_2(\mathbf{y}) - \mathbf{y}|^2 + (L_2 - L_1)^2}} \begin{pmatrix} \mathbf{m}_2(\mathbf{y}) - \mathbf{y} \\ L_2 - L_1 \end{pmatrix}. \tag{45}$$

The projection of $\mathbf{t} \in \mathbb{R}^3$ onto the first target plane is called the optical momentum $\mathbf{p}_t \in \mathbb{R}^2$.

The optical path length V_1 from the source to the first target is used as a Hamilton characteristic function. It satisfies $\partial V_1/\partial \mathbf{x} = \mathbf{0}$ and $\partial V_1/\partial \mathbf{y} = \mathbf{p}_t$, see [62, Sec. 2] for details. The first equation states that V_1 does not depend on $\mathbf{x}$, so $V_1 = V_1(\mathbf{y})$. We can solve the second equation to find V_1. We prescribe the average value of V_1 to make it unique.

We use the following parameterizations for the reflectors. The first reflector is given by its height $u(\mathbf{x})$ measured from the source. The second reflector is described by the distance $w(\mathbf{y})$ along a ray from the second reflector to the first target. This leads to

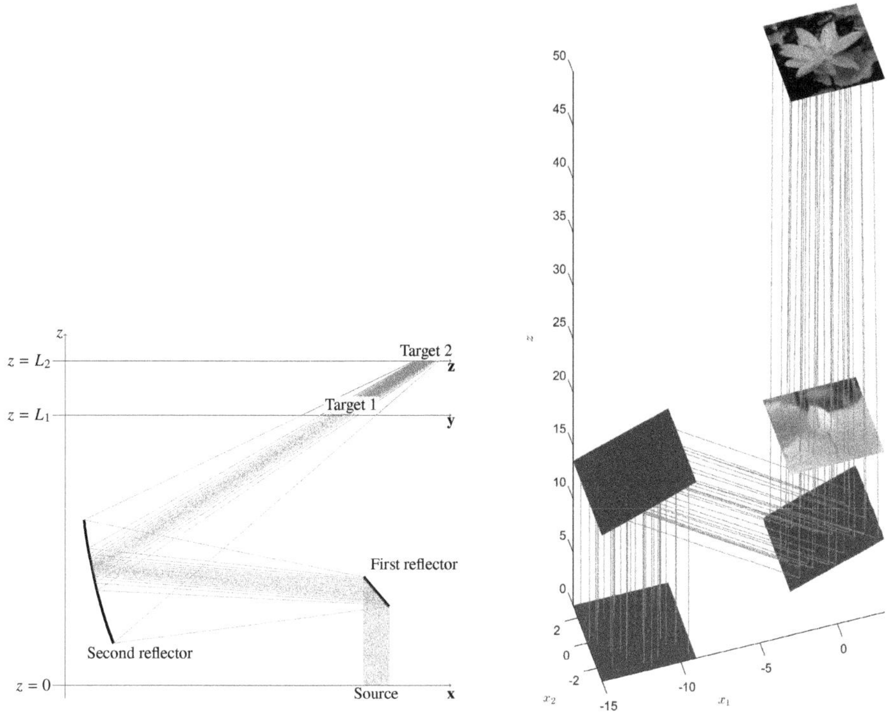

(a) 2D system

(b) 3D system

Figure 14 Reflector systems with a parallel source and two targets.

$$\mathbf{r}_1(\mathbf{x}) = \begin{pmatrix} \mathbf{x} \\ u(\mathbf{x}) \end{pmatrix}, \tag{46a}$$

$$\mathbf{r}_2(\mathbf{y}) = \begin{pmatrix} \mathbf{y} \\ L_1 \end{pmatrix} - w(\mathbf{y})\mathbf{t}. \tag{46b}$$

Expressing V_1 in terms of the unknown functions u and w, imposing conservation of luminous flux and transport boundary conditions leads to a generated Jacobian equation for the optical mapping $\mathbf{m}_1$. From $\mathbf{m}_1$ we can compute both u and w, which gives the shapes of the reflectors. More details on the mathematical model can be found in [12].

3 Numerical Method

Our numerical method for finding the optical surface(s) is an iterative least-squares solver. The implementation is based on the following breakdown in substeps. We compute $\mathbf{P}$, $\mathbf{b}$ and $\mathbf{m}$ such that, cf. Eqs. (16), (40) and (41),

$$\mathbf{C}\,\mathrm{D}\mathbf{m} = \mathbf{P}, \tag{47a}$$
$$\det(\mathbf{P}) = \det(\mathbf{C}) \cdot F(\mathbf{x}, \mathbf{m}(\mathbf{x})), \tag{47b}$$
$$\mathbf{b}(\mathbf{x}) = \mathbf{m}(\mathbf{x}), \qquad \mathbf{x} \in \partial S. \tag{47c}$$

Here, **b** maps the boundary of the source domain to the boundary of the target domain.

Our solution strategy is to minimize the functionals

$$J_I(\mathbf{m}, \mathbf{P}) = \frac{1}{2} \iint_S \|\mathbf{C}\,\mathrm{D}\mathbf{m} - \mathbf{P}\|_F^2 \, d\mathbf{x}, \tag{48a}$$
$$J_B(\mathbf{m}, \mathbf{b}) = \frac{1}{2} \oint_{\partial S} \|\mathbf{m} - \mathbf{b}\|^2 \, ds, \tag{48b}$$
$$J(\mathbf{m}, \mathbf{P}, \mathbf{b}) = \alpha\, J_I(\mathbf{m}, \mathbf{P}) + (1 - \alpha)\, J_B(\mathbf{m}, \mathbf{b}). \tag{48c}$$

In Eq. (48a), $\|\cdot\|_F$ is the Frobenius norm. The weight $\alpha \in (0, 1)$ in Eq. (48c) balances accuracy of the mapping inside the domain and at the boundary.

The functionals are minimized in an iterative procedure: We let $k = 0$ and choose an initial guess $\mathbf{m}^0$. Next, we compute

$$\mathbf{P}^{k+1} = \operatorname*{argmin}_{\mathbf{P} \in \mathcal{P}} J_I(\mathbf{m}^k, \mathbf{P}), \tag{49a}$$
$$\mathbf{b}^{k+1} = \operatorname*{argmin}_{\mathbf{b} \in \mathcal{B}} J_B(\mathbf{m}^k, \mathbf{b}), \tag{49b}$$
$$\mathbf{m}^{k+1} = \operatorname*{argmin}_{\mathbf{m} \in \mathcal{M}} J(\mathbf{m}, \mathbf{P}^{k+1}, \mathbf{b}^{k+1}), \tag{49c}$$

where the function spaces $\mathcal{P}$, $\mathcal{B}$ and $\mathcal{M}$ are defined as

$$\mathcal{P} = \left\{\mathbf{P} \in C^1(S)^{2\times 2} \,\middle|\, \mathbf{P} = \mathbf{P}^T, \det(\mathbf{P}) = \det(\mathbf{C}) \cdot F(\mathbf{x}, \mathbf{m}(\mathbf{x}))\right\}, \tag{50a}$$
$$\mathcal{B} = \left\{\mathbf{b} \in C(\partial S)^2 \,\middle|\, \mathbf{b}(\mathbf{x}) \in \partial T \text{ for all } \mathbf{x} \in \partial S\right\}, \tag{50b}$$
$$\mathcal{M} = C^2(S)^2. \tag{50c}$$

The minimization of $J_I(\mathbf{m}^k, \mathbf{P})$ to compute $\mathbf{P}^{k+1}$ can be performed pointwise because the integrand does not depend on any derivatives of $\mathbf{P}^{k+1}$. It requires the solution of a constrained minimization problem. We discretize the source domain S and denote the grid points by $\mathbf{x}_{ij}$. We approximate $\mathrm{D}\mathbf{m}$ with central differences at each grid point. Next we determine the elements of the matrix $\mathbf{P}^{k+1}$ by minimizing $\frac{1}{2}\|\mathbf{C}\,\mathrm{D}\mathbf{m} - \mathbf{P}\|_F^2$ under the constraint $\det(\mathbf{P}) = \det(\mathbf{C}) \cdot F(\mathbf{x}_{ij}, \mathbf{m}(\mathbf{x}_{ij}))$. Details are given in [49, Sec. 8.3]. The minimization of $J_B(\mathbf{m}^k, \mathbf{b})$ to compute $\mathbf{b}^{k+1}$ can also be performed pointwise. For each point $\mathbf{x}_{ij}$ on the boundary of the source domain, we project $\mathbf{m}^k(\mathbf{x}_{ij})$ on the boundary of the target domain to find $\mathbf{b}^{k+1}(\mathbf{x}_{ij})$. For the minimization of $J(\mathbf{m}, \mathbf{P}^{k+1}, \mathbf{b}^{k+1})$ to compute $\mathbf{m}^{k+1}$, we require the first variation with respect to $\mathbf{m}$ to vanish:

$$
\begin{aligned}
&\delta J(\mathbf{m}, \mathbf{P}^{k+1}, \mathbf{b}^{k+1})(\boldsymbol{\eta}) \\
&= \lim_{\epsilon \to 0} \frac{1}{\epsilon} \Big(J(\mathbf{m} + \epsilon\boldsymbol{\eta}, \mathbf{P}^{k+1}, \mathbf{b}^{k+1}) - J(\mathbf{m}, \mathbf{P}^{k+1}, \mathbf{b}^{k+1}) \Big) \\
&= \alpha \iint_S ((\mathbf{C}\,\mathrm{D}\mathbf{m} - \mathbf{P}^{k+1}) : \mathbf{C}\mathrm{D}\boldsymbol{\eta})\, d\mathbf{x} + (1-\alpha) \oint_{\partial S} (\mathbf{m} - \mathbf{b}^{k+1}) \cdot \boldsymbol{\eta}\, ds = 0,
\end{aligned}
$$

for all $\boldsymbol{\eta} \in \mathcal{M}$. Here, $\mathbf{A} : \mathbf{B}$ denotes the Frobenius inner product of matrices $\mathbf{A}$ and $\mathbf{B}$. Using Gauss's theorem and the fundamental lemma of calculus of variations [19, p. 185], we find the coupled boundary value problem

$$
\nabla \cdot (\mathbf{C}^T \mathbf{C}\,\mathrm{D}\mathbf{m}) = \nabla \cdot (\mathbf{C}^T \mathbf{P}^{k+1}), \qquad \mathbf{x} \in S, \tag{51a}
$$

$$
\alpha(\mathbf{C}^T \mathbf{C}\,\mathrm{D}\mathbf{m}) \cdot \mathbf{n} + (1-\alpha)\mathbf{m} = \alpha \mathbf{C}^T \mathbf{P}^{k+1}\mathbf{n} + (1-\alpha)\mathbf{b}^{k+1}, \quad \mathbf{x} \in \partial S, \tag{51b}
$$

where $\mathbf{n}$ is the unit outward normal on ∂S. The divergence of a 2×2-matrix function $\mathbf{A}$ is defined as $\nabla \cdot \mathbf{A} = \begin{pmatrix} \nabla \cdot \mathbf{b}_1 \\ \nabla \cdot \mathbf{b}_2 \end{pmatrix}$ where $\mathbf{B} = \mathbf{A}^T = (\mathbf{b}_1, \mathbf{b}_2)$. We discretize Eq. (51) using a standard finite volume method [53, App. B].

The required smoothness for $\mathbf{P}$, $\mathbf{b}$ and $\mathbf{m}$ follows from Eq. (51). In Eq. (51a), we need $\mathbf{m}$ to be at least twice differentiable and $\mathbf{P}$ to be once differentiable. From Eq. (51b), it follows that $\mathbf{b}$ needs to be at least once differentiable, because the boundary condition contains first-order derivatives of $\mathbf{m}$.

For Models 1 and 2, we repeat the steps (49a)–(49c) until convergence. Upon convergence of the iterative procedure, the location of the surface is calculated from the optical map using the zero-gradient condition Eq. (11) (for Model 1) or Eq. (26) (for Model 2). We minimize the functional

$$
I(u) = \tfrac{1}{2} \iint_S \|\nabla_{\mathbf{x}} c(\cdot, \mathbf{m}) - \nabla u\|^2\, d\mathbf{x}. \tag{52}
$$

Again, we require the first variation $\delta I(u)(w)$ to vanish for all $w \in C^2(S)$. A similar derivation as above for Eq. (51) leads to the following Neumann problem for u

$$
\Delta u = \nabla \cdot \nabla_{\mathbf{x}} c(\cdot, \mathbf{m}), \quad \mathbf{x} \in S, \tag{53a}
$$

$$
\nabla u \cdot \mathbf{n} = \nabla_{\mathbf{x}} c(\cdot, \mathbf{m}) \cdot \mathbf{n}, \quad \mathbf{x} \in \partial S. \tag{53b}
$$

We discretize Eq. (53) using standard finite differences. Because u is determined up to an additive constant, we prescribe either the distance from the source to the first optical surface along a specific ray or the average value of u.

For Model 3, we need to do an additional step during the iteration to compute u. To do so, we use Eq. (37) for computing the surface. Dividing Eq. (37) by $H_v(\mathbf{x}, \mathbf{m}, u)$ gives

$$
\frac{\nabla_{\mathbf{x}} H(\mathbf{x}, \mathbf{m}, u)}{H_v(\mathbf{x}, \mathbf{m}, u)} + \nabla u = \mathbf{0}.
$$

Setting $\mathbf{f} = (\nabla_{\mathbf{x}} H)/H_v$, we minimize the functional

$$I(u, \mathbf{m}) = \frac{1}{2} \iint_S \|\mathbf{f} + \nabla u\|^2 \, d\mathbf{x}. \tag{54}$$

We compute

$$u^{k+1} = \underset{u \in \mathcal{U}}{\operatorname{argmin}}\, I(u, \mathbf{m}^{k+1}), \qquad \mathcal{U} = C^2(S), \tag{55}$$

by requiring the first variation $\delta I(u, \mathbf{m}^{k+1})(w)$ to vanish for all $w \in \mathcal{U}$. We obtain the boundary value problem

$$\nabla \cdot (\mathbf{f} + \nabla u) = \mathbf{f}_v \cdot (\mathbf{f} + \nabla u), \quad \mathbf{x} \in S, \tag{56a}$$

$$(\mathbf{f} + \nabla u) \cdot \mathbf{n} = 0, \quad \mathbf{x} \in \partial S. \tag{56b}$$

We solve Eq. (56) and enforce a unique solution in the same way as we did for Eq. (53).

For details on the numerical algorithm for SMA and GMA equations, see [57]. A full description of the generalization to GJEs can be found in [54]. Table 4 presents an overview of papers in which numerical results have been computed with the iterative least-squares solver. The table illustrates the general applicability of our framework, showing solutions for twelve of the sixteen basic optical systems in Fig. 1. The four missing systems can be solved using the same techniques but have not yet been treated, as they are of limited practical relevance.

Table 4 Overview of papers on basic optical systems solved using the iterative least-squares method.

Reflector systems	Parallel target	Point target	Near-field target	Far-field target
Parallel source	[76, 77]	[66]	[55, 60, 62]	[3, 49, 50, 53]
Point source	[66]	[11]	[59]	[57, 63]
Lens systems				
Parallel source	[64, 75, 76, 78]			[49, 63, 67]
Point source			[61]	[55, 56, 58, 62, 67, 68]

4 Numerical Results

In this section we tried capturing the growth and blossoming of the Woudschoten conference in a numerical experiment. We have computed reflectors that map a uniform emittance into illuminances that correspond to an image of a lotus bud at the first target and an image of a lotus in full bloom at the second. The optical system is sketched in Fig. 14b. The prescribed illuminances and the optical mappings are shown in Fig. 15. The numerical results have been computed using a 501×501 grid and 50 iterations of the least-squares solver.

(a) Target distribution at T_1

(b) Target distribution at T_2

(c) Optical map $\mathbf{m}_1 : S \to T_1$

(d) Optical map $\mathbf{m}_2 : T_1 \to T_2$

(e) Ray-traced illumination at T_1

(f) Ray-traced illumination at T_2

Figure 15 Numerical results for the 3D parallel-to-two-targets reflector system.

The parameter α in the least-squares algorithm was set to 0.5 for computing the mapping from source to first target and to 0.9992 for computing the mapping from first to second target. The source domain is $-15 \leq x_1 \leq -9$, $-3 \leq x_2 \leq 3$, $z = 0$, Target 1 is $-3 \leq y_1, y_2 \leq 3$, $z = 15$ and Target 2 is $-3 \leq z_1, z_2 \leq 3$, $z = 50$. The average optical path length from source to first target is 40. The average distance from source to first reflector is 15. A Quasi-Monte Carlo ray-tracing verification is done with five million light rays. We see that the ray-traced illuminances in Figs. 15e and 15f show good agreement with the complicated target distributions that were prescribed. Only 50 iteration steps were needed, showing the fast convergence of the method.

5 Conclusions

In this chapter we presented mathematical models for a parallel-to-parallel reflector system, a parallel-to-point reflector system and a parallel-to-near-field lens system. The approach can be generalized to sixteen basic optical systems, leading to an equation of standard Monge-Ampère type (Model 1), generalized Monge-Ampère type (Model 2) or generated Jacobian type (Model 3). The exact classification is shown in Fig. 1. The models are proper generalizations in the sense that Model 1 systems can also be described by Model 2, and Model 2 systems can be formulated using generating functions (Model 3) too.

We outlined a numerical method to solve the nonlinear partial differential equations and presented a challenging numerical example with complicated target light distributions.

Acknowledgements This work was supported by the NWO Perspectief program *Optical coherence; optimal delivery and positioning* (OPTIC, P21-20).

References

1. S. B. Angenent, S. Haker, and A. R. Tannenbaum. Minimizing flows for the Monge–Kantorovich problem. *SIAM Journal on Mathematical Analysis*, 35(1):61–97, 2003.
2. M. J. H. Anthonissen, L. B. Romijn, J. H. M. ten Thije Boonkkamp, and W. L. IJzerman. Unified mathematical framework for a class of fundamental freeform optical systems. *Optics Express*, 29(20):31650–31664, 2021.
3. R. Beltman, J. ten Thije Boonkkamp, and W. IJzerman. A least-squares method for the inverse reflector problem in arbitrary orthogonal coordinates. *Journal of Computational Physics*, 367:347–373, 2018.
4. J.-D. Benamou, G. Carlier, M. Cuturi, L. Nenna, and G. Peyré. Iterative Bregman projections for regularized transportation problems. *SIAM Journal on Scientific Computing*, 37(2):A1111–A1138, 2015.
5. J.-D. Benamou, B. D. Froese, and A. M. Oberman. Two numerical methods for the elliptic Monge–Ampère equation. *ESAIM: Mathematical Modelling and Numerical Analysis*, 44(4):737–758, 2010.

6. J.-D. Benamou, B. D. Froese, and A. M. Oberman. Numerical solution of the optimal transportation problem using the Monge–Ampère equation. *Journal of Computational Physics*, 260:107–126, 2014.
7. M. W. M. C. Bertens, M. J. H. Anthonissen, J. H. M. ten Thije Boonkkamp, and W. L. IJzerman. An iterative least-squares method for the hyperbolic Monge–Ampère equation with transport boundary condition. *SIAM Journal on Scientific Computing*, 47(3):B735–B761, 2025.
8. M. Born and E. Wolf. *Principles of Optics: 60th Anniversary Edition.* Cambridge University, 7 edition, 2019.
9. C. Bösel and H. Gross. Ray mapping approach for the efficient design of continuous freeform surfaces. *Optics Express*, 24(13):14271–14282, 2016.
10. C. Bösel and H. Gross. Double freeform illumination design for prescribed wavefronts and irradiances. *Journal of the Optical Society of America A*, 35(2):236–243, 2018.
11. P. A. Braam, J. H. M. ten Thije Boonkkamp, M. J. H. Anthonissen, R. Beltman, and W. L. IJzerman. Mathematical model for inverse freeform design of a point-to-point two-reflector system. *Journal of the Optical Society of America A, Optics, Image Science and Vision*, 41(11):2156–2162, 2024.
12. P. A. Braam, J. H. M. ten Thije Boonkkamp, M. J. H. Anthonissen, K. Mitra, R. Beltman, and W. L. IJzerman. Inverse freeform design of a parallel-to-two-target reflector system. *Journal of the Optical Society of America A, Optics, Image Science and Vision*, 42(8):1133–1143, 2025.
13. K. Brix, Y. Hafizogullari, and A. Platen. Designing illumination lenses and mirrors by the numerical solution of Monge–Ampère equations. *Journal of the Optical Society of America A*, 32(11):2227–2236, 2015.
14. K. Brix, Y. Hafizogullari, and A. Platen. Solving the Monge–Ampère equations for the inverse reflector problem. *Mathematical Models and Methods in Applied Sciences*, 25(5):803–837, 2015.
15. D. A. Bykov, L. L. Doskolovich, A. A. Mingazov, E. A. Bezus, and N. L. Kazanskiy. Linear assignment problem in the design of freeform refractive optical elements generating prescribed irradiance distributions. *Optics Express*, 26(21):27812–27825, 2018.
16. A. Caboussat, R. Glowinski, and D. Gourzoulidis. A least-squares/relaxation method for the numerical solution of the three-dimensional elliptic Monge–Ampère equation. *Journal of Scientific Computing*, 77(1):53–78, 2018.
17. A. Caboussat, R. Glowinski, and D. C. Sorensen. A least-squares method for the numerical solution of the Dirichlet problem for the elliptic Monge–Ampère equation in dimension two. *ESAIM: Control, Optimisation and Calculus of Variations*, 19(3):780–810, 2013.
18. S. Chang, R. Wu, L. An, and Z. Zheng. Design beam shapers with double freeform surfaces to form a desired wavefront with prescribed illumination pattern by solving a Monge–Ampère type equation. *Journal of Optics*, 18(12):125602, 2016.
19. R. Courant and D. Hilbert. *Methods of Mathematical Physics*, volume 2. John Wiley & Sons, Ltd, 1989.
20. E. J. Dean and R. Glowinski. Numerical solution of the two-dimensional elliptic Monge–Ampère equation with Dirichlet boundary conditions: An augmented Lagrangian approach. *Comptes Rendus Mathématique*, 336(9):779–784, 2003.
21. E. J. Dean and R. Glowinski. Numerical solution of the two-dimensional elliptic Monge–Ampère equation with Dirichlet boundary conditions: A least-squares approach. *Comptes Rendus Mathématique*, 339(12):887–892, 2004.
22. E. J. Dean and R. Glowinski. Numerical methods for fully nonlinear elliptic equations of the Monge–Ampère type. *Computer Methods in Applied Mechanics and Engineering*, 195(13):1344–1386, 2006.
23. L. L. Doskolovich, D. A. Bykov, E. S. Andreev, E. A. Bezus, and V. I. Oliker. Designing double freeform surfaces for collimated beam shaping with optimal mass transportation and linear assignment problems. *Optics Express*, 26(19):24602–24613, 2018.
24. L. L. Doskolovich, D. A. Bykov, A. A. Mingazov, and E. A. Bezus. Optimal mass transportation and linear assignment problems in the design of freeform refractive optical elements generating far-field irradiance distributions. *Optics Express*, 27(9):13083–13097, 2019.
25. Z. Feng, D. Cheng, and Y. Wang. Iterative wavefront tailoring to simplify freeform optical design for prescribed irradiance. *Optics Letters*, 44(9):2274–2277, 2019.

26. Z. Feng, B. D. Froese, C.-Y. Huang, D. Ma, and R. Liang. Creating unconventional geometric beams with large depth of field using double freeform-surface optics. *Applied Optics*, 54(20):6277–6281, 2015.
27. Z. Feng, L. Huang, G. Jin, and M. Gong. Designing double freeform optical surfaces for controlling both irradiance and wavefront. *Optics Express*, 21(23):28693–28701, 2013.
28. B. D. Froese. A numerical method for the elliptic Monge–Ampère equation with transport boundary conditions. *SIAM Journal on Scientific Computing*, 34(3):A1432–A1459, 2012.
29. B. D. Froese and A. M. Oberman. Convergent finite difference solvers for viscosity solutions of the elliptic Monge–Ampère equation in dimensions two and higher. *SIAM Journal on Numerical Analysis*, 49(4):1692–1714, 2011.
30. B. D. Froese and A. M. Oberman. Fast finite difference solvers for singular solutions of the elliptic Monge–Ampère equation. *Journal of Computational Physics*, 230(3):818–834, 2011.
31. B. D. Froese and A. M. Oberman. Convergent filtered schemes for the Monge–Ampère partial differential equation. *SIAM Journal on Numerical Analysis*, 51(1):423–444, 2013.
32. T. Glimm and V. I. Oliker. Optical design of single reflector systems and the Monge–Kantorovich mass transfer problem. *Journal of Mathematical Sciences*, 117(3):4096–4108, 2003.
33. E. Haber, T. Rehman, and A. R. Tannenbaum. An efficient numerical method for the solution of the L_2 optimal mass transfer problem. *SIAM Journal on Scientific Computing*, 32(1):197–211, 2010.
34. E. Hecht. *Optics*. Pearson Education, 2017.
35. E. L. Kawecki, O. Lakkis, and T. Pryer. A finite element method for the Monge–Ampère equation with transport boundary conditions. *arXiv preprint arXiv:1807.03535*, 2018.
36. S. A. Kochengin and V. I. Oliker. Determination of reflector surfaces from near-field scattering data. *Inverse Problems*, 13(2):363–373, 1997.
37. S. A. Kochengin and V. I. Oliker. Determination of reflector surfaces from near-field scattering data II. Numerical solution. *Numerische Mathematik*, 79(4):553–568, 1998.
38. S. A. Kochengin and V. I. Oliker. Computational algorithms for constructing reflectors. *Computing and Visualization in Science*, 6(1):15–21, 2003.
39. O. Lakkis and T. Pryer. A finite element method for second order nonvariational elliptic problems. *SIAM Journal on Scientific Computing*, 33(2):786–801, 2011.
40. O. Lakkis and T. Pryer. A finite element method for nonlinear elliptic problems. *SIAM Journal on Scientific Computing*, 35(4):A2025–A2045, 2013.
41. G. Monge. *Mémoire sur la théorie des déblais et des remblais*. De l'Imprimerie Royale, 1781.
42. A. M. Oberman. Convergent difference schemes for degenerate elliptic and parabolic equations: Hamilton–Jacobi equations and free boundary problems. *SIAM Journal on Numerical Analysis*, 44(2):879–895, 2006.
43. A. M. Oberman. Wide stencil finite difference schemes for the elliptic Monge–Ampère equation and functions of the eigenvalues of the Hessian. *Discrete and Continuous Dynamical Systems - B*, 10(1):221, 2008.
44. V. I. Oliker. On reconstructing a reflecting surface from the scattering data in the geometric optics approximation. *Inverse Problems*, 5(1):51–65, 1989.
45. V. I. Oliker. Optical design of freeform two-mirror beam-shaping systems. *Journal of the Optical Society of America A*, 24(12):3741–3752, 2007.
46. V. I. Oliker. Controlling light with freeform multifocal lens designed with supporting quadric method (SQM). *Optics Express*, 25(4):A58–A72, 2017.
47. V. I. Oliker, L. L. Doskolovich, and D. A. Bykov. Beam shaping with a plano-freeform lens pair. *Optics Express*, 26(15):19406–19419, 2018.
48. V. I. Oliker, J. Rubinstein, and G. M. Wolansky. Supporting quadric method in optical design of freeform lenses for illumination control of a collimated light. *Advances in Applied Mathematics*, 62:160–183, 2015.
49. C. R. Prins. *Inverse Methods for Illumination Optics*. PhD thesis, Eindhoven University of Technology, Eindhoven, the Netherlands, 2014.
50. C. R. Prins, R. Beltman, J. H. M. ten Thije Boonkkamp, W. L. IJzerman, and T. W. Tukker. A least-squares method for optimal transport using the Monge–Ampère equation. *SIAM Journal on Scientific Computing*, 37(6):B937–B961, 2015.

51. C. R. Prins, J. H. M. ten Thije Boonkkamp, J. van Roosmalen, W. L. IJzerman, and T. W. Tukker. A Monge–Ampère-solver for free-form reflector design. *SIAM Journal on Scientific Computing*, 36(3):B640–B660, 2014.
52. H. Ries and A. Rabl. Edge-ray principle of nonimaging optics. *Journal of the Optical Society of America A, Optics, Image Science and Vision*, 11(10):2627–2632, 1994.
53. L. B. Romijn. *Generated Jacobian Equations in Freeform Optical Design*. PhD thesis, Eindhoven University of Technology, Eindhoven, the Netherlands, Oct. 2021.
54. L. B. Romijn, J. H. M. ten Thije Boonkkamp, M. J. H. Anthonissen, and W. L. IJzerman. Generating-function approach for double freeform lens design. *Journal of the Optical Society of America A*, 38(3):356–368, 2021.
55. L. B. Romijn, J. H. M. ten Thije Boonkkamp, M. J. H. Anthonissen, and W. L. IJzerman. An iterative least-squares method for generated Jacobian equations in freeform optical design. *SIAM Journal on Scientific Computing*, 43(2):B298–B322, 2021.
56. L. B. Romijn, J. H. M. ten Thije Boonkkamp, and W. L. IJzerman. Freeform lens design for a point source and far-field target. *Journal of the Optical Society of America A, Optics, Image Science and Vision*, 36(11):1926–1939, Nov. 2019.
57. L. B. Romijn, J. H. M. ten Thije Boonkkamp, and W. L. IJzerman. Inverse reflector design for a point source and far-field target. *Journal of Computational Physics*, 408, May 2020.
58. L. B. Romijn, J. H. M. ten Thije Boonkkamp, and W. L. IJzerman. A Monge–Ampère least-squares solver for the design of a freeform lens. In F. J. Vermolen and C. Vuik, editors, *Numerical Mathematics and Advanced Applications ENUMATH 2019*, pages 833–840, Cham, 2021. Springer International Publishing.
59. J. H. M. ten Thije Boonkkamp, M. J. H. Anthonissen, P. A. Braam, and W. L. IJzerman. Generating function formulation for inverse freeform design. In T. E. Kidger, S. David, and T. Schupp, editors, *Illumination Optics VII*, volume 13022, page 130220H. International Society for Optics and Photonics, SPIE, 2024.
60. J. H. M. ten Thije Boonkkamp, K. Mitra, M. J. H. Anthonissen, L. Kusch, P. A. Braam, and W. L. IJzerman. Inverse methods for freeform optical design. In Y. Wang, T. E. Kidger, and R. Wu, editors, *Optical Design and Testing XIV*, volume 13237, page 132370P. International Society for Optics and Photonics, SPIE, 2024.
61. J. H. M. ten Thije Boonkkamp, K. Mitra, M. J. H. Anthonissen, L. Kusch, P. A. Braam, and W. L. IJzerman. Freeform optical design: mathematical models and numerical methods. In Y. Wang, T. E. Kidger, and R. Wu, editors, *Optical Design and Testing XV*, volume 13716, page 137160M. International Society for Optics and Photonics, SPIE, 2025.
62. J. H. M. ten Thije Boonkkamp, K. Mitra, M. J. H. Anthonissen, L. Kusch, P. A. Braam, and W. L. IJzerman. Inverse freeform design in non-imaging optics: Hamilton's theory of geometrical optics, optimal transport, and least-squares solvers. *Frontiers in Physics*, 13, 2025.
63. J. H. M. ten Thije Boonkkamp, L. B. Romijn, and W. L. IJzerman. Generalized Monge–Ampère equations for illumination freeform design. In Y. Wang, P. Benítez, and O. Matoba, editors, *Optical Design and Testing IX*, volume 11185, page 1118504. International Society for Optics and Photonics, SPIE, 2019.
64. J. H. M. ten Thije Boonkkamp, L. B. Romijn, N. K. Yadav, and W. L. IJzerman. Monge–Ampère type equations for freeform illumination optics. In *Illumination Optics V*, Proceedings of SPIE, United States, Jan. 2018. SPIE. Illumination Optics V 2018 ; Conference date: 14-05-2018 Through 16-05-2018.
65. N. S. Trudinger. On the local theory of prescribed Jacobian equations. *Discrete and Continuous Dynamical Systems*, 34(4):1663–1681, 2014.
66. A. H. van Roosmalen, M. J. H. Anthonissen, W. L. IJzerman, and J. H. M. ten Thije Boonkkamp. Design of a freeform two-reflector system to collimate and shape a point source distribution. *Optics Express*, 29(16):25605–25625, Aug 2021.
67. A. H. van Roosmalen, M. J. H. Anthonissen, W. L. IJzerman, and J. H. M. ten Thije Boonkkamp. Fresnel reflections in inverse freeform lens design. *Journal of the Optical Society of America A, Optics, Image Science and Vision*, 39(6):1045–1052, Jun 2022.
68. A. H. van Roosmalen, M. J. H. Anthonissen, W. L. IJzerman, and J. H. M. ten Thije Boonkkamp. Least-squares method for generated Jacobian equations in freeform optical design. In A. Sequeira,

A. Silvestre, S. S. Valtchev, and J. Janela, editors, *Numerical Mathematics and Advanced Applications ENUMATH 2023, Volume 2*, pages 302–309, Cham, 2025. Springer Nature Switzerland.
69. S. Verma. *Inverse Methods for the Optical Design of Freeform Imaging Systems*. PhD thesis, Eindhoven University of Technology, Eindhoven, the Netherlands, 2026.
70. C. Villani. *Optimal Transport: Old and New*, volume 338 of *Grundlehren der mathematischen Wissenschaften*. Springer, Berlin, Heidelberg, 2009.
71. R. Wu and H. Hua. Direct design of aspherical lenses for extended non-Lambertian sources in three-dimensional rotational geometry. *Optics Express*, 24(2):1017–1030, 2016.
72. R. Wu, C. Y. Huang, X. Zhu, H.-N. Cheng, and R. Liang. Direct three-dimensional design of compact and ultra-efficient freeform lenses for extended light sources. *Optica*, 3(8):840–843, 2016.
73. R. Wu, P. Liu, Y. Zhang, Z. Zheng, H. Li, and X. Liu. A mathematical model of the single freeform surface design for collimated beam shaping. *Optics Express*, 21(18):20974–20989, 2013.
74. R. Wu, L. Xu, P. Liu, Y. Zhang, Z. Zheng, H. Li, and X. Liu. Freeform illumination design: A nonlinear boundary problem for the elliptic Monge–Ampère equation. *Optics Letters*, 38(2):229–231, 2013.
75. N. Yadav, J. ten Thije Boonkkamp, and W. IJzerman. Computation of double freeform optical surfaces using a Monge–Ampère solver: Application to beam shaping. *Optics Communications*, 439:251–259, 2019.
76. N. K. Yadav. *Monge–Ampère problems with non-quadratic cost function. Application to freeform optics*. PhD thesis, Eindhoven University of Technology, Eindhoven, the Netherlands, Sept. 2018.
77. N. K. Yadav, L. B. Romijn, J. H. M. ten Thije Boonkkamp, and W. L. IJzerman. A least-squares method for the design of two-reflector optical systems. *Journal of Physics: Photonics*, 1(3), July 2019.
78. N. K. Yadav, J. H. M. ten Thije Boonkkamp, and W. L. IJzerman. A Monge–Ampère problem with non-quadratic cost function to compute freeform lens surfaces. *Journal of Scientific Computing*, 80(1):475–499, July 2019.
79. L. Yang, Y. Liu, Z. Ding, J. Zhang, X. Tao, Z. Zheng, and R. Wu. Design of freeform lenses for illuminating hard-to-reach areas through a light-guiding system. *Optics Express*, 28(25):38155–38168, 2020.
80. Y. Zhang, R. Wu, P. Liu, Z. Zheng, H. Li, and X. Liu. Double freeform surfaces design for laser beam shaping with Monge–Ampère equation method. *Optics Communications*, 331:297–305, 2014.

Computing for Life: Modeling and Simulation of Electrodiffusion in Dense Reconstructions of Cerebral Tissue

Halvor Herlyng, Marius Causemann, Gaute T. Einevoll, Ada J. Ellingsrud, Geir Halnes and Marie E. Rognes

Abstract Excitable tissue is fundamental to brain function, yet its study is complicated by extreme morphological complexity and the physiological processes governing its dynamics. Consequently, detailed computational modeling of this tissue represents a formidable task, requiring both efficient numerical methods and robust implementations. Meanwhile, efficient and robust methods for image segmentation and meshing are needed to provide realistic geometries for which numerical solutions are tractable. Here, we present a computational framework that models electrodiffusion in excitable cerebral tissue, together with realistic geometries generated from electron microscopy data. To demonstrate a possible application of the framework, we simulate electrodiffusive dynamics in cerebral tissue during neuronal activity. Our results and findings highlight the numerical and computational challenges associated with modeling and simulation of electrodiffusion and other multiphysics in dense reconstructions of cerebral tissue.

1 Introduction

Our brains are composed of complex cellular tissue consisting of nerve and glial cells, separated by the tortuous extracellular space, and with vasculature running throughout [40]. Brain tissue is excitable: cells communicate through electromagnetic, chemical

Halvor Herlyng, Marius Causemann, Ada J. Ellingsrud
Simula Research Laboratory, Kristian Augusts gate 23, 0164 Oslo, Norway, hherlyng@simula.no, mariusca@simula.no, ada@simula.no

Marie E. Rognes
Simula Research Laboratory, Kristian Augusts gate 23, 0164 Oslo, Norway and K. G. Jebsen Centre for Brain Fluid Research, Oslo, Norway, meg@simula.no

Gaute T. Einevoll
Norwegian University of Life Sciences, Ås, Norway and University of Oslo, Oslo, Norway, gaute.einevoll@nmbu.no

Geir Halnes
Norwegian Artificial Intelligence Research Consortium, Oslo, Norway, geir.halnes@nora.ai

B. Koren and D. Roose (eds.), *Numerical Analysis and Scientific Computing: an Anthology*, Lecture Notes in Computational Science and Engineering 156,
https://doi.org/10.1007/978-3-032-33428-2_14

and mechanical signals, with ion dynamics playing a central role. Signals can be sent directly from cell to cell through synapses via electric potential spikes known as action potentials (APs), while electrochemical changes in the extracellular environment may also excite or inhibit signals.

Understanding the physical mechanisms underlying brain function is key to better understand neurodegenerative and other neurological disorders. A 2024 study estimated that over one-third of the global population suffers or has suffered from a neurological condition, such as stroke, epilepsy and dementia [73]. Computational modeling of neuronal biophysics offers a route to improved understanding: models can be tailored to study specific physical mechanisms, such as the response of a cell to an action potential, or how changes in action potential firing frequencies affect ion concentrations in the surrounding cerebral tissue. Moreover, simulations are flexible because the choice of parameters, initial and boundary conditions can often be altered swiftly at low or no cost. However, an inherent modelling challenge is the range of spatiotemporal scales involved in brain biophysics [10, 40]. In space, these processes occur over centimeters at the organ level to sub-nanometers at the atomic level inside cells (Figure 1). In time, processes occurring over seconds and over hours may influence each other.

Traditionally, computational modeling of brain tissue has focused on electrophysiology. An important reason for this is the impressively accurate model of a giant squid axon potential developed by Hodgkin and Huxley in 1952 [33, 31, 20]. Analogies between electrical circuits and neural tissue have resulted in volume conductor theory models, compartmental models and network models, where extra- and intracellular potentials are calculated based on modeling neurons as compartments in a volume conductor or as components of electrical circuits. These models often account for the geometry implicitly, by including homogenized parameters such as volume fractions and tortuosity. Multi-compartmental modeling based on cable theory has developed into the standard framework for mechanistic modeling of neurons [29], including detailed whole neuron models [50]. Over the last decades, there has been increasing interest in models accounting for geometry and morphology explicitly [1, 23]. This development has taken place in tandem with modeling cardiac cells and tissue, another example of excitable tissues [79].

While most computational modeling of cerebral tissue has focused on neurons, which are the main signaling units of the nervous system [40], in recent years, glial cells have received an increasing amount of attention [17]. Glial cells constitute at least half of human brain tissue and are now considered to play an important role in cerebral dynamics. For example, the glia known as astrocytes respond to neurotransmitters, partake in modulation and extracellular metabolism of adenosine triphosphate (ATP), and are involved in neuronal signaling pathways through receptor dynamics [26, 45, 14]. Astrocytes also regulate brain function through ion concentrations, with examples including potassium buffering [15] and the dynamic role of astrocytic chloride concentrations in volume regulation and cell proliferation [80]. The ion concentrations both in the extracellular and intracellular spaces of the brain play an indispensable role in brain functioning, with up to two orders of magnitude differences between physiological and pathological concentrations [60].

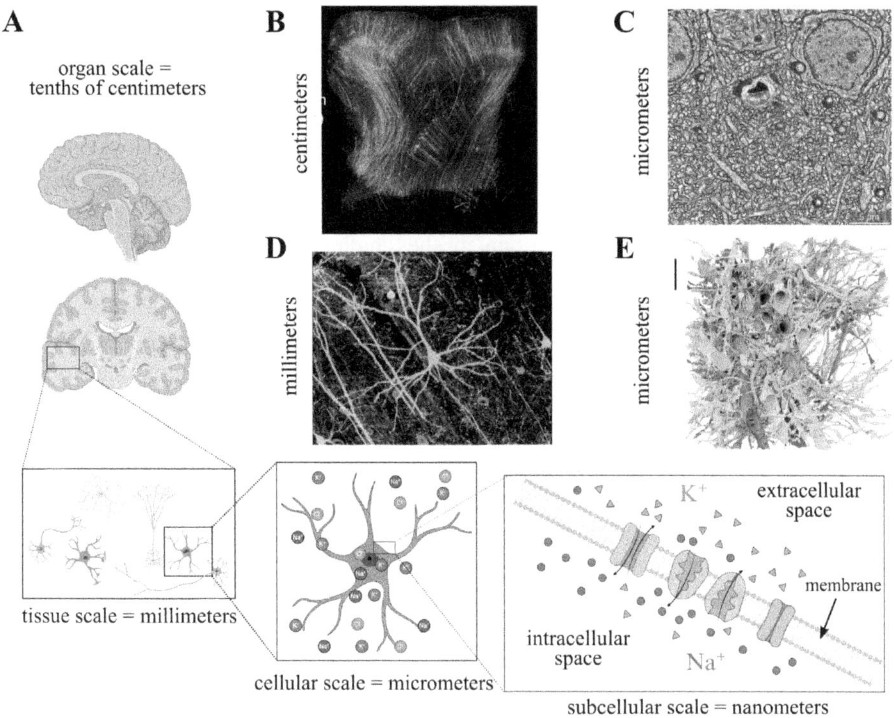

Figure 1 A. The physical scales of the brain, from organ scale to subcellular scale. At the subcellular scale, ions are transported across the cellular membrane through ion channels. The ion channels illustrated on the membrane are, from left to right: open generic ion channel, Na^+/K^+-ATPase pump importing potassium ions, Na^+/K^+-ATPase pump exporting sodium ions, and a closed generic ion channel. The illustrations in this panel were created using Biorender. **B.** Plane illumination microscopy of macaque primary motor cortex tissue at centimeter scale [25] **C.** Confocal microscopy image of a neuron (green) in rodent visual cortex tissue. Image spans half a millimeter [44]. **D.** Light-microscopy based reconstructed cross section of mouse hippocampus tissue at the micrometer scale [76]. **E.** 3D rendering of mouse hippocampus tissue, scale bar 3 μm [76]. The images in panels **B–E** are reproduced under the terms of the Creative Commons Attribution License.

An open question relates to the role of electrodiffusion in brain tissue under physiological and pathological conditions and the coupled response of electric potentials and ion concentrations. Many electrodiffusion models homogenize or idealize the brain tissue, often considering zero-dimensional cellular compartments or rectangular or cylindrical geometries [30, 71, 66, 67, 68, 69]. These models are particularly suitable for studying the effect of ionic fluxes. For example, Halnes *et al.* [30] studied a one-dimensional model of an astrocyte coupled with the extracellular space, stimulating the system by adding potassium and removing sodium ions in the astrocyte. Sætra *et al.* [67] used a similar stimulus for an electrodiffusive compartmental model with neuronal, glial and extracellular spaces. There has also been development of electrodiffusive models that explicitly represent the cellular geometries [49, 56, 47, 23, 46]. These partial differential equation (PDE)-based models are typically solved numerically using finite difference, finite element, or finite volume methods. The explicit geometry representation of individual cells and membranes allows for studying the interplay between cellular shape and function, a key tenet of cellular biology, and for studying spatially varying membrane mechanisms such as distributed ion channels, pumps or cotransporters.

Cell-by-cell models of ionic electrodiffusion are highly demanding to construct and solve computationally [78], for several reasons. First, the morphology of neurons, astrocytes and the extracellular space is highly complex. This makes the generation of well-defined, conforming computational meshes challenging in its own right. Moreover, the fine geometric detail calls for high-resolution meshes, and the large number of mesh vertices and cells in turn increases computational cost. Second, electrodiffusion involves systems of PDEs with strongly coupled variables, whose dynamics evolve on multiple spatial and temporal scales. Development of efficient and scalable numerical solution algorithms for such coupled problems is an active field of research [62, 7, 21].

Over the last three decades, advances in bioimaging have revolutionized our ability to study the structure and function of living systems. At the nano- to microscale, the new capabilities of cellular microscopy have resulted in the generation and wide availability of large-scale high-resolution imaging data, as strikingly exemplified by increasingly complex dense reconstructions of cortical tissue. Through open distribution of such high-quality imaging data, there now exist several datasets with extremely detailed image data of brain tissue [16]. Together with modern image-processing software and powerful hardware, the image data enables generation of realistic computational geometries of brain tissue. These convergent developments give new opportunities for detailed simulations of brain electrophysiology, and brain multiphysics more generally. Examples include simulations of ionic electrodiffusion using realistic, three-dimensional geometries representing single neurons [21] and geometries with a few neuronal and glial cells [7], surrounded by, extracellular space. Moreover, very detailed geometries have been used in other computational studies of the brain, e.g. in simulating interstitial solute transport in the neuropil [34] and mechanics at the gliovascular interface [13].

In this chapter, we present a computational framework for numerically solving ionic electrodiffusion PDEs defined over extremely detailed geometries of mouse visual cortex tissue. We highlight different aspects of the computational pipeline: from the geometric properties of these brain tissues to the numerical properties of the resulting finite element discretization.

2 Dense Reconstructions of Brain Tissue from Cell Imaging

Brain imaging has been fundamental to our understanding of brain structure and function [86]. Today, the combination of highly detailed cell imaging data and computational processing power allows for a new level of detail in terms of digital representations of brain tissue. Neurons, glial cells and the extracellular space are however characterized by tortuous, meandering structures making meshing of such high-resolution digital reconstructions of brain tissue challenging. One aspect is the computational demand in processing the images, generating surfaces and volumetric meshes. Another key element is that the generated computational mesh must be of sufficient quality for numerical

PDE solution algorithms. Here, we illustrate how these challenges can be addressed by current meshing technology by presenting a pipeline generating conforming tetrahedral meshes from dense image reconstructions.

2.1 From Segmented Cell Imaging to Conforming Finite Element Meshes

The pipeline we present is implemented as the open-source software EMI-Meshing [12]. All software tools mentioned in the following text are third-party components integrated in EMI-Meshing. The pipeline consists of the following steps (see also Figure 2):

1. Select and extract segmented image data.
2. Filter and/or process the segmentation.
3. Construct triangulated surfaces representing the interface between intracellular and extracellular spaces.
4. Generate a tetrahedral mesh of the complete volume conforming to the interface surfaces.

First, the raw segmentation data is downloaded at a voxel resolution of $32 \times 32 \times 40$ nm. This data is subsequently resampled to an isotropic resolution of 20 nm. To reduce noise and focus on significant biological structures, we remove small cellular fragments with volumes below 5000 voxels (approximately 0.04 μm^3) and retain only the N largest biological cells.

We then apply a sequence of morphological operations to process the cellular shapes, utilizing the GPU-accelerated library *pyclesperanto* [27]. The cells are first dilated into the background, followed by morphological opening and closing operations, which effectively smooth the cellular boundaries. Crucially, all cells are then eroded; this step ensures a physiological volume fraction for the extracellular space and prevents contact between adjacent cells. Following these morphological operations, any remaining disconnected components smaller than the 5000 voxel threshold are removed.

Surface meshes are extracted from the processed voxel data using the marching cubes algorithm implemented in *PyVista* [75]. To ensure computational tractability, the resolution of these surface meshes is reduced by a factor of ten using surface remeshing via the *pyacvd* library [41]. We mitigate staircase artifacts arising from the voxel approximation by applying five iterations of Taubin smoothing in *PyVista*. Next, a cubic surface of edge length L is generated to define the outer boundary of the extracellular space. The parameters L and the number of biological cells N are important, as they together determine the size and biological complexity of the generated mesh. We will in our numerical experiments vary L and N to study performance of the computational framework we present on meshcs of varying size and complexity.

Finally, we generate a volumetric mesh of the complete domain, conforming to the triangulated interface surfaces to accurately represent both the intracellular regions and the extracellular space. This step is performed using the fTetWild software [36] on the eX3 high-performance computing infrastructure [70]. The resulting mesh resolution is governed by the *envelope size* and *ideal edge length* parameters. The *envelope size* represents the maximal deviation from the input surface, while the *ideal edge length*

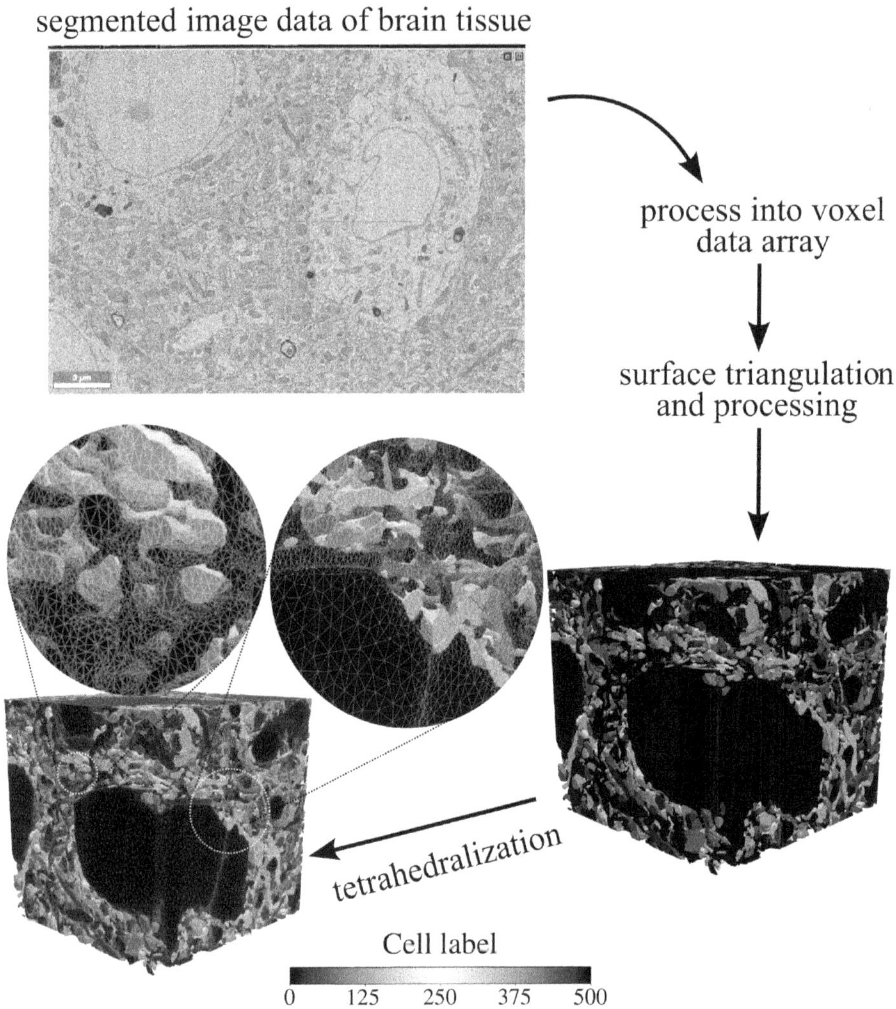

Figure 2 Illustration of the computational meshing pipeline. Segmented electron microscopy image data, here shown in the MICrONS explorer interface [16], is downloaded and processed into a voxel data array. The voxel data is next converted into triangulated surfaces representing the interfaces between the biological cells and extracellular space. These surfaces go through a sequence of morphological processing steps to ensure sufficient mesh quality downstream. Finally, a tetrahedral mesh is generated conforming to the interface surfaces.

defines the target size of the generated tetrahedral mesh cells. The ideal edge length should not be confused with the side length L of the cube of tissue, though setting the ideal edge length as a fraction of the size of the tissue cube may be a natural choice. Here, using a small *envelope size* of 18 nm and a larger *ideal edge length* (1–1.5 % of the diagonal of the box bounding the cube of side lengths L), the mesh resolution naturally adapts to the shape and morphology of the cellular regions, such that larger intracellular volumes are covered by larger mesh cells than the narrow extracellular space gaps. To ensure element quality suitable for numerical simulation, the mesh is optimized until a maximal conformal AMIPS energy [24] of 10 is reached, thereby eliminating severely distorted tetrahedra. The segmentation labels are transferred from the image data to the volumetric mesh, ensuring that all biological cells and interface surfaces are uniquely labeled in the final computational mesh. These labels or mesh tags facilitate modelling different properties, membrane mechanisms, and initial and boundary conditions for the different biological cells.

We next apply this meshing pipeline to generate computational meshes from the *IARPA MICrONS Cortical mm*^{3} dataset published by the MICrONS consortium [16]. This open dataset includes 3D images spanning a 1.4mm × 0.87mm × 0.84mm volume of mouse visual cortex obtained using two-photon microscopy, microCT and serial electron microscopy (Figure 2). The image data are already segmented, i.e., each connected component within the 3D image has been identified and assigned a unique label, and can be viewed using the web-visualization tool Neuroglancer [48]. We consider a series of cubical volumes of reconstructed tissue of size $L \times L \times L$ with center point $(225182, 107314, 22000)$ and with $L = 5, 10, 20, 30$ μm. For each of these cubical volumes, we generate a series of computational meshes including the N largest biological cells with $N = 100, 200, 300, 400, 500$ (see Figure 3). The remaining volume is relabeled as extracellular space.

2.2 Most Mesh Vertices Populate Cellular Membranes

We next analyze and present quantitative mesh characteristics for the series of cubical meshes of the mouse visual cortex with varying number of biological cells N and varying cube side lengths L. When including an increasing number of (biological) cells, the general trend for the computational meshes is an increase in the number of computational cells and vertices (Figure 4A, B). However, the increase is not uniform, for instance the number of mesh cells and vertices is lower for the case $N = 400$ than for $N = 300$ when $L = 10$. Notably, the ratio of vertices located on the cell membranes to the total number of vertices generally increases as the meshes become denser. For the largest and densest meshes, over 90% of the mesh vertices are on cellular membranes (Figure 4C). Moreover, the volume occupied by the extracellular space is reduced as more biological cells are packed into the geometries (Figure 4D). This also results in a narrower and more tortuous extracellular space (ECS). The ECS fraction of the densest meshes are between 17–25%. In real brain tissue, the fraction of extracellular space is commonly reported as around 20% [15]. Note that for the mesh with $L = 5$ μm, we only include the $N = 100$ and $N = 200$ meshes, as the maximum number of cells in

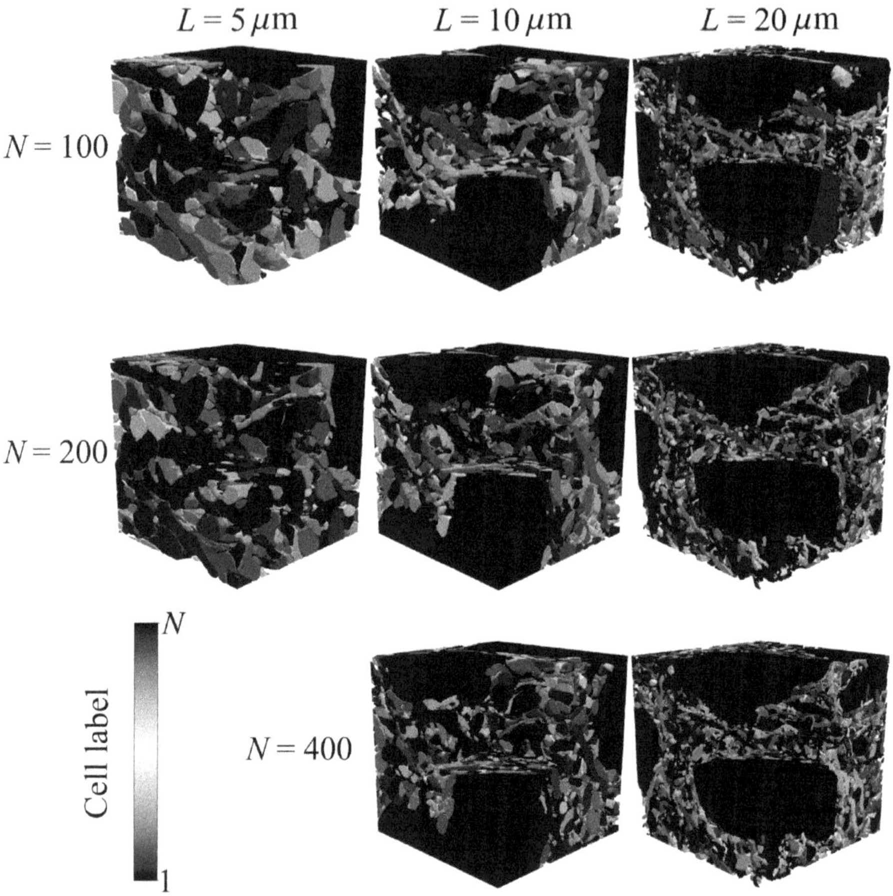

Figure 3 A sample of the computational meshes generated using the EMI-Meshing pipeline [12] of size $L \times L \times L$ containing N biological cells surrounded by extracellular space.

this cube of tissue, given the processing parameters we employ, is just under 200 cells. For the mesh with size $L = 10$ μm, we further present detailed statistics including the number of ICS and ECS cells, and maximum and minimum edge lengths (Table 1).

3 Modeling Electrodiffusion at the Cellular Scale in Explicit Geometries

With computational cellular geometries available, we next turn to mathematical modeling of electrodiffusion. In a domain $\Omega \subset \mathbb{R}^3$ representing cerebral tissue, we model ion concentrations $c_k = c_k(\boldsymbol{x}, t)$ with $k \in K$ for a set of ions $K = \{\mathrm{Na}^+, \mathrm{K}^+, \mathrm{Cl}^-\}$ and the electric potential $\phi = \phi(\boldsymbol{x}, t)$, for $\boldsymbol{x} \in \Omega$ and times $t \in [0, T]$. We will for notational

Table 1 Statistics for the meshes with size $L = 10\ \mu\text{m}$ containing N biological cells ($N = 100, 200, 300, 400, 500$. ICS: intracellular space, ECS: extracellular space.

Quantity	$N = 100$	200	300	400	500
Number of cells	3 292 043	4 489 915	6 693 005	5 625 770	8 061 336
Number of cells in the ICS	1 560 517	2 023 846	2 960 339	2 478 330	3 523 077
Number of cells in the ECS	1 731 526	2 364 127	3 732 666	3 147 440	4 538 259
ECS volume fraction [%]	26.2	20.7	18.4	17.6	17.1
Vertices	574 314	759 501	1 158 188	968 230	1 389 747
Membrane vertices	465 906	661 225	1 073 278	881 889	1 292 304
Membrane vertices fraction [%]	81.1	87.1	92.7	91.1	93.0
Max edge length [nm]	456.1	435.6	449.8	422.6	450.9
Min edge length [nm]	3.6	2.3	5.3	4.5	3.9

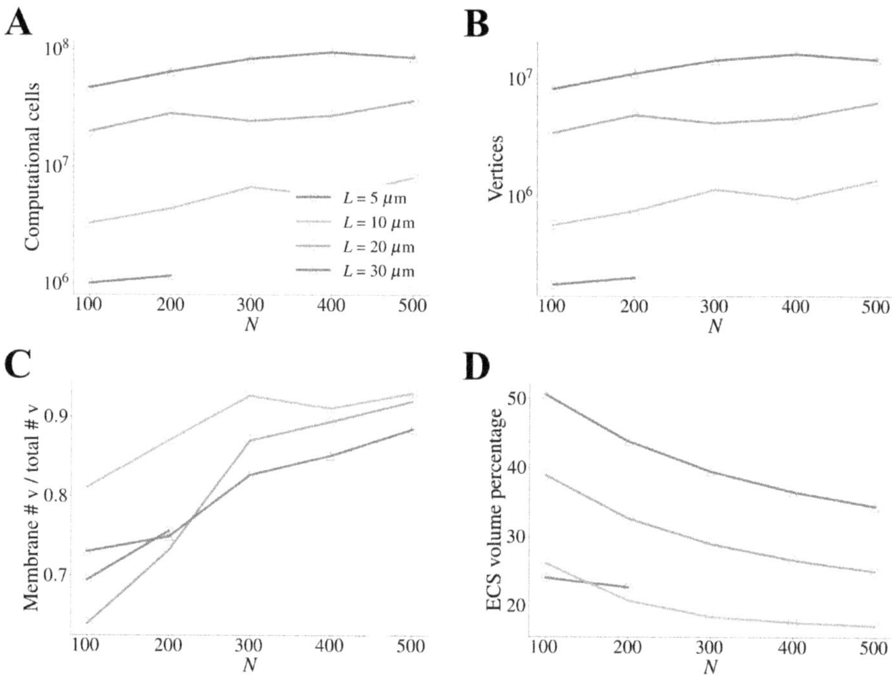

Figure 4 Statistics of cellular geometries of mouse visual cortex tissue. (**A**) Total number of computational cells in the mesh. (**B**) Total number of mesh vertices. (**C**) The ratio of total number of membrane vertices to total number of mesh vertices. (**D**) Percentage of the total mesh volume that is occupied by extracellular space. Common legend for all subfigures.

brevity drop the valence of the ions in the ion subscripts, i.e. we will use Na, K and Cl when explicitly subscripting the three ions K. At the cellular scale, brain tissue is subdivided into intracellular space, extracellular space, and the cellular membranes that separate these spaces (Figure 5) [2]. We first state the general governing equations of electrodiffusion in cerebral tissue, before extending the equations to intracellular and extracellular spaces.

3.1 Governing Equations of Electrodiffusion

In the absence of fluid convection, the movement of ions in brain tissue is described by the Nernst-Planck equation, which states that the total flux $\boldsymbol{J}_k = \boldsymbol{J}_k(\boldsymbol{x}, t)$ of an ion k is a combination of molecular diffusion and electrical drift [74]:

$$\boldsymbol{J}_k = \boldsymbol{J}_{k,\text{diffusion}} + \boldsymbol{J}_{k,\text{drift}} = -D_k \nabla c_k - \frac{z_k D_k F}{RT} c_k \nabla \phi. \tag{1}$$

Here D_k and z_k are the free diffusion coefficient and valence of ion k, while $R = 8.314\ \text{Jmol}^{-1}\text{K}^{-1}$ is the universal gas constant, $T = 300$ K is absolute temperature and $F = 96485\ \text{Cmol}^{-1}$ is Faraday's constant. By assuming conservation of ions in the domain Ω, it follows that

$$\frac{\partial c_k}{\partial t} + \nabla \cdot \boldsymbol{J}_k = f_k \quad \text{in } \Omega, \tag{2}$$

where $f_k = f_k(\boldsymbol{x}, t)$ is the rate of addition or removal of ions.

An equation for the electric potential is derived by assuming bulk electroneutrality in the domain Ω. This is equivalent to no charge separation, meaning that the total sum of ion charges equals zero:

$$F \sum_{k=1}^{K} z_k c_k = 0. \tag{3}$$

One can derive an alternative form of (3) by differentiating the sum with respect to time, invoking (2), and assuming that the source terms f_k are electroneutral. This gives

$$F \sum_{k=1}^{K} z_k \nabla \cdot \boldsymbol{J}_k = 0. \tag{4}$$

In a discrete setting, the ellipticity of (4) can make this formulation favorable compared to (3). Coupling (3) or (4) with (1) and (2) has been studied in several works under several names [49, 28, 62, 23, 71]. Here we refer to this system of equations as the Kirchhoff–Nernst–Planck (KNP) equations.

3.2 Governing Equations in Multiple Cells

Equations (1), (2) and (4) describe electroneutral electrodiffusion in the bulk of a domain Ω. We now turn to modeling excitable tissue where multiple cells communicate through the extracellular space across the cellular membranes. The model is based on the Extracellular-Membrane-Intracellular (EMI) framework, which considers separate domains for the extracellular space, the intracellular spaces, and the cellular membranes [79]. In principle, there can be N intracellular domains $\Omega_{i,j}$ for $j = 1, 2 \ldots, N$ representing biological cells with a common extracellular domain Ω_e, with cellular membranes Γ_j denoting the interface between $\Omega_{i,j}$ and Ω_e. In this section, we describe the case with one biological cell ($N = 1$, Figure 5). The generalization to N non-intersecting biological cells is immediate.

Let $c_{k,r} = c_{k,r}(\boldsymbol{x}, t)$ be the ion concentrations and $\phi_r = \phi_r(\boldsymbol{x}, t)$ the electric potential in a domain Ω_r, where $r = \{i, e\}$ denotes the intra- or extracellular space. We continue to consider the ion set $K = \{\mathrm{Na}^+, \mathrm{K}^+, \mathrm{Cl}^-\}$. Equations (1), (2) and (4) describe electrodiffusion in the bulk of the domains Ω_r:

$$\frac{\partial c_{k,r}}{\partial t} + \nabla \cdot \boldsymbol{J}_{k,r} = f_{k,r} \quad \text{in } \Omega_r, \tag{5}$$

$$F \sum_{k=1}^{K} z_k \nabla \cdot \boldsymbol{J}_{k,r} = 0 \quad \text{in } \Omega_r, \tag{6}$$

with the flux $\boldsymbol{J}_{k,r}$ given by the Nernst-Planck equation (1). As we have coupled the KNP equations with the EMI framework, we refer to (5) and (6) as the KNP-EMI equations.

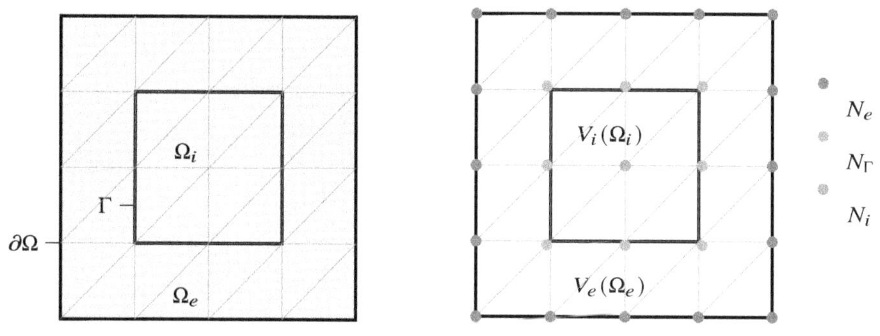

Figure 5 Conceptual illustration of the geometry and meshes. Left: idealized geometry, illustrating the intracellular (Ω_i) and extracellular (Ω_e) spaces separated by a membrane Γ. The boundary of the domain is $\partial\Omega$. Right: The degree of freedom layout of the idealized cellular geometry when discretized with linear continuous Lagrange elements. The function spaces $V_i(\Omega_i)$ and $V_e(\Omega_e)$ share the degrees of freedom on Γ.

The intra- and extracellular dynamics are coupled at the cellular membrane Γ that separates the intra- and extracellular spaces. Cellular membranes in cerebral tissue are not freely permeable to ions, but lined with selective ion channels that only permit transport of select ions under specific circumstances. Because of this, there are spatial differences in the intra- and extracellular ion concentrations and thus a charge difference across the membrane. The concentration differences induce an electric potential difference across the membrane, labeled the membrane potential $\phi_m = \phi_i - \phi_e$, and the membrane acts like a capacitor storing electric energy [74]. On the cellular membrane Γ, we assume continuity in the membrane current density I_m:

$$I_m = F \sum_{k \in K} z_k \boldsymbol{J}_{k,i} \cdot \boldsymbol{n}_i = -F \sum_{k \in K} z_k \boldsymbol{J}_{k,e} \cdot \boldsymbol{n}_e, \quad \text{on } \Gamma \tag{7}$$

where the normal vector $\boldsymbol{n}_r = \boldsymbol{n}_r(\boldsymbol{x})$ is directed out of the domain Ω_r.

We will here consider the *single-dimensional* form of the KNP-EMI equations, where the membrane current density I_m that exists on a lower-dimensional manifold is eliminated from the system of equations [79, 7, 23]. We assume that the membrane current I_m is the sum of an ionic current I_{ion}, whose model depends on the physiological scenario considered, and a capacitive current I_{cap}, which reflects the electric energy stored in the membrane. Assuming that the membrane is a parallel plate capacitor with capacitance C_m, the capacitive current is

$$I_{\text{cap}} = C_m \frac{\partial \phi_m}{\partial t}. \tag{8}$$

We assume that both the ionic current and the capacitive current have ion-specific contributions:

$$I_{\text{ion}} + I_{\text{cap}} = \sum_{k \in K} I_{\text{ion},k} + \sum_{k \in K} I_{\text{cap},k}, \tag{9}$$

with

$$I_{\text{cap},k} = \alpha_{k,r} I_{\text{cap}} \tag{10}$$

weighted by the coefficients

$$\alpha_{k,r} = \frac{z_k^2 D_{k,r} c_{k,r}}{\sum_{l \in K} z_l^2 D_{l,r} c_{l,r}}. \tag{11}$$

By definition $\sum_{k \in K} \alpha_{k,r} = 1$. Using (11) and (10) with the relation (9) and the definition of the membrane current (7), the flux of an ion k is expressed as a function of the ionic currents and the membrane potential

$$\boldsymbol{J}_{k,r} \cdot \boldsymbol{n}_r = \pm_r \frac{I_{\text{ion},k} + I_{\text{cap},k}}{F z_k} = \pm_r \frac{I_{\text{ion},k} + \alpha_{k,r} C_m \frac{\partial \phi_m}{\partial t}}{F z_k} \tag{12}$$

for $r \in \{i, e\}$, with the signs $\pm_i = 1$ and $\pm_e = -1$.

3.3 Membrane Mechanisms and Stimulus

Cellular membrane mechanisms such as ion channels, pumps and cotransporters are central to excitable tissue dynamics. The membrane mechanisms determine how cells respond to external stimuli such as synaptic signaling, or to changes in the extracellular environment. When modelling membrane mechanisms, one considers the cell types and biological scenarios at hand. For example, neurons and astrocytes have different membrane mechanisms [2, 30]. Here, we let neuronal membrane dynamics be governed by the Hodgkin–Huxley model [33], an Na^+/K^+-ATPase pump [39] and the two cotransporters KCC2 and NKCC1 [84, 22]. Glial membrane dynamics are governed by an ATP pump [30], passive leak channels with an inward-rectifying potassium current [30], and the KCC1 and NKCC1 cotransporters [53].

The Hodgkin–Huxley model [33] enables simulating action potentials, the communicating signal of neurons. The model includes ion channel currents that are voltage-gated; i.e., whether they are open or closed depends on the difference between a resting membrane potential ϕ_{rest} and an ion-specific reversal potential E_k. The Hodgkin–Huxley model includes passive (leak) channels and active channels. Letting $g_{k,\text{leak}}$ be the conductance of the leak channel of ion k, the general expression for a neuronal leak channel current is

$$I_{\text{leak},k,n} = g_{\text{leak},k,n}(\phi_m - E_k).$$

The reversal potential

$$E_k = \frac{RT}{z_k F} \log \frac{c_{k,e}}{c_{k,i}},$$

follows from the Nernst-Planck equation, considering a steady state where the diffusive flux balances the electrical drift across the cellular membrane [74].

The active channels include gating variables n, m and h that reflect the probability that an ion channel is open or closed. For sodium and potassium, the active channel currents are

$$I_{\text{active,Na}} = \overline{g}_{\text{Na}} m^3 h(\phi_m - E_{\text{Na}}),$$
$$I_{\text{active,K}} = \overline{g}_{\text{K}} n^4 (\phi_m - E_{\text{K}}),$$

where $\overline{g}_k$ is the maximum conductivity of the channel for ion k. There is no active channel for chloride. Time evolution of the gating variables n, m, h is governed by ordinary differential equations (ODEs) of the general form

$$\frac{\partial \zeta}{\partial t} = \alpha_\zeta (1 - \zeta) - \beta_\zeta \zeta, \tag{13}$$

where $\zeta = (n(t), m(t), h(t))$. The coefficients α_ζ and β_ζ depend on the difference $\phi_m - \phi_{\text{resting}}$. The ODEs (13) are equipped with appropriate initial conditions $\zeta(t = 0) = \zeta_0 = (n_0, m_0, h_0)$. See [33] for full comprehensive details of the model and its coefficients.

In addition to passive and active ion channels, which are driven by the difference $\phi_m - E_k$, we model membrane mechanisms that are driven by differences in intra- and extracellular concentrations. First, we consider the sodium-potassium pump Na^+/K^+-ATPase. This pump uses ATP (adenosine triphosphate) as energy to pump sodium out of a cell and potassium into the cell. For every ATP molecule consumed, three sodium ions are exported and two potassium ions are imported, meaning the pump is a net exporter of electric charge. The membrane current density associated with the ATP pump on neuronal cell membranes is modeled by

$$I_{\mathrm{ATP},n} = \frac{S_{\mathrm{ATP},n}}{(1 + P_{\mathrm{K}}/c_{\mathrm{K},e})^2(1 + P_{\mathrm{Na}}/c_{\mathrm{Na},i})^3},$$

where $S_{\mathrm{ATP},n} = 0.25\ \mathrm{Am}^{-2}$ is the pump strength, while $P_{\mathrm{K}} = 1.5$ mM and $P_{\mathrm{Na}} = 10$ mM are threshold values that determine whether the pump strength increases or decreases [55, 39].

For the neuronal membranes, we also consider the two cotransporters NKCC1 and KCC2. The NKCC1 cotransporter transports sodium, potassium and chloride into the cell, while the KCC2 cotransporter transports potassium and chloride out of cells [84, 22, 55]. The membrane current densities associated with the cotransporters are defined as

$$I_{\mathrm{NKCC1},n} = \delta S_{\mathrm{NKCC1},n} \ln \frac{c_{\mathrm{K},e} c_{\mathrm{Cl},e}}{c_{\mathrm{K},i} c_{\mathrm{Cl},i}},$$

$$I_{\mathrm{KCC2}} = S_{\mathrm{KCC2}} \ln \frac{c_{\mathrm{Na},e} c_{\mathrm{K},e} c^2_{\mathrm{Cl},e}}{c_{\mathrm{Na},i} c_{\mathrm{K},i} c^2_{\mathrm{Cl},i}},$$

where $S_{\mathrm{NKCC1},n} = 0.0023\ \mathrm{Am}^{-2}$ and $S_{\mathrm{KCC2}} = 0.0068\ \mathrm{Am}^{-2}$ are the maximum strengths of the cotransporters. The function $\delta = \delta(c^0_{\mathrm{K},e}, c_{\mathrm{K},e})$ is defined as

$$\delta(c^0_{\mathrm{K},e}, c_{\mathrm{K},e}) = \frac{1.0}{(1.0 + (0.03/(c_{\mathrm{K},e} - c^0_{\mathrm{K},e}))^{10})} \tag{14}$$

if $c_{\mathrm{K},e} \in [3.0, c^0_{\mathrm{K},e}]$, else $\delta = 0$ to suppress activity of the NKCC1 cotransporter [55].

For glial cells, we consider mechanisms described in previous astrocyte modeling studies [53, 30, 55]. We define the glial leak channel currents with different conductances than for the neuronal cells

$$I_{\mathrm{leak},k,g} = g_{\mathrm{leak},k,g} f_{\mathrm{Kir-Na},k}(\phi_m - E_k),$$

where

$$f_{\mathrm{Kir-Na},k} = \begin{cases} 1.0, & \text{for } k \in \{\mathrm{Na}, \mathrm{Cl}\}, \\ \sqrt{\frac{c_{\mathrm{K},e}}{c^0_{\mathrm{K},e}}} \frac{AB}{CD}, & \text{for } k = \mathrm{K}. \end{cases} \tag{15}$$

This function is used to model the potassium leak channel as inward-rectifying [30]. In Equation (15), $c^0_{\mathrm{K},e}$ is the initial extracellular potassium concentration, while the coefficients A, B, C and D are defined as

$$\begin{aligned} A &= 1 + e^{0.433}, \\ B &= 1 + e^{\frac{-(0.1186 + E^0_{\mathrm{K}})}{0.0441}}, \\ C &= 1 + e^{\frac{(\phi m - E_{\mathrm{K}} + 0.0185)}{0.0425}}, \\ D &= 1 + e^{\frac{-(0.1186 + \phi m)}{0.0441}}, \end{aligned}$$

where E^0_{K} is the initial potassium Nernst potential.

We model a $\mathrm{Na}^+/\mathrm{K}^+$-ATPase pump current and cotransporters for glia, similar to the approach for neurons. The glial ATP pump current density is expressed as

$$I_{\mathrm{ATP},g} = \frac{S_{\mathrm{ATP},g}}{(1 + P_{\mathrm{K}}/c_{\mathrm{K},e})(1 + P_{\mathrm{Na}}/c_{\mathrm{Na},i})^{3/2}},$$

where $S_{\mathrm{ATP},g} = 0.12\ \mathrm{Am}^{-2}$, $P_{\mathrm{Na}} = 10$ mM and $P_{\mathrm{K}} = 1.5$ mM are the maximum pump strength, sodium and potassium pump thresholds, respectively [53, 30].

For glia, we consider the NKCC1 and KCC1 cotransporters [53, 55]. The membrane current densities associated with these cotransporters are

$$\begin{aligned} I_{\mathrm{NKCC1},g} &= \delta S_{\mathrm{NKCC1},g} \ln \frac{c_{\mathrm{K},e} c_{\mathrm{Cl},e}}{c_{\mathrm{K},i} c_{\mathrm{Cl},i}}, \\ I_{\mathrm{KCC1}} &= S_{\mathrm{KCC1}} \ln \frac{c_{\mathrm{Na},e} c_{\mathrm{K},e} c^2_{\mathrm{Cl},e}}{c_{\mathrm{Na},i} c_{\mathrm{K},i} c^2_{\mathrm{Cl},i}}, \end{aligned}$$

where the maximum cotransporter strengths are $S_{\mathrm{NKCC1},g} = 5.2 \times 10^{-4}\ \mathrm{Am}^{-2}$ and $S_{\mathrm{KCC1}} = 0.018\ \mathrm{Am}^{-2}$, and δ is given by Equation (14).

We stimulate the brain tissue by adding a stimulus current density to the sodium ionic current density. We model an exponentially decreasing stimulus with the expression

$$I_{\mathrm{stim}} = \overline{g}_{\mathrm{stim}} e^{(-t \bmod T_{\mathrm{stim}})/a_{\mathrm{stim}}}. \qquad (16)$$

Here $\overline{g}_{\mathrm{stim}}$ is the maximum stimulus conductance. The exponential factor in Equation (16) makes I_{stim} decay exponentially with a time constant $a_{\mathrm{stim}} = 0.5$ ms, and the modulus operation is used to turn the current on periodically with a period T_{stim}.

To summarize, the ionic current densities for neuronal cell membranes are

$$\begin{aligned} I_{\mathrm{ion,Na},n} &= I_{\mathrm{leak,Na},n} + I_{\mathrm{active,Na}} + I_{\mathrm{stim}} + 3I_{\mathrm{ATP},n} - I_{\mathrm{NKCC1},n}, \\ I_{\mathrm{ion,K},n} &= I_{\mathrm{leak,K}} + I_{\mathrm{active,K}} - 2I_{\mathrm{ATP},n} - I_{\mathrm{NKCC1},n} - I_{\mathrm{KCC2}}, \\ I_{\mathrm{ion,Cl},n} &= I_{\mathrm{leak,Cl}} + 2I_{\mathrm{NKCC1},n} + I_{\mathrm{KCC2}}, \end{aligned}$$

while in the cases where we consider glial cells, the glial ionic membrane current densities are

$$
\begin{aligned}
I_{\text{ion,Na},g} &= I_{\text{leak,Na},g} + I_{\text{active,Na}} + 3I_{\text{ATP},g} - I_{\text{NKCC1},g}, \\
I_{\text{ion,K},g} &= I_{\text{leak,K}} - 2I_{\text{ATP},g} - I_{\text{NKCC1},g} - I_{\text{KCC1}}, \\
I_{\text{ion,Cl},g} &= I_{\text{leak,Cl},g} + 2I_{\text{NKCC1},g} + I_{\text{KCC1}}.
\end{aligned}
$$

3.4 Boundary Conditions

With the membrane currents defined, it remains to determine boundary and initial conditions to close the KNP-EMI equation system. As boundary conditions, we consider a no-flux condition on all boundaries $\partial\Omega_r$ exterior to the domain Ω_r for all times $t > 0$:

$$
\boldsymbol{J}_{k,r} \cdot \boldsymbol{n}_r = 0, \quad \forall k \quad \text{on } \partial\Omega_r, \quad \forall r. \tag{17}
$$

With the boundary condition (17) enforced on all boundaries, the electric potentials are only determined up to a constant. We return to this point in Section 4.3.

3.5 Initial Conditions and Defining Resting States

As initial conditions, we set the initial concentrations $c_{k,r}(\boldsymbol{x}, 0) = c^0_{k,r}$ and electric potentials $\phi_r(\boldsymbol{x}, 0) = \phi^0_r$.

To define initial conditions that constitute a resting state with no changes in ionic concentrations or electrical potentials, we solve a system of ordinary differential equations (ODEs) that represent the dynamics in a single spatial point. When only considering neurons, we collapse the intracellular spaces and the extracellular space into two separate compartments. When modeling both neuronal and glial cells, we consider two separate intracellular compartments, one for each cell type, and a third compartment for the extracellular space. In these compartmental models, the change in the membrane potential at steady state is given by the total ionic current I_{ion}, since then the total membrane current density $I_m = I_{\text{ion}} + I_{\text{cap}} = 0$. The ion concentration changes are in turn given by the molar fluxes associated with the total ionic current densities of the respective ions. The resulting steady-state ODE system for the three-compartment model is

$$
\begin{aligned}
\frac{\mathrm{d}\phi_{m,l}}{\mathrm{d}t} &= -\frac{1}{C_m} I_{\text{ion},l}, \\
\frac{\mathrm{d}c_{k,i,l}}{\mathrm{d}t} &= -\frac{I_{\text{ion},k,l}}{z_k F}\frac{|\Gamma|_l}{|\Omega|_{i,l}}, \\
\frac{\mathrm{d}c_{k,e}}{\mathrm{d}t} &= \frac{1}{z_k F}\left(I_{\text{ion},k,n}\frac{|\Gamma|_n}{|\Omega|_e} + I_{\text{ion},k,g}\frac{|\Gamma|_g}{|\Omega|_e}\right), \\
\frac{\mathrm{d}\zeta}{\mathrm{d}t} &= \alpha_\zeta(1-\zeta) - \beta_\zeta\zeta.
\end{aligned} \tag{18}
$$

Table 2 Material and model parameters. (*) Used to determine initial conditions with a steady-state approximation. (†) Calibrated based on reference.

Parameter	Symbol	Value	Unit	Ref.
Universal gas constant	R	8.314	$JK^{-1}mol^{-1}$	
Faraday constant	F	96485	$Cmol^{-1}$	
Temperature	T	300	K	
Free diffusion coefficient sodium	D_{r,Na^+}	1.33×10^{-9}	m^2s^{-1}	[32]
Free diffusion coefficient potassium	D_{r,K^+}	1.96×10^{-9}	m^2s^{-1}	[32]
Free diffusion coefficient chloride	D_{r,Cl^-}	2.03×10^{-9}	m^2s^{-1}	[32]
Initial neuronal sodium concentration (ICS)	$[Na^+]_{i,n}^0$	10	mM	[57]*
Initial glial sodium concentration (ICS)	$[Na^+]_{i,g}^0$	15	mM	[55]*
Initial sodium concentration (ECS)	$[Na^+]_e^0$	145	mM	[55]*
Initial neuronal potassium concentration (ICS)	$[K^+]_{i,n}^0$	130	mM	[57]*
Initial glial potassium concentration (ICS)	$[K^+]_{i,g}^0$	100	mM	[55]*
Initial potassium concentration (ECS)	$[K^+]_e^0$	3	mM	[55]*
Initial neuronal chloride concentration (ICS)	$[Cl^-]_{i,n}^0$	5	mM	[57]*
Initial glial chloride concentration (ICS)	$[Cl^-]_{i,g}^0$	5	mM	[55]*
Initial chloride concentration (ECS)	$[Cl^-]_e^0$	134	mM	[55]*
Initial neuronal membrane potential	$\phi_{m,n}^0$	−70	mV	[74]*
Initial glial membrane potential	$\phi_{m,g}^0$	−85	mV	[55]*
Resting membrane potential	$\phi_{resting}$	−65	mV	
Membrane capacitance	C_m	0.01	Fm^{-2}	[55]
Initial potassium activation probability	n_0	0.244		*
Initial sodium activation probability	m_0	0.029		*
Initial sodium inactivation probability	h_0	0.754		*
Sodium maximum conductivity	$\overline{g}_{Na}$	1200	Sm^{-2}	[33]
Potassium maximum conductivity	$\overline{g}_K$	360	Sm^{-2}	[33]
Neuronal sodium leak conductivity	$g_{Na,leak,n}$	0.3	Sm^{-2}	[7]†
Glial sodium leak conductivity	$g_{Na,leak,g}$	1	Sm^{-2}	[53]
Neuronal potassium leak conductivity	$g_{K,leak,n}$	0.1	Sm^{-2}	[55]†
Glial potassium leak conductivity	$g_{K,leak,g}$	16.96	Sm^{-2}	[53]
Neuronal chloride leak conductivity	$g_{Cl,leak,n}$	0.25	Sm^{-2}	[53]
Glial chloride leak conductivity	$g_{Cl,leak,g}$	2.0	Sm^{-2}	[53]†
Neuronal ATP pump maximum strength	$S_{ATP,n}$	0.25	Am^{-2}	[39, 21]†
Glial ATP pump maximum strength	$S_{ATP,g}$	0.12	Am^{-2}	[30]†
ATP pump sodium threshold	P_{Na}	10	mM	[30]
ATP pump potassium threshold	P_K	1.5	mM	[30]
Neuronal NKCC1 maximum strength	$S_{NKCC1,n}$	0.0023	Am^{-2}	[22]†
Glial NKCC1 maximum strength	$S_{NKCC1,g}$	5.2×10^{-4}	Am^{-2}	[53]
KCC2 maximum strength	S_{KCC2}	0.0068	Am^{-2}	[22]†
KCC1 maximum strength	S_{KCC1}	0.018	Am^{-2}	[53]
Stimulus time constant	a_{stim}	5×10^{-4}	s	
Maximum stimulus conductance	$\overline{g}_{stim}$	60–500	Sm^{-2}	

This ODE system respects our convention of considering currents out of a cell as negative fluxes. The subscript $l \in \{n, g\}$ denotes neuronal or glial cell, $|\Gamma|_l$ is the total surface area of the cellular membranes separating the compartment of cell type l from the extracellular space, $|\Omega|_{i,l}$ is the total volume of all intracellular compartments of cell type l, and $|\Omega|_e$ is the volume of the extracellular space. The two-compartment system results from reducing (18) by considering only $l = n$. A steady-state solution to the system (18) will constitute a steady state for the full PDE system, given that we set the initial condition constant in space and that the membrane mechanisms are homogeneous in space.

To determine initial conditions for the PDE system by solving the ODE system (18), we consider a set of initial conditions for the ODE system. The initial membrane potentials used to solve the ODE system are −70 mV and −85 mV for the neurons and astrocytes, respectively (taken from [7, 55, 54]). The initial concentrations are based on physiological concentrations in neurons and glial cells (taken from [57, 54, 55, 53]). Finally, the initial conditions for the gating variables are set by solving (13) at steady state with $\phi_m = -70$ mV. All the membrane mechanism parameters and physical constants in the KNP-EMI model used in this work are listed in Table 2. We note that some of the membrane mechanism parameters were calibrated to balance the membrane mechanisms.

4 Discretization and Numerical Solution of the KNP-EMI Equations

The KNP-EMI model equations can be summarized as follows: find the ionic concentrations $c_{k,r} = c_{k,r}(\boldsymbol{x}, t)$ and electric potentials $\phi_r = \phi_r(\boldsymbol{x}, t)$ in the domains Ω_r such that for all $t \in [0, T]$

$$\frac{\partial c_{k,r}}{\partial t} + \nabla \cdot \boldsymbol{J}_{k,r} = f_{k,r} \qquad \text{in } \Omega_r, \tag{19a}$$

$$F \sum_{k=1}^{K} z_k \nabla \cdot \boldsymbol{J}_{k,r} = 0 \qquad \text{in } \Omega_r, \tag{19b}$$

$$\boldsymbol{J}_{k,r} \cdot \boldsymbol{n}_r = \pm_r \frac{I_{\text{ion},k} + \alpha_{k,r} C_m \frac{\partial \phi_m}{\partial t}}{F z_k} \qquad \text{on } \Gamma, \tag{19c}$$

$$\boldsymbol{J}_{k,r} = -D_{k,r} \nabla c_{k,r} - \frac{z_{k,r} D_{k,r} F}{RT} c_{k,r} \nabla \phi_r \qquad \text{in } \Omega_r, \tag{19d}$$

$$c_{k,r}(\boldsymbol{x}, 0) = c_{k,r}^0 \qquad \text{in } \Omega_r, \tag{19e}$$

$$\phi_r(\boldsymbol{x}, 0) = \phi_r^0 \qquad \text{in } \Omega_r, \tag{19f}$$

$$\boldsymbol{J}_{k,r} \cdot \boldsymbol{n}_r = 0 \qquad \text{on } \partial\Omega_r. \tag{19g}$$

We solve these equations using the solution algorithm presented by Benedusi et al. [7]. The scheme is based on a finite difference method in time, an operator splitting scheme decoupling the ODEs from the PDEs, a finite element spatial discretization, and finally an iterative, preconditioned solver for the resulting linear system. For completeness, we describe the solution steps in detail below.

4.1 Operator Splitting and Time Discretization

The KNP-EMI equations are nonlinear, involving both a bilinear coupling of each concentration field with the electric potential in each bulk domain Ω_r cf. (19d) and nonlinear dynamics at the membrane cf. (19c). The bulk and membrane dynamics occur on different timescales, with much faster membrane dynamics. To decouple across these different scales, we consider an implicit-explicit operator splitting scheme [23, 7]. To this end, we first partition the time interval $[0, T]$ uniformly into times[1] $t^n = n\Delta t$ with $n = 1, \dots, N_t$ for N_t timesteps with timestep size Δt. At each timestep t^n, given the membrane potential at the previous timestep, we first solve the gating variable ODEs (13) to obtain an approximation of the ionic currents $I_{\mathrm{ion},k}$. Specifically, we solve the ODEs from t^{n-1} to t^n using m steps of a Rush–Larsen method [64] with a local timestep $\Delta t_{\mathrm{ODE}} = \Delta t/m$ (we let $m = 25$). Next, we integrate (19a) and (19c) in time using an implicit Euler method, and solve (19a)–(19d) by evaluating the just-computed ionic currents directly. We linearize the bilinear term by evaluating the concentration in the electrical drift term and in the coefficients $\alpha_{k,r}$ at the previous timestep.

4.2 Finite Element Spatial Discretization

Our computational meshes are composed of tessellated intracellular domains $\mathcal{T}_i$ and extracellular space $\mathcal{T}_e$ (Figure 5). In this section, we present the finite element discretization for just one connected intracellular domain. This extends to N disjoint intracellular domains readily, as will be demonstrated in Section 5. Relative to each mesh domain $\mathcal{T}_r$, we define the finite element space of continuous piecewise linear polynomials $V(\mathcal{T})$. The fully discrete weak form of the "PDE part" of the KNP-EMI problem then reads: at each time t^n, find the concentration $c_{k,r}^{n,h} \in V(\mathcal{T}_r)$ for $k \in K$ and electric potentials $\phi_r^{n,h} \in V(\mathcal{T}_r)$ such that

$$\int_{\Omega_r} \left(c_{k,r}^{n,h} v_{k,r} - \Delta t \boldsymbol{J}_{k,r}^{n,h} \cdot \nabla v_{k,r} \right) \mathrm{d}x \pm_r \frac{\alpha_{k,r}^n C_m}{z_k F} \int_\Gamma \phi_m^{n,h} v_{k,r} \, \mathrm{d}s \\ = \int_{\Omega_r} \left((c_{k,r}^{n-1,h} + \Delta t f_{k,r}^n) v_{k,r} \right) \mathrm{d}x - \int_\Gamma g_{k,r}^{n-1,h} v_{k,r} \, \mathrm{d}s \quad (20)$$

for all $v_{k,r} \in V(\mathcal{T}_r), k \in K, r \in \{i, e\}$, and

$$- \Delta t \sum_{k \in K} z_k \int_{\Omega_r} \boldsymbol{J}_{k,r}^{n,h} \cdot \nabla w_r \, \mathrm{d}x \pm_r \frac{C_m}{F} \int_\Gamma \phi_m^{n,h} w_r \, \mathrm{d}s = - \int_\Gamma q_r^{n-1,h} w_r \, \mathrm{d}s \quad (21)$$

[1] Since the unknown variables already have an ion index k and a domain index r, we denote the timestep n by a superscript to avoid a triple subscript.

for all $w_r \in V(\mathcal{T}_r)$. Here, we have introduced the short-hand

$$g_{k,r}^{n,h} = \pm_r \frac{1}{z_k F}\left(\Delta t I_{\mathrm{ion},k}^{n,h} - \alpha_{k,r}^{n} C_m \phi_m^{n,h}\right) \text{ and } q_r^{n,h} = \pm_r \left(\Delta t I_{\mathrm{ion},k}^{n,h} - C_m \phi_m^{n,h}\right)/F. \tag{22}$$

The discrete weak form induces a linear system $A^n x^n = b^n$ to be solved at every timestep n. The system matrix A^n has the following block structure:

$$A^n = \begin{bmatrix} A_{c,i} & \tilde{B}_i & 0 & C_{i,e} \\ B_i & A_{\phi,i} & 0 & D_{i,e} \\ 0 & C_{e,i} & A_{c,e} & \tilde{B}_e \\ 0 & D_{e,i} & B_e & A_{\phi,e} \end{bmatrix}^n \tag{23}$$

where $A_{f,r}$ are the concentration and potential blocks in domain r, B_r and $\tilde{B}_r$ are blocks coupling the potentials and the concentrations in r, C_{r_1,r_2} is the concentration-potential coupling block between space r_1 and r_2, and D_{r_1,r_2} are the blocks coupling the potentials in spaces r_1 and r_2. Note that $A_{c,r}$ are block matrices with each of the individual $A_{c_k,i}$ block matrices on the diagonal. The blocks in (23) coupling the intra- and extracellular spaces arise from the shared degrees of freedom on the cellular membranes (Figure 5). For a comprehensive exposition of the block structure, we refer to Benedusi *et al.* [7].

4.3 Solving the KNP-EMI Equations in Dense Cortical Tissue Geometries

We solve the linear KNP-EMI PDE system (23) in a monolithic fashion using an iterative method as previously described [7]. The iterative solver uses a Generalized Minimal Residual Method (GMRES) [65] and is preconditioned with an Algebraic Multigrid (AMG) method [85, 18]. We use

$$P = \begin{bmatrix} A_{c,i} & 0 & 0 & 0 \\ 0 & A_{\phi,i} & 0 & 0 \\ 0 & 0 & A_{c,e} & 0 \\ 0 & 0 & 0 & A_{\phi,e} \end{bmatrix}$$

as a preconditioner for the linear system, where all of the stiffness matrices are evaluated with the initial conditions. We noted previously that our choice of boundary conditions leaves the electric potentials only determined up to a constant. Thus, without additional constraints, the matrix A^n (23) is singular. We address this by providing the iterative solver with the nullspace of A^n, which is the set of all constants. We also orthogonalize b^n with respect to the nullspace.

The computational model is implemented in DOLFINx [5], which is interfaced with the linear algebra library PETSc. We use the functionality of the library multiphenicsx [4] to discretize the intra- and extracellular spaces separately. We note that the method results in a doubling of the membrane degrees of freedom in Figure 5, because the discrete intra- and extracellular spaces both have degrees of freedom on the cellular membranes. The software is parallelized with MPI communication. The model implementation was verified in two ways (Section 6): (i) a convergence study on idealized geometries; (ii) a convergence study on the mouse visual cortex meshes. The implementation is openly available at GitHub: https://github.com/hherlyng/knp-emi-cgx.

The key question now is how this numerical solution strategy performs in realistic or near-realistic simulation scenarios defined over dense cortical tissue geometries. To study this, we simulate the firing of an action potential within the cortical tissue geometries generated in Section 2. In each geometry, the axon of one of the neurons is stimulated by a stimulus current density as introduced in Section 3.3. As an example of the setup, Figure 6A shows the geometry for the mesh with $L = 5\ \mu$m and $N = 100$, with the stimulated axon in pink. For simplicity in this performance test case, we consider all other intracellular domains within the tissue cube to also be neurons, by employing the neuronal membrane mechanisms (as introduced in Section 3.3).

The simulations are run for a total of 200 timesteps with $\Delta t = 2.5 \times 10^{-5}$ s to a final time of 5 ms. The total stimulus current applied to the neuronal membrane (Figure 6B) makes the membrane potential cross the threshold required for firing a single action potential (Figure 6C). We note that the simulations are run with a varying number of processes for varying L to achieve reasonable runtimes as the problem sizes increase (Table 3). All simulations were run using the high-performance computing architecture eX3 [70] on a node with two AMD EPYC Genoa-X 9684X 96-Core dual processors.

Table 3 Problem sizes (numbers of degrees of freedom) on the mouse visual cortex meshes with varying domain boundary side lengths L and total number of biological cells N. The – entries signify that there are no meshes for $L = 5\ \mu$m and $N > 200$.

L [μm]	$N = 100$	$N = 200$	$N = 300$	$N = 400$	$N = 500$
5	1 219 172	1 452 136	–	–	–
10	4 161 248	5 683 448	8 926 676	7 401 376	10 729 424
20	22 892 896	34 403 796	31 837 296	35 739 444	48 709 344
30	57 002 756	78 171 324	105 776 564	121 689 728	111 075 140

We first examine the assembly times – the total time spent assembling the system matrix, right-hand side vector and preconditioner matrix over the whole simulation time (Table 4). As expected, these times increase with increasing problem size in a controlled fashion. A general trend is that assembly times remain fairly constant when concomitantly doubling the number of MPI cores and the number of degrees of freedom, indicating scalable assembly.

In the per-timestep linear solver, the number of iterations until convergence is highest during the action potential when the variables rapidly change with time (Figure 7). After the membrane potential peaks around timestep 25, iteration counts remain fairly constant until it slightly increases, concurrent with the membrane potential kink at around timestep 100. For the last 100 iterations the iteration counts fall at first before stabilizing, as a result of the membrane potential being fairly constant.

The solver performance in terms of average number of iterations is stable for the smaller geometries with $L = 5\ \mu$m and $L = 10\ \mu$m, and for the $L = 20\ \mu$m geometries with up to $N = 300$ biological cells, while the iteration counts start increasing from

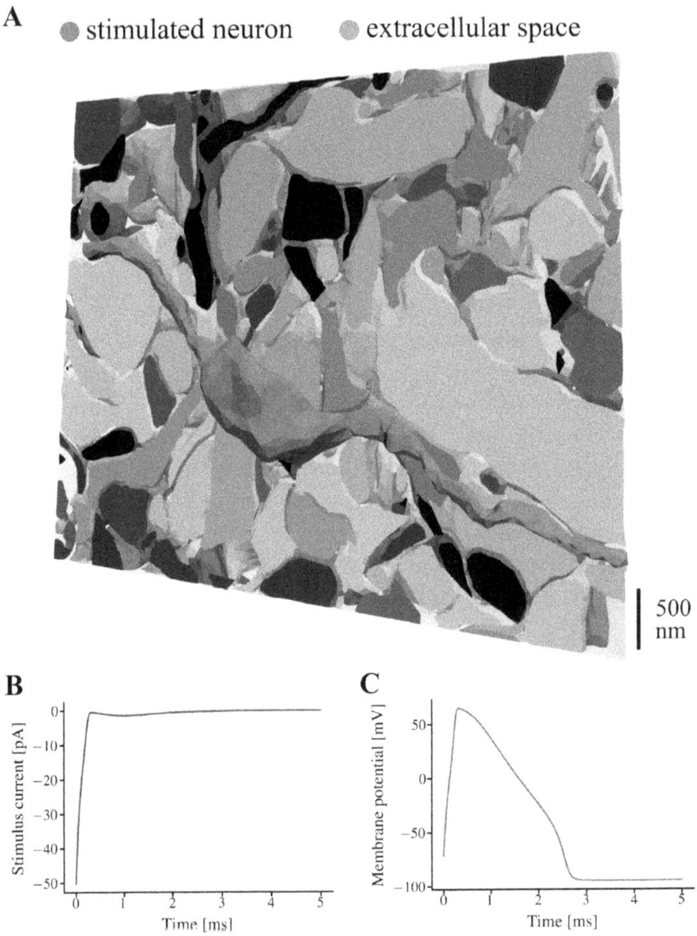

Figure 6 Simulation setup to test the performance of the iterative linear solver. **A.** A clip of the $L = 5$ μm, $N = 100$ geometry with a central axon highlighted in pink. The extracellular space is shown in light blue, while the remaining cells or cell fragments are shown in separate, unique colors. **B.** The total stimulus current applied to the membrane of the axon by adding a sodium stimulus current to the ionic membrane current density. **C.** The membrane potential measured in a point on the axon's membrane ($\boldsymbol{x} = (1186, 3241, 3594)$ nm).

Table 4 Assembly times when simulating an action potential on the mouse visual cortex meshes with varying domain boundary side lengths L and total number of biological cells N. The – entries signify that there are no meshes for $L = 5$ μm and $N > 200$.

L [μm]	$N = 100$	$N = 200$	$N = 300$	$N = 400$	$N = 500$
5 (16 cores)	1390 s	1783 s	–	–	–
10 (32 cores)	2269 s	3459 s	5543 s	5414 s	8314 s
20 (64 cores)	7219 s	10749 s	10260 s	12751 s	16355 s
30 (128 cores)	7597 s	14828 s	21652 s		

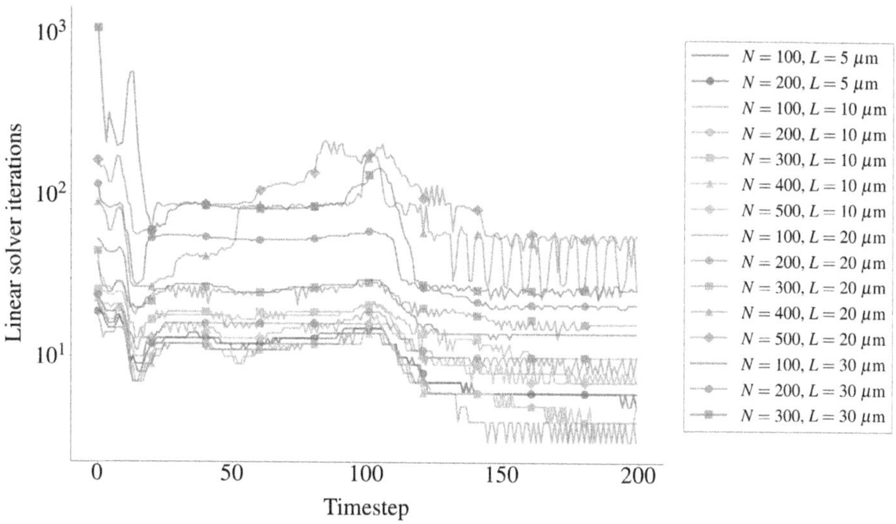

Figure 7 Iteration counts for 200 timesteps when solving the KNP-EMI equations on the various mouse visual cortex geometries during a simulation of an action potential. Note the logarithmic scaling of the vertical axis.

$N = 300$ towards $N = 500$ (Table 5). For the $L = 30\ \mu m$ geometries, iteration counts are higher than for the smaller geometries and increase with increasing N. For $N = 400$ and $N = 500$, the linear solver reaches the prescribed maximum number of iterations per timestep (5000). The total time spent solving the linear system generally increases as problem sizes increase.

Table 5 Solver times (in seconds) and average iteration counts (in brackets) when simulating an action potential on the mouse visual cortex meshes with varying domain boundary side lengths L and total number of biological cells N. The – entries signify that there are no meshes for $L = 5\ \mu m$ and $N > 200$. Max its. means the simulation exceeded the prescribed maximum number of linear solver iterations for a single timestep (5000).

L [μm]	$N = 100$	$N = 200$	$N = 300$	$N = 400$	$N = 500$
5 (16 cores)	175 s [10]	203 s [10]	–	–	–
10 (32 cores)	435 s [14]	443 s [9]	1103 s [16]	648 s [9]	1263 s [12]
20 (64 cores)	1105 s [9]	2316 s [13]	3258 s [22]	10400 s [62]	21172 s [100]
30 (128 cores)	3206 s [22]	10901 s [40]	30127 s [86]	max its.	max its.

We now look at strong scaling of the solver by comparing performance on the $L = 10\ \mu m$ geometries when using 16 and 32 cores. Assembly times are roughly cut in half when doubling the number of cores (Table 6). The solver times are halved for the $N = 100$ and $N = 400$ geometries, and even more than halved for the $N = 200$ geometry. Mostly, the average number of iterations is similar when running on 16 and

32 cores. An interesting exception is the $N = 300$ geometry, for which the average iteration count almost doubles when doubling the number of cores.

Table 6 Solver times and average iteration counts when simulating an action potential on the mouse visual cortex meshes with the $L = 10$ μm geometries for varying number of biological cells N.

Metric (# Cores)	$N = 100$	$N = 200$	$N = 300$	$N = 400$	$N = 500$
Assembly (16)	4796 s	7621 s	12354 s	10745 s	15819 s
Assembly (32)	2269 s	3459 s	5543 s	5414 s	8314 s
Solve (16)	844 s	1283 s	1443 s	1204 s	2578 s
Solve (32)	435 s	443 s	1103 s	648 s	1263 s
Average iterations (16)	13	13	9	9	15
Average iterations (32)	14	9	16	9	12

In summary, these findings indicate that the solution strategy performs well for reasonably complex meshes, but also that there are clear opportunities for further research in solution algorithms for simulation of electrodiffusion in dense tissue reconstructions.

5 Electrodiffusive Interplay in Geometrically Detailed Brain Tissue During Low- and High-Frequency Firing

Neuronal activity is governed by ionic movement across cellular membranes. A key change is the release of potassium from excited neurons into the extracellular space (ECS), which, if not cleared efficiently, can enhance neuronal excitability and disrupt neuronal function [3, 52, 81]. The restoration of extracellular ionic levels, via e.g. glial uptake or homeostatic mechanisms such as the Na^+/K^+-ATPase sodium-potassium pump, following neuronal activity is thus critical for healthy brain function [42, 83]. During sustained high-frequency firing, the homeostatic mechanisms are not able to keep up, leading to larger, persistent ionic shifts that will alter excitability and firing patterns [72]. The intricate morphology of individual cells and the surrounding ECS may further affect local electrodiffusive dynamics, a factor often neglected in simplified models.

To demonstrate the capabilities of the electrodiffusive simulation framework detailed in the present work, we consider the two scenarios (i) low-frequency (normal) neuronal activity and (ii) high-frequency firing, in a geometrically detailed domain with neurons, glial cells and ECS. We analyze and compare the two scenarios and point to observations where shape affects the dynamics (and thus potentially function). Specifically, we consider the mesh introduced in Section 2 with a domain length $L = 5$ μm and $N = 100$ biological cells. The membrane model is described in Section 3.3 and notably includes homeostatic mechanisms. We further apply the stimulus current (16) and consider two different stimuli with periods $T_{\text{stim}} = 1$ s and $T_{\text{stim}} = 20$ ms, which correspond to firing frequencies of 1 and 50 Hz, respectively. The stimulus current is applied at all mesh facets of the axon where $x \in (1.15, 1.25)$ μm. The tissue domain is illustrated in Figure 8.

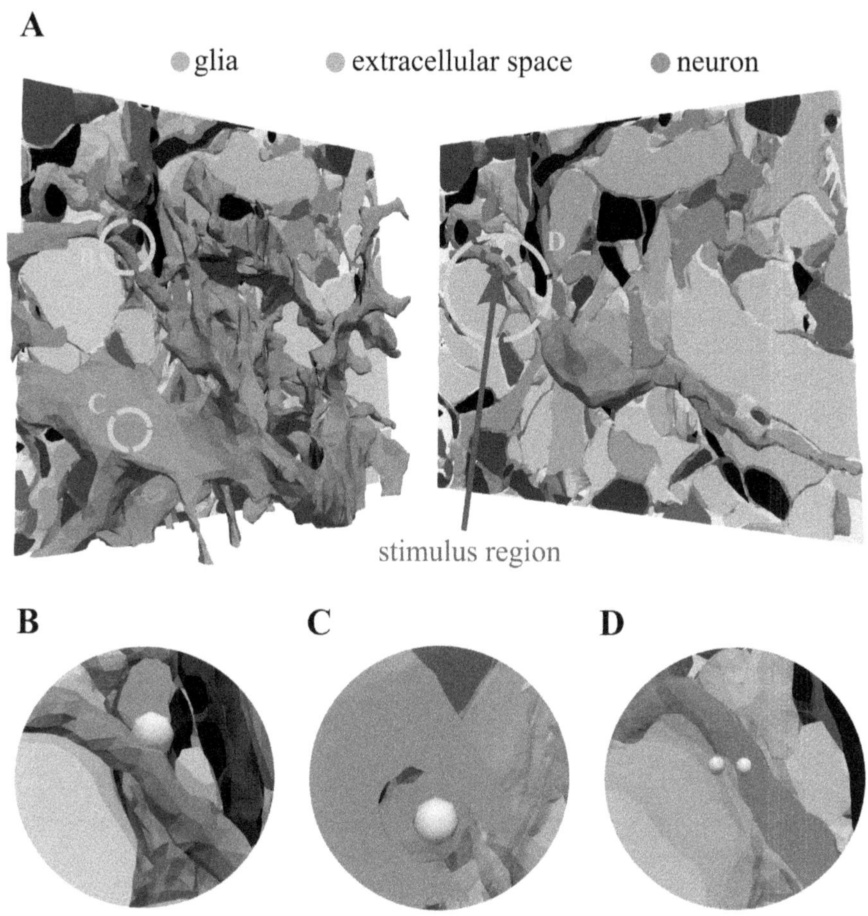

Figure 8 A. Illustration of the computational mesh with $L = 5$ μm and $N = 100$ with an astrocyte (green) partly wrapping around the stimulated axon (pink). The dashed blue ellipse indicates the stimulus region. The extracellular space is colored opaque light blue. All other colored structures are biological cells, with a unique color for every unique cell. **B.** Gray spherical glyph denoting point where glial intracellular concentrations are measured. **C.** Gray spherical glyph denoting point where distal extracellular concentrations are measured. **D.** Gray spherical glyphs denoting points where proximal extracellular concentrations (left) and neuronal intracellular concentrations (right) are measured.

5.1 Membrane Potentials and Ion Concentrations Develop Over Different Time Scales During Low Neuronal Activity

We start by analyzing the temporal dynamics of the system during a low firing frequency of 1 Hz over a period of 250 ms. We observe that an action potential fires at the onset of the simulation as a result of the stimulus current applied to the axon (Figure 9A). After the action potential, the membrane potential returns to a resting state at −72.5 mV, which is a 1.4% change from the initial state of −71.5 mV (Figure 9A). Meanwhile, the glial membrane potential initially increases rapidly during the action potential, by up to 0.6 mV from the initial value of −85 mV, before it steadily decreases over time reaching −84.6 mV at 250 ms (Figure 9B).

The action potential initiates rapid, coupled ionic shifts in both the intracellular and extracellular compartments. The neuronal Na^+ and K^+ concentrations respectively increase and decrease by approximately 4 mM during the action potential, before the concentrations slowly go back toward baseline levels as the homeostatic mechanisms (pumps, co-transporters and glial uptake) are activated. We observe that the neuronal Cl^- concentration rapidly increases as the neuron depolarizes, and decreases as the neuron repolarizes. Moreover, the Cl^- concentration drops below the initial state, indicating a total loss of neuronal Cl^-. Note that the changes in Cl^- are two orders of magnitude smaller than those observed for Na^+ and K^+ (Figure 9C).

In the extracellular space proximal to the stimulated region, all ion concentrations change abruptly when the stimulus current is applied (Figure 9D). In contrast to the neuronal concentrations, the shifts in the extracellular concentrations are of similar magnitude for all species. After the action potential is fired, both extracellular Na^+ and Cl^- concentrations proximal to the stimulated neuron settle at elevated values (around 1 mM higher than the initial concentrations). The extracellular K^+ concentration increases during the action potential, before decreasing as the homeostatic mechanisms are activated, and is still decreasing at the end of the simulation. Distal to the neuron, the extracellular Na^+ and Cl^- concentrations decrease by 0.017 and 0.012 mM, respectively, an order of magnitude less than the shifts of these ions proximal to the neuron. The distal K^+ concentration behaves similarly to the proximal concentration, but the increase from the initial K^+ concentration to the peak concentration value in the distal point is half of that observed in the proximal point (Figure 9D).

The glial ion concentrations develop steadily over time throughout the entire simulation, with a decreasing Na^+ concentration while K^+ and Cl^- concentrations increase (Figure 9E). The ion concentration changes of the three species are similar in magnitude; by the end of the simulation, the Na^+ concentration has decreased by 0.026 mM, while K^+ and Cl^- concentrations have increased by 0.055 and 0.029 mM, respectively. In general, the ion concentrations stabilize over a much slower time scale than the membrane potentials. Both the membrane potentials and the concentrations converge to a different set of steady-state values than the initial conditions (Figure 9C–E).

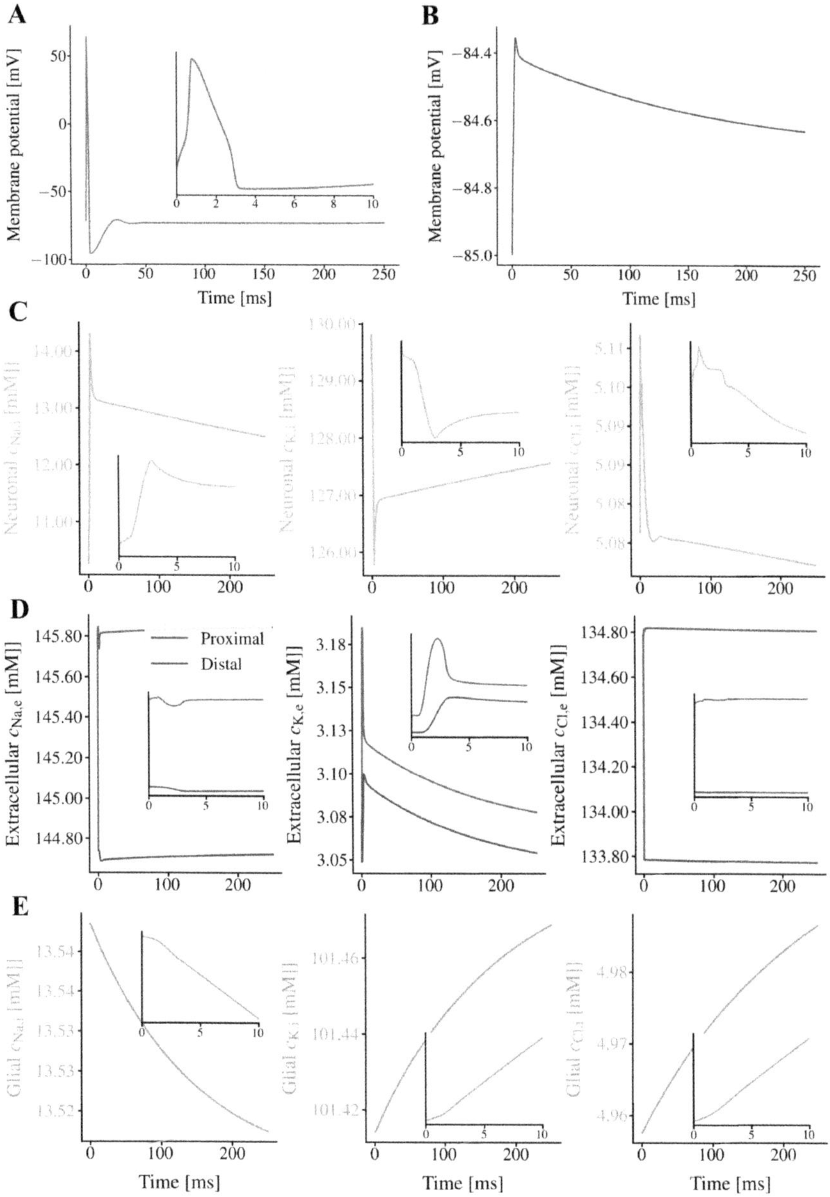

Figure 9 Neuronal membrane potential (**A**), glial membrane potential (**B**), neuronal ion concentrations (**C**), extracellular ion concentrations (**D**) and glial ion concentrations (**E**) when simulating stimulation of an axon at a physiological frequency of 1 Hz on the $L = 5$ μm mouse visual cortex mesh with $N = 100$ biological cells. The insets show the graphs during the first 10 ms. The insets have separate axes limits for the ICS and ECS concentrations (like in the original plots). In panel **C**, proximal and distal refer to the ECS points shown in Figure 8D and C, respectively.

5.2 Sustained High-Frequency Firing Induces a Burst-Firing Mode

During high-frequency firing, where the axon is stimulated at a frequency of 50 Hz, the neuron initially fires a train of action potentials for approximately 400 ms (Figure 10A). At $t = 400$ ms, the neuron transitions into a new firing mode, characterized by bursts of two action potentials separated by silent periods that last around 180 ms. We observe that the amplitude of the membrane depolarization decreases over the first 400 ms; the initial AP peaks at 65 mV, while the last AP before the neuron transitions into a new firing mode peaks at less than 28 mV. We further observe an alteration of the hyperpolarization phase for each AP during the initial firing phase. Correspondingly, the glial membrane potential exhibits a slow, gradual depolarization over the 1000 ms simulation time, overlaid with small, rapid oscillations corresponding to the neuronal APs (Figure 10B). Notably, the initial AP depolarizes the glial membrane more heavily than the following APs.

The transition in neuronal firing pattern is underpinned by significant ionic shifts in both the intracellular and extracellular compartments. We observe that the neuronal Na^+ concentration increases rapidly while the K^+ concentration decreases rapidly during the initial firing phase. Both concentrations change by almost 40 mM during the sustained train of APs in the first 400 ms of the simulation. Following the transition to the burst-firing mode, the ion concentration trends reverse during the silenced periods, indicating that the homeostatic mechanisms are activated. The neuronal Cl^- concentration displays significant oscillations, peaking near the onset of the second firing phase (Figure 10C). The Cl^- concentration changes are two orders of magnitude smaller than Na^+ and K^+ concentration changes.

We further observe that the extracellular K^+ concentration increases rapidly during the initial firing phase, reaching a peak of approximately 3.40 mM and 3.33 mM in the proximal and distal points, respectively, before decreasing (with notable peaks during the bursts) in the second firing phase (Figure 10D). The extracellular Na^+ and Cl^- concentrations decrease consistently over the simulation period in the distal point, whereas in the proximal point both concentrations increase rapidly at the onset of the simulation before Na^+ slowly increases while Cl^- decreases.

In contrast to the spiking behavior observed in the neuron and the ECS, the glial ion concentrations change gradually and more smoothly over time during both the AP train and the burst-firing mode (Figure 10E). The Na^+ concentration decreases, while both K^+ and Cl^- increase. The rates of change decrease after the onset of the burst-firing mode, and all ion concentrations appear to saturate.

5.3 Cellular Morphology Determines How Ion Concentrations and Electric Potentials Develop in Space During Neuronal Activity

We now turn to the spatial distribution of ion concentrations and electric potentials. During firing of a single action potential, the extracellular K^+ concentration is fairly constant in the initial phase when the membrane potential rises (Figure 11A, B). Subse-

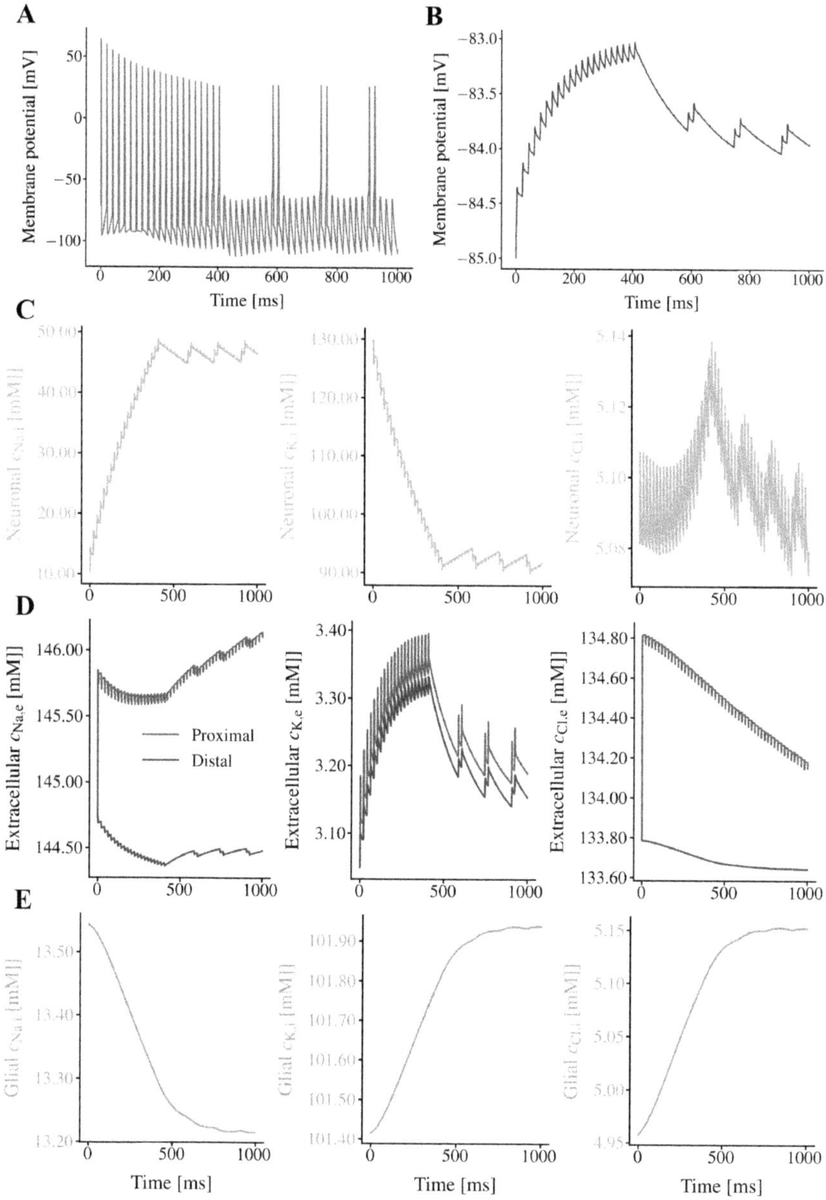

Figure 10 Neuronal membrane potential (**A**), glial membrane potential (**B**), neuronal ion concentrations (**C**), extracellular ion concentrations (**D**) and glial ion concentrations (**E**) when simulating stimulation of an axon at a frequency of 50 Hz on the $L = 5$ μm mouse visual cortex mesh with $N = 100$ biological cells. The insets show the graphs during the first 10 ms. The insets have separate axes limits for the ICS and ECS concentrations (like in the original plots). In panel **C**, proximal and distal refer to the ECS points shown in Figure 8**D** and **C**, respectively.

quent to the membrane potential peak, the increase in extracellular K^+ spreads through the extracellular space rapidly as the voltage-gate K^+ channels open (Figure 11C–E). During the action potential, the extracellular Na^+ concentration decreases gradually proximal to the neuron, with less notable changes in distal locations (Figure 11F–J).

There are notable local differences in the neuronal concentrations, with distinctly greater K^+ and lower Na^+ concentrations in the axonal varicosity (widened section) of the stimulated neuron after the action potential, compared to the slender structures of the axon. (Figure 11E, J). At time $t = 2.5$ ms, the Na^+ concentration is 2-3 mM higher in the slender structures than in the varicosity.

At a given time instant, there are also small spatial differences in the electric potentials in the stimulated neuron and the extracellular space. After the peak of the first action potential at time $t = 2.5$ ms, the extracellular potential ϕ_e is most negative close to the neuron, especially near the axonal varicosity (Figure 12A). The intracellular potential ϕ_i is also lowest on the membrane of the varicosity. After $t = 1000$ ms, subsequent to sustained stimulus with a frequency of 50 Hz, the situation is reversed for the extracellular potential (Figure 12B); the potential is less negative close to the neuron. Also, the maximum spatial differences in ϕ_e at time $t = 1000$ ms are around 0.14 mV in magnitude, more than an order greater than the maximum difference of 0.008 mV at $t = 2.5$ ms.

In the astrocyte wrapping around the stimulated neuron, the intracellular K^+ at $t = 2.5$ ms increases most closely to the neuron in slender structures (Figure 13A). The spatial variation in glial K^+ concentration is as low as 0.002 mM, whereas in the extracellular space, the concentrations differ by around 0.14 mM. Later at $t = 1000$ ms, the intracellular K^+ differs by around 0.016 mM (Figure 13B). The extracellular K^+ concentration is essentially uniform in space, except for locally elevated values in the stimulus region (Figure 13B inset).

6 Discussion and Outlook

We have presented a computational framework for simulation of electrodiffusion in cellular geometries based on a finite element method for the KNP-EMI equations. We have also demonstrated a pipeline for generating realistic computational meshes from electron microscopy data that represents dense reconstructions of cerebral tissue. Using such a geometry of mouse visual cortex tissue, we applied our computational framework to simulate electrodiffusive dynamics during neuronal activity at low- and high-frequency firing.

The local structure of the extracellular space (ECS) is a key determinant of concentration gradients and temporal dynamics, and can lead to local extracellular accumulation, depletion, and ion trapping in narrow extracellular confines [35, 51]. Localized ion concentration dynamics like this affect membrane reversal potentials and set the balance between transmembrane fluxes and diffusive clearance. In line with this, our results showed that morphology had clear consequences for local ion concentrations: wider and more slender axonal segments developed systematically different intra- and extra-

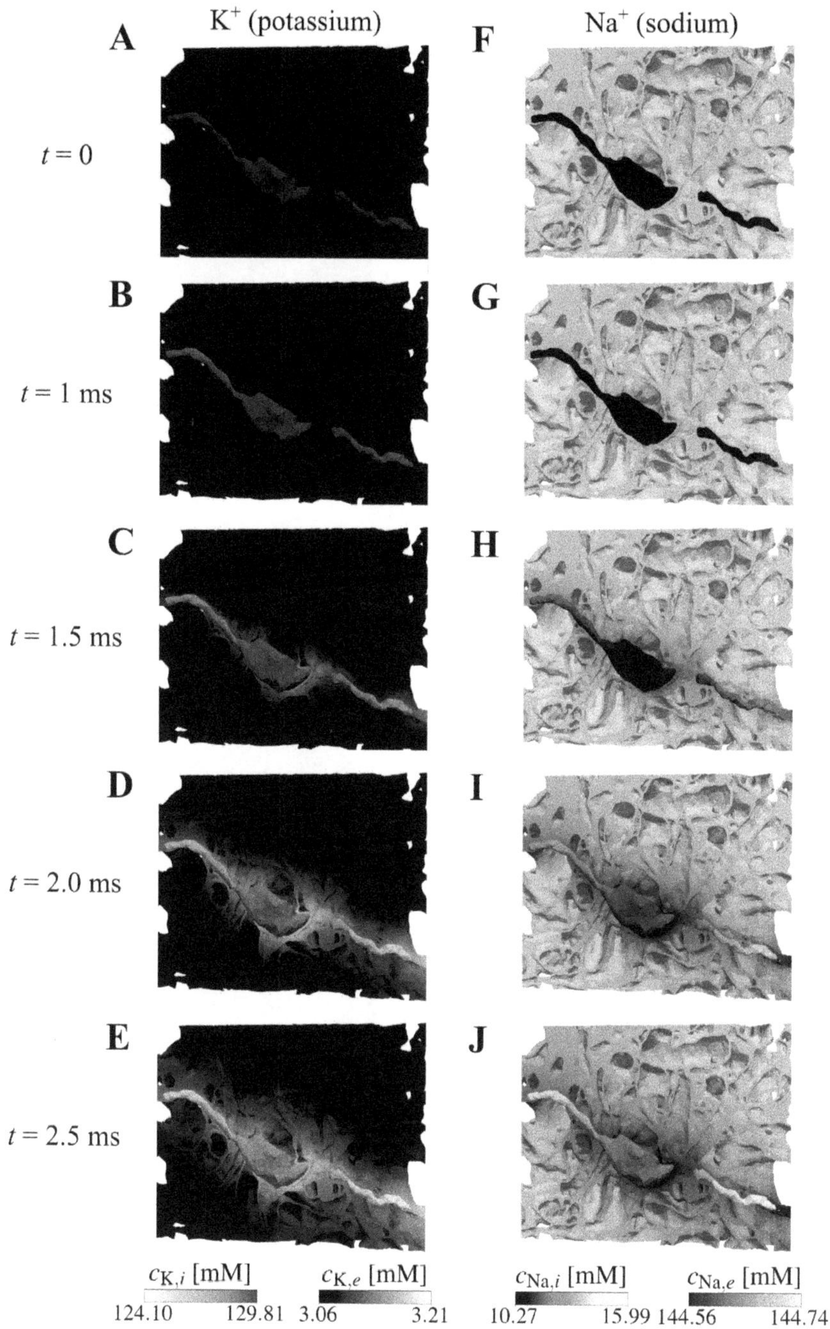

Figure 11 Potassium (**A–E**) and sodium (**F–J**) concentrations in the stimulated neuron and the extracellular space over time when simulating the firing of an action potential using the $L = 5\ \mu$m, $N = 100$ mouse visual cortex geometry. The ECS is clipped and no other cells than the stimulated neuron are shown.

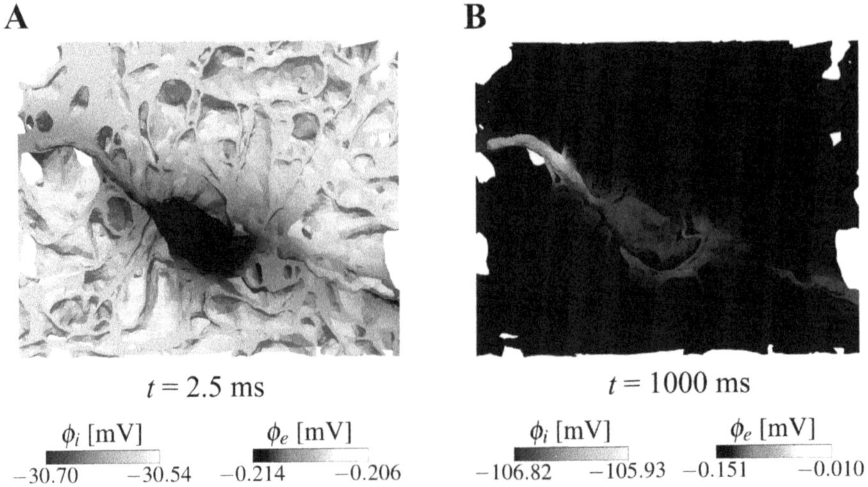

Figure 12 Electric potentials in the stimulated neuron and the extracellular space at times $t = 2.5$ ms (**A**) and $t = 1000$ ms (**B**). The ECS is clipped and the only cell shown is the neuron that is stimulated. Note the individual colorbars for each of the subpanels.

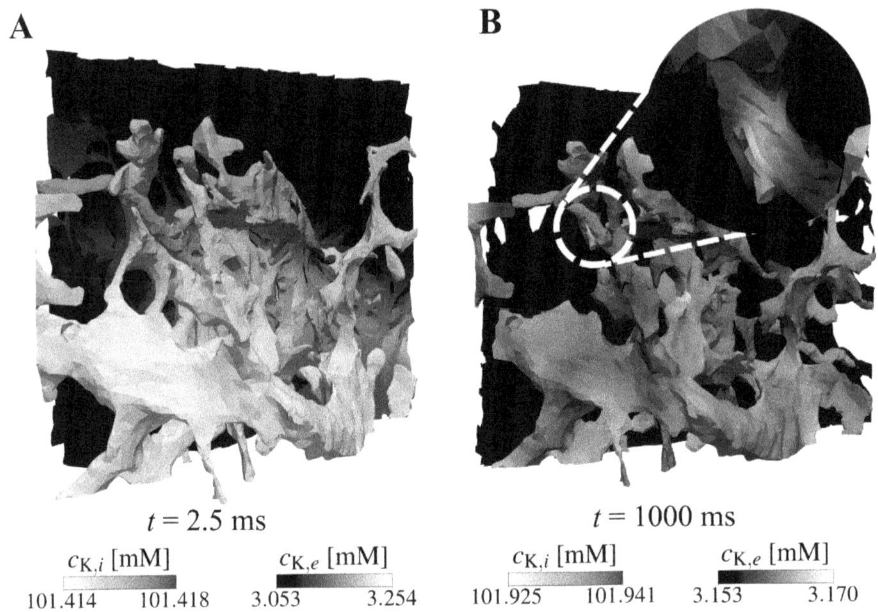

Figure 13 Potassium concentrations in an astrocyte (glial cell) and the extracellular space at times $t = 2.5$ ms (**A**) and $t = 1000$ ms (**B**) when simulating a neuron (pink) with a frequency of 50 Hz. The ECS is clipped. Note the individual colorbars for each of the subpanels.

cellular ion concentration profiles due to differences in local membrane area and the surrounding ECS geometry (See e.g., Figure 11).

By explicitly resolving the ECS, our framework is suited to capture morphology-dependent effects on ion concentration dynamics on a fine scale. As such, it lends itself to simulating local modulation of reversal potentials in regions with high-density clusters of ion channels, such as nodes of Ranvier, the axon initial segment, or other specific dendritic and axonal regions [11, 59, 77, 82]. Also, if equipping the membranes with synapses (not done in the present study), the framework would be suited to simulate effects of synaptic spillover on nearby membrane receptors [63, 6], effects of such spillovers on synaptic plasticity [43], or effects of localized ion concentration dynamics on synaptic plasticity [58, 19].

Effects like those discussed above are inaccessible to standard HH-type models which typically do not account for extracellular variations. Some of these effects can be captured qualitatively in simpler morphologies extended with an explicit representation of the ECS (e.g. [23]), but realistic, complex morphologies are crucial for correctly predicting where local gradients arise, how long they persist, and how they interact across neighboring structures in a dense neuropil. In that sense, the detailed geometry does not only add spatial "decoration" but fundamentally constrains the electrodiffusive dynamics that can occur [51].

We also reported on the characteristics of dense reconstructions of cerebral tissue by generating a series of meshes with varying domain sizes and number of biological cells. We found that the number of membrane points in the geometries are high compared to the total number of points, especially for geometries with a high number of biological cells. This poses a challenge for strongly coupled problems such as the electrodiffusive KNP-EMI equations, with ion concentrations and electric potentials developing over varying temporal and spatial scales. Since theory based on spectral analysis and simulations based on the EMI equations suggest that the preconditioner we employ will struggle when most of the mesh vertices are located on cellular membranes [8, 9, 7], we assessed how the iterative solver performs in this setting. For the smaller geometries, the solver performed consistently in spite of high membrane-point ratios, but performance deteriorated with increasingly complex geometries. It would be interesting to investigate how PDE splitting strategies, contrasting our monolithic approach, would perform on these geometries.

In terms of limitations, two important considerations are the following. First, there are uncertainties in the membrane mechanism parameters that are adopted from simpler models. This calls for validation by comparing simulations with physical experiments. Another alternative is an uncertainty quantification study. Second, the geometries used have narrow extracellular structures where the distance separating two cells may approach the limit of the electroneutrality assumption in the KNP equations. The KNP approximation has shown to be excellent on the micrometer scale [28]. When modeling dynamics very close to the cellular membranes at the nanometer scale, the small spatial and temporal scales require solving a Poisson equation for the electric potentials, forming the Poisson-Nernst-Planck (PNP) equations to fully incorporate the rapid dynamics [37]. A comparison study of the two approaches (KNP and PNP) on the mouse visual cortex geometries could shed light on this matter.

In conclusion, the contents of this chapter exemplify the current computational state of the art in terms of numerical simulations of ionic electrodiffusion in image-based, realistic brain tissue geometries. Indeed, we argue that the techniques presented have reached a level of maturity and scalability that allows for in-silico studies of fully coupled electrophysiology and electrodiffusion in dense reconstructions of neuropil on the order of tens of micrometers. Yet, open methodological questions remain in particular with regard to scalable and accurate solution algorithms in the presence of highly convoluted geometries and longer timescales. Moreover, we envision that fast solvers and surrogate models can be built from solution techniques that leverage the regular structure of the imaging data to a greater extent and other representation architectures.

Acknowledgements We are grateful to Pietro Benedusi for interesting discussions and for providing input to Figure 5. The research presented in this chapter has benefited from the Experimental Infrastructure for Exploration of Exascale Computing (eX3), a computational infrastructure that is financially supported by the Research Council of Norway under contract #270053. G. T. E. acknowledges funding from the European Union Horizon 2020 Research and Innovation Programme under Grant Agreement No. 101147319 (EBRAINS 2.0). M. C. acknowledges funding from the European Research Council under grant No. 101141807 (aCleanBrain). M.E.R. acknowledges support from Stiftelsen Kristian Gerhard Jebsen via the K. G. Jebsen Centre for Brain Fluid Research and Wellcome via Award 313298/Z/24/Z (Next-generation simulation and learning in imaging-based biomedicine).

Competing Interests The authors have no conflicts of interest to declare that are relevant to the content of this chapter.

Appendix

Numerical verification of the iterative numerical method with idealized geometries

Here, we verify spatial convergence properties of the iterative numerical method by performing a convergence study with two idealized problems. The first problem considers the domain $\Omega = [0,1]^2$ μm^2 with boundary $\partial\Omega$, with an inner square $\Omega_i = [0.25, 0.75]^2$ μm^2 as intracellular space and $\Omega_e = \Omega \setminus \Omega_i$ as extracellular space. The interface that separates Ω_i and Ω_e is the cellular membrane Γ. The second problem is similar, and considers the cube $\Omega = [0,1]^3$ μm^3 with an inner cube $\Omega_i = [0.25, 0.75]^3$ μm^3 as intracellular space and $\Omega_e = \Omega \setminus \Omega_i$ as extracellular space. The cellular membrane Γ is also here defined as the interface between Ω_i and Ω_e. The two problem domains are meshed with triangles and tetrahedra with mesh edge lengths h. To assess convergence, we consider a series of increasingly refined meshes. The coarsest mesh edge length of the two problems is denoted h_0.

Convergence is verified by manufacturing solutions for both problems [61]. The parameter values of R, F, T and C_m are all set to 1. We solve the problems for one timestep using $\Delta t = 1 \times 10^{-5}$, and compute the L^2-error for the concentrations and potentials Tables 7 and 8. Although a priori error estimates for the KNP-EMI problem are not yet well-established, the convergence rate when using continuous Lagrange elements of order p to approximate a finite element solution is bounded from above by $p + 1$ [38]. We observe convergence rates of $p + 1$, both for the 2D and the 3D test problems.

Table 7 Error norms and convergence rates for the two-dimensional verification problem on meshes of increasing resolution characterized by the edge lengths h. The coarsest mesh has mesh edge length h_0.

h/h_0	$\|c_{Na,i} - c_{Na,i,exact}\|_{L^2}$	$\|c_{K,i} - c_{K,i,exact}\|_{L^2}$	$\|c_{Cl,i} - c_{Cl,i,exact}\|_{L^2}$	$\|\phi_i - \phi_{i,exact}\|_{L^2}$
1	9.01×10^{-3} (–)	9.00×10^{-3} (–)	1.80×10^{-2} (–)	9.26×10^{-2} (–)
1/2	2.33×10^{-3} (1.95)	2.33×10^{-3} (1.95)	4.67×10^{-3} (1.95)	2.48×10^{-2} (1.90)
1/4	5.88×10^{-4} (1.99)	5.88×10^{-4} (1.99)	1.17×10^{-3} (1.99)	6.30×10^{-3} (1.98)
1/8	1.47×10^{-4} (2.00)	1.47×10^{-4} (2.00)	2.94×10^{-4} (2.00)	1.58×10^{-3} (2.00)
1/16	3.68×10^{-5} (2.00)	3.63×10^{-5} (2.02)	7.26×10^{-5} (2.02)	3.95×10^{-4} (2.00)
	$\|c_{Na,e} - c_{Na,e,exact}\|_{L^2}$	$\|c_{K,e} - c_{K,e,exact}\|_{L^2}$	$\|c_{Cl,e} - c_{Cl,e,exact}\|_{L^2}$	$\|\phi_e - \phi_{e,exact}\|_{L^2}$
1	3.12×10^{-2} (–)	1.04×10^{-2} (–)	4.16×10^{-2} (–)	6.13×10^{-2} (–)
1/2	8.08×10^{-3} (1.95)	2.69×10^{-3} (1.95)	1.08×10^{-3} (1.95)	1.67×10^{-2} (1.88)
1/4	2.04×10^{-3} (1.99)	6.79×10^{-4} (1.99)	2.72×10^{-3} (1.99)	4.25×10^{-3} (1.97)
1/8	5.09×10^{-4} (2.00)	1.70×10^{-4} (2.00)	6.78×10^{-4} (2.00)	1.07×10^{-3} (1.99)
1/16	1.26×10^{-4} (2.00)	4.19×10^{-5} (2.02)	1.68×10^{-4} (2.02)	2.68×10^{-4} (2.00)

Table 8 Error norms and convergence rates for the three-dimensional verification problem on meshes of increasing resolution characterized by the edge lengths h. The coarsest mesh has mesh edge length h_0.

h/h_0	$\|c_{Na,i} - c_{Na,i,exact}\|_{L^2}$	$\|c_{K,i} - c_{K,i,exact}\|_{L^2}$	$\|c_{Cl,i} - c_{Cl,i,exact}\|_{L^2}$	$\|\phi_i - \phi_{i,exact}\|_{L^2}$
1	6.70×10^{-3} (–)	6.70×10^{-3} (–)	1.34×10^{-2} (–)	6.82×10^{-2} (–)
1/2	1.79×10^{-3} (1.90)	1.79×10^{-3} (1.90)	3.58×10^{-3} (1.90)	1.93×10^{-2} (1.82)
1/4	4.54×10^{-4} (1.98)	4.55×10^{-4} (1.98)	9.09×10^{-4} (1.98)	4.99×10^{-3} (1.95)
1/8	1.14×10^{-4} (2.00)	1.14×10^{-4} (2.00)	2.28×10^{-4} (2.00)	1.26×10^{-3} (1.99)
1/16	2.83×10^{-5} (2.01)	2.83×10^{-5} (2.01)	5.66×10^{-5} (2.01)	3.16×10^{-4} (2.00)
	$\|c_{Na,e} - c_{Na,e,exact}\|_{L^2}$	$\|c_{K,e} - c_{K,e,exact}\|_{L^2}$	$\|c_{Cl,e} - c_{Cl,e,exact}\|_{L^2}$	$\|\phi_e - \phi_{e,exact}\|_{L^2}$
1	3.55×10^{-2} (–)	1.12×10^{-2} (–)	4.73×10^{-2} (–)	6.74×10^{-2} (–)
1/2	9.47×10^{-3} (1.90)	3.16×10^{-3} (1.90)	1.26×10^{-3} (1.90)	1.96×10^{-2} (1.78)
1/4	2.41×10^{-3} (1.98)	8.02×10^{-4} (1.98)	3.21×10^{-3} (1.98)	5.11×10^{-3} (1.94)
1/8	6.03×10^{-4} (2.00)	2.01×10^{-4} (2.00)	8.04×10^{-4} (2.00)	1.29×10^{-3} (1.98)
1/16	1.50×10^{-4} (2.01)	4.99×10^{-5} (2.01)	2.00×10^{-4} (2.01)	3.24×10^{-4} (2.00)

Numerical verification with a realistic cerebral tissue geometry

With the spatial convergence properties verified on idealized geometries, we here verify that the numerical solution of the KNP-EMI model converges on the mouse visual cortex tissue mesh. We perform a convergence study by solving the KNP-EMI equations on the original $L = 5\ \mu$m, $N = 100$ mesh (generated with EMI-Meshing as introduced in Section 2), and two uniformly refined versions of the mesh. We run a similar simulation as described in Section 5; an axon is stimulated with the stimulus current density defined in Equation (16) with $T_{\text{stim}} = 1$ s. The simulation is run for 800 timesteps with $\Delta t = 2.5 \times 10^{-5}$ to a final time of $T = 10$ ms. We evaluate the membrane potential in the point $\boldsymbol{x} = (1186, 3241, 3594)$ nm on the stimulated neuron. The neuronal intracellular sodium concentration, the extracellular potassium concentration proximal to the neuron, and the glial intracellular Cl^- concentration are all evaluated in the same points as described in Section 5, see Figure 8. The solutions obtained are similar to the results observed in Section 5, and the results agree well across the different mesh versions (Figure 14).

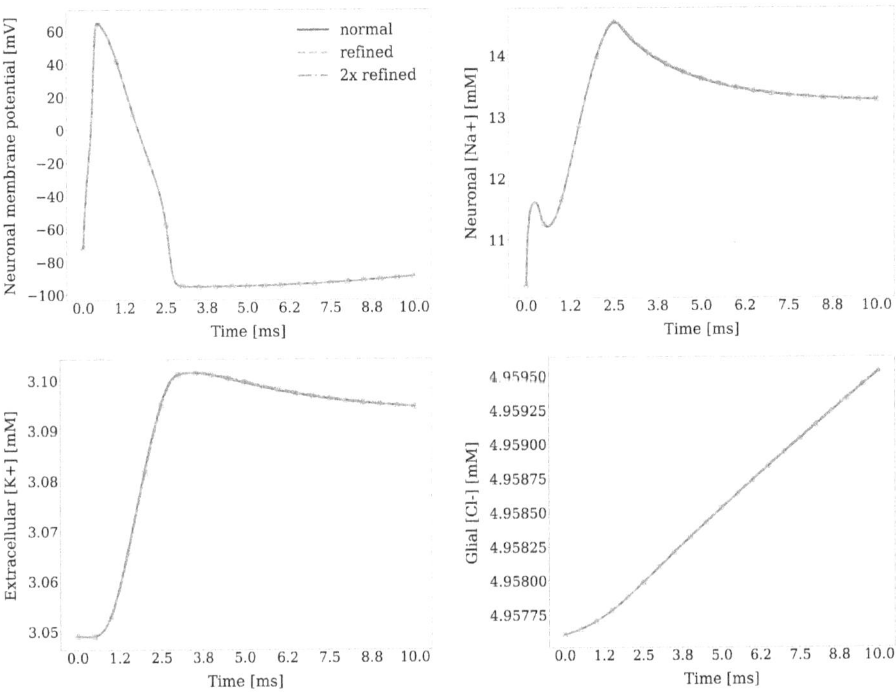

Figure 14 Convergence of solution under mesh refinement on the $L = 5\ \mu$m, $N = 100$ mouse visual cortex meshes. Three meshes of increasing refinement are considered, and we simulate an action potential as described in Section 5.1. The membrane potential is evaluated on the neuronal membrane at $\boldsymbol{x} = (1186, 3241, 3594)$ nm, while the concentrations are evaluated in the points described in Figure 8. Common legend for all subfigures.

References

1. A. Agudelo-Toro and A. Neef. Computationally efficient simulation of electrical activity at cell membranes interacting with self-generated and externally imposed electric fields. *Journal of Neural Engineering*, 10(2):026019, 2013.
2. D. J. Aidley. *The physiology of excitable cells*. Cambridge University Press, 1998.
3. P. Aitken and G. Somjen. The sources of extracellular potassium accumulation in the CA1 region of hippocampal slices. *Brain Research*, 369(1-2):163–167, 1986.
4. F. Ballarin. multiphenicsx - easy prototyping of multiphysics problems in FEniCSx, 2020.
5. I. A. Baratta, J. P. Dean, J. S. Dokken, M. Habera, J. S. Hale, C. N. Richardson, M. E. Rognes, M. W. Scroggs, N. Sime, and G. N. Wells. DOLFINx: The next generation FEniCS problem solving environment, Dec. 2023.
6. B. Barbour. An evaluation of synapse independence. *Journal of Neuroscience*, 21(20):7969–7984, 2001.
7. P. Benedusi, A. J. Ellingsrud, H. Herlyng, and M. E. Rognes. Scalable approximation and solvers for ionic electrodiffusion in cellular geometries. *SIAM Journal on Scientific Computing*, 46(5):B725–B751, 2024.
8. P. Benedusi, P. Ferrari, M. Causemann, and S. Serra-Capizzano. Dense cell-by-cell systems of PDEs: approximation, spectral analysis, and preconditioning. *arXiv preprint arXiv:2409.13432*, 2024.
9. P. Benedusi, P. Ferrari, M. E. Rognes, and S. Serra-Capizzano. Modeling excitable cells with the EMI equations: Spectral analysis and iterative solution strategy. *Journal of Scientific Computing*, 98(3):58, 2024.
10. R. F. Betzel and D. S. Bassett. Multi-scale brain networks. *Neuroimage*, 160:73–83, 2017.
11. J. H. Caldwell, K. L. Schaller, R. S. Lasher, E. Peles, and S. R. Levinson. Sodium channel Nav1.6 is localized at nodes of Ranvier, dendrites, and synapses. *Proceedings of the National Academy of Sciences*, 97(10):5616–5620, 2000.
12. M. Causemann. Emi-meshing: High-quality extracellular-membrane-intracellular meshes of the mouse visual cortex, 8 2024.
13. M. Causemann, R. Enger, and M. E. Rognes. Stretch and flow at the gliovascular interface: high-fidelity modelling of the mechanics of astrocyte endfeet. *bioRxiv*, pages 2025–05, 2025.
14. A. B. Chen, M. Duque, A. Rymbek, M. Dhanasekar, V. M. Wang, X. Mi, L. Tocquer, S. Narayan, E. M. Legorreta, M. Eddison, G. Yu, C. Wyart, D. A. Prober, F. Engert, and M. B. Ahrens. Norepinephrine changes behavioral state through astroglial purinergic signaling. *Science*, 388(6748):769–775, 2025.
15. K. C. Chen and C. Nicholson. Spatial buffering of potassium ions in brain extracellular space. *Biophysical Journal*, 78(6):2776–2797, 2000.
16. T. M. Consortium. Functional connectomics spanning multiple areas of mouse visual cortex. *Nature*, 640(8058):435–447, 2025.
17. M. De Pittà and H. Berry. *Computational glioscience*. Springer, 2019.
18. H. P. Developers. *hypre*: High performance preconditioners. `https://llnl.gov/casc/hypre`, `https://github.com/hypre-space/hypre`.
19. N. Doyon, L. Vinay, S. A. Prescott, and Y. De Koninck. Chloride regulation: a dynamic equilibrium crucial for synaptic inhibition. *Neuron*, 89(6):1157–1172, 2016.
20. G. T. Einevoll. Mathematical modeling of neural activity. In *Dynamics of complex interconnected systems: Networks and bioprocesses*, pages 127–145. Springer, 2006.
21. A. J. Ellingsrud, P. Benedusi, and M. Kuchta. A splitting, discontinuous Galerkin solver for the cell-by-cell electroneutral Nernst–Planck framework. *SIAM Journal on Scientific Computing*, 47(2):B477–B504, 2025.
22. A. J. Ellingsrud, C. Daversin-Catty, and M. E. Rognes. A cell-based model for ionic electrodiffusion in excitable tissue. *Modeling Excitable Tissue: The EMI Framework*, 7:14–27, 2021.
23. A. J. Ellingsrud, A. Solbrå, G. T. Einevoll, G. Halnes, and M. E. Rognes. Finite element simulation of ionic electrodiffusion in cellular geometries. *Frontiers in Neuroinformatics*, 14:11, 2020.
24. X.-M. Fu, Y. Liu, and B. Guo. Computing locally injective mappings by advanced mips. *ACM Transactions on Graphics (TOG)*, 34(4):1–12, 2015.

25. A. Glaser, J. Chandrashekar, S. Vasquez, C. Arshadi, R. Javeri, N. Ouellette, X. Jiang, J. Baka, G. Kovacs, M. Woodard, et al. Expansion-assisted selective plane illumination microscopy for nanoscale imaging of centimeter-scale tissues. *Elife*, 12:RP91979, 2025.
26. K. A. Guttenplan, I. Maxwell, E. Santos, L. A. Borchardt, E. Manzo, L. Abalde-Atristain, R. D. Kim, and M. R. Freeman. GPCR signaling gates astrocyte responsiveness to neurotransmitters and control of neuronal activity. *Science*, 388(6748):763–768, 2025.
27. R. Haase, P. Rajasekhar, T. Lambert, J. Nunez-Iglesias, C. Caporal, C. Avenel, and T. Korten. clesperanto/pyclesperanto_prototype. https://github.com/clEsperanto/pyclesperanto_prototype, feb 21 2025.
28. G. Halnes, T. Mäki-Marttunen, D. Keller, K. H. Pettersen, O. A. Andreassen, and G. T. Einevoll. Effect of ionic diffusion on extracellular potentials in neural tissue. *PLOS Computational Biology*, 12(11):e1005193, 2016.
29. G. Halnes, T. V. Ness, S. Næss, E. Hagen, K. H. Pettersen, and G. T. Einevoll. *Electric brain signals: foundations and applications of biophysical modeling*. Cambridge University Press, 2024.
30. G. Halnes, I. Østby, K. H. Pettersen, S. W. Omholt, and G. T. Einevoll. Electrodiffusive model for astrocytic and neuronal ion concentration dynamics. *PLOS Computational Biology*, 9(12):1–16, 12 2013.
31. M. Häusser. The Hodgkin-Huxley theory of the action potential. *Nature Neuroscience*, 3(11):1165–1165, 2000.
32. B. Hille. Ion channels of excitable membranes, 2001. Sinauer, Sunderland, MA.
33. A. L. Hodgkin and A. F. Huxley. A quantitative description of membrane current and its application to conduction and excitation in nerve. *The Journal of Physiology*, 117(4):500, 1952.
34. K. E. Holter, B. Kehlet, A. Devor, T. J. Sejnowski, A. M. Dale, S. W. Omholt, O. P. Ottersen, E. A. Nagelhus, K.-A. Mardal, and K. H. Pettersen. Interstitial solute transport in 3D reconstructed neuropil occurs by diffusion rather than bulk flow. *Proceedings of the National Academy of Sciences*, 114(37):9894–9899, 2017.
35. S. Hrabeťová, J. Hrabe, and C. Nicholson. Dead-space microdomains hinder extracellular diffusion in rat neocortex during ischemia. *Journal of Neuroscience*, 23(23):8351–8359, 2003.
36. Y. Hu, T. Schneider, B. Wang, D. Zorin, and D. Panozzo. Fast tetrahedral meshing in the wild. *ACM Trans. Graph.*, 39(4), 8 2020.
37. K. H. Jæger, E. Ivanovic, J. P. Kucera, and A. Tveito. Nano-scale solution of the Poisson-Nernst-Planck (PNP) equations in a fraction of two neighboring cells reveals the magnitude of intercellular electrochemical waves. *PLOS Computational Biology*, 19(2):e1010895, 2023.
38. C. Johnson. *Numerical solution of partial differential equations by the finite element method*. Courier Corporation, 2009.
39. H. Kager, W. J. Wadman, and G. G. Somjen. Simulated seizures and spreading depression in a neuron model incorporating interstitial space and ion concentrations. *Journal of Neurophysiology*, 84(1):495–512, 2000.
40. E. R. Kandel, J. H. Schwartz, T. M. Jessell, S. Siegelbaum, A. J. Hudspeth, S. Mack, et al. *Principles of Neural Science*, volume 4. McGraw-Hill New York, 2000.
41. A. Kaszynski. pyvista/pyacvd. https://github.com/pyvista/pyacvd, oct 10 2025.
42. P. Kofuji and E. Newman. Potassium buffering in the central nervous system. *Neuroscience*, 129(4):1043–1054, 2004.
43. D. M. Kullmann and F. Asztely. Extrasynaptic glutamate spillover in the hippocampus: evidence and implications. *Trends in Neurosciences*, 21(1):8–14, 1998.
44. W.-C. A. Lee, H. Huang, G. Feng, J. R. Sanes, E. N. Brown, P. T. So, and E. Nedivi. Dynamic remodeling of dendritic arbors in GABAergic interneurons of adult visual cortex. *PLoS Biology*, 4(2):e29, 2006.
45. K. B. Lefton, Y. Wu, Y. Dai, T. Okuda, Y. Zhang, A. Yen, G. M. Rurak, S. Walsh, R. Manno, B.-E. Myagmar, J. D. Dougherty, V. K. Samineni, P. C. Simpson, and T. Papouin. Norepinephrine signals through astrocytes to modulate synapses. *Science*, 388(6748):776–783, 2025.
46. C. L. Lopreore, T. M. Bartol, J. S. Coggan, D. X. Keller, G. E. Sosinsky, M. H. Ellisman, and T. J. Sejnowski. Computational modeling of three-dimensional electrodiffusion in biological systems: application to the node of Ranvier. *Biophysical Journal*, 95(6):2624–2635, 2008.

47. B. Lu, M. J. Holst, J. A. McCammon, and Y. Zhou. Poisson–Nernst–Planck equations for simulating biomolecular diffusion–reaction processes I: Finite element solutions. *Journal of Computational Physics*, 229(19):6979–6994, 2010.
48. MICrONS Consortium. Neuroglancer. Accessed: 11.08.2025.
49. Y. Mori and C. Peskin. A numerical method for cellular electrophysiology based on the electrodiffusion equations with internal boundary conditions at membranes. *Communications in Applied Mathematics and Computational Science*, 4(1):85–134, 2009.
50. A. Nandi, T. Chartrand, W. Van Geit, A. Buchin, Z. Yao, S. Y. Lee, Y. Wei, B. Kalmbach, B. Lee, E. Lein, et al. Single-neuron models linking electrophysiology, morphology, and transcriptomics across cortical cell types. *Cell Reports*, 40(6), 2022.
51. C. Nicholson and S. Hrabeťová. Brain extracellular space: the final frontier of neuroscience. *Biophysical Journal*, 113(10):2133–2142, 2017.
52. C. Nicholson, G. Ten Bruggencate, H. Stockle, and R. Steinberg. Calcium and potassium changes in extracellular microenvironment of cat cerebellar cortex. *Journal of Neurophysiology*, 41(4):1026–1039, 1978.
53. I. Østby, L. Øyehaug, G. T. Einevoll, E. A. Nagelhus, E. Plahte, T. Zeuthen, C. M. Lloyd, O. P. Ottersen, and S. W. Omholt. Astrocytic mechanisms explaining neural-activity-induced shrinkage of extraneuronal space. *PLOS Computational Biology*, 5(1):e1000272, 2009.
54. L. Øyehaug. Slow ion concentration oscillations and multiple states in neuron–glia interaction—insights gained from reduced mathematical models. *Frontiers in Network Physiology*, 3:1189118, 2023.
55. L. Øyehaug, I. Østby, C. M. Lloyd, S. W. Omholt, and G. T. Einevoll. Dependence of spontaneous neuronal firing and depolarisation block on astroglial membrane transport mechanisms. *Journal of Computational Neuroscience*, 32(1):147–165, 2012.
56. J. Pods, J. Schönke, and P. Bastian. Electrodiffusion models of neurons and extracellular space using the Poisson-Nernst-Planck equations—numerical simulation of the intra-and extracellular potential for an axon model. *Biophysical Journal*, 105(1):242–254, 2013.
57. D. Purves, G. J. Augustine, and D. Fitzpatrick. *Neuroscience*. Sunderland (MA): Sinauer Associates, 2nd edition, 2001.
58. J. V. Raimondo, R. J. Burman, A. A. Katz, and C. J. Akerman. Ion dynamics during seizures. *Frontiers in Cellular Neuroscience*, 9:419, 2015.
59. M. N. Rasband. The axon initial segment and the maintenance of neuronal polarity. *Nature Reviews Neuroscience*, 11(8):552–562, 2010.
60. R. Rasmussen, J. O'Donnell, F. Ding, and M. Nedergaard. Interstitial ions: A key regulator of state-dependent neural activity? *Progress in Neurobiology*, 193, 10 2020.
61. P. J. Roache. Code verification by the method of manufactured solutions. *Journal of Fluids Engineering*, 124(1):4–10, 1 2001.
62. T. Roy, J. Andrej, and V. A. Beck. A scalable DG solver for the electroneutral Nernst-Planck equations. *Journal of Computational Physics*, 475:111859, 2023.
63. D. A. Rusakov and D. M. Kullmann. Extrasynaptic glutamate diffusion in the hippocampus: ultrastructural constraints, uptake, and receptor activation. *Journal of Neuroscience*, 18(9):3158–3170, 1998.
64. S. Rush and H. Larsen. A practical algorithm for solving dynamic membrane equations. *IEEE Transactions on Biomedical Engineering*, pages 389–392, 2007.
65. Y. Saad and M. H. Schultz. GMRES: A generalized minimal residual algorithm for solving nonsymmetric linear systems. *SIAM Journal on Scientific and Statistical Computing*, 7(3):856–869, 1986.
66. M. J. Sætra, G. T. Einevoll, and G. Halnes. An electrodiffusive, ion conserving Pinsky-Rinzel model with homeostatic mechanisms. *PLOS Computational Biology*, 16(4):e1007661, 2020.
67. M. J. Sætra, G. T. Einevoll, and G. Halnes. An electrodiffusive neuron-extracellular-glia model for exploring the genesis of slow potentials in the brain. *PLoS Computational Biology*, 17(7):e1008143, 2021.
68. M. J. Sætra, A. J. Ellingsrud, and M. E. Rognes. Neural activity induces strongly coupled electro-chemo-mechanical interactions and fluid flow in astrocyte networks and extracellular space—-A computational study. *PLoS Computational Biology*, 19(7):e1010996, 2023.

69. M. J. Sætra and Y. Mori. An electrodiffusive network model with multicompartmental neurons and synaptic connections. *PLOS Computational Biology*, 20(11):e1012114, 2024.
70. Simula Research Laboratory. Experimental infrastructure for exploration of exascale computing. Accessed: 30.07.2025.
71. A. Solbrå, A. W. Bergersen, J. van den Brink, A. Malthe-Sørenssen, G. T. Einevoll, and G. Halnes. A Kirchhoff-Nernst-Planck framework for modeling large scale extracellular electrodiffusion surrounding morphologically detailed neurons. *PLOS Computational Biology*, 14(10):e1006510, 2018.
72. G. G. Somjen. Ion regulation in the brain: implications for pathophysiology. *The Neuroscientist*, 8(3):254–267, 2002.
73. J. D. Steinmetz, K. M. Seeher, N. Schiess, E. Nichols, B. Cao, C. Servili, V. Cavallera, E. Cousin, H. Hagins, M. E. Moberg, et al. Global, regional, and national burden of disorders affecting the nervous system, 1990–2021: a systematic analysis for the Global Burden of Disease Study 2021. *The Lancet Neurology*, 23(4):344–381, 2024.
74. D. Sterratt, B. Graham, A. Gillies, G. Einevoll, and D. Willshaw. *Principles of computational modelling in neuroscience*. Cambridge University Press, 2023.
75. B. Sullivan and A. Kaszynski. PyVista: 3D plotting and mesh analysis through a streamlined interface for the Visualization Toolkit (VTK). *Journal of Open Source Software*, 4(37):1450, May 2019.
76. M. R. Tavakoli, J. Lyudchik, M. Januszewski, V. Vistunou, N. Agudelo Dueñas, J. Vorlaufer, C. Sommer, C. Kreuzinger, B. Oliveira, A. Cenameri, et al. Light-microscopy-based connectomic reconstruction of mammalian brain tissue. *Nature*, pages 1–13, 2025.
77. J. S. Trimmer and K. J. Rhodes. Localization of voltage-gated ion channels in mammalian brain. *Annual Review of Physiology*, 66(1):477–519, 2004.
78. A. Tveito, K. H. Jæger, G. T. Lines, Ł. Paszkowski, J. Sundnes, A. G. Edwards, T. Māki-Marttunen, G. Halnes, and G. T. Einevoll. An evaluation of the accuracy of classical models for computing the membrane potential and extracellular potential for neurons. *Frontiers in Computational Neuroscience*, 11:27, 2017.
79. A. Tveito, K.-A. Mardal, and M. E. Rognes. *Modeling excitable tissue: The EMI framework*. Springer Nature, 2021.
80. V. Untiet. Astrocytic chloride regulates brain function in health and disease. *Cell Calcium*, 118:102855, 2024.
81. D. Utzschneider, J. Kocsis, and M. Devor. Mutual excitation among dorsal root ganglion neurons in the rat. *Neuroscience Letters*, 146(1):53–56, 1992.
82. H. Vacher, D. P. Mohapatra, and J. S. Trimmer. Localization and targeting of voltage-dependent ion channels in mammalian central neurons. *Physiological Reviews*, 88(4):1407–1447, 2008.
83. W. Walz. Role of astrocytes in the clearance of excess extracellular potassium. *Neurochemistry International*, 36(4-5):291–300, 2000.
84. Y. Wei, G. Ullah, and S. J. Schiff. Unification of neuronal spikes, seizures, and spreading depression. *Journal of Neuroscience*, 34(35):11733–11743, 2014.
85. U. M. Yang et al. BoomerAMG: A parallel algebraic multigrid solver and preconditioner. *Applied Numerical Mathematics*, 41(1):155–177, 2002.
86. C. Yen, C.-L. Lin, and M.-C. Chiang. Exploring the frontiers of neuroimaging: a review of recent advances in understanding brain functioning and disorders. *Life*, 13(7):1472, 2023.

www.ingramcontent.com/pod-product-compliance
Ingram Content Group UK Ltd.
Pitfield, Milton Keynes, MK11 3LW, UK
UKHW020935180926
475692UK00003B/47
* 9 7 8 3 0 3 2 3 3 4 2 7 5 *